WORLD CHECKLIST AND BIBLIOGRAPHY OF

Euphorbiaceae 1

(with Pandaceae)

WORLD CHECKLIST AND BIBLIOGRAPHY OF

Euphorbiaceae 1

(with Pandaceae)

Introduction
Euphorbiaceae: general references
Euphorbiaceae:
Aalius – Crossophora

Rafaël Govaerts, David G. Frodin
and
Alan Radcliffe-Smith

(assisted by Susan Carter, Mike Gilbert and Victor Steinmann for Euphorbiinae, Hans-Jürgen Esser for Hippomaneae, and Petra Hoffman for *Antidesma*)

First published March 2000

World Checklists and Bibliographies, 4.
[The first three in this series were published respectively as Magnoliaceae (October 1996), Fagales (May 1998) and Coniferae (August 1998)].

Address of the principal authors:
Herbarium, Royal Botanic Gardens, Kew, Richmond, Surrey TW9 3AE, United Kingdom

ISBN 1 900347 83 0

Cover photograph: *Antidesma montanum* Blume.
Kebun Raya Indonesia, Bogor, 1996, *P. Hoffman*

Cover design by Jeff Eden for Media Resources, Information Services Department, Royal Botanic Gardens, Kew

Page make-up by Media Resources, Information Services Department, Royal Botanic Gardens, Kew, from text generated by David G. Frodin using Microsoft Access 2.0® and Microsoft Word 6.0®

Printed in The European Union by Redwood Books Limited, Trowbridge, Wiltshire, UK.

Published by
The Royal Botanic Gardens, Kew
2000

Summary

Volume 1

Volume 2

Volume 3

Volume 4

Contents, Volume 1

Euphorbiaceae: Aalius – Crossophora

List of Colour Plates

Preface

This *World Checklist and Bibliography of Euphorbiaceae (with Pandaceae)* is the fourth in a series intended in the first instance to document families and other plant groups of particular interest to the programmes of the Royal Botanic Gardens, Kew and, at the same time, to contribute to current international efforts to record the world's biota (Reichhardt 1999). The series was initiated in 1996 with publication of *World Checklist and Bibliography of Magnoliaceae* (Frodin & Govaerts 1996).

The history of previous worldwide and regional checklists was outlined in the preface to the Magnoliaceae Checklist and is not here repeated. The present series runs in parallel with four other international programmes: 1) the Global Plant Checklist of the International Organization for Plant Information (IOPI)(Burnett 1994); 2) Species 2000 (Bisby 1994, Bisby & Smith 1996); 3) the Species Plantarum Project of IOPI (Orchard 1999); and 4) the *World Checklist of Seed Plants* (Govaerts 1995-). All have different aims: the Global Plant Checklist has concentrated on the 'merger' of national and regional floristic databases and rendering them accessible on the World Wide Web; Species 2000 is developing a system for simultaneous Web access to existing and developing systematic databases of organisms, living and fossil; the Species Plantarum Project aims at the production of a world Flora published in family or infrafamilial fascicles in the tradition of *Das Pflanzenreich* (Engler *et al.*, 1900-); and the *World Checklist of Seed Plants* is in effect a taxonomically validated corollary to *Index Kewensis* and, like it, is arranged by genera. The *World Checklists and Bibliographies* differ in their focus on families (or groups thereof), the inclusion of annotated source references, and summary accounts of genera and (where appropriate) family subdivisions as well as families. We hope the present volume, on Euphorbiaceae and Pandaceae, will fill a long-standing need for complete summaries of their genera, among them *Acalypha, Croton, Jatropha* and *Phyllanthus* as well as the very large *Euphorbia*, the last with 1836 species here listed.

D. G. Frodin

Royal Botanic Gardens, Kew
June 1999

Introduction

Aims and scope

The work presented here is a checklist of species and infraspecific taxa, with synonymy, places of publication, and indication of distribution and habit with occasional notes, of the Euphorbiales of the world (here comprising the Euphorbiaceae and Pandaceae, corresponding to part of the Malpighiales of the Angiosperm Phylogeny Group (1998)). Integrated with it are bibliographic references on these families with, as far as possible, at least one citation for every accepted genus. It is the fourth of a planned series initiated in 1996 with the *World Checklist and Bibliography of Magnoliaceae* (Frodin & Govaerts 1996) and continues our aim of furnishing synonymised and validated accounts of selected plant families.

Compilation of the two families was initiated in 1994. A first draft was run in 1997 and this was revised during 1997-98, with further minor revisions in 1999 before final printing. There are over 1700 bibliographic entries while in the checklist are names of 10 148 accepted taxa and 14 568 synonyms below generic rank (with altogether 966 generic names), a ratio of 1.43. 326 genera with 8935 species are recognised (322 genera, 8916 species in Euphorbiaceae; 4 genera, 19 species in Pandaceae).[1] 108 binomials have been excluded from the order. Of the principal authors, Rafaël Govaerts is responsible for the checklist, David Frodin has undertaken the general editing as well as the bibliographies and commentaries, and Alan Radliffe-Smith has acted as overall taxonomic referee. Hans-Jürgen Esser has overseen the genera of the difficult tribe Hippomaneae (in Euphorbioideae) and in addition has advised on Thailand. Advice has also been given by Christine Barker, Susan Carter (Euphorbiinae), Mike Gilbert (*Euphorbia*, especially in Africa), Petra Hoffmann (especially for *Antidesma*), Victor Steinmann (American *Euphorbia*), Peter van Welzen (Malesian genera), and Tim Whitmore and Stuart Davies (*Macaranga*). Once again we would like to thank Mr Gilbert for help with respect to *Flora of China*. The second author also wishes to express thanks to Dr van Welzen for enabling participation in two successive workshops on Malesian Euphorbiaceae.

The Euphorbiaceae and Pandaceae have been for this checklist circumscribed in accordance with current Kew practice (Brummitt 1992). Although recognised in essence since the time of Linnaeus,[2] its delimitation remains controversial with many alternative concepts on offer. *Daphniphyllum* (now in Daphniphyllaceae) was definitively removed only in 1919 although Jean Müller had first proposed family status fifty years before, while separation of *Centroplacus, Galearia, Microdesmis* and *Panda* as a distinct family (Pandaceae) was first advocated in 1913 but even now is not universally accepted. Proposals for removing some additional genera currently included in Euphorbiaceae to further segregate families have existed since the first half of the nineteenth century; certain of these have over time enjoyed support here and there but hitherto have not found lasting or widespread favour (Webster 1987; Meeuse 1990). Current research, however, indicates that there may be a good case for one such proposal: separation of *Drypetes* and allied genera (as Putranjivaceae; Tokuoka & Tobe 1999).[3] All this lends weight to arguments that Euphorbiaceae, while a concept of long standing, are – given a historical basis largely in features of ovary and fruit – less securely founded than such a family as Apiaceae *sensu stricto*, with a well-correlated, relatively catholic set of characters. Without certain adjustments, such families may prove hard to accomodate in a monophyletically based phylogenetic classification. Contemporary research in Euphorbiaceae is, however, not sufficiently far advanced for a new consensus to have emerged (cf. Angiosperm Phylogeny Group 1998: 534).

Arrangement of volumes

As the size of Euphorbiaceae requires that the work be in four volumes, the division has been made as follows: 1, Introduction, family, regional and subfamily references in Euphorbiaceae, and genera *Aalius-Crossophora*; 2, *Croton-Excoecariopsis*; 3, *Fahrenheitia-Oxydectes*; 4, *Pachystemon-Zygospermum*, Pandaceae, and unplaced and excluded taxa. We prefer that the three largest genera, especially *Euphorbia*, be not spread over more than one volume; hence they are not uniform in size.

Bibliography

The bibliography is selective rather than exhaustive. For each genus, where available, only the more important references are given: monographs, complete or partial revisions, synopses and other relatively significant contributions. A selection of 'plant portraits' is also listed, the major sources being *Botanical Magazine*, *Icones Plantarum* and *Flowering Plants of Africa*. As far as possible, all references have been annotated with regard to their contents and particular features. The language(s) of each paper are given in abbreviated form. Cross-references may sometimes be given, either to papers listed under another genus or under tribes or geographical regions. Contributions on the family as a whole, along with those on particular geographical regions (where all or a part of the family is covered) and individual tribes, are listed in a block preceding the first genus. Papers thought to be of particular significance are bulleted (•). Compilation was initially carried out with the use of R:base® version 4.5, a PC database program for personal computers from Abacus/Microrim (available in MS-DOS, OS/2 and Windows® formats), but final additions and editing were done within Microsoft Access® version 2.0 for Windows. Frequent reference is made in generic headings to Webster's *Synopsis of the genera and suprageneric taxa of Euphorbiaceae* (1994; see **General**); it appears therein as 'Synopsis, 1994'. Other general works not cited among generic references are given in full.

Checklist

Structure

The checklist is derived from a database encompassing 24 fields and complying with the data standards proposed by *Species Plantarum/Flora of the World*, several of them in association with the Taxonomic Databases Working Group (TDWG). Compilation of the database was effected using Foxbase®, a Dbase-class database program for personal computers.

Names

Names of accepted genera and their species and infraspecific taxa are listed alphabetically. Synonymous genera (and species) are intercalated. For each accepted taxon, associated synonyms are listed chronologically if heterotypic, with any homotypic

Notes

[1] The current figure for species compares with an estimate of 4000 made in 1892 for the first edition of Engler's *Syllabus der Pflanzenfamilien*. By 1904 this had increased to 4500 (which remained through the ninth/tenth editions of 1924); by 1936 (following publication of Pax and Hoffmann's account in *Die natürlichen Pflanzenfamilien*), to 7200; and for the 12th edition in 1964, to 7500 (in 290 genera). For the first edition of Mabberley's *Plant-book* (1987) a figure of 7750 (in 329 genera) was given.

[2] His natural order no. 38, 'Tricoccae' (Linnaeus 1764 & 1792 were left out of the references and have been added). An earlier version of the *Ordines* appears in *Philosophia botanica* (Linnaeus 1751: 27-36), where 'Tricocca' appears as no. 47. Yet 'Tricoccae' has even older antecedents; it is, for example, no. 8 of the classes in *Florae leydensis prodromus* (van Royen 1740). Eight of the twelve genera included (pp. 193-203) correspond to those in modern Euphorbiaceae, including the 'Big Four': *Acalypha, Croton, Euphorbia* and *Phyllanthus*.

[3] This is in fact a revival of a proposal first made by Endlicher in 1841 (*Enchir. Bot.*: 174).

synonyms following in a given lead; in addition, all synonyms in an accepted genus are listed alphabetically at the end of that genus. Doubtful and excluded taxa are likewise summarised in an appendix following the last genus of Pandaceae.[4] Place and date of publication of all names are given. Citation of authors follows *Authors of Plant Names* (Brummitt & Powell 1992); for book abbreviations, the standard is *Taxonomic Literature*, 2nd edn. (Stafleu & Cowan 1976-88; supplements, 1992-); and periodicals are abbreviated according to *Botanico-Periodicum-Huntianum/ Supplementum* (Bridson 1991). A question mark (?) following a name and author indicates that a place of publication has yet to be established. Names of nothospecies ('hybrids') are preceded by a multiplication sign (×), with the place of publication being followed by the names of the parents if known. Basionyms or replaced synonyms of accepted names are designated by an asterisk (*). For genera, the number of accepted species and the geographical distribution are furnished together with general comments and the suprageneric taxa as used at Kew. Type species are not indicated, but reference may be made to Farr *et al.* (1979, 1986) or Greuter *et al.* (1993).

Acceptance of taxa

Acceptance of species and infraspecific taxa is based on assessments of literature and common practice with reference to specialist advice and, where necessary, to the herbarium. At species level, there have historically been in some genera differences of opinion with respect to limits as well as what infraspecific taxa (if any) should be recognised. These have

Notes

[4] Many if not most of these names have never been re-evaluated since their publication and their disposition cannot yet be established. In addition, the status and the placement given for some synonymy in this work requires further investigation.

[5] The family was in that work never completely revised; there are, for instance, no accounts of *Croton, Euphorbia, Glochidion* or *Phyllanthus* or their immediate allies such as *Breynia, Reverchonia, or Synadenium*. Partial surveys of these genera did appear, of course, in Pax and Hoffmann's account of the family for *Die natürlichen Pflanzenfamilien* (1931) and have long been standard; on the other hand, for *Croton* and *Phyllanthus* the infrageneric classification adopted was on the whole little different from that of Müller in 1866. The largest genus ever revised by the Breslau team was *Acalypha* (1924), which appeared shortly before Pax's formal retirement from his professorship. Key recent revisions, all cited under their respective genera, include Bally (1961) on *Monadenium*, R. Dressler (1957) on *Pedilanthus*, S. Dressler (1996) on Asia/Pacific *Bridelia*, Esser (1997) on Malesian *Homalanthus*, Gillespie (1993) on *Plukenetia*, Henderson (1992) on *Amperea*, Ingram (1957, 1980) on *Argythamnia* and *Chiropetalum*, Kruijt (1996) on *Sapium* and allies, Pojarkova (1960) on *Andrachne*, Rogers (1951) on *Stillingia*, Rogers & Appun (1973) on *Manihot*, Schaeffer (1971) on *Endospermum*, Schultes (1940s through 1990) on *Hevea*, Webster (1964, 1979, 1984) respectively on *Margaritaria, Meineckia* and *Flueggea*, and Webster & Armbruster (1991) on neotropical *Dalechampia*. Dehgan and Webster (1979) and Webster (1993) have presented infrageneric accounts respectively for *Jatropha* and *Croton*, McVaugh (1944) for *Cnidoscolus*, and Webster (1956-57) a full revision of West Indian *Phyllanthus* with a partial review of its difficult infrageneric classification. An informal but useful overview of *Euphorbia* appears in the *Royal Horticultural Society Dictionary of Gardening* (1992); with two large groups of enthusiasts the genus continues to draw many contributions and some partial revisions. Accounts for *Antidesma* by Petra Hoffmann, *Aporusa* by Anne Schot and *Baccaurea* by Raoul Haegens (all prepared under the aëgis of the *Flora Malesiana* Euphorbiaceae programme) are expected. Major studies by Esser (1994) on *Mabea* and other Hippomaneae, Brunel (1987) on African *Phyllanthus* and allies, and Gillespie (1988) on *Omphalea* circulate as formally unpublished dissertations. A number of floras fully cover some genera and in addition provide major accounts for others, such as *Phyllanthus* in New Caledonia where Schmid (1987) recognised 111 native species. The many larger and smaller studies by H. K. Airy-Shaw (with T. C. Whitmore for *Macaranga* and G. Webster for *Phyllanthus*) cover a very substantial proportion of the Malesian species, while Jean Léonard and Alan Radcliffe-Smith between them have over more than four decades contributed substantially to our knowledge of non-euphorbiid African genera. Peter van Welzen and his collaborators are well into revisions for accounts for *Flora Malesiana* and *Flora of Thailand*, N. P. Balakrishnan, T. Chakraborty and others have gradually worked towards a family treatment in the new *Flora of India*, while Paul Forster and others are active in work towards *Flora of Australia*. There is considerable activity in Madagascar and the Americas, where our knowledge is less well consolidated. Yet on the whole our current documentation is far from effectively superseding the accounts in *Das Pflanzenreich* (and *Die natürlichen Pflanzenfamilien*).

[6] These systems, all cited in the main text under **General**, were reviewed in Webster (1987).

been most pronounced in the northern hemisphere; examples include *Mercurialis* and *Euphorbia* subgen. *Chamaesyce* and *Esula* and within widespread species such as the Mediterranean *E. characias*. Elsewhere, our knowledge of species is still largely at the alpha-taxonomic level. *Euphorbia* (along with satellites such as *Monadenium* and *Pedilanthus*), parts of *Phyllanthus*, and economically significant genera such as *Hevea, Manihot,* and *Vernicia* and its relatives have enjoyed the most attention, but in general there have been few revisions of medium-sized and larger genera since the accounts of Pax, Hoffmann, Grüning and Jablonski in *Das Pflanzenreich* (1910-24).[5] Considerable reliance has therefore had to be placed on subsequent flora treatments and regional surveys in arriving at current practice; reference to these is made under appropriate headings.

At higher taxonomic levels, we have for the genera of Euphorbiaceae by and large adopted the system of Webster (1994), the best and most recent available pending publication of the third author's *Genera Euphorbiacearum*. Those of Pax and Hoffmann in 1931 and (less completely) Hurusawa in 1954 were still much indebted to that of Müller in 1866, notably in the arrangement of tribes, while that of Hutchinson in 1969 has never found favour.[6] Our infrafamilial headings in the family-bibliographical section of the main text (pp. 13-42) correspond to the five subfamilies of Webster except that the tribe Euphorbieae is for convenience given its own heading. As already noted, Pandaceae are covered separately following Euphorbiaceae. This follows Kew practice and is thus contrary to Webster (1994) where the family is not distinguished.

A few necessary new names and combinations have been made in the text. A goodly number of other taxa presently are in need of transfer but require restudy. These have for convenience been grouped with other unplaced names at the end of the present work.

Geographical Distribution

Distributions of **species** and taxa of lower rank are furnished in two ways: firstly by a generalised statement in narrative form, and secondly as TDWG geographical codes (Hollis & Brummitt, 1992) expressed to that system's third level. Examples of the former include:

E. & C. U.S.A.
Texas to C. America
Mexico (Veracruz)
Europe to Iran
E. Himalaya, Tibet, China (W. Yunnan)
Philippines (Luzon)
Irian Jaya (Watjetoni Mts.)
Cult.

When the presence of a taxon in a given region or location is not certainly known, a question mark is used, e.g. New Ireland ?; when an exact location within a country or is not known, a question mark within brackets is used, e.g. Mexico (?). Distributions of **genera** are furnished in a relatively simplified form, any special features being given within brackets.

With respect to the TDWG codes, the **region** is indicated by the two-digit number (representative of the first two levels), the first digit also indicating the continent. The letter codes following the digits, when given, represent the third-level **unit** (a country, state or other comparable area). They usually are the first three letters of a given unit's name, but sometimes are contractions. 'ALL' is used if the species is known to occur in every area of a given region. If the country code is not known, '+' is used. If the taxon is extinct in a given region, '–' is used after the country code. Naturalisation is expressed by putting the third-level codes in lower case and, if in a second-level region all occurrences are the result of naturalisation, the code number for the region is placed in brackets. The application of question marks is as indicated above for geographical regions. Examples include:

12 SPA–	[SW. Europe: Spain, where extinct]
32 +	[C. Asia (more exact distribution not known)]
36 CHN 38 JAP KOR	[China and Eastern Asia; China, Japan, Korea]
51 NZN NZS	[New Zealand: North and South Islands]
76 ARI 77 NWM TEX 79 ALL 80 GUA HON?	[SW. & SC. U.S.A., Mexico & C. America: Arizona, New Mexico, Texas, Mexico, Guatemala and perhaps Honduras]
77 TEX	[SC. U.S.A.: Texas]
(10) (11) 42 50 51 60 85	[of a genus: New Guinea, Australia, SW. Pacific Islands, New Zealand, and S. America; naturalized in NW. and C. Europe]

The units are based on polities or islands or island groups.

Life-forms

The terminology for *life-forms,* definitions of which follow, is based on the system of Raunkiær (1934, especially chapters 1 and 2) with modifications derived from *Nouvelle flore de la Belgique, du Grand-Duché de Luxembourg, du Nord de la France et des régions voisines* (Lambinon *et al.*, 1992, pp. xx-xxi). Examples are taken as far as possible from Euphorbiaceae and Pandaceae.

Main categories

Phanerophyte (phan.)
Stems: woody and indefinitely persistent
Buds: normally 3 m or more above ground
e.g.: very large shrubs; small, medium and large trees: *Bischofia javanica, Euphorbia haeleeleana, Hevea brasiliensis, Hura crepitans, Panda oleosa, Ricinus communis*

Nanophanerophyte (nanophan.)
Stems: woody and indefinitely persistent
Buds: above soil level but normally less than 3 m above ground
e.g. shrubs such as *Acalypha concinna, Andrachne chinensis, Cnidoscolus urens, Euphorbia pulcherrima, Koilodepas wallichianum*

Herbaceous phanerophyte (herb. phan.)
Stems: herbaceous and persisting for several years
Buds: above soil level
e.g. bananas and plantains such as *Musa basjoo*; none known in Euphorbiaceae or Pandaceae

Chamaephyte (cham.)
Stems: herbaceous and/or woody and persistent
Buds: on or just above soil level but never above 0.50 m from ground
e.g. Amperea xiphoclada, Andrachne fedtschenkoi, Clutia alaternoides, Euphorbia handeniensis, Jatropha horizontalis, Manihot stipularis, Oreoporanthera alpina

Hemicryptophyte (hemicr.)
Stems: herbaceous, often dying back after the growing season but with buds or growth at soil level surviving
Buds: just on or below soil level
e.g. Acalypha fredericii, Cnidoscolus texanus, Jatropha scaposa, Microstachys ditassoides, Tragia urens

Geophyte [not abbreviated]
Hemicryptophytes which survive unfavouable seasons in the form of a rhizome, bulb, tuber or root bud.
Buds: below soil level
e.g. Euphorbia arrecta

Therophyte (ther.)
Plants surviving unfavourable seasons in the form of seeds, completing their life-cycle during the favourable season.
e.g. annuals such as *Euphorbia hirta, E. ocellata* and *E. thymifolia*

Aquatic plants

Helophyte (hel.)
Hemicryptophytes growing in soil saturated with water or in water and with leaf- and flower-bearing shoots held above water
Buds: on or below soil level
e.g. Echinodorus, Oryza, Spartina, Typha; Haematostemon coriaceus

Hydrophyte [inclusive of the three following subcategories]
Plants with stems and vegetative shoots entirely in water, the leaves usually submerged and/or floating; flower-bearing parts may emerge above the water
Buds: permanently or temporarily on the bottom of the water

Hydrohemicryptophyte (hydrohemicr.): aquatic hemicryptophytes
e.g. Stratiotes

Hydrogeophyte: aquatic geophytes [not abbreviated]
e.g. Nuphar, Nymphaea, Nymphoides

Hydrotherophyte (hydrother.): aquatic therophytes
e.g. Lemna; Utricularia vulgaris

Others

Bamboo
Biennial

Supplementary information

Climbing plants (cl.), including:
Climbing phanerophytes (cl. phan.); *e.g. Omphalea megacarpa*
Climbing chamaephytes (cl. cham.); *e.g. Dalechampia pentaphylla, Tragia balfourii*
Climbing nanophanerophytes (cl. nanophan.); *e.g. Dalechampia canescens, Manihot peruviana, Plukenetia volubilis, Tragia bailloniana*
Climbing hemicryptophytes (cl. hemicr.); *e.g. Tragia adenanthera, T. glabrescens*
Climbing tuberous geophytes (cl. tuber geophyte); *e.g. Tropaeolum tuberosum*
Climbing therophytes (cl. ther.); *e.g. Tragia muelleriana*
Sometimes climbing nanophanerophytes ((cl.) nanophan.), *e.g. Omphalea queenslandiae*
Sometimes climbing chamaephytes ((cl.) cham.), *e.g. Tragia dodecandra, T. pinnata*
Sometimes climbing hemicryptophytes ((cl.) hemicr.), *e.g. Tragia betonicifolia*

Succulent plants (succ.), including:

Succulent phanerophytes (succ. phan.); *e g. Euphorbia candelabrum, E. tanaensis*

Succulent nanophanerophytes (succ. nanophan.); *e.g.*: *Euphorbia vulcanorum, Jatropha glandulifera*

Succulent hemicryptophytes (succ. hemicr.); *e.g. Euphorbia eyassiana*

Succulent chamaephytes (succ. cham.); *e.g. Euphorbia baioensis, E. obesa, Jatropha spicata, Monademium ritchieii*

Succulent therophytes (succ. ther.); *e.g. Sedum azureum*

Climbing succulent nanophanerophytes (cl. succ. nanophan.); *e.g. Euphorbia fluminis*

Tuber plants (tuber), including:

Tuber nanophanerophytes (tuber nanophan.); *e g.Euphorbia elliotii, Manihot pringlei*

Tuber chamaephytes (tuber cham.); *e.g.*: *Jatropha prunifolia,Monadenium laeve*

Tuber geophyte (tuber geophyte); *e g. Cnidoscolus maculatus, Euphorbia davyi, Jatropha macrorhiza, Monadenium pedunculatum*

Succulent tuber chamaephytes (succ. tuber cham.); *e.g. Euphorbia suzannae-marnieriae*

Parasitic plants (par.), including

Hemiparasitic plants (hemipar.), i.e. parasitic plants with a continued ability to photosynthesize; *e.g. Cassytha filiformis* (hemipar. cl. nanophan.), *Viscum orientale* (hemipar. nanophan.); rare or absent in Euphorbiaceae and Pandaceae.

Holoparasitic plants (holopar.), i.e. parasitic plants fully dependent upon their host; *e.g. Orobanche ramosa* (holopar. ther.), *Rafflesia tuan-mudae* (holopar. geophyte); rare or absent in Euphorbiaceae and Pandaceae.

Note: **Holomycotrophs** are organisms dependent wholly on dead as opposed to living organic matter, but the distinction is not a sharp one. Vascular holomycotrophs are usually herbaceous and feature roots associated with ecto- or endotrophic root-fungi (mycorrhizae). The fungi draw nutrients from the organic matter. Examples include *Monotropa uniflora* and *Neottia nidus-avis*. None is known in Euphorbiaceae and Pandaceae.

Report Generation

Consolidation of output from the two databases and generation of raw reports with formatting instructions is accomplished with the use of Microsoft Access® version 2.0. Using a 'generic/text only' driver, the reports are then printed to a disk file. Final processing is carried out in Microsoft Word® version 6.0 with the aid of specially written macros. Both packages operate under Microsoft Windows® NT 4.0. Selected illustrations are incorporated into the print version of this publication.

Illustrations

Almost all have been taken from the Kew collections. Some have not before been published.

Electronic Generation

Current plans include distribution of the database to the International Organization for Plant Information for dissemination on their Global Plant Checklist Web page and, in due course, to Species 2000. Other options are currently under study.

Acknowledgments

We wish to acknowledge with thanks the help of our internal and external advisers as well as other Kew staff, particularly Sally Hinchcliffe and (through 1997) Milan Svanderlik (Information Services Department). We also would like to thank our external reviewer, Grady Webster, for his contributions. The second author would like moreover to thank the Library for obtaining materials not held at Kew.

References

Angiosperm Phylogeny Group (1998). An ordinal classification for the families of flowering plants. *Ann. Missouri Bot. Gard.* **85**: 531-553.

Bisby, F. A. (1994). Global master species databases and biodiversity. *Biology International* **29**: 33-40.

Bisby, F. A. & Smith, P. (1996). *Species 2000 Project Plan (version 3.1).* [2], 42 pp. Southampton: Species 2000 Secretariat, University of Southampton. [www.sp2000.org]

Bridson, G., comp. & ed. (1991). *Botanico-Periodicum Huntianum/Supplementum.* Pittsburgh: Hunt Institute for Botanical Documentation.

Burnett, J. (1994). IOPI and the Global Plant Checklist project. *Biology International* **29**: 40-44.

Brummitt, R. K. (1992). *Vascular Plant Families and Genera.* 804 pp. Kew: Royal Botanic Gardens.

Brummitt, R. K. & Powell, C. E. (1992). *Authors of Plant Names.* 732 pp. Kew: Royal Botanic Gardens.

Engler, A. *et al.*, eds (1900-). *Das Pflanzenreich. Regni vegetabilis conspectus.* Berlin: Engelmann (for Preussisches Akademie der Wissenschaften) (nos. 1-106, 1900-43); Akademie-Verlag (nos. 106 (reissue), 107-108, 1953-68).

Farr, E. R., Leussink, J. A. & Stafleu, F. A., eds. (1979). *Index Nominum Genericorum (Plantarum).* 3 vols. Utrecht: Bohn, Scheltema & Holkema. (Regnum Vegetabile 100-102).

Farr, E. R., Leussink, J. A. & Zijlstra, G., eds. (1986). *Index Nominum Genericorum (Plantarum): Supplementum I.* xv, 126 pp. The Hague: Bohn, Scheltema & Holkema. (Regnum Vegetabile 113.)

Frodin, D.G. & Govaerts, R. (1996). *World Checklist and Bibliography of Magnoliaceae.* vii, 72 pp. Kew: Royal Botanic Gardens.

Govaerts, R. (1995-). *World Checklist of Seed Plants.* Antwerp: MIM (letter 'A'); Continental Publishing (letter 'B', 1996, 'C', 1999). [In continuation.]

Greuter, W. *et al.* (1993). *Names in Current Use for Extant Plant Genera* (Names in current use, 3). xxvii, 1464 pp. Koenigstein: Koeltz. (Regnum Vegetabile 129.)

Hollis, S. & Brummitt, R. K. (1992). *World Geographical Scheme for Recording Plant Distributions.* ix, 105 pp. Pittsburgh, Penna: Hunt Institute for Botanical Documentation, Carnegie-Mellon University. (for the International Working Group on Taxonomic Databases for Plant Sciences). (Plant Taxonomic Database Standards, 2: version 1.0.)

Lambinon, J. *et al.* (1992). *Nouvelle Flore de la Belgique, du Grand-Duché de Luxembourg, du Nord de la France et des Régions Voisines.* 4th edn. cxx, 1092 pp., illus., map. Meise: Éditions du Patrimoine, Jardin Botanique National de Belgique.

Linnaeus, C. (1794). *Genera plantarum.* 6th edn. xx, 580 pp., index. Stockholm: Salvius.

Linnaeus, C. (1792). *Praelectiones in ordines naturales plantarum proprio et J.C. Fab*ricii (ed. P.D. Giseke). 1, 662 pp., illus. Hamburg: B.G. Hoffmann.

Orchard, A. E., ed. (1999). *Species plantarum/Flora of the world: introduction to the series.* iv, 90 pp. Canberra: ABRS, Environment Australia.

Raunkiær, C. (1934). *The life forms of plants and statistical plant geography.* xvi, 632 pp., illus. London: Oxford University Press, London.

Reichhardt, A. (1999). It's sink or swim as a tidal wave of data approaches (with sidebar: 'The catalogue of life could become reality'). *Nature* **399** (10 June 1999): 517-520.

Royen, A. van (1740). *Florae leydensis prodromus.* Leiden: Luchtmans.

Stafleu, F. & Cowan, R. S. (1976-88). *Taxonomic Literature: A Selective Guide to Botanical Publications and Collections with Dates, Commentaries and Types*. 2nd edn. 7 vols. Utrecht: Bohn, Scheltema & Holkema. (Regnum Vegetabile 94, 98, 105, 110, 112, 115, 116.) Continued as Stafleu, F. *et al.* (1992-). *Taxonomic Literature, Supplement.* Vols. 1- . Koenigstein: Koeltz. (Regnum Vegetabile 125- , *passim*. As of 1999 five volumes published.)

Tokuoka, T. & Tobe, H. (1999). Embryology of tribe Drypeteae, an enigmatic taxon of Euphorbiaceae. *Pl. Syst. Evol.* **215**: 189-208.

Webster, G. (1987). The saga of the spurges: a review of classification and relationships in the Euphorbiales. In S. L. Jury *et al.* (eds), *The Euphorbiales – chemistry, taxonomy and economic botany*: 3-46. London: Linnean Society of London. (Also in *Bot. J. Linn. Soc.* **94**: 3-46).

Webster, G. (1994). Synopsis of the genera and suprageneric taxa of Euphorbiaceae. *Ann. Missouri Bot. Gard.* **81**: 33-144.

Abbreviations

al.	alii: others
archip.	archipelago
auct.	auctoris: of author
C.	central
Ch.	Chinese
cham.	chamaephyte [life-form]
cit.	citations; cited
cl.	climber
cl. phan.	climbing phanerophyte [life-form]
cl. hemicr.	climbing hemicryptophyte [life-form]
cl. nanophan.	climbing nanophanerophyte [life-form]
cl. ther.	climbing therophyte [life-form]
cl. tuber geophyte	climbing tuberous geophyte [life-form]
Co.	comitas: county or (China) *hsien*
comb.	combinatio: combination
cons.	conservandus: conserved
cult.	cultus: cultivated
cv.	cultivarietas: cultivar
descr.	descriptio: description
Distr.	District
Du.	Dutch
E.	east(ern)
En.	English
etc.	et cetera: and the rest
e.g.	exempli gratia: for example
Fr.	French
Ge	German
hel.	helophyte [life-form]
hemicr.	hemicryptophyte
hemipar.	hemiparasitic
herb.	herbaceous
herb. phan.	herbaceous phanerophyte
holopar.	holoparasitic
hort.	hortorum: of gardens *or* hortulanorum: of horticulturalists
hydrohemicr.	hydrohemicryptophyte
hydrother.	hydrotherophyte
I./Is.	island/islands
ICBN	International Code of Botanical Nomenclature
i.e.	id est: that is
ign.	ignotus: unknown
in litt.	in litteris: in correspondence
ined.	ineditus: unpublished, provisional name
inq.	inquilinus: naturalised
i.q.	idem quod: the same as
La.	Latin
Medit.	Mediterranean
Mt./Mts.	mountain/mountains
N.	north(ern)
nanophan.	nanophanerophyte [life-form]
nom. cons.	nomen conservandum: conserved name [ICBN]
nom. illeg.	nomen illegitimum: illegitimate name [ICBN]
nom. inval.	nomen invalidum: invalidly published name [ICBN]
nom. nud.	nomen nudum: name without a description [ICBN]

nom. rejic.	nomen rejiciendum: rejected name [ICBN]
nom. superfl.	nomen superfluum: name superfluous when published [ICBN]
nov.	novus: new
par.	parasitic
Pen.	peninsula(r)
phan.	phanerophyte (life-form)
p.p.	pro parte: partly
prov.	province
q.e.	quod est: which is
q.v.	quod vide: which see
reg.	regio: region
Rep.	republic
Ru.	Russian
S.	south(ern)
seq.	sequens: following
s.l.	sensu lato: in the broad sense
Sp.	Spanish (in literature references)
sp.	species
sphalm.	sphalmate: by mistake
s.s.	sensu stricto
st.	status
subtrop.	subtropical
syn.	synonymon: synonym
temp.	temperate
ther.	therophyte [life-form]
trop.	tropical
viz.	videlicet: namely
W.	west(ern)
?	not known, doubtful (all contexts)
*	basionym or replaced synonym (before a name)
×	nothotaxon (before a genus or species name)
+	range more than as indicated but not certainly known
–	extinct (after TDWG unit)
•	bibliographic references of particular significance

Euphorbiaceae

322 genera, 8910 species. Nearly worldwide, although predominately in warmer parts. The family has long attracted scientific and popular attention on account of its sheer diversity of forms; moreover, some of its members are of great economic importance. It has been circumscribed more or less within its present compass since Adrien de Jussieu's *De Euphorbiacearum generibus medisque earundum viribus tentamen* (1824), with the most acceptable modern major subdivisions being the subfamilies of Webster in a system first proposed in 1975. A great proportion of our present understanding of the family rests on the work of Johannes Mueller (Mueller Argoviensis) as published in the de Candolle *Prodromus* (1866); this followed the treatment of Euphorbieae alone for the same work by Boissier (1862). Important general surveys covering geography, relationships and phylogeny were published by Bentham (1878) and Pax (1924). The perception of certain internal discontinuities along with the wide range of forms have, however, from time to time given rise to calls for segregation of one or more 'satellite' families or even dismemberment (for discussion see Radcliffe-Smith (1987) and Meeuse (1990; for both, see **General**)). In addition, there have been disagreements with respect to the infrafamilial arrangement along with new insights derived from additional lines of evidence. Controversy has also existed concerning the place of the family in the system of angiosperms; several workers have in the past admitted multiple affinities or even that it was polyphyletic. Evidence from palynology, seed morphology and anatomy, serology and molecular biology has been seen as suggesting that the Euphorbiaceae s.l. are indeed heterogeneous, with not only a major division represented in the distinction between biovulate (Phyllanthoideae, Oldfieldioideae) and uniovulate (Acalyphoideae, Crotonoideae and Euphorbioideae) members – each with distinct affinities within the angiosperms – but also a greater range of divergence among the former, with separate family status or outright exclusion likely for some genera or groups of genera. Molecular systematic research at Kew and elsewhere is indeed pointing in this direction, with the most serious consequence being removal of Phyllanthoideae and Oldfieldioideae – and especially the former's tribe Drypeteae – to a distinct family or families. The uniovulate representatives form a more coherent group and for this reason calls have been heard for their amalgamation into a single subfamily; conversely, there is good reason to separate Euphorbieae from other Euphorbioideae. By contrast, however, there remains a strong school of advocates of a diverse but essentially coherent Euphorbiaceae, albeit with one major internal subdivision; what is actually at stake is our understanding of the system of angiosperms and the possible or even likelihood of the unnaturalness of the Cronquistian subclasses Dilleniidae and Rosidae. It is currently thought premature, however, to proceed to serious development of a new system of 'euphorbioid' families; most authorities call for retention of the family in its traditional sense, sometimes with exclusion of Pandaceae. There is also extensive support for the Webster system of subdivision into subfamilies and tribes, which has taken account of much modern evidence and, now that a fully worked-out version is available, is seen as more satisfactory than the long-standard arrangement of Pax and Hoffmann. The genera adopted here will as far as possible conform to those adopted for the forthcoming *Genera Euphorbiacearum* by Alan Radcliffe-Smith. At genus and species level, coverage of the family since the great (though incomplete) series of revisions by Ferdinand Pax and his associates in *Das Pflanzenreich* (1910-24) is improving but still patchy; there are many notable modern accounts in regional floras as well as several good regional and generic revisions. Nevertheless, there are many areas where an effective delineation of genera remains problematic; in addition, the real worth of many characters remains to be tested. Indeed, for some – notably the truly protean *Euphorbia* – we may never reach a consensus; human understanding may simply resist any attempts at 'rational' solutions following one or another body of formal logic.

General

Included here are the major systems (Jussieu 1823, 1824; Baillon 1858, 1874; Boissier & Mueller 1862-66; Bentham 1878, Pax 1890, Pax & Hoffmann 1931, Hurusawa 1954, Hutchinson 1969, Webster 1975, 1994) along with major essays by Croizat (1973), Meeuse (1990) and Webster (1994) and a historical survey by Webster (1987).

- Jussieu, A. de (1823). Considérations sur la famille des Euphorbiacées. Mem. Mus. Natl. Hist. Nat. 10: 317-355. (Also publ. separately, 39 pp., 1824, Paris. Summary in Ann. Sci. Nat., 1: 136-146. 1824.) Fr. — Review of characters; summary of author's views on the circumscription of the family and of its 'sections' (pp. 348-355; pp. 352-354 on *Euphorbia*); no formal synopsis. [Six sections (i.e. tribes) recognised.]
- Jussieu, A. de (1824). De Euphorbiacearum generibus medisque earundum viribus tentamen. 118 pp., 18 pls. Paris. (Summary in Ann. Sci. Nat. 1: 146-167. 1824.) Fr. — First overall revision of family in the era of 'natural' classification; a generic monograph with descriptions and commentary (the genera arranged into six unnamed 'sections'); conspectus on pp. 63-71. Following the formal treatment is a treatise on medicinal uses and properties of various genera, and an index to all generic and vernacular names in both parts. [At the time already 300 species of *Euphorbia* were known. Also of interest is the inclusion of *Dalechampia* with section 6 which otherwise is limited to the modern Euphorbieae; Hurusawa (1954) retained a similar disposition in his survey of the family. His sections 1-2 correspond to Phyllanthoideae and Oldfieldioideae; sections 3-5 approximately to Crotonoideae, Acalyphoideae and Sapioideae (Euphorbioideae without Euphorbieae).]
- Baillon, H. (1858). Étude générale du groupe des Euphorbiacées. 684 pp.; atlas of 52 pp, 27 pls. Paris: Masson. Fr. — A 'genera Euphorbiacearum' with, firstly, a conspectus of major groups and genera (pp. 3-44); this is followed by a systematic organography, remarks on distribution and affinities of the family, a survey of its classification (pp. 255-280), and then a systematic treatment of 235 genera with 'synoptic' keys, descriptions, commentary, lists of species with specimens seen (not, however, in *Euphorbia* where only the names of nine sections are given), and references to floras and other works; excluded genera, bibliography, table of contents, and complete index at end prior to the atlas. [Presents an alternative view of the family to that of Johannes Mueller (Argoviensis), but one without a hierarchy; 14 'series' in 2 major subdivisions (Uniovulées and Biovulées) recognized with reference to 'archetypes' in the opening conspectus. A revised and abridged treatment by the author may be found in his *Histoire des plantes* 5: 105-256 (1874). The lack of a well-developed hierarchy rendered this system relatively uninformative with respect to relationships]

  Hasskarl, J. K. (1859). Clavis analytica generum Euphorbiacearum secundum Baillon. Flora 42: 721-746. La. — Analytical keys to the genera in Baillon (1858), the leads not always dichotomous.

  Klotzsch, J. F. (1859). Linné's natürliche Pflanzenklasse Tricoccae des Berliner Herbarium's im Allgemeinen und die natürliche Ordnung Euphorbiaceae insbesondere. Monatsber. Königl. Preuss. Akad. Wiss. Berlin (1859): 236-254, 2 pls. (Separately reprinted with unchanged pagination.) Ge. — Precursor (and nomenclaturally prior) to fuller treatment (1860, q.v.); includes a historical survey of the Tricoccae along with discussion of the merits or otherwise of different proposed systems. [For fuller notes see **Euphorbieae**.]

  Klotzsch, J. F. (1860). Linné's natürliche Pflanzenklasse Tricoccae des Berliner Herbarium's im Allgemeinen und die natürliche Ordnung Euphorbiaceae insbesondere. Abhandl. Königl. Preuss. Akad. Wiss. Berlin (1859), Phys. Abhandl.: 1-108, 2 pls. (Also separately published, Berlin; ii, 108 pp., 2 pls.) Ge. — Includes a history of the 'Tricoccae' since 1751 along with arguments for its segregation into 6 'orders' (foreshadowing the modern subfamilies) of which Euphorbiaceae s.s. (as treated here, corresponding to Euphorbieae in modern systems) is one. [For fuller notes see **Euphorbieae**.]

- Boissier, E. and J. Mueller (Argoviensis) (1862-66). Euphorbiaceae. In A. de Candolle (ed.), Prodromus systematis naturalis regni vegetabilis, 15(2): 1-1286. Paris: Masson. La. — Complete concise treatment including synonymy, descriptions, distribution, selected exsiccatae and commentary along with synoptic separators but no analytical keys. [Last nominally complete treatment of family. Euphorbieae by Edmond Boissier (pp. 3-188, 1261-1269); remainder by Johannes Mueller (pp. 189-1261, 1269-1286).]

  Mueller (Argoviensis), J. (1863-65). Vorläufige Mittheilungen aus dem für De Candolle's Prodromus bestimmten Manuscript über diese Familie. Linnaea 32: 1-126; 34: 1-224. Ge. — Partial keys as well as descriptions and notes; some infrageneric synopsis presented (e.g. in *Acalypha* and *Phyllanthus*).

  Mueller (Argoviensis), J. (1866). Euphorbiaceae [except Euphorbieae]. See Boissier & Mueller (1862-66).

  Baillon, H. E. (1873). Nouvelles observations sur les Euphorbiacées. Adansonia 11: 72-138. Fr. — Supplementary to the Étude Générale of 1858; also a precursory to treatment of family in vol. 5 of his *Histoire des plantes* (1874). Includes further contributions on New Caledonian species as well as a discussion of the author's philosophy of classification in the family and a defence of his broad circumscriptions of, among others, *Antidesma*, *Phyllanthus* and *Excoecaria*.

- Baillon, H. E. (1874). Euphorbiacées. In *idem*, Histoire des plantes 5: 105-256, illus. Paris. Fr. — Generic encyclopaedia, with descriptions and commentary. [Reflects the author's system of 1858 with revisions in the same way that Pax's *Pflanzenfamilien* treatment of 1890 reflects Mueller's.]
- Bentham, G. (1878). Notes on Euphorbiaceae. J. Linn. Soc. Bot. 17: 185-267. En. — Review of family including history, nomenclature, classification (with examination of individual tribes), origin and geographical distribution (with the belief that 'the most ancient home of the Order was in the Old World, whence it spread in very remote times to America'; within the former there were two centres, eastern tropical Africa and 'towards the eastern extremity'. [A companion to the author's formal treatment of the family in vol. 3 of *Genera plantarum* (1880).]
- Pax, F. (1890). Euphorbiaceae. In A. Engler & K. Prantl (eds), Die natürlichen Pflanzenfamilien, III, 3: 1-119, 456-458, illus. Leipzig. La/Ge. — Generic encyclopaedia, with keys, concise descriptions, and distribution; includes general introduction to family. [Not considered to be that markedly different from Mueller's system, but regarded as better than that of 1931.]
- Pax, F. and K. Hoffmann (Bridelieae by E. Jablonski (Jablonszky), Stenolobieae by G. Gruening) (1910-24). Euphorbiaceae. In A. Engler (ed.), Das Pflanzenreich, IV 147. Illus. Berlin: Engelmann (im Auftrage der Preuss. Akademie der Wissenschaften). (Heft 42, 44, 47, 52, 57, 63, 65, 68, 81, 85.) La/Ge. — Concise monographic treatments with keys, descriptions, synonymy, references, exsiccatae, indication of distribution, and commentary. [Incomplete. Euphorbieae as well as *Croton*, *Phyllanthus* and certain other smaller genera were never published.]

  Pax, F. and K. Hoffmann (1919). Euphorbiaceae-Additamentum VI. In A. Engler (ed.), Das Pflanzenreich, IV 147 XIV: 1-81. Berlin: Engelmann (im Auftrage der Preuss. Akademie der Wissenschaften). (Heft 68, pars.) La/Ge. — Additions and corrections to previous contributions; includes additional new genera. Pp. 64-76 comprise a list of all collections seen for the tribes and subtribes treated in Heft 68 as well as the Additamenta.

  Pax, F. (1924). Die Phylogenie der Euphorbiaceae. Bot. Jahrb. Syst. 59: 129-182. Ge. — Phytogeography and phylogeny, with also reference to individual tribes and subtribes. The major spread of the family was achieved through intercontinental and archipelagic connections, with 'continental drift' not rejected. 52 references cited (without, however, mention of Bentham, although relatively similar conclusions were reached).

  Pax, F. and K. Hoffmann (1924). Euphorbiaceae-Additamentum VII. In A. Engler (ed.), Das Pflanzenreich, IV 147 XVII: 179-204. Berlin: Engelmann (im Auftrage der Preuss. Akademie der Wissenschaften). (Heft 85, pars.) La/Ge. — Additions and corrections to previous contributions; includes additional new genera. Pp. 205-231 comprise an index of collections seen as well as an index to all botanical names for both parts of Heft 85.

- Pax, F. and K. Hoffmann (1931). Euphorbiaceae. In A. Engler (ed.), Die natürlichen Pflanzenfamilien, 2. Aufl., 19c: 11-233, illus. Leipzig. Ge. — Generic encyclopaedia with keys, descriptions, indication of distribution, special features, and literature references. [A standard reference, used as a framework for many taxonomic and subdisciplinary studies. Now out of date but no equivalent has appeared to succeed it.]
- Hurusawa, I. (1954). Eine nochmalige Durchsicht des herkömmlichen Systems der Euphorbiaceen in weiteren Sinne. J. Fac. Sci. Univ. Tokyo, III, 6: 209-342, illus., 4 pls. Ge. — This study has two aims: a revision of Euphorbiaceae and its allies (including Antidesmataceae, corresponding to Pax's Phyllanthoideae) for Japan and former Japanese territories in east Asia and a higher-level systematic study of this array of families. In his restricted sense of Euphorbiaceae (corresponding to Pax's Crotonoideae) the author recognises four subfamilies: Euphorbioideae (limited to Euphorbieae and *Dalechampia*), Acalyphoideae, Crotonoideae (limited to Crotoneae) and Sapioideae (proposed on p. 310 and including much of what Webster assigns to Crotonoideae). Keys to suprageneric taxa (in Latin) are given but only irregularly to lower taxa (save where significant taxonomic changes are being made). Entries include, for species, types, distribution and synonymy and for supraspecific taxa, type species. In addition to Euphorbiaceae s.s. and Antidesmataceae, the author included in his order Euphorbiales the 'stenoloboid' genera as families Porantheraceae and Ricinocarpaceae. Daphniphyllaceae was definitively separated into its own order. In Euphorbiales several new tribes were proposed outright or by re-ranking; some have been taken up by later workers. [With respect to Euphorbieae *Euphorbia* s.l. is segregated, with generic distinction given to *Chamaesyce*, *Esula* (as *Galarhoeus*), *Agaloma* and others (see key, p. 230). The author's justification for this is based on growth-forms (see pp. 228-230 and particularly the figure on p. 229).]

Airy-Shaw, H. K. (1965). Diagnoses of new families, new names, etc., for the seventh edition of Willis's 'Dictionary'. Kew Bull. 18: 249-273. En. — Includes several taxonomic and nomenclatural revisions relating to Euphorbiaceae s.l.; several segregate families proposed, with discussion (in Euphorbiaceae the Androstachydaceae, Bischofiaceae, Hymenocardiaceae, and Uapacaceae).

Hutchinson, J. (1969). Tribalism in the family Euphorbiaceae. Amer. J. Bot. 56: 738-758, illus. En. — A proposed classification of the family, with descriptions of tribes, numerous representative illustrations, and enumeration of included genera. [Precursory to a projected account (never published) in the author's *Genera of flowering plants*. Strong reliance was placed on habit ('Gestalt') and gross morphology without correlation from studies in anatomy, cytology or palynology; Webster (1987; see below) believes the author's recognised intuition was here not very successful.]

Croizat, L. (1973). Les Euphorbiacées vues en elles-mêmes, et dans leurs rapports envers l'angiospermie en général. 206 pp. (Mem. Soc. Brot. 23). Coimbra. Fr. — An extended essay on morphology, the reproductive unit (especially in *Euphorbia*), phylogeny, and panbiogeography. 'Les genres contemporains sont l'issue en 3ème ou 4ème génération de formes anciennes ayant évolué chacun à soi en différents domaines géographiques' (p. 63). [The last major contribution on the family by the author; apart from ideas on the reproductive unit, philosophically I (D. Frodin) think this to be generally on the mark.]

Webster, G. L. (1975). Conspectus of a new classification of the Euphorbiaceae. Taxon 24: 593-601. En. — A new system of the family proposed; key to 5 subfamilies and synopsis of 52 tribes (and many additional subtribes) with included genera. The 'Stenolobieae' of Mueller are dismembered. Pollen morphology was seen as especially useful in suprageneric classification. [For historical discussion, see Webster (1987).]

Jury, S. L., T. Reynolds, D. F. Cutler and F. J. Evans (eds) (1987). The Euphorbiales: chemistry, taxonomy and economic botany. [iv], 326 pp., illus. London: Academic Press. (Repr. from Bot. J. Linn. Soc. 94(1/2).) En. — The first 2 articles are separately accounted for here.

Radcliffe-Smith, A. (1987). Segregate families from the Euphorbiaceae. In S. L. Jury et al. (eds), The Euphorbiales: 47-66. London. En. — Discussion of 11 older and more recent segregates as well as the tribe Paivaeusinae of Pax and Hoffmann. [All Paivaeusinae now in Oldfieldioideae. Of the segregate families, Buxaceae, Aextoxicaceae, Didymelaceae, Daphniphyllaceae and Pandaceae are definitely excluded, but the last-named is retained within Euphorbiales.]

• Webster, G. L. (1987). The saga of the spurges: a review of classification and relationships in the Euphorbiales. In S. L. Jury et al. (eds), The Euphorbiales: 3-46. London. En. — Historical survey and review, including recent systems of classification; recognition of certain segregate families not advocated. Table 1 gives summary characters of the subfamilies of Euphorbiaceae and the tribes of Pandaceae.

Meeuse, A. D. J. (1990). The Euphorbiaceae auct. plur.: an unnatural taxon. 38 pp. Delft: Eburon. En. — Presents arguments for a Euphorbiaceae *exclusive* of Webster's subfamilies Oldfieldioideae (here called Paivaeusaceae) and Phyllanthoideae (here Phyllanthaceae and Putranjivaceae). *Androstachys, Antidesma, Bischofia, Hymenocardia, Pera,* and *Uapaca* are moreover also given or restored to family rank along with the tribe Galearieae (= Pandaceae). [Some support for these views is found in evidence from more recent molecular systematic studies, notably family status for Phyllanthoideae. Pandaceae have already been accepted at Kew as a distinct family. For an alternative position, see Radcliffe-Smith (1987).]

• Webster, G. L. (1994). Classification of the Euphorbiaceae. Ann. Missouri Bot. Gard. 81: 3-32, maps. En. — A general review of the systematics of the family, with discussion of principal characters and their presumed sequences grouped under five main headings; review of palynology, dispersal (including sexual expression), ecology, and geographical distribution (with maps of representative infrafamilial taxa and their genera); list of references. The Phyllanthoideae retain their position as the least advanced subfamily (p. 4); the potential significance of the dichotomy between biovulate and uniovulate genera is also there acknowledged.

• Webster, G. L. (1994). Synopsis of the genera and suprageneric taxa of Euphorbiaceae. Ann. Missouri Bot. Gard. 81: 33-144. En. — A formal system of the family covering all genera; includes keys along with chromosome numbers and evidence from anatomy and morphology. [317 genera in 49 tribes in 5 subfamilies. Suprageneric taxa are described, while for genera are given distribution and pertinent commentary.]

Special

The main early general accounts of anatomy are those of Pax (1884) and Gaucher (1902). Relatively detailed studies pertaining to parts of the family are those of Rittershausen (1892; see **Acalyphoideae**), Rothdauscher (1896; see **Phyllanthoideae**), Froembling (1896; see **Crotonoideae**), and Herbert (1897; see **Euphorbioideae**). Since 1918 there have been studies on tree architecture, micromorphology, wood anatomy, pollen, embryology, karyology, phytochemistry, and serology. For some, the evidence suggests that the family as traditionally conceived may be heterogeneous; for Vogel (1986), Jensen et al. (1994) and Seigler (1994), by contrast, the family - despite its diversity - remains a coherent concept but instead raises questions about the worth of the recently established and widely accepted subclasses Dilleniidae and Rosidae as well as the relationships of families in Ehrendorfer's angiosperm 'Mittelbau'. This recalls the thoughts of Croizat (1973; see **General**).

Baillon, H. (1858). Recherches sur l'organisation des Euphorbiacées. 241 pp. Paris. (Thèse, Paris.) Fr. — Forms the first part of the author's *Étude générale* (see **General**).

Pax, F. (1884). Die Anatomie der Euphorbiaceen in ihrer Bezeihung zum System derselben. Bot. Jahrb. Syst. 5: 384-421, 2 pls. Ge. — A general survey of anatomical features with suggestions as to their use in classification. Major suites found useful included laticifers, phloem and trichomes.

Gaucher, L. (1902). Recherches anatomiques sur les Euphorbiacées. Ann. Sci. Nat., Bot., VIII, 15: 161-309, illus. Fr. — General anatomy of stem and leaf and of laticifers; systematic descriptions of tribes and subtribes; conclusions, with classification of laticiferous taxa; list of species studied with vouchers. [375 species in all covered in 25 (of 26) tribes and 126 (of 208) genera.Written with little reference to the work of Pax (1884) or Radlkofer's students in Bavaria (see references under subfamilies). Bibliographic references are very scanty.]

Arnoldi, W. (1912). Zur Embryologie einiger Euphorbiaceen. Trav. Mus. Bot. Acad. Sci. St.-Pétersb. 9: 136-154, illus. Ge. — Descriptive account of species from nine genera.

Michaelis, P. (1924). Blütenmorphologische Untersuchungen an den Euphorbiaceen. 150 pp., 41 pls. (Bot. Abhandl. (Jena), 3). Jena: Fischer. Ge. — A response to a then-current debate on the evolution of the angiosperms ('pseudanthy' of Wettstein vs. 'euanthy' of Hallier and Arber and Parkin); systematic survey of many genera; general analysis, organ by organ (pp. 99-122). The author concluded that its place was not in the 'Monochlamydeae' but was better near the 'Columniferae' (i.e. Malvales) though he admitted the floral structure was reminiscent of Urticales. No vouchers are given.

Mandl, K. (1926). Beitrag zur Kenntnis der Anatomie der Samen mehrerer Euphorbiaceen-Arten. Österr. Bot. Zeitschr. 75: 1-17, 4 pls. Ge. — Anatomy of selected species with descriptions and figures; remarks on developmental trends.

Janssonius, H.K. (1929). A contribution to the natural classification of the Euphorbiaceae. Trop. Woods 19: 8-10. En. — Report of consistency of wood structure of *Putranjiva*, *Aporusa*, *Baccaurea* and *Cyclostemon* (=*Drypetes*) in comparison with other members of Phyllanthoideae; affinities of these latter (among others Bixales and Flacourtiaceae were suggested); disagreement of *Acalypha* in relation to the Crotonoideae of Pax and a suggestion that it was nearer Phyllanthoideae; differences between the two major subfamilies; need for exclusion of *Daphniphyllum* from the family.

Janssonius, H.K. (1934). Mikrographie des Holzes der auf Java vorkommenden Baumarten, 5. Monochlamydeae, I. 835 pp., illus. Leiden. Ge. — Euphorbiaceae, pp. 442-812; the most detailed study available of mature woods in the family.

Dehay, C. (1935). L'appareil libéro-ligneux foliaire des Euphorbiacées. Ann. Sci. Nat., Bot., X, 17: 147-290, illus., 4 pls. Fr. — Systematic anatomy of petiolar and laminar vasculature in a wide range of genera and species, with figures and descriptions of their arrangement and histology; summary and general conclusions (the family was considered related to Urticales).

Croizat, L. (1940). On the phylogeny of the Euphorbiaceae and some of their presumed allies. Revista Universitaria Chile 25: 205-220. En. — Comprises a discussion of different systems of classification (particularly Bentham & Hooker and Engler & Prantl, with divergent views on the nature of apetaly) and of the place within them of the Euphorbiaceae. The author favours an affinity with Malvales and Sapindales. A consideration of the cyathium follows, here viewed as comprising androecial and gynoecial 'glomerules'.

Perry, B. A. (1943). Chromosome number and phylogenetic relationships in the Euphorbiaceae. Amer. J. Bot. 30: 527-543, illus. En. — Background to study (with note that the small size of the chromosomes and problems with latex made fixation difficult); reports from a range of genera and species (including data published elsewhere); discussion of evolutionary trends. [Some 50% of the 109 species (in 22 genera) investigated were found to be polyploids; in addition, perennials were more likely to be polyploid than annuals. A very wide range of forms and numbers was in particular found in *Euphorbia*, prompting the suggestion that segregate genera should be recognised.]

De Wildeman, É. (1944). Les latex des Euphorbiacées, I. Considérations générales. 68 pp. (Mém. Inst. Colon. Belge, Sci. Nat. Méd., Collect. 8°, 12(4)). Brussels. Fr. — A rather rambling review, with descriptions of laticifers in various species, summaries of laticifer classification in various genera (especially *Euphorbia*), and remarks on chemical and toxic properties; speculations on the function of latex (pp. 50 ff.); uses and risks of chemical characters in taxonomy. No summary is provided; there are, however, many references in footnotes. [The author makes clear that there is a great diversity of laticifer structure and latex composition in the family, but does not make significant attempts to correlate it or to bring it to bear on classification.]

- Punt, W. (1962). Pollen morphology of the Euphorbiaceae with special reference to taxonomy. 116 pp., illus. (Wentia 7). Amsterdam: North-Holland. En. — Introduction, methodology and glossary; systematic palynological treatment (the suprageneric taxa and selected genera arranged following Pax & Hoffmann 1931) with descriptions and diagnoses of types and subtypes and lists of taxa studied with vouchers and specific measurements; general taxonomic commentaries on Phyllanthoideae (pp. 46-47) and Crotonoideae (pp. 105-106); summary (pp. 106-107), list of references and index to all names. [Pollen features are considered 'exceedingly useful' in Phyllanthoideae and in Crotonoideae point to a need for complete taxonomic revision. There is a marked distinction between the pollens of the two subfamilies; this would support those favouring family rank for Phyllanthoideae. It may be noted here that *Hymenocardia* and *Bischofia*, problematic with respect to other characters, could not, on the basis of pollen data, readily be placed with other Phyllanthoideae.]

Bernhard, F. (1966). Contribution à l'étude des glandes foliaires chez les Crotonoïdées (Euphorbiacées). Mém. Inst. Fond. Afrique Noire 75: 71-156, illus. (Based on a Thèse du Doctorat de 3eme cycle, Univ. Paris, 1964.) Fr. — Systematic morphological study; includes chapters on euphorbiaceous glands in genera, morphology of glands in Crotonoideae (sensu Pax), and anatomy and development in selected species (all material being derived from the Muséum National d'Histoire Naturelle, Paris, but not vouchered). References, pp. 154-156. [The glands were found to be structurally relatively uniform but possessing a great diversity of external form; there were possibly two lines of historical development. They were of value in taxonomy at specific and generic level. No homology with Malvales was detected.]

Hallé, F. (1971). Architecture and growth of tropical trees exemplified by the Euphorbiaceae. Biotropica 3: 56-62, illus. En. — A study of tree models with reference to those featured in the family; future directions for study suggested. [In this respect the family is quite rich; however, no phylogenetic inferences were drawn. The author indicates that studies had to be extended to its non-arborescent members in order to understand the processes of miniaturisation (and, I might add, precocity and neoteny).]

Venkata Rao, C. (1971). Anatomy of the inflorescence of some Euphorbiaceae. Bot. Notis. 124: 39-64, illus. En. — Studies of selected genera and species; evidence suggests that evolution has proceeded mainly by reduction and specialisation from less advanced 'types' (represented in particular by *Jatropha curcas*). *Dalechampia* is considered relatively advanced. A reconstruction of an inflorescence type ancestral to the cyathium of Euphorbieae is also put forward; among the species examined this comes closest to *Acalypha indica*. [The author also would support a break-up of *Euphorbia* into segregate genera; as part of this he accepts Schmidt's *Diplocyathium*.]

- Hans, A. S. (1973). Chromosomal conspectus of the Euphorbiaceae. Taxon 22: 591-636. En. — Review and index of chromosome counts (covering some 350 species); arrangement follows Bentham (1880). In his conclusions, the author drew particular attention to *Bischofia javanica* and suggested that at n=98 it was a relictual polyploid. [Genera from the former tribe Buxeae were also included; these are now referred to Buxaceae or Simmondsiaceae in Buxales.]

Urbatsch, L. E. et al. (1975). Chromosome numbers for North American Euphorbiaceae. Amer. J. Bot. 62: 494-500, illus. En. — Presentation of new and reviewed data (table, pp. 498-499); few if any inferences. [By the time of writing of the paper counts existed for only 6% of the then-estimated 7000 species in the family.]

Uhlarz, H. (1978). Über die Stipularorgane der Euphorbiaceae, unter besonderer Berücksichtigung ihrer Rudimentation. 65 pp., illus. (Trop. Subtrop. Pflanzenw. 23). Mainz. Ge. — Systematic observations of selected genera (arranged following the Pax & Hoffmann 1931 system); four trends related to rudimentation recognised (pp. 61-62), including glandularisation and spinification, or combinations of these.

- Vogel, C. (1986). Phytoserologische Untersuchungen zur Systematik der Euphorbiaceae. 124 pp., illus. Berlin: Cramer/Borntraeger. (Dissertationes Botanicae 98.) Ge. — Centers around serological studies on 4 genera in 3 (uniovulate) subfamilies (for antisera) and a selection of 30 additional genera (for antigens). The results and discussion include 'phylogenetic' analyses (see particularly fig. 17, p. 62) and considerations on infrafamilial relationships. The author suggests that the uniovulate subfamilies are mutually closely enough related to warrant union as Crotonoideae. The position of the genera *Euphorbia* and *Mallotus* could not be ascertained. Her proposals also call for far fewer tribes than presently recognised. Altogether there is some distinctly radical thought but its full development is not here undertaken.
- Punt, W. (1987). A survey of pollen morphology in Euphorbiaceae with special reference to *Phyllanthus*. Bot. J. Linn. Soc. 94: 127-142, illus. En. — Introduction with review of scope and quality of sources; survey of pollen morphology in relation to the system of Webster (1975); systematic survey of pollen in *Phyllanthus* in relation to the subgeneric scheme of Webster.

Rudall, P. J. (1987). Laticifers in Euphorbiaceae: a conspectus. Bot. J. Linn. Soc. 94: 143-163, illus. En. — Review of past work (with terminology an impediment to understanding; an important early contribution was that of Pax (1884)); examination of a large range of genera and species. A table of forms and occurrences of laticifers is presented on pp. 146-149. Laticifer evidence is also examined in relation to classification and evolution. [Particular attention was paid to the problem of non-articulated and articulated states (with *Jatropha* found to have both forms) with the conclusion that their ancestry was common. Articulated laticifers were found to be restricted to Crotonoideae.]

Jensen, U., I, Vogel-Bauer & M. Nitschke (1994). Leguminlike proteins and the systematics of the Euphorbiaceae. Ann. Missouri Bot. Gard. 81: 160-179, illus. En. — Survey of a wide range of the Euphorbiaceae with reference to five serological standard-species (for their mutual distances, see fig. 2 on p. 170), using the so-called leguminlike seed proteins (widespread in flowering plants); comparisons with other families. The results suggest strong affinities with the superorders Malviflorae, Rutiflorae and Violiflorae; together these 'bridge' the major subclasses Dilleniidae and Rosidae of the widely used Cronquist system. Within the Euphorbiaceae, viewed by the authors as coherent, division into five subfamilies was not supported; rather, only two should be accepted, the basically biovulate Phyllanthoideae and the basically uniovulate Euphorbioideae.

Kapil, R. N. & A. K. Bhatnagar (1994). The contribution of embryology to the systematics of the Euphorbiaceae. Ann. Missouri Bot. Gard. 81: 145-159. En. — Embryological features seen as more conservative and useful in the delineation and understanding of relationships. The survey includes consideration of several families included at one or another time in Euphorbiales; the authors conclude that the order should be limited to Euphorbiaceae s.l. Internally, there was support for Webster's division of the family into 5 subfamilies. Corner's proposal (in *Seeds of Dicotyledons*, 1976) that the biovulate and uniovulate genera respectively had different affinities was rejected, with in particular *Bischofia* showing an intermediate seed structure. Good support was found for a relationship with Malvales. Only 16 of the 50 tribes were, however, examined.

Seigler, D. S. (1994). Phytochemistry and systematics of the Euphorbiaceae. Ann. Missouri Bot. Gard. 81: 380-401. En. — Description of the range of secondary metabolites; review of potential systematic value of secondary characters. Support was found for an affinity of the family with Geraniales and Malvales as well as for the idea that the distinction between Dilleniidae and Rosidae in the Cronquist system represents an exaggeration; moreover, the Euphorbiaceae may have arisen from ancestors intermediate to Geraniales and Malvales.

Sutter, D. & P. K. Endress (1995). Aspects of gynoecium structure and macrosystematics in Euphorbiaceae. Bot. Jahrb. Syst. 116: 517-536, illus. En. — A study of the ovule and other parts of the gynoecium in 17 species in four subfamilies (Webster system); discussion with respect to internal similarities, ovule structure in related families, and two recent classifications (Meeuse and Huber). [Designed as a test of renewed arguments over the monophyly of the family. No support was found for Meeuse's proposal to exclude Phyllanthoideae and Oldfieldioideae from the rest of the family (as two distinct families) nor for Huber's view of a linalean relationship for Phyllanthoideae/Oldfieldioideae and a violalean/malvalean one for the remainder (based respectively on 2 vs. 1 ovules/carpel). Nevertheless, in the authors' words (p. 518): 'The question remains whether the Euphorbiaceae s.l. are monophyletic or polyphyletic'.]

Tokuoka, T. & H. Tobe (1995). Embryology and systematics of Euphorbiaceae sens. lat.: a review and perspective. J. Pl. Res. 108: 97-106. En. — A literature survey, with discussion of the individual subfamilies; the authors conclude that available embryological evidence does *not* support monophyly in the family. Only 5 of 50 characters examined were found to be potentially useful; all were associated with mature ovules and seeds allowing use of herbarium material. References, pp. 102-105 (inclusive of the 110 publications specifically in embryology, for which there is a classified index, pp. 105-106). [*Euphorbia* has been studied to a far greater extent than have other genera. The authors' conclusion is of interest given Hutchinson's opinion that the family had multiple origins, a view also supported by seed structure (cf. Corner, *The seeds of dicotyledons*, 1976).]

Africa

The basic treatments for tropical and southern Africa are those of Brown (1911-13, 1915) and Pax (1921). More recent regional revisions include those in *Flora of West Tropical Africa, Flora of Ethiopia and Eritrea* (neither separately accounted for here), *Flora of Tropical East Africa, Flora Zambesiaca* and *Flore d'Afrique Centrale* (the latter two still in progress). Morocco was covered by Vindt (1955-60) but the only overall coverage of North Africa in more recent years is in *Med-Checklist* (see **Eurasia**).

Baillon, H. E. (1860-63). Species Euphorbiacearum, A: Euphorbiacées africaines. Adansonia 1: 58-87, 139-173, 251-286, pl. 5; 2: 27-55; 3: 133-166. Fr. — Synoptic treatments with synonymy, references, descriptions of novelties and exsiccatae; descriptions of novelties included. [Basic for knowledge of African Euphorbiaceae.]

Pax, F. (1893-1900). Euphorbiaceae africanae. Bot. Jahrb. Syst. 15: 522-535; 19: 76-127, illus.; 23: 518-536; 26: 325-329; 28: 18-27. (Beiträge zur Flora von Afrika.) Ge. — Novelties and notes.

Brown, N. E. (1911-13). Euphorbiaceae. Flora of tropical Africa 6(1): 441-1020, 1034-1059. London. (Pp. 441-576, 1911; 577-960, 1912; 961-1020 and 1034-1059, 1913.) En. — Flora treatment, with keys. [The major synthesis prior to 1914; not yet entirely supplanted though now out of date.]

Brown, N. E. (1915-25). Euphorbiaceae. Flora Capensis, 5(2): 216-516, 585-586. London. (Pp. 216-384, 1915; 385-516, 1920; 585-586, 1925.) En. — Flora treatment, with keys. [Remains the most recent overall for southern Africa, although now out of date.]

Pax, F. (1921). Geraniales-Euphorbiaceae. In A. Engler (ed.), Die Pflanzenwelt Afrikas, 3(2): 1-168, illus. Leipzig: Engelmann. (Die Vegetation der Erde, 9.) Ge. — Generic-level survey with coverage of infrageneric taxa, annotated lists of representative species, and some keys; index at end of part-volume. [For some genera, notably *Euphorbia*, not yet superseded in terms of continental coverage.]

Vindt, J. (1953-60). Monographie des Euphorbiacées du Maroc, I-II. Trav. Inst. Sci. Chérif. 6: i-xx, 1-217, pls. 1-3, fold. map; 19: i-xxix, 219-533. Rabat. (I, Revision et systématique; II, Anatomie.) Fr. — Part I comprises an amply descriptive illustrated revision with keys, synonymy, references, literature citations, localities with exsiccatae (keyed to folding map), indication of habitat and overall distribution, and critical notes; gazetteer of localities and keys to *Euphorbia* respectively on vegetative features, seeds and capsules, and index. Part II covers systematic anatomy, supplements to the revision (pp. 439-470), general conclusions (pp. 473-480), an extensive source bibliography, and (at end) a full table of contents.

Léonard, J. (1955). Notulae systematicae XVIII. Euphorbiaceae africanae novae. Bull. Jard. Bot. État 25: 281-301. Fr. — Includes among notes on several genera a description of *Cyttaranthus*; a precursor to *Flora du Congo Belge.*

Léonard, J. (1958). Notes sur diverses Euphorbiacées africains des genres *Croton, Crotonogyne, Dalechampia, Grossera* et *Thecacoris*. Bull. Jard. Bot. État 28: 111-122. Fr-. — Precursor to *Flore du Congo Belge* treatment.

Léonard, J. (1959). Notes sur diverses Euphorbiacées africaines des genres *Bridelia, Croton, Grossera, Maprounea* et *Tetrorchidium*. Bull. Jard. Bot. État 29: 195-204. Fr. — Precursor to *Flore du Congo Belge* treatment.

Léonard, J. (1961). Euphorbiaceae africanae novae (*Bridelia, Croton, Jatropha, Mildbraedia*). Bull. Jard. Bot. État 31: 55-67. Fr. — Precursor to *Flore du Congo Belge* treatment.

Léonard, J. (1961). Observations sur des espèces africaines de *Clutia, Ricinodendron,* et *Sapium* (Euphorbiacées). Bull. Jard. Bot. État 31: 391-406. Fr. — Precursor to *Flore du Congo Belge* treatment.

- Léonard, J. (1962). Euphorbiaceae, [1]. 214 pp., 13 pls. (Fl. Congo: Spermatophytes 8(1)). Brussels. Fr. — Flora treatment; encompasses the Paxian tribes Bridelieae, Crotoneae, Clutieae, Gelonieae, Hippomaneae, Chrozophoreae and Dalechampieae. [For continuation, see Léonard 1995, 1996.]

Radcliffe-Smith, A. (1972-). Notes on African Euphorbiaceae, I- . Kew Bull. 27, *passim*. En. — Mainly precursory to floras; continuing as of 1998. Some keys and small revisions are included. [Where appropriate invididual parts are cited under genera.]

Carter, S. et al. (1984). A collection of and notes on Euphorbiaceae from the Cape Verde Islands. Senckenbergiana, Biol. 64: 429-451. En. — Based on large collections by Werner Lobin; includes keys, distribution and exsiccatae.

Brunel, J. F. (1987). Sur le genre *Phyllanthus* L. et quelques genres voisins de la tribu des Phyllantheae Dumort. (Euphorbiaceae, Phyllantheae) en Afrique intertropicale et à Madagascar. xiii, 472, 196 pp., illus. Strasbourg. (Thèse, Université de Strasbourg I). Fr. — The main part includes a historical survey, character analysis, and suprageneric revision; an annex, a catalogue of *Phyllanthus* in infratropical Africa. [Also listed under individual genera.]

- Radcliffe-Smith, A. (part 2 by S. Carter and A. Radcliffe-Smith) (1987-88). Euphorbiaceae, 1-2. Flora of Tropical East Africa, Euphorbiaceae. 599 pp., illus. Rotterdam: Balkema. En. — Flora treatment, with keys, descriptions, synonymy, localities with exsiccatae and notes. Part 1, genera except those in Euphorbieae; part 2, Euphorbieae. [See also Radcliffe-Smith, 1995.]
- Léonard, J. (1995). Euphorbiaceae, II. 115 pp., illus., maps (Flore d'Afrique centrale (Zaïre-Rwanda-Burundi): Spermatophytes). Brussels. Fr. — Flora treatment, with keys, descriptions, and localities with exsiccatae; covers *Antidesma, Cyathogyne* (*=Thecacoris*), *Maesobotrya, Martretia, Protomegabaria, Spondianthus* and *Thecacoris* (all in Phyllantheae-Antidesminae of Pax & Hoffmann, 1931).

Radcliffe Smith, A. (1995). Additions and corrections to "Euphorbiaceae" for *Flora of Tropical East Africa*. Kew Bull. 50: 809-816. En. — Includes some revised keys.

- Léonard, J. (1996). Euphorbiaceae, III. 74 pp., illus., maps (Flore d'Afrique centrale (Zaïre-Rwanda-Burundi): Spermatophytes). Meise. Fr. — Flora treatment with keys, descriptions, and localities with exsiccatae; covers *Argomuellera, Crotonogynopsis, Discoglypremna, Mareya, Mareyopsis* (*=Mareya*), *Necepsia* and *Pycnocoma* (all in the first two series of Acalypheae-Mercurialinae of Pax & Hoffmann, 1931).
- Radcliffe-Smith, A. (1996). Euphorbiaceae [except Euphorbieae]. Fl. Zambesiaca 9(4): 1-337. Kew. En. — Flora treatment with keys, descriptions, synonymy, references and literature citations, localities with exsiccatae, indication of overall distribution and habitat, and commentary; index to all names at end. [Does not cover Euphorbieae.]

Americas

Modern treatments of the family appear for the most part in floras; there have been few separate area revisions. A primary survey was made by Baillon in 1864-65; this was followed in 1873-74 by Mueller's revision of the family for *Flora brasiliensis*. In the first half of the twentieth century there appeared treatments for Surinam by Lanjouw (1931), a series for Argentina (Lourteig & O'Donell, 1942, 1943; O'Donell & Lourteig 1942, 1942), and an extension to Reiche's incomplete Chilean flora (Behn, 1942-43). A patchwork of further contributions has characterised the remainder of the twentieth century; the most substantial of these have been for Panama (Webster & Burch, 1967; Webster & Huft, 1988), Costa Rica (Burger & Huft, 1995) and the southeastern United States (Webster, 1967).

- Baillon, H. E. (1864-65). Species Euphorbiacearum: Euphorbiacées américaines. Adansonia 4: 257-377; 5: 221-240, 305-360. Fr. — Synoptic treatment with descriptions of novelties, synonymy, localities with exsiccatae, and commentary.
- Mueller (Argoviensis), J. (1873-74). Euphorbiaceae. Fl. Brasiliensis 11(2): 1-726, pls. 1-104. La. — Flora treatment of 869 species (856 in Brazil) in 62 genera. An index for the part-volume follows (pp. 728-752). [Remains standard as no comparable successor is available.]

Millspaugh, C. F. (1909). Praenunciae bahamenses, II. Publ. Field Mus. Nat. Hist., Bot. Ser. 2: 289-318. Euphorbieae, pp. 299-306. En. — Precursory to *Bahama flora* by Britton and Millspaugh (1920). Includes description and synopsis of *Euphorbiodendron* (now *Euphorbia* subgen. *Esula* sect. *Adenorima*).

Millspaugh, C. F. (1913). The genera *Pedilanthus* and *Cubanthus*, and other American Euphorbiaceae. Publ. Field Mus. Nat. Hist., Bot. Ser. 2: 353-397. En. — Includes description and synopsis of *Cubanthus*, a segregate from *Pedilanthus*, as well as *Dendrocousinia* (a segregate of *Sebastiania*, Crotonoideae), and additions in Cuba to *Euphorbiodendron* (now part of *Euphorbia*) along with other notes.

- Lanjouw, J. (1931). The Euphorbiaceae of Suriname. 195 pp., 5 pls, 1 folding map. Utrecht. (Ph.D. dissertation, University of Utrecht.) En. — The second of the two parts of this work is a preprint from the *Flora of Suriname* treatment (publ. 1932 as vol. 2(1): 1-101); the first, 'New and critical species', is more analytical. [For additions and corrections, see *idem*, *Fl. Suriname* 2(1): 457-470. 1939; A. R. A. Görts-van Rijn in *ibid.*, 2(2): 387-424. 1976.]

Wheeler, L. C. (1939). A miscellany of New World Euphorbiaceae, [I]-II. Contr. Gray Herb. 124: 35-42; 127: 48-78, pls. 3-4. En. — Descriptions of novelties and miscellaneous notes with some keys. [Arrangement of genera follows Pax and Hoffmann (1931).]

Lourteig, A. and C. A. O'Donell (1942). Acalypheae argentinae (Euphorbiaceae). Lilloa 8: 273-333, pls. 1-9, map. Sp. — Descriptive revision with keys, synonymy, references, types, indication of distribution, habitat, and origin (if relevant), and localities with exsiccatae; map showing southern distributional limits of the 6 genera covered (p. 331); literature at end.

O'Donell, C. A. and A. Lourteig (1942). Chrozophoreae argentinae. Lilloa 8: 37-81, illus., maps, pls. 1-7. Sp. — Descriptive revision with keys, synonymy, references, types, indication of distribution, habitat, and origin (if relevant), localities with exsiccatae, figures and distribution maps; literature at end. [Covers 4 genera, 3 of them with species mapped. *Aonikena* is combined here with *Chiropetalum*.]

O'Donell, C. A. and A. Lourteig (1942). Hippomaneae argentinae. Lilloa 8: 545-592, illus., maps, pls. 1-6. Sp. — Descriptive revision with keys, synonymy, references, types, indication of distribution, habitat, and origin (if relevant), localities with exsiccatae, and illustrations; 2 maps showing southern distributional limits of the 5 genera covered (pp. 580 and 591); literature at end.

- Behn, K. (1942-43(1944)). Flora de Chile: las familias Euforbiaceas, Aextoxicaceas y Calitriquaceas. Revista Chilena Hist. Nat. Pura Aplic. 46/47: 145-196, 4 pls. Sp. — Compiled revisions (Euphorbiaceae, pp. 145-188) in the style of a flora account, with keys, descriptions, references and citations, synonymy, vernaucular names, derivations of scientific names, and indication of distribution and habitat. [Written as an extension to the incomplete *Flora de Chile* of Reiche.)

Croizat, L. (1943). Notes on American Euphorbiaceae, with descriptions of eleven new species. J. Washington Acad. Sci. 33: 11-20. En. — Notes and novelties; includes representatives of *Phyllanthus*.

Lourteig, A. and C. A. O'Donell (1943). Euphorbiaceae argentinae: Phyllantheae, Dalechampieae, Cluytieae, Manihoteae. Lilloa 9: 77-173, illus., pls. 1-18. Sp. — Descriptive revision with keys, synonymy, references, types, indication of distribution, habitat, and origin (if relevant), localities with exsiccatae, and illustrations; 2 maps showing southern distributional limits of the genera covered (pp. 104 and 150); literature and list of all names, abbreviations and collections seen in the four installments of this revision at end.

• Lourteig, A. and C. A. O'Donell (1943). Euphorbiaceae. In H. Descole (ed.), *Genera et species plantarum argentinarum*, 1: 145-317, pls. 63-138 (incl. maps). Buenos Aires: Kraft. La/Sp. — Sumptuously formatted illustrated descriptive flora with keys (based on the authors' precursary revisions); references, pp. 301-303. [Omits Euphorbieae and does not cover Crotoneae save to generic rank. The maps show generic ranges grouped by tribes).]

Croizat, L. (1945). New or critical Euphorbiaceae from the Americas. J. Arnold Arbor. 26: 181-196. En. — Includes among other notes a description of *Moacroton* and a synopsis.

Martínez, M. (1955). Familia de la Euforbiáceas del Estado de México. 19 pp., illus. Toluca, Méx.: Dirección de Agricultura y Ganaderia, Edo. de México. Sp. — Flora treatment, with descriptions and keys.

Rambo, B. (1960). Euphorbiaceae riograndenses. Pesquisas (Rio Grande do Sul), Bot. 9: 1-78. Lu. — Extensive treatment, with many exsiccatae.

McVaugh, R. (1961). Euphorbiaceae novae novo-galicianae. Brittonia 13: 145-205. En. — Novelties and notes; precursor to projected treatment in *Flora novo-galiciana*. [For further contributions see McVaugh 1993, 1995.]

Moreira, E. A. and G. Hatschbach (1964). Lista das Euphorbiáceas do Estado do Paraná. Bol. Inst. Hist. Nat., Bot. 5: 1-16. Lu. — State list, with exsiccatae. [Not seen.]

Jablonski, E. (1965). Euphorbiaceae. In B. Maguire and collaborators, The botany of the Guayana Highland VI (Mem. New York Bot. Gard. 12(3): 150-178). New York. En. — Separate entries for this work appear under all genera.

Jablonski, E. (1967). Euphorbiaceae. In B. Maguire and collaborators, The botany of the Guayana Highland, VII (Mem. N.Y. Bot. Gard. 17(1): 80-190). New York. En. — Separate entries for this work appear under all genera.

• Webster, G. L. (1967). The genera of Euphorbiaceae in the southeastern United States. J. Arnold Arbor. 48: 303-430. London. En. — Descriptive treatment, with key and commentaries, of 22 genera native in the United States, of which 18 in the Southeast; six additional genera naturalized. 115 species known in region. Detailed family and generic bibliographies!

• Webster, G. L. and D. Burch (1967(1968)). Euphorbiaceae. Ann. Missouri Bot. Gard. 54: 211-350, illus. (Flora of Panama, VI: family 97). En. — Flora treatment, with keys, descriptions, synonymy, localities with exsiccatae, indication of distribution and habitat, and commentary. [For revision, see Webster and Huft (1988, under **Americas**).]

Webster, G. L. (1970). Notes on Galapagos Euphorbiaceae. Madroño 20: 257-263. En. — Notes on *Croton scouleri* and various species of *Acalypha*; no keys. A precursor to the flora treatment by Wiggins and Porter (1971). [The variation in *C. scouleri* was considered as still not well understood.]

Urbatsch, L. E. et al. (1975). Chromosome numbers for North American Euphorbiaceae. See **Special**.

Secco, R. S. (1985). Revisão taxonômica dos gêneros *Anomalocalyx* Ducke, *Dodecastigma* Ducke, *Pausandra* Radlk., *Pogonophora* Miers ex Benth. e *Sagotia* Baill. (Euphorbiaceae). Manaus. (Unpubl. Tese de Mestrado, INPA.) Lu. — Parts have appeared in Boletim do Museu Paraense 'Emilio Goeldi'.

Richardson, J. W., D. Burch and T. S. Cochrane (1987(1988)). Euphorbiaceae. Trans. Wisconsin Acad. Sci. Arts Letters 75: 97-129, illus., maps (Flora of Wisconsin, Preliminary Report 69). En. — Detailed background study for a projected state flora; includes keys, descriptions, indication of overall and local distribution as well as habitats and biology, and taxonomic commentary; figures and dot distribution maps for each species (the latter including phenology). [*Chamaesyce* is here segregated from *Euphorbia*.]

- Webster, G. L. and M. J. Huft (1988). Revised synopsis of Panamanian Euphorbiaceae. Ann. Missouri Bot. Gard. 75: 1087-1144. En. — Successor to 1968 flora treatment, similar in style and content and with added exsiccatae and revised keys where required; separate index at end.

Cardiel Sanz, J. M. & P. Franco Roselli (1992). Euphorbiaceae. Fl. Real Exped. Bot. Nuevo Reyno de Granada 23: 13-118, col. pls. Madrid. Sp. — Folio plates with accompanying text including modern identifications; many literature references under the family (pp. 17-20) as well as individual genera.

- Gillespie, L. J. (1993). Euphorbiaceae of the Guianas: annotated species checklist and key to the genera. Brittonia 45(1): 56-94. En. — Introduction; key to genera; concise checklist with synonymy, typification and indication of distribution. Asterisks indicate if a species is elsewhere found only in the Guayana Region or is only in cultivation (or adventive). Excluded species, references and an index appear on pp. 93-94. Totals of 61 genera and 261 species are accounted for.

McVaugh, R. (1993). Euphorbiaceae novo-galicianae revisae. Contr. Univ. Michigan Herb. 19: 207-239. En. — Additions and revisions to the author's 1961 paper; includes among other notes comments on *Chamaesyce* vs. *Euphorbia* (with a somewhat pragmatic decision in favour of separation).

- Burger, W. & M. Huft (1995). Euphorbiaceae. Fl. Costaricensis, family 113 (Fieldiana, Bot. n.s. 36): 119 pp., illus. Chicago. En. — Flora treatment with keys, descriptions, synonymy, references, localities with exsiccatae, indication of distribution, and notes.

McVaugh, R. (1995). Euphorbiacearum sertum Novo-Galicianarum revisarum. Contr. Univ. Michigan Herb., 20: 173-215, illus. En. — Further contributions, precursory to treatment in *Flora Novo-Galiciana*.

Murillo Aldana, J. & P. Franco Rosselli (1995). Las Euforbiáceas de la región de Araracuara. 191 pp., 48 text-fig. (incl. map). Bogotá: Instituto de Ciencias Naturales, Universidad Nacional de Colombia. (Estudios en la Amazonia Colombiana, 9.) En. — Descriptive treatment (64 species in 36 genera, some 56% of the species known from Colombian Amazonia) with analytical (Lamarckian) and multiaccess keys, vernacular names, synonymy, indication of distribution and habitat, and notes on properties and uses. Standard references, if available, appear under genera. At least one species in each genus is illustrated. References, lists of specimens seen with localities, and indexes are given at the end. The introduction contains the plan of the work and its basis, along with floristic considerations and general analytical and multiaccess keys with figures of diagnostic characters.

Gillespie, L. J. & W. S. Armbruster (1997). A contribution to the Guianan flora: *Dalechampia, Haematostemon, Omphalea, Pera, Plukenetia* and *Tragia* (Euphorbiaceae) with notes on subfamily Acalyphoideae. See **Acalyphoideae**.

- Steinmann, V. W. & R. S. Felger (1997). The Euphorbiaceae of Sonora, Mexico. Aliso 16: 1-71, illus., map. En. — Descriptive revision, with keys, of 19 genera and 143 species; includes synonymy, types, localities with exsiccatae, indication of phenology, distribution, habitat and altitudinal range, and commentary. The revision proper is preceded by a general floristic, ecological and biogeographical account.

Asia

Notable treatments include revisions for NE. Asia by Hayata (1904) and Hurusawa (1954), for former Indochina by Gagnepain and Beille (1925-27), for Taiwan by Keng (1955), for Thailand by Airy-Shaw (1971, with revisions in 1977), and for the Andaman and Nicobar Islands by Chakraborty and Balakrishnan (1992). Records for Vietnam were brought up to date by Nguyen Nghia Thin (1995), and since 1994 all Chinese species have been covered by P.-T. Li in vol. 44 of *Flora Reipublicae Popularis Sinicae* (the English version is currently in preparation). Several of Airy-Shaw's contributions on Malesian Euphorbiaceae also cover species in mainland Asia apart from Thailand. A further review for Vietnam appeared in 1998, along with keys for Nepal and Thailand.

Hayata, B. (1904). Revisio Euphorbiacearum et Buxacearum Japonicarum. 92 pp., 6 pls. (J. Coll. Sci. Imp. Univ. Tokyo, 20(3)). Tokyo. La. — Regional revision with keys, descriptions, synonymy, references and citations, vernacular names, and indication of distribution along with exsiccatae seen; no commentary. The plates are all of botanical details. [A predecessor of Hurusawa's 1954 revision.]

Gagnepain, F. and L. Beille (1925-27). Euphorbiaceae. Fl. Gén. Indo-Chine 5: 229-673, illus. Paris. Fr. — Flora treatment with keys, descriptions, synonymy, localities with exsiccatae, indication of distribution, and notes. [Remains a standard account in SE Asia; should be used along with the works of Airy-Shaw and (for Vietnam) the 1980s list by Nguyen Nghia Thin.]

Croizat, L. (1940). New and critical Euphorbiaceae from eastern tropical Asia. J. Arnold Arbor. 21: 490-510. En-. — Covers *Actephila, Securinega, Glochidion, Breynia, Sauropus* (including *Breyniopsis*), *Antidesma, Croton, Mallotus, Cleidion, Epiprinus, Sapium, Euphorbia*, and *Chamaesyce* (the last-named here separate from *Euphorbia*). Regarding separation or not of the last two there is an extensive discussion.

Croizat, L. (1941). The tribe Plukenetiinae [sic!] of the Euphorbiaceae in eastern tropical Asia. J. Arnold Arbor. 22: 417-431. En. — Contains numerous additions and revisions to the Pax and Hoffmann subtribe, encompassing *Tragia, Cnesmone, Megistostigma* and *Plukenetia* as well as additions to *Sphaerostylis* and a note on *Ramelia* (=*Bocquillonia*). A key to the first four genera is presented.

Croizat, L. (1942). New and critical Euphorbiaceae from the tropical Far East. J. Arnold Arbor. 23: 495-508. En. — Contributions in *Croton* (extensive), *Claoxylon, Ostodes, Cheilosa* and *Sapium*

Croizat, L. (1942). On certain Euphorbiaceae from the tropical Far East. J. Arnold Arbor. 23: 29-54. En. — Contributions in *Actephila, Phyllanthus, Securinega, Phyllanthodendron* (extensive treatment), *Dicoelia, Cleistanthus, Croton* (extensive additions), *Alchornea, Ptychopyxis, Koilodepas, Macaranga, Epiprinus* (incl. *Symphyllia*; reduction not subsequently accepted by some) and *Trigonostemon*.

• Hurusawa, I. (1954). Eine nochmalige Durchsicht des herkömmlichen Systems der Euphorbiaceen in weiteren Sinne. J. Fac. Sci. Univ. Tokyo, III, 6: 209-342, illus., 4 pls. Ge. — Includes a revision of Euphorbiaceae and its allies (including Antidesmataceae, corresponding to Pax's Phyllanthoideae) for Japan and former Japanese territories in east Asia, in succession to that of Hayata (1904). [For further details, see corresponding entry under **General**.]

Keng, H. (1955). The Euphorbiaceae of Taiwan. Taiwania, 6: 27-66. En. — Synoptic revision. For *Euphorbia* sens. lat. succeeded by contributions from Lin and Hsieh (1991) and Lin, Chaw and Hsieh (1991; for both, see that genus).

Airy-Shaw, H. K. (1969). Notes on Malesian and other Asiatic Euphorbiaceae, CXII. Notes on the subtribe Plukenetiinae Pax. Kew Bull. 23: 114-121. En. — Notes on *Tragia, Pachystylidium, Cnesmone* and *Megistostigma*, with key to all species of the last-named. In effect an extension to Croizat (1941).

Airy-Shaw, H. K. (1971). Notes on Malesian and other Asiatic Euphorbiaceae, CXXIV. A key to the Securineginae and some deceptively similar Phyllanthinae of tropical and temperate Asia. Kew Bull. 25: 491-493. En. — Analytical key covering *Richeriella, Margaritaria, Securinega* (and *Flueggea*), and parts of *Phyllanthus*. [*Cicca* is here kept separate from *Phyllanthus*.]

• Airy-Shaw, H. K. (1971). The Euphorbiaceae of Siam. Kew Bull. 26: 191-363. En. — Concise regional revision, with keys. See also his Additions and corrections .. Kew Bull. 32: 69-83 (1977).

Airy-Shaw, H. K. (1977). Additions and corrections to Euphorbiaceae of Siam. Kew Bull. 32: 69-83. En. — Includes (p. 79-80) a new genus *Myladenia*.

- Chakraborty, T. and N. P. Balakrishnan (1992). The family Euphorbiaceae of Andaman and Nicobar Islands. 122 pp., illus. (J. Econ. Tax. Bot., Addit. Ser. 9). Jodhpur. En. — Area revision with keys, descriptions, synonymy, references and citations, vernacular names, indication of phenology, distribution, habitat, uses and properties, representative exsiccatae, and some commentary; complete index at end. [Covers 110 species (of which 13 cultivated) in 40 genera; 1 genus (*Sphyranthera*) and 18 species endemic.]

Singh, M. (1994). Succulent Euphorbiaceae of India. 55 pp., illus., map. Delhi. En. — Brief introduction; illustrated descriptive popular treatments of *Euphorbia*, *Jatropha* and *Synadenium* (the last-named represented by the naturalised *S. grantii*). Each species occupies a two-page spread with description, reference citations, distribution and commentary. In *Euphorbia* are two in the Tirucalli group of subgen. *Esula* (one introduced), several from subgen. *Euphorbia*, and four geophytic species including *E. fusiformis* and one not yet described (comparative table on pp. 40-41). *Jatropha* is represented here by seven species of which six are native. Literature, p. 54; no index.

Tripathi, A. K., B. K. Shukla & V. Mudgal (1994). Floral elements of Madhya Pradesh (Acanthaceae and Euphorbiaceae). viii, 249 pp., illus. New Delhi: Ashish Publishing House. En. — Pp. 141-234 cover Euphorbiaceae; these incorporate a descriptive treatment with keys, synonymy, notes on habitat, distribution, and phenology, and distribution maps (with districts shaded where a species is present). The revision is followed by a list of literature consulted for both families (pp. 235-237) and complete index.

- Li P.-T. et al. (1994-97). Euphorbiaceae, 1-3. Fl. Reipublicae Popularis Sinicae, 44(1-3). Illus. Beijing. Ch. — Flora treatment with keys, descriptions, synonymy, references and citations, indication of distribution and habitat, and notes. Part 1 covers Phyllanthoideae; part 2, Acalyphoideae and Crotonoideae; and part 3, Euphorbioideae.

Nguyen Nghia Thin (1995). Euphorbiaceae of Vietnam. 50 pp. Hanoi: 'Agriculture' Publishing House. En. — Synoptic list of genera and species, with key to subfamilies and tribes; review of phylogeny, photogeography, and properties and uses; list of literature consulted and index to all names. An appendix (pp. 46-50) furnishes descriptions of some new species. [Should be used together with Gagnepain and Beille's account in the *Flore générale* (1925-27).]

Kurosawa, T. (1998). Tentative keys for Nepalese Euphorbiaceae. Newsl. Himalayan Bot. 22: 12-26. En. — Keys to genera and species; miscellaneous taxonomic notes. [Inclusive of some species not identified or described *Chamaesyce* is considered as distinct from *Euphorbia*.]

Nguyen Nghia Thin (1998). Taxonomy and phylogeny of Euphorbiaceae in Vietnam. In A. L. Zheng *et al.* (eds), *Floristic characteristics and diversity of East Asian plants: proceedings of the first international symposium* (Kunming, 1996): 367-383. Beijing: China Higher Education Press; Berlin: Springer. En. — Includes basic statistics and key to tribes along with discussions of character trends and internal and external affinities (with a ground plan, p. 379) along with conclusions and references. [Phyllanthoideae and, within that subfamily, Wielandieae considered most primitive. The family was estimated as having 6000-8000 species.]

Welzen, P. C. van (1998). Analytical key to the genera of Thai Euphorbiaceae. Thai Forest Bull., Bot. 26: 1-17. En. — Keys to genera (including *Galearia*) and some species; 3 references. [Incorporates results of changes to Thai Euphorbiaceae since publication of Air-Shaw's account.]

Nguyen Nghia Thin (n.d. (after 1988)). Taxonomy of the Euphorbiaceae of the Vietnamese flora. 44 pp., illus. Hanoi. En. — Includes synoptic list of genera, species (422) and additional varieties (21); key to tribes. [A preliminary version of Nguyen Nghia Thin 1995.]

Australasia

Basic knowledge was synthesised by Baillon in two contributions in the 1860s and by Bentham in *Flora australiensis*. A further important contribution is the *Partial Synopsis* of Airy-Shaw (1980). The 'Stenolobieae' of Johannes Mueller have been shown to be a phylogenetic grade or homoplasy and the characteristic heath-like genera formerly included therein are now dispersed among Phyllanthoideae, Oldfieldioideae, Acalyphoideae and Crotonoideae.

Baillon, H. E. (1866). Species Euphorbiacearum: Euphorbiacées australiennes. Adansonia 6: 282-345. Fr. — Synoptic treatment with descriptions of novelties, synonymy, localities with exsiccatae, and commentary. [Basic to later knowledge for the continent and Tasmania.]

Baillon, H. E. (1866-67). Species Euphorbiacearum: recherches complémentaires sur les Euphorbiacées australiennes. Adansonia 7: 352-360, 2 pls. Fr. — Additions to the author's review of 1865-66.

Airy-Shaw, H. K. (1971). New or noteworthy Euphorbiaceae-Ricinocarpoideae from Western Australia. Kew Bull. 26: 67-71. En. — Novelties and notes.

Airy-Shaw, H. K. (1976). New or noteworthy Australian Euphorbiaceae, I. Kew Bull. 31: 341-398, illus. En. — Notes and novelties precursory to 1980 revision of 'Platylobeae'; includes figures of *Margaritaria dubium-traceyi* and *Whyanbeelia terrae-reginae* (the latter genus first described here).

Airy-Shaw, H. K. (1980). New or noteworthy Australian Euphorbiaceae, II. Muelleria 4: 207-241. En. — Additions to Australian 'Platylobeae' (the first of the two subdivisions of Mueller (Argoviensis)). Key genera: *Meoroepera*, pp. 217-218; *Petalostigma*, pp. 218-219; *Dissiliaria*, p. 220; *Choriceras*, p. 220-221.

- Airy-Shaw, H. K. (1980a). An alphabetical check-list of native Australian Euphorbiaceae (excluding *Phyllanthus*, *Euphorbia* and the Stenolobeae). Muelleria 4: 243-245. En. — Checklist; precursor to descriptive treatment (Airy-Shaw 1980 in Kew Bulletin 35).
- Airy-Shaw, H. K. (1980b). A partial synopsis of the Euphorbiaceae-Platylobeae of Australia (excluding *Phyllanthus*, *Euphorbia* and *Calycopeplus*. Kew Bull. 35: 577-700. En. — Complete treatment of family except for the second of the two subdivisions of Mueller (Argoviensis), the 'Stenolobeae' (all Australasian!), along with the cited 'platyloboid' genera.

Eurasia

The whole of Europe and northern Eurasia has been documented in floristic publications since 1945. Only *Euphorbia* is at all well represented; however, for this genus species concepts vary among the major floras and no overall synthesis has been essayed. Kuzmanov (1964) contributed a paper on the biogeography and possible evolution of *Euphorbia* subgen. *Esula* in Europe with Macaronesia notable for the presence of certain of its lesser advanced groups. *Mercurialis* is also well-represented, and *Chrozophora* is found in southern areas.

Zimmermann, W. G., G. Hegi & H. Beger (1923). Euphorbiaceae. Illustrierte Flora von Mitteleuropa 5(1): 113-190, illus., col. pls. 177-178. Munich. Ge. — Detailed descriptive illustrated flora treatment with a substantial general introduction to the family (including original morphological observations by Zimmermann), keys, indication of distribution and habitat, vernacular names, and extensive biological, economic and other commentary. For additions and corrections, see *idem*, 1968. Nachträge, Berichtigungen und Ergänzungen zu den unveräncerten Nachdrucken der Bände IV/3 and V/1 bis V/4 mit Verzeichnissen der lateinischen und deutschen Pflanzennamen. Munich (Euphorbiaceae, 5(1), pp. 671b, 671c, 671d and 672).

Pojarkova, A. I., B. K. Shishkin and Ya. I. Prokhanov (1949). Euphorbiaceae. Flora SSSR 14: 267-495. Moscow/Leningrad. Ru. — The largest part of this chapter comprises Prokhanov's treatment of *Euphorbia*. [Whole text and keys also available in English in the Israel Program for Scientific Translations edition of this work.]

Tutin, T. G. et al. (eds) (1968). Euphorbiaceae. Fl. Europaea 2: 211-226. Cambridge. En. — Concise flora treatment. [By far the largest genus is *Euphorbia*, for which no other modern treatment encompassing all Europe is available.]

Greuter, W. et al. (eds) (1986). Euphorbiaceae. Med-Checklist 3: 205-224. Geneva/Berlin. En. — Tabular checklist. [By far the largest genus is *Euphorbia*, for which no other modern regional treatment is available.]

Malagassia

Basic revisions exist for *Euphorbia* (Denis 1921; there since have been many additions), *Croton* (q.v.) and the Phyllanthoideae. From 1935 Leandri followed up his Malagasy enumeration with many papers (partly also listed under the subfamilies) as well as a flora treatment (1958) for Phyllanthoideae and *Androstachys* (=*Stachyandra*), now in Oldfieldioideae. At the present time research is in progress towards a new synthesis, with studies by Alan Radcliffe-Smith, Gordon McPherson, Petra Hoffmann and others. Revisions of several smaller genera have been published. Mascarene representatives were fully revised by Coode and collaborators for *Flore des Mascareignes* (1982).

- Denis, M. (1921). Les Euphorbiées des îles australes d'Afrique. 153 pp., illus. Nemours. (Thèse, Nemours, 1921. Reissued in Rev. Gén. Bot. 34: 1-64, 96-123, 170-177, 214-236, 287-299, 346-366. 1922.) Fr. — Mostly on *Euphorbia*; for fuller description see under that genus.
- Leandri, J. (1935). Euphorbiaceae. 51 pp. Tananarive: Colonie de Madagascar et Dépendances/Paris: Société d'Éditions Géographiques, Maritimes et Coloniales. (Académie Malgache: Catalogue des plantes de Madagascar.) Fr. — Concise systematically arranged enumeration with references, synonymy, indication of distribution, habitat and phenology, and diagnostic features; no index. [The most recent nominally complete enumeration of family for Madagascar and the Comoros; superseded for Phyllanthoideae (sensu Pax and Hoffmann). Extended by the author's Contributions à l'étude des Euphorbiacées de Madagascar (1937-47; see below).]

Leandri, J. (1937). Sur la distribution et las affinités des Phyllanthées de Madagascar. See **Phyllanthoideae**.

Leandri, J. (1937-47). Contributions à l'étude des Euphorbiacées de Madagascar, I-XI. Notul. Syst. (Paris) 6-13, illus., *passim*. Fr. — Contains treatments in various tribes, preparatory to the *Flore de Madagascar* account. [Most separately accounted for under individual genera or subfamilies.]

Leandri, J. (1939). Euphorbiacées malgaches nouvelles recoltées par M. H. Perrier de la Bâthie. Bull. Soc. Bot. France 85: 523-533. Fr. — Includes among other notes descriptions of *Benoistia*, *Cladogelonium*, *Claoxylopsis*, *Danguyodrypetes* (now reduced to *Lingelsheimia*) and *Paragelonium* (now in *Aristogeitonia*).

Leandri, J. (1948). Sur les affinités des Euphorbiacées de Madagascar. Mém. Acad. Malg., hors-série: 163-167. (In memoriam M. Dr. Maurice Fontoynont.) Fr. — Statistical table covering all genera in Madagascar with number of Malagasy species, the world total, and world distribution; discussion and summary with general speculations on possible sources of the flora. [The so-called 'archaic' elements apparently retained a strong affinity with South America in spite of the elapse of time since fragmentation of 'Africa-Brazil'. A total of 410 species (360 endemic) in 53 genera were by then known for the territory, about a quarter more than in 1935.]

Leandri, J. (1952). Les arbres et grands arbustes malgaches de la famille des Euphorbiacées. Natural. Malgache 4: 47-82, illus. Fr. — Systematic enumeration, with keys, of genera (34) and species (117) with vernacular names and distribution. [A useful introductory account. Includes some fine illustrations by Mlle. J. Vesque and the author.]

Leandri, J. (1957). Notes systématiques sur les Euphorbiacées-Phyllanthées de Madagascar. Mém. Inst. Sci. Madagascar, B (Biol. Sci.), 8: 205-261. Fr. — Novelties and notes (the largest part on *Phyllanthus*); precursory to *Flore de Madagascar* treatment. A historical review also covering individual genera is included.

• Leandri, J. (1958). Euphorbiacées, I. Flore de Madagascar et des Comores (plantes vasculaires), famille 111: 1-209, 35 text-fig. Tananarive: Gouvernment Général de Madagascar. Fr. — Flora treatment, with keys and illustrations; covers 18 genera of Phyllanthoideae (sensu Pax and Hoffmann 1931, including *Androstachys* which now is usually placed in Oldfieldioideae). An index is included but there is no general part apart from the key to genera. [The remaining subfamilies (Acalyphoideae, Crotonoideae and Euphorbioideae) have yet to be published for this flora.]

Leandri, J. (1962). Notes sur les Euphorbiacées malgaches. Adansonia, II, 2: 216-223. Fr. — Novelties and notes; includes description of *Bossera*.

Coode, M. J. E. (with A. Radcliffe-Smith and A. J. Scott) (1982). 160. Euphorbiacées. 117 pp. (in J. Bosser et al. (eds), Flore des Mascareignes, 153. Lauracées à 160. Euphorbiacées). Mauritius: Govt. Printer (for Sugar Industry Research Institute *et al.*) Fr. — Flora treatment, with only a few exsiccatae cited. For additional details, see M. J. E. Coode, 1979-80. Notes on Euphorbiaceae in the Mascarene Islands, I-II. Kew Bull. 33: 109-120; 34: 39-48. For *Euphorbia*, see also Denis (1921) under Euphorbieae.

Brunel, J. F. (1987). Sur le genre *Phyllanthus* L. et quelques genres voisins de la tribu des Phyllantheae Dumort. (Euphorbiaceae, Phyllantheae) en Afrique intertropicale et à Madagascar. xiii, 472, 196 pp., illus. Strasbourg. (Thèse, Université de Strasbourg I). Fr. — The main part includes a historical survey, character analysis, and suprageneric revision. [The catalogue of *Phyllanthus* in the annex is not directly applicable to Madagascar.]

Radcliffe-Smith, A. (1988-). Notes on Madagascan Euphorbiaceae, I- . Kew Bull. 43: 625-647, *passim*. En. — Four parts to date (1993), also cited under individual genera.

Malesia

The whole region was reviewed or restudied by H. K. Airy-Shaw in a long series of papers and several island treatments from 1960 through 1983. In 1993, a new programme for Malesian Euphorbiaceae was undertaken at the Rijksherbarium, Leiden, under the direction of Peter van Welzen; its principal goal is a family treatment in *Flora Malesiana*. Genera were allocated to specialists and students; already some revisions have been published. *Malesian Euphorbiaceae Newsletter* was instituted in 1993 (with issue 9 in March 1999) and a short symposium on progress was held at the 3rd Flora Malesiana Symposium in 1995 (a further workshop also took place at the 4th Symposium in 1998). Whitmore (1997) in his phytogeographic review gives figures of 91 genera and 1254 species as being present in Malesia.

Robinson, C. B. (1909). Philippine Phyllantheae. Philip. J. Sci. 4, Bot.: 71-105. En. — 35 species, of which 19 in *Phyllanthus*; keys, descriptions, citations of exsiccatae and commentary. Also includes *Cicca* (=*Phyllanthus acidus*), *Securinega* (=*Flueggea* with respect to local species), *Flueggea*, *Sauropus*, *Glochidion* and *Breynia*.

Smith, J. J. (1910). Euphorbiaceae. In Koorders, S. H. and T. Valeton, Bijdrage tot de kennis der boomsoorten op Java 12: 9-637. Buitenzorg. Meded. Dept. Landb. 10. La/Du. — A 'very important contribution' (van Steenis).

Merrill, E. D. (1912-20). Notes on Philippine Euphorbiaceae, I-III. Philipp. J. Sci., Bot. 7: 379-410; 9: 461-493 (1914); Philipp. J. Sci. 16: 539-579. En. — A long series of novelties and notes, without keys. [Some 100 new species were described in these three papers, a major basis for the treatment of the family in the author's *Enumeration of Philippine flowering plants* (1923-26).]

Airy-Shaw, H. K. (1960). Notes on Malaysian Euphorbiaceae, II-XV, XVI-XIX. Kew Bull. 14: 353-397, 469-475. En. — For contents, see individual genera.

Airy-Shaw, H. K. (1963). Notes on Malaysian and other Asiatic Euphorbiaceae, XX-XLVIII. Kew Bull. 16: 341-372. En. — For contents, see individual genera.

Airy-Shaw, H. K. (1965). Notes on Malaysian and other Asiatic Euphorbiaceae, XLIX-LV. Kew Bull. 19: 299-328. En. — For contents, see individual genera.

Airy-Shaw, H. K. (1966-67). Notes on Malaysian and other Asiatic Euphorbiaceae, LVI-LXVI, LXVII-LXXXII. Kew Bull. 20: 25-49, 379-415. En. — For contents, see individual genera.

Airy-Shaw, H. K. (1968). Notes on Malesian and other Asiatic Euphorbiaceae, LXXXIII-XCVI. Kew Bull. 21: 353-418. En. — For contents, see individual genera.

Airy-Shaw, H. K. (1969). Notes on Malesian and other Asiatic Euphorbiaceae, CXII. Notes on the subtribe Plukenetiinae Pax. Kew Bull. 23: 114-121. En. — Notes on *Tragia*, *Pachystylidium*, *Cnesmone* and *Megistostigma*, with key to all species of the last-named. In effect an extension to Croizat (1941).

Airy-Shaw, H. K. (1969). Notes on Malesian and other Asiatic Euphorbiaceae, XCVII-CXIX. Kew Bull. 23: 1-131. En. — For contents, see individual genera.

Airy-Shaw, H. K. (1971). Notes on Malesian and other Asiatic Euphorbiaceae, CXX-CXLVIII. Kew Bull. 25: 473-553. En. — For contents, see individual genera.

Airy-Shaw, H. K. (1972). Notes on Malesian and other Asiatic Euphorbiaceae, CXLIX-CLXIX. Kew Bull. 27: 3-93. En. — For contents, see individual genera.

Airy-Shaw, H. K. (1974). Notes on Malesian and other Asiatic Euphorbiaceae, CLXX-CLXXXVI. Kew Bull. 29: 281-331. En. — For contents, see individual genera.

Airy-Shaw, H. K. (1974). Noteworthy Euphorbiaceae from tropical Asia (Burma to New Guinea). Ic. Pl. 38: pl. 3701-3725. En. — A suite of 25 species of particular interest, featuring drawings, descriptions and commentary.

• Airy-Shaw, H. K. (1975). The Euphorbiaceae of Borneo. 245 pp. (Kew Bull. Addit. Ser. 4). London: HMSO. En. — Concise regional monograph, with keys, synonymy, representative exsiccatae (relatively few), and indication of distribution; index. [In parts now superseded.]

Airy-Shaw, H. K. (1978). Notes on Malesian and other Asiatic Euphorbiaceae, CLXXXVII-CCVI. Kew Bull. 32: 361-418. En. — For contents, see individual genera.

Airy-Shaw, H. K. (1978-79). Notes on Malesian and other Asiatic Euphorbiaceae, CCVII-CCXXI, CCXXII-CCXXVIII. Kew Bull. 33: 25-77, 529-538. En. — For contents, see individual genera.

Airy-Shaw, H. K. (1980). New Euphorbiaceae from New Guinea. Kew Bull. 34: 591-598. En. — Notes and novelties, preparatory to *The Euphorbiaceae of New Guinea.*

Airy-Shaw, H. K. (1980). Notes on Euphorbiaceae from Indomalesia, Australia and the Pacific, CCXXIX-CCXLI. Kew Bull. 35: 383-399. En. — Notes and novelties; for contents see individual genera.

• Airy-Shaw, H. K. (1980). The Euphorbiaceae of New Guinea. 243 pp., illus. (Kew Bull. Addit. Ser. 8.) London: HMSO. En. — Concise regional monograph, with keys, synonymy, representative exsiccatae (relatively few), indication of distribution, and some illustrations; index. [In parts now superseded.]

Airy-Shaw, H. K. (1981). Notes on Asiatic, Malesian and Melanesian Euphorbiaceae, CCXLII-CCLII. Kew Bull. 36: 599-612. En. — For contents, see individual genera.

• Airy-Shaw, H. K. (1981). The Euphorbiaceae of Sumatra. Kew Bull. 36: 239-374, illus. En. — Concise area revision with keys, citations, limited synonymy, distribution and commentary. Genera, alphabetically arranged, are described, and a separate index to all names is furnished at end.

Airy-Shaw, H. K. (1982). New Euphorbiaceae from Sumatra, New Guinea, Australia and New Caledonia. Kew Bull. 37: 377-381. En. — Notes and novelties.

• Airy-Shaw, H. K. (1982). The Euphorbiaceae of Central Malesia (Celebes, Moluccas, Lesser Sunda Is.). Kew Bull. 37: 1-40. En. — Concise enumeration with descriptions of novelties, literature citations and summary of distribution; no keys or *exsiccatae*.

- Airy-Shaw, H. K. (1983). An alphabetical enumeration of the Euphorbiaceae of the Philippine Islands. 56 pp. Kew: the author. En. — Concise enumeration with synonymy and exsiccatae represented at Kew; no keys. [A work of limited circulation; intended as a partial revision of the account in *Enumeration of Philippine Flowering Plants*.]

Whitmore, T. C. (1997(1998)). The phytogeography of Malesian Euphorbiaceae. In J. Dransfield, M. J. E. Coode & D. A. Simpson (eds), *Plant diversity in Malesia*, III: *proceedings of the third international Flora Malesiana symposium* (Kew, July 1995): 389-404, maps. Kew: Royal Botanic Gardens. En. — 'Narrative' overview based on a relatively well-revised base of knowledge, with several maps each divided into four primary units (West, East, Central and South Malesia); results broadly suggest that there has been strong speciation in the perhumid parts of West and East Malesia, and that the Philippines is currently the least well-understood area.

Oceania

Flora treatments have been published in recent years for New Caledonia (see below), French Polynesia (see below) and Fiji (in *Flora vitiensis nova*).

Baillon, H. E. (1862). Species Euphorbiacearum: Euphorbiaceae neo-caledonicae. Adansonia 2: 211-242. Fr. — Synoptic treatment with descriptions of novelties, synonymy, localities with exsiccatae, and commentary; covers 18 genera and 58 species. [The first description of what is now *Neoguillauminia cleopatra* appears on p. 213; Baillon described it in *Euphorbia* but created for it a new section *Decadenia*. *E. vieillardii*, now part of *E. tannensis*, is also described.]

Guillaumin, A. (1929). Révision des Euphorbiacées de la Nouvelle-Calédonie. 48 pp. (Arch. Bot. (Caen), Mém. 2(3)). Caen. (Materiaux pur la Flore de la Nouvelle-Calédonie XXVI.) Fr. — Additions and corrections to Baillon's treatment, with analytical keys; 31 genera and 171 species now recorded. General comments appear on pp. 45-47. [The family was reckoned to be the fourth largest among vascular plants in the territory. The treatment includes *Trisyngyne*, now excluded from Euphorbiaceae and referred to *Nothofagus* (Fagaceae or Nothofagaceae). The whole is now superseded.]

- McPherson, G., C. Tirel and M. Schmid (1987-91). Euphorbiacées, I-II. Flore de la Nouvelle-Calédonie et dépendances, 14, 17. Illus., maps. Paris. Fr. — Complete treatment; Phyllanthoideae (part 2) quite the largest of the 5 subfamilies represented (123 species, with 111 native *Phyllanthi*). Other subfamilies comprise part 1 (Euphorbioideae, 11 species; Crotonoideae, 26; Acalyphoideae, 36; Oldfieldioideae, 16).

Florence, J. (1996). Gallicae Polynesiae florae praecursores, 1. Nouveautes taxonomiques dans les Euphorbiaceae, Piperaceae et Urticaceae. Bull. Mus. Natl. Hist. Nat., IV, B (Adansonia), 18: 239-274, illus. Fr. — Includes, in Euphorbiaceae, notes and novelties in *Euphorbia* (as *Chamaesyce*), *Glochidion* and *Macaranga*. [See also family account in *Flore de la Polynésie française*, 1 (1997).]

- Florence, J. (1997). Euphorbiaceae. Fl. Polynésie Française 1: 27-141, illus. Paris: ORSTOM. Fr. — Flora treatment with keys, descriptions, synonymy, types, indication of vernacular names, phenology, distribution, ecology and uses, localities with exsiccatae, and critical notes; listings of cultivated, doubtful and excluded species. [A very careful treatment.]

1. Phyllanthoideae

A great proportion of this subfamily was revised by Pax and his collaborators for *Das Pflanzenreich* but not *Phyllanthus, Glochidion* and their near relatives. Several genera were removed to the new subamily Oldfieldioideae following the work of Köhler (1965). Porantthereae, formerly in the 'Stenolobieae', are now included here. On the other hand, there are persistent hints that the subfamily is heterogeneous; several genera have at one or another time been segregated into distinct families and, more significantly, its core may have nothing to do with the rest of the family. *Bischofia* – by some also given family rank – is seen by advocates of unity as a 'link' between its two main lines. A goodly number of studies are now available covering anatomy, morphology, pollen and seed structure. More are to come using the methods of molecular systematics; preliminary results suggest that the subfamily, along with the Oldfieldioideae, should indeed be removed to a separate family (M. Chase, personal communication, July 1998). The results also confirm ideas about the distinctness of Drypeteae, with family rank (as Putranjivaceae) also likely (cf. Tokuoka & Tobe, 1999).

Froembling, W. (1896). Anatomisch-systematische Untersuchung von Blatt und Axe der Crotoneen und Euphyllantheen. Bot. Centralbl. 17(65): 129-139, 177-192, 241-249, 289-297, 321-329, 369-378, 403-411, 433-441; 2 pls. (Diss., Univ. München; reprinted, 76 pp., 2 pls., Cassel.) Ge. — A systematic-anatomical study of *Croton* and its immediate allies as well as of *Phyllanthus* and its immediate allies. Most of the paper is on *Croton* where a great diversity of structure was observed; such did not obtain in *Phyllanthus* for which coverage is limited to the last installment (with some attention given to the phenomenon of phylloclady in part of the genus). All species studied are listed, and keys to the internal structures in *Croton* are spread over much of the work. The captions to the two plates appear in the last installment. [Prepared under the direction of L. Radlkofer.]

Rothdauscher, H. (1896). Ueber die anatomischen Verhältnisse von Blatt und Axe der Phyllantheen (mis Ausschluss der Euphyllantheen). Bot. Centralbl. 17(68): 65-79, 97-108, 129-136, 161-169, 193-203, 248-253, 280-285, 305-315, 338-346, 385-393. (Diss., Univ. München; reprinted, 89 pp., Cassel.) Ge. — A systematic-anatomical study of Phyllanthoideae (except *Phyllanthus* and immediate relatives); general part in first two installments with summary of the distribution of anatomical features among genera on pp.104-108; special part from third installment onwards with descriptions of the anatomical features within individual genera (all vouchered!). [Prepared under the direction of L. Radlkofer.]

Gruening, G. (1913). Euphorbiaceae-Porantheroideae et Ricinocarpoideae. In A. Engler (ed.), Das Pflanzenreich, IV 147 [St]. 97 pp., illus. Berlin: Engelmann (im Auftrage der Preuss. Akademie der Wissenschaften). (Heft 58.) La/Ge. — The Porantheroideae here encompass 4 genera, *Poranthera* (including *Oreoporanthera*) (tribe Porantthereae) and *Micrantheum, Pseudanthus* and *Stachystemon* (tribe Caletieae). [In Webster's system the Porantthereae are a subtribe of Antidesmeae in the Phyllanthoideae and the Caletieae are classed with Oldfieldioideae.]

Jablonski, E. (as E. Jablonszky) (1915). Euphorbiaceae-Phyllanthoideae-Bridelieae. In A. Engler (ed.), Das Pflanzenreich, IV 147 VIII. 98 pp., illus. Berlin: Engelmann (im Auftrage der Preuss. Akademie der Wissenschaften). (Heft 65.) La/Ge. — Includes *Cleistanthus* (130, Africa to Asia-Pacific) and *Bridelia* (over 60, Africa to Asia-Pacific). [*Dendrophyllanthus*, from New Caledonia, and *Godefroya*, from Cambodia, not covered here, were by 1931 added to the tribe; the former is now part of *Phyllanthus* and the latter part of *Cleistanthus*.]

Pax, F. and K. Hoffmann (1922). Euphorbiaceae-Phyllanthoideae-Phyllantheae. In A. Engler (ed.), Das Pflanzenreich, IV 147 XV. 349 pp., illus. Berlin: Engelmann (im Auftrage der Preuss. Akademie der Wissenschaften). (Heft 81.) La/Ge. — Subtribes from Antidesminae to Bischoffiinae (except Phyllanthinae and Glochidiinae). Of entry 17 (p. 2), *Cubincola* is in *Cneorum* (Cneoraceae) and *Riseleya* is in *Drypetes*).

Leandri, J. (1937). Contributions à l'étude des Euphorbiacées de Madagascar, I-II: Phyllanthées. Notul. Syst. (Paris) 6: 11-35; 7: 185-199. Fr. — Precursory notes on Phyllanthoideae, preparatory to an account in *Flore de Madagascar et des Comores* (see Leandri, 1958, under **Malagassia**).

Leandri, J. (1937). Sur la distribution et les affinités des Phyllanthées de Madagascar. Bull. Soc. Bot. France 84: 61-72, 93-98, illus., maps. Fr. — An account of the geographical and ecological distribution of the group and its species, with notes on possible origins (including maps) as well as methods of propagule dispersal; putative history (pp. 96-98); list of references (p. 98). [The author, cautiously, supports continental separation as an explanation for the isolation of Madagascar.]

Leandri, J. (1944). Contribution à l'étude des Euphorbiacées de Madagascar, VIII: Bridéliées, Géloniées. Notul. Syst. (Paris) 11: 151-159. Fr. — Novelties and notes.

Leandri, J. (1957). Notes systématiques sur les Euphorbiacées-Phyllanthées de Madagascar. Mém. Inst. Sci. Madagascar, B (Biol. Sci.), 8: 205-261. Fr. — Novelties and notes (the largest part on *Phyllanthus*); precursory to *Flore de Madagascar* treatment. A historical review also covering individual genera is included.

• Köhler, E. (1965). Die Pollenmorphologie der biovulaten Euphorbiaceae und ihre Bedeutung für die Taxonomie. Grana Palynol. 6: 26-120, illus. (Based on the author's 1962 dissertation of the same title, Universität Jena.) Ge. — General introduction; methodology and taxa studied (231 species in 75 genera, arranged following Pax 1925 and Pax & Hoffmann 1931; see **General**); description of pollen types; discussion (pp. 62-92) of taxonomic value and of 27 perceived pollen types and (pp. 92 ff.) of pollen evolutionary trends (diagrams, pp. 96-97); proposal for a new classification (pp. 98-101). [In this paper subfamily Oldfieldioideae is proposed for the first time; it was formalized in Webster 1967 (see **Americas**).]

Airy-Shaw, H. K. (1971). Notes on Malesian and other Asiatic Euphorbiaceae, CXXIV. A key to the Securineginae and some deceptively similar Phyllanthinae of tropical and temperate Asia. Kew Bull. 25: 491-493. En. — Analytical key covering *Richeriella, Margaritaria, Securinega* (and *Flueggea*), and parts of *Phyllanthus*. [*Cicca* is here kept separate from *Phyllanthus*.]

Levin, G. A. (1986). Systematic foliar morphology of Phyllanthoideae (Euphorbiaceae), I-III. Ann. Missouri Bot. Gard. 73: 29-85, 86-98, illus.; Syst. Bot. 11: 515-530. En. — The first part, 'Conspectus', comprises a character table and systematic presentation of data, with illustrations, covering 259 species in 49 genera (all vouchered). Parts 2, 'Phenetic analysis' and 3, 'Cladistic analysis', are attempts at resolution of the data and include speculations on putative relationships of the family. The author in part 1 concludes that leaf architecture provides useful characters at higher levels and that the family is dilleniid with an affinity in the vicinity of the Violales. Earlier suggestions of a relationship with Celastrales (in the Rosidae) were more or less rejected. In part 2 he notes that 'at least in the Phyllanthoideae, the amount of parallelism and convergence in leaf architectural and cuticular characters is small enough that overall similarity usually reflects presumed evolutionary relationship'. The cladistic analyses of part 3 were even more congruent with earlier classifications.

Mennega, A.M.W. (1987). Wood anatomy of the Euphorbiaceae, in particular of the subfamily Phyllanthoideae. Bot. J. Linn. Soc. 94: 111-126, illus. En. — Includes systematic descriptions of 116 species of 35 genera in 13 tribes following Webster's classification. Wood-anatomical evidence supports two groups of tribes: an *Aporusa*-type (less advanced) and a *Glochidion*-type (more advanced). *Uapaca* was found to be quite distinctive, supporting Airy-Shaw's suggestions for family status.

Simpson, M. C. & G. A. Levin (1994). Pollen ultrastructure of the biovulate Euphorbiaceae. Int. J. Pl. Sci. 155: 313-341, illus. En. — Phyllanthoideae and Oldfieldioideae; systematic investigation of a range of genera with descriptions of pollen from representative species; table of species and characters; cladistic analyses. [Retention of *Hymenocardia* in the subfamily is supported.]

Stuppy, W. (1995). Systematische Morphologie und Anatomie der Samen der biovulaten Euphorbiaceen. iii, 364 pp., illus. Kaiserslautern. (Doctoral dissertation, Universität Kaiserslautern.) Ge. — A study of the surface morphology and histology of seeds from 163 species in 81 genera (51 in Phyllanthoideae, 22 in Oldfieldioideae) building on Corner's treatment in *Seeds of Dicotyledons* (1976). The author shows there to be large differences between the uni- and biovulate branches of the family, with the former more homogeneous (and perhaps warranting treatment as a single subfamily). The biovulate genera by contrast include several worthy of elevation or restoration to family rank; indeed, the whole group is viewed as potentially polyphyletic. Nevertheless, their origins might point to relatively close relationships among their progenitors. Specific recommentations regarding taxonomic position are made with respect to some genera.

Govaerts, R. & A. Radcliffe-Smith (1996). New names and combinations in Euphorbiaceae: Phyllanthoideae. Kew Bull. 51: 175-178. En. — Nomenclatural, precursory to *World Checklist and Bibliography of Euphorbiaceae* (this volume). Includes taxa in *Aporusa, Drypetes, Glochidion* and *Phyllanthus*.

Radcliffe-Smith, A. & R. Govaerts (1997). New names and combinations in Euphorbiaceae-Phyllanthoïdeae. Kew Bull. 51: 175-178. En. — Miscellaneous names and combinations, many in *Phyllanthus*.

Tokuoka, T. & H. Tobe (1999). Embryology of tribe Drypeteae, an enigmatic taxon of Euphorbiaceae. Pl. Syst. Evol. 215: 189-208, illus. En. — Review of previous evidence from ovule and seed morphology, chemistry and gene sequences as well as phyllanthoid embryology (a poorly documented area), along with new research; findings suggest that tribe is misplaced in Euphorbiaceae (and not even closely related), thus confirming the suggestion of Meeuse (1990) that separate family status (as Putranjivaceae) was in order. Alternative affinities were explored, with the best placement thought to be near Erythroxylaceae. A cladogram is presented (without Phyllanthoideae!) on p. 205. [*Drypetes* is notably the only genus in Euphorbiaceae producing glucasinolates (mustard-oil); this suggested to Meeuse a place in Capparales. Other supporting evidence for this conclusion was found, however, to be wanting.]

2. Oldfieldioideae

The first recognition of part of this subfamily came in 1965 with the publication of Androstachydaceae by Airy-Shaw (1965) for the Afro-Malagasy *Androstachys* s.l. In the same year, as part of his extensive palynological study, Köhler (1965) proposed the name Oldfieldoideae. Meeuse (1990; see **General**) somewhat tentatively proposed that the subfamily be reorganised as 2 families, Androstachydaceae (monotypic; now 2 genera) and Paivaeusaceae (all the rest). Although there are only some 100 species – making it by far the smallest subfamily – 28 genera (in four tribes) were accepted by Webster. They are for the greater part found in the southern hemisphere, and many appear relictual. The most coherent group appears to be the Caletieae, mainly in Australasia (and formerly in the 'Stenolobieae'). Hayden (1994) has suggested an origin in the primitive phyllanthoid complex, with particular reference to Wielandiinae. To the writers they appear rather heterogeneous. One of the few constant characters distinguishing the group from Phyllanthoideae appears to be spiny pollen; in addition, most genera are carunculate. An extensive comparative palynological study was produced by Levin & Simpson (1994). More are to come using the methods of molecular systematics; preliminary results suggest that the subfamily, along with the Phyllanthoideae, should be removed to a separate family (M. Chase, personal communication, July 1998).

Gruening, G. (1913). Euphorbiaceae-Porantheroideae et Ricinocarpoideae. In A. Engler (ed.), Das Pflanzenreich, IV 147 [St]. 97 pp., illus. Berlin: Engelmann (im Auftrage der Preuss. Akademie der Wissenschaften). (Heft 58.) La/Ge. — The Porantheroideae here encompass 4 genera, *Poranthera* (including *Oreoporanthera*) (tribe Poranthereae) and *Micrantheum, Pseudanthus* and *Stachystemon* (tribe Caletieae). [In Webster's system the Porantheraeae are a subtribe of Antidesmeae in the Phyllanthoideae and the Caletieae are classed with Oldfieldioideae; this follows the work of Köhler 1965.]

- Köhler, E. (1965). Die Pollenmorphologie der biovulaten Euphorbiaceae und ihre Bedeutung für die Taxonomie. Grana Palynol. 6: 26-120, illus. (Based on the author's 1962 dissertation of the same title, Universität Jena.) Ge. — General introduction; methodology and taxa studied (231 species in 75 genera, arranged following Pax 1925 and Pax & Hoffmann 1931; see **General**); description of pollen types; discussion (pp. 62-92) of taxonomic value and of 27 perceived pollen types and (pp. 92 ff.) of pollen evolutionary trends (diagrams, pp. 96-97); proposal for a new classification (pp. 98-101). [In this paper subfamily Oldfieldioideae is proposed for the first time; it was formalized in Webster 1967 (see **Americas**).]

Hayden, W. J. (1980). Systematic anatomy of Oldfieldioideae (Euphorbiaceae). College Park, Md. (Ph.D. dissertation, Univ. of Maryland, College Park.) En. — For published version (in part) see Hayden (1994).

Hayden, W. J. (1994). Systematic anatomy of Euphorbiaceae subfamily Oldfieldioideae, 1. Overview. Ann. Missouri Bot. Gard. 81: 180-202, illus. En. — Treatment of leaf and wood anatomy and karyology; considerations on possible internal relationships; phylogram (p. 182). [The phylogram also suggests that Drypetinae and Petalostigmatinae are closer to Oldfieldioideae than to the rest of the Phyllanthoideae s.l. Both subtribes have also been at one or another time been ranked as distinct families.]

- Levin, G. A. & M. G. Simpson (1994). Phylogenetic implications of pollen ultrastructure in the Oldfieldioideae (Euphorbiaceae). Ann. Missouri Bot. Gard. 81: 203-238, illus. En. — Detailed comparative study; cladistic essays (overall strict consensus trees, p. 225-226 and preferred diagrams for each tribe following); proposed classification (p. 230) and supporting cladogram (p. 231). [The diagrams suggest greater coherence among the Oldfieldioideae than the Phyllanthoideae. The 'basal' genus is *Croizatia*.]

Simpson, M. C. & G. A. Levin (1994). Pollen ultrastructure of the biovulate Euphorbiaceae. Int. J. Pl. Sci. 155: 313-341, illus. En. — Phyllanthoideae and Oldfieldioideae; systematic investigation of a range of genera with descriptions of pollen from representative species; table of species and characters; cladistic analyses.

Stuppy, W. (1995). Systematische Morphologie und Anatomie der Samen der biovulaten Euphorbiaceen. See **Phyllanthoideae**. — Worthy of particular note is a proposal for exclusion of *Oldfieldia* from the family.

3. Acalyphoideae

Most of the subfamily was revised for *Das Pflanzenreich* by Pax and K. Hoffmann, although their arrangement of genera bears only a partial relationship to that currently accepted. A basic anatomical study was essayed by Rittershausen (1892). No recent overall survey is available but in more general contributions on the family suggestions have been made that its delimitation from Crotonoideae (and Euphorbioideae) is not sharp.

Rittershausen, P. (1892). Anatomisch-systematische Untersuchung von Blatt und Achse der Acalypheen. xv, 123 pp., illus., 1 leaf of plates. Munich. (Diss., Univ. Erlangen.) Ge. — [Not seen; reference from Froembling 1896.]

Pax, F. (1910). Euphorbiaceae-Crotonoideae-Adrianeae. In A. Engler (ed.), Das Pflanzenreich, IV 147 II. 111 pp., illus. Berlin: Engelmann (im Auftrage der Preuss. Akademie der Wissenschaften). (Heft 44.) La/Ge. — 8 genera, including *Manihot*. [As conceived here the tribe encompassed unrelated elements. The circumscription of Adrianeae was reduced in 1931 with the exclusion of *Manihot* and its allies.]

Pax, F. and K. Hoffmann (1911). Euphorbiaceae-Crotonoideae-Cluytieae. In A. Engler (ed.), Das Pflanzenreich, IV 147 III. 124 pp., illus. Berlin: Engelmann (im Auftrage der Preuss. Akademie der Wissenschaften). (Heft 47.) La/Ge. — Subtribes Codeiaeinae (genera 1-12), Ricinodendrinae (genera 13-15), Cluytiinae (genera 16-20) and Galeariinae (genera 21-24). [No. 9, *Erismanthus*, no. 16, *Clutia* and nos. 20-24 now in Acalyphoideae; remainder in Crotonoideae except no. 17, *Schistostigma*, now part of *Cleistanthus* (Phyllanthoideae), and no. 19, *Uranthera*, now in *Phyllanthus* via *Phyllanthodendron* (Phyllanthoideae).]

Pax, F. and K. Hoffmann (1912). Euphorbiaceae-Crotonoideae-Acalypheae-Chrozophorinae. In A. Engler (ed.), Das Pflanzenreich, IV 147 VI. 142 pp., illus. Berlin: Engelmann (im Auftrage der Preuss. Akademie der Wissenschaften). (Heft 57.) La/Ge. — 20 genera in two series, Regulares and Irregulares. [Taxon raised to tribal status by 1931.]

Pax, F. and K. Hoffmann (1912). Euphorbiaceae-Crotonoideae-Gelonieae. In A. Engler (ed.), Das Pflanzenreich, IV 147 IV. 41 pp., illus. Berlin: Engelmann (im Auftrage der Preuss. Akademie der Wissenschaften). (Heft 52, pars.) La/Ge. — Of the three subtribes accepted, only the genera of Chaetocarpinae, *Chaetocarpus* (incl. *Mettenia*) and *Cheilosa*, may be referred to Acalyphoideae.

Gruening, G. (1913). Euphorbiaceae-Porantheroideae et Ricinocarpoideae. In A. Engler (ed.), Das Pflanzenreich, IV 147 [St]. 97 pp., illus. Berlin: Engelmann (im Auftrage der Preuss. Akademie der Wissenschaften). (Heft 58.) La/Ge. — The Ricinocarpoideae here encompasses *Ricinocarpos*, *Bertya* and *Beyeria* (tribe Ricinocarpeae) along with *Monotaxis* and *Amperea* (Ampereae). [In Webster's system the former tribe is classed with Crotonoideae, the latter, Acalyphoideae.]

Prain, D. (1913). Mercurialineae and Adenoclineae of South Africa. Ann. Bot. 27: 371-410. En. — Mostly on *Adenocline*, *Leidesia* and *Seidelia* (see there); these are entirely or largely S African.

Pax, F. and K. Hoffmann (1914). Euphorbiaceae-Crotonoideae-Acalypheae-Mercurialinae. In A. Engler (ed.), Das Pflanzenreich, IV 147 VII. 473 pp., illus. Berlin: Engelmann (im Auftrage der Preuss. Akademie der Wissenschaften). (Heft 63.) La/Ge. — Includes, among other genera, treatments of *Alchornea*, *Claoxylon*, *Macaranga*, *Mallotus* and *Melanolepis*.

Pax, F. and K. Hoffmann (1919). Euphorbiaceae-Crotonoideae-Acalypheae-Epiprininae. In A. Engler (ed.), Das Pflanzenreich, IV 147 X. pp. 109-111, illus. Berlin: Engelmann (im Auftrage der Preuss. Akademie der Wissenschaften). (Heft 68, pars.) La/Ge. — 1 genus, *Epiprinus*, SE Asia.

Pax, F. and K. Hoffmann (1919). Euphorbiaceae-Crotonoideae-Acalypheae-Plukenetiinae. In A. Engler (ed.), Das Pflanzenreich, IV 147 IX. pp. 1-108, illus. Berlin: Engelmann (im Auftrage der Preuss. Akademie der Wissenschaften). (Heft 68, pars.) La/Ge. — 19 genera, including among others *Tragia* and *Plukenetia*.

Pax, F. and K. Hoffmann (1919). Euphorbiaceae-Crotonoideae-Acalypheae-Ricininae. In A. Engler (ed.), Das Pflanzenreich, IV 147 XI. pp. 112-134, illus. Berlin: Engelmann (im Auftrage der Preuss. Akademie der Wissenschaften). (Heft 68, pars.) La/Ge. — 3 genera, *Homonoia* (incl. *Spathiostemon*), *Lasiococca* (1, Himalaya) and *Ricinus* (1, Africa/Asia but now widely spread).

Pax, F. and K. Hoffmann (1919). Euphorbiaceae-Crotonoideae-Dalechampieae. In A. Engler (ed.), Das Pflanzenreich, IV 147 XII. 59 pp., illus. Berlin: Engelmann (im Auftrage der Preuss. Akademie der Wissenschaften). (Heft 68, pars.) La/Ge. — Only the genus *Dalechampia* (100, mostly American).

Pax, F. and K. Hoffmann (1919). Euphorbiaceae-Crotonoideae-Pereae. In A. Engler (ed.), Das Pflanzenreich, IV 147 XIII. 14 pp., illus. Berlin: Engelmann (im Auftrage der Preuss. Akademie der Wissenschaften). (Heft 68, pars.) La/Ge. — Only the genus *Pera* (35, Americas).

Pax, F. and K. Hoffmann (1924). Euphorbiaceae-Crotonoideae-Acalypheae-Acalyphinae. In A. Engler (ed.), Das Pflanzenreich, IV 147 XVI: pp. 1-178, illus. Berlin: Engelmann (im Auftrage der Preuss. Akademie der Wissenschaften). (Heft 85, pars.) La/Ge. — 2 genera, *Acalypha* and *Acalyphopsis* (the latter now reduced to *Acalypha*).

Croizat, L. (1941). The tribe Plukenetiinae [sic!] of the Euphorbiaceae in eastern tropical Asia. J. Arnold Arbor. 22: 417-431. En. — Contains numerous additions and revisions to the Pax and Hoffmann subtribe, encompassing *Tragia*, *Cnesmone*, *Megistostigma* and *Plukenetia* as well as additions to *Sphaerostylis* and a note on *Ramelia* (=*Bocquillonia*). A key to the first four genera is presented.

Leandri, J. (1941). Contribution à l'étude des Euphorbiacées de Madagascar, IV: Acalyphées. Notul. Syst. (Paris) 9: 156-188, illus. Fr. — Novelties and notes.

Leandri, J. (1943). Contribution à l'étude des Euphorbiacées de Madagascar, VI: Acalyphées. Notul. Syst. (Paris) 10: 252-291, illus. Fr. — Novelties and notes, in continuation of Leandri (1941); mostly on *Acalypha* and including key to the 22 known species (pp. 288-290).

Airy-Shaw, H. K. (1969). Notes on Malesian and other Asiatic Euphorbiaceae, CXII. Notes on the subtribe Plukenetiinae Pax. Kew Bull. 23: 114-121. En. — Notes on *Tragia*, *Pachystylidium*, *Cnesmone* and *Megistostigma*, with key to all species of the last-named. In effect an extension to Croizat (1941).

Gillespie, L. J. (1994). Pollen morphology and phylogeny of the tribe Plukenetieae (Euphorbiaceae). Ann. Missouri Bot. Gard. 81: 317-348, illus. En. — Pollen morphology and systematics.

Gillespie, L. J. & W. S. Armbruster (1997). A contribution to the Guianan flora: *Dalechampia*, *Haematostemon*, *Omphalea*, *Pera*, *Plukenetia* and *Tragia* (Euphorbiaceae) with notes on subfamily Acalyphoideae. iv, 48 pp., illus. (Smithsonian Contr. Bot. 86). Washington. En. — Comprises illustrated descriptive regional revisions of six genera with keys, synonymy, types, indication of vernacular names, distribution and ecology, localities with exsiccatae, and commentary; key references at end of each genus. A list of taxa and collections, literature, and index conclude the work. The opening headings contain keys to the subfamilies of Euphorbiaceae and a key to all genera of Acalyphoideae in the Guianas. [The generic accounts are not separately accounted for under generic headings; none of the genera has a significant percentage of its species in the region.]

Nowicke, J. W., M. Takahashi & G. L. Webster (1998-99). Pollen morphology, exine structure and systematics of Acalyphoideae (Euphorbiaceae), I-II. Rev. Palaeobot. Palynol. 102: 115-152; 105: 1-62, illus. En. — Systematic-palynological studies of the genera (with one or more representative species) of Clutieae, Pogonophoreae, Chaetocarpeae, Pereae, Cheiloseae, Erismantheae (in part), Dicoelieae, Galearieae, and Ampereae (in part I) and Agrostistachydeae, Chrozophoreae, Caryodendreae, Bernardieae and Pycnocomeae (in Part II), the tribes arranged and circumscribed following Webster (Synopsis, 1994). Light, scanning electron and transmission electron microscopic approaches have all been used and each genus is described and discussed. Vouchers are listed at the beginning of each paper. The authors suggest that in general the results – although arranged according to a different system of the family – strongly support the findings of Punt (1962; see **Special**). They also intended that the papers be essentially documentary. [The genera covered correspond to nos. 89-134 in Webster's system. An additional paper will conclude the survey.]

4. Crotonoideae

The subfamily was anatomically surveyed by Froembling (1896); pollen and laticifers have been studied respectively by Nowicke (1994) and Rudall (1994). Most genera – but not *Croton* or its immediate allies – were revised for *Das Pflanzenreich* from 1910 to 1913.

Froembling, W. (1896). Anatomisch-systematische Untersuchung von Blatt und Axe der Crotoneen und Euphyllantheen. Bot. Centralbl. 17(65): 129-139, 177-192, 241-249, 289-297, 321-329, 369-378, 403-411, 433-441; 2 pls. (Diss., Univ. München; reprinted, 76 pp., 2 pls., Cassel.) Ge. — A systematic-anatomical study of *Croton* and its immediate allies as well as of *Phyllanthus* and its immediate allies. Most of the paper is on *Croton* where a great diversity of structure was observed; such did not obtain in *Phyllanthus* for which coverage is limited to the last installment (with some attention given to the phenomenon of phylloclady in part of the genus). All species studied are listed, and keys to the internal structures in *Croton* are spread over much of the work. The captions to the two plates appear in the last installmentlment. [Prepared under the direction of L. Radlkofer.]

Pax, F. (1910). Euphorbiaceae-Crotonoideae-Adrianeae. In A. Engler (ed.), Das Pflanzenreich, IV 147 II. 111 pp., illus. Berlin: Engelmann (im Auftrage der Preuss.

Akademie der Wissenschaften). (Heft 44.) La/Ge. — 8 genera, including *Manihot.* [As conceived here the tribe encompassed unrelated elements. The circumscription of Adrianeae was reduced in 1931 with the exclusion of *Manihot* and its allies.]

Pax, F. (1910). Euphorbiaceae-Crotonoideae-Jatropheae. In A. Engler (ed.), Das Pflanzenreich, IV 147 [I]. 148 pp., illus. Berlin: Engelmann (im Auftrage der Preuss. Akademie der Wissenschaften). (Heft 42.) La/Ge. — Included here are Micrandrinae and Jatrophinae (the latter also including what later became Joannesieae). Genera covered include, among others, *Jatropha* (and *Cnidoscolus*), *Hevea* and *Aleurites* (sens. lat.). Jatropheae were later by Pax and Hoffmann (1931) ranked as a subtribe of Cluytieae.

Pax, F. and K. Hoffmann (1911). Euphorbiaceae-Crotonoideae-Cluytieae. In A. Engler (ed.), Das Pflanzenreich, IV 147 III. 124 pp., illus. Berlin: Engelmann (im Auftrage der Preuss. Akademie der Wissenschaften). (Heft 47.) La/Ge. — Subtribes Codeiaeinae (genera 1-12), Ricinodendrinae (genera 13-15), Cluytiinae (genera 16-20) and Galeariinae (genera 21-24). [No. 9, *Erismanthus*, no. 16, *Clutia* and nos. 20-24 now in Acalyphoideae; remainder in Crotonoideae except no. 17, *Schistostigma*, now part of *Cleistanthus* (Phyllanthoideae), and no. 19, *Uranthera*, now in *Phyllanthus* via *Phyllanthodendron* (Phyllanthoideae).]

Pax, F. and K. Hoffmann (1912). Euphorbiaceae-Crotonoideae-Gelonieae. In A. Engler (ed.), Das Pflanzenreich, IV 147 IV. 41 pp., illus. Berlin: Engelmann (im Auftrage der Preuss. Akademie der Wissenschaften). (Heft 52, pars.) La/Ge. — Of the three subtribes covered, the genera in Chaetocarpinae are now referred to Acalyphoideae; those in Geloniinae and Tetrorchidiinae remain in Crotonoideae.

Gruening, G. (1913). Euphorbiaceae-Porantheroideae et Ricinocarpoideae. In A. Engler (ed.), Das Pflanzenreich, IV 147 [St]. 97 pp., illus. Berlin: Engelmann (im Auftrage der Preuss. Akademie der Wissenschaften). (Heft 58.) La/Ge. — The Ricinocarpoideae here encompasses *Ricinocarpos*, *Bertya* and *Beyeria* (tribe Ricinocarpeae) along with *Monotaxis* and *Amperea* (Ampereae). [In Webster's system the former tribe is classed with Crotonoideae, the latter, Acalyphoideae.]

Leandri, J. (1944). Contribution à l'étude des Euphorbiacées de Madagascar, VIII: Bridéliées, Géloniées. Notul. Syst. (Paris) 11: 151-159. Fr. — Novelties and notes; includes keys, notably to *Gelonium* (=*Suregada*).

Nowicke, J. W. (1994). A palynological study of Crotonoideae (Euphorbiaceae). Ann. Missouri Bot. Gard. 81: 245-269, illus. En. — Treatment of 69 species from 34 genera in 12 (of 13) tribes (sensu Webster) using LM, SEM and TEM methodology; descriptions of pollen (with vouchers) arranged by tribes; discussion (pollen diversity is not the same from one subfamily to another). The Thymeleaceae are considered closely related to Euphorbiaceae.

Rudall, P. (1994). Laticifers in Crotonoideae (Euphorbiaceae): homology and evolution. Ann. Missouri Bot. Gard. 81: 270-282, illus. En. — Review of previous work and reports of species investigated by the author; discrimination of laticifer origin in Crotonoideae as compared with the rather more extensively investigated Euphorbioideae; phylogenetic considerations (with expression of the idea that *Omphalea* is misplaced in Acalyphoideae). The developmental relationship of laticifers and sclereids is also discussed.

Radcliffe-Smith, A. & R. Govaerts (1997). New names and combinations in the Crotonoïdeae. Kew Bull. 52: 183-189. En. — Miscellaneous names and combinations, many in *Croton*.

5. Euphorbioideae (except Euphorbieae)

The Hippomaneae were anatomically surveyed by Herbert (1897). Taxonomically they remain a very difficult tribe; nevertheless, the studies of Esser (1994) and Kruijt (1996), both including phylogenetic analyses, represent serious attempts at some resolution (with one response being the recognition of several segregates from *Sapium*, *Sebastiania* and other genera). There is increasing support for separate subfamily status for Euphorbieae (see Gilbert 1994 under that heading).

Herbert, H. (1897). Anatomische Untersuchung von Blatt und Axe der Hippomaneen. Munich. (Diss., Univ. München.) Ge. — [Not seen; reference from Webster 1994.]

Pax, F. and K. Hoffmann (1912). Euphorbiaceae-Hippomaneae. In A. Engler (ed.), Das Pflanzenreich, IV 147 V. 319 pp., illus. Berlin: Engelmann (im Auftrage der Preuss. Akademie der Wissenschaften). (Heft 52, pars.) La/Ge. — 26 genera in nine subtribes. [All now in Euphorbioideae except subtribe 1, Omphaleinae (now in Acalyphoideae) and 4, Trisyngyninae (=*Nothofagus*, Fagaceae or Nothofagaceae).]

Esser, H.-J. (1994). Systematische Studien an den Hippomaneae Adr. Jussieu ex Bartling (Euphorbiaceae) insbesondere den Mabeinae Pax et K. Hoffm. 305 pp., pp., illus., maps. Hamburg. (Unpubl. Ph.D. dissertation, Univ. of Hamburg.) Ge. — Detailed descriptive treatment with special reference to *Senefelderopsis*, *Dendrothrix*, *Mabea*, *Senefeldera*, *Pseudosenefeldera* and *Rhodothyrsus*, the key to genera on pp. 74-75. A key to all genera in Hippomaneae appears on pp. 24-31 followed by a list of genera, p. 32, and conspectus, pp. 33-67 (31 genera and 4 unplaced species). The tribal key and conspectus are preceded by reviews of taxonomic history and characters and the limits and internal classification of the tribe. Pp. 286-305 encompass bibliography, acknowledgments, index to all names, and an addendum, while appendices A-C include an index to specimens seen, collections in herb. HBG, and illustrations, cladograms and distribution maps.

• Kruijt, R. C. (1996). A taxonomic monograph of *Sapium* Jacq., *Anomostachys* (Baill.) Hurus., *Duvigneaudia* J. Léonard and *Sclerocroton* Hochst. (Euphorbiaceae tribe Hippomaneae). 109 pp., illus. (Biblioth. Bot. 146). Stuttgart. En. — A major revision with a markedly narrower circumscription of *Sapium*; includes in addition to the generic treatments sections on taxonomic philosophy (a clear distinction is drawn between morpholologically and phylogenetically based approaches), a phylogenetic analysis of *Sapium* and its allies, and (p. 8) a key to these genera. A provisional treatment of *Shirakia* is also furnished, along with an extensive list of references and index to collections seen.

• Esser, H.-J., P. van Welzen & T. Djarwaningsih (1997(1998)). A phylogenetic classification of the Malesian Hippomaneae (Euphorbiaceae). Syst. Bot. 22: 617-628. En. — Review of genera; discussion of characters (see also Appendix 1, p. 627), optimisation, coding and cladistic analysis (main cladogram, p. 623); conclusions and new classification (but no formal synopsis). [Among proposed changes are the union of *Sebastiania* with *Excoecaria* (and *Glyphostylus*) and the transfer of two *Sebastiania* species to and expanded *Gymnanthes*. Two new genera were also recognised but not named.]

6. Euphorbieae

Key early studies are those of Klotzsch (1859, 1860), an advocate of separate family status for this tribe as well as for a multigeneric *Euphorbia*. The whole tribe was also revised by Boissier (1862); his broad concept of *Euphorbia* has generally prevailed although there is an increasing body of opinion at least favouring separation of *Chamaesyce*. Gilbert (1994) has argued that the tribe is less closely related to the rest of Euphorbioideae than usually thought; all other tribes hitherto included therein should be excluded to a separate subfamily. In contrast to the other major groups of Euphorbiaceae, the Euphorbieae have an enormous horticultural following; essentially enthusiast publications have been a regular feature in the literature since the little handbook of Berger (1907). The ten volumes of *Euphorbia Journal* contain documentation and iconography of hundreds of taxa, while Turner (1995; see *Euphorbia*) is a useful introduction to hardy temperate species. A first modern checklist for the whole tribe was that of Oudejans (1990; supplement, 1993).

Klotzsch, J. F. (1859). Linné's natürliche Pflanzenklasse Tricoccae des Berliner Herbarium's im Allgemeinen und die natürliche Ordnung Euphorbiaceae insbesondere. Monatsber. Königl. Preuss. Akad. Wiss. Berlin (1859): 236-254, 2 pls. (Separately reprinted with unchanged pagination.) Ge. — Precursor (and nomenclaturally prior) to fuller treatment (1860, q.v.); includes a historical survey along with discussion of the merits or otherwise of different proposed systems. An outline of the author's segregate genera with their sections and representative species follows. [Klotzsch was an advocate of many segregate genera within *Euphorbia* and of the tribe Euphorbieae as a distinct family; see in particular pp. 245-247.]

Klotzsch, J. F. (1860). Linné's natürliche Pflanzenklasse Tricoccae des Berliner Herbarium's im Allgemeinen und die natürliche Ordnung Euphorbiaceae insbesondere. Abhandl. Königl. Preuss. Akad. Wiss. Berlin (1859), Phys. Abhandl.: 1-108, 2 pls. (Also separately published, Berlin; ii, 108 pp., 2 pls.) Ge. — History of the 'Tricoccae' since 1751 and arguments for its segregation into 6 'orders' (foreshadowing the modern subfamilies) of which Euphorbiaceae s.s. (as treated here) is one; proposal for division of the restricted family into three tribes, Euphorbieae, Pedilantheae and Anthostemeae (formally described on p. 20) with in the first of these two subtribes, Anisophyllae and Tithymalae (p. 13) encompassing respectively 8 and 7 genera. The formal treatment (written in association with C.A.F. Garcke) of genera and species in Euphorbieae and Pedilantheae follows with synonymy, references, descriptions of novelties, and summary of distribution (with some exsiccatae cited). [The fullest coverage of *Euphorbia* and its immediate allies prior to that of Boissier. The taxonomic philosophy adopted here contrasts strongly with that of Boissier which continues to guide modern practice. It may be worthy of note that Klotzsch and Garcke's subtribe Anisophyllae corresponds to *Euphorbia* subgen. *Chamaesyce* and *Agaloma*, with the latter divided into 7 genera, while the Tithymalae encompass the remaining modern subgenera. This corroborates with continuing suggestions that the current conception of subgen. *Agaloma* is too broad.]

• Boissier, E. (1862). Euphorbiaceae subordo I. Euphorbieae. In A. de Candolle (ed.), Prodromus 15(2): 3-188, 1261-1269. Paris. La. — Last nominally complete treatment of *Euphorbia* and its immediate relatives.

Berger, A. (1907). Sukkulente Euphorbien. 135 pp., illus. Stuttgart: Ulmer. Ge. — Mostly selected species of *Euphorbia*; descriptions, illustrations and cultural notes.

Millspaugh, C. F. (1913). The genera *Pedilanthus* and *Cubanthus*, and other American Euphorbiaceae. Publ. Field Mus. Nat. Hist., Bot. Ser. 2: 353-397. En. — Includes description and synopsis of *Cubanthus*, a segregate from *Pedilanthus*, as well as *Dendrocousinia* (a segregate of *Sebastiania*, Crotonoideae), and additions in Cuba to *Euphorbiodendron* (now part of *Euphorbia*) along with other notes.

• Pax, F. and K. Hoffmann (1931). Euphorbieae. In A. Engler (ed.), Die natürlichen Pflanzenfamilien, 2. Aufl., 19c: 207-223. Leipzig. Ge. — References, synoptic subdivision, representative species; encompasses *Euphorbia* (then estimated at 1600 species) along with 'satellite' and related genera. Remains the most recent tribal 'overview', but the authors' subdivision of *Euphorbia* is now out of date.

• White, A., R. A. Dyer and B. L. Sloane (1941). The succulent Euphorbieae (southern Africa). 2 vols. Illus. (part col.). Pasadena, Calif.: Abbey Garden Press. En. — The standard work, not yet superseded although much added to. [More fully described under *Euphorbia* (193 species represented) but also includes regional coverage of *Monadenium* and *Synadenium*.]

• Wheeler, L. C. (1943). The genera of living Euphorbieae. Amer. Midl. Nat. 3: 456-503. En. — A very detailed but tedious synopsis and nomenclatural review; mostly on *Euphorbia* (see there).

LaFon, R., H. Schwartz and D. Koutnik (eds) (1983-96). Euphorbia Journal. Vols. 1-10, illus., maps. Mill Valley, Calif.: Strawberry Press. En. — A well-illustrated plant-enthusiasts' journal, with articles by growers, specialists and others; ten volumes published, with general index, by 1996. [Should be consulted for additional information on a wide range of the more or less succulent species, including those in other genera of Euphorbieae; regional surveys have also been published.]

- Oudejans, R. C. H. M. (1990). World catalogue of species names published in the tribe Euphorbieae (Euphorbiaceae) with their geographical distribution. 444 pp. Utrecht. En. — Includes places of publication, accepted equivalents (where relevant), detailed distribution by geographical units (following Frodin, *Guide to standard floras of the world*, 1984), and some derivations of names. [For additions and corrections, see *idem*, 1993. World catalogue of species names published in the tribe Euphorbieae (Euphorbiaceae) with their geographical distribution: cumulative supplement 1. 78 pp. Scherpenzeel, the Netherlands.]

Gilbert, M. G. (1994). The relationships of the Euphorbieae (Euphorbiaceae). Ann. Missouri Bot. Gard. 81: 283-288, illus. En. — A partial critique of the system of Webster (1994; see **General**) with respect to the Euphorbioideae and more specifically the Euphorbieae. The author calls for the application of more rigorous systematic analyses although he does not question the homogeneity of the Euphorbieae (p. 284). Differences in inflorescence structure as well as the uniqueness of the cyathium are such that the tribe is probably not directly related to the Hippomaneae and related tribes. Two suggestions are made: a full union of Euphorbioideae s.l. with Crotonoideae or establishment of Hippomaneae (with related tribes) and Euphorbieae as mutually distinct subfamilies.

Singh, M. (1994). Succulent Euphorbiaceae of India. 55 pp., illus. New Delhi: Meena Singh. En. — National treatment; includes key. (For more details see *Euphorbia*.)

Aalius

Synonyms:
Aalius Rumph. ex Lam. === **Sauropus** Blume
Aalius androgyna (L.) Kuntze === **Sauropus androgynus** (L.) Merr.
Aalius assimilis (Thwaites) Kuntze === **Sauropus assimilis** Thwaites
Aalius brevipes (Müll.Arg.) Kuntze === **Sauropus brevipes** Müll.Arg.
Aalius ceratogynum Kuntze === **Sauropus quadrangularis** (Willd.) Müll.Arg.
Aalius compressa (Müll.Arg.) Kuntze === **Sauropus quadrangularis** (Willd.) Müll.Arg.
Aalius forcipata (Hook.f.) Kuntze === **Sauropus macranthus** Hassk.
Aalius lanceolata (Hook.f.) Kuntze === **Sauropus lanceolatus** Hook.f.
Aalius macrantha (Hassk.) Kuntze === **Sauropus macranthus** Hassk.
Aalius macrophylla (Hook.f.) Kuntze === **Sauropus macranthus** Hassk.
Aalius oblongifolia (Hook.f.) Kuntze === **Sauropus oblongifolius** Hook.f.
Aalius pubescens (Hook.f.) Kuntze === **Sauropus quadrangularis** (Willd.) Müll.Arg.
Aalius quadrangularis (Müll.Arg.) Kuntze === **Sauropus quadrangularis** (Willd.) Müll.Arg.
Aalius retroversa (Wight) Kuntze === **Sauropus retroversus** Wight
Aalius rhamnodes (Blume) Kuntze === **Sauropus rhamnoides** Blume
Aalius rigida (Thwaites) Kuntze === **Sauropus rigidus** Thwaites
Aalius rostrata (Miq.) Kuntze === **Sauropus rostratus** Miq.
Aalius spectabilis (Miq.) Kuntze === **Sauropus macranthus** Hassk.
Aalius stipitata (Hook.f.) Kuntze === **Sauropus stipitatus** Hook.f.
Aalius sumatrana (Miq.) Kuntze === **Sauropus androgynus** (L.) Merr.
Aalius trinervia (Hook.f. & Thomson ex Müll.Arg.) Kuntze === **Sauropus trinervius** Hook.f. & Thomson ex Müll.Arg.

Aapaca

Synonyms:
Aapaca Metzdorff === **Uapaca** Baill.

Acalypha

462 species, tropics but extending into warm-temperate regions (save in Europe); with *Acalyphopsis* and *Corythea*, the fourth-largest genus in the family. Shrubs or less usually small trees or perennial or annual herbs (rarely lianas), usually found in more open habitats. The greater part of the species is American, with also a substantial number in Africa and in Madagascar (c. 22) but rather fewer from Asia to the Pacific (in Malesia, c. 25). Pax & Hoffmann (1924, p. 6-11) tabulated species numbers by area and for each also listed the 'sections' (in their sense) and species present. Supraspecific classification in this genus has not been examined in depth for more than a century, although a start was made by Seberg (1984). The first major scheme, by Müller (1866; see **General**), was based strongly on inflorescence characters (since shown to be less useful at higher levels); two subgenera (*Linostachys* and *Euacalypha*) and many 'Gruppe' (the latter designated merely by an § sign) were recognised. That by Pax and Hoffmann (1924), though featuring an additional subgenus *Androcephala* (monotypic, comprising the Malagasy *A. diminuata*), was otherwise only slightly revised from its predecessor and has largely been followed since. However, Müller's 'Gruppe' were therein arbitrarily reranked as sections, with those in subgen. *Acalypha* (the vast majority) additionally grouped into several intermediate 'series' (here used at a level contrary to modern practice). This classification passed into the authors' account of the genus for *Die natürlichen Pflanzenfamilien* (1931); here, 3 subgenera, 8 'series' (in subgen. *Acalypha*) and 40 'sections' were accounted for. Hurusawa (1954; see **General**), recognised 6 subgenera. Webster (1967: 371; see **Americas**) argued against too many subgenera and moreover suggested that in contemporary terms Muller's 'Gruppe' more nearly corresponded to series rather than sections; formal ranking was therein given to ten with representatives in North America north of Mexico. Seberg (1984), following extensive reviews of old and new characters, showed that the Müllerian arrangement

and its derivatives were to a large extent artificial; he furthermore indicated that the genus as a whole was distinguished by just one synapomorphy (a distinctive anther morphology). The two largest subgenera each were found to be distinguished by two synapomorphies (based on female flowers pedicellate or not and the size of their subtending bracts). At national or regional level, there have been several studies since 1931 (some only in floras or family-wide studies, not here separately cited), but collectively their coverage is inevitably patchy. Current work includes a review of Malesian species by Esti Munawaroh (Bogor) as a contribution towards a *Flora Malesiana* account. The widely cultivated *A. wilkesiana*, of which there are several 'fancy' forms, is correctly *A. amentacea* subsp. *wilkesiana*. This is nominally distributed from the Bismarck Archipelago to Fiji but should perhaps be regarded as a prehistoric cultigen. *A. hispida*, similarly of Melanesian origin, is also widely cultivated. (Acalyphoideae)

Pax, F. (with K. Hoffmann) (1912). *Corythea*. In A. Engler (ed.), Das Pflanzenreich, IV 147 V (Euphorbiaceae-Hippomaneae): 156. Berlin. (Heft 52.) La/Ge. — 1 species, Mexico. [Included at present in *Acalypha*.]

Nitschke, R. (1923). Die geographische Verbreitung der Gattung *Acalypha*. Bot. Arch. 4: 277-317, map. Ge. — Tedious account and analysis of geographical distribution along with an ecological table and evolutionary speculation; list of 358 established and some additional doubtful or incompletely known species. [Prepared in conjunction with Pax and Hoffmann's revision in *Das Pflanzenreich*.]

• Pax, F. & K. Hoffmann (1924). *Acalypha*. In A. Engler (ed.), Das Pflanzenreich, IV 147 XVI (Euphorbiaceae-Crotonoideae-Acalypheae-Acalyphinae): 12-177. Berlin. (Heft 85.) La/Ge. — Nearly 400 species in 3 subgenera, the third (*Acalypha*) much the largest, with 8 suprasectional 'series'. (For geographical breakdown see introduction to the subtribe, pp. 7-11.) There is a distinct radiation in the Galápagos (nos. 278-287). *A. wilkesiana*, well-known in cultivation, is no. 329. [The last overall revision of the genus.]

Pax, F. & K. Hoffmann (1924). *Acalyphopsis*. In A. Engler (ed.), Das Pflanzenreich, IV 147 XVI (Euphorbiaceae-Crotonoideae-Acalypheae-Acalyphinae): 178. Berlin. (Heft 85.) La/Ge. — 1 species, Sulawesi (now reduced to *Acalypha*).

Staner, P. (1938). Révision des espèces congolaises du genre *Acalypha* L. Bull. Jard. Bot. État 15: 132-146, 1 fig. Fr. — Synoptic revision with key, descriptions of novelties, synonymy, references and citations, localities with exsiccatae, and commentary; covers 20 species and some additional infraspecific taxa. [The stems of one species, *A. senensis*, were reported as being used for fiber in the Kasenga region in the east.]

Leandri, J. (1943). Contribution à l'étude des Euphorbiacées de Madagascar, VI: Acalyphées. Notul. Syst. (Paris) 10: 252-291. Fr. — Most of the paper comprises novelties and notes in *Acalypha*; a key to all 22 known species of Malagasy *Acalyphae* is included (pp. 288-290).

• Hurusawa, I. (1954). Eine nochmalige Durchsicht des herkömmlichen Systems der Euphorbiaceen in weiteren Sinne. J. Fac. Sci. Univ. Tokyo III, 6: 209-342, illus., 4 pls. Ge. — Contains (pp. 295-301) a revised system of subgenera and sections in *Acalypha* with key, type species, and synonymy but little discussion; inflorescence diagrams, p. 299.

Airy-Shaw, H. K. (1966). Notes on Malaysian and other Asiatic Euphorbiaceae, LXXVII. New species of *Acalypha* L. Kew Bull. 20: 406-408. En. — 2 novelties from New Guinea.

Jablonski, E. (1967). *Acalypha*. Euphorbiaceae, Guayana Highland (Mem. New York Bot. Gard. 17(1)): 140-142. New York. En. — 5 species, none new; the cultivated *A. wilkesiana* is included.

Radcliffe-Smith, A. (1973). Notes on African Euphorbiaceae, III. *Acalypha*. Kew Bull. 28: 287-295. En. — For *Flora of Tropical East Africa*. Further notes in idem, IV (ibid.): 521-523.

Airy-Shaw, H. K. (1978). Notes on Malesian and other Asiatic Euphorbiaceae, CCXIX. *Acalypha* L. Kew Bull. 33: 71-74. En. — Descriptions of 4 new species from New Guinea.

• Seberg, O. (1984). Taxonomy and phylogeny of the genus *Acalypha* (Euphorbiaceae) in the Galapagos Archipelago. Nordic J. Bot. 4: 159-190, illus., map. En. — A study of the distinctive radiation of the genus in these islands with in addition a preliminary phylogenetic analysis for the whole of the Americas (based on 42 taxa; list, p. 161, and tree, p. 165); includes (pp. 175-188) a formal taxonomic treatment (4 species) with key, synonymy with references and citations, lengthy descriptions, indication of distribution and habitat, and concise specific commentary; separate index at end. [The Galápagos species fall within nos. 39-43 (in sect. 5, *Pheloideae*) and 278-287 (in sect. 27, *Cuspidata*) in the Pax & Hoffmann arrangement; however, they are all considered derived from the mainland *A. cuspidata* complex. This, along with the more general study of mainland taxa, shows clearly the artificiality of the Mueller infrageneric scheme. The author was, however, apparently unaware of the reranking of the Mueller 'series' (used in Pax & Hoffmann 1924) to sections (as used in Pax & Hoffmann 1931).]

Radcliffe-Smith, A. (1989). Notes on African Euphorbiaceae, XX. *Acalypha* (ii), etc. Kew Bull. 44: 439-454. En. — Includes notes on other genera; precursory to *Flora Zambesiaca*.

Radcliffe-Smith, A. (1990). Notes on Australian Euphorbiaceae, I. New records of *Acalypha* in Australia. Kew Bull. 45: 677-679. En. — Report of *A. pubiflora*, an African species, in Australia (associated with *Adansonia gregrorii*, the Australian baobab). [In sect. 30, *Paucibracteatae*, following Pax & Hoffmann 1931. Also parallels *Euphorbia sarcostemmoides*, whose relatives are in Africa.]

Cardiel, J. M. (1994). A synopsis of the Colombian species of *Acalypha* subgenus *Linostachys* (Euphorbiaceae). Brittonia 46: 200-207, illus. En. — Revision (7 species) with key, descriptions of three novelties, synonymy, references, types, and commentary including distribution, habitat, and taxonomic notes; list of literature at end. [Subgenus 1 in Pax & Hoffmann 1931.]

Cardiel, J. M. (1994). Las especies herbaceas de *Acalypha* (Euphorbiaceae) de Colombia. An. Jard. Bot. Madrid 52: 151-157, illus. Sp. — Revision (5 species) with key, descriptions of novelties, synonymy, references, types, commentary, and indication of distribution and habitat; literature list at end. [All these species are in subgen. *Acalypha*.]

Forster, P. I. (1994). A taxonomic revision of *Acalypha* L. (Euphorbiaceae) in Australia. Austrobaileya 4: 209-226, illus. En. — Regional revision (8 species of which 6 native (with 1 new, *A. lyonsii*) and 2 naturalised); includes key, descriptions, synonymy, references, types, indication of distribution and habitat, localities with (sometimes selected) exsiccatae, phenology, and commentary along with conservation status; literature list but no separate index. [The native species, some more or less ruderal, are of scattered affinity in the Mueller and Pax & Hoffmann systems.]

Levin, G. A. (1994(1995)). Systematics of the *Acalypha californica* complex (Euphorbiaceae). Madroño 41: 254-265, illus./map. En. — Biosystematic analysis; only one variable species worthy of formal recognition. A map with records in the form of varying-size pie charts is included; vouchers follow the list of references. [In sect. 36, *Betulinae* Muell. Arg., of Pax & Hoffmann 1931.]

Levin, G. A. (1995). Euphorbiaceae, spurge family: 1. *Acalypha* and *Cnidoscolus*. J. Ariz.-Nev. Acad. Sci. 29: 18-24, illus., maps. En. — Revision of Arizonan species of *Acalypha* precursory to treatment of the genus in a new state flora.

- McVaugh, R. (1995). Euphorbiacearum sertum Novo-Galicianarum revisarum. Contr. Univ. Michigan Herb. 20: 173-215, illus. En. — Revision precursory to treatment in *Flora Novo-Galiciana*; includes (pp. 173-188) treatment of *Acalypha* (8 species) with partial keys.
- Cardiel, J. M. (1996). Tipificación de las especies de *Acalypha* L. (Euphorbiaceae) descritas por Jacquin. An. Jard. Bot. Madrid 54: 230-233. Sp. — Typification of Jacquin's 7 species in this genus, with general discussion.

Acalypha L., Sp. Pl.: 1003 (1753).

Trop. & Subtrop. 21 22 23 24 25 26 27 28 29 31 35 36 38 40 41 42 50 60 61 62 (63) 71 72 74 75 76 77 78 79 80 81 82 83 84 85.

Mercuriastrum Heist. ex Fabr., Enum.: 202 (1759).

Cupameni Adans., Fam. Pl. 2: 356 (1763).

Caturus L., Mant. Pl. 1: 19 (1767).

Galurus Spreng., Anleit. Kenntn. Gew. 2(2): 364 (1817).

Usteria Dennst., Schlüssel Hortus Malab.: 31 (1818).

Cupamenis Raf., Sylva Tellur.: 67 (1838).

Acalyphes Hassk., Cat. Horto Bogor.: 235 (1844), orth. var.

Linostachys Klotzsch ex Schltdl., Linnaea 19: 235 (1847).

Odonteilema Turcz., Bull. Soc. Imp. Naturalistes Moscou 21(1): 587 (1848).
Paracelsea Zoll., Natuurk. Tijdschr. Ned.-Indië 14(4): 171 (1857), nom. illeg.
Calyptrospatha Klotzsch ex Baill., Étude Euphorb.: 440 (1858).
Gymnalypha Griseb., Bonplandia 6: 2 (1858).
Corythea S.Watson, Proc. Amer. Acad. Arts 22: 451 (1887).
Ricinocarpus Burm. ex Kuntze, Revis. Gen. Pl. 2: 615 (1891).
Acalyphopsis Pax & K.Hoffm. in H.G.A.Engler, Pflanzenr., IV, 147, XVI: 178 (1924).

Acalypha abingdonii Seberg, Nordic J. Bot. 4: 178 (1984).
Galapagos (Marchena I., Pinta I.). 83 GAL. Cham.

Acalypha acapulcensis Fernald, Proc. Amer. Acad. Arts 33: 87 (1897).
Mexico (Guerrero). 79 MXS. Nanophan.

Acalypha accedens Müll.Arg., Linnaea 24: 35 (1865). *Acalypha accedens* var. *genuina* Müll.Arg. in C.F.P.von Martius, Fl. Bras. 11(2): 361 (1874), nom. inval. *Ricinocarpus accedens* (Müll.Arg.) Kuntze, Revis. Gen. Pl. 2: 617 (1891).
Brazil (Rio de Janeiro). 84 BZL. Nanophan.
Acalypha brachyandra Baill., Adansonia 5: 232 (1865). *Acalypha accedens* var. *brachyandra* (Baill.) Müll.Arg. in A.P.de Candolle, Prodr. 15(2): 362 (1866). *Ricinocarpus brachyandrus* (Baill.) Kuntze, Revis. Gen. Pl. 2: 617 (1891).
Acalypha dupraeana Baill., Adansonia 5: 229 (1865).

Acalypha acmophylla Hemsl., J. Linn. Soc., Bot. 26: 436 (1894).
China (Hubei). 36 CHC. Nanophan.

Acalypha acrogyna Pax, Bot. Jahrb. Syst. 43: 323 (1909).
Zaire (Kivu), Uganda, NW. Tanzania, Sudan, Ethiopia, Zambia, Zimbabwe. 23 ZAI 24 ETH SUD 25 TAN 26 ZAM ZIM. Nanophan. or phan.

Acalypha acuminata Benth., Hooker's J. Bot. Kew Gard. Misc. 6: 329 (1854). *Ricinocarpus acuminatus* (Benth.) Kuntze, Revis. Gen. Pl. 2: 617 (1891).
Brazil (Amazonas). 84 BZN. Nanophan.

Acalypha adenostachya Müll.Arg., Linnaea 34: 21 (1865). *Ricinocarpus adenostachyus* (Müll.Arg.) Kuntze, Revis. Gen. Pl. 2: 617 (1891).
Mexico. 79 MXC MXE MXN MXS. Nanophan.

Acalypha akoensis Hayata, J. Coll. Sci. Imp. Univ. Tokyo 30: 266 (1911).
C. & S. Taiwan. 38 TAI. Nanophan.

Acalypha alchorneoides Rusby, Bull. New York Bot. Gard. 8: 101 (1912).
Bolivia. 83 BOL. Nanophan.

Acalypha aldabrica Pax & K.Hoffm. in H.G.A.Engler, Pflanzenr., IV, 147, XVI: 136 (1924).
Aldabra. 29 ALD. Nanophan.

Acalypha alexandrii Urb., Symb. Antill. 5: 387 (1908).
Jamaica. 81 JAM. Nanophan.
Acalypha hernandifolia Griseb., Fl. Brit. W. I.: 47 (1859), nom. illeg.

Acalypha aliena Brandegee, Proc. Calif. Acad. Sci., II, 3: 172 (1891).
Mexico to Nicaragua. 79 MXE MXN MXS MXT 80 NIC. Ther.
Acalypha simplicissima Millsp., Publ. Field Mus. Nat. Hist., Bot. Ser. 2: 417 (1916).

Acalypha allenii Hutch., Bull. Misc. Inform. Kew 1911: 229 (1911).
SC. Trop. Africa to Botswana. 26 MLW MOZ ZAM ZIM 27 BOT. Hemicr.

Acalypha alnifolia Klein ex Willd., Sp. Pl. 4: 525 (1805). *Ricinocarpus alnifolius* (Klein ex Willd.) Kuntze, Revis. Gen. Pl. 2: 617 (1891).
S. India. 40 IND. Cham.
Acalypha capitata Willd., Sp. Pl. 4: 525 (1805). *Ricinocarpus capitatus* (Willd.) Kuntze, Revis. Gen. Pl. 2: 617 (1891).
Acalypha alnifolia Wall., Numer. List: 7752A (1847), nom. inval.
Acalypha capitata var. *ambigua* Müll.Arg., Linnaea 34: 27 (1865).

Acalypha alopecuroides Jacq., Icon. Pl. Rar. 3: 19 (1790). *Ricinocarpus alopecuroides* (Jacq.) Kuntze, Revis. Gen. Pl. 2: 617 (1891).
Mexico to Peru, Caribbean. 79 MXE MXS 80 COS PAN 81 BAH BER CUB DOM HAI 82 VEN 83 CLM PER. Ther. or cham.
Acalypha alopecuroides lusus *glanduligera* Klotzsch in B.Seemann, Bot. Voy. Herald: 102 (1853).
Acalypha alopecuroides f. *polycephala* Müll.Arg., Linnaea 34: 50 (1865).
Acalypha alopecuroides lusus *hispida* Müll.Arg., Linnaea 34: 50 (1865).

Acalypha ambigua Pax, Bot. Jahrb. Syst. 19: 96 (1894).
Burundi to Namibia. 23 BUR ZAI 25 TAN 26 ANG ZAM ZIM 27 NAM. Hemicr.
Acalypha polymorpha var. *angustifolia* Müll.Arg., J. Bot. 1: 335 (1864).
Acalypha polymorpha var. *depauperata* Müll.Arg., J. Bot. 1: 335 (1864).
Acalypha dumetorum Pax in O.Warburg (ed.), Kunene-Sambesi Exped.: 283 (1903), nom. illeg.

Acalypha amblyodonta (Müll.Arg.) Müll.Arg. in C.F.P.von Martius, Fl. Bras. 11(2): 365 (1874).
SE. Brazil, NE. Argentina, Paraguay. 84 BZL 85 AGE PAR. Nanophan.
Acalypha dupraeana var. *gaudichaudii* Baill., Adansonia 5: 330 (1865). *Acalypha amblyodonta* var. *gaudichaudii* (Baill.) Müll.Arg. in C.F.P.von Martius, Fl. Bras. 11(2): 366 (1874).
* *Acalypha cuspidata* var. *amblyodonta* Müll.Arg., Linnaea 34: 37 (1866). *Ricinocarpus amblyodontus* (Müll.Arg.) Kuntze, Revis. Gen. Pl. 2: 618 (1891).
Acalypha amblyodonta var. *hispida* Müll.Arg. in C.F.P.von Martius, Fl. Bras. 11(2): 366 (1874).
Acalypha amblyodonta var. *repanda* Müll.Arg. in C.F.P.von Martius, Fl. Bras. 11(2): 366 (1874).
Acalypha amblyodonta var. *villosa* Müll.Arg. in C.F.P.von Martius, Fl. Bras. 11(2): 366 (1874).

Acalypha amentacea Roxb., Fl. Ind. ed. 1832, 3: 676 (1832).
Taiwan to Caroline Is. 38 TAI 42 BIS BOR MOL PHI SUL 60 FIJ 62 CRL. Nanophan.

subsp. **amentacea**
Borneo (Sabah), Philippines, Sulawesi, Maluku, Caroline Is. 42 BOR MOL PHI SUL 62 CRL. Nanophan.
Acalypha glandulosa Blanco, Fl. Filip.: 749 (1837). *Ricinocarpus blancoanus* Kuntze, Revis. Gen. Pl. 2: 617 (1891).
Acalypha affinis Klotzsch, Nova Acta Acad. Caes. Leop-Carol. German. Nat. Cur. 19(Suppl. 1): 416 (1843).
Acalypha amboinensis Benth., Hooker's J. Bot. Kew Gard. Misc. 2: 233 (1843). *Acalypha grandis* var. *amboinensis* (Benth.) Müll.Arg., Linnaea 34: 10 (1865).
Acalypha stipulacea Klotzsch, Nova Acta Acad. Caes. Leop-Carol. German. Nat. Cur. 19(Suppl. 1): 416 (1843). *Ricinocarpus stipulaceus* (Klotzsch) Kuntze, Revis. Gen. Pl. 2: 618 (1891).
Ricinocarpus philippinensis Kuntze, Revis. Gen. Pl. 2: 617 (1891).
Acalypha luzonica Pax & K.Hoffm. in H.G.A.Engler, Pflanzenr., IV, 147, XVI: 153 (1924).
Acalypha philippinensis Müll.Arg. in H.G.A.Engler, Pflanzenr., IV, 147, XVI: 152 (1924).

var. **grandis** (Benth.) Fosberg, Smithsonian Contr. Bot. 45: 8 (1980).
Fiji. 60 FIJ. Nanophan.
* *Acalypha grandis* Benth., Hooker's J. Bot. Kew Gard. Misc. 2: 232 (1843). *Acalypha grandis* var. *genuina* Müll.Arg., Linnaea 34: 10 (1865), nom. inval. *Ricinocarpus grandis* (Benth.) Kuntze, Revis. Gen. Pl. 2: 618 (1891).
Acalypha exaltata Baill., Adansonia 2: 225 (1862). *Ricinocarpus exaltatus* (Baill.) Kuntze, Revis. Gen. Pl. 2: 618 (1891).
Acalypha grandis var. *villosa* Müll.Arg., Flora 47: 441 (1864).
Acalypha consimilis Müll.Arg. in A.P.de Candolle, Prodr. 15(2): 807 (1866). *Ricinocarpus consimilis* (Müll.Arg.) Kuntze, Revis. Gen. Pl. 2: 617 (1891).
Acalypha finitima S.Moore, J. Linn. Soc., Bot. 45: 403 (1921).
Acalypha grandis f. *atropurpurea* Gilli, Ann. Naturhist. Mus. Wien 83: 436 (1979 publ. 1980), nom. illeg.

var. **heterotricha** Fosberg, Smithsonian Contr. Bot. 45: 9 (1980).
Caroline Is. (Babeldaob I.). 62 CRL. Nanophan.

var. **palauensis** Fosberg, Smithsonian Contr. Bot. 45: 9 (1980).
Caroline Is. (Palau). 62 CRL. Nanophan.

var. **trukensis** (Pax & K.Hoffm.) Fosberg, Smithsonian Contr. Bot. 45: 10 (1980).
Caroline Is. 62 CRL. Nanophan.
* *Acalypha trukensis* Pax & K.Hoffm. in H.G.A.Engler, Pflanzenr., IV, 147, XVI: 151 (1924).

var. **velutina** (Müll.Arg.) Fosberg, Smithsonian Contr. Bot. 45: 10 (1980).
Philippines, Taiwan. 38 TAI 42 PHI. Nanophan.
Acalypha angatensis Blanco, Fl. Filip.: 750 (1837). *Ricinocarpus angatensis* (Blanco) Kuntze, Revis. Gen. Pl. 2: 617 (1891).
Acalypha tomentosa Blanco, Fl. Filip.: 750 (1837).
* *Acalypha grandis* var. *velutina* Müll.Arg., Flora 47: 441 (1864).
Acalypha formosana Hayata, J. Coll. Sci. Imp. Univ. Tokyo 30: 267 (1911).

subsp. **wilkesiana** (Müll.Arg.) Fosberg, Smithsonian Contr. Bot. 45: 10 (1980).
Bismarck Archip., Solomon Is., Vanuatu, Fiji. 42 BIS 60 FIJ SOL VAN. Nanophan. or phan. – Many cultivars widely cultivated.
Acalypha circinata A.Gray ex Seem., Syn. Pl. Vit.: 11 (1862), pro syn.
* *Acalypha wilkesiana* Müll.Arg. in A.P.de Candolle, Prodr. 15(2): 817 (1866). *Ricinocarpus wilkesianus* (Müll.Arg.) Kuntze, Revis. Gen. Pl. 2: 618 (1891).
Acalypha wilkesiana f. *circinata* Müll.Arg. in A.P.de Candolle, Prodr. 15(2): 817 (1866). *Acalypha amentacea* f. *circinata* (Müll.Arg.) Fosberg, Smithsonian Contr. Bot. 45: 11 (1980).
Acalypha tricolor Seem., Fl. Vit.: 225 (1867).
Acalypha wilkesiana var. *marginata* Mill., Ann. Bot. Hort. 36: 157 (1876). *Acalypha marginata* (Mill.) J.J.Sm., Meded. Dept. Landb. Ned.-Indië 10: 19 (1910), nom. illeg.
Acalypha macrophylla Veitch, Ill. Hort. 24: 59 (1877), nom. illeg. *Acalypha wilkesiana* f. *macrophylla* J.J.Sm., Meded. Dept. Landb. Ned.-Indië 10: 20 (1910).
Acalypha musaica auct., Gard. Chron., II, 1877: 520 (1877).
Acalypha macafeeana Veitch, Cat.: 25 (1878).
Acalypha triumphans L.Linden & Rodig, Ill. Hort.: 55 (1888). *Acalypha wilkesiana* f. *triumphans* (L.Linden & Rodig) J.J.Sm., Meded. Dept. Landb. Ned.-Indië 10: 20 (1910).
Acalypha hamiltoniana Briant, Cat.(1893).
Acalypha godseffiana Mast., Gard. Chron., III, 23: 241 (1898).
Acalypha wilkesiana f. *illustris* J.J.Sm., Meded. Dept. Landb. Ned.-Indië 10: 20 (1910).
Acalypha wilkesiana f. *monstrosa* J.J.Sm., Meded. Dept. Landb. Ned.-Indië 10: 18 (1910).
Acalypha illustris Pax & K.Hoffm. in H.G.A.Engler, Pflanzenr., IV, 147, XVI: 154 (1924), nom. inval.
Acalypha torta Pax & K.Hoffm. in H.G.A.Engler, Pflanzenr., IV, 147, XVI: 154 (1924), pro syn.
Acalypha compacta Guilf. ex C.T.White, Gard. Chron., III, 94: 343 (1933).

Acalypha amplexicaulis A.C.Sm., J. Arnold Arbor. 33: 400 (1952).
Fiji (N. & W. Viti Levu). 60 FIJ. Nanophan. or phan.

Acalypha ampliata Pax & K.Hoffm. in H.G.A.Engler, Pflanzenr., IV, 147, XVI: 138 (1924).
Brazil (Rio de Janeiro). 84 BZL. Nanophan.

Acalypha anadenia Standl., Publ. Field Mus. Nat. Hist., Bot. Ser. 22: 33 (1940).
Mexico (Guerrero). 79 MXS.

Acalypha andina Müll.Arg., Flora 55: 26 (1872).
Ecuador. 83 ECU. Nanophan.

Acalypha andringitrensis Leandri, Notul. Syst. (Paris) 10: 277 (1942).
Madagascar. 29 MDG.

Acalypha anemioides Kunth in F.W.H.von Humboldt, A.J.A.Bonpland & C.S.Kunth, Nov. Gen. Sp. 2: 94 (1817). *Acalypha anemioides* var. *genuina* Müll.Arg. in A.P.de Candolle, Prodr. 15(2): 886 (1866), nom. inval. *Ricinocarpus anemioides* (Kunth) Kuntze, Revis. Gen. Pl. 2: 617 (1891).
NE. Mexico (San Luis Potosí). 79 MXE. Hemicr.
Acalypha monostachya Benth., Pl. Hartw.: 15 (1839), nom. illeg.
Acalypha agrimonioides D.Dietr., Syn. Pl. 5: 377 (1852).
Acalypha anemioides var. *eglandulosa* Müll.Arg. in A.P.de Candolle, Prodr. 15(2): 886 (1866).

Acalypha angustata Sond., Linnaea 23: 115 (1850). *Acalypha peduncularis* var. *angustata* (Sond.) Müll.Arg., Linnaea 34: 28 (1865).
S. Africa. 27 NAT OFS SWZ TVL. Hemicr.
Acalypha angustata var. *glabra* Sond., Linnaea 23: 115 (1850).
Acalypha schinzii var. *denticulata* Pax, Bull. Herb. Boissier 6: 734 (1898).

Acalypha angustifolia Sw., Prodr.: 99 (1788). *Acalypha angustifolia* var. *genuina* Müll.Arg., Linnaea 34: 22 (1865), nom. inval. *Ricinocarpus angustifolius* (Sw.) Kuntze, Revis. Gen. Pl. 2: 617 (1891).
Hispaniola. 81 DOM HAI. (Cl.) nanophan.
Acalypha carpinifolia Poir. in J.B.A.M.de Lamarck, Encycl. 6: 203 (1804). *Acalypha carpinifolia* var. *genuina* Müll.Arg., Linnaea 34: 22 (1865), nom. inval. *Ricinocarpus carpinifolius* (Poir.) Kuntze, Revis. Gen. Pl. 2: 617 (1891).
Acalypha laevigata Willd., Sp. Pl. 4: 527 (1805).
Acalypha domingensis Spreng., Syst. Veg. 3: 880 (1826). *Acalypha carpinifolia* var. *domingensis* (Spreng.) Müll.Arg., Linnaea 34: 22 (1865).
Acalypha angustifolia var. *glabrata* Müll.Arg., Linnaea 34: 22 (1865).
Acalypha arcuata Urb., Repert. Spec. Nov. Regni Veg. 18: 188 (1922).

Acalypha angustissima Pax, Bot. Jahrb. Syst. 19: 99 (1894).
Zaire. 23 ZAI. Hemicr.

Acalypha annobonae Pax & K.Hoffm. in H.G.A.Engler, Pflanzenr., IV, 147, XVI: 50 (1924).
Annobon. 23 GGI. Nanophan.

Acalypha apetiolata Allem & J.L.Wächt., Revista Brasil. Biol. 37: 85 (1977).
Brazil (Rio Grande do Sul). 84 BZS. Cham. or nanophan.

Acalypha apodanthes Standl. & L.O.Williams, Ceiba 1: 241 (1951).
N. & C. Costa Rica. 80 COS. Nanophan. or phan.
* *Acalypha ferdinandii* var. *pubescens* K.Hoffm. in H.G.A.Engler, Pflanzenr., IV, 147, XVI: 64 (1924).

Acalypha arciana (Baill.) Müll.Arg. in C.F.P.von Martius, Fl. Bras. 11(2): 362 (1874).
Brazil (Bahia). 84 BZE. Nanophan.
Acalypha brasiliensis var. *mollis* Müll.Arg., Linnaea 34: 37 (1865).
* *Acalypha dupraeana* var. *arciana* Baill., Adansonia 5: 230 (1865). *Ricinocarpus arcianus* (Baill.) Kuntze, Revis. Gen. Pl. 2: 618 (1891).

Acalypha argomuelleri Briq., Annuaire Conserv. Jard. Bot. Genève 4: 229 (1900).
Peru. 83 PER. Nanophan.

Acalypha aristata Kunth in F.W.H.von Humboldt, A.J.A.Bonpland & C.S.Kunth, Nov. Gen. Sp. 2: 93 (1817). *Ricinocarpus aristatus* (Kunth) Kuntze, Revis. Gen. Pl. 2: 617 (1891).
Mexico, Trop. America. 79 MXE MXG 80 COS GUA HON PAN 81 WIN 82 FRG SUR VEN 83 BOL CLM ECU PER 84 BZN. Ther. or cham.
Acalypha arvensis Poepp. & Endl., Nov. Gen. Sp. Pl. 3: 21 (1841). *Acalypha arvensis* var. *genuina* Müll.Arg. in A.P.de Candolle, Prodr. 15(2): 881 (1866), nom. inval. *Ricinocarpus arvensis* (Poepp. & Endl.) Kuntze, Revis. Gen. Pl. 2: 617 (1891).
Acalypha pavoniana Müll.Arg., Linnaea 34: 50 (1865). *Acalypha arvensis* var. *pavoniana* (Müll.Arg.) Müll.Arg. in A.P.de Candolle, Prodr. 15(2): 881 (1866).
Acalypha hystrix Balb. ex Spreng., Syst. Veg. 3: 883 (1891), nom. nud.
Acalypha arvensis var. *belangeri* Briq., Annuaire Conserv. Jard. Bot. Genève 4: 229 (1900).
Acalypha capitellata Brandegee, Univ. Calif. Publ. Bot. 6: 183 (1915).

Acalypha aronioides Pax & K.Hoffm. in H.G.A.Engler, Pflanzenr., IV, 147, XVI: 113 (1924).
Peru. 83 PER. Nanophan.

Acalypha aspericocca Pax & K.Hoffm. in H.G.A.Engler, Pflanzenr., IV, 147, XVI: 112 (1924).
Brazil (Rio de Janeiro). 84 BZL. Nanophan.

Acalypha australis L., Sp. Pl.: 1004 (1753). *Acalypha gemina* var. *genuina* Müll.Arg., Linnaea 34: 41 (1865), nom. inval. *Ricinocarpus australis* (L.) Kuntze, Revis. Gen. Pl. 2: 617 (1891).
Temp. & Subtrop. E. Asia. 36 CHM CHS 38 JAP TAI 42 PHI (50) nsw qld. Ther.
Acalypha virgata Thunb., Fl. Jap.: 268 (1784).
Acalypha sessilis Poir. in J.B.A.M.de Lamarck, Encycl. 6: 205 (1804).
Acalypha pauciflora Hornem., Hort. Bot. Hafn.: 909 (1815).
Acalypha gemina Spreng., Syst. Veg. 3: 880 (1826).
Acalypha chinensis Roxb., Fl. Ind. ed. 1832, 3: 677 (1832).
Acalypha lanceolata Wall., Numer. List: 7789 (1847), nom. inval.
Acalypha gemina var. *lanceolata* Hayata, J. Coll. Sci. Imp. Univ. Tokyo 20: 51 (1904).

Acalypha bakeriana Baill., Bull. Mens. Soc. Linn. Paris 2: 1180 (1895).
C. Madagascar. 29 MDG. Nanophan.

Acalypha balansae Guillaumin, Arch. Bot. Mém. 2(3): 38 (1929).
New Caledonia (Nouméa Reg.). 60 NWC. Nanophan. or phan.

Acalypha baronii Baker, J. Linn. Soc., Bot. 20: 254 (1884).
C. Madagascar. 29 MDG. Nanophan. or phan.

Acalypha benensis Britton, Bull. Torrey Bot. Club 28: 304 (1901).
Bolivia, Peru. 83 BOL PER. Nanophan.
Acalypha macrostachya Rusby, Mem. Torrey Bot. Club 4: 57 (1895), nom. illeg.
Acalypha tomentosula Ule, Verh. Bot. Vereins Prov. Brandenburg 50: 79 (1908).

Acalypha benguelensis Müll.Arg., J. Bot. 2: 335 (1864). *Ricinocarpus benguelensis* (Müll.Arg.) Kuntze, Revis. Gen. Pl. 2: 617 (1891).
Angola. 26 ANG. Hemicr.

Acalypha benguelensis var. *adenogyne* Müll.Arg., J. Bot. 1: 335 (1864).
Acalypha benguelensis var. *trichogyne* Müll.Arg., J. Bot. 1: 335 (1864).
Acalypha teusczii Pax, Bot. Jahrb. Syst. 19: 98 (1894).

Acalypha berteroana Müll.Arg., Linnaea 34: 33 (1865). *Ricinocarpus berteroanus* (Müll.Arg.) Kuntze, Revis. Gen. Pl. 2: 617 (1891).
Puerto Rico. 81 PUE. Nanophan.

Acalypha bipartita Müll.Arg., Flora 47: 538 (1864). *Ricinocarpus bipartitus* (Müll.Arg.) Kuntze, Revis. Gen. Pl. 2: 617 (1891).
Zaire (Orientale, Kivu), S. Sudan, Uganda, Rwanda, Burundi, Kenya, Tanzania. 23 BUR RWA ZAI 24 SUD 25 KEN TAN UGA. Cl. cham. or cl. nanophan.

Acalypha bisetosa Bertol. ex Spreng., Syst. Veg. 3: 879 (1826). *Ricinocarpus bisetosus* (Bertol. ex Spreng.) Kuntze, Revis. Gen. Pl. 2: 617 (1891).
Puerto Rico. 81 PUE. Nanophan.
Acalypha cuspidata Griseb., Fl. Brit. W. I.: 48 (1859).

Acalypha boinensis Leandri, Notul. Syst. (Paris) 10: 268 (1942).
Madagascar. 29 MDG.

Acalypha boiviniana Baill., Adansonia 1: 272 (1861). *Ricinocarpus boivinianus* (Baill.) Kuntze, Revis. Gen. Pl. 2: 617 (1891).
Tanzania (Zanzibar). 25 TAN. Nanophan. – Close to *A. acrogyna*.

Acalypha boliviensis Müll.Arg., Linnaea 34: 162 (1865). *Ricinocarpus boliviensis* (Müll.Arg.) Kuntze, Revis. Gen. Pl. 2: 617 (1891).
Bolivia. 83 BOL. Ther.

Acalypha bopiana Rusby, Mem. New York Bot. Gard. 7: 287 (1927).
Bolivia. 83 BOL.

Acalypha botteriana Müll.Arg., Linnaea 34: 46 (1865). *Ricinocarpus botterianus* (Müll.Arg.) Kuntze, Revis. Gen. Pl. 2: 617 (1891). *Acalypha botteriana* var. *genuina* Müll.Arg., Linnaea 34: 46 (1965), nom. inval.
Mexico (Veracruz). 79 MXG.
Acalypha botteriana var. *pubescens* Müll.Arg., Linnaea 34: 46 (1865).

Acalypha brachiata Krauss, Flora 28: 83 (1845). – FIGURE, p. 52.
Trop. & S. Africa. 22 GNB GUI IVO MLI NGA SEN 23 CAF CMN CON ZAI 24 ETH SUD 25 KEN TAN UGA 26 ANG MLW MOZ ZAM ZIM 27 BOT CPP NAM NAT SZW TVL. Hemicr. or cham.
Acalypha languida E.Mey. in J.F.Drège, Zwei Pflanzengeogr. Dokum.: 161 (1843), nom. nud. *Ricinocarpus languidus* Kuntze, Revis. Gen. Pl. 2: 618 (1891).
Acalypha petiolaris Hochst. ex C.Krauss, Flora 28: 83 (1845). *Ricinocarpus petiolaris* (Hochst. ex C.Krauss) Kuntze, Revis. Gen. Pl. 2: 618 (1891).
Acalypha hirsuta Hochst. ex A.Rich., Tent. Fl. Abyss. 2: 248 (1850).
Acalypha sidifolia A.Rich., Tent. Fl. Abyss. 2: 249 (1850), nom. illeg. *Ricinocarpus sidifolius* (A.Rich.) Kuntze, Revis. Gen. Pl. 2: 618 (1891).
Acalypha villicaulis Hochst. ex A.Rich., Tent. Fl. Abyss. 2: 248 (1850). *Ricinocarpus villicaulis* (Hochst. ex A.Rich.) Kuntze, Revis. Gen. Pl. 2: 618 (1891).
Acalypha senensis Klotzsch in W.C.H.Peters, Naturw. Reise Mossambique: 96 (1861). *Ricinocarpus senensis* (Klotzsch) Kuntze, Revis. Gen. Pl. 2: 618 (1891).
Acalypha zambesica Müll.Arg., Flora 47: 440 (1864). *Ricinocarpus zambesicus* (Müll.Arg.) Kuntze, Revis. Gen. Pl. 2: 618 (1891).

Acalypha brachiata Krauss (as *A. villicaulis* Hochst. ex A. Rich.)
Artist: W.H. Fitch
Trans. Linn. Soc. 29: pl. 98 (1875)
KEW ILLUSTRATIONS COLLECTION

Acalypha tenuis Müll.Arg., Linnaea 34: 30 (1865). *Ricinocarpus tenuis* (Müll.Arg.) Kuntze, Revis. Gen. Pl. 2: 618 (1891).
Acalypha tenuis var. *eglandulosa* Müll.Arg., Linnaea 34: 30 (1865).
Acalypha tenuis var. *glandulosa* Müll.Arg., Linnaea 34: 30 (1865).
Acalypha villicaulis Müll.Arg. in A.P.de Candolle, Prodr. 15(2): 845 (1866), nom. illeg.
Acalypha villicaulis var. *minor* Müll.Arg., Abh. Naturwiss. Vereine Bremen 7: 26 (1880).
Acalypha haplostyla Pax, Bot. Jahrb. Syst. 19: 98 (1894). *Acalypha senensis* var. *haplostyla* (Pax) Hutch. in D.Oliver, Fl. Trop. Afr. 6(1): 889 (1912).

Acalypha rehmannii Pax, Bull. Herb. Boissier 6: 733 (1898).
Acalypha haplostyla var. *longifolia* De Wild., Études Fl. Bas- Moyen-Congo 1: 277 (1906).
Acalypha chariensis Beille, Bull. Soc. Bot. France 55(8): 80 (1908). *Acalypha senensis* var. *chariensis* (Beille) Hutch. in D.Oliver, Fl. Trop. Afr. 6(1): 888 (1912).
Acalypha zambesica var. *brevistyla* Beille, Bull. Soc. Bot. France 8: 81 (1908).
Acalypha senegalensis Pax & K.Hoffm. in H.G.A.Engler, Pflanzenr., IV, 147, XVI: 79 (1924).

Acalypha brachyclada Müll.Arg. in A.P.de Candolle, Prodr. 15(2): 862 (1866). *Ricinocarpus brachycladus* (Müll.Arg.) Kuntze, Revis. Gen. Pl. 2: 617 (1891).
Peru. 83 PER. Nanophan.

Acalypha brasiliensis Müll.Arg., Linnaea 34: 37 (1865). *Ricinocarpus brasiliensis* (Müll.Arg.) Kuntze, Revis. Gen. Pl. 2: 617 (1891).
Brazil, NE. Argentina. 84 BZC BZE BZL BZS 85 AGE. Nanophan.
Acalypha brasiliensis f. *obtusa* Müll.Arg., Linnaea 34: 37 (1865). *Acalypha brasiliensis* var. *obtusa* (Müll.Arg.) Müll.Arg. in C.F.P.von Martius, Fl. Bras. 11(2): 363 (1874).
Acalypha brasiliensis f. *cordata* Müll.Arg., Linnaea 11(2): 363 (1874). *Acalypha brasiliensis* var. *cordata* (Müll.Arg.) Müll.Arg. in C.F.P.von Martius, Fl. Bras. 11(2): 363 (1874).
Acalypha brasiliensis var. *brevipes* Müll.Arg. in C.F.P.von Martius, Fl. Bras. 11(2): 363 (1874).
Acalypha brasiliensis var. *glabrata* Müll.Arg. in C.F.P.von Martius, Fl. Bras. 11(2): 364 (1874).
Acalypha brasiliensis var. *longipes* Müll.Arg. in C.F.P.von Martius, Fl. Bras. 11(2): 363 (1874).
Acalypha brasiliensis var. *maxima* Müll.Arg. in C.F.P.von Martius, Fl. Bras. 11(2): 364 (1874).
Acalypha brasiliensis var. *angustifolia* Pax & K.Hoffm. in H.G.A.Engler, Pflanzenr., IV, 147, XVI: 117 (1924).

Acalypha brevibracteata Müll.Arg. in A.P.de Candolle, Prodr. 15(2): 855 (1866). *Ricinocarpus brevibracteatus* (Müll.Arg.) Kuntze, Revis. Gen. Pl. 2: 617 (1891).
Brazil (Rio de Janeiro). 84 BZL. Nanophan. ?

Acalypha brevicaulis Müll.Arg., Linnaea 34: 13 (1865). *Ricinocarpus brevicaulis* (Müll.Arg.) Kuntze, Revis. Gen. Pl. 2: 617 (1891).
Mexico (Hidalgo). 79 MXE. Hemicr.
Acalypha crassicaulis Klotzsch ex Pax & K.Hoffm. in H.G.A.Engler, Pflanzenr., IV, 147, XVI: 159 (1924).

Acalypha brittonii Rusby, Bull. Torrey Bot. Club 28: 303 (1901).
Bolivia. 83 BOL. Nanophan.

Acalypha buchtienii Pax, Repert. Spec. Nov. Regni Veg. 5: 227 (1908).
Bolivia. 83 BOL. Nanophan. or phan.
Acalypha mollissima Rusby ex Pax & K.Hoffm. in H.G.A.Engler, Pflanzenr., IV, 147, XVI: 68 (1924), pro syn.

Acalypha buddleifolia Pax & K.Hoffm. in H.G.A.Engler, Pflanzenr., IV, 147, XVI: 113 (1924).
Peru. 83 PER. Nanophan.

Acalypha bullata Müll.Arg., Linnaea 34: 17 (1865). *Ricinocarpus bullatus* (Müll.Arg.) Kuntze, Revis. Gen. Pl. 2: 617 (1891).
Peru. 83 PER.
Acalypha rugosa Klotzsch ex Pax & K.Hoffm. in H.G.A.Engler, Pflanzenr., IV, 147, XVI: 66 (1924), pro syn.

Acalypha burquezii V.W.Steinm. & Felger, Aliso 16: 11 (1997).
Mexico (Sonora, Chihuahua). 79 MXE MXN. Ther.

Acalypha bussei Hutch. in D.Oliver, Fl. Trop. Afr. 6(1): 1056 (1913).
Kenya, Tanzania. 25 KEN TAN. Ther.

Acalypha caeciliae Pax & K.Hoffm. in H.G.A.Engler, Pflanzenr., IV, 147, XVI: 29 (1924).
Missouri. 74 MSO.

Acalypha californica Benth., Bot. Voy. Sulphur: 51 (1844). *Ricinocarpus californicus* (Benth.) Kuntze, Revis. Gen. Pl. 2: 617 (1891).
Mexico (Baja California, NW. Sonora, NW. Sinaloa), S. California, SW. Arizona. 76 ARI CAL 79 MXN. Nanophan.
Acalypha pringlei S.Watson, Proc. Amer. Acad. Arts 20: 373 (1885).
Acalypha stokesiae Pax & K.Hoffm. in H.G.A.Engler, Pflanzenr., IV, 147, XVI: 138 (1924).

Acalypha capensis (L.f.) Prain, Bull. Misc. Inform. Kew 1913: 15 (1913).
Cape Prov. 27 CPP. Nanophan.
* *Urtica capensis* L.f., Suppl. Pl.: 417 (1781). *Mercurialis capensis* (L.f.) Spreng. ex Sond., Linnaea 23: 111 (1850). *Leidesia capensis* (L.f.) Müll.Arg. in A.P.de Candolle, Prodr. 15(2): 793 (1866).
Tragia villosa Thunb., Prodr. Fl. Cap.: 14 (1794). *Acalypha decumbens* var. *villosa* (Thunb.) Müll.Arg. in A.P.de Candolle, Prodr. 15(2): 864 (1866). *Acalypha capensis* var. *villosa* (Thunb.) Pax & K.Hoffm. in H.G.A.Engler, Pflanzenr., IV, 147, XVI: 127 (1924).
Acalypha cordata Thunb., Prodr. Fl. Cap.: 117 (1800). *Acalypha decumbens* var. *cordata* (Thunb.) Müll.Arg. in A.P.de Candolle, Prodr. 15(2): 864 (1866). *Acalypha capensis* var. *cordata* (Thunb.) Prain, Bull. Misc. Inform. Kew 1913: 16 (1913).
Acalypha decumbens Thunb., Prodr. Fl. Cap.: 117 (1800). *Acalypha decumbens* var. *genuina* Müll.Arg. in A.P.de Candolle, Prodr. 15(2): 864 (1866), nom. inval. *Ricinocarpus decumbens* (Thunb.) Kuntze, Revis. Gen. Pl. 2: 617 (1891). *Acalypha capensis* var. *decumbens* (Thunb.) Prain, Bull. Misc. Inform. Kew 1913: 16 (1913).
Acalypha discolor E.Mey. in J.F.Drège, Zwei Pflanzengeogr. Dokum.: 129 (1843).
Acalypha kraussiana Buchinger ex Krauss., Flora 27: 84 (1845).
Acalypha lamiifolia Scheele, Linnaea 25: 587 (1852). *Acalypha capensis* f. *lamiifolia* (Scheele) Prain, Bull. Misc. Inform. Kew 1913: 16 (1913).
Acalypha discolor var. *major* Baill., Adansonia 3: 158 (1863).
Acalypha decumbens Müll.Arg. in A.P.de Candolle, Prodr. 15(2): 864 (1866), nom. illeg.
Acalypha grandidentata Müll.Arg. in A.P.de Candolle, Prodr. 15(2): 823 (1866). *Ricinocarpus grandidentatus* (Müll.Arg.) Kuntze, Revis. Gen. Pl. 2: 618 (1891). *Acalypha capensis* f. *grandidentata* (Müll.Arg.) Prain, Bull. Misc. Inform. Kew 1913: 16 (1913).
Acalypha prostrata Zeyh. ex Pax. & K.Hoffm. in H.G.A.Engler, Pflanzenr., IV, 147, XVI: 127 (1924).

Acalypha caperonioides Baill., Adansonia 3: 157 (1863). *Acalypha peduncularis* var. *caperonioides* (Baill.) Müll.Arg. in A.P.de Candolle, Prodr. 15(2): 846 (1866).
Malawi, Mozambique, Zimbabwe, S. Africa. 26 MLW MOZ ZIM 27 CPP NAT OFS SWZ TVL. Hemicr.
Acalypha peduncularis var. *glabrata* Sond., Linnaea 23: 115 (1850).
Acalypha peduncularis Pax in H.G.A.Engler, Pflanzenw. Ost-Afrikas C: 239 (1895), nom. illeg.
Acalypha caperonioides var. *galpinii* Prain, Bull. Misc. Inform. Kew 1913: 23 (1913).

Acalypha capillipes Müll.Arg., Linnaea 34: 40 (1865). *Acalypha eremorum* var. *capillipeda* Baill., Adansonia 6: 317 (1866). *Ricinocarpus capillipes* (Müll.Arg.) Kuntze, Revis. Gen. Pl. 2: 617 (1891).
SE. Queensland, New South Wales. 50 NSW QLD. Nanophan.

Acalypha carrascoana Cardiel, Anales Jard. Bot. Madrid 52: 153 (1994 publ. 1995).
Colombia, Venezuela. 82 VEN 83 CLM. Ther.

PLATE 1

A. *Acalypha hispida* B. *Alchornea schomburgki*

PLATE 2

A. *Antidesma bunius*

B. *Baccaurea parviflora*

C. *Cheilosa montana*

PLATE 3

Acalypha supera

PLATE 4

Adenopeltis serrata

PLATE 5

Anthostema senegalensis

PLATE 6

Breynia fruticosa

PLATE 7

Cleistanthus oblongifolius

PLATE 8

Cnidoscolus urens

Acalypha carthagenensis Jacq., Enum. Syst. Pl.: 32 (1760). *Ricinocarpus carthagenensis* (Jacq.) Kuntze, Revis. Gen. Pl. 2: 617 (1891).
N. Colombia. 83 CLM. Nanophan. – Provisionally accepted.

Acalypha castroviejoi Cardiel, Brittonia 46: 205 (1994).
Colombia (Santander). 83 CLM. Cham. or nanophan.

Acalypha caturus Blume, Bijdr.: 629 (1826). *Ricinocarpus caturus* (Blume) Kuntze, Revis. Gen. Pl. 2: 617 (1891).
Taiwan (Langu I.), Malesia, Caroline Is. (Ponape). 38 TAI 42 BOR JAW LSI MOL PHI SUL SUM 62 CRL. Nanophan. or phan.
* *Caturus spiciflorus* L., Mant. Pl. 1: 127 (1767). *Ricinocarpus spiciflorus* (L.) Kuntze, Revis. Gen. Pl. 2: 618 (1891).
Acalypha minahassae Koord., Meded. Lands Plantentuin 19: 579 (1898).
Acalypha similis Koord., Meded. Lands Plantentuin 19: 579 (1898).
Acalypha cardiophylla Merr., Philipp. J. Sci. 1(Suppl.): 80 (1906).
Acalypha caturus f. *angustifolia* J.J.Sm., Meded. Dept. Landb. Ned.-Indië 10: 511 (1910).
Acalypha subcinerea Elmer, Leafl. Philipp. Bot. 7: 2631 (1915).
Acalypha ponapensis Kaneh. & Hatus., Bot. Mag. (Tokyo) 54: 434 (1940). *Acalypha cardiophylla* var. *ponapensis* (Kaneh. & Hatus.) Fosberg, Smithsonian Contr. Bot. 45: 11 (1980).

Acalypha centromalayca Pax & K.Hoffm. in H.G.A.Engler, Pflanzenr., IV, 147, XVI: 150 (1924).
Sulawesi, Maluku (Bacan), Irian Jaya. 42 MOL NWG SUL. Nanophan. or phan.

Acalypha ceraceopunctata Pax, Bot. Jahrb. Syst. 45: 238 (1910).
Ghana, Nigeria, Cameroon. 22 GHA NGA 23 CMN. Cham. or nanophan.

Acalypha chamaedrifolia (Lam.) Müll.Arg. in A.P.de Candolle, Prodr. 15(2): 879 (1866).
S. Florida, Caribbean. 78 FLA 81 CUB DOM HAI JAM LEE PUE WIN. Hemicr.
* *Croton chamaedrifolius* Lam., Encycl. 2: 214 (1786). *Acalypha chamaedrifolia* var. *genuina* Müll.Arg., Linnaea 34: 48 (1865), nom. inval. *Ricinocarpus chamaedrifolius* (Lam.) Kuntze, Revis. Gen. Pl. 2: 617 (1891).
Acalypha reptans Sw., Prodr.: 99 (1788).
Acalypha corchorifolia Willd., Sp. Pl. 4: 524 (1805).
Acalypha adscendens Hornem., Hort. Bot. Hafn., Suppl.: 108 (1819).
Acalypha chamaedrifolia var. *brevipes* Müll.Arg., Linnaea 34: 48 (1865).
Acalypha hispaniolae Urb., Repert. Spec. Nov. Regni Veg. 15: 410 (1919).
Acalypha hotteana Urb., Ark. Bot. 20A(15): 59 (1926).

Acalypha chiapensis Brandegee, Univ. Calif. Publ. Bot. 10: 411 (1924).
Mexico (Chiapas). 79 MXT.

Acalypha chibomboa Baill., Adansonia 1: 269 (1861).
Comoros. 29 COM. Nanophan.

Acalypha chirindica S.Moore, J. Linn. Soc., Bot. 40: 199 (1911).
Zaire (Shaba), Tanzania, Mozambique, Malawi, Zambia, Zimbabwe. 23 ZAI 25 TAN 26 MLW MOZ ZAM ZIM. Nanophan.

Acalypha chlorocardia Standl., Publ. Field Mus. Nat. Hist., Bot. Ser. 8: 18 (1930).
Belize. 80 BLZ.

Acalypha chocoana Cardiel, Brittonia 46: 201 (1994).
Colombia (Chocó). 83 CLM. Nanophan. or phan.

Acalypha chordantha F.Seym., Phytologia 43: 161 (1979).
Nicaragua. 80 NIC.

Acalypha chorisandra Baill., Adansonia 5: 235 (1865). *Ricinocarpus chorisandrus* (Baill.) Kuntze, Revis. Gen. Pl. 2: 617 (1891).
S. Brazil. 84 BZS. Nanophan.

Acalypha ciliata Forssk., Fl. Aegypt.-Arab.: 162 (1775). *Acalypha ciliata* var. *genuina* Müll.Arg., Linnaea 34: 44 (1865), nom. inval. *Ricinocarpus ciliatus* (Forssk.) Kuntze, Revis. Gen. Pl. 2: 617 (1891).
Africa, SW. Yemen, NE. Pakistan, W. India, Sri Lanka. 21 CVI 22 BEN GHA GNB GUI IVO LBR MLI NGA SEN SIE TOG 23 CAF CMN CON GAB GGI ZAI 24 ETH SUD 24 KEN TAN UGA 26 ANG MLW MOZ ZAM ZIM 27 BOT NAM TVL 35 YEM 40 IND PAK SRL. Ther.
Acalypha rubra Wight ex Wall., Numer. List: 7781 (1847), nom. inval.
Acalypha ciliata var. *trichophora* Müll.Arg., Linnaea 34: 44 (1865).

Acalypha cincta Müll.Arg., Linnaea 34: 20 (1865). *Ricinocarpus cinctus* (Müll.Arg.) Kuntze, Revis. Gen. Pl. 2: 617 (1891).
Mexico. 79 MXE MXN MXS. Nanophan.
Acalypha gentryi Standl., Publ. Field Mus. Nat. Hist., Bot. Ser. 22: 34 (1940).

Acalypha cinerea Pax & K.Hoffm. in H.G.A.Engler, Pflanzenr., IV, 147, XVI: 102 (1924).
Mexico (?). 79 MXE. Nanophan.

Acalypha cinnamomifolia Pax & K.Hoffm. in H.G.A.Engler, Pflanzenr., IV, 147, XVI: 142 (1924).
NE. Sulawesi, Maluku, New Guinea. 42 MOL NWG SUL. Phan.

var. **cinnamomifolia**
NE. Sulawesi, Maluku, New Guinea. 42 MOL NWG SUL. Phan.

var. **induta** Airy Shaw, Kew Bull., Addit. Ser. 8: 15 (1980).
Papua New Guinea. 42 NWG. Phan.

Acalypha claoxyloides Hutch., Bull. Misc. Inform. Kew 1918: 205 (1918).
Aldabra. 29 ALD. Nanophan.

Acalypha claussenii (Turcz.) Müll.Arg., Linnaea 34: 51 (1865).
WC. & SE. Brazil. 84 BZC BZL. Hemicr.
* *Odonteilema claussenii* Turcz., Bull. Soc. Imp. Naturalistes Moscou 21(1): 588 (1848).
Ricinocarpus claussenii (Turcz.) Kuntze, Revis. Gen. Pl. 2: 617 (1891).

Acalypha clutioides Radcl.-Sm., Kew Bull. 28: 287 (1873).
S. Zaire, Zambia. 23 ZAI 26 ZAM. Hemicr.

Acalypha codonocalyx Baill., Adansonia 1: 271 (1861). *Ricinocarpus codonocalyx* (Baill.) Kuntze, Revis. Gen. Pl. 2: 617 (1891).
Comoros (Mohili). 29 COM. Nanophan.

Acalypha coleispica Pax & K.Hoffm. in H.G.A.Engler, Pflanzenr., IV, 147, XVI: 27 (1924).
Mexico (?). 79 +. Hemicr.

Acalypha colombiana Cardiel, Anales Jard. Bot. Madrid 48: 21 (1995).
Colombia (Cundinamarca, Meta). 83 CLM. Nanophan. or phan.

Acalypha communis Müll.Arg., Linnaea 34: 23 (1865). *Ricinocarpus communis* (Müll.Arg.) Kuntze, Revis. Gen. Pl. 2: 617 (1891).

C. & S. Brazil, Bolivia, Paraguay, Uruguay, N. Argentina. 83 BOL 84 BZC BZL BZS 85 AGE AGW PAR URU. Nanophan.

Acalypha hirta Spreng., Syst. Veg. 4(2): 315 (1827), nom. illeg. *Acalypha communis* var. *hirta* (Spreng.) Müll.Arg., Linnaea 34: 24 (1865).

Acalypha virgata Vell., Fl. Flumin.: 10 (1829), nom. illeg.

Acalypha betulodes Klotzsch ex Baill., Adansonia 5: 228 (1865).

Acalypha betuloides Klotzsch ex Baill., Adansonia 5: 228 (1865).

Acalypha communis var. *puberula* Müll.Arg., Linnaea 34: 24 (1865).

Acalypha communis var. *tomentella* Müll.Arg., Linnaea 34: 23 (1865).

Acalypha communis var. *tomentosa* Müll.Arg., Linnaea 34: 23 (1865).

Acalypha pilifera Klotzsch ex Baill., Adansonia 5: 227 (1865).

Acalypha urticoides Klotzsch ex Baill., Adansonia 5: 227 (1865).

Acalypha variabilis Klotzsch ex Baill., Adansonia 5: 226 (1865).

Acalypha variabilis var. *albescens* Baill., Adansonia 5: 227 (1865).

Acalypha communis var. *intermedia* Müll.Arg. in C.F.P.von Martius, Fl. Bras. 11(2): 350 (1874).

Acalypha communis var. *obscura* Müll.Arg. in C.F.P.von Martius, Fl. Bras. 11(2): 350 (1874).

Acalypha communis var. *pallida* Müll.Arg. in C.F.P.von Martius, Fl. Bras. 11(2): 349 (1874).

Acalypha cordobensis Müll.Arg., J. Bot. 12: 228 (1874). *Ricinocarpus cordobensis* (Müll.Arg.) Kuntze, Revis. Gen. Pl. 3(2): 291 (1898).

Acalypha cordoviensis Müll.Arg., J. Bot. 12: 228 (1874).

Acalypha communis var. *rotundata* Griseb., Abh. Königl. Ges. Wiss. Göttingen 24: 59 (1879).

Acalypha gracilis Griseb., Abh. Königl. Ges. Wiss. Göttingen 24: 59 (1879).

Acalypha agrestis Morong ex Britton, Ann. New York Acad. Sci. 7: 225 (1893). *Acalypha communis* var. *agrestis* (Morong ex Britton) Chodat & Hassl., Bull. Herb. Boissier, II, 1: 397 (1901).

Acalypha apicalis N.E.Br., Trans. & Proc. Bot. Soc. Edinburgh 20: 70 (1894).

Acalypha amphigyne Moore, Trans. Linn. Soc. London, Bot. 4: 467 (1895).

Acalypha communis f. *grandifolia* Chodat & Hassl., Bull. Herb. Boissier, II, 5: 604 (1905).

Acalypha communis var. *guaranitica* Chodat & Hassl., Bull. Herb. Boissier, II, 5: 605 (1905).

Acalypha communis var. *tomentosa* Chodat & Hassl., Bull. Herb. Boissier, II, 5: 604 (1905).

Acalypha paraguariensis Chodat & Hassl., Bull. Herb. Boissier, II, 5: 606 (1905).

Acalypha humilis Pax & K.Hoffm., Repert. Spec. Nov. Regni Veg. 8: 162 (1910).

Acalypha communis var. *hirtiformis* Pax & K.Hoffm. in H.G.A.Engler, Pflanzenr., IV, 147, XVI: 39 (1924).

Acalypha communis var. *hispida* Müll.Arg. in H.G.A.Engler, Pflanzenr., IV, 147, XVI: 39 (1924).

Acalypha communis var. *salicifolia* Pax & K.Hoffm. in H.G.A.Engler, Pflanzenr., IV, 147, XVI: 39 (1924).

Acalypha communis var. *saltensis* Pax & K.Hoffm. in H.G.A.Engler, Pflanzenr., IV, 147, XVI: 39 (1924).

Acalypha montevidensis Klotzsch ex Pax & K.Hoffm. in H.G.A.Engler, Pflanzenr., IV, 147, XVI: 39 (1924), pro syn.

Acalypha comonduana Millsp., Proc. Calif. Acad. Sci., II, 2: 222 (1889).

Mexico (Baja California). 79 MXN. Nanophan.

Acalypha comorensis Pax, Bot. Jahrb. Syst. 19: 95 (1894).

Comoros. 29 COM. Nanophan.

Acalypha concinna Airy Shaw, Kew Bull. 33: 74 (1978).

Papua New Guinea. 42 NWG. Nanophan.

Acalypha confertiflora Pax & K.Hoffm. in H.G.A.Engler, Pflanzenr., IV, 147, XVI: 53 (1924).
Mexico (Oaxaca). 79 MXS. Nanophan.

Acalypha conspicua Müll.Arg. in A.P.de Candolle, Prodr. 15(2): 832 (1866). *Ricinocarpus conspicuus* (Müll.Arg.) Kuntze, Revis. Gen. Pl. 2: 617 (1891).
Mexico (Veracruz). 79 MXG.

Acalypha contermina Müll.Arg., Linnaea 34: 46 (1865). *Ricinocarpus conterminus* (Müll.Arg.) Kuntze, Revis. Gen. Pl. 2: 617 (1891).
Peru. 83 PER.

Acalypha controversa (Kuntze) K.Schum., Just's Bot. Jahresber. 26: 348 (1898).
Bolivia. 83 BOL. Phan.
* *Ricinocarpus controversus* Kuntze, Revis. Gen. Pl. 3(2): 290 (1898).

Acalypha costaricensis (Kuntze) Knobl., Just's Bot. Jahresber. 19: 337 (1894).
S. Mexico to Panama. 79 MXT 80 COS PAN. Nanophan.
* *Ricinocarpus costaricensis* Kuntze, Revis. Gen. Pl. 2: 616 (1891).

Acalypha crenata Hochst. ex A.Rich., Tent. Fl. Abyss. 2: 245 (1850). *Ricinocarpus crenatus* (Hochst. ex A.Rich.) Kuntze, Revis. Gen. Pl. 2: 617 (1891).
Trop. Africa. 21 CVI 22 GHA GUI IVO LBR NGA SEN 23 CMN ZAI 24 ETH SOM SUD 25 KEN TAN UGA 26 MLW MOZ ZAM ZIM. Ther.
Acalypha abortiva Hochst. ex Baill., Étude Euphorb.: 443 (1858). *Acalypha indica* lusus *abortiva* (Hochst. ex Baill.) Müll.Arg., Linnaea 34: 42 (1865).
Acalypha vahliana Oliv., Trans. Linn. Soc. London 29: 147 (1875). *Ricinocarpus vahlianus* (Oliv.) Kuntze, Revis. Gen. Pl. 2: 618 (1891).

Acalypha × cristata Radcl.-Sm., Kew Bull. 44: 443 (1989). A. ciliata × A. fimbriata.
SE. Kenya, Tanzania. 25 KEN TAN. Ther.

Acalypha crockeri Fosberg, Lloydia 3: 114 (1940).
Solomon Is. (Rennell). 60 SOL.

Acalypha cubensis Urb., Repert. Spec. Nov. Regni Veg. 28: 225 (1930).
W. & C. Cuba. 81 CUB. Nanophan.

Acalypha cuneata Poepp. & Endl., Nov. Gen. Sp. Pl. 3: 22 (1845). *Acalypha cuneata* var. *genuina* Müll.Arg., Linnaea 34: 11 (1865), nom. inval. *Ricinocarpus cuneatus* (Poepp. & Endl.) Kuntze, Revis. Gen. Pl. 2: 617 (1891).
Panama to N. Brazil. 80 PAN 82 VEN 83 CLM ECU 84 BZN. Nanophan. or phan.
Acalypha obovata Benth., Bot. Voy. Sulphur: 163 (1846). *Acalypha cuneata* var. *obovata* (Benth.) Müll.Arg., Linnaea 34: 11 (1865).
Acalypha longifolia Baill., Étude Euphorb.: 443 (1858), nom. nud.
Acalypha klotzschii Baill., Adansonia 5: 231 (1865).
Acalypha eggersii Pax, Bot. Jahrb. Syst. 26: 505 (1899).
Acalypha castaneifolia Poepp. ex Pax & K.Hoffm. in H.G.A.Engler, Pflanzenr., IV, 147, XVI: 163 (1924), nom. nud.

Acalypha cuprea Herzog, Repert. Spec. Nov. Regni Veg. 7: 60 (1909).
Bolivia. 83 BOL. Nanophan.

Acalypha cuspidata Jacq., Pl. Hort. Schoenbr. 2: 63 (1797). *Acalypha cuspidata* var. *genuina* Müll.Arg., Linnaea 34: 37 (1865), nom. inval. *Ricinocarpus cuspidatus* (Jacq.) Kuntze, Revis. Gen. Pl. 2: 617 (1891).

Caribbean to Ecuador. 79 + 80 + 81 CUB DOM HAI JAM LEE NLA WIN 82 VEN 83 CLM ECU. Nanophan.
Acalypha vestita Benth., Bot. Voy. Sulphur: 164 (1846).
Acalypha asterifolia Rusby, Descr. S. Amer. Pl.: 48 (1920).
Acalypha santae-martae Pax & K.Hoffm. in H.G.A.Engler, Pflanzenr., IV, 147, XVI: 121 (1924).

Acalypha dalzellii Hook.f., Fl. Brit. India 5: 414 (1887).
India. 40 IND. Nanophan.

Acalypha decaryana Leandri, Notul. Syst. (Paris) 10: 284 (1942).
Madagascar. 29 MDG.

Acalypha delgadoana McVaugh, Contr. Univ. Michigan Herb. 20: 173 (1995).
Mexico (Jalisco). 79 MXS.

Acalypha delpyana Gagnep., Bull. Soc. Bot. France 70: 872 (1923 publ. 1924).
C. Thailand, Cambodia, Laos. 41 CBD LAO THA. (Cl.) nanophan.

Acalypha deltoidea Robyns & Lawalrée, Bull. Jard. Bot. État 18: 267 (1947).
Zaire. 23 ZAI.

Acalypha depauperata Müll.Arg., Linnaea 34: 160 (1865). *Ricinocarpus depauperatus* (Müll.Arg.) Kuntze, Revis. Gen. Pl. 2: 617 (1891).
Mexico (Oaxaca). 79 MXS. Nanophan.

Acalypha depressa Sessé & Moç., Pl. Nov. Hisp.: 165 (1890).
Mexico. 79 +.

Acalypha depressinervia (Kuntze) K.Schum., Just's Bot. Jahresber. 26: 348 (1901).
S. Africa. 27 CPP NAT OFS SWZ TVL. Hemicr.
Acalypha schinzii Pax, Bull. Herb. Boissier 6: 734 (1898).
* *Ricinocarpus depressinervius* Kuntze, Revis. Gen. Pl. 3(2): 291 (1898).
Acalypha oweniae Harv. ex Pax & K.Hoffm. in H.G.A.Engler, Pflanzenr., IV, 147, XVI: 78 (1924), pro syn.

Acalypha dewevrei Pax, Bot. Jahrb. Syst. 28: 24 (1899).
Zaire. 23 ZAI. Nanophan.

Acalypha dictyoneura Müll.Arg., Linnaea 34: 12 (1865). *Ricinocarpus dictyoneurus* (Müll.Arg.) Kuntze, Revis. Gen. Pl. 2: 617 (1891).
Ecuador, Peru. 83 ECU PER. Nanophan.

f. **dictyoneura**
Ecuador, Peru. 83 ECU PER. Nanophan.

f. **reducta** Müll.Arg., Linnaea 34: 12 (1865).
Peru. 83 PER. Nanophan.

Acalypha digynostachya Baill., Adansonia 5: 233 (1865). *Ricinocarpus digynostachyus* (Baill.) Kuntze, Revis. Gen. Pl. 2: 617 (1891).
S. Brazil. 84 BZS. Nanophan.

Acalypha dikuluwensis P.A.Duvign. & Dewit, Bull. Soc. Roy. Bot. Belgique 96: 121 (1963).
Zaire, Zambia. 23 ZAI 26 ZAM.
Acalypha cupricola Robyns, Natuurw. Tijdschr. Ned.-Indië 14(Congr.Num.): 103 (1932), nom. nud.

Acalypha diminuata Baill. in A.Grandidier, Hist. Phys. Madagascar, Atlas: 194 (1891).
Madagascar. 29 MDG. Nanophan.

Acalypha dimorpha Müll.Arg. in C.F.P.von Martius, Fl. Bras. 11(2): 354 (1874). *Ricinocarpus dimorphus* (Müll.Arg.) Kuntze, Revis. Gen. Pl. 2: 618 (1891).
SE. Brazil. 84 BZL. Nanophan.

Acalypha dioica S.Watson, Proc. Amer. Acad. Arts 25: 162 (1890).
Mexico (Nuevo León). 79 MXE. Hemicr.
Acalypha dissitiflora S.Watson, Proc. Amer. Acad. Arts 26: 148 (1891).

Acalypha distans Müll.Arg. in A.P.de Candolle, Prodr. 15(2): 820 (1866). *Ricinocarpus distans* (Müll.Arg.) Kuntze, Revis. Gen. Pl. 2: 617 (1891).
W. & C. Cuba. 81 CUB. Nanophan.

Acalypha divaricata Müll.Arg., Linnaea 34: 34 (1865). *Ricinocarpus divaricatus* (Müll.Arg.) Kuntze, Revis. Gen. Pl. 2: 617 (1891).
Peru. 83 PER. Nanophan.

Acalypha diversifolia Jacq., Pl. Hort. Schoenbr. 2: 63 (1797). *Acalypha diversifolia* var. *genuina* Müll.Arg. in A.P.de Candolle, Prodr. 15(2): 854 (1866), nom. inval. *Ricinocarpus diversifolius* (Jacq.) Kuntze, Revis. Gen. Pl. 2: 617 (1891).
Mexico to Bolivia. 79 MXG MXT 80 COS GUA HON PAN 82 FRG GUY SUR VEN 83 BOL CLM PER 84 BZC BZL BZN. Nanophan. or phan.
Acalypha leptostachya Kunth in F.W.H.von Humboldt, A.J.A.Bonpland & C.S.Kunth, Nov. Gen. Sp. 2: 96 (1817). *Acalypha leptostachya* var. *genuina* Müll.Arg., Linnaea 34: 34 (1865), nom. inval. *Acalypha diversifolia* var. *leptostachya* (Kunth) Müll.Arg. in A.P.de Candolle, Prodr. 15(2): 853 (1866).
Acalypha popayanensis Kunth in F.W.H.von Humboldt, A.J.A.Bonpland & C.S.Kunth, Nov. Gen. Sp. 7: 173 (1825). *Acalypha leptostachya* var. *popayanensis* (Kunth) Müll.Arg., Linnaea 34: 34 (1865). *Acalypha diversifolia* var. *popayanensis* (Kunth) Müll.Arg. in A.P.de Candolle, Prodr. 15(2): 854 (1866).
Acalypha microgyna Poepp. & Endl., Nov. Gen. Sp. Pl. 3: 21 (1845).
Acalypha samydifolia Poepp. & Endl., Nov. Gen. Sp. Pl. 3: 21 (1845). *Ricinocarpus samydifolius* (Poepp. & Endl.) Kuntze, Revis. Gen. Pl. 2: 618 (1891).
Acalypha ulmifolia Benth., Pl. Hartw.: 252 (1846).
Acalypha betuloides Pav. ex Klotzsch in B.Seemann, Bot. Voy. Herald: 101 (1853).
Acalypha billbergiana Klotzsch in B.Seemann, Bot. Voy. Herald: 101 (1853).
Acalypha carpinifolia Poepp. ex Seem., Bot. Voy. Herald: 101 (1853). *Acalypha leptostachya* var. *carpinifolia* (Poepp. ex Seem.) Müll.Arg., Linnaea 34: 35 (1865). *Acalypha diversifolia* var. *carpinifolia* (Poepp. ex Seem.) Müll.Arg. in A.P.de Candolle, Prodr. 15(2): 854 (1866).
Acalypha panamensis Klotzsch in B.Seemann, Bot. Voy. Herald: 101 (1853).
Acalypha hartwegiana Benth. ex Baill., Étude Euphorb.: 442 (1858).
Acalypha diversifolia var. *squarrosa* Müll.Arg. in C.F.P.von Martius, Fl. Bras. 11(2): 358 (1874).
Acalypha diversifolia Rusby, Mem. Torrey Bot. Club 4: 257 (1895), nom. illeg.
Acalypha salicioides Rusby, Descr. S. Amer. Pl.: 46 (1920).
Acalypha callosa var. *glabra* Britton ex Pax & K.Hoffm. in H.G.A.Engler, Pflanzenr., IV, 147, XVI: 108 (1924).
Acalypha diversifolia var. *claoneura* Pax & K.Hoffm. in H.G.A.Engler, Pflanzenr., IV, 147, XVI: 108 (1924).
Acalypha tabascensis Lundell, Lloydia 4: 51 (1941).

Acalypha douilleana Rusby, Mem. New York Bot. Gard. 7: 285 (1927).
Bolivia. 83 BOL.

Acalypha dregei Gand., Bull. Soc. Bot. France 60: 27 (1913).
S. Africa. 27 CPP.

Acalypha dumetorum Müll.Arg., J. Bot. 2: 334 (1864). *Ricinocarpus dumetorum* (Müll.Arg.) Kuntze, Revis. Gen. Pl. 2: 617 (1891).
Angola. 26 ANG. Nanophan.

Acalypha echinus Pax & K.Hoffm. in H.G.A.Engler, Pflanzenr., IV, 147, XVI: 20 (1924).
SE. Kenya, E. Tanzania. 25 KEN TAN. Nanophan.

Acalypha ecklonii Baill., Adansonia 3: 158 (1863). *Ricinocarpus ecklonii* (Baill.) Kuntze, Revis. Gen. Pl. 2: 617 (1891).
Namibia, Cape Prov. 27 CPP NAM. Ther.
Acalypha brachiata E.Mey. in J.F.Drège, Zwei Pflanzengeogr. Dokum.: 129 (1843), nom. nud.
Acalypha cordata E.Mey. in J.F.Drège, Zwei Pflanzengeogr. Dokum.: 161 (1843), nom. nud.

Acalypha ecuadorica Pax & K.Hoffm. in H.G.A.Engler, Pflanzenr., IV, 147, XVI: 68 (1924).
Ecuador. 83 ECU. Nanophan.

Acalypha elizabethiae R.A.Howard, Phytologia 61: 1 (1986).
Windward Is. 81 WIN.

Acalypha elliptica Sw., Prodr.: 99 (1788). *Ricinocarpus ellipticus* (Sw.) Kuntze, Revis. Gen. Pl. 2: 617 (1891).
Jamaica. 81 JAM. Nanophan.

Acalypha elskensii De Wild., Pl. Bequaert. 3: 488 (1926).
Zaire. 23 ZAI.

Acalypha emirnensis Baill., Adansonia 1: 270 (1861). *Ricinocarpus emirnensis* (Baill.) Kuntze, Revis. Gen. Pl. 2: 617 (1891).
Madagascar. 29 MDG. Nanophan.

Acalypha engleri Pax, Bot. Jahrb. Syst. 34: 372 (1904).
SE. Kenya, Tanzania (incl. Zanzibar). 25 KEN TAN. Nanophan.
Acalypha sigensis Pax & K.Hoffm. in H.G.A.Engler, Pflanzenr., IV, 147, XVI: 92 (1924).

Acalypha erecta Paul G.Wilson, Kew Bull. 13: 170 (1958).
Mexico (México State). 79 MXC.

Acalypha eremorum Müll.Arg., Flora 47: 440 (1864). *Acalypha eremorum* var. *sessilis* Baill., Adansonia 6: 317 (1866). *Ricinocarpus eremorum* (Müll.Arg.) Kuntze, Revis. Gen. Pl. 2: 617 (1891).
C. & S. Queensland, NE. New South Wales. 50 NSW QLD. Nanophan.

Acalypha eriophylla Hutch., Bull. Misc. Inform. Kew 1911: 185 (1911).
Angola. 26 ANG. Hemicr.

Acalypha eriophylloides S.Moore, J. Bot. 57: 250 (1919).
Angola. 26 ANG.

Acalypha erosa Rusby, Bull. Torrey Bot. Club 28: 305 (1901).
Bolivia. 83 BOL.

Acalypha erubescens Rob. & Greenm., Proc. Amer. Acad. Arts 29: 393 (1894).
Mexico (San Luis Potosí). 79 MXE. Ther.

Acalypha eugeniifolia Rusby, Bull. New York Bot. Gard. 4: 443 (1907).
Bolivia. 83 BOL. Nanophan.

Acalypha euphrasiostachys Bartlett, Proc. Amer. Acad. Arts 43: 55 (1907).
Guatemala. 80 GUA. Nanophan.

Acalypha explorationis Airy Shaw, Kew Bull. 33: 71 (1978).
Irian Jaya. 42 NWG. Nanophan.

Acalypha fasciculata Müll.Arg., Linnaea 34: 31 (1865). *Ricinocarpus fasciculatus* (Müll.Arg.) Kuntze, Revis. Gen. Pl. 2: 618 (1891).
Réunion. 29 REU. Nanophan.

Acalypha ferdinandii K.Hoffm. in H.G.A.Engler, Pflanzenr., IV, 147, XVI: 63 (1924).
C. America. 80 COS GUA HON. Phan.

Acalypha filiformis Poir. in J.B.A.M.de Lamarck, Encycl. 6: 205 (1804).
St. Helena, W. Indian Ocean. 28 STH 29 COM MAU MDG REU. Nanophan.

subsp. **filiformis**
Mauritius, Réunion, Madagascar. 29 MAU MDG REU. Nanophan.
Acalypha filiformis var. *arborea* Poir. in J.B.A.M.de Lamarck, Encycl. 6: 205 (1804).
Acalypha reticulata var. *arborea* (Poir.) Müll.Arg., Linnaea 34: 32 (1865).
Tragia reticulata Poir. in J.B.A.M.de Lamarck, Encycl. 7: 725 (1806). *Claoxylon reticulatum* (Poir.) Bojer, Hortus Maurit.: 285 (1837). *Acalypha reticulata* var. *genuina* Müll.Arg., Linnaea 34: 32 (1865), nom. inval. *Acalypha reticulata* (Poir.) Müll.Arg., Linnaea 34: 32 (1865). *Ricinocarpus reticulatus* (Poir.) Kuntze, Revis. Gen. Pl. 2: 618 (1891).
Acalypha lantanifolia Bojer, Hortus Maurit.: 286 (1837).
Tragia arborea Comm. ex Baill., Adansonia 1: 267 (1861), pro syn.

var. **goudotiana** (Baill.) Govaerts in R.Govaerts, D.G.Frodin & A.Radcliffe-Smith, World Checklist Bibliogr. Euphorbiaceae: 62 (2000).
Madagascar. 29 MDG. Nanophan.
* *Acalypha goudotiana* Baill., Adansonia 1: 268 (1861). *Acalypha reticulata* var. *goudotiana* (Baill.) Müll.Arg., Linnaea 34: 32 (1865).

var. **ovalifolia** (Baill.) Govaerts in R.Govaerts, D.G.Frodin & A.Radcliffe-Smith, World Checklist Bibliogr. Euphorbiaceae: 62 (2000).
Comoros (incl. Mayotte). 29 COM. Nanophan.
* *Acalypha ovalifolia* Baill., Adansonia 1: 269 (1861). *Acalypha reticulata* var. *ovalifolia* (Baill.) Müll.Arg., Linnaea 34: 32 (1865).

var. **pervilleana** (Baill.) Govaerts in R.Govaerts, D.G.Frodin & A.Radcliffe-Smith, World Checklist Bibliogr. Euphorbiaceae: 62 (2000).
Madagascar (Nosi Bé I.). 29 MDG. Nanophan.
* *Acalypha pervilleana* Baill., Adansonia 1: 273 (1861). *Acalypha reticulata* var. *pervilleana* (Baill.) Müll.Arg., Linnaea 34: 32 (1865).

subsp. **rubra** (Müll.Arg.) Govaerts in R.Govaerts, D.G.Frodin & A.Radcliffe-Smith, World Checklist Bibliogr. Euphorbiaceae: 62 (2000).
St. Helena. 28 STH. Nanophan. – Extinct since 1843.
* *Acalypha rubra* Roxb. in A.Beaston, Tracts St. Helena, App. 1: 295 (1816), nom. illeg. *Acalypha reticulata* var. *rubra* Müll.Arg. in A.P.de Candolle, Prodr. 15(2): 851 (1866). *Acalypha rubrinervis* Cronk, Bull. Nat. Hist. Mus. (Bot.) 25: 98 (1995).

var. **urophylla** (Baill.) Govaerts in R.Govaerts, D.G.Frodin & A.Radcliffe-Smith, World Checklist Bibliogr. Euphorbiaceae: 63 (2000).
Madagascar (Nosi Bé I.). 29 MDG. Nanophan.
* *Acalypha urophylla* Baill., Adansonia 1: 273 (1861). *Acalypha reticulata* var. *urophylla* (Baill.) Müll.Arg., Linnaea 34: 32 (1865).

var. **urophylloides** (Pax & K.Hoffm.) Govaerts in R.Govaerts, D.G.Frodin & A.Radcliffe-Smith, World Checklist Bibliogr. Euphorbiaceae: 63 (2000).
Comoros, Madagascar (Sacatia I.). 29 COM MDG. Nanophan.
* *Acalypha urophylla* Pax, Bot. Jahrb. Syst. 19: 96 (1894), nom. illeg. *Acalypha paxii* Palacky, Cat. Pl. Madagasc. 2: 25 (1907). *Acalypha reticulata* var. *urophylloides* Pax & K.Hoffm. in H.G.A.Engler, Pflanzenr., IV, 147, XVI: 105 (1924).

Acalypha filipes (S.Watson) McVaugh, Brittonia 13: 149 (1961).
Mexico. 79 MXC MXE MXN MXS. Nanophan.
* *Corythea filipes* S.Watson, Proc. Amer. Acad. Arts 22: 451 (1887).
Acalypha coryloides Rose, Contr. U. S. Natl. Herb. 1: 357 (1895).

Acalypha fimbriata Schumach. & Thonn. in C.F.Schumacher, Beskr. Guin. Pl.: 409 (1827).
Acalypha vahliana Müll.Arg., Linnaea 34: 43 (1865).
Trop. & S. Africa. 22 GAN GHA GNB GUI IVO LBR MLI NGA SEN SIE TOG 23 BUR CAF CMN CON GGI ZAI 24 SUD 25 TAN UGA 26 ANG MLW MOZ ZAM ZIM 27 NAM TVL. Ther.
Acalypha fimbriata Hochst. ex A.Rich., Tent. Fl. Abyss. 2: 251 (1850).

Acalypha firmula Müll.Arg., Linnaea 34: 21 (1865). *Ricinocarpus firmulus* (Müll.Arg.) Kuntze, Revis. Gen. Pl. 2: 618 (1891).
El Salvador. 80 ELS.

Acalypha fissa (Müll.Arg.) Hutch., Bull. Misc. Inform. Kew 1913: 27 (1913).
Cuba. 81 CUB. Hemicr.
* *Acalypha chamaedrifolia* var. *fissa* Müll.Arg. in A.P.de Candolle, Prodr. 15(2): 879 (1866).

Acalypha flaccida Hook.f., Trans. Linn. Soc. London 20: 186 (1847). *Acalypha parvula* var. *flaccida* (Hook.f.) Müll.Arg. in A.P.de Candolle, Prodr. 15(2): 878 (1866).
Galapagos (San Salvador). 83 GAL. Ther.

Acalypha flagellata Millsp., Publ. Field Mus. Nat. Hist., Bot. Ser. 2: 417 (1916).
Mexico (Yucatán Pen.). 79 MXT. Nanophan.

Acalypha flavescens S.Watson, Proc. Amer. Acad. Arts 26: 149 (1891).
Mexico (San Luis Potosí: Tamasopa canyon). 79 MXE. Nanophan.

Acalypha forbesii S.Moore, J. Bot. 53: 336 (1914).
Peru. 83 PER. Hemicr. ?

Acalypha forsteriana Müll.Arg. in A.P.de Candolle, Prodr. 15(2): 807 (1866). *Ricinocarpus forsterianus* (Müll.Arg.) Kuntze, Revis. Gen. Pl. 2: 618 (1891).
Vanuatu. 60 VAN. Nanophan. or phan.
* *Acalypha virgata* G.Forst., Fl. Ins. Austr.: 67 (1786), nom. illeg.

Acalypha fournieri Müll.Arg., Linnaea 34: 162 (1865). *Ricinocarpus fournieri* (Müll.Arg.) Kuntze, Revis. Gen. Pl. 2: 618 (1891).
Mexico. 79 MXE MXG. Hemicr.

Acalypha fragilis Pax & K.Hoffm., Repert. Spec. Nov. Regni Veg. 41: 226 (1937).
Brazil (Ceará). 84 BZE.

Acalypha fredericii Müll.Arg. in A.P.de Candolle, Prodr. 15(2): 828 (1866).
Mexico (Veracruz). 79 MXG. Hemicr.

Acalypha friesii Pax & K.Hoffm. in H.G.A.Engler, Pflanzenr., IV, 147, XVI: 50 (1924).
Argentina (Jujuy). 85 AGW. Nanophan.

Acalypha fruticosa Forssk., Fl. Aegypt.-Arab.: 161 (1775). *Ricinocarpus fruticosus* (Forssk.) Kuntze, Revis. Gen. Pl. 2: 618 (1891).
Namibia, E. Africa, Yemen, S. India, Sri Lanka, Burma. 23 BUR 24 ETH SOM SUD 25 KEN TAN UGA 26 MLW MOZ 27 BOT NAM 35 YEM 40 IND SRL 41 BMA. Nanophan. or phan.

var. **eglandulosa** Radcl.-Sm., Kew Bull. 28: 288 (1973).
Kenya, Sudan, Ethiopia, Uganda, Burundi, Tanzania. 23 BUR 24 ETH SUD 25 KEN TAN UGA. Nanophan. or phan.
Acalypha holtzii Pax & K.Hoffm. in H.G.A.Engler, Pflanzenr., IV, 147, XVI: 130 (1924).
Acalypha kilimandscharica Volkens ex Pax & K.Hoffm. in H.G.A.Engler, Pflanzenr., IV, 147, XVI: 130 (1924).

var. **fruticosa**
Namibia, E. Africa, Yemen, S. India, Sri Lanka, Burma. 23 BUR 24 ETH SOM SUD 25 KEN TAN UGA 26 MLW MOZ 27 BOT NAM 35 YEM 40 IND SRL 41 BMA. Nanophan. or phan.
Acalypha betulina Retz., Observ. Bot. 5: 30 (1789).
Acalypha capitata Wall., Numer. List: 7783A, B (1847), nom. inval.
Acalypha fruticosa var. *villosa* Hutch. in D.Oliver, Fl. Trop. Afr. 6(1): 896 (1912).
Acalypha paxiana Dinter ex Pax & K.Hoffm. in H.G.A.Engler, Pflanzenr., IV, 147, XVI: 169 (1924), pro syn.
Acalypha chrysadenia Suess. & Friedrich, Mitt. Bot. Staatssamml. München 8: 333 (1953).

Acalypha fryeri Hutch., Bull. Misc. Inform. Kew 1918: 206 (1918).
Aldabra. 29 ALD.

Acalypha fulva I.M.Johnst., Contr. Gray Herb. 75: 29 (1925).
Peru. 83 PER.

Acalypha fuscescens Müll.Arg. in A.P.de Candolle, Prodr. 15(2): 821 (1866). *Ricinocarpus fuscescens* (Müll.Arg.) Kuntze, Revis. Gen. Pl. 2: 618 (1891).
Angola, Zambia. 26 ANG ZAM. Hemicr. or nanophan.

Acalypha garnieri Standl. & L.O.Williams, Ceiba 1: 147 (1950).
Nicaragua. 80 NIC.

Acalypha gaumeri Pax & K.Hoffm. in H.G.A.Engler, Pflanzenr., IV, 147, XVI: 173 (1924).
Mexico (Yucatán). 79 MXT. Nanophan.

Acalypha gigantesca McVaugh, Contr. Univ. Michigan Herb. 20: 175 (1995).
Mexico (Jalisco). 79 MXS.

Acalypha gillmanii Radcl.-Sm., Kew Bull. 30: 675 (1975 publ. 1976).
E. & SE. Tanzania. 25 TAN. Nanophan.

Acalypha glabrata Thunb., Prodr. Pl. Cap.: 117 (1800). *Acalypha glabrata* var. *genuina* Müll.Arg., Linnaea 34: 36 (1865), nom. inval. *Ricinocarpus glabratus* (Thunb.) Kuntze, Revis. Gen. Pl. 2: 618 (1891).
Zimbabwe, Mozambique, S. Africa. 26 MOZ ZIM 27 BOT CPP NAT SWZ TVL. Nanophan.

f. **glabrata**
Zimbabwe, Mozambique, S. Africa. 26 MOZ ZIM 27 BOT CPP NAT SWZ TVL. Nanophan.
Acalypha velutina E.Mey. in J.F.Drège, Zwei Pflanzengeogr. Dokum.: 161 (1843), nom. nud.
Acalypha betulina Sond., Linnaea 23: 116 (1850).
Acalypha betulina var. *latifolia* Sond., Linnaea 23: 117 (1850). *Acalypha glabrata* var. *latifolia* (Sond.) Müll.Arg., Linnaea 34: 36 (1865). *Ricinocarpus glabratus* var. *latifolius* (Sond.) Kuntze, Revis. Gen. Pl. 3(2): 291 (1898).

f. **pilosior** (Kuntze) Prain & Hutch., Bull. Misc. Inform. Kew 1913: 13 (1913).
S. Africa. 27 BOT CPP NAT TVL. Nanophan.
Acalypha glabrata var. *pilosa* Pax, Bull. Herb. Boissier 6: 733 (1898).
* *Ricinocarpus glabratus* f. *pilosior* Kuntze, Revis. Gen. Pl. 3(2): 291 (1898).

Acalypha glandulifolia Buchinger & Meisn. ex Krauss, Flora 28: 83 (1845). *Acalypha peduncularis* var. *glandulifolia* (Buchinger & Meisn. ex Krauss) Müll.Arg., Linnaea 34: 28 (1865).
S. Africa. 27 NAT SWZ TVL. Hemicr.
Acalypha entumenica Prain, Bull. Misc. Inform. Kew 1913: 22 (1913).

Acalypha glandulosa Cav., Anales Hist. Nat. 2: 141 (1800).
Mexico (S. Guanajuato), Colombia (Boyacá). 79 MXE 83 CLM. Cham. or nanophan.

Acalypha glechomifolia A.Rich. in R.de la Sagra, Hist. Fis. Cuba, Bot. 2: 205 (1850). *Acalypha reptans* var. *glechomifolia* (A.Rich.) Müll.Arg., Linnaea 34: 48 (1865). *Acalypha chamaedrifolia* var. *glechomifolia* (A.Rich.) Müll.Arg. in A.P.de Candolle, Prodr. 15(2): 879 (1866).
Cuba, Hispaniola. 81 CUB DOM HAI. Hemicr.

Acalypha gossweileri S.Moore, J. Bot. 57: 250 (1919).
Angola. 26 ANG.

Acalypha gracilis Spreng., Syst. Veg. 4(2): 315 (1827). *Acalypha gracilis* var. *genuina* Müll.Arg., Linnaea 34: 24 (1865), nom. inval. *Ricinocarpus gracilis* (Spreng.) Kuntze, Revis. Gen. Pl. 2: 618 (1891).
S. Brazil. 84 BZS. Nanophan.
Acalypha divaricata Baill., Adansonia 5: 234 (1865). *Acalypha gracilis* var. *divaricata* (Baill.) Pax & K.Hoffm. in H.G.A.Engler, Pflanzenr., IV, 147, XVI: 84 (1924).
Acalypha gracilis var. *fruticulosa* Müll.Arg., Linnaea 34: 24 (1865).
Acalypha gracilis var. *pubescens* Müll.Arg. in C.F.P.von Martius, Fl. Bras. 11(2): 352 (1874).
Acalypha diversifolia var. *leptostachya* Glaz., Mém. Soc. Bot. France 3: 623 (1912).
Acalypha striolata Lingelsh., Mitth. Thüring. Bot. Vereins, n.s., 29: 48 (1912).

Acalypha grandibracteata Merr., Philipp. J. Sci., C 5: 191 (1910).
Philippines. 42 PHI. Nanophan. or phan.

Acalypha grandispicata Rusby, Bull. Torrey Bot. Club 28: 304 (1901).
Bolivia. 83 BOL. Nanophan.

Acalypha grisea Pax & K.Hoffm. in H.G.A.Engler, Pflanzenr., IV, 147, XVI: 56 (1924).
Mexico (Jalisco). 79 MXS. Nanophan.

Acalypha grisebachiana (Kuntze) Pax & K.Hoffm. in H.G.A.Engler, Pflanzenr., IV, 147, XVI: 148 (1924).
Trinidad. 81 TRT. Nanophan.
* *Ricinocarpus grisebachianus* Kuntze, Revis. Gen. Pl. 2: 616 (1891).

Acalypha grueningiana Pax & K.Hoffm. in H.G.A.Engler, Pflanzenr., IV, 147, XVI: 79 (1924).
Angola. 26 ANG. Hemicr.

Acalypha guatemalensis Pax & K.Hoffm. in H.G.A.Engler, Pflanzenr., IV, 147, XVI: 27 (1924).
Guatemala. 80 GUA. Hemicr.

Acalypha gummifera Lundell, Contr. Univ. Michigan Herb. 4: 10 (1940).
Honduras. 80 HON.

Acalypha hainanensis Merr. & Chun, Sunyatsenia 5: 91 (1940).
Hainan. 36 CHH.

Acalypha haploclada Pax & K.Hoffm. in H.G.A.Engler, Pflanzenr., IV, 147, XVI: 51 (1924).
Mexico (Veracruz, Oaxaca). 79 MXG MXS. Hemicr.

Acalypha harmandiana Gagnep., Bull. Soc. Bot. France 70: 873 (1923 publ. 1924).
Cambodia. 41 CBD.

Acalypha hassleriana Chodat, Bull. Herb. Boissier, II, 5: 606 (1905). *Acalypha hassleriana* var. *genuina* Pax & K.Hoffm. in H.G.A.Engler, Pflanzenr., IV, 147, XVI: 41 (1924), nom. inval.
Paraguay. 85 PAR. Nanophan.
Ricinocarpus glandulosus Kuntze, Revis. Gen. Pl. 2: 618 (1891). *Acalypha glandulosa* (Kuntze) Chodat & Hassl., Bull. Herb. Boissier, II, 5: 605 (1905), nom. illeg. *Acalypha hassleriana* var. *glandulosa* (Kuntze) Pax & K.Hoffm. in H.G.A.Engler, Pflanzenr., IV, 147, XVI: 41 (1924).
Acalypha glandulosa var. *brevistachya* Chodat & Hassl., Bull. Herb. Boissier, II, 5: 606 (1905).

Acalypha havanensis Müll.Arg., Linnaea 34: 49 (1865). *Ricinocarpus havanensis* (Müll.Arg.) Kuntze, Revis. Gen. Pl. 2: 618 (1891).
W. & WC. Cuba. 81 CUB. Ther. or cham.

Acalypha helenae Buscal. & Muschl., Bot. Jahrb. Syst. 49: 477 (1913).
Zambia. 26 ZAM. Nanophan.

Acalypha hellwigii Warb., Bot. Jahrb. Syst. 18: 198 (1894).
Sulawesi, New Guinea. 42 NWG SUL. (Cl.) nanophan. or phan.

var. **hellwigii**
New Guinea. 42 NWG. (Cl.) nanophan. or phan.
Acalypha scandens var. *glabra* Warb., Bot. Jahrb. Syst. 13: 359 (1891). *Acalypha hellwigii* var. *glabra* (Warb.) K.Schum. & Lauterb., Fl. Schutzgeb. Südsee: 402 (1900).

var. **mollis** (Warb.) K.Schum. & Lauterb., Fl. Schutzgeb. Südsee: 402 (1900).
Sulawesi, New Guinea. 42 NWG SUL. Nanophan. or phan.
* *Acalypha scandens* var. *mollis* Warb., Bot. Jahrb. Syst. 13: 360 (1891).
Acalypha sogerensis S.Moore, J. Bot. 61(Suppl.): 47 (1923).

Acalypha hernandiifolia Sw., Prodr.: 99 (1788). *Acalypha hernandiifolia* var. *genuina* Müll.Arg., Linnaea 34: 11 (1865), nom. inval. *Ricinocarpus hernandiifolius* (Sw.) Kuntze, Revis. Gen. Pl. 2: 616 (1891).
Jamaica. 81 JAM. Nanophan. or phan.

var. **hernandiifolia**
Jamaica. 81 JAM. Nanophan. or phan.

var. **pubescens** Müll.Arg., Linnaea 34: 11 (1865).
Jamaica. 81 JAM. Nanophan. or phan.

Acalypha herzogiana Pax & K.Hoffm., Meded. Rijks-Herb. 40: 24 (1921).
Bolivia. 83 BOL. Hemicr.

Acalypha heteromorpha Rusby, Mem. New York Bot. Gard. 7: 286 (1927).
Bolivia. 83 BOL.

Acalypha hibiscifolia Britton, Mem. Torrey Bot. Club 9: 257 (1895).
Bolivia. 83 BOL. Nanophan.

Acalypha hildebrandtii Baill., Bull. Mens. Soc. Linn. Paris 2: 1005 (1892).
Madagascar. 29 MDG. Nanophan.

Acalypha hispida Burm.f., Fl. Indica: 303 (1768). *Ricinocarpus hispidus* (Burm.f.) Kuntze, Revis. Gen. Pl. 2: 618 (1891).
Bismarck Archip. 42 BIS. Nanophan. – Widely cultivated elsewere.
Acalypha densiflora Blume, Bijdr.: 628 (1826).
Acalypha sanderi N.E.Br., Gard. Chron., III, 1896: 392 (1896). *Acalypha hispida* var. *sanderi* (N.E.Br.) J.J.Sm., Meded. Dept. Landb. Ned.-Indië 10: 19 (1910).
Acalypha sanderiana K.Schum., Notizbl. Bot. Gart. Berlin-Dahlem 2: 127 (1898).

Acalypha hochstetteriana Müll.Arg., Linnaea 34: 39 (1865).
Ethiopia. 24 ETH. Ther.
* *Mercurialis alternifolia* Hochst. ex Baill., Étude Euphorb.: 490 (1858). *Ricinocarpus alternifolius* (Hochst. ex Baill.) Kuntze, Revis. Gen. Pl. 2: 617 (1891).

Acalypha hoffmanniana Hurus., J. Fac. Sci. Univ. Tokyo, Sect. 3, Bot. 6: 297 (1954).
S. Sulawesi. 42 SUL. Nanophan.
* *Acalyphopsis celebica* Pax & K.Hoffm. in H.G.A.Engler, Pflanzenr., IV, 147, XVI: 178 (1924).

Acalypha hologyna Baker, J. Linn. Soc., Bot. 21: 441 (1885).
Madagascar. 29 MDG. Nanophan.

Acalypha homblei De Wild., Repert. Spec. Nov. Regni Veg. 13: 145 (1914).
Zaire. 23 ZAI. Hemicr.

Acalypha hontauyuensis H.Keng, J. Wash. Acad. Sci. 41: 204 (1951).
Taiwan (Lanyu I.). 38 TAI. Nanophan.

Acalypha huillensis Pax & K.Hoffm. in H.G.A.Engler, Pflanzenr., IV, 147, XVI: 70 (1924).
Angola. 26 ANG. Nanophan.

Acalypha humbertii Leandri, Notul. Syst. (Paris) 10: 274 (1942).
Madagascar. 29 MDG.

Acalypha humboltiana Baill. in A.Grandidier, Hist. Phys. Madagascar, Atlas: 190 (1891).
Madagascar. 29 MDG. Nanophan.

Acalypha hutchinsonii Britton, Mem. Torrey Bot. Club 16: 77 (1920).
SC. Cuba. 81 CUB. Nanophan.

Acalypha hypogaea S.Watson, Proc. Amer. Acad. Arts 22: 451 (1887).
Mexico (Jalisco). 79 MXS. Ther.

Acalypha inaequalis Rusby, Bull. Torrey Bot. Club 28: 303 (1901).
Bolivia. 83 BOL. Nanophan.

Acalypha inaequilqtera Cardiel, Brittonia 46: 203 (1994).
Colombia (Boyacá). 83 CLM. Nanophan. or phan.

Acalypha indica L., Sp. Pl.: 1003 (1753). *Ricinocarpus indicus* (L.) Kuntze, Revis. Gen. Pl. 2: 618 (1891).
Trop. Old World. 22 ALL 23 ZAI 24 ETH SOM 25 KEN TAN UGA 26 ALL 27 BOT CPP NAM NAT SWZ TVL 29 MAU MDG 38 NNS TAI 40 IND PAK SRL 41 MLY THA 42 JAW LSI NWG PHI SUM. Ther. or cham.
Acalypha decidua Forssk., Fl. Aegypt.-Arab.: 161 (1775). *Ricinocarpus deciduus* (Forssk.) Kuntze, Revis. Gen. Pl. 2: 617 (1891).
Acalypha spicata Forssk., Fl. Aegypt.-Arab.: 161 (1775).
Acalypha caroliniana Blanco, Fl. Filip.: 748 (1837), nom. illeg.
Acalypha canescens Wall., Numer. List: 7785 (1847), nom. inval.
Acalypha ciliata Wall., Numer. List: 7779J (1847), nom. inval.
Acalypha chinensis Benth., Fl. Hongk.: 303 (1861).
Acalypha fimbriata Baill., Adansonia 1: 272 (1861).
Acalypha bailloniana Müll.Arg., Linnaea 34: 44 (1865). *Ricinocarpus baillonianus* (Müll.Arg.) Kuntze, Revis. Gen. Pl. 2: 617 (1891). *Acalypha indica* var. *bailloniana* (Müll.Arg.) Hutch. in D.Oliver, Fl. Trop. Afr. 6(1): 904 (1912).
Acalypha somalium Müll.Arg., Bremen Abh. 7: 27 (1880).
Acalypha somalensis Pax, Bot. Jahrb. Syst. 19: 100 (1894).
Acalypha cupamenii Dragend., Heilpfl.: 380 (1898).
Acalypha minima H.Keng, Taiwania 6: 32 (1955). *Acalypha indica* var. *minima* (H.Keng) S.F.Huang & T.C.Huang, Taiwania 36: 83 (1991).

Acalypha infesta Poepp. & Endl., Nov. Gen. Sp. Pl. 3: 22 (1845). *Ricinocarpus infestus* (Poepp. & Endl.) Kuntze, Revis. Gen. Pl. 2: 617 (1891).
Ecuador, Peru. 83 ECU PER. Ther.
Acalypha rotundifolia Vahl ex Baill., Étude Euphorb.: 442 (1858). *Acalypha infesta* var. *rotundifolia* (Vahl ex Baill.) Müll.Arg., Linnaea 34: 23 (1865).
Acalypha infesta var. *stenoloba* Müll.Arg., Linnaea 34: 23 (1865).
Acalypha infestans Müll.Arg., Linnaea 34: 23 (1865).

Acalypha insulana Müll.Arg., Flora 4: 39 (1864). *Ricinocarpus insulanus* (Müll.Arg.) Kuntze, Revis. Gen. Pl. 2: 618 (1891).
Fiji, Samoa. 60 FIJ SAM. Cl. nanophan. or phan.

var. **anisodonta** (Müll.Arg.) Govaerts in R.Govaerts, D.G.Frodin & A.Radcliffe-Smith, World Checklist Bibliogr. Euphorbiaceae: 68 (2000).
Fiji. 60 FIJ. Nanophan. or phan.
* *Acalypha anisodonta* Müll.Arg. in A.P.de Candolle, Prodr. 15(2): 819 (1866). *Ricinocarpus anisodontus* (Müll.Arg.) Kuntze, Revis. Gen. Pl. 2: 617 (1891).
Acalypha anisodonta var. *subvillosa* Müll.Arg. in A.P.de Candolle, Prodr. 15(2): 818 (1866). *Acalypha insulana* var. *subvillosa* (Müll.Arg.) A.C.Sm., J. Arnold Arbor. 33: 397 (1952).

var. **flavicans** Müll.Arg. in A.P.de Candolle, Prodr. 15(2): 818 (1866).
Fiji. 60 FIJ. Cl. nanophan. or phan.

var. **insulana**
Fiji, Samoa. 60 FIJ SAM. Cl. nanophan. or phan.
Acalypha insulana var. *pubescens* Müll.Arg. in A.P.de Candolle, Prodr. 15(2): 818 (1866).
Acalypha insulana var. *stipularis* Müll.Arg. in A.P.de Candolle, Prodr. 15(2): 818 (1866). *Acalypha stipularis* (Müll.Arg.) Engl., Bot. Jahrb. Syst. 7: 462 (1886).
Acalypha insulana var. *villosa* Müll.Arg. in A.P.de Candolle, Prodr. 15(2): 818 (1866).
Acalypha latifolia Müll.Arg. in A.P.de Candolle, Prodr. 15(2): 817 (1866). *Ricinocarpus latifolius* (Müll.Arg.) Kuntze, Revis. Gen. Pl. 2: 618 (1891).

Acalypha scandens Warb., Bot. Jahrb. Syst. 13: 359 (1891).
Acalypha stipulacea K.Schum. & Lauterb., Fl. Schutzgeb. Südsee: 403 (1900).
Acalypha weinlandii K.Schum. ex Pax & K.Hoffm. in H.G.A.Engler, Pflanzenr., IV, 147, XVI: 166 (1924), pro syn.

Acalypha integrifolia Willd., Sp. Pl. 4: 530 (1805). *Ricinocarpus integrifolius* (Willd.) Kuntze, Revis. Gen. Pl. 2: 617 (1891).
Madagascar, Mauritius, Réunion. 29 MAU MDG REU. Nanophan.

var. **crateriana** Coode, Kew Bull. 34: 44 (1979).
Mauritius. 29 MAU. Nanophan.
Acalypha commersoniana f. *unicolor* Müll.Arg. in A.P.de Candolle, Prodr. 15(2): 850 (1866).

var. **gracilipes** (Baill.) Müll.Arg. in A.P.de Candolle, Prodr. 15(2): 850 (1866).
Madagascar. 29 MDG. Nanophan.
* *Acalypha gracilipes* Baill., Adansonia 1: 273 (1861).

subsp. **integrifolia**
Mauritius, Réunion. 29 MAU REU. Nanophan.
Tragia colorata Poir. in J.B.A.M.de Lamarck, Encycl. 7: 725 (1806). *Acalypha colorata* (Poir.) Spreng., Syst. Veg. 3: 879 (1826). *Acalypha commersoniana* Baill. ex Müll.Arg. in A.P.de Candolle, Prodr. 15(2): 849 (1866). *Acalypha integrifolia* var. *colorata* (Poir.) Pax & K.Hoffm. in H.G.A.Engler, Pflanzenr., IV, 147, XVI: 106 (1924).
Acalypha discolor Bojer, Hortus Maurit.: 286 (1837), nom. nud.
Acalypha tomentosa Bojer, Hortus Maurit.: 285 (1837).
Acalypha odorata Steud., Nomencl. Bot., ed. 2, 2: 696 (1841).
Tragia odorata Steud., Nomencl. Bot., ed. 2, 2: 696 (1841).
Tragia lobata Wall., Numer. List: 7796 (1847), nom. nud.
Acalypha commersoniana var. *concolor* Baill., Adansonia 1: 267 (1861), nom. nud.
Acalypha commersoniana var. *discolor* Baill., Adansonia 1: 267 (1861), nom. nud.
Tragia fruticosa Comm. ex Baill., Adansonia 1: 267 (1861).
Tragia obtusata Vahl ex Baill., Adansonia 1: 267 (1861).
Acalypha commersoniana f. *concolor* Müll.Arg. in A.P.de Candolle, Prodr. 15(2): 849 (1866). *Acalypha integrifolia* var. *concolor* (Müll.Arg.) Pax & K.Hoffm. in H.G.A.Engler, Pflanzenr., IV, 147, XVI: 106 (1924).
Acalypha commersoniana f. *discolor* Müll.Arg. in A.P.de Candolle, Prodr. 15(2): 850 (1866).
Acalypha commersoniana f. *purpurea* Müll.Arg. in A.P.de Candolle, Prodr. 15(2): 849 (1866).
Acalypha commersoniana f. *purpureomarginata* Müll.Arg. in A.P.de Candolle, Prodr. 15(2): 849 (1866).
Acalypha commersoniana var. *acutifolia* Müll.Arg. in A.P.de Candolle, Prodr. 15(2): 849 (1866).
Acalypha commersoniana var. *longifolia* Müll.Arg. in A.P.de Candolle, Prodr. 15(2): 850 (1866).
Acalypha commersoniana var. *obtusifolia* Müll.Arg. in A.P.de Candolle, Prodr. 15(2): 850 (1866).
Tragia integrifolia Willd. ex Müll.Arg. in A.P.de Candolle, Prodr. 15(2): 948 (1866).

var. **longifolia** (Müll.Arg.) Coode, Kew Bull. 34: 41 (1979).
Mauritius. 29 MAU. Nanophan.
Acalypha reticulata f. *aberrans* Müll.Arg., Linnaea 34: 32 (1865).
* *Acalypha reticulata* var. *longifolia* Müll.Arg., Linnaea 34: 32 (1865).

subsp. **marginata** (Poir.) Coode, Kew Bull. 34: 42 (1979).
Mauritius. 29 MAU. Nanophan.
* *Tragia marginata* Poir. in J.B.A.M.de Lamarck, Encycl. 7: 725 (1806). *Acalypha marginata* (Poir.) Spreng., Syst. Veg. 3: 879 (1826). *Ricinocarpus marginatus* (Poir.) Kuntze, Revis. Gen. Pl. 2: 618 (1891).
Tragia castaneifolia Juss. ex Baill., Adansonia 1: 267 (1861).

subsp. **panduriformis** Coode, Kew Bull. 34: 42 (1979).
Réunion. 29 REU. Nanophan.

var. **parvifolia** (Baill. ex Müll.Arg.) Pax & K.Hoffm. in H.G.A.Engler, Pflanzenr., IV, 147, XVI: 106 (1924).
Mauritius. 29 MAU. Nanophan.
Acalypha commersoniana var. *brevifolia* Müll.Arg. in A.P.de Candolle, Prodr. 15(2): 850 (1866).
* *Acalypha commersoniana* var. *parvifolia* Baill. ex Müll.Arg. in A.P.de Candolle, Prodr. 15(2): 850 (1866).

var. **saltuum** Coode, Kew Bull. 34: 43 (1979).
Mauritius. 29 MAU. Nanophan.

Acalypha intermedia De Wild., Pl. Bequaert. 3: 490 (1926).
Zaire. 23 ZAI.

Acalypha jerzedowskii Calderón, Acta Ci. Potos. 7: 312 (1978).
Mexico (México State). 79 MXC. Ther.

Acalypha jubifera Rusby, Descr. S. Amer. Pl.: 48 (1920).
Bolivia. 83 BOL.

Acalypha juliflora Pax, Bot. Jahrb. Syst. 19: 95 (1894).
NW. Madagascar. 29 MDG. Nanophan.
Acalypha polynema Baill., Bull. Mens. Soc. Linn. Paris 2: 1197 (1895).

Acalypha juruana Ule, Verh. Bot. Vereins Prov. Brandenburg 50: 78 (1908 publ. 1909).
Brazil (Amazonas). 84 BZN. Nanophan.

Acalypha karwinskii Müll.Arg., Flora 55: 41 (1872).
Mexico. 79 +. Hemicr.

Acalypha katharinae Pax in H.G.A.Engler, Pflanzenr., IV, 147, XVI: 63 (1924).
Mexico (Oaxaca). 79 MXS. Hemicr.

Acalypha kerrii Craib, Bull. Misc. Inform. Kew 1911: 465 (1911).
Thailand, Burma, Vietnam. 41 BMA THA VIE. Nanophan. or phan.
Acalypha lacei Hutch., Bull. Misc. Inform. Kew 1914: 381 (1914).
Acalypha heterostachya Gagnep., Bull. Soc. Bot. France 70: 874 (1923 publ. 1924).
Acalypha siamensis Gagnep., Bull. Soc. Bot. France 70: 874 (1923 publ. 1924), nom. illeg. *Acalypha gagnepainii* Merr., J. Arnold Arbor. 19: 39 (1938).

Acalypha × koraensis Radcl.-Sm., Kew Bull. 41: 451 (1986). A. crenata × A. indica.
Kenya. 25 KEN. Ther.

Acalypha kotoensis Hayata, Icon. Pl. Formosan. 9: 99 (1920).
Taiwan. 38 TAI.

Acalypha laevigata Sw., Prodr.: 99 (1788). *Ricinocarpus laevigatus* (Sw.) Kuntze, Revis. Gen. Pl. 2: 618 (1891).
W. Jamaica. 81 JAM. Nanophan.

Acalypha lagascana Müll.Arg., Flora 55: 27 (1872).
Mexico (?). 79 +. Nanophan.

Acalypha lagoensis Müll.Arg. in C.F.P.von Martius, Fl. Bras. 11(2): 367 (1874). *Ricinocarpus lagoensis* (Müll.Arg.) Kuntze, Revis. Gen. Pl. 2: 618 (1891).
Brazil (Minas Gerais). 84 BZL. Nanophan.

Acalypha lagopus McVaugh, Contr. Univ. Michigan Herb. 20: 176 (1995).
Mexico (Michoacán). 79 MXS.

Acalypha lanceolata Willd., Sp. Pl. 4: 524 (1805). *Acalypha wightiana* var. *lanceolata* (Willd.) Müll.Arg., Linnaea 34: 43 (1865). *Ricinocarpus lanceolatus* (Willd.) Kuntze, Revis. Gen. Pl. 2: 617 (1891).
Trop. Africa to Polynesia. 22 + 23 ZAI 24 ETH 25 ALL 26 MLW MOZ 36 CHS 40 CKI IND SRL 41 BMA MLY THA 42 BIS BOR? JAW LSI MOL NWG PHI SUL SUM 50 QLD WAU 60 FIJ SAM SOL 61 COO. Ther.

var. **glandulosa** (Müll.Arg.) Radcl.-Sm., Kew Bull. 44: 444 (1989).
Trop. Africa. 22 + 23 ZAI 24 ETH 25 ALL 26 MLW MOZ. Ther.
* *Acalypha crenata* var. *glandulosa* Müll.Arg., Linnaea 34: 43 (1865). *Acalypha boehmerioides* var. *glandulosa* (Müll.Arg.) Pax & K.Hoffm. in H.G.A.Engler, Pflanzenr., IV, 147, XVI: 97 (1924).
Acalypha glomerata Hutch., Bull. Misc. Inform. Kew 1911: 229 (1911).

var. **lanceolata**
India to S. Pacific. 36 CHS 40 CKI IND SRL 41 BMA MLY THA 42 BIS BOR? JAW LSI MOL NWG PHI SUL SUM 50 QLD WAU 60 FIJ SAM SOL 61 COO. Ther.
Acalypha hispida Blume, Bijdr.: 628 (1826).
Acalypha zeylanica Raf., New Fl. 1: 46 (1836).
Acalypha hispida var. *pubescens* Hook. & Arn., Bot. Beechey Voy.: 213 (1837).
Acalypha flexuosa Wight ex Steud., Nomencl. Bot., ed. 2, 1: 9 (1840).
Acalypha hispida Wall., Numer. List: 7780 (1847), nom. inval.
Acalypha virginica Wall., Numer. List: 7779 (1847), nom. inval.
Acalypha corchorifolia Vahl ex Baill., Étude Euphorb.: 443 (1858).
Acalypha boehmerioides Miq., Fl. Ned. Ind., Eerste Bijv.: 459 (1861). *Acalypha boehmerioides* var. *genuina* Pax & K.Hoffm. in H.G.A.Engler, Pflanzenr., IV, 147, XVI: 96 (1924), nom. inval.
Acalypha fallax Müll.Arg., Linnaea 34: 43 (1865). *Ricinocarpus fallax* (Müll.Arg.) Kuntze, Revis. Gen. Pl. 2: 616 (1891).
Acalypha wightiana Müll.Arg., Linnaea 34: 43 (1865). *Acalypha wightiana* var. *genuina* Müll.Arg., Linnaea 34: 43 (1865), nom. inval.
Acalypha albicans B.Heyne ex Hook.f., Fl. Brit. India 5: 417 (1887), pro syn.
Acalypha collina Hayne ex Hook.f., Fl. Brit. India 5: 417 (1887).
Acalypha floribunda B.Heyne ex Hook.f., Fl. Brit. India 5: 417 (1887), pro syn.
Acalypha indica K.Schum. & Hollr., Fl. Kais. Wilh. Land: 75 (1889).
Acalypha indica var. *australis* F.M.Bailey, Queensland Dept. Agric. Div. Entomol. Bull. 9: 16 (1891).

Acalypha lancetillae Standl., Publ. Field Mus. Nat. Hist., Bot. Ser. 4: 312 (1929).
Honduras. 80 HON.

Acalypha langiana Müll.Arg., Linnaea 34: 159 (1865). *Ricinocarpus langianus* (Müll.Arg.) Kuntze, Revis. Gen. Pl. 2: 618 (1891).
W. Mexico. 79 MXN? MXS. Nanophan.

var. **langiana**
Mexico (Oaxaca). 79 MXN? MXS. Nanophan.

var. **rigens** McVaugh, Contr. Michigan Herb. 20: 179 (1995).
Mexico (Jalisco). 79 MXS.

Acalypha laxiflora Müll.Arg., Linnaea 34: 19 (1865). *Ricinocarpus laxiflorus* (Müll.Arg.) Kuntze, Revis. Gen. Pl. 2: 618 (1891).
Cuba, Mexico (Veracruz). 79 MXG 81 CUB. Nanophan.
Acalypha leptostachya A.Rich. in R.de la Sagra, Hist. Fis. Cuba, Bot. 11: 205 (1850), nom. illeg.

Acalypha lechleri Rusby, Bull. Torrey Bot. Club 28: 304 (1901).
Bolivia. 83 BOL.

Acalypha leicesterfieldiensis Radcl.-Sm. & Govaerts, Kew Bull. 52: 477 (1997).
Jamaica. 81 JAM.
* *Acalypha jamaicensis* Britton, Bull. Torrey Bot. Club 39: 7 (1912), nom. illeg.

Acalypha leonii Baill., Bull. Mens. Soc. Linn. Paris 2: 1197 (1895).
Madagascar. 29 MDG. Nanophan.

Acalypha lepidopagensis Leandri, Notul. Syst. (Paris) 10: 280 (1942).
Madagascar. 29 MDG.

Acalypha lepinei Müll.Arg., Linnaea 34: 14 (1865). *Ricinocarpus lepinei* (Müll.Arg.) Kuntze, Revis. Gen. Pl. 2: 618 (1891).
Society Is. (Tahiti, Bora Bora, Raiatea). 61 SCI. Nanophan. or phan.

Acalypha leptoclada Benth., Bot. Voy. Sulphur: 164 (1846). *Ricinocarpus leptocladus* (Benth.) Kuntze, Revis. Gen. Pl. 2: 618 (1891).
Mexico (between San Blas and Tepic). 79 MXE. Nanophan.

Acalypha leptomyura Baill., Bull. Mens. Soc. Linn. Paris 2: 1004 (1892).
W. Madagascar. 29 MDG. Nanophan.
Euphorbia leptomyura Baill. ex Poiss., Rech. Fl. Mérid. Madagascar: 62 (1912), sphalm.

Acalypha leptopoda Müll.Arg., Linnaea 34: 39 (1865). *Ricinocarpus leptopodus* (Müll.Arg.) Kuntze, Revis. Gen. Pl. 2: 618 (1891).
Mexico to W. Panama. 79 MXS 80 COS GUA PAN. Nanophan.
Acalypha leptopoda var. *glabrescens* Müll.Arg. in A.P.de Candolle, Prodr. 15(2): 824 (1866).
Acalypha leptopoda var. *mollis* Müll.Arg. in A.P.de Candolle, Prodr. 15(2): 824 (1866).
Acalypha lotsii Donn.Sm., Bot. Gaz. 10: 544 (1895).

Acalypha leptorhachis Müll.Arg., Linnaea 34: 7 (1865). *Ricinocarpus leptorhachis* (Müll.Arg.) Kuntze, Revis. Gen. Pl. 2: 618 (1891).
W. Cuba. 81 CUB. Nanophan.

Acalypha liebmanniana Müll.Arg., Linnaea 34: 161 (1865).
Mexico (Veracruz). 79 MXG. Nanophan.
Acalypha liebmannii Müll.Arg. in A.P.de Candolle, Prodr. 15(2): 829 (1866). *Ricinocarpus liebmannii* (Müll.Arg.) Kuntze, Revis. Gen. Pl. 2: 618 (1891).

Acalypha lignosa Brandegee, Univ. Calif. Publ. Bot. 6: 184 (1915).
Mexico (Oaxaca). 79 MXS. Nanophan.

Acalypha lindeniana Müll.Arg. in A.P.de Candolle, Prodr. 15(2): 827 (1866). *Ricinocarpus lindenianus* (Müll.Arg.) Kuntze, Revis. Gen. Pl. 2: 618 (1891).
Mexico (Puebla, Veracruz). 79 MXC MXG. Nanophan.

Acalypha linearifolia Leandri, Notul. Syst. (Paris) 10: 275 (1942).
Madagascar. 29 MDG.

Acalypha longiacuminata Hayata, Icon. Pl. Formosan. 9: 100 (1920).
S. Taiwan. 38 TAI. Nanophan.

Acalypha longipes S.Watson, Proc. Amer. Acad. Arts 26: 149 (1891).
Mexico (San Luis Potosí). 79 MXE. Nanophan.

Acalypha longispica Warb., Bot. Jahrb. Syst. 18: 197 (1894).
New Guinea (incl. D'Entrecasteaux Is.), Bismarck Archip. 42 BIS NWG. Phan.
Acalypha grandis K.Schum. & Lauterb., Fl. Schutzgeb. Südsee: 401 (1900), nom. illeg.
Acalypha caturoides K.Schum. & Lauterb., Fl. Schutzgeb. Südsee, Nachtr.: 298 (1905).
Acalypha protracta S.Moore, J. Bot. 61(Suppl.): 47 (1923).

Acalypha longispicata Müll.Arg., Linnaea 34: 163 (1865). *Ricinocarpus longispicatus* (Müll.Arg.) Kuntze, Revis. Gen. Pl. 2: 618 (1891).
Mexico (San Luis Potosí). 79 MXE. Hemicr.

Acalypha longistipularis Müll.Arg., Linnaea 34: 51 (1865). *Ricinocarpus longistipularis* (Müll.Arg.) Kuntze, Revis. Gen. Pl. 2: 618 (1891).
Mexico (Oaxaca). 79 MXS. Nanophan.

Acalypha lovelandii (McVaugh) McVaugh, Contr. Univ. Michigan Herb. 20: 180 (1995).
Mexico (Nayarit, Jalisco). 79 MXS.
* *Acalypha subviscida* var. *lovelandii* McVaugh, Brittonia 13: 153 (1961).

Acalypha lucida Rusby, Bull. New York Bot. Gard. 4: 444 (1907).
Bolivia. 83 BOL. Nanophan.

Acalypha lyallii Baker, J. Linn. Soc., Bot. 20: 255 (1884).
C. Madagascar. 29 MDG. Hemicr.

Acalypha lycioides Pax & K.Hoffm., Meded. Rijks-Herb. 40: 24 (1921).
N. Argentina, Bolivia. 83 BOL 85 AGE. Nanophan.
Acalypha divaricata Griseb., Abh. Königl. Ges. Wiss. Göttingen 24: 60 (1879), nom. illeg.

Acalypha lyonsii P.I.Forst., Austrobaileya 4: 216 (1994).
Queensland (Cook). 50 QLD. Nanophan. or phan.

Acalypha macbridei I.M.Johnst., Contr. Gray Herb. 75: 28 (1925).
Peru. 83 PER.

Acalypha macrodonta Müll.Arg., Linnaea 34: 51 (1865). *Ricinocarpus macrodontus* (Müll.Arg.) Kuntze, Revis. Gen. Pl. 2: 618 (1891).
Peru. 83 PER. Nanophan.

Acalypha macrostachya Jacq., Pl. Hort. Schoenbr. 2: 63 (1797). *Acalypha macrostachya* var. *genuina* Müll.Arg. in C.F.P.von Martius, Fl. Bras. 11(2): 345 (1874), nom. inval. *Ricinocarpus macrostachyus* (Jacq.) Kuntze, Revis. Gen. Pl. 2: 618 (1891).
Mexico, Trop. America. 79 MXS MXT 80 COS CPI GUA HON PAN 81 WIN 82 GUY SUR VEN 83 BOL CLM ECU PER 84 BZL. Nanophan. or phan.
Acalypha hirsutissima Willd., Sp. Pl. 4: 528 (1805). *Acalypha macrostachya* var. *hirsutissima* (Willd.) Müll.Arg., Linnaea 34: 11 (1865).
Acalypha macrostachya Willd., Sp. Pl. 4: 521 (1805), nom. illeg.
Acalypha cucullata Poir. in J.B.A.M.de Lamarck, Encycl., Suppl. 4: 683 (1816).
Acalypha caudata Kunth in F.W.H.von Humboldt, A.J.A.Bonpland & C.S.Kunth, Nov. Gen. Sp. 2: 95 (1817).

Acalypha macrophylla Kunth in F.W.H.von Humboldt, A.J.A.Bonpland & C.S.Kunth, Nov. Gen. Sp. 2: 96 (1817). *Acalypha macrostachya* f. *macrophylla* (Kunth) Müll.Arg. in A.P.de Candolle, Prodr. 15(2): 810 (1866). *Acalypha macrostachya* var. *macrophylla* (Kunth) Müll.Arg. in C.F.P.von Martius, Fl. Bras. 11(2): 345 (1874).
Acalypha sidifolia Kunth in F.W.H.von Humboldt, A.J.A.Bonpland & C.S.Kunth, Nov. Gen. Sp. 2: 95 (1817). *Acalypha macrostachya* var. *sidifolia* (Kunth) Müll.Arg., Linnaea 34: 11 (1865).
Acalypha tristis Poepp. & Endl., Nov. Gen. Sp. Pl. 3: 22 (1841). *Acalypha macrostachya* f. *tristis* (Poepp. & Endl.) Müll.Arg. in A.P.de Candolle, Prodr. 15(2): 811 (1866). *Acalypha macrostachya* var. *tristis* (Poepp. & Endl.) Müll.Arg. in C.F.P.von Martius, Fl. Bras. 11(2): 345 (1874).
Acalypha callosa Benth., Pl. Hartw.: 252 (1846). *Ricinocarpus callosus* (Benth.) Kuntze, Revis. Gen. Pl. 2: 617 (1891).
Acalypha seemannii Klotzsch in B.Seemann, Bot. Voy. Herald: 102 (1853).
Acalypha cancana Müll.Arg., Flora 47: 438 (1864). *Ricinocarpus cancanus* (Müll.Arg.) Kuntze, Revis. Gen. Pl. 2: 617 (1891).
Acalypha heterodonta Müll.Arg., Linnaea 34: 12 (1865). *Ricinocarpus heterodontus* (Müll.Arg.) Kuntze, Revis. Gen. Pl. 2: 618 (1891).
Acalypha heterodonta var. *hirsuta* Müll.Arg., Linnaea 34: 12 (1865).
Acalypha heterodonta var. *psilocarpa* Müll.Arg., Linnaea 34: 12 (1865).
Acalypha heterodonta var. *trichoclada* Müll.Arg., Linnaea 34: 12 (1865).
Acalypha neogranatensis Müll.Arg., Linnaea 34: 15 (1865). *Ricinocarpus neogranatensis* (Müll.Arg.) Kuntze, Revis. Gen. Pl. 2: 618 (1891).
Acalypha macrostachya f. *puberula* Müll.Arg. in A.P.de Candolle, Prodr. 15(2): 810 (1866).
Acalypha tarapotensis Müll.Arg. in A.P.de Candolle, Prodr. 15(2): 808 (1866). *Ricinocarpus tarapotensis* (Müll.Arg.) Kuntze, Revis. Gen. Pl. 2: 618 (1891).
Acalypha lehmanniana Pax, Bot. Jahrb. Syst. 26: 505 (1899).
Acalypha foliosa Rusby, Bull. New York Bot. Gard. 4: 443 (1907).
Acalypha amplifolia Rusby, Descr. S. Amer. Pl.: 46 (1920).
Acalypha caucana Müll.Arg. in H.G.A.Engler, Pflanzenr., IV, 147, XVI: 149 (1924).
Acalypha obtusifolia Pax & K.Hoffm. in H.G.A.Engler, Pflanzenr., IV, 147, XVI: 147 (1924).
Acalypha piperoides Klotzsch ex Pax & K.Hoffm. in H.G.A.Engler, Pflanzenr., IV, 147, XVI: 145 (1924), pro syn.
Acalypha pittieri Pax & K.Hoffm. in H.G.A.Engler, Pflanzenr., IV, 147, XVI: 18 (1924).
Acalypha hicksii Riley, Bull. Misc. Inform. Kew 1927: 126 (1927).
Acalypha fertilis Standl. & L.O.Williams, Ceiba 1: 146 (1950).

Acalypha macrostachyoides Müll.Arg. in A.P.de Candolle, Prodr. 15(2): 809 (1866).
Ricinocarpus macrostachyoides (Müll.Arg.) Kuntze, Revis. Gen. Pl. 2: 618 (1891).
Mexico (Veracruz). 79 MXG. Nanophan.

Acalypha macularis Pax & K.Hoffm. in H.G.A.Engler, Pflanzenr., IV, 147, XVI: 138 (1924).
Brazil (Rio de Janeiro). 84 BZL. Nanophan.

Acalypha madagascariensis Pax & K.Hoffm. in H.G.A.Engler, Pflanzenr., IV, 147, XVI: 162 (1924).
Madagascar. 29 MDG. Nanophan.

Acalypha madreporica Baill. in A.Grandidier, Hist. Phys. Madagascar, Atlas: 196 (1891).
Madagascar. 29 MDG. Phan.

Acalypha maestrensis Urb., Repert. Spec. Nov. Regni Veg. 28: 224 (1930).
Cuba (Sierra Maestra). 81 CUB. Nanophan.

Acalypha mairei (H.Lév.) C.K.Schneid. in C.S.Sargent, Pl. Wilson. 3: 301 (1916).
SC. China, N. Thailand. 36 CHC 41 THA. Nanophan.
* *Morus mairei* H.Lév., Repert. Spec. Nov. Regni Veg. 13: 265 (1914).

f. **mairei**
SC. China, N. Thailand. 36 CHC 41 THA. Nanophan.
Acalypha szechuanensis Hutch. in C.S.Sargent, Pl. Wilson. 2: 524 (1916).

f. **schneideriana** (Pax & K.Hoffm.) W.T.Wang, Vasc. Pl. Henqduan Mt. 1: 1057 (1993).
China (Yunnan), N. Thailand. 36 CHC 41 THA. Nanophan.
* *Acalypha schneideriana* Pax & K.Hoffm. in H.G.A.Engler, Pflanzenr., IV, 147, XVI: 138 (1924).

Acalypha malabarica Müll.Arg., Linnaea 34: 42 (1865). *Ricinocarpus malabaricus* (Müll.Arg.) Kuntze, Revis. Gen. Pl. 2: 618 (1891).
India. 40 IND. Ther.

Acalypha × **malawiensis** Radcl.-Sm., Kew Bull. 44: 444 (1989). A. chirindica × A. ornata.
Malawi. 26 MLW. Nanophan.

Acalypha mandonii Müll.Arg., Linnaea 34: 162 (1865). *Ricinocarpus mandonii* (Müll.Arg.) Kuntze, Revis. Gen. Pl. 2: 618 (1891).
Bolivia. 83 BOL. Nanophan.
Acalypha mollis Rusby, Mem. Torrey Bot. Club 6: 119 (1876).

Acalypha manniana Müll.Arg., Flora 47: 441 (1864). *Ricinocarpus mannianus* (Müll.Arg.) Kuntze, Revis. Gen. Pl. 2: 618 (1891).
W. Trop. Africa to Uganda. 22 BEN GHA NGA 23 CMN BUR GGI RWA ZAI 25 UGA. Nanophan.

Acalypha mapirensis Pax, Repert. Spec. Nov. Regni Veg. 7: 110 (1909).
Peru, Bolivia. 83 BOL PER. Nanophan.
Acalypha mapirensis Pax in O.Buchtien, Contr. Fl. Bolivia: 125 (1910), nom. nud.
Acalypha mapirensis var. *pubescens* Pax & K.Hoffm. in H.G.A.Engler, Pflanzenr., IV, 147, XVI: 65 (1924).
Acalypha mapirensis var. *scabra* Pax & K.Hoffm. in H.G.A.Engler, Pflanzenr., IV, 147, XVI: 65 (1924).

Acalypha marissima M.G.Gilbert, Kew Bull. 42: 360 (1987).
Ethiopia. 24 ETH.

Acalypha martiana Müll.Arg. in C.F.P.von Martius, Fl. Bras. 11(2): 359 (1874). *Ricinocarpus martianus* (Müll.Arg.) Kuntze, Revis. Gen. Pl. 2: 618 (1891).
Brazil. 84 BZL. Nanophan. or phan.

Acalypha matsudai Hayata, Icon. Pl. Formosan. 9: 100 (1920).
Taiwan (Hengchun Pen.). 38 TAI. Cham.

Acalypha medibracteata Radcl.-Sm. & Govaerts, Kew Bull. 52: 477 (1997).
Madagascar. 29 MDG.
* *Acalypha gagnepainii* Leandri, Notul. Syst. (Paris) 10: 274 (1942), nom. illeg.

var. **calcicola** (Leandri) Radcl.-Sm. & Govaerts, Kew Bull. 52: 477 (1997).
Madagascar. 29 MDG.
* *Acalypha gagnepainii* var. *calcicola* Leandri, Not. Syst. (Paris) 10: 275 (1942).

var. **medibracteata**
Madagascar. 29 MDG.

Acalypha meiodonta Baill., Bull. Mens. Soc. Linn. Paris 2: 1197 (1893).
C. Madagascar. 29 MDG. Nanophan.

Acalypha melochiifolia Müll.Arg. in A.P.de Candolle, Prodr. 15(2): 821 (1866). *Ricinocarpus melochiifolius* (Müll.Arg.) Kuntze, Revis. Gen. Pl. 2: 618 (1891).
Mexico (Veracruz). 79 MXG.

Acalypha membranacea A.Rich. in R.de la Sagra, Hist. Fis. Cuba, Bot. 2: 204 (1850). *Ricinocarpus membranaceus* (A.Rich.) Kuntze, Revis. Gen. Pl. 2: 618 (1891).
W. & C. Cuba, Colombia ? 81 CUB 83 CLM ? Nanophan.
Acalypha adenophora Griseb., Nachr. Königl. Ges. Wiss. Georg-Augusts-Univ. 1: 175 (1865).
Acalypha macrosperma Müll.Arg., Linnaea 34: 18 (1865).
Acalypha squarrosa Klotzsch ex Pax & K.Hoffm. in H.G.A.Engler, Pflanzenr., IV, 147, XVI: 50 (1924).

Acalypha mentiens Gand., Bull. Soc. Bot. France 60: 27 (1913).
SW. Cape Prov. 27 CPP.

Acalypha mexicana Müll.Arg., Linnaea 34: 41 (1865). *Ricinocarpus mexicanus* (Müll.Arg.) Kuntze, Revis. Gen. Pl. 2: 618 (1891). *Acalypha indica* var. *mexicana* (Müll.Arg.) Pax & K.Hoffm. in H.G.A.Engler, Pflanzenr., IV, 147, XVI: 35 (1924).
SE. Arizona to C. America. 76 ARI 79 MXE MXS MXT 80 COS GUA. Ther.
Acalypha tenuis Klotzsch ex Pax & K.Hoffm. in H.G.A.Engler, Pflanzenr., IV, 147, XVI: 25 (1924).

Acalypha meyeri Pax & K.Hoffm. in H.G.A.Engler, Pflanzenr., IV, 147, XVI: 165 (1924).
Philippines (Luzon). 42 PHI. Nanophan. or phan. – Perhaps abnormal *A. amentacea*.

Acalypha michoacanensis Sessé & Moç., Fl. Mexic., ed. 2: 221 (1894).
Mexico (Michoacán). 79 MXS.

Acalypha microcephala Müll.Arg., Linnaea 34: 160 (1865). *Ricinocarpus microcephalus* (Müll.Arg.) Kuntze, Revis. Gen. Pl. 2: 618 (1891).
Mexico (Oaxaca). 79 MXS. Nanophan.

Acalypha microphylla Klotzsch in B.Seemann, Bot. Voy. Herald: 278 (1856).
Mexico, Nicaragua. 79 MXN MXS MXT 80 NIC. Hemicr. or cham.

var. **interior** McVaugh, Contr. Univ. Michigan Herb. 20: 183 (1995).
Mexico (Sinaloa, Jalisco, Colima, Michoacán, Nayarit). 79 MXN MXS. Ther. or hemicr.

var. **microphylla**
Mexico (Jalisco, Sinaloa, Oaxaca, Colima, Chiapas), Nicaragua. 79 MXN MXS MXT 80 NIC. Hemicr. or cham.
Acalypha parvifolia Müll.Arg., Linnaea 34: 161 (1865). *Ricinocarpus parvifolius* (Müll.Arg.) Kuntze, Revis. Gen. Pl. 2: 618 (1891).
Acalypha nicaraguensis Pax & K.Hoffm. in H.G.A.Engler, Pflanzenr., IV, 147, XVI: 54 (1924).

Acalypha mogotensis Urb., Repert. Spec. Nov. Regni Veg. 28: 226 (1930).
W. Cuba. 81 CUB. Hemicr. or cham.

Acalypha mollis Kunth in F.W.H.von Humboldt, A.J.A.Bonpland & C.S.Kunth, Nov. Gen. Sp. 2: 94 (1817). *Ricinocarpus mollis* (Kunth) Kuntze, Revis. Gen. Pl. 2: 618 (1891).
Mexico (México State, San Luis Potosí). 79 MXC MXE. Nanophan.
Acalypha microstachya Benth., Pl. Hartw.: 71 (1840).
Acalypha mollis var. *polystachya* Müll.Arg., Linnaea 34: 19 (1865).

Acalypha monostachya Cav., Anales Hist. Nat. 2: 138 (1800). *Ricinocarpus monostachyus* (Cav.) Kuntze, Revis. Gen. Pl. 2: 618 (1891).
N. Mexico, New Mexico, Texas. 77 NWM TEX 79 MXE. Ther. or hemicr.

Acalypha hederacea Torr. in W.H.Emory, Rep. U.S. Mex. Bound. 2(1): 200 (1858).
Acalypha hederacea var. *genuina* Müll.Arg. in A.P.de Candolle, Prodr. 15(2): 885 (1866), nom. inval.
Acalypha hederacea var. *oligodonta* Müll.Arg., Linnaea 34: 53 (1865).
Acalypha hederacea var. *orbicularis* Müll.Arg. in A.P.de Candolle, Prodr. 15(2): 886 (1866).
Ricinocarpus hederaceus (Torr.) Kuntze, Revis. Gen. Pl. 2: 618 (1891).

Acalypha mortoniana Lundell, Bull. Torrey Bot. Club 64: 552 (1937).
Guatemala. 80 COS? GUA. Nanophan.

Acalypha muelleriana Urb., Symb. Antill. 1: 338 (1899).
Costa Rica, Venezuela. 80 COS 82 VEN. Nanophan.
* *Acalypha bisetosa* Müll.Arg. in A.P.de Candolle, Prodr. 15(2): 801 (1866), nom. illeg.

Acalypha multicaulis Müll.Arg., Linnaea 34: 53 (1865). *Acalypha multicaulis* var. *genuina* Müll.Arg. in C.F.P.von Martius, Fl. Bras. 11(2): 354 (1874), nom. inval. *Ricinocarpus multicaulis* (Müll.Arg.) Kuntze, Revis. Gen. Pl. 2: 618 (1891).
Brazil, Paraguay, NE. Argentina. 84 BZE BZL BZS 85 AGE PAR. Nanophan.
Acalypha tenuicaulis Baill., Adansonia 5: 234 (1865).
Acalypha multicaulis var. *tomentella* Müll.Arg. in C.F.P.von Martius, Fl. Bras. 11(2): 354 (1874).
Acalypha hederaceus Kuntze, Revis. Gen. Pl. 2: 292 (1891).
Acalypha pruriens Chodat & Hassl., Bull. Herb. Boissier, II, 5: 604 (1905), nom. illeg.
Acalypha multicaulis var. *glabrescens* Pax & K.Hoffm. in H.G.A.Engler, Pflanzenr., IV, 147, XVI: 88 (1924).
Acalypha multicaulis var. *tenuispica* Pax & K.Hoffm. in H.G.A.Engler, Pflanzenr., IV, 147, XVI: 88 (1924).

Acalypha multifida N.E.Br., Bull. Misc. Inform. Kew 1921: 296 (1921).
Zaire. 23 ZAI.

Acalypha multiflora (Standl.) Radcl.-Sm., Kew Bull. 31: 226 (1976).
Mexico (Nayarit). 79 MXS.
* *Corythea multiflora* Standl., Contr. U. S. Natl. Herb. 23: 649 (1923).

Acalypha multispicata S.Watson, Proc. Amer. Acad. Arts 26: 148 (1891).
Mexico (Jalisco). 79 MXS. Hemicr.

Acalypha mutisii Cardiel, Anales Jard. Bot. Madrid 48: 17 (1990).
Colombia. 83 CLM. Nanophan.

Acalypha nana (Müll.Arg.) Griseb. ex Hutch., Bull. Misc. Inform. Kew 1913: 27 (1913).
W. Cuba. 81 CUB. Hemicr.
Acalypha pygmaea Griseb., Nachr. Ges. Wiss. Göttingen Jahresber. 1865: 176 (1865), nom. illeg.
* *Acalypha chamaedrifolia* var. *nana* Müll.Arg. in A.P.de Candolle, Prodr. 15(2): 880 (1866).

Acalypha nematorhachis K.Schum. & Lauterb., Fl. Schutzgeb. Südsee: 402 (1900).
NE. Papua New Guinea. 42 NWG. Nanophan.

Acalypha nemorum F.Muell. ex Müll.Arg., Linnaea 34: 38 (1865). *Ricinocarpus nemorum* (F.Muell. ex Müll.Arg.) Kuntze, Revis. Gen. Pl. 2: 618 (1891).
C. & S. Queensland, NE. New South Wales. 50 NSW QLD. Nanophan.
Acalypha cunninghamii Müll.Arg., Linnaea 34: 35 (1865). *Ricinocarpus cunninghamii* (Müll.Arg.) Kuntze, Revis. Gen. Pl. 2: 617 (1891).

Acalypha neomexicana Müll.Arg., Linnaea 34: 19 (1865). *Ricinocarpus neomexicanus* (Müll.Arg.) Kuntze, Revis. Gen. Pl. 2: 618 (1891).
New Mexico, Arizona, W. Texas, Mexico. 76 ARI 77 NWM TEX 79 MXE MXN MXS. Ther.

Acalypha neptunica Müll.Arg., Bremen Abh. 7: 26 (1882). *Acalypha neptunica* var. *genuina* Pax & K.Hoffm. in H.G.A.Engler, Pflanzenr., IV, 147, XVI: 109 (1924), nom. inval.
Trop. Africa. 22 GHA NGA 23 CMN ZAI 24 ETH SUD 25 KEN TAN UGA 26 MLW. Nanophan. or phan.

var. **neptunica**
Trop. Africa. 22 GHA NGA 23 CMN ZAI 24 SUD 25 KEN TAN UGA. Nanophan. or phan.
Mallotus brevipes Pax ex Engl., Pflanzenw. Ost-Afrikas A: 18 (1895), nom. nud.
Acalypha mildbraediana Pax, Bot. Jahrb. Syst. 43: 323 (1909).
Acalypha mildbraediana var. *glabrescens* Pax, Bot. Jahrb. Syst. 43: 223 (1909). *Acalypha neptunica* var. *glabrescens* (Pax) Pax & K.Hoffm. in H.G.A.Engler, Pflanzenr., IV, 147, XVI: 109 (1924).
Acalypha subsessilis Hutch., Bull. Misc. Inform. Kew 1911: 231 (1911). *Acalypha subsessilis* var. *glabra* Pax & K.Hoffm. in H.G.A.Engler, Pflanzenr., IV, 147, XVI: 110 (1924), nom. illeg.

var. **pubescens** (Pax) Hutch. in D.Oliver, Fl. Trop. Afr. 6(1): 908 (1912).
Trop. Africa. 22 GHA NGA 23 CMN ZAI 24 ETH SUD 25 KEN TAN UGA 26 MLW. Nanophan. or phan.
* *Acalypha mildbraediana* var. *pubescens* Pax, Bot. Jahrb. Syst. 43: 323 (1909).
Acalypha subsessilis var. *mollis* Hutch., Bull. Misc. Inform. Kew 1911: 231 (1911).
Acalypha neptunica var. *vestita* Pax & K.Hoffm. in H.G.A.Engler, Pflanzenr., IV, 147, XVI: 109 (1924).

Acalypha nervulosa Airy Shaw, Kew Bull. 18: 407 (1966).
Irian Jaya. 42 NWG. Nanophan. or phan.

Acalypha nitschkeana Pax & K.Hoffm. in H.G.A.Engler, Pflanzenr., IV, 147, XVI: 88 (1924).
Paraguay. 85 PAR. Nanophan.
Acalypha dimorpha Chodat & Hassl., Bull. Herb. Boissier, II, 5: 604 (1905), nom. illeg.
Acalypha multicaulis Chodat & Hassl., Bull. Herb. Boissier, II, 5: 605 (1905), nom. illeg.

Acalypha noronhae Ridl., J. Linn. Soc., Bot. 27: 59 (1890).
Brazil (Fernando de Noronha). 84 BZE. Nanophan.

Acalypha novoguineensis Warb., Bot. Jahrb. Syst. 13: 359 (1891).
Papua New Guinea. 42 NWG. Phan.
Acalypha longispica K.Schum. & Lauterb., Fl. Schutzgeb. Südsee: 401 (1900).

Acalypha nubicola McVaugh, Brittonia 13: 150 (1961).
Mexico (SW. Michoacán). 79 MXT. Nanophan.

Acalypha nyasica Hutch. in D.Oliver, Fl. Trop. Afr. 6(1): 894 (1912).
S. Tanzania, Mozambique, Malawi. 25 TAN 26 MLW MOZ. Ther.

Acalypha oblancifolia Lundell, Wrightia 5: 243 (1976).
Guatemala. 80 GUA.

Acalypha obscura Müll.Arg., Linnaea 34: 163 (1865). *Ricinocarpus obscurus* (Müll.Arg.) Kuntze, Revis. Gen. Pl. 2: 618 (1891).
Mexico (San Luis Potosí). 79 MXE. Hemicr.

Acalypha ocymoides Kunth in F.W.H.von Humboldt, A.J.A.Bonpland & C.S.Kunth, Nov. Gen. Sp. 2: 93 (1817). *Ricinocarpus ocymoides* (Kunth) Kuntze, Revis. Gen. Pl. 2: 618 (1891).
Mexico (Michoacán: Volcan de Jarullo). 79 MXS. Ther.

Acalypha oligantha Müll.Arg., Linnaea 34: 159 (1865). *Ricinocarpus oliganthus* (Müll.Arg.) Kuntze, Revis. Gen. Pl. 2: 618 (1891).
Mexico (Veracruz). 79 MXG. Nanophan.

Acalypha oligodonta Müll.Arg. in A.P.de Candolle, Prodr. 15(2): 831 (1866). *Ricinocarpus oligodontus* (Müll.Arg.) Kuntze, Revis. Gen. Pl. 2: 618 (1891).
Mexico (Sierra de San Pedro Nolasco). 79 MXS. Ther.

Acalypha omissa Pax & K.Hoffm. in H.G.A.Engler, Pflanzenr., IV, 147, XVI: 111 (1924).
S. Brazil. 84 BZS. Nanophan.
Acalypha brasiliensis f. *microphylla* Müll.Arg. ex Pax & K.Hoffm. in H.G.A.Engler, Pflanzenr., IV, 147, XVI: 110 (1924), pro syn.

Acalypha oreopola Greenm., Proc. Amer. Acad. Arts 39: 82 (1903).
Mexico (Guerrero). 79 MXS.

Acalypha ornata Hochst. ex A.Rich., Tent. Fl. Abyss. 2: 247 (1850). *Acalypha ornata* var. *genuina* Müll.Arg., Linnaea 34: 19 (1865), nom. inval. *Ricinocarpus ornatus* (Hochst. ex A.Rich.) Kuntze, Revis. Gen. Pl. 2: 618 (1891).
Trop. & S. Africa. 22 NGA 23 CAF CMN ZAI 24 ETH SUD 25 KEN TAN UGA 26 ANG MLW MOZ ZAM ZIM 27 BOT NAM SWZ TVL. Cham. or nanophan.
Acalypha adenotricha A.Rich., Tent. Fl. Abyss. 2: 248 (1850). *Ricinocarpus adenotrichus* (A.Rich.) Kuntze, Revis. Gen. Pl. 2: 617 (1891).
Acalypha livingstoniana Müll.Arg., Flora 47: 440 (1864). *Ricinocarpus livingstonianus* (Müll.Arg.) Kuntze, Revis. Gen. Pl. 2: 618 (1891).
Acalypha nigritiana Müll.Arg., Flora 47: 440 (1864). *Ricinocarpus nigritianus* (Müll.Arg.) Kuntze, Revis. Gen. Pl. 2: 618 (1891).
Acalypha ornata var. *bracteosa* Müll.Arg., Flora 47: 443 (1864).
Acalypha ornata var. *pilosa* Müll.Arg., Flora 47: 443 (1864).
Acalypha ornata var. *glandulosa* Müll.Arg., Linnaea 34: 19 (1865).
Acalypha grantii Baker & Hutch., Bull. Misc. Inform. Kew 1911: 230 (1911).
Acalypha swynnertonii S.Moore, J. Linn. Soc., Bot. 40: 200 (1911).
Acalypha moggii Compton, J. S. African Bot. 41: 48 (1975).

Acalypha ostryifolia Riddell, Syn. Fl. West. States: 33 (1835).
E. & SE. U.S.A. to C. America, Caribbean. 74 KAN 75 NWJ 78 ALA FLA GEO KTY LOU 77 TEX 79 MXE MXN 80 GUA 81 CUB DOM HAI JAM PUE. Ther.
Acalypha caroliniana Elliott, Sketch Bot. S. Carolina 2: 645 (1824), nom. illeg. *Ricinocarpus carolinianus* Kuntze, Revis. Gen. Pl. 2: 617 (1891).
Acalypha corchorifolia A.Rich. in R.de la Sagra, Hist. Fis. Cuba, Bot. 3: 203 (1850). *Acalypha persimilis* var. *corchorifolia* (A.Rich.) Müll.Arg. in A.P.de Candolle, Prodr. 15(2): 842 (1866).
Acalypha polystachya Griseb., Fl. Brit. W. I.: 48 (1859), nom. illeg.
Acalypha persimilis Müll.Arg., Linnaea 34: 25 (1865). *Ricinocarpus persimilis* (Müll.Arg.) Kuntze, Revis. Gen. Pl. 2: 618 (1891).
Acalypha persimilis var. *scabra* Müll.Arg., Linnaea 34: 25 (1865).
Acalypha setosa Bello, Anales Soc. Esp. Hist. Nat. 12: 111 (1883), nom. illeg.
Acalypha pedunculata Klotzsch ex Pax & K.Hoffm. in H.G.A.Engler, Pflanzenr., IV, 147, XVI: 43 (1924), pro syn.

Acalypha oxyodonta (Müll.Arg.) Müll.Arg. in C.F.P.von Martius, Fl. Bras. 11(2): 367 (1874).
Ricinocarpus oxyodontus (Müll.Arg.) Kuntze, Revis. Gen. Pl. 2: 618 (1891).
Brazil (Minas Gerais). 84 BZL. Cham. or nanophan.
* *Acalypha cuspidata* var. *oxyodonta* Müll.Arg., Linnaea 34: 37 (1865).
Acalypha dupraeana var. *hilarii* Baill., Adansonia 5: 230 (1865).
Acalypha dupraeana var. *sylvicola* Baill., Adansonia 5: 230 (1865).

Acalypha padifolia Kunth in F.W.H.von Humboldt, A.J.A.Bonpland & C.S.Kunth, Nov. Gen. Sp. 2: 97 (1817). *Ricinocarpus padifolius* (Kunth) Kuntze, Revis. Gen. Pl. 2: 618 (1891).
W. South America. 83 BOL CLM ECU PER. Nanophan.
Acalypha erythrostachya Müll.Arg., Linnaea 34: 51 (1865). *Ricinocarpus erythrostachyus* (Müll.Arg.) Kuntze, Revis. Gen. Pl. 2: 617 (1891).
Acalypha coriifolia Pax & K.Hoffm. in H.G.A.Engler, Pflanzenr., IV, 147, XVI: 63 (1924).

Acalypha pallescens Urb., Repert. Spec. Nov. Regni Veg. 28: 24 (1930).
Jamaica. 81 JAM.

Acalypha palmeri Pax & K.Hoffm. in H.G.A.Engler, Pflanzenr., IV, 147, XVI: 157 (1924).
SW. Mexico. 79 MXS. Nanophan.

Acalypha pancheriana Baill., Adansonia 2: 225 (1862). *Ricinocarpus pancherianus* (Baill.) Kuntze, Revis. Gen. Pl. 2: 618 (1891).
New Caledonia (incl. Loyalty Is.). 60 NWC. Nanophan. or phan.
Acalypha neocaledonica Müll.Arg. in A.P.de Candolle, Prodr. 15(2): 812 (1866). *Ricinocarpus neocaledonicus* (Müll.Arg.) Kuntze, Revis. Gen. Pl. 2: 618 (1891).
Acalypha schlechteri Pax & K.Hoffm. in H.G.A.Engler, Pflanzenr., IV, 147, XVI: 155 (1924), nom. illeg.

Acalypha papillosa Rose, Contr. U. S. Natl. Herb. 1: 358 (1895).
Mexico (Sonora, Sinaloa, Chihuahua). 79 MXE MXN. Nanophan.

Acalypha parvula Hook.f., Trans. Linn. Soc. London 20: 185 (1847). *Acalypha parvula* var. *genuina* Müll.Arg. in A.P.de Candolle, Prodr. 15(2): 878 (1866), nom. inval. *Ricinocarpus parvulus* (Hook.f.) Kuntze, Revis. Gen. Pl. 2: 618 (1891).
Galapagos. 83 GAL. Ther. or cham.

var. **chathamensis** (Rob.) G.L.Webster, Madroño 20: 363 (1970).
Galapagos (San Cristóbal). 83 GAL.
* *Acalypha chathamensis* Rob., Proc. Amer. Acad. Arts 38: 163 (1902).

var. **parvula**
Galapagos. 83 GAL. Ther. or cham.
Acalypha cordifolia Hook.f., Trans. Linn. Soc. London 20: 186 (1847). *Acalypha parvula* var. *cordifolia* (Hook.f.) Müll.Arg. in A.P.de Candolle, Prodr. 15(2): 877 (1866). *Acalypha hookeri* J.F.Macbr., Publ. Field Mus. Nat. Hist., Bot. Ser. 11: 26 (1931), nom. illeg.
Acalypha cordifolia Andersson, Galapagos Veg.: 103 (1854).
Acalypha diffusa Andersson, Kongl. Vetensk. Acad. Handl. 1854: 240 (1855). *Acalypha parvula* f. *diffusa* (Andersson) Müll.Arg. in A.P.de Candolle, Prodr. 15(2): 878 (1866).
Acalypha spicata Andersson, Kongl. Vetensk. Acad. Handl. 1854: 239 (1855), nom. illeg.
Acalypha parvula var. *procumbens* Müll.Arg. in A.P.de Candolle, Prodr. 15(2): 878 (1866).
Acalypha parvula var. *pubescens* Müll.Arg. in A.P.de Candolle, Prodr. 15(2): 877 (1866).
Acalypha albemarlensis Rob., Proc. Amer. Acad. Arts 38: 163 (1902).

var. **reniformis** (Hook.f.) Müll.Arg. in A.P.de Candolle, Prodr. 15(2): 878 (1866).
Galapagos. 83 GAL. Ther.
* *Acalypha reniformis* Hook.f., Trans. Linn. Soc. London 20: 187 (1847).
Acalypha adamsii Rob., Proc. Amer. Acad. Arts 38: 161 (1902).

var. **strobilifera** (Hook.f.) Müll.Arg. in A.P.de Candolle, Prodr. 15(2): 877 (1866).
Galapagos. 83 GAL. Ther.
* *Acalypha strobilifera* Hook.f., Trans. Linn. Soc. London 20: 187 (1847).

Acalypha patens Müll.Arg. in A.P.de Candolle, Prodr. 15(2): 848 (1866). *Ricinocarpus patens* (Müll.Arg.) Kuntze, Revis. Gen. Pl. 2: 618 (1891).
Cape Prov. 27 CPP.

Acalypha paucifolia Baker & Hutch., Bull. Misc. Inform. Kew 1911: 230 (1911).
Mozambique, Zambia, Tanzania, Malawi. 25 TAN 26 MLW MOZ ZAM. Hemicr. or cham.

Acalypha paupercula Pax & K.Hoffm., Meded. Rijks-Herb. 40: 24 (1921).
Bolivia. 83 BOL. Hemicr.

Acalypha peckoltii Müll.Arg. in C.F.P.von Martius, Fl. Bras. 11(2): 365 (1874). *Ricinocarpus peckoltii* (Müll.Arg.) Kuntze, Revis. Gen. Pl. 2: 618 (1891).
Brazil (Rio de Janeiro). 84 BZL. Nanophan.

Acalypha peduncularis Meisn. ex C.Krauss, Flora 28: 82 (1845). *Acalypha peduncularis* var. *genuina* Müll.Arg., Linnaea 34: 29 (1865), nom. inval. *Ricinocarpus peduncularis* (Meisn. ex C.Krauss) Kuntze, Revis. Gen. Pl. 2: 618 (1891).
S. Africa. 27 CPP LES NAT OFS SWZ TVL. Hemicr.
Acalypha peduncularis var. *ferox* Pax ex F.Wilms, Exsicc.: 2265 .
Acalypha crassa Buchinger ex C.Krauss, Flora 28: 83 (1845). *Acalypha peduncularis* var. *crassa* (Buchinger ex C.Krauss) Müll.Arg., Linnaea 34: 28 (1865).
Acalypha zeyheri var. *pubescens* Müll.Arg., Linnaea 34: 29 (1865).

Acalypha pendula C.Wright ex Griseb., Nachr. Ges. Wiss. Göttingen Jahresber. 1865: 176 (1865). *Acalypha chamaedrifolia* var. *pendula* (C.Wright ex Griseb.) Müll.Arg. in A.P.de Candolle, Prodr. 15(2): 879 (1866).
W. & C. Cuba, Hispaniola. 81 CUB DOM HAI. Hemicr.

Acalypha perrieri Leandri, Notul. Syst. (Paris) 10: 273 (1942).
Madagascar. 29 MDG.

Acalypha peruviana Müll.Arg., Linnaea 34: 17 (1865). *Ricinocarpus peruvianus* (Müll.Arg.) Kuntze, Revis. Gen. Pl. 2: 618 (1891).
Peru. 83 PER. Nanophan.

Acalypha phleoides Cav., Anales Hist. Nat. 2: 139 (1801). *Acalypha phleoides* var. *genuina* Müll.Arg. in A.P.de Candolle, Prodr. 15(2): 876 (1866), nom. inval.
S. U.S.A. to C. America. 76 ARI 77 TEX 79 MXE MXG MXN MXS 80 GUA. Hemicr.
Acalypha hirta Cav., Anales Hist. Nat. 2: 141 (1801). *Acalypha phleoides* var. *hirta* (Cav.) Müll.Arg. in A.P.de Candolle, Prodr. 15(2): 876 (1866).
Acalypha pastoris DC. ex Willd., Enum. Pl.: 993 (1809).
Acalypha rubra Willd., Enum. Pl.: 992 (1809).
Acalypha prunifolia Kunth in F.W.H.von Humboldt, A.J.A.Bonpland & C.S.Kunth, Nov. Gen. Sp. 2: 92 (1817).
Acalypha pastoris Schrank, Denkschr. Bayer. Bot. Ges. Regensburg 2: 39 (1822).
Acalypha phleoides Torr. in W.H.Emory, Rep. U.S. Mex. Bound. 2(1): 199 (1858), nom. illeg. *Ricinocarpus phleoides* (Torr.) Kuntze, Revis. Gen. Pl. 2: 618 (1891).
Acalypha lindheimeri Müll.Arg., Linnaea 34: 47 (1865). *Ricinocarpus lindheimeri* (Müll.Arg.) Kuntze, Revis. Gen. Pl. 2: 618 (1891).
Acalypha ehretiifolia Klotzsch ex Pax & K.Hoffm. in H.G.A.Engler, Pflanzenr., IV, 147, XVI: 27 (1924), nom. illeg.
Acalypha lindheimeri var. *major* Pax & K.Hoffm. in H.G.A.Engler, Pflanzenr., IV, 147, XVI: 26 (1924).

Acalypha phyllonomifolia Airy Shaw, Kew Bull. 20: 406 (1966).
Papua New Guinea. 42 NWG. Nanophan.

Acalypha pilocardia Gilli, Repert. Spec. Nov. Regni Veg. 92: 678 (1981).
Ecuador. 83 ECU.

Acalypha pilosa Cav., Anales Hist. Nat. 2: 136 (1800). *Ricinocarpus pilosus* (Cav.) Kuntze, Revis. Gen. Pl. 2: 618 (1891).
Mexico, C. America. 79 MXS 80 PAN. Ther.

Acalypha pippenii McVaugh, Brittonia 13: 152 (1961).
Mexico (Michoacán). 79 MXS. Ther.

Acalypha platyphylla Müll.Arg., Linnaea 34: 6 (1865). *Ricinocarpus platyphyllus* (Müll.Arg.) Kuntze, Revis. Gen. Pl. 2: 618 (1891).
Colombia to Peru. 83 CLM ECU PER. Nanophan. or phan.
Acalypha subandina Ule, Verh. Bot. Vereins Prov. Brandenburg 50: 77 (1908 publ. 1909).

Acalypha pleiogyne Airy Shaw, Kew Bull. 32: 69 (1977).
Thailand. 41 THA.

Acalypha plicata Müll.Arg. in A.P.de Candolle, Prodr. 15(2): 855 (1866). *Ricinocarpus plicatus* (Müll.Arg.) Kuntze, Revis. Gen. Pl. 2: 618 (1891).
Colombia, Bolivia to N. Argentina. 83 BOL CLM 85 AGE AGW PAR. Cham. or nanophan.
Acalypha cordifolia Griseb., Abh. Königl. Ges. Wiss. Göttingen 19: 49 (1874).
Acalypha cordifolia var. *polyadenia* Griseb., Abh. Königl. Ges. Wiss. Göttingen 24: 60 (1879).
Acalypha plicata Griseb., Abh. Königl. Ges. Wiss. Göttingen 24: 60 (1879), nom. illeg.
Acalypha lagoensis var. *grandifolia* Chodat & Hassl., Bull. Herb. Boissier, II, 5: 604 (1895).
Acalypha flabellifera Rusby, Mem. Torrey Bot. Club 6: 119 (1896).

Acalypha pohliana Müll.Arg. in C.F.P.von Martius, Fl. Bras. 11(2): 360 (1874). *Ricinocarpus pohlianus* (Müll.Arg.) Kuntze, Revis. Gen. Pl. 2: 618 (1891).
SE. Brazil. 84 BZL. Nanophan.

Acalypha poiretii Spreng., Syst. Veg. 3: 879 (1826). *Ricinocarpus poiretii* (Spreng.) Kuntze, Revis. Gen. Pl. 2: 618 (1891).
Mexico, Trop. America. 79 MXE MXE 80 GUA 81 DOM HAI 82 FRG GUY? SUR? 84 BZC BZE BZL BZN BZS 85 AGW. Ther.
Acalypha alnifolia Poir. in J.B.A.M.de Lamarck, Encycl. 6: 203 (1804), nom. illeg.
Acalypha macrostachyos Poir. in J.B.A.M.de Lamarck, Encycl. 6: 208 (1804), nom. illeg.
Acalypha hispida Willd., Sp. Pl. 4: 523 (1805), nom. illeg.
Acalypha indica Vell., Fl. Flumin.: 10 (1829).
Acalypha cylindrica Roxb., Fl. Ind. ed. 1832, 3: 678 (1832).
Acalypha villosa Vahl ex Baill., Étude Euphorb.: 442 (1858), nom. illeg.
Acalypha rhombifolia Baill., Adansonia 5: 230 (1865).

Acalypha polymorpha Müll.Arg., J. Bot. 2: 335 (1864). *Ricinocarpus polymorphus* (Müll.Arg.) Kuntze, Revis. Gen. Pl. 2: 618 (1891).
C. & E. Trop. Africa. 23 BUR RWA ZAI 25 KEN TAN UGA 26 ANG MLW MOZ ZAM ZIM. Hemicr.
Acalypha polymorpha var. *elliptica* Müll.Arg., J. Bot. 2: 335 (1864).
Acalypha polymorpha var. *oblongifolia* Müll.Arg., J. Bot. 2: 335 (1864).
Acalypha polymorpha var. *sericea* Müll.Arg., J. Bot. 2: 335 (1864).
Acalypha crotonoides Pax, Bot. Jahrb. Syst. 19: 97 (1894).
Acalypha stuhlmannii Pax, Bot. Jahrb. Syst. 19: 99 (1894).

Acalypha crotonoides var. *caudata* Hutch. ex R.E.Fr., Wiss. Erg. Schwed. Rhod.-Kongo Exped. 1(1): 123 (1914).
Acalypha goetzei Pax & K.Hoffm. in H.G.A.Engler, Pflanzenr., IV, 147, XVI: 78 (1924).
Acalypha shirensis Hutch. ex Pax & K.Hoffm. in H.G.A.Engler, Pflanzenr., IV, 147, XVI: 32 (1924).

Acalypha polystachya Jacq., Pl. Hort. Schoenbr. 2: 64 (1797). *Ricinocarpus polystachyus* (Jacq.) Kuntze, Revis. Gen. Pl. 2: 618 (1891).
Mexico to Ecuador. 79 MXE MXN MXS MXT 80 COS 83 ECU. Ther.
Acalypha filifera S.Watson, Proc. Amer. Acad. Arts 22: 451 (1887).
Acalypha polystachya Sessé & Moç., Fl. Mexic., ed. 2: 221 (1894), nom. illeg.
Acalypha matudai Lundell, Contr. Univ. Michigan Herb. 4: 10 (1940).

Acalypha porcina Standl. & L.O.Williams, Ceiba 3: 208 (1953).
Nicaragua. 80 NIC.

Acalypha porphyrantha Standl., J. Arnold Arbor. 11: 32 (1930).
Honduras. 80 HON.

Acalypha portoricensis Müll.Arg., Linnaea 34: 22 (1865). *Ricinocarpus portoricensis* (Müll.Arg.) Kuntze, Revis. Gen. Pl. 2: 618 (1891).
Puerto Rico. 81 PUE. Nanophan.

Acalypha pruinosa Urb., Symb. Antill. 5: 388 (1908).
Jamaica. 81 JAM. Nanophan.
* *Acalypha elliptica* Griseb., Fl. Brit. W. I.: 47 (1859), nom. illeg.

Acalypha prunifolia Nees & Mart., Nova Acta Acad. Caes. Leop-Carol. German. Nat. Cur. 11: 37 (1823). *Ricinocarpus prunifolius* (Nees & Mart.) Kuntze, Revis. Gen. Pl. 2: 618 (1891).
S. Brazil. 84 BZS. Nanophan.
Acalypha klotzschiana Baill., Adansonia 5: 231 (1865).

Acalypha pruriens Nees & Mart., Nova Acta Acad. Caes. Leop-Carol. German. Nat. Cur. 11: 36 (1823). *Ricinocarpus pruriens* (Nees & Mart.) Kuntze, Revis. Gen. Pl. 2: 618 (1891).
E. Brazil. 84 BZE BZL. Nanophan.

Acalypha pseudalopecuroides Pax & K.Hoffm. in H.G.A.Engler, Pflanzenr., IV, 147, XVI: 86 (1924).
Mexico to Nicaragua. 79 MXC MXN MXS 80 NIC. Ther.

Acalypha pseudovagans Pax & K.Hoffm. in H.G.A.Engler, Pflanzenr., IV, 147, XVI: 52 (1924).
Mexico (San Luis Potosí). 79 MXE. Hemicr.

Acalypha psilostachya Hochst. ex A.Rich., Tent. Fl. Abyss. 2: 246 (1850). *Ricinocarpus psilostachyus* (Hochst. ex A.Rich.) Kuntze, Revis. Gen. Pl. 2: 618 (1891).
C., NE. & E. Trop. Africa. 23 BUR ZAI 24 ETH SUD 25 KEN TAN UGA 26 ANG MLW MOZ ZAM. Hemicr. or nanophan.

var. **glandulosa** Hutch. in D.Oliver, Fl. Trop. Afr. 6(1): 900 (1912).
C. & E. Trop. Africa. 23 BUR ZAI 25 KEN TAN UGA 26 MLW MOZ ZAM. Hemicr. or nanophan.

var. **psilostachya**
C., NE. & E. Trop. Africa. 23 BUR ZAI 24 ETH SUD 25 KEN TAN UGA 26 ANG MLW MOZ ZAM. Hemicr. or nanophan.
Acalypha johnstonii Pax in A.G.H.Engler, Hochgebirgsfl. Afrika: 283 (1892).
Acalypha bequaertii Staner, Bull. Jard. Bot. État 15: 141 (1938).

Acalypha pubiflora (Klotzsch) Baill., Adansonia 1: 268 (1861).
SC. Trop. & S. Africa, Madagascar, N. Australia. 26 MLW MOZ ZIM 27 BOT TVL 29 MDG 50 WAU. Nanophan.
* *Calyptrospatha pubiflora* Klotzsch in W.C.H.Peters, Naturw. Reise Mossambique: 97 (1861). *Ricinocarpus pubiflorus* (Klotzsch) Kuntze, Revis. Gen. Pl. 2: 618 (1891).

subsp. **australica** Radcl.-Sm., Kew Bull. 45: 678 (1990).
N. Western Australia. 50 WAU. Nanophan.

subsp. **pubiflora**
SC. Trop. & S. Africa, Madagascar. 26 MLW MOZ ZIM 27 BOT TVL 29 MDG. Nanophan.

Acalypha pulchrespicata Däniker, Vierteljahrschr. Naturf. Ges. Zürich 77(19): 228 (1932).
New Caledonia (Loyalty Is.). 60 NWC.

Acalypha punctata Meisn. ex C.Krauss, Flora 28: 83 (1845). *Acalypha peduncularis* var. *punctata* (Meisn. ex C.Krauss) Müll.Arg., Linnaea 34: 28 (1865).
S. Africa. 27 CPP LES NAT TVL. Hemicr.
Acalypha punctata var. *longifolia* Prain, Bull. Misc. Inform. Kew 1913: 24 (1913).
Acalypha punctata var. *rogersii* Prain, Bull. Misc. Inform. Kew 1913: 24 (1913).
Acalypha longifolia E.Mey. ex Pax. & K.Hoffm. in H.G.A.Engler, Pflanzenr., IV, 147, XVI: 75 (1924).

Acalypha purpurascens Kunth in F.W.H.von Humboldt, A.J.A.Bonpland & C.S.Kunth, Nov. Gen. Sp. 2: 97 (1817). *Acalypha purpurascens* var. *genuina* Müll.Arg., Linnaea 34: 52 (1865), nom. inval. *Ricinocarpus purpurascens* (Kunth) Kuntze, Revis. Gen. Pl. 2: 618 (1891).
Mexico (San Luis Potosí). 79 MXE. Hemicr.
Acalypha purpurascens var. *eglandulosa* Müll.Arg., Linnaea 34: 52 (1865).

Acalypha purpusii Brandegee, Univ. Calif. Publ. Bot. 6: 53 (1914).
Mexico (Oaxaca). 79 MXS. Hemicr.

Acalypha pycnantha Urb., Ark. Bot. 10A(15): 59 (1926).
Haiti. 81 HAI. Cham. or nanophan.

Acalypha pygmaea A.Rich. in R.de la Sagra, Hist. Fis. Cuba, Bot. 3: 205 (1853). *Acalypha reptans* var. *pygmaea* (A.Rich.) Müll.Arg., Linnaea 34: 49 (1865). *Acalypha chamaedrifolia* var. *pygmaea* (A.Rich.) Müll.Arg. in A.P.de Candolle, Prodr. 15(2): 880 (1866).
W. & C. Cuba. 81 CUB. Hemicr.

Acalypha racemosa Wall. ex Baill., Étude Euphorb.: 443 (1858).
Trop. Africa, India, Sri Lanka, Jawa, Lesser Sunda Is. (Flores). 22 GHA IVO NGA TOG 23 CAF CON CMN GGI ZAI 24 CHA ETH SUD 25 KEN TAN UGA 26 ANG MLW MOZ ZIM 40 IND SRL 42 JAW LSI. Hemicr. or nanophan.
Acalypha paniculata Miq., Fl. Ned. Ind. 1(2): 406 (1859).
Acalypha wallichiana Thwaites, Enum. Pl. Zeyl.: 271 (1861).
Acalypha paniculata f. *depauperata* Müll.Arg. in A.P.de Candolle, Prodr. 15(2): 802 (1866).

Acalypha radians Torr. in W.H.Emory, Rep. U.S. Mex. Bound. 2(1): 200 (1858). *Acalypha radians* var. *genuina* Müll.Arg., Linnaea 34: 52 (1865), nom. inval. *Ricinocarpus radians* (Torr.) Kuntze, Revis. Gen. Pl. 2: 618 (1891).
New Mexico, Texas. 77 NWM TEX. Hemicr.
Acalypha radians var. *geraniifolia* Müll.Arg., Linnaea 34: 52 (1865).

Acalypha radicans Müll.Arg., Linnaea 34: 39 (1865). *Ricinocarpus radicans* (Müll.Arg.) Kuntze, Revis. Gen. Pl. 2: 618 (1891).
SE. Brazil. 84 BZL. Nanophan.

Acalypha radinostachya Donn.Sm., Bot. Gaz. 54: 243 (1912).
NC. Costa Rica. 80 COS. Nanophan.

Acalypha radula Baill. in A.Grandidier, Hist. Phys. Madagascar, Atlas: 193 (1891).
Madagascar. 29 MDG.

Acalypha rafaelensis Standl., Contr. U. S. Natl. Herb. 23: 629 (1923).
Mexico (San Luis Potosí: Agua del Medio at Minas de San Rafael). 79 MXE. Nanophan.

Acalypha raivavensis F.Br., Bernice P. Bishop Mus. Bull. 130: 145 (1935).
Tubuai Is. (Raivavae, Tubuai). 61 TUB. Nanophan. or phan.
Acalypha tubuaiensis H.St.John, Nordic J. Bot. 3: 449 (1983).

Acalypha rapensis F.Br., Bernice P. Bishop Mus. Bull. 130: 146 (1935).
Tubuai Is. (Rapa I.). 61 TUB. Nanophan. or phan.
Acalypha stokesii F.Br., Bernice P. Bishop Mus. Bull. 130: 146 (1935), nom. illeg.
Acalypha polynesiaca Radcl.-Sm. & Govaerts, Kew Bull. 52: 477 (1997).

Acalypha reflexa Müll.Arg., Linnaea 34: 33 (1865). *Ricinocarpus reflexus* (Müll.Arg.) Kuntze, Revis. Gen. Pl. 2: 618 (1891).
Peru. 83 PER. Nanophan.

Acalypha repanda Müll.Arg., Flora 47: 439 (1864). *Ricinocarpus repandus* (Müll.Arg.) Kuntze, Revis. Gen. Pl. 2: 618 (1891).
Fiji, Tonga. 60 FIJ TON. Nanophan. or phan.

var. **denudata** (Müll.Arg.) A.C.Sm., J. Arnold Arbor. 33: 399 (1952).
Fiji. 60 FIJ. Nanophan. or phan.
* *Acalypha denudata* Müll.Arg. in A.P.de Candolle, Prodr. 15(2): 819 (1866).
Acalypha laevifolia Müll.Arg. in A.P.de Candolle, Prodr. 15(2): 853 (1866). *Ricinocarpus laevifolius* (Müll.Arg.) Kuntze, Revis. Gen. Pl. 2: 618 (1891).

var. **repanda**
Fiji, Tonga. 60 FIJ TON. Nanophan. or phan.
Acalypha anisodonta var. *subsericea* Müll.Arg. in A.P.de Candolle, Prodr. 15(2): 818 (1866).
Acalypha insulana var. *glabrescens* Müll.Arg. in A.P.de Candolle, Prodr. 15(2): 818 (1866).

Acalypha retifera Standl. & L.O.Williams, Ceiba 3: 209 (1953).
Honduras. 80 HON.

Acalypha rhombifolia Schltdl., Linnaea 7: 382 (1832). *Ricinocarpus rhombifolius* (Schltdl.) Kuntze, Revis. Gen. Pl. 2: 618 (1891).
Mexico (Veracruz). 79 MXG. Hemicr.

Acalypha richardiana Baill., Adansonia 1: 268 (1861). *Ricinocarpus richardianus* (Baill.) Kuntze, Revis. Gen. Pl. 2: 618 (1891).
Madagascar. 29 MDG. Nanophan.

Acalypha riedeliana Baill., Adansonia 5: 231 (1865). *Ricinocarpus riedelianus* (Baill.) Kuntze, Revis. Gen. Pl. 2: 618 (1891).
S. Brazil. 84 BZS. Nanophan.

Acalypha rivularis Seem., Bonplandia 9: 258 (1861). *Ricinocarpus rivularis* (Seem.) Kuntze, Revis. Gen. Pl. 2: 618 (1891).
Fiji. 60 FIJ. Nanophan. or phan.

Acalypha rottleroides Baill., Adansonia 1: 270 (1861). *Ricinocarpus rottleroides* (Baill.) Kuntze, Revis. Gen. Pl. 2: 618 (1891).
Madagascar. 29 MDG. Nanophan.

Acalypha rubroserrata Pax & K.Hoffm. in H.G.A.Engler, Pflanzenr., IV, 147, XVI: 28 (1924).
C. Mexico. 79 MXC. Nanophan.

Acalypha ruderalis Mart. ex Britton, Ann. New York Acad. Sci. 7: 225 (1893).
Paraguay. 85 PAR.

Acalypha ruiziana Müll.Arg., Linnaea 34: 16 (1865). *Ricinocarpus ruizianus* (Müll.Arg.) Kuntze, Revis. Gen. Pl. 2: 618 (1891).
Peru, Ecuador. 83 ECU PER. Nanophan.
Acalypha granulata Ruiz ex Pax & K.Hoffm. in H.G.A.Engler, Pflanzenr., IV, 147, XVI: 67 (1924), nom. inval.

Acalypha rupestris Urb., Repert. Spec. Nov. Regni Veg. 28: 223 (1930).
Cuba (Sierra Maestra). 81 CUB. Nanophan.

Acalypha rusbyi Dorr, Brittonia 43: 226 (1991).
Bolivia. 83 BOL.
* *Acalypha williamsii* Rusby, Descr. S. Amer. Pl.: 47 (1920).

Acalypha sabulicola Brandegee, Univ. Calif. Publ. Bot. 6: 183 (1915).
Mexico (?). 79 +. Nanophan.

Acalypha salicifolia Müll.Arg., Flora 47: 438 (1864). *Ricinocarpus salicifolius* (Müll.Arg.) Kuntze, Revis. Gen. Pl. 2: 618 (1891).
Ecuador. 83 ECU. Nanophan.

Acalypha salvadorensis Standl., J. Wash. Acad. Sci. 14: 96 (1924).
SW. Mexico, El Salvador. 79 MXS 80 ELS. Ther.
Acalypha neomexicana var. *jaliscana* McVaugh, Contr. Univ. Michigan Herb. 20: 186 (1995).

Acalypha salviifolia Baill., Étude Euphorb.: 443 (1858).
Madagascar. 29 MDG. Phan.
Tragia salviifolia Bojer ex Baill., Étude Euphorb.: 443 (1858).
Acalypha radula Baker, J. Linn. Soc., Bot. 20: 254 (1883).

Acalypha saxicola Wiggins, Contr. Dudley Herb. 3: 70 (1940).
Mexico (Baja California). 79 MXN.

Acalypha scabrosa Sw., Prodr.: 100 (1788). *Acalypha scabrosa* var. *genuina* Müll.Arg., Linnaea 34: 33 (1865), nom. inval. *Ricinocarpus scabrosus* (Sw.) Kuntze, Revis. Gen. Pl. 2: 618 (1891).
Jamaica. 81 JAM. Nanophan.
Acalypha betulifolia Sw., Prodr.: 99 (1788). *Acalypha scabrosa* var. *betulifolia* (Sw.) Müll.Arg., Linnaea 34: 33 (1865).
Acalypha scabrosa var. *ovata* Griseb., Fl. Brit. W. I.: 47 (1859).
Acalypha scabrosa var. *elongata* Urb., Symb. Antill. 5: 389 (1908).

Acalypha scandens Benth., Hooker's J. Bot. Kew Gard. Misc. 6: 329 (1854). *Ricinocarpus scandens* (Benth.) Kuntze, Revis. Gen. Pl. 2: 618 (1891).
Guianas, Venezuela, Brazil (Amazonas), SE. Colombia, Peru. 82 GUY SUR VEN 83 CLM PER 84 BZN. Cl. nanophan.

Acalypha schiedeana Schltdl., Linnaea 7: 384 (1832). *Acalypha schiedeana* var. *genuina* Müll.Arg., Linnaea 7: 384 (1832), nom. inval. *Ricinocarpus schiedeanus* (Schltdl.) Kuntze, Revis. Gen. Pl. 2: 618 (1891).
Mexico, C. America, Colombia, Venezuela. 79 MXE MXG 80 COS GUA 82 VEN 83 CLM. Nanophan.
Acalypha deppeana Schltdl., Linnaea 7: 385 (1832). *Acalypha schiedeana* var. *deppeana* (Schltdl.) Müll.Arg., Linnaea 7: 384 (1832).
Acalypha schiedeana f. *angustifolia* Müll.Arg., Linnaea 7: 384 (1832). *Acalypha schiedeana* var. *angustifolia* (Müll.Arg.) Pax & K.Hoffm. in H.G.A.Engler, Pflanzenr., IV, 147, XVI: 55 (1924).
Acalypha schiedeana f. *latifoliam* Müll.Arg., Linnaea 7: 384 (1832).
Acalypha schiedeana var. *macrodonta* Müll.Arg., Linnaea 7: 384 (1832).
Acalypha schiedeana var. *purpusiana* Pax & K.Hoffm. in H.G.A.Engler, Pflanzenr., IV, 147, XVI: 55 (1924).

Acalypha schimpffii Diels, Biblioth. Bot. 24(116): 103 (1937).
Ecuador. 83 ECU.

Acalypha schlechtendaliana Müll.Arg., Linnaea 34: 6 (1865). *Acalypha schlechtendaliana* var. *genuina* Müll.Arg., Linnaea 34: 6 (1865), nom. inval. *Ricinocarpus schlechtendalianus* (Müll.Arg.) Kuntze, Revis. Gen. Pl. 2: 618 (1891).
Mexico (Veracruz). 79 MXG. Nanophan.
* *Acalypha filiformis* Klotzsch ex Schltdl., Linnaea 19: 235 (1847), nom. illeg.
Acalypha schlechtendaliana var. *mollis* Müll.Arg., Linnaea 34: 6 (1865).

Acalypha schlechteri Gand., Bull. Soc. Bot. France 60: 27 (1913).
Northern Prov. 27 TVL.

Acalypha schlumbergeri Müll.Arg. in A.P.de Candolle, Prodr. 15(2): 861 (1866). *Ricinocarpus schlumbergeri* (Müll.Arg.) Kuntze, Revis. Gen. Pl. 2: 618 (1891).
Mexico (?). 79 +. Nanophan.

Acalypha schreiteri Lillo ex Lourteig & O'Donell, Lilloa 8: 327 (1942).
N. Argentina. 85 AGE.

Acalypha schultesii Cardiel, Anales Jard. Bot. Madrid 52: 155 (1994 publ. 1995).
Colombia (Amazonas). 83 CLM. Ther.

Acalypha segetalis Müll.Arg., J. Bot. 2: 336 (1864). *Ricinocarpus segetalis* (Müll.Arg.) Kuntze, Revis. Gen. Pl. 2: 618 (1891).
Trop. & S. Africa. 22 BEN GHA IVO NGA SEN SIE TOG 23 CMN CON ZAI 24 ETH SUD 25 KEN UGA 26 ANG MLW MOZ ZAM ZIM 27 BOT CPP NAT OFS TVL. Ther.
Acalypha sessilis var. *brevibracteata* Müll.Arg., Flora 47: 465 (1864). *Acalypha gemina* var. *brevibracteata* (Müll.Arg.) Müll.Arg. in A.P.de Candolle, Prodr. 15(2): 866 (1866).
Acalypha sessilis var. *exserta* Müll.Arg., Flora 47: 465 (1864). *Acalypha gemina* var. *exserta* (Müll.Arg.) Müll.Arg. in A.P.de Candolle, Prodr. 15(2): 866 (1866).
Acalypha sessilis De Wild. & T.Durand, Bull. Herb. Boissier, II, 1: 47 (1900), nom. illeg.
Mercurialis cucullata Dinter ex Pax in H.G.A.Engler, Pflanzenr., IV, 147, XVI: 36 (1924), pro syn.

Acalypha sehnemii Allem & Irgang, Bol. Soc. Argent. Bot. 17: 305 (1976).
Brazil (Rio Grande do Sul). 84 BZS. Cham.

Acalypha seleriana Greenm., Publ. Field Columbian Mus., Bot. Ser. 2: 254 (1907).
Mexico. 79 MXG MXT. Nanophan.
Acalypha mollis Millsp., Publ. Field Columbian Mus., Bot. Ser. 1: 302 (1896).

Acalypha seminuda Müll.Arg. in C.F.P.von Martius, Fl. Bras. 11(2): 360 (1874). *Ricinocarpus seminudus* (Müll.Arg.) Kuntze, Revis. Gen. Pl. 2: 618 (1891).
SE. Brazil. 84 BZL. Nanophan.

Acalypha senilis Baill., Adansonia 5: 228 (1865). *Ricinocarpus senilis* (Baill.) Kuntze, Revis. Gen. Pl. 2: 618 (1891).
Uruguay. 85 URU. Nanophan.
Acalypha rotundifolia Herter, Ananes Mus. Nac. Montevideo 2: 80 (1910).

Acalypha septemloba Müll.Arg., Flora 55: 27 (1872).
C. America. 80 COS GUA? PAN. Hemicr.
Ricinocarpus irazuensis Kuntze, Revis. Gen. Pl. 2: 616 (1891). *Acalypha irazuensis* (Kuntze) Pax & K.Hoffm. in H.G.A.Engler, Pflanzenr., IV, 147, XVI: 53 (1924).

Acalypha sericea Andersson, Kongl. Vetensk. Acad. Handl. 1854: 238 (1855). *Acalypha parvula* f. *sericea* (Andersson) Müll.Arg., Linnaea 34: 47 (1865).
Galapagos. 83 GAL. Cham.

var. **baurii** (B.L.Rob. & Greenm.) G.L.Webster, Madroño 20: 261 (1970).
Galapagos (SW. San Cristobal). 83 GAL.
* *Acalypha baurii* B.L.Rob. & Greenm., Amer. J. Sci. 3(50): 144 (1895).

var. **indefessus** G.L.Webster, Madroño 20: 261 (1970).
Galapagos (Santa Cruz). 83 GAL.

var. **sericea**
Galapagos. 83 GAL. Cham.

Acalypha sessilifolia S.Watson, Proc. Amer. Acad. Arts 22: 450 (1887).
Mexico (Jalisco). 79 MXS. Hemicr.

Acalypha setosa A.Rich. in R.de la Sagra, Hist. Fis. Cuba, Bot. 2: 204 (1850). *Ricinocarpus setosus* (A.Rich.) Kuntze, Revis. Gen. Pl. 2: 618 (1891).
Caribbean, Mexico to Ecuador. 79 MXG MXT 80 COS GUA 81 BAH CUB DOM HAI PUE WIN 83 CLM ECU. Ther.
Acalypha polystachya Spreng., Syst. Veg. 3: 883 (1826), nom. illeg.
Acalypha scleroloba Müll.Arg. in H.G.A.Engler, Pflanzenr., IV, 147, XVI: 43 (1924), pro syn.

Acalypha siamensis Oliv. ex Gage, Rec. Bot. Surv. India 9: 238 (1922).
Burma, Pen. Malaysia, Thailand, Laos, Vietnam. 41 BMA LAO THA VIE 42 MLY. Nanophan. or phan.

var. **denticulata** Airy Shaw, Kew Bull. 32: 70 (1977).
Thailand. 41 THA.

var. **siamensis**
Burma, Pen. Malaysia, Thailand, Laos, Vietnam. 41 BMA LAO THA VIE 42 MLY. Nanophan. or phan.
Acalypha evrardii Gagnep., Bull. Soc. Bot. France 70: 871 (1923 publ. 1924).
Acalypha sphenophylla Pax & K.Hoffm. in H.G.A.Engler, Pflanzenr., IV, 147, XVI: 110 (1924).

Acalypha skutchii I.M.Johnst., J. Arnold Arbor. 19: 120 (1938).
Guatemala, Mexico (Chiapas). 79 MXT 80 GUA. Phan.

Acalypha sonderi Gand., Bull. Soc. Bot. France 60: 27 (1913).
SW. Cape Prov. 27 CPP.

Acalypha sonderiana Müll.Arg., Linnaea 34: 9 (1865). *Ricinocarpus sonderianus* (Müll.Arg.) Kuntze, Revis. Gen. Pl. 2: 618 (1891).
S. Mozambique, Kwazulu-Natal. 26 MOZ 27 NAT. Nanophan. or phan.
* *Acalypha petiolaris* Sond., Linnaea 23: 117 (1850), nom. illeg.

Acalypha soratensis Pax & K.Hoffm. in H.G.A.Engler, Pflanzenr., IV, 147, XVI: 126 (1924).
Bolivia. 83 BOL. Nanophan.

Acalypha spachiana Baill., Adansonia 1: 272 (1861). *Ricinocarpus spachianus* (Baill.) Kuntze, Revis. Gen. Pl. 2: 618 (1891).
Madagascar. 29 MDG. Nanophan.
Acalypha spachiana var. *acutifolia* Baill., Adansonia 1: 272 (1861).
Acalypha spachiana var. *latifolia* Baill., Adansonia 1: 272 (1861).
Acalypha spachiana var. *minor* Baill., Adansonia 1: 272 (1861).
Acalypha buchenavii Müll.Arg., Abh. Naturwiss. Vereine Bremen 7: 27 (1880).
Tragia saxatilis Bojer ex Pax in H.G.A.Engler, Pflanzenr., IV, 147, XVI: 33 (1924), nom. illeg.

Acalypha spectabilis Airy Shaw, Kew Bull. 33: 71 (1978).
NE. Papua New Guinea. 42 NWG. Nanophan.

Acalypha spinescens Benth., Hooker's Icon. Pl. 13: t. 1291 (1871).
Sulawesi. 42 SUL. Nanophan. – Provisionally accepted.

Acalypha squarrosa Pax, Bot. Jahrb. Syst. 19: 91 (1894).
Madagascar. 29 MDG. Nanophan.

Acalypha stachyura Pax, Repert. Spec. Nov. Regni Veg. 7: 110 (1909).
W. South America. 83 BOL CLM ECU PER. Nanophan. or phan.
Acalypha macrophylla Ule, Verh. Bot. Vereins Prov. Brandenburg 50: 79 (1908), nom. illeg. *Acalypha ulei* Radcl.-Sm. & Govaerts, Kew Bull. 52: 478 (1997).

Acalypha stellipila Pax & K.Hoffm. in H.G.A.Engler, Pflanzenr., IV, 147, XVI: 49 (1924).
Ecuador. 83 ECU. Nanophan.

Acalypha stenoloba Müll.Arg., Flora 55: 41 (1872).
Peru, Bolivia. 83 BOL PER. Nanophan.
Acalypha capillaris Rusby, Mem. Torrey Bot. Club 4: 257 (1895).
Acalypha baenitzii Pax, Repert. Spec. Nov. Regni Veg. 5: 227 (1908).
Acalypha ovata Pax & K.Hoffm., Meded. Rijks-Herb. 40: 23 (1921).

Acalypha stenophylla K.Schum., Bot. Jahrb. Syst. 9: 206 (1888).
NE. Papua New Guinea. 42 NWG. Nanophan.

Acalypha stricta Poepp. & Endl., Nov. Gen. Sp. Pl. 3: 21 (1845). *Ricinocarpus strictus* (Poepp. & Endl.) Kuntze, Revis. Gen. Pl. 2: 618 (1891).
Peru. 83 PER. Nanophan.
Acalypha mollissima Klotzsch ex Pax & K.Hoffm. in H.G.A.Engler, Pflanzenr., IV, 147, XVI: 64 (1924), pro syn.

Acalypha subbullata Pax & K.Hoffm. in H.G.A.Engler, Pflanzenr., IV, 147, XVI: 67 (1924).
? 8 +. Nanophan.

Acalypha subcastrata Aresch., Freg. Eugenies Resa Bot.: 137 (1910).
Ecuador, Peru. 83 ECU PER. Ther.

Acalypha subintegra Airy Shaw, Kew Bull. 33: 73 (1978).
New Guinea (Louisiade Archip., Trobriand Is.). 42 NWG. Nanophan. or phan.

Acalypha subsana Mart. ex Colla, Herb. Pedem. 5: 113 (1836).
Brazil. 84 BZL.

Acalypha subscandens Rusby, Descr. S. Amer. Pl.: 47 (1920).
Colombia. 83 CLM.

Acalypha subterranea Paul G.Wilson, Hooker's Icon. Pl. 36: t. 3588 (1962).
Mexico (México State, Guerrero). 79 MXC MXS. Hemicr. or cham.

Acalypha subtomentosa Lag., Gen. Sp. Pl.: 21 (1816). *Ricinocarpus subtomentosus* (Lag.) Kuntze, Revis. Gen. Pl. 2: 618 (1891).
Mexico. 79 +.

Acalypha subvillosa Müll.Arg. in C.F.P.von Martius, Fl. Bras. 11(2): 341 (1874). *Ricinocarpus subvillosus* (Müll.Arg.) Kuntze, Revis. Gen. Pl. 2: 618 (1891).
Venezuela, Brazil. 82 VEN 84 BZN. Nanophan.

Acalypha subviscida S.Watson, Proc. Amer. Acad. Arts 21: 440 (1886).
Mexico to Guatemala. 79 MXE MXN 80 GUA. Cham. or nanophan.

Acalypha suirebiensis Yamam., J. Soc. Trop. Agric. 5: 178 (1933).
E. Taiwan. 38 TAI. Cham.

Acalypha supera Forssk., Fl. Aegypt.-Arab.: 162 (1775).
Trop. Africa to Trop. Asia. 23 CMN ZAI 24 ETH 25 KEN TAN UGA 26 MLW MOZ 35 YEM 36 CHC 40 ASS IND PAK SRL 42 JAW LSI SUL SUM. Ther.
Acalypha brachystachya Hornem., Hort. Bot. Hafn.: 909 (1815). *Ricinocarpus brachystachyus* (Hornem.) Kuntze, Revis. Gen. Pl. 2: 617 (1891).
Acalypha conferta Roxb., Fl. Ind. ed. 1832, 3: 677 (1832).
Acalypha calyciformis Wight ex Wall., Numer. List: 7786 (1847), nom. inval.
Acalypha fissa Wall., Numer. List: 7786B (1847), nom. inval.
Tragia tenuis Wall., Numer. List: 7787 (1847), nom. inval.
Acalypha elegantula Hochst. ex A.Rich., Tent. Fl. Abyss. 2: 246 (1850).

Acalypha swallowensis Fosberg, Lloydia 3: 114 (1940).
Santa Cruz Is. 60 SCZ.

Acalypha synoica Pax & K.Hoffm. in H.G.A.Engler, Pflanzenr., IV, 147, XVI: 156 (1924).
Mexico (Puebla). 79 MXC. Nanophan.

Acalypha tacanensis Lundell, Contr. Univ. Michigan Herb. 4: 11 (1940).
Mexico (Chiapas). 79 MXT.

Acalypha tamaulipasensis Lundell, Wrightia 5: 244 (1976).
Mexico (Tamaulipas). 79 MXE.

Acalypha tenuicauda Pax & K.Hoffm. in H.G.A.Engler, Pflanzenr., IV, 147, XVI: 149 (1924).
Guatemala. 80 GUA. Nanophan.

Acalypha tenuifolia Müll.Arg. in A.P.de Candolle, Prodr. 15(2): 863 (1866). *Ricinocarpus tenuifolius* (Müll.Arg.) Kuntze, Revis. Gen. Pl. 2: 618 (1891).
Venezuela. 82 VEN. Nanophan.

Acalypha tenuipes Pax & K.Hoffm. in H.G.A.Engler, Pflanzenr., IV, 147, XVI: 122 (1924).
Ecuador. 83 ECU. Nanophan.

Acalypha tenuiramea Müll.Arg. in A.P.de Candolle, Prodr. 15(2): 858 (1866). *Ricinocarpus tenuirameus* (Müll.Arg.) Kuntze, Revis. Gen. Pl. 2: 618 (1891).
Brazil (Rio de Janeiro). 84 BZL. Nanophan.

Acalypha tomentosa Sw., Prodr.: 99 (1788). *Ricinocarpus tomentosus* (Sw.) Kuntze, Revis. Gen. Pl. 2: 618 (1891).
Dominican Rep., Haiti. 81 DOM HAI. Nanophan.
Acalypha multipartita Moench., Suppl. Meth.: 122 (1802).
Acalypha platyodonta Urb., Ark. Bot. 17(7): 37 (1922).

Acalypha totonaca Sessé & Moç., Fl. Mexic., ed. 2: 221 (1894).
Mexico (?). 79 +. – Provisionally accepted.

Acalypha tracheliifolia Pax & K.Hoffm. in H.G.A.Engler, Pflanzenr., IV, 147, XVI: 41 (1924).
NE. Argentina. 85 AGE. Hemicr.

Acalypha trachyloba Müll.Arg., Flora 55: 25 (1872).
Mexico (Oaxaca). 79 MXS. Nanophan.
Acalypha glandulifera Rob. & Greenm., Amer. J. Sci. 50: 164 (1895).

Acalypha transvaalensis Gand., Bull. Soc. Bot. France 60: 27 (1913).
Northern Prov. 27 TVL.

Acalypha tricholoba Müll.Arg., Linnaea 34: (1865). *Ricinocarpus tricholobus* (Müll.Arg.) Kuntze, Revis. Gen. Pl. 2: 618 (1891).
S. Mexico, Guatemala. 79 MXT 80 GUA. Nanophan.

Acalypha trilaciniata Paul G.Wilson, Kew Bull. 13: 170 (1958).
Mexico (Michoacán). 79 MXS.

Acalypha triloba Müll.Arg., Linnaea 34: 23 (1865). *Ricinocarpus trilobus* (Müll.Arg.) Kuntze, Revis. Gen. Pl. 2: 618 (1891).
Mexico, C. America. 79 + 80 COS GUA. Hemicr.

Acalypha tunguraguae Pax & K.Hoffm. in H.G.A.Engler, Pflanzenr., IV, 147, XVI: 66 (1924).
Ecuador. 83 ECU. Nanophan.

Acalypha uleana L.B.Sm. & Downs, Phytologia 22: 90 (1971).
Brazil (Santa Catarina). 84 BZS. Cham.

Acalypha umbrosa Brandegee, Erythea 7: 7 (1899).
Revillagigedo Is. (Socorro). 79 MXI.

Acalypha unibracteata Müll.Arg., Linnaea 34: 160 (1865). *Ricinocarpus unibracteatus* (Müll.Arg.) Kuntze, Revis. Gen. Pl. 2: 618 (1891).
Mexico (Veracruz). 79 MXG. Nanophan.

Acalypha urostachya Baill., Adansonia 5: 229 (1865). *Ricinocarpus urostachyus* (Baill.) Kuntze, Revis. Gen. Pl. 2: 618 (1891).
Brazil (?). 84 +. Nanophan.

Acalypha vagans Cav., Anales Hist. Nat. 2: 139 (1800). *Acalypha vagans* var. *genuina* Müll.Arg., Linnaea 34: 161 (1865), nom. inval. *Ricinocarpus vagans* (Cav.) Kuntze, Revis. Gen. Pl. 2: 618 (1891).
Mexico. 79 MXE MXG MXS. Nanophan.
Acalypha vagans var. *glandulosa* Müll.Arg., Linnaea 34: 161 (1865).
Acalypha glutinosa Klotzsch ex Pax & K.Hoffm. in H.G.A.Engler, Pflanzenr., IV, 147, XVI: 60 (1924), pro syn.
Acalypha scutellariifolia Klotzsch ex Pax & K.Hoffm. in H.G.A.Engler, Pflanzenr., IV, 147, XVI: 59 (1924).

Acalypha vallartae McVaugh, Contr. Univ. Michigan Herb. 20: 187 (1995).
Mexico (Jalisco, Nayarit). 79 MXS. Ther. or hemicr.

Acalypha variegata Rusby, Mem. New York Bot. Gard. 7: 285 (1927).
Bolivia. 83 BOL.

Acalypha vellamea Baill., Adansonia 5: 228 (1865). *Ricinocarpus vellameus* (Baill.) Kuntze, Revis. Gen. Pl. 2: 618 (1891).
WC. Brazil, Paraguay. 84 BZC 85 PAR. Cham.
Acalypha communis var. *brevipes* Müll.Arg., Linnaea 34: 21 (1865). *Acalypha brevipes* (Müll.Arg.) Müll.Arg. in C.F.P.von Martius, Fl. Bras. 11(2): 348 (1874), nom. illeg. *Ricinocarpus brevipes* (Müll.Arg.) Kuntze, Revis. Gen. Pl. 2: 618 (1891).
Acalypha communis var. *brevipetiolata* Chodat & Hassl., Bull. Herb. Boissier, II, 5: 605 (1905).
Acalypha goyazensis Glaz., Bull. Soc. Bot. France 59(3): 623 (1912 publ. 1913), nom. nud.
Acalypha serratifolia Klotzsch ex Pax & K.Hoffm. in H.G.A.Engler, Pflanzenr., IV, 147, XVI: 40 (1924), pro syn.

Acalypha velutina Hook.f., Trans. Linn. Soc. London 20: 186 (1847). *Acalypha parvula* f. *velutina* (Hook.f.) Müll.Arg., Linnaea 34: 48 (1865).
Galapagos (Santa Maria). 83 GAL. Hemicr.
Acalypha velutina var. *minor* Hook.f., Trans. Linn. Soc. London 20: 186 (1847).

Acalypha verbenacea Standl., Contr. Dudley Herb. 1: 75 (1927).
Mexico (Nayarit). 79 MXS. Cham. or nanophan.

Acalypha vermifera Rusby, Mem. New York Bot. Gard. 7: 286 (1927).
Bolivia. 83 BOL.

Acalypha veronicoides Pax & K.Hoffm. in H.G.A.Engler, Pflanzenr., IV, 147, XVI: 89 (1924).
Mexico (San Luis Potosí: Minas de San Rafael). 79 MXE. Hemicr.

Acalypha villosa Jacq., Select. Stirp. Amer. Hist.: 254 (1763). *Acalypha villosa* var. *genuina* Müll.Arg., Linnaea 34: 8 (1865), nom. inval. *Ricinocarpus villosus* (Jacq.) Kuntze, Revis. Gen. Pl. 2: 618 (1891).
Mexico to Paraguay. 79 MXT 80 COS HON NIC PAN 82 GUY VEN 83 BOL CLM ECU PER 84 BZC BZE BZL BZN 85 AGE PAR. Nanophan.
Acalypha linostachys Baill., Adansonia 5: 235 (1865).
Acalypha villosa var. *intermedia* Müll.Arg., Linnaea 34: 8 (1865).
Acalypha villosa var. *tomentosa* Müll.Arg., Linnaea 34: 8 (1865).
Acalypha villosa f. *paniculata* Müll.Arg. in A.P.de Candolle, Prodr. 15(2): 802 (1866). *Acalypha villosa* var. *paniculata* (Müll.Arg.) Pax & K.Hoffm. in H.G.A.Engler, Pflanzenr., IV, 147, XVI: 17 (1924).
Acalypha villosa var. *trichopoda* Müll.Arg. in C.F.P.von Martius, Fl. Bras. 11(2): 340 (1874).
Acalypha karsteniana Pax & K.Hoffm. in H.G.A.Engler, Pflanzenr., IV, 147, XVI: 19 (1924).

Acalypha muricata Klotzsch ex Pax & K.Hoffm. in H.G.A.Engler, Pflanzenr., IV, 147, XVI: 17 (1924), pro syn.
Acalypha villosa var. *latiuscula* Pax & K.Hoffm. in H.G.A.Engler, Pflanzenr., IV, 147, XVI: 17 (1924).

Acalypha vincentina Urb., Repert. Spec. Nov. Regni Veg. 28: 222 (1930).
St. Vincent. 81 WIN.

Acalypha virgata L., Amoen. Acad. 5: 410 (1760). *Ricinocarpus virgatus* (L.) Kuntze, Revis. Gen. Pl. 2: 618 (1891).
Jamaica. 81 JAM. Nanophan.

Acalypha virginica L., Sp. Pl.: 1003 (1753). *Ricinocarpus virginicus* (L.) Kuntze, Revis. Gen. Pl. 2: 618 (1891).
SE. Canada, C. & E. U.S.A. 72 ONT 74 ILL IOW MSO 75 INI NWJ NWY OHI PEN 78 ALA FLA GEO KTY NCA SCA TEN VRG 77 TEX. Ther.

var. **deamii** Weath., Rhodora 29: 197 (1927). *Acalypha rhomboidea* var. *deamii* (Weath.) Weath., Rhodora 39: 16 (1937). *Acalypha deamii* (Weath.) Ahles in G.N.Jones & G.D.Fuller, Vasc. Pl. Illinois: 301 (1955).
NE. U.S.A. 75 INI OHI. Ther.

var. **gracilens** (A.Gray) Müll.Arg., Linnaea 34: 45 (1865).
SE. U.S.A. 78 ALA FLA GEO NCA SCA. Ther.
Acalypha virginica Michx., Fl. Bor.-Amer. 2: 215 (1803).
* *Acalypha gracilens* A.Gray., Manual: 408 (1848).
Acalypha gracilens var. *monococca* Engelm. ex A.Gray, Manual, ed. 2: 390 (1856). *Acalypha monococca* (Engelm. ex A.Gray) Lill.W.Mill. & Gandhi, Sida 13: 123 (1988).
Acalypha virginica var. *fraseri* Müll.Arg., Linnaea 34: 44 (1865).
Acalypha virginica var. *gracilescens* Müll.Arg., Linnaea 34: 44 (1865).
Acalypha virginica var. *intermedia* Müll.Arg., Linnaea 34: 44 (1865).

var. **rhomboidea** (Raf.) Cooperr., Michigan Bot. 23: 165 (1984).
NC. & E. U.S.A. 74 KAN NEB OKL 75 NWJ OHI 78 MRY NCA SCA. Ther.
* *Acalypha rhomboidea* Raf., New Fl. 1: 45 (1836).

var. **virginica**
SE. Canada, C. & E. U.S.A. 72 ONT 74 ILL IOW MSO 75 INI NWJ NWY OHI PEN 78 KTY NCA SCA TEN VRG 77 TEX. Ther.
Acalypha caroliniana Walter, Fl. Carol.: 238 (1788).
Acalypha brevipes Raf., New Fl. 1: 44 (1836).
Acalypha crenulata Raf., New Fl. 1: 44 (1836).
Acalypha echinata Raf., New Fl. 1: 45 (1836).
Acalypha urticifolia Raf., New Fl. 1: 45 (1836). *Ricinocarpus urticifolius* (Raf.) Kuntze, Revis. Gen. Pl. 2: 618 (1891).
Acalypha virginea Dragend., Heilpfl.: 380 (1898).
Acalypha virginica f. *purpurea* Farw., Pap. Michigan Acad. Sci. 2: 27 (1923).

Acalypha volkensii Pax in H.G.A.Engler, Pflanzenw. Ost-Afrikas C: 239 (1895).
S. Sudan, Ethiopia, Somalia, Uganda, Kenya, Tanzania. 24 ETH SOM SUD 25 KEN TAN UGA. Cl. cham. or cl. nanophan.
Acalypha psilostachyoides Pax, Annuario Reale Ist. Bot. Roma 6: 183 (1896).

Acalypha vulneraria Baill., Bull. Mens. Soc. Linn. Paris 2: 1180 (1895).
Madagascar. 29 MDG. Nanophan.

Acalypha warburgii Pax & K.Hoffm. in H.G.A.Engler, Pflanzenr., IV, 147, XVI: 155 (1924).
Philippines (Luzon). 42 PHI. Nanophan. or phan.

Acalypha weddelliana Baill., Adansonia 5: 232 (1865). *Acalypha weddelliana* var. *genuina* Müll.Arg. in C.F.P.von Martius, Fl. Bras. 11(2): (1874), nom. inval.
E. Brazil. 84 BZE BZL. Nanophan.
Acalypha brasiliensis var. *psilophylla* Müll.Arg., Linnaea 34: 37 (1865).
Acalypha estrellana Baill., Adansonia 5: 237 (1865).
Acalypha major Baill., Adansonia 5: 236 (1865). *Acalypha weddelliana* var. *major* (Baill.) Müll.Arg. in C.F.P.von Martius, Fl. Bras. 11(2): 364 (1874).
Ricinocarpus weddellianus (Baill.) Kuntze, Revis. Gen. Pl. 2: 618 (1891).
Acalypha weddelliana var. *janeirensis* Pax & K.Hoffm. in H.G.A.Engler, Pflanzenr., IV, 147, XVI: 122 (1924).

Acalypha welwitschiana Müll.Arg., J. Bot. 2: 334 (1864). *Ricinocarpus welwitschianus* (Müll.Arg.) Kuntze, Revis. Gen. Pl. 2: 618 (1891).
Burundi to SC. Trop. Africa. 23 BUR 25 TAN 26 ANG MLW MOZ ZAM. Cham.
Acalypha angolensis Müll.Arg., J. Bot. 2: 335 (1864). *Ricinocarpus angolensis* (Müll.Arg.) Kuntze, Revis. Gen. Pl. 2: 617 (1891).

Acalypha wigginsii G.L.Webster, Madroño 20: 261 (1970).
Galapagos (Santa Cruz I.). 83 GAL. Cham. or nanophan.

Acalypha wilderi Merr., Bernice P. Bishop Mus. Bull. 86: 64 (1931).
Cook Is. (Rarotonga). 61 COO. Nanophan. or phan.

Acalypha williamsii Rusby, Bull. New York Bot. Gard. 8(28): 101 (1912).
Bolivia. 83 BOL.

Acalypha wilmsii Pax ex Prain & Hutch., Bull. Misc. Inform. Kew 1913: 24 (1913).
S. Africa. 27 CPP NAT OFS SWZ TVL. Hemicr.

Acalypha wui H.S.Kiu, J. Trop. Subtrop. Bot. 3: 17 (1995).
China (W. Guangdong). 36 CHS. Nanophan.

Acalypha yucatanensis Millsp., Publ. Field Columbian Mus., Bot. Ser. 1: 371 (1898).
SE. Mexico. 79 MXT.

Acalypha zeyheri Baill., Adansonia 3: 156 (1863). *Ricinocarpus zeyheri* (Baill.) Kuntze, Revis. Gen. Pl. 2: 618 (1891).
Cape Prov. 27 CPP. Hemicr.
Acalypha peduncularis var. *psilogyne* Müll.Arg., Linnaea 34: 28 (1865).
Acalypha zeyheri var. *glabrata* Müll.Arg., Linnaea 34: 29 (1865).

Acalypha zollingeri Müll.Arg., Linnaea 34: 40 (1865). *Ricinocarpus zollingeri* (Müll.Arg.) Kuntze, Revis. Gen. Pl. 2: 617 (1891).
Lesser Sunda Is. (Sumbawa). 42 LSI. Nanophan.

Synonyms:

Acalypha abortiva Hochst. ex Baill. === **Acalypha crenata** Hochst. ex A.Rich.
Acalypha accedens var. *brachyandra* (Baill.) Müll.Arg. === **Acalypha accedens** Müll.Arg.
Acalypha accedens var. *genuina* Müll.Arg. === **Acalypha accedens** Müll.Arg.
Acalypha acuminata Baill. === **Cleidion spiciflorum** (Burm.f.) Merr. var. **spiciflorum**
Acalypha acuminata Vahl ex Baill. === **Cleidion spiciflorum** (Burm.f.) Merr. var. **spiciflorum**
Acalypha acuta Thunb. === **Adenocline acuta** (Thunb.) Baill.
Acalypha adamsii Rob. === **Acalypha parvula** var. **reniformis** (Hook.f.) Müll.Arg.
Acalypha adenophora Griseb. === **Acalypha membranacea** A.Rich.
Acalypha adenotricha A.Rich. === **Acalypha ornata** Hochst. ex A.Rich.

Acalypha adscendens Hornem. === **Acalypha chamaedrifolia** (Lam.) Müll.Arg.
Acalypha affinis Klotzsch === **Acalypha amentacea** Roxb. subsp. **amentacea**
Acalypha agrestis Morong ex Britton === **Acalypha communis** Müll.Arg.
Acalypha agrimonioides D.Dietr. === **Acalypha anemioides** Kunth
Acalypha albemarlensis Rob. === **Acalypha parvula** Hook.f. var. **parvula**
Acalypha albicans B.Heyne ex Hook.f. === **Acalypha lanceolata** Willd. var. **lanceolata**
Acalypha alnifolia Poir. === **Acalypha poiretii** Spreng.
Acalypha alnifolia Wall. === **Acalypha alnifolia** Klein ex Willd.
Acalypha alopecuroides lusus *glanduligera* Klotzsch === **Acalypha alopecuroides** Jacq.
Acalypha alopecuroides lusus *hispida* Müll.Arg. === **Acalypha alopecuroides** Jacq.
Acalypha alopecuroides f. *polycephala* Müll.Arg. === **Acalypha alopecuroides** Jacq.
Acalypha amblyodonta var. *gaudichaudii* (Baill.) Müll.Arg. === **Acalypha amblyodonta** (Müll.Arg.) Müll.Arg.
Acalypha amblyodonta var. *hispida* Müll.Arg. === **Acalypha amblyodonta** (Müll.Arg.) Müll.Arg.
Acalypha amblyodonta var. *repanda* Müll.Arg. === **Acalypha amblyodonta** (Müll.Arg.) Müll.Arg.
Acalypha amblyodonta var. *villosa* Müll.Arg. === **Acalypha amblyodonta** (Müll.Arg.) Müll.Arg.
Acalypha amboinensis Benth. === **Acalypha amentacea** Roxb. subsp. **amentacea**
Acalypha amentacea f. *circinata* (Müll.Arg.) Fosberg === **Acalypha amentacea** subsp. **wilkesiana** (Müll.Arg.) Fosberg
Acalypha amphigyne Moore === **Acalypha communis** Müll.Arg.
Acalypha amplifolia Rusby === **Acalypha macrostachya** Jacq.
Acalypha anemioides var. *eglandulosa* Müll.Arg. === **Acalypha anemioides** Kunth
Acalypha anemioides var. *genuina* Müll.Arg. === **Acalypha anemioides** Kunth
Acalypha angatensis Blanco === **Acalypha amentacea** var. **velutina** (Müll.Arg.) Fosberg
Acalypha angolensis Müll.Arg. === **Acalypha welwitschiana** Müll.Arg.
Acalypha angustata var. *glabra* Sond. === **Acalypha angustata** Sond.
Acalypha angustifolia var. *genuina* Müll.Arg. === **Acalypha angustifolia** Sw.
Acalypha angustifolia var. *glabrata* Müll.Arg. === **Acalypha angustifolia** Sw.
Acalypha anisodonta Müll.Arg. === **Acalypha insulana** var. **anisodonta** (Müll.Arg.) Govaerts
Acalypha anisodonta var. *subsericea* Müll.Arg. === **Acalypha repanda** Müll.Arg. var. **repanda**
Acalypha anisodonta var. *subvillosa* Müll.Arg. === **Acalypha insulana** var. **anisodonta** (Müll.Arg.) Govaerts
Acalypha apicalis N.E.Br. === **Acalypha communis** Müll.Arg.
Acalypha arcuata Urb. === **Acalypha angustifolia** Sw.
Acalypha arvensis Poepp. & Endl. === **Acalypha aristata** Kunth
Acalypha arvensis var. *belangeri* Briq. === **Acalypha aristata** Kunth
Acalypha arvensis var. *genuina* Müll.Arg. === **Acalypha aristata** Kunth
Acalypha arvensis var. *pavoniana* (Müll.Arg.) Müll.Arg. === **Acalypha aristata** Kunth
Acalypha asterifolia Rusby === **Acalypha cuspidata** Jacq.
Acalypha baenitzii Pax === **Acalypha stenoloba** Müll.Arg.
Acalypha bailloniana Müll.Arg. === **Acalypha indica** L.
Acalypha baurii B.L.Rob. & Greenm. === **Acalypha sericea** var. **baurii** (B.L.Rob. & Greenm.) G.L.Webster
Acalypha benguelensis var. *adenogyne* Müll.Arg. === **Acalypha benguelensis** Müll.Arg.
Acalypha benguelensis var. *trichogyne* Müll.Arg. === **Acalypha benguelensis** Müll.Arg.
Acalypha bequaertii Staner === **Acalypha psilostachya** Hochst. ex A.Rich. var. **psilostachya**
Acalypha betulifolia Sw. === **Acalypha scabrosa** Sw.
Acalypha betulina Sond. === **Acalypha glabrata** Thunb. f. **glabrata**
Acalypha betulina Schweinf. === **Cephalocroton cordofanus** Hochst.
Acalypha betulina Retz. === **Acalypha fruticosa** Forssk. var. **fruticosa**
Acalypha betulina var. *latifolia* Sond. === **Acalypha glabrata** Thunb. f. **glabrata**
Acalypha betulodes Klotzsch ex Baill. === **Acalypha communis** Müll.Arg.
Acalypha betuloides Klotzsch ex Baill. === **Acalypha communis** Müll.Arg.
Acalypha betuloides Pav. ex Klotzsch === **Acalypha diversifolia** Jacq.
Acalypha billbergiana Klotzsch === **Acalypha diversifolia** Jacq.
Acalypha bisetosa Müll.Arg. === **Acalypha muelleriana** Urb.

Acalypha boehmerioides Miq. === **Acalypha lanceolata** Willd. var. **lanceolata**
Acalypha boehmerioides var. *genuina* Pax & K.Hoffm. === **Acalypha lanceolata** Willd. var. **lanceolata**
Acalypha boehmerioides var. *glandulosa* (Müll.Arg.) Pax & K.Hoffm. === **Acalypha lanceolata** var. **glandulosa** (Müll.Arg.) Radcl.-Sm.
Acalypha botteriana var. *genuina* Müll.Arg. === **Acalypha botteriana** Müll.Arg.
Acalypha botteriana var. *pubescens* Müll.Arg. === **Acalypha botteriana** Müll.Arg.
Acalypha brachiata E.Mey. === **Acalypha ecklonii** Baill.
Acalypha brachyandra Baill. === **Acalypha accedens** Müll.Arg.
Acalypha brachystachya Hornem. === **Acalypha supera** Forssk.
Acalypha bracteata Miq. === ? [LSI]
Acalypha brasiliensis var. *angustifolia* Pax & K.Hoffm. === **Acalypha brasiliensis** Müll.Arg.
Acalypha brasiliensis var. *brevipes* Müll.Arg. === **Acalypha brasiliensis** Müll.Arg.
Acalypha brasiliensis var. *cordata* (Müll.Arg.) Müll.Arg. === **Acalypha brasiliensis** Müll.Arg.
Acalypha brasiliensis f. *cordata* Müll.Arg. === **Acalypha brasiliensis** Müll.Arg.
Acalypha brasiliensis var. *glabrata* Müll.Arg. === **Acalypha brasiliensis** Müll.Arg.
Acalypha brasiliensis var. *longipes* Müll.Arg. === **Acalypha brasiliensis** Müll.Arg.
Acalypha brasiliensis var. *maxima* Müll.Arg. === **Acalypha brasiliensis** Müll.Arg.
Acalypha brasiliensis f. *microphylla* Müll.Arg. ex Pax & K.Hoffm. === **Acalypha omissa** Pax & K.Hoffm.
Acalypha brasiliensis var. *mollis* Müll.Arg. === **Acalypha arciana** (Baill.) Müll.Arg.
Acalypha brasiliensis var. *obtusa* (Müll.Arg.) Müll.Arg. === **Acalypha brasiliensis** Müll.Arg.
Acalypha brasiliensis f. *obtusa* Müll.Arg. === **Acalypha brasiliensis** Müll.Arg.
Acalypha brasiliensis var. *psilophylla* Müll.Arg. === **Acalypha weddelliana** Baill.
Acalypha brevipes Raf. === **Acalypha virginica** L. var. **virginica**
Acalypha brevipes (Müll.Arg.) Müll.Arg. === **Acalypha vellamea** Baill.
Acalypha buchenavii Müll.Arg. === **Acalypha spachiana** Baill.
Acalypha callosa Benth. === **Acalypha macrostachya** Jacq.
Acalypha callosa var. *glabra* Britton ex Pax & K.Hoffm. === **Acalypha diversifolia** Jacq.
Acalypha calyciformis Wight ex Wall. === **Acalypha supera** Forssk.
Acalypha cancana Müll.Arg. === **Acalypha macrostachya** Jacq.
Acalypha canescens Wall. === **Acalypha indica** L.
Acalypha capensis D.Dietr. === **Bernardia corensis** (Jacq.) Müll.Arg.
Acalypha capensis var. *cordata* (Thunb.) Prain === **Acalypha capensis** (L.f.) Prain
Acalypha capensis var. *decumbens* (Thunb.) Prain === **Acalypha capensis** (L.f.) Prain
Acalypha capensis f. *grandidentata* (Müll.Arg.) Prain === **Acalypha capensis** (L.f.) Prain
Acalypha capensis f. *lamiifolia* (Scheele) Prain === **Acalypha capensis** (L.f.) Prain
Acalypha capensis var. *villosa* (Thunb.) Pax & K.Hoffm. === **Acalypha capensis** (L.f.) Prain
Acalypha caperonioides var. *galpinii* Prain === **Acalypha caperonioides** Baill.
Acalypha capillaris Rusby === **Acalypha stenoloba** Müll.Arg.
Acalypha capitata Wall. === **Acalypha fruticosa** Forssk. var. **fruticosa**
Acalypha capitata Willd. === **Acalypha alnifolia** Klein ex Willd.
Acalypha capitata var. *ambigua* Müll.Arg. === **Acalypha alnifolia** Klein ex Willd.
Acalypha capitellata Brandegee === **Acalypha aristata** Kunth
Acalypha cardiophylla Merr. === **Acalypha caturus** Blume
Acalypha cardiophylla var. *ponapensis* (Kaneh. & Hatus.) Fosberg === **Acalypha caturus** Blume
Acalypha caroliniana Blanco === **Acalypha indica** L.
Acalypha caroliniana Walter === **Acalypha virginica** L. var. **virginica**
Acalypha caroliniana Elliott === **Acalypha ostryifolia** Riddell
Acalypha carpinifolia Poepp. ex Seem. === **Acalypha diversifolia** Jacq.
Acalypha carpinifolia Poir. === **Acalypha angustifolia** Sw.
Acalypha carpinifolia var. *domingensis* (Spreng.) Müll.Arg. === **Acalypha angustifolia** Sw.
Acalypha carpinifolia var. *genuina* Müll.Arg. === **Acalypha angustifolia** Sw.
Acalypha castaneifolia Poepp. ex Pax & K.Hoffm. === **Acalypha cuneata** Poepp. & Endl.
Acalypha caturoides K.Schum. & Lauterb. === **Acalypha longispica** Warb.
Acalypha caturus f. *angustifolia* J.J.Sm. === **Acalypha caturus** Blume

Acalypha caucana Müll.Arg. === **Acalypha macrostachya** Jacq.
Acalypha caudata Kunth === **Acalypha macrostachya** Jacq.
Acalypha celebica Koord. === ?
Acalypha chamaedrifolia var. *brevipes* Müll.Arg. === **Acalypha chamaedrifolia** (Lam.) Müll.Arg.
Acalypha chamaedrifolia var. *fissa* Müll.Arg. === **Acalypha fissa** (Müll.Arg.) Hutch.
Acalypha chamaedrifolia var. *genuina* Müll.Arg. === **Acalypha chamaedrifolia** (Lam.) Müll.Arg.
Acalypha chamaedrifolia var. *glechomifolia* (A.Rich.) Müll.Arg. === **Acalypha glechomifolia** A.Rich.
Acalypha chamaedrifolia var. *nana* Müll.Arg. === **Acalypha nana** (Müll.Arg.) Griseb. ex Hutch.
Acalypha chamaedrifolia var. *pendula* (C.Wright ex Griseb.) Müll.Arg. === **Acalypha pendula** C.Wright ex Griseb.
Acalypha chamaedrifolia var. *pygmaea* (A.Rich.) Müll.Arg. === **Acalypha pygmaea** A.Rich.
Acalypha chariensis Beille === **Acalypha brachiata** Krauss
Acalypha chathamensis Rob. === **Acalypha parvula** var. **chathamensis** (Rob.) G.L.Webster
Acalypha chinensis Benth. === **Acalypha indica** L.
Acalypha chinensis Roxb. === **Acalypha australis** L.
Acalypha chrysadenia Suess. & Friedrich === **Acalypha fruticosa** Forssk. var. **fruticosa**
Acalypha ciliata Wall. === **Acalypha indica** L.
Acalypha ciliata var. *genuina* Müll.Arg. === **Acalypha ciliata** Forssk.
Acalypha ciliata var. *trichophora* Müll.Arg. === **Acalypha ciliata** Forssk.
Acalypha circinata A.Gray ex Seem. === **Acalypha amentacea** subsp. **wilkesiana** (Müll.Arg.) Fosberg
Acalypha collina Hayne ex Hook.f. === **Acalypha lanceolata** Willd. var. **lanceolata**
Acalypha colorata Baker === **Acalypha integrifolia** Willd. subsp. **integrifolia**
Acalypha colorata (Poir.) Spreng. === **Acalypha integrifolia** Willd. subsp. **integrifolia**
Acalypha commersoniana Baill. ex Müll.Arg. === **Acalypha integrifolia** Willd. subsp. **integrifolia**
Acalypha commersoniana var. *acutifolia* Müll.Arg. === **Acalypha integrifolia** Willd. subsp. **integrifolia**
Acalypha commersoniana var. *brevifolia* Müll.Arg. === **Acalypha integrifolia** var. **parvifolia** (Baill. ex Müll.Arg.) Pax & K.Hoffm.
Acalypha commersoniana f. *concolor* Müll.Arg. === **Acalypha integrifolia** Willd. subsp. **integrifolia**
Acalypha commersoniana var. *concolor* Baill. === **Acalypha integrifolia** Willd. subsp. **integrifolia**
Acalypha commersoniana var. *discolor* Baill. === **Acalypha integrifolia** Willd. subsp. **integrifolia**
Acalypha commersoniana f. *discolor* Müll.Arg. === **Acalypha integrifolia** Willd. subsp. **integrifolia**
Acalypha commersoniana var. *longifolia* Müll.Arg. === **Acalypha integrifolia** Willd. subsp. **integrifolia**
Acalypha commersoniana var. *obtusifolia* Müll.Arg. === **Acalypha integrifolia** Willd. subsp. **integrifolia**
Acalypha commersoniana var. *parvifolia* Baill. ex Müll.Arg. === **Acalypha integrifolia** var. **parvifolia** (Baill. ex Müll.Arg.) Pax & K.Hoffm.
Acalypha commersoniana f. *purpurea* Müll.Arg. === **Acalypha integrifolia** Willd. subsp. **integrifolia**
Acalypha commersoniana f. *purpureomarginata* Müll.Arg. === **Acalypha integrifolia** Willd. subsp. **integrifolia**
Acalypha commersoniana f. *unicolor* Müll.Arg. === **Acalypha integrifolia** var. **crateriana** Coode
Acalypha communis var. *agrestis* (Morong ex Britton) Chodat & Hassl. === **Acalypha communis** Müll.Arg.
Acalypha communis var. *brevipes* Müll.Arg. === **Acalypha vellamea** Baill.
Acalypha communis var. *brevipetiolata* Chodat & Hassl. === **Acalypha vellamea** Baill.
Acalypha communis f. *grandifolia* Chodat & Hassl. === **Acalypha communis** Müll.Arg.
Acalypha communis var. *guaranitica* Chodat & Hassl. === **Acalypha communis** Müll.Arg.
Acalypha communis var. *hirta* (Spreng.) Müll.Arg. === **Acalypha communis** Müll.Arg.
Acalypha communis var. *hirtiformis* Pax & K.Hoffm. === **Acalypha communis** Müll.Arg.

Acalypha communis var. *hispida* Müll.Arg. === **Acalypha communis** Müll.Arg.
Acalypha communis var. *intermedia* Müll.Arg. === **Acalypha communis** Müll.Arg.
Acalypha communis var. *obscura* Müll.Arg. === **Acalypha communis** Müll.Arg.
Acalypha communis var. *pallida* Müll.Arg. === **Acalypha communis** Müll.Arg.
Acalypha communis var. *puberula* Müll.Arg. === **Acalypha communis** Müll.Arg.
Acalypha communis var. *rotundata* Griseb. === **Acalypha communis** Müll.Arg.
Acalypha communis var. *salicifolia* Pax & K.Hoffm. === **Acalypha communis** Müll.Arg.
Acalypha communis var. *saltensis* Pax & K.Hoffm. === **Acalypha communis** Müll.Arg.
Acalypha communis var. *tomentella* Müll.Arg. === **Acalypha communis** Müll.Arg.
Acalypha communis var. *tomentosa* Müll.Arg. === **Acalypha communis** Müll.Arg.
Acalypha communis var. *tomentosa* Chodat & Hassl. === **Acalypha communis** Müll.Arg.
Acalypha compacta Guilf. ex C.T.White === **Acalypha amentacea** subsp. **wilkesiana** (Müll.Arg.) Fosberg
Acalypha conferta Roxb. === **Acalypha supera** Forssk.
Acalypha consimilis Müll.Arg. === **Acalypha amentacea** var. **grandis** (Benth.) Fosberg
Acalypha corchorifolia Vahl ex Baill. === **Acalypha lanceolata** Willd. var. **lanceolata**
Acalypha corchorifolia Willd. === **Acalypha chamaedrifolia** (Lam.) Müll.Arg.
Acalypha corchorifolia A.Rich. === **Acalypha ostryifolia** Riddell
Acalypha cordata Thunb. === **Acalypha capensis** (L.f.) Prain
Acalypha cordata E.Mey. === **Acalypha ecklonii** Baill.
Acalypha cordifolia Andersson === **Acalypha parvula** Hook.f. var. **parvula**
Acalypha cordifolia Griseb. === **Acalypha plicata** Müll.Arg.
Acalypha cordifolia Hook.f. === **Acalypha parvula** Hook.f. var. **parvula**
Acalypha cordifolia var. *polyadenia* Griseb. === **Acalypha plicata** Müll.Arg.
Acalypha cordobensis Müll.Arg. === **Acalypha communis** Müll.Arg.
Acalypha cordoviensis Müll.Arg. === **Acalypha communis** Müll.Arg.
Acalypha corensis Jacq. === **Bernardia corensis** (Jacq.) Müll.Arg.
Acalypha coriifolia Pax & K.Hoffm. === **Acalypha padifolia** Kunth
Acalypha coryloides Rose === **Acalypha filipes** (S.Watson) McVaugh
Acalypha crassa Buchinger ex C.Krauss === **Acalypha peduncularis** Meisn. ex C.Krauss
Acalypha crassicaulis Klotzsch ex Pax & K.Hoffm. === **Acalypha brevicaulis** Müll.Arg.
Acalypha crenata var. *glandulosa* Müll.Arg. === **Acalypha lanceolata** var. **glandulosa** (Müll.Arg.) Radcl.-Sm.
Acalypha crenulata Raf. === **Acalypha virginica** L. var. **virginica**
Acalypha crotonoides Pax === **Acalypha polymorpha** Müll.Arg.
Acalypha crotonoides var. *caudata* Hutch. ex R.E.Fr. === **Acalypha polymorpha** Müll.Arg.
Acalypha cucullata Poir. === **Acalypha macrostachya** Jacq.
Acalypha cuneata var. *genuina* Müll.Arg. === **Acalypha cuneata** Poepp. & Endl.
Acalypha cuneata var. *obovata* (Benth.) Müll.Arg. === **Acalypha cuneata** Poepp. & Endl.
Acalypha cunninghamii Müll.Arg. === **Acalypha nemorum** F.Muell. ex Müll.Arg.
Acalypha cupamenii Dragend. === **Acalypha indica** L.
Acalypha cupricola Robyns === **Acalypha dikuluwensis** P.A.Duvign. & Dewit
Acalypha cuspidata Griseb. === **Acalypha bisetosa** Bertol. ex Spreng.
Acalypha cuspidata var. *amblyodonta* Müll.Arg. === **Acalypha amblyodonta** (Müll.Arg.) Müll.Arg.
Acalypha cuspidata var. *genuina* Müll.Arg. === **Acalypha cuspidata** Jacq.
Acalypha cuspidata var. *oxyodonta* Müll.Arg. === **Acalypha oxyodonta** (Müll.Arg.) Müll.Arg.
Acalypha cylindrica Roxb. === **Acalypha poiretii** Spreng.
Acalypha deamii (Weath.) Ahles === **Acalypha virginica** var. **deamii** Weath.
Acalypha decidua Forssk. === **Acalypha indica** L.
Acalypha decumbens Thunb. === **Acalypha capensis** (L.f.) Prain
Acalypha decumbens Müll.Arg. === **Acalypha capensis** (L.f.) Prain
Acalypha decumbens var. *cordata* (Thunb.) Müll.Arg. === **Acalypha capensis** (L.f.) Prain
Acalypha decumbens var. *genuina* Müll.Arg. === **Acalypha capensis** (L.f.) Prain
Acalypha decumbens var. *villosa* (Thunb.) Müll.Arg. === **Acalypha capensis** (L.f.) Prain
Acalypha densiflora Blume === **Acalypha hispida** Burm.f.

Acalypha dentata Schumach. & Thonn. === **Mallotus oppositifolius** (Geiseler) Müll.Arg. var. **oppositifolius**
Acalypha denudata Müll.Arg. === **Acalypha repanda** var. **denudata** (Müll.Arg.) A.C.Sm.
Acalypha deppeana Schltdl. === **Acalypha schiedeana** Schltdl.
Acalypha diffusa Andersson === **Acalypha parvula** Hook.f. var. **parvula**
Acalypha digyneia Raf. === ?
Acalypha dimorpha Chodat & Hassl. === **Acalypha nitschkeana** Pax & K.Hoffm.
Acalypha discolor Bojer === **Acalypha integrifolia** Willd. subsp. **integrifolia**
Acalypha discolor E.Mey. === **Acalypha capensis** (L.f.) Prain
Acalypha discolor var. *major* Baill. === **Acalypha capensis** (L.f.) Prain
Acalypha dissitiflora S.Watson === **Acalypha dioica** S.Watson
Acalypha divaricata Baill. === **Acalypha gracilis** Spreng.
Acalypha divaricata Raf. === ?
Acalypha divaricata Griseb. === **Acalypha lycioides** Pax & K.Hoffm.
Acalypha diversifolia Rusby === **Acalypha diversifolia** Jacq.
Acalypha diversifolia var. *carpinifolia* (Poepp. ex Seem.) Müll.Arg. === **Acalypha diversifolia** Jacq.
Acalypha diversifolia var. *claoneura* Pax & K.Hoffm. === **Acalypha diversifolia** Jacq.
Acalypha diversifolia var. *genuina* Müll.Arg. === **Acalypha diversifolia** Jacq.
Acalypha diversifolia var. *leptostachya* Glaz. === **Acalypha gracilis** Spreng.
Acalypha diversifolia var. *leptostachya* (Kunth) Müll.Arg. === **Acalypha diversifolia** Jacq.
Acalypha diversifolia var. *popayanensis* (Kunth) Müll.Arg. === **Acalypha diversifolia** Jacq.
Acalypha diversifolia var. *squarrosa* Müll.Arg. === **Acalypha diversifolia** Jacq.
Acalypha domingensis Spreng. === **Acalypha angustifolia** Sw.
Acalypha dumetorum Pax === **Acalypha ambigua** Pax
Acalypha dupraeana Baill. === **Acalypha accedens** Müll.Arg.
Acalypha dupraeana var. *arciana* Baill. === **Acalypha arciana** (Baill.) Müll.Arg.
Acalypha dupraeana var. *gaudichaudii* Baill. === **Acalypha amblyodonta** (Müll.Arg.) Müll.Arg.
Acalypha dupraeana var. *hilarii* Baill. === **Acalypha oxyodonta** (Müll.Arg.) Müll.Arg.
Acalypha dupraeana var. *sylvicola* Baill. === **Acalypha oxyodonta** (Müll.Arg.) Müll.Arg.
Acalypha echinata Raf. === **Acalypha virginica** L. var. **virginica**
Acalypha eggersii Pax === **Acalypha cuneata** Poepp. & Endl.
Acalypha ehretiifolia Klotzsch ex Pax & K.Hoffm. === **Acalypha phleoides** Cav.
Acalypha elegantula Hochst. ex A.Rich. === **Acalypha supera** Forssk.
Acalypha elliptica Griseb. === **Acalypha pruinosa** Urb.
Acalypha entumenica Prain === **Acalypha glandulifolia** Buchinger & Meisn. ex Krauss
Acalypha eremorum var. *capillipeda* Baill. === **Acalypha capillipes** Müll.Arg.
Acalypha eremorum var. *sessilis* Baill. === **Acalypha eremorum** Müll.Arg.
Acalypha erythrostachya Müll.Arg. === **Acalypha padifolia** Kunth
Acalypha estrellana Baill. === **Acalypha weddelliana** Baill.
Acalypha evrardii Gagnep. === **Acalypha siamensis** Oliv. ex Gage var. **siamensis**
Acalypha exaltata Baill. === **Acalypha amentacea** var. **grandis** (Benth.) Fosberg
Acalypha fallax Müll.Arg. === **Acalypha lanceolata** Willd. var. **lanceolata**
Acalypha ferdinandii var. *pubescens* K.Hoffm. === **Acalypha apodanthes** Standl. & L.O.Williams
Acalypha fertilis Standl. & L.O.Williams === **Acalypha macrostachya** Jacq.
Acalypha filifera S.Watson === **Acalypha polystachya** Jacq.
Acalypha filiformis Klotzsch ex Schltdl. === **Acalypha schlechtendaliana** Müll.Arg.
Acalypha filiformis var. *arborea* Poir. === **Acalypha filiformis** Poir. subsp. **filiformis**
Acalypha fimbriata Baill. === **Acalypha indica** L.
Acalypha fimbriata Hochst. ex A.Rich. === **Acalypha fimbriata** Schumach. & Thonn.
Acalypha finitima S.Moore === **Acalypha amentacea** var. **grandis** (Benth.) Fosberg
Acalypha fissa Wall. === **Acalypha supera** Forssk.
Acalypha flabellifera Rusby === **Acalypha plicata** Müll.Arg.
Acalypha flexuosa Wight ex Steud. === **Acalypha lanceolata** Willd. var. **lanceolata**
Acalypha floribunda B.Heyne ex Hook.f. === **Acalypha lanceolata** Willd. var. **lanceolata**
Acalypha foliosa Rusby === **Acalypha macrostachya** Jacq.

Acalypha formosana Hayata === **Acalypha amentacea** var. **velutina** (Müll.Arg.) Fosberg
Acalypha fruticosa var. *villosa* Hutch. === **Acalypha fruticosa** Forssk. var. **fruticosa**
Acalypha fruticulosa Raf. === ?
Acalypha gagnepainii Merr. === **Acalypha kerrii** Craib
Acalypha gagnepainii Leandri === **Acalypha medibracteata** Radcl.-Sm. & Govaerts
Acalypha gagnepainii var. *calcicola* Leandri === **Acalypha medibracteata** var. **calcicola** (Leandri) Radcl.-Sm. & Govaerts
Acalypha gemina Spreng. === **Acalypha australis** L.
Acalypha gemina var. *brevibracteata* (Müll.Arg.) Müll.Arg. === **Acalypha segetalis** Müll.Arg.
Acalypha gemina var. *exserta* (Müll.Arg.) Müll.Arg. === **Acalypha segetalis** Müll.Arg.
Acalypha gemina var. *genuina* Müll.Arg. === **Acalypha australis** L.
Acalypha gemina var. *lanceolata* Hayata === **Acalypha australis** L.
Acalypha gentryi Standl. === **Acalypha cincta** Müll.Arg.
Acalypha giraldii Pax ex Diels === **Discocleidion rufescens** (Franch.) Pax & K.Hoffm.
Acalypha giraldii Pamp. === **Alchornea davidii** Franch.
Acalypha glabrata Vahl ex A.Juss. === **Thecacoris madagascariensis** A.Juss. var. **madagascariensis**
Acalypha glabrata var. *genuina* Müll.Arg. === **Acalypha glabrata** Thunb.
Acalypha glabrata var. *latifolia* (Sond.) Müll.Arg. === **Acalypha glabrata** Thunb. f. **glabrata**
Acalypha glabrata var. *pilosa* Pax === **Acalypha glabrata** f. **pilosior** (Kuntze) Prain & Hutch.
Acalypha glandulifera Rob. & Greenm. === **Acalypha trachyloba** Müll.Arg.
Acalypha glandulosa (Kuntze) Chodat & Hassl. === **Acalypha hassleriana** Chodat
Acalypha glandulosa Blanco === **Acalypha amentacea** Roxb. subsp. **amentacea**
Acalypha glandulosa var. *brevistachya* Chodat & Hassl. === **Acalypha hassleriana** Chodat
Acalypha glomerata Hutch. === **Acalypha lanceolata** var. **glandulosa** (Müll.Arg.) Radcl.-Sm.
Acalypha glutinosa Klotzsch ex Pax & K.Hoffm. === **Acalypha vagans** Cav.
Acalypha godseffiana Mast. === **Acalypha amentacea** subsp. **wilkesiana** (Müll.Arg.) Fosberg
Acalypha goetzei Pax & K.Hoffm. === **Acalypha polymorpha** Müll.Arg.
Acalypha goudotiana Baill. === **Acalypha filiformis** var. **goudotiana** (Baill.) Govaerts
Acalypha goyazensis Glaz. === **Acalypha vellamea** Baill.
Acalypha gracilens A.Gray. === **Acalypha virginica** var. **gracilens** (A.Gray) Müll.Arg.
Acalypha gracilens var. *monococca* Engelm. ex A.Gray === **Acalypha virginica** var. **gracilens** (A.Gray) Müll.Arg.
Acalypha gracilipes Baill. === **Acalypha integrifolia** var. **gracilipes** (Baill.) Müll.Arg.
Acalypha gracilis Griseb. === **Acalypha communis** Müll.Arg.
Acalypha gracilis var. *divaricata* (Baill.) Pax & K.Hoffm. === **Acalypha gracilis** Spreng.
Acalypha gracilis var. *fruticulosa* Müll.Arg. === **Acalypha gracilis** Spreng.
Acalypha gracilis var. *genuina* Müll.Arg. === **Acalypha gracilis** Spreng.
Acalypha gracilis var. *pubescens* Müll.Arg. === **Acalypha gracilis** Spreng.
Acalypha grandidentata Müll.Arg. === **Acalypha capensis** (L.f.) Prain
Acalypha grandifolia Poir. === **Claoxylon grandifolium** (Poir.) Müll.Arg.
Acalypha grandis K.Schum. & Lauterb. === **Acalypha longispica** Warb.
Acalypha grandis Benth. === **Acalypha amentacea** var. **grandis** (Benth.) Fosberg
Acalypha grandis var. *amboinensis* (Benth.) Müll.Arg. === **Acalypha amentacea** Roxb. subsp. **amentacea**
Acalypha grandis f. *atropurpurea* Gilli === **Acalypha amentacea** var. **grandis** (Benth.) Fosberg
Acalypha grandis var. *genuina* Müll.Arg. === **Acalypha amentacea** var. **grandis** (Benth.) Fosberg
Acalypha grandis var. *velutina* Müll.Arg. === **Acalypha amentacea** var. **velutina** (Müll.Arg.) Fosberg
Acalypha grandis var. *villosa* Müll.Arg. === **Acalypha amentacea** var. **grandis** (Benth.) Fosberg
Acalypha grantii Baker & Hutch. === **Acalypha ornata** Hochst. ex A.Rich.
Acalypha granulata Ruiz ex Pax & K.Hoffm. === **Acalypha ruiziana** Müll.Arg.
Acalypha hamiltoniana Briant === **Acalypha amentacea** subsp. **wilkesiana** (Müll.Arg.) Fosberg
Acalypha haplostyla Pax === **Acalypha brachiata** Krauss
Acalypha haplostyla var. *longifolia* De Wild. === **Acalypha brachiata** Krauss
Acalypha hartwegiana Benth. ex Baill. === **Acalypha diversifolia** Jacq.

Acalypha hassleriana var. *genuina* Pax & K.Hoffm. === **Acalypha hassleriana** Chodat
Acalypha hassleriana var. *glandulosa* (Kuntze) Pax & K.Hoffm. === **Acalypha hassleriana** Chodat
Acalypha hederacea Torr. === **Acalypha monostachya** Cav.
Acalypha hederacea var. *genuina* Müll.Arg. === **Acalypha monostachya** Cav.
Acalypha hederacea var. *oligodonta* Müll.Arg. === **Acalypha monostachya** Cav.
Acalypha hederacea var. *orbicularis* Müll.Arg. === **Acalypha monostachya** Cav.
Acalypha hederaceus Kuntze === **Acalypha multicaulis** Müll.Arg.
Acalypha hellwigii var. *glabra* (Warb.) K.Schum. & Lauterb. === **Acalypha hellwigii** Warb. var. **hellwigii**
Acalypha hernandifolia Griseb. === **Acalypha alexandrii** Urb.
Acalypha hernandiifolia var. *genuina* Müll.Arg. === **Acalypha hernandiifolia** Sw.
Acalypha heterodonta Müll.Arg. === **Acalypha macrostachya** Jacq.
Acalypha heterodonta var. *hirsuta* Müll.Arg. === **Acalypha macrostachya** Jacq.
Acalypha heterodonta var. *psilocarpa* Müll.Arg. === **Acalypha macrostachya** Jacq.
Acalypha heterodonta var. *trichoclada* Müll.Arg. === **Acalypha macrostachya** Jacq.
Acalypha heterostachya Gagnep. === **Acalypha kerrii** Craib
Acalypha hicksii Riley === **Acalypha macrostachya** Jacq.
Acalypha hirsuta Hochst. ex A.Rich. === **Acalypha brachiata** Krauss
Acalypha hirsuta Mart. ex Colla === ?
Acalypha hirsutissima Willd. === **Acalypha macrostachya** Jacq.
Acalypha hirta Cav. === **Acalypha phleoides** Cav.
Acalypha hirta Spreng. === **Acalypha communis** Müll.Arg.
Acalypha hispaniolae Urb. === **Acalypha chamaedrifolia** (Lam.) Müll.Arg.
Acalypha hispida Willd. === **Acalypha poiretii** Spreng.
Acalypha hispida Wall. === **Acalypha lanceolata** Willd. var. **lanceolata**
Acalypha hispida Blume === **Acalypha lanceolata** Willd. var. **lanceolata**
Acalypha hispida var. *pubescens* Hook. & Arn. === **Acalypha lanceolata** Willd. var. **lanceolata**
Acalypha hispida var. *sanderi* (N.E.Br.) J.J.Sm. === **Acalypha hispida** Burm.f.
Acalypha holtzii Pax & K.Hoffm. === **Acalypha fruticosa** var. **eglandulosa** Radcl.-Sm.
Acalypha hookeri J.F.Macbr. === **Acalypha parvula** Hook.f. var. **parvula**
Acalypha hotteana Urb. === **Acalypha chamaedrifolia** (Lam.) Müll.Arg.
Acalypha humilis Pax & K.Hoffm. === **Acalypha communis** Müll.Arg.
Acalypha hystrix Balb. ex Spreng. === **Acalypha aristata** Kunth
Acalypha illustris Pax & K.Hoffm. === **Acalypha amentacea** subsp. **wilkesiana** (Müll.Arg.) Fosberg
Acalypha indica Vell. === **Acalypha poiretii** Spreng.
Acalypha indica K.Schum. & Hollr. === **Acalypha lanceolata** Willd. var. **lanceolata**
Acalypha indica lusus *abortiva* (Hochst. ex Baill.) Müll.Arg. === **Acalypha crenata** Hochst. ex A.Rich.
Acalypha indica var. *australis* F.M.Bailey === **Acalypha lanceolata** Willd. var. **lanceolata**
Acalypha indica var. *bailloniana* (Müll.Arg.) Hutch. === **Acalypha indica** L.
Acalypha indica var. *mexicana* (Müll.Arg.) Pax & K.Hoffm. === **Acalypha mexicana** Müll.Arg.
Acalypha indica var. *minima* (H.Keng) S.F.Huang & T.C.Huang === **Acalypha indica** L.
Acalypha infesta var. *rotundifolia* (Vahl ex Baill.) Müll.Arg. === **Acalypha infesta** Poepp. & Endl.
Acalypha infesta var. *stenoloba* Müll.Arg. === **Acalypha infesta** Poepp. & Endl.
Acalypha infestans Müll.Arg. === **Acalypha infesta** Poepp. & Endl.
Acalypha insulana var. *glabrescens* Müll.Arg. === **Acalypha repanda** Müll.Arg. var. **repanda**
Acalypha insulana var. *pubescens* Müll.Arg. === **Acalypha insulana** Müll.Arg. var. **insulana**
Acalypha insulana var. *stipularis* Müll.Arg. === **Acalypha insulana** Müll.Arg. var. **insulana**
Acalypha insulana var. *subvillosa* (Müll.Arg.) A.C.Sm. === **Acalypha insulana** var. **anisodonta** (Müll.Arg.) Govaerts
Acalypha insulana var. *villosa* Müll.Arg. === **Acalypha insulana** Müll.Arg. var. **insulana**
Acalypha integrifolia var. *colorata* (Poir.) Pax & K.Hoffm. === **Acalypha integrifolia** Willd. subsp. **integrifolia**

Acalypha integrifolia var. *concolor* (Müll.Arg.) Pax & K.Hoffm. === **Acalypha integrifolia** Willd. subsp. **integrifolia**
Acalypha interrupta Schltdl. === **Bernardia dodecandra** (Sessé ex Cav.) Govaerts
Acalypha irazuensis (Kuntze) Pax & K.Hoffm. === **Acalypha septemloba** Müll.Arg.
Acalypha jamaicensis Britton === **Acalypha leicesterfieldiensis** Radcl.-Sm. & Govaerts
Acalypha jamaicensis Raf. === ?
Acalypha japonica Houtt. ex Steud. === **Boehmeria platyphylla** D.Don (Urticaceae)
Acalypha jardinii Müll.Arg. === ? (close to *A. cuspidata*)
Acalypha johnstonii Pax === **Acalypha psilostachya** Hochst. ex A.Rich. var. **psilostachya**
Acalypha karsteniana Pax & K.Hoffm. === **Acalypha villosa** Jacq.
Acalypha kilimandscharica Volkens ex Pax & K.Hoffm. === **Acalypha fruticosa** var. **eglandulosa** Radcl.-Sm.
Acalypha klotzschiana Baill. === **Acalypha prunifolia** Nees & Mart.
Acalypha klotzschii Baill. === **Acalypha cuneata** Poepp. & Endl.
Acalypha kraussiana Buchinger ex Krauss. === **Acalypha capensis** (L.f.) Prain
Acalypha lacei Hutch. === **Acalypha kerrii** Craib
Acalypha laevifolia Müll.Arg. === **Acalypha repanda** var. **denudata** (Müll.Arg.) A.C.Sm.
Acalypha laevigata Willd. === **Acalypha angustifolia** Sw.
Acalypha lagoensis var. *grandifolia* Chodat & Hassl. === **Acalypha plicata** Müll.Arg.
Acalypha lamiifolia Scheele === **Acalypha capensis** (L.f.) Prain
Acalypha lanceolata Wall. === **Acalypha australis** L.
Acalypha languida E.Mey. === **Acalypha brachiata** Krauss
Acalypha lantanifolia Bojer === **Acalypha filiformis** Poir. subsp. **filiformis**
Acalypha latifolia Müll.Arg. === **Acalypha insulana** Müll.Arg. var. **insulana**
Acalypha lehmanniana Pax === **Acalypha macrostachya** Jacq.
Acalypha leonensis Benth. === **Mareya micrantha** (Benth.) Müll.Arg.
Acalypha leptopoda var. *glabrescens* Müll.Arg. === **Acalypha leptopoda** Müll.Arg.
Acalypha leptopoda var. *mollis* Müll.Arg. === **Acalypha leptopoda** Müll.Arg.
Acalypha leptostachya A.Rich. === **Acalypha laxiflora** Müll.Arg.
Acalypha leptostachya Kunth === **Acalypha diversifolia** Jacq.
Acalypha leptostachya var. *carpinifolia* (Poepp. ex Seem.) Müll.Arg. === **Acalypha diversifolia** Jacq.
Acalypha leptostachya var. *genuina* Müll.Arg. === **Acalypha diversifolia** Jacq.
Acalypha leptostachya var. *popayanensis* (Kunth) Müll.Arg. === **Acalypha diversifolia** Jacq.
Acalypha liebmannii Müll.Arg. === **Acalypha liebmanniana** Müll.Arg.
Acalypha lindheimeri Müll.Arg. === **Acalypha phleoides** Cav.
Acalypha lindheimeri var. *major* Pax & K.Hoffm. === **Acalypha phleoides** Cav.
Acalypha linostachys Baill. === **Acalypha villosa** Jacq.
Acalypha livingstoniana Müll.Arg. === **Acalypha ornata** Hochst. ex A.Rich.
Acalypha longifolia E.Mey. ex Pax. & K.Hoffm. === **Acalypha punctata** Meisn. ex C.Krauss
Acalypha longifolia Baill. === **Acalypha cuneata** Poepp. & Endl.
Acalypha longispica K.Schum. & Lauterb. === **Acalypha novoguineensis** Warb.
Acalypha lotsii Donn.Sm. === **Acalypha leptopoda** Müll.Arg.
Acalypha luzonica Pax & K.Hoffm. === **Acalypha amentacea** Roxb. subsp. **amentacea**
Acalypha macafeeana Veitch === **Acalypha amentacea** subsp. **wilkesiana** (Müll.Arg.) Fosberg
Acalypha macrophylla Ule === **Acalypha stachyura** Pax
Acalypha macrophylla Veitch === **Acalypha amentacea** subsp. **wilkesiana** (Müll.Arg.) Fosberg
Acalypha macrophylla Kunth === **Acalypha macrostachya** Jacq.
Acalypha macrosperma Müll.Arg. === **Acalypha membranacea** A.Rich.
Acalypha macrostachya Rusby === **Acalypha benensis** Britton
Acalypha macrostachya Willd. === **Acalypha macrostachya** Jacq.
Acalypha macrostachya var. *genuina* Müll.Arg. === **Acalypha macrostachya** Jacq.
Acalypha macrostachya var. *hirsutissima* (Willd.) Müll.Arg. === **Acalypha macrostachya** Jacq.
Acalypha macrostachya var. *macrophylla* (Kunth) Müll.Arg. === **Acalypha macrostachya** Jacq.
Acalypha macrostachya f. *macrophylla* (Kunth) Müll.Arg. === **Acalypha macrostachya** Jacq.
Acalypha macrostachya f. *puberula* Müll.Arg. === **Acalypha macrostachya** Jacq.

Acalypha macrostachya var. *sidifolia* (Kunth) Müll.Arg. === **Acalypha macrostachya** Jacq.
Acalypha macrostachya f. *tristis* (Poepp. & Endl.) Müll.Arg. === **Acalypha macrostachya** Jacq.
Acalypha macrostachya var. *tristis* (Poepp. & Endl.) Müll.Arg. === **Acalypha macrostachya** Jacq.
Acalypha macrostachyos Poir. === **Acalypha poiretii** Spreng.
Acalypha major Baill. === **Acalypha weddelliana** Baill.
Acalypha mapirensis Pax === **Acalypha mapirensis** Pax
Acalypha mapirensis var. *pubescens* Pax & K.Hoffm. === **Acalypha mapirensis** Pax
Acalypha mapirensis var. *scabra* Pax & K.Hoffm. === **Acalypha mapirensis** Pax
Acalypha mappa (L.) Willd. === **Macaranga mappa** (L.) Müll.Arg.
Acalypha marginata (Mill.) J.J.Sm. === **Acalypha amentacea** subsp. **wilkesiana** (Müll.Arg.) Fosberg
Acalypha marginata (Poir.) Spreng. === **Acalypha integrifolia** subsp. **marginata** (Poir.) Coode
Acalypha matudai Lundell === **Acalypha polystachya** Jacq.
Acalypha micrantha Benth. === **Mareya micrantha** (Benth.) Müll.Arg.
Acalypha microgyna Poepp. & Endl. === **Acalypha diversifolia** Jacq.
Acalypha microstachya Benth. === **Acalypha mollis** Kunth
Acalypha mildbraediana Pax === **Acalypha neptunica** Müll.Arg. var. **neptunica**
Acalypha mildbraediana var. *glabrescens* Pax === **Acalypha neptunica** Müll.Arg. var. **neptunica**
Acalypha mildbraediana var. *pubescens* Pax === **Acalypha neptunica** var. **pubescens** (Pax) Hutch.
Acalypha minahassae Koord. === **Acalypha caturus** Blume
Acalypha minima H.Keng === **Acalypha indica** L.
Acalypha moggii Compton === **Acalypha ornata** Hochst. ex A.Rich.
Acalypha mollis Millsp. === **Acalypha seleriana** Greenm.
Acalypha mollis Rusby === **Acalypha mandonii** Müll.Arg.
Acalypha mollis var. *polystachya* Müll.Arg. === **Acalypha mollis** Kunth
Acalypha mollissima Klotzsch ex Pax & K.Hoffm. === **Acalypha stricta** Poepp. & Endl.
Acalypha mollissima Rusby ex Pax & K.Hoffm. === **Acalypha buchtienii** Pax
Acalypha monococca (Engelm. ex A.Gray) Lill.W.Mill. & Gandhi === **Acalypha virginica** var. **gracilens** (A.Gray) Müll.Arg.
Acalypha monostachya Benth. === **Acalypha anemioides** Kunth
Acalypha montevidensis Klotzsch ex Pax & K.Hoffm. === **Acalypha communis** Müll.Arg.
Acalypha multicaulis Chodat & Hassl. === **Acalypha nitschkeana** Pax & K.Hoffm.
Acalypha multicaulis var. *genuina* Müll.Arg. === **Acalypha multicaulis** Müll.Arg.
Acalypha multicaulis var. *glabrescens* Pax & K.Hoffm. === **Acalypha multicaulis** Müll.Arg.
Acalypha multicaulis var. *tenuispica* Pax & K.Hoffm. === **Acalypha multicaulis** Müll.Arg.
Acalypha multicaulis var. *tomentella* Müll.Arg. === **Acalypha multicaulis** Müll.Arg.
Acalypha multipartita Moench. === **Acalypha tomentosa** Sw.
Acalypha muralis Zipp. ex Span. === ?
Acalypha muricata Klotzsch ex Pax & K.Hoffm. === **Acalypha villosa** Jacq.
Acalypha musaica auct. === **Acalypha amentacea** subsp. **wilkesiana** (Müll.Arg.) Fosberg
Acalypha neocaledonica Müll.Arg. === **Acalypha pancheriana** Baill.
Acalypha neogranatensis Müll.Arg. === **Acalypha macrostachya** Jacq.
Acalypha neomexicana var. *jaliscana* McVaugh === **Acalypha salvadorensis** Standl.
Acalypha neptunica var. *genuina* Pax & K.Hoffm. === **Acalypha neptunica** Müll.Arg.
Acalypha neptunica var. *glabrescens* (Pax) Pax & K.Hoffm. === **Acalypha neptunica** Müll.Arg. var. **neptunica**
Acalypha neptunica var. *vestita* Pax & K.Hoffm. === **Acalypha neptunica** var. **pubescens** (Pax) Hutch.
Acalypha nicaraguensis Pax & K.Hoffm. === **Acalypha microphylla** Klotzsch var. **microphylla**
Acalypha nigritiana Müll.Arg. === **Acalypha ornata** Hochst. ex A.Rich.
Acalypha obovata Benth. === **Acalypha cuneata** Poepp. & Endl.
Acalypha obtusa Thunb. === **Leidesia procumbens** (L.) Prain
Acalypha obtusata Spreng. ex Steud. === **Adenocline violifolia** (Kunze) Prain
Acalypha obtusifolia Pax & K.Hoffm. === **Acalypha macrostachya** Jacq.
Acalypha odorata Steud. === **Acalypha integrifolia** Willd. subsp. **integrifolia**

Acalypha ornata var. *bracteosa* Müll.Arg. === **Acalypha ornata** Hochst. ex A.Rich.
Acalypha ornata var. *genuina* Müll.Arg. === **Acalypha ornata** Hochst. ex A.Rich.
Acalypha ornata var. *glandulosa* Müll.Arg. === **Acalypha ornata** Hochst. ex A.Rich.
Acalypha ornata var. *pilosa* Müll.Arg. === **Acalypha ornata** Hochst. ex A.Rich.
Acalypha ovalifolia Baill. === **Acalypha filiformis** var. **ovalifolia** (Baill.) Govaerts
Acalypha ovata Pax & K.Hoffm. === **Acalypha stenoloba** Müll.Arg.
Acalypha oweniae Harv. ex Pax & K.Hoffm. === **Acalypha depressinervia** (Kuntze) K.Schum.
Acalypha panamensis Klotzsch === **Acalypha diversifolia** Jacq.
Acalypha paniculata Miq. === **Acalypha racemosa** Wall. ex Baill.
Acalypha paniculata f. *depauperata* Müll.Arg. === **Acalypha racemosa** Wall. ex Baill.
Acalypha paraguariensis Chodat & Hassl. === **Acalypha communis** Müll.Arg.
Acalypha parvifolia Müll.Arg. === **Acalypha microphylla** Klotzsch var. **microphylla**
Acalypha parvula var. *cordifolia* (Hook.f.) Müll.Arg. === **Acalypha parvula** Hook.f. var. **parvula**
Acalypha parvula f. *diffusa* (Andersson) Müll.Arg. === **Acalypha parvula** Hook.f. var. **parvula**
Acalypha parvula var. *flaccida* (Hook.f.) Müll.Arg. === **Acalypha flaccida** Hook.f.
Acalypha parvula var. *genuina* Müll.Arg. === **Acalypha parvula** Hook.f.
Acalypha parvula var. *procumbens* Müll.Arg. === **Acalypha parvula** Hook.f. var. **parvula**
Acalypha parvula var. *pubescens* Müll.Arg. === **Acalypha parvula** Hook.f. var. **parvula**
Acalypha parvula f. *sericea* (Andersson) Müll.Arg. === **Acalypha sericea** Andersson
Acalypha parvula f. *velutina* (Hook.f.) Müll.Arg. === **Acalypha velutina** Hook.f.
Acalypha pastoris Schrank === **Acalypha phleoides** Cav.
Acalypha pastoris DC. ex Willd. === **Acalypha phleoides** Cav.
Acalypha pauciflora Hornem. === **Acalypha australis** L.
Acalypha pavoniana Müll.Arg. === **Acalypha aristata** Kunth
Acalypha paxiana Dinter ex Pax & K.Hoffm. === **Acalypha fruticosa** Forssk. var. **fruticosa**
Acalypha paxii Palacky === **Acalypha filiformis** var. **urophylloides** (Pax & K.Hoffm.) Govaerts
Acalypha peduncularis Pax === **Acalypha caperonioides** Baill.
Acalypha peduncularis var. *angustata* (Sond.) Müll.Arg. === **Acalypha angustata** Sond.
Acalypha peduncularis var. *caperonioides* (Baill.) Müll.Arg. === **Acalypha caperonioides** Baill.
Acalypha peduncularis var. *crassa* (Buchinger ex C.Krauss) Müll.Arg. === **Acalypha peduncularis** Meisn. ex C.Krauss
Acalypha peduncularis var. *ferox* Pax ex F.Wilms === **Acalypha peduncularis** Meisn. ex C.Krauss
Acalypha peduncularis var. *genuina* Müll.Arg. === **Acalypha peduncularis** Meisn. ex C.Krauss
Acalypha peduncularis var. *glabrata* Sond. === **Acalypha caperonioides** Baill.
Acalypha peduncularis var. *glandulifolia* (Buchinger & Meisn. ex Krauss) Müll.Arg. === **Acalypha glandulifolia** Buchinger & Meisn. ex Krauss
Acalypha peduncularis var. *psilogyne* Müll.Arg. === **Acalypha zeyheri** Baill.
Acalypha peduncularis var. *punctata* (Meisn. ex C.Krauss) Müll.Arg. === **Acalypha punctata** Meisn. ex C.Krauss
Acalypha pedunculata Klotzsch ex Pax & K.Hoffm. === **Acalypha ostryifolia** Riddell
Acalypha persimilis Müll.Arg. === **Acalypha ostryifolia** Riddell
Acalypha persimilis var. *corchorifolia* (A.Rich.) Müll.Arg. === **Acalypha ostryifolia** Riddell
Acalypha persimilis var. *scabra* Müll.Arg. === **Acalypha ostryifolia** Riddell
Acalypha pervilleana Baill. === **Acalypha filiformis** var. **pervilleana** (Baill.) Govaerts
Acalypha petiolaris Sond. === **Acalypha sonderiana** Müll.Arg.
Acalypha petiolaris Hochst. ex C.Krauss === **Acalypha brachiata** Krauss
Acalypha philippinensis Müll.Arg. === **Acalypha amentacea** Roxb. subsp. **amentacea**
Acalypha phleoides Torr. === **Acalypha phleoides** Cav.
Acalypha phleoides var. *genuina* Müll.Arg. === **Acalypha phleoides** Cav.
Acalypha phleoides var. *hirta* (Cav.) Müll.Arg. === **Acalypha phleoides** Cav.
Acalypha pilifera Klotzsch ex Baill. === **Acalypha communis** Müll.Arg.
Acalypha pinnata Poir. === **Tragia pinnata** (Poir.) A.Juss.
Acalypha piperoides Klotzsch ex Pax & K.Hoffm. === **Acalypha macrostachya** Jacq.
Acalypha pittieri Pax & K.Hoffm. === **Acalypha macrostachya** Jacq.
Acalypha platyodonta Urb. === **Acalypha tomentosa** Sw.
Acalypha plicata Griseb. === **Acalypha plicata** Müll.Arg.

Acalypha polymorpha var. *angustifolia* Müll.Arg. === **Acalypha ambigua** Pax
Acalypha polymorpha var. *depauperata* Müll.Arg. === **Acalypha ambigua** Pax
Acalypha polymorpha var. *elliptica* Müll.Arg. === **Acalypha polymorpha** Müll.Arg.
Acalypha polymorpha var. *oblongifolia* Müll.Arg. === **Acalypha polymorpha** Müll.Arg.
Acalypha polymorpha var. *sericea* Müll.Arg. === **Acalypha polymorpha** Müll.Arg.
Acalypha polynema Baill. === **Acalypha juliflora** Pax
Acalypha polynesiaca Radcl.-Sm. & Govaerts === **Acalypha rapensis** F.Br.
Acalypha polystachya Griseb. === **Acalypha ostryifolia** Riddell
Acalypha polystachya Spreng. === **Acalypha setosa** A.Rich.
Acalypha polystachya Sessé & Moç. === **Acalypha polystachya** Jacq.
Acalypha ponapensis Kaneh. & Hatus. === **Acalypha caturus** Blume
Acalypha popayanensis Kunth === **Acalypha diversifolia** Jacq.
Acalypha pringlei S.Watson === **Acalypha californica** Benth.
Acalypha prostrata Zeyh. ex Pax. & K.Hoffm. === **Acalypha capensis** (L.f.) Prain
Acalypha protracta S.Moore === **Acalypha longispica** Warb.
Acalypha prunifolia Kunth === **Acalypha phleoides** Cav.
Acalypha pruriens Chodat & Hassl. === **Acalypha multicaulis** Müll.Arg.
Acalypha psilostachyoides Pax === **Acalypha volkensii** Pax
Acalypha punctata var. *longifolia* Prain === **Acalypha punctata** Meisn. ex C.Krauss
Acalypha punctata var. *rogersii* Prain === **Acalypha punctata** Meisn. ex C.Krauss
Acalypha purpurascens var. *eglandulosa* Müll.Arg. === **Acalypha purpurascens** Kunth
Acalypha purpurascens var. *genuina* Müll.Arg. === **Acalypha purpurascens** Kunth
Acalypha pygmaea Griseb. === **Acalypha nana** (Müll.Arg.) Griseb. ex Hutch.
Acalypha radians var. *genuina* Müll.Arg. === **Acalypha radians** Torr.
Acalypha radians var. *geraniifolia* Müll.Arg. === **Acalypha radians** Torr.
Acalypha radula Baker === **Acalypha salviifolia** Baill.
Acalypha rehmannii Pax === **Acalypha brachiata** Krauss
Acalypha reniformis Hook.f. === **Acalypha parvula** var. **reniformis** (Hook.f.) Müll.Arg.
Acalypha reptans Sw. === **Acalypha chamaedrifolia** (Lam.) Müll.Arg.
Acalypha reptans var. *glechomifolia* (A.Rich.) Müll.Arg. === **Acalypha glechomifolia** A.Rich.
Acalypha reptans var. *pygmaea* (A.Rich.) Müll.Arg. === **Acalypha pygmaea** A.Rich.
Acalypha reticulata (Poir.) Müll.Arg. === **Acalypha filiformis** Poir. subsp. **filiformis**
Acalypha reticulata f. *aberrans* Müll.Arg. === **Acalypha integrifolia** var. **longifolia** (Müll.Arg.) Coode
Acalypha reticulata var. *arborea* (Poir.) Müll.Arg. === **Acalypha filiformis** Poir. subsp. **filiformis**
Acalypha reticulata var. *genuina* Müll.Arg. === **Acalypha filiformis** Poir. subsp. **filiformis**
Acalypha reticulata var. *goudotiana* (Baill.) Müll.Arg. === **Acalypha filiformis** var. **goudotiana** (Baill.) Govaerts
Acalypha reticulata var. *longifolia* Müll.Arg. === **Acalypha integrifolia** var. **longifolia** (Müll.Arg.) Coode
Acalypha reticulata var. *ovalifolia* (Baill.) Müll.Arg. === **Acalypha filiformis** var. **ovalifolia** (Baill.) Govaerts
Acalypha reticulata var. *pervilleana* (Baill.) Müll.Arg. === **Acalypha filiformis** var. **pervilleana** (Baill.) Govaerts
Acalypha reticulata var. *rubra* Müll.Arg. === **Acalypha filiformis** subsp. **rubra** (Müll.Arg.) Govaerts
Acalypha reticulata var. *urophylla* (Baill.) Müll.Arg. === **Acalypha filiformis** var. **urophylla** (Baill.) Govaerts
Acalypha reticulata var. *urophylloides* Pax & K.Hoffm. === **Acalypha filiformis** var. **urophylloides** (Pax & K.Hoffm.) Govaerts
Acalypha rhombifolia Baill. === **Acalypha poiretii** Spreng.
Acalypha rhomboidea Raf. === **Acalypha virginica** var. **rhomboidea** (Raf.) Cooperr.
Acalypha rhomboidea var. *deamii* (Weath.) Weath. === **Acalypha virginica** var. **deamii** Weath.
Acalypha rotundifolia Herter === **Acalypha senilis** Baill.
Acalypha rotundifolia Vahl ex Baill. === **Acalypha infesta** Poepp. & Endl.
Acalypha rubra Roxb. === **Acalypha filiformis** subsp. **rubra** (Müll.Arg.) Govaerts

Acalypha rubra Willd. === **Acalypha phleoides** Cav.
Acalypha rubra Wight ex Wall. === **Acalypha ciliata** Forssk.
Acalypha rubrinervis Cronk === **Acalypha filiformis** subsp. **rubra** (Müll.Arg.) Govaerts
Acalypha ruderalis Mart. ex Colla === ?
Acalypha rugosa Klotzsch ex Pax & K.Hoffm. === **Acalypha bullata** Müll.Arg.
Acalypha salicioides Rusby === **Acalypha diversifolia** Jacq.
Acalypha samydifolia Poepp. & Endl. === **Acalypha diversifolia** Jacq.
Acalypha sanderi N.E.Br. === **Acalypha hispida** Burm.f.
Acalypha sanderiana K.Schum. === **Acalypha hispida** Burm.f.
Acalypha santae-martae Pax & K.Hoffm. === **Acalypha cuspidata** Jacq.
Acalypha scabra Vahl ex Baill. === **Claoxylon parviflorum** A.Juss.
Acalypha scabrosa var. *betulifolia* (Sw.) Müll.Arg. === **Acalypha scabrosa** Sw.
Acalypha scabrosa var. *elongata* Urb. === **Acalypha scabrosa** Sw.
Acalypha scabrosa var. *genuina* Müll.Arg. === **Acalypha scabrosa** Sw.
Acalypha scabrosa var. *ovata* Griseb. === **Acalypha scabrosa** Sw.
Acalypha scandens Warb. === **Acalypha insulana** Müll.Arg. var. **insulana**
Acalypha scandens var. *glabra* Warb. === **Acalypha hellwigii** Warb. var. **hellwigii**
Acalypha scandens var. *mollis* Warb. === **Acalypha hellwigii** var. **mollis** (Warb.) K.Schum. & Lauterb.
Acalypha schiedeana f. *angustifolia* Müll.Arg. === **Acalypha schiedeana** Schltdl.
Acalypha schiedeana var. *angustifolia* (Müll.Arg.) Pax & K.Hoffm. === **Acalypha schiedeana** Schltdl.
Acalypha schiedeana var. *deppeana* (Schltdl.) Müll.Arg. === **Acalypha schiedeana** Schltdl.
Acalypha schiedeana var. *genuina* Müll.Arg. === **Acalypha schiedeana** Schltdl.
Acalypha schiedeana f. *latifoliam* Müll.Arg. === **Acalypha schiedeana** Schltdl.
Acalypha schiedeana var. *macrodonta* Müll.Arg. === **Acalypha schiedeana** Schltdl.
Acalypha schiedeana var. *purpusiana* Pax & K.Hoffm. === **Acalypha schiedeana** Schltdl.
Acalypha schinzii Pax === **Acalypha depressinervia** (Kuntze) K.Schum.
Acalypha schinzii var. *denticulata* Pax === **Acalypha angustata** Sond.
Acalypha schlechtendaliana var. *genuina* Müll.Arg. === **Acalypha schlechtendaliana** Müll.Arg.
Acalypha schlechtendaliana var. *mollis* Müll.Arg. === **Acalypha schlechtendaliana** Müll.Arg.
Acalypha schlechteri Pax & K.Hoffm. === **Acalypha pancheriana** Baill.
Acalypha schneideriana Pax & K.Hoffm. === **Acalypha mairei** f. **schneideriana** (Pax & K.Hoffm.) W.T.Wang
Acalypha scleroloba Müll.Arg. === **Acalypha setosa** A.Rich.
Acalypha scleropumila A.Chev. === ?
Acalypha scutellariifolia Klotzsch ex Pax & K.Hoffm. === **Acalypha vagans** Cav.
Acalypha seemannii Klotzsch === **Acalypha macrostachya** Jacq.
Acalypha senegalensis Pax & K.Hoffm. === **Acalypha brachiata** Krauss
Acalypha senensis Klotzsch === **Acalypha brachiata** Krauss
Acalypha senensis var. *chariensis* (Beille) Hutch. === **Acalypha brachiata** Krauss
Acalypha senensis var. *haplostyla* (Pax) Hutch. === **Acalypha brachiata** Krauss
Acalypha serratifolia Klotzsch ex Pax & K.Hoffm. === **Acalypha vellamea** Baill.
Acalypha sessilis De Wild. & T.Durand === **Acalypha segetalis** Müll.Arg.
Acalypha sessilis Poir. === **Acalypha australis** L.
Acalypha sessilis var. *brevibracteata* Müll.Arg. === **Acalypha segetalis** Müll.Arg.
Acalypha sessilis var. *exserta* Müll.Arg. === **Acalypha segetalis** Müll.Arg.
Acalypha setosa Bello === **Acalypha ostryifolia** Riddell
Acalypha shirensis Hutch. ex Pax & K.Hoffm. === **Acalypha polymorpha** Müll.Arg.
Acalypha siamensis Gagnep. === **Acalypha kerrii** Craib
Acalypha sidifolia Kunth === **Acalypha macrostachya** Jacq.
Acalypha sidifolia A.Rich. === **Acalypha brachiata** Krauss
Acalypha sigensis Pax & K.Hoffm. === **Acalypha engleri** Pax
Acalypha silvestri Pamp. === **Alchornea davidii** Franch.
Acalypha similis Koord. === **Acalypha caturus** Blume
Acalypha simplicissima Millsp. === **Acalypha aliena** Brandegee

Acalypha sogerensis S.Moore === **Acalypha hellwigii** var. **mollis** (Warb.) K.Schum. & Lauterb.
Acalypha somalensis Pax === **Acalypha indica** L.
Acalypha somalium Müll.Arg. === **Acalypha indica** L.
Acalypha spachiana var. *acutifolia* Baill. === **Acalypha spachiana** Baill.
Acalypha spachiana var. *latifolia* Baill. === **Acalypha spachiana** Baill.
Acalypha spachiana var. *minor* Baill. === **Acalypha spachiana** Baill.
Acalypha sphenophylla Pax & K.Hoffm. === **Acalypha siamensis** Oliv. ex Gage var. **siamensis**
Acalypha spicata Andersson === **Acalypha parvula** Hook.f. var. **parvula**
Acalypha spicata Forssk. === **Acalypha indica** L.
Acalypha spiciflora Poir. === **Claoxylon parviflorum** A.Juss.
Acalypha spiciflora Burm.f. === **Cleidion spiciflorum** (Burm.f.) Merr.
Acalypha spiciflora Thouars ex Baill. === **Claoxylon linostachys** Baill. subsp. **linostachys**
Acalypha spicigera Klotzsch === **Cleidion spiciflorum** (Burm.f.) Merr. var. **spiciflorum**
Acalypha squarrosa Klotzsch ex Pax & K.Hoffm. === **Acalypha membranacea** A.Rich.
Acalypha stipulacea K.Schum. & Lauterb. === **Acalypha insulana** Müll.Arg. var. **insulana**
Acalypha stipulacea Klotzsch === **Acalypha amentacea** Roxb. subsp. **amentacea**
Acalypha stipularis (Müll.Arg.) Engl. === **Acalypha insulana** Müll.Arg. var. **insulana**
Acalypha stokesiae Pax & K.Hoffm. === **Acalypha californica** Benth.
Acalypha stokesii F.Br. === **Acalypha rapensis** F.Br.
Acalypha striolata Lingelsh. === **Acalypha gracilis** Spreng.
Acalypha strobilifera Hook.f. === **Acalypha parvula** var. **strobilifera** (Hook.f.) Müll.Arg.
Acalypha stuhlmannii Pax === **Acalypha polymorpha** Müll.Arg.
Acalypha subandina Ule === **Acalypha platyphylla** Müll.Arg.
Acalypha subcinerea Elmer === **Acalypha caturus** Blume
Acalypha subsessilis Hutch. === **Acalypha neptunica** Müll.Arg. var. **neptunica**
Acalypha subsessilis var. *glabra* Pax & K.Hoffm. === **Acalypha neptunica** Müll.Arg. var. **neptunica**
Acalypha subsessilis var. *mollis* Hutch. === **Acalypha neptunica** var. **pubescens** (Pax) Hutch.
Acalypha subviscida var. *lovelandii* McVaugh === **Acalypha lovelandii** (McVaugh) McVaugh
Acalypha swynnertonii S.Moore === **Acalypha ornata** Hochst. ex A.Rich.
Acalypha szechuanensis Hutch. === **Acalypha mairei** (H.Lév.) C.K.Schneid. f. **mairei**
Acalypha tabascensis Lundell === **Acalypha diversifolia** Jacq.
Acalypha tarapotensis Müll.Arg. === **Acalypha macrostachya** Jacq.
Acalypha tenera Royle === ?
Acalypha tenuicaulis Baill. === **Acalypha multicaulis** Müll.Arg.
Acalypha tenuis Müll.Arg. === **Acalypha brachiata** Krauss
Acalypha tenuis Klotzsch ex Pax & K.Hoffm. === **Acalypha mexicana** Müll.Arg.
Acalypha tenuis var. *eglandulosa* Müll.Arg. === **Acalypha brachiata** Krauss
Acalypha tenuis var. *glandulosa* Müll.Arg. === **Acalypha brachiata** Krauss
Acalypha teusczii Pax === **Acalypha benguelensis** Müll.Arg.
Acalypha tiliifolia Poir. === **Acalypha sp.**
Acalypha tomentosa Bojer === **Acalypha integrifolia** Willd. subsp. **integrifolia**
Acalypha tomentosa Blanco === **Acalypha amentacea** var. **velutina** (Müll.Arg.) Fosberg
Acalypha tomentosula Ule === **Acalypha benensis** Britton
Acalypha torta Pax & K.Hoffm. === **Acalypha amentacea** subsp. **wilkesiana** (Müll.Arg.) Fosberg
Acalypha tricolor Seem. === **Acalypha amentacea** subsp. **wilkesiana** (Müll.Arg.) Fosberg
Acalypha tristis Poepp. & Endl. === **Acalypha macrostachya** Jacq.
Acalypha triumphans L.Linden & Rodig === **Acalypha amentacea** subsp. **wilkesiana** (Müll.Arg.) Fosberg
Acalypha trukensis Pax & K.Hoffm. === **Acalypha amentacea** var. **trukensis** (Pax & K.Hoffm.) Fosberg
Acalypha tubuaiensis H.St.John === **Acalypha raivavensis** F.Br.
Acalypha ulei Radcl.-Sm. & Govaerts === **Acalypha stachyura** Pax
Acalypha ulmifolia Benth. === **Acalypha diversifolia** Jacq.
Acalypha urophylla Pax === **Acalypha filiformis** var. **urophylloides** (Pax & K.Hoffm.) Govaerts

Acalypha urophylla Baill. === **Acalypha filiformis** var. **urophylla** (Baill.) Govaerts
Acalypha urticifolia Raf. === **Acalypha virginica** L. var. **virginica**
Acalypha urticifolia Poir. === **Acalypha sp.**
Acalypha urticoides Klotzsch ex Baill. === **Acalypha communis** Müll.Arg.
Acalypha vagans var. *genuina* Müll.Arg. === **Acalypha vagans** Cav.
Acalypha vagans var. *glandulosa* Müll.Arg. === **Acalypha vagans** Cav.
Acalypha vahliana Oliv. === **Acalypha crenata** Hochst. ex A.Rich.
Acalypha vahliana Müll.Arg. === **Acalypha fimbriata** Schumach. & Thonn.
Acalypha variabilis Klotzsch ex Baill. === **Acalypha communis** Müll.Arg.
Acalypha variabilis var. *albescens* Baill. === **Acalypha communis** Müll.Arg.
Acalypha vedeliana Baill. === **Macaranga vedeliana** (Baill.) Müll.Arg.
Acalypha velutina E.Mey. === **Acalypha glabrata** Thunb. f. **glabrata**
Acalypha velutina var. *minor* Hook.f. === **Acalypha velutina** Hook.f.
Acalypha venosa Poir. === **Leptonema venosum** (Poir.) A.Juss.
Acalypha vestita Benth. === **Acalypha cuspidata** Jacq.
Acalypha villicaulis Hochst. ex A.Rich. === **Acalypha brachiata** Krauss
Acalypha villicaulis Müll.Arg. === **Acalypha brachiata** Krauss
Acalypha villicaulis var. *minor* Müll.Arg. === **Acalypha brachiata** Krauss
Acalypha villosa Vahl ex Baill. === **Acalypha poiretii** Spreng.
Acalypha villosa var. *genuina* Müll.Arg. === **Acalypha villosa** Jacq.
Acalypha villosa var. *intermedia* Müll.Arg. === **Acalypha villosa** Jacq.
Acalypha villosa var. *latiuscula* Pax & K.Hoffm. === **Acalypha villosa** Jacq.
Acalypha villosa f. *paniculata* Müll.Arg. === **Acalypha villosa** Jacq.
Acalypha villosa var. *paniculata* (Müll.Arg.) Pax & K.Hoffm. === **Acalypha villosa** Jacq.
Acalypha villosa var. *tomentosa* Müll.Arg. === **Acalypha villosa** Jacq.
Acalypha villosa var. *trichopoda* Müll.Arg. === **Acalypha villosa** Jacq.
Acalypha virgata Thunb. === **Acalypha australis** L.
Acalypha virgata Vell. === **Acalypha communis** Müll.Arg.
Acalypha virgata G.Forst. === **Acalypha forsteriana** Müll.Arg.
Acalypha virginea Dragend. === **Acalypha virginica** L. var. **virginica**
Acalypha virginica Wall. === **Acalypha lanceolata** Willd. var. **lanceolata**
Acalypha virginica Michx. === **Acalypha virginica** var. **gracilens** (A.Gray) Müll.Arg.
Acalypha virginica var. *fraseri* Müll.Arg. === **Acalypha virginica** var. **gracilens** (A.Gray) Müll.Arg.
Acalypha virginica var. *gracilescens* Müll.Arg. === **Acalypha virginica** var. **gracilens** (A.Gray) Müll.Arg.
Acalypha virginica var. *intermedia* Müll.Arg. === **Acalypha virginica** var. **gracilens** (A.Gray) Müll.Arg.
Acalypha virginica f. *purpurea* Farw. === **Acalypha virginica** L. var. **virginica**
Acalypha wallichiana Thwaites === **Acalypha racemosa** Wall. ex Baill.
Acalypha weddelliana var. *genuina* Müll.Arg. === **Acalypha weddelliana** Baill.
Acalypha weddelliana var. *janeirensis* Pax & K.Hoffm. === **Acalypha weddelliana** Baill.
Acalypha weddelliana var. *major* (Baill.) Müll.Arg. === **Acalypha weddelliana** Baill.
Acalypha weinlandii K.Schum. ex Pax & K.Hoffm. === **Acalypha insulana** Müll.Arg. var. **insulana**
Acalypha wightiana Müll.Arg. === **Acalypha lanceolata** Willd. var. **lanceolata**
Acalypha wightiana var. *genuina* Müll.Arg. === **Acalypha lanceolata** Willd. var. **lanceolata**
Acalypha wightiana var. *lanceolata* (Willd.) Müll.Arg. === **Acalypha lanceolata** Willd.
Acalypha wilkesiana Müll.Arg. === **Acalypha amentacea** subsp. **wilkesiana** (Müll.Arg.) Fosberg
Acalypha wilkesiana f. *circinata* Müll.Arg. === **Acalypha amentacea** subsp. **wilkesiana** (Müll.Arg.) Fosberg
Acalypha wilkesiana f. *illustris* J.J.Sm. === **Acalypha amentacea** subsp. **wilkesiana** (Müll.Arg.) Fosberg
Acalypha wilkesiana f. *macrophylla* J.J.Sm. === **Acalypha amentacea** subsp. **wilkesiana** (Müll.Arg.) Fosberg

Acalypha wilkesiana var. *marginata* Mill. === **Acalypha amentacea** subsp. **wilkesiana** (Müll.Arg.) Fosberg
Acalypha wilkesiana f. *monstrosa* J.J.Sm. === **Acalypha amentacea** subsp. **wilkesiana** (Müll.Arg.) Fosberg
Acalypha wilkesiana f. *triumphans* (L.Linden & Rodig) J.J.Sm. === **Acalypha amentacea** subsp. **wilkesiana** (Müll.Arg.) Fosberg
Acalypha williamsii Rusby === **Acalypha rusbyi** Dorr
Acalypha zambesica Müll.Arg. === **Acalypha brachiata** Krauss
Acalypha zambesica var. *brevistyla* Beille === **Acalypha brachiata** Krauss
Acalypha zeyheri var. *glabrata* Müll.Arg. === **Acalypha zeyheri** Baill.
Acalypha zeyheri var. *pubescens* Müll.Arg. === **Acalypha peduncularis** Meisn. ex C.Krauss
Acalypha zeylanica Raf. === **Acalypha lanceolata** Willd. var. **lanceolata**

Acalyphes

Synonyms:
Acalyphes Hassk. === **Acalypha** L.

Acalyphopsis

Synonyms:
Acalyphopsis Pax & K.Hoffm. === **Acalypha** L.
Acalyphopsis celebica Pax & K.Hoffm. === **Acalypha hoffmanniana** Hurus.

Acanthocaulon

Synonyms:
Acanthocaulon Klotzsch === **Platygyna** Mercier

Acantholoma

Synonyms:
Acantholoma Gaudich. ex Baill. === **Pachystroma** Müll.Arg.

Acanthopyxis

Synonyms:
Acanthopyxis Miq. ex Lanj. === **Caperonia** A.St.-Hil.

Accia

Synonyms:
Accia A.St.-Hil. === **Plukenetia** L.

Acidocroton

11 species, W. Indies (Cuba, Hispaniola) and S. America (Colombia); microphyllous, mostly thorny shrubs or dry forest or scrub. *A. ekmanii* in Cuba inhabits hard limestone *mogotes*, while in Haiti *A. montanus* is similarly a limestone-dweller. Most species are relatively local; those in Cuba and Hispaniola appear mutually rather closely related. Pax's 1910 treatment is now obsolete on account of the many subsequent additions, but there has yet been no reassessment. Contrary to Webster (Synopsis, 1994) *Ophellantha* has been maintained as distinct (q.v.). (Crotonoideae)

Pax, F. (1910). *Acidocroton*. In A. Engler (ed.), Das Pflanzenreich, IV 147 [I] (Euphorbiaceae-Jatropheae): 13-14. Berlin. (Heft 42.) La/Ge. — 1 species, Cuba.

Pax, F. (with K. Hoffmann) (1914). *Acidocroton*. In A. Engler (ed.), Das Pflanzenreich, IV 147 VII [Euphorbiaceae-Additamentum V]: 397. Berlin. (Heft 63.) La/Ge. — 1 sp., *A. verrucosus*, Jamaica.

Fernández Alonso, J. L. & R. Jaramillo Mejia (1995). Hallazgo del género *Acidocroton* Griseb. (Euphorbiaceae) en Suramerica, en un bosque seco de Colombia. Caldasia 17(82-85): 389-394, illus. Sp. — First record of genus in South America, recorded from dry forest; description of a new species, *A. gentryi*.

Acidocroton Griseb., Fl. Brit. W. I.: 42 (1859).
Caribbean, Colombia. 81 83.

Acidocroton acunae Borhidi & O.Muñiz, Acta Bot. Acad. Sci. Hung. 22: 305 (1976 publ. 1977).
Cuba. 81 CUB.

Acidocroton adelioides Griseb., Fl. Brit. W. I.: 42 (1859).
W. & C. Cuba. 81 CUB. Nanophan.

Acidocroton ekmanii Urb., Symb. Antill. 9: 210 (1924).
E. Cuba. 81 CUB. Nanophan.

Acidocroton gentryi Fern.Alonso & R.Jaram., Caldasia 17: 389 (1995).
Colombia (Cundinamarca). 83 CLM. Nanophan. or phan.

Acidocroton horridus Urb. & Ekman, Ark. Bot. 20A(15): 64 (1926).
Haiti. 81 HAI. Nanophan.

Acidocroton litoralis Urb. & Ekman, Ark. Bot. 20A(15): 62 (1926).
Haiti. 81 HAI. Nanophan.

Acidocroton lobulatus Urb., Symb. Antill. 9: 209 (1924).
SE. Cuba. 81 CUB. Nanophan.

Acidocroton montanus Urb. & Ekman, Ark. Bot. 20A(15): 63 (1926).
Hispaniola. 81 DOM HAI. Nanophan.

Acidocroton oligostemon Urb., Symb. Antill. 9: 208 (1924).
C. & E. Cuba. 81 CUB. NAnophan.

Acidocroton trichophyllus Urb., Symb. Antill. 9: 211 (1924).
Cuba. 81 CUB. Nanophan.

subsp. **pilosulus** (Urb.) Borhidi, Acta Bot. Hung. 29: 183 (1983).
Cuba (Sierra de Nipe). 81 CUB. Nanophan.
* *Acidocroton pilosulus* Urb., Repert. Spec. Nov. Regni Veg. 28: 227 (1930).

subsp. **trichophyllus**
Cuba (Holguín). 81 CUB. Nanophan.

Acidocroton verrucosus Urb., Symb. Antill. 7: 513 (1913).
Jamaica. 81 JAM.

Synonyms:
Acidocroton pilosulus Urb. === **Acidocroton trichophyllus** subsp. **pilosulus** (Urb.) Borhidi
Acidocroton spinosus (Standl.) G.L.Webster === **Ophellantha spinosa** Standl.
Acidocroton steyermarkii (Standl.) G.L.Webster === **Ophellantha steyermarkii** Standl.

Acidoton

6 species, West Indies, Central America and N. tropical South America; includes *Gitara*. Only the mainland *A. nicaraguensis*, a shrub or small tree, is widely distributed; the others, all shrubs, are local in the Greater Antilles. Two sections were recognised by Pax & Hoffmann (1919), one not spiny and with distichously arranged leaves (*A. urens* in Jamaica), the other with spines and spirally arranged leaves. Their account, however, does not include *A. nicaraguensis* which was later assigned to *Gitara* (Pax & Hoffmann 1924). The genus is allied to *Tragia* but is dioecious. (Acalyphoideae)

Pax, F. & K. Hoffmann (1919). *Acidoton*. In A. Engler (ed.), Das Pflanzenreich, IV 147 IX (Euphorbiaceae-Acalypheae-Plukenetiinae): 24-26. Berlin. (Heft 68.) La/Ge. — 2 species, Antilles.

Pax, F. & K. Hoffmann (1924). *Gitara*. In A. Engler (ed.), Das Pflanzenreich, IV 147 XVII (Euphorbiaceae-Additamentum VII): 187-188. Berlin. (Heft 85.) La/Ge. — 1 species, Venezuela. [Now combined with *Acidoton*.]

Acidoton Sw., Prodr.: 83 (1788).
Trop. America. 81 82 83.
Durandeeldea Kuntze, Revis. Gen. Pl. 2: 603 (1891).
Gitara Pax & K.Hoffm. in H.G.A.Engler, Pflanzenr., IV, 147, XVII: 187 (1924).

Acidoton haitiensis Alain, Brittonia 20: 154 (1968).
Haiti. 81 HAI. Nanophan.

Acidoton lanceolatus Urb. & Ekman, Ark. Bot. 20A(15): 61 (1926).
Haiti. 81 HAI. Nanophan.

Acidoton microphyllus Urb., Symb. Antill. 3: 302 (1913).
Hispaniola. 81 DOM HAI. Nanophan.

Acidoton nicaraguensis (Hemsl.) G.L.Webster, Ann. Missouri Bot. Gard. 54: 191 (1967).
C. America, Colombia, Venezuela, Peru. 80 COS GUA NIC PAN 82 VEN 83 CLM PER. Nanophan. or phan.
* *Cleidion nicaraguense* Hemsl., Biol. Cent.-Amer., Bot. 3: 130 (1883).
Gitara venezolana Pax & K.Hoffm. in H.G.A.Engler, Pflanzenr., IV, 147, XVII: 187 (1924). *Acidoton venezolanus* (Pax & K.Hoffm.) G.L.Webster, Ann. Missouri Bot. Gard. 54: 191 (1967).

Acidoton urens Sw., Prodr.: 83 (1788).
Jamaica. 81 JAM. Nanophan.
Acidoton innocuus Baill., Étude Euphorb.: 402 (1858).

Acidoton variifolius Urb. & Ekman, Ark. Bot. 20A(15): 60 (1926).
Hispaniola. 81 DOM HAI. Nanophan.

Synonyms:
Acidoton acidothamnus (Griseb.) Kuntze === **Flueggea acidoton** (L.) G.L.Webster
Acidoton baillonianus (Müll.Arg.) Kuntze === **Margaritaria discoidea** var. **triplosphaera** Radcl.-Sm.
Acidoton buxifolius (Reut.) Kuntze === **Flueggea tinctoria** (L.) G.L.Webster
Acidoton congestus (Benth. ex Müll.Arg.) Kuntze === **Jablonskia congesta** (Benth. ex Müll.Arg.) G.L.Webster
Acidoton durissimus (J.F.Gmel.) Kuntze === **Securinega durissima** J.F.Gmel.
Acidoton ellipticus (Spreng.) Kuntze === **Flueggea elliptica** (Spreng.) Baill.
Acidoton flexuosus (Müll.Arg.) Kuntze === **Flueggea flexuosa** Müll.Arg.

Acidoton flueggeoides (Müll.Arg.) Kuntze === **Flueggea suffruticosa** (Pall.) Baill.
Acidoton griseus (Müll.Arg.) Kuntze === **Flueggea virosa** (Roxb. ex Willd.) Voigt
Acidoton hilarianus (Müll.Arg.) Kuntze === **Meineckia neogranatensis** subsp. **hilariana** (Baill.) G.L.Webster
Acidoton innocuus Baill. === **Acidoton urens** Sw.
Acidoton leucopyrus (Willd.) Kuntze === **Flueggea leucopyrus** Willd.
Acidoton obovatus (Willd.) Kuntze === **Flueggea virosa** (Roxb. ex Willd.) Voigt
Acidoton phyllanthoides (Baill.) Kuntze === **Flueggea virosa** (Roxb. ex Willd.) Voigt
Acidoton ramiflorus (Aiton) Kuntze === **Flueggea suffruticosa** (Pall.) Baill.
Acidoton schuechianus (Müll.Arg.) Kuntze === **Flueggea schuechiana** (Müll.Arg.) G.L.Webster
Acidoton trichogynus (Baill.) Kuntze === **Meineckia trichogynis** (Baill.) G.L.Webster
Acidoton venezolanus (Pax & K.Hoffm.) G.L.Webster === **Acidoton nicaraguensis** (Hemsl.) G.L.Webster
Acidoton virosus (Roxb. ex Willd.) Kuntze === **Flueggea virosa** (Roxb. ex Willd.) Voigt

Acidoton

Rejected against *Acidoton* Sw.

Synonyms:
Acidoton P.Browne === **Flueggea** Willd.

Aconceveibum

Synonyms:
Aconceveibum Miq. === **Mallotus** Lour.

Actephila

21 species, Asia (India and China), Malesia (6), Australia and the southwestern Pacific; shrubs (many of them small) or small trees. *A. excelsa* is widely distributed, with several formal varieties, and *A. lindleyi* (formerly *A. mooriana*) is found over most of the eastern part of the range; other species are more or less local. Various relationships have been suggested but Webster has settled for the relatively plesiomorphic Wielandieae following Mennega (1987; see **Phyllanthoideae**). No overall revision has been published since 1922 and the genus remains poorly understood. (Phyllanthoideae)

Pax, F. & K. Hoffmann (1922). *Actephila*. In A. Engler (ed.), Das Pflanzenreich, IV 147 XV (Euphorbiaceae-Phyllanthoideae-Phyllantheae): 191-195. Berlin. (Heft 81.) La/Ge. — 8 species, Asia, Malesia, Australia and the Pacific.

Airy-Shaw, H. K. (1971). Notes on Malesian and other Asiatic Euphorbiaceae, CXXVII. The genus *Actephila* Bl. in Eastern Malesia and Australia. Kew Bull. 25: 496-500. En. — 1 new combination, 1 new species. *A. lindleyi* (formerly *A. nitida*) is considered to have been only cultivated in Tahiti; it is native from the Moluccas through New Guinea to the Solomons and eastern Australia.

Airy-Shaw, H. K. (1978). Notes on Malesian and other Asiatic Euphorbiaceae, CXCII. A new species of *Actephila* Bl. from New Guinea. Kew Bull. 32: 380. En. — Description of *A. dolichopoda* from the Port Moresby region.

Actephila Blume, Bijdr.: 581 (1826).
S. China, Trop. Asia, N. Australia, SW. Pacific. 36 40 41 42 50 60.
Lithoxylon Endl., Gen. Pl.: 1122 (1840).
Anomospermum Dalzell, Hooker's J. Bot. Kew Gard. Misc. 3: 228 (1851).

Actephila albidula Gagnep., Bull. Soc. Bot. France 71: 566 (1924).
Vietnam. 41 VIE.

Actephila anthelminthica Gagnep., Bull. Soc. Bot. France 72: 458 (1925).
Vietnam. 41 VIE.

Actephila aurantiaca Ridl., Bull. Misc. Inform. Kew 1923: 360 (1923).
Pen. Malaysia. 42 MLY.

Actephila collinsiae Hunter ex Craib, Bull. Misc. Inform. Kew 1923: 96 (1924).
Thailand. 41 THA. Nanophan.
Actephila siamensis Pierre ex Gagnep., Bull. Soc. Bot. France 71: 568 (1924).

Actephila daii Yhin, J. Biol. (Vietnam) 9: 39 (1987).
Vietnam. 41 VIE.

Actephila dolichopoda Airy Shaw, Kew Bull. 32: 380 (1978).
Papua New Guinea. 42 NWG. Nanophan.

Actephila excelsa (Dalzell) Müll.Arg., Linnaea 32: 78 (1863).
S. China, Trop. Asia. 36 CHC CHS 40 ASS IND 41 AND BMA THA VIE 42 BOR JAW MLY NWG PHI SUM. Nanophan. or phan.
* *Anomospermum excelsum* Dalzell, Hooker's J. Bot. Kew Gard. Misc. 3: 228 (1851).
Actephila excelsa var. *genuinum* Pax & K.Hoffm. in H.G.A.Engler, Pflanzenr., IV, 147, XV: 192 (1922), nom. inval.

var. **acuminata** Airy Shaw, Kew Bull. 26: 209 (1972).
Thailand, Vietnam, Pen. Malaysia, Burma ?, Assam ? 40 ASS? 41 BMA? THA VIE 42 MLY. Nanophan.

var. **excelsa**
China (Yunnan, Guangxi), India, Assam, Burma, Thailand, Pen. Malaysia ? 36 CHC CHS 40 ASS IND 41 BMA THA 42 MLY? Nanophan. or phan.
Actephila neilgherrensis Wight, Icon. Pl. Ind. Orient. 5: t. 1910 (1852).
Savia zeylanica Baill., Étude Euphorb.: 571 (1858). *Actephila zeylanica* (Baill.) Müll.Arg., Linnaea 34: 77 (1865). *Actephila excelsa* var. *zeylanica* (Baill.) Pax & K.Hoffm. in H.G.A.Engler, Pflanzenr., IV, 147, XV: 192 (1922).
Actephila bantamensis Miq., Fl. Ned. Ind. 1(2): 356 (1859).
Actephila thomsonii Müll.Arg., Linnaea 34: 65 (1865). *Actephila excelsa* var. *thomsonii* (Müll.Arg.) Pax & K.Hoffm. in H.G.A.Engler, Pflanzenr., IV, 147, XV: 192 (1922).
Actephila dolichantha Croizat, J. Arnold Arbor. 23: 30 (1942).

var. **javanica** (Miq.) Pax & K.Hoffm. in H.G.A.Engler, Pflanzenr., IV, 147, XV: 192 (1922).
Thailand, Malesia. 41 THA 42 BOR JAW MLY NWG PHI SUM. Nanophan. or phan.
Savia actephila Hassk., Cat. Hort. Bot. Bogor.: 243 (1844).
* *Actephila javanica* Miq., Fl. Ned. Ind. 1(2): 356 (1859).
Actephila major Müll.Arg., Linnaea 32: 77 (1863).
Actephila gigantifolia Koord., Meded. Lands Plantentuin 19: 625 (1897).
Actephila minahassae Koord., Meded. Lands Plantentuin 19: 625 (1897).
Pimelodendron dispersum Elmer, Leafl. Philipp. Bot. 1: 308 (1908). *Actephila dispersa* (Elmer) Merr., Philipp. J. Sci., C 4: 276 (1909).
Actephila gitingensis Elmer, Leafl. Philipp. Bot. 3: 903 (1910).
Actephila magnifolia Elmer, Leafl. Philipp. Bot. 3: 904 (1910).

var. **puberula** (Kurz) Pax & K.Hoffm. in H.G.A.Engler, Pflanzenr., IV, 147, XV: 193 (1922).
Andaman Is. 41 AND. Nanophan. or phan.
* *Actephila puberula* Kurz, J. Asiat. Soc. Bengal, Pt. 2, Nat. Hist. 42(2): 236 (1873).

Actephila foetida Domin, Biblioth. Bot. 89: 315 (1927).
Queensland. 50 QLD. Nanophan.

Actephila latifolia Benth., Fl. Austral. 6: 89 (1873).
Maluku (Key I.), Queensland. 42 MOL 50 QLD. Nanophan.

Actephila lindleyi (Steud.) Airy Shaw, Kew Bull. 25: 496 (1971).
Maluku, Lesser Sunda Is., New Guinea, Solomon Is. to N. & E. Australia. 42 LSI MOL NWG 50 NSW QLD 60 SOL. Nanophan. or phan.
Securinega nitida W.T.Aiton, Hortus Kew. 5: 383 (1813), nom. illeg. *Lithoxylon lindleyi* Steud., Nomencl. Bot., ed. 2, 2: 57 (1841). *Lithoxylon nitidum* Müll.Arg. in A.P.de Candolle, Prodr. 15(2): 232 (1866). *Actephila nitida* (Müll.Arg.) Benth. & Hook. ex Drake, Ill. Fl. Ins. Pacif.: 177, 286 (1892).
Lithoxylon grandifolium Müll.Arg., Linnaea 34: 65 (1865). *Actephila grandifolia* (Müll.Arg.) Baill., Adansonia 6: 330 (1866).
Actephila mooriana Baill., Adansonia 6: 330 (1866).
Actephila petiolaris Benth., Fl. Austral. 6: 89 (1873).

Actephila longipedicellata (Merr.) Croizat, J. Arnold Arbor. 23: 29 (1942).
Indo-China. 41 VIE.
* *Cleistanthus longipedicellatus* Merr., Univ. Calif. Publ. Bot. 10: 425 (1924).

Actephila macrantha Gagnep., Bull. Soc. Bot. France 72: 459 (1925).
Vietnam. 41 VIE.

Actephila mearsii C.T.White, Proc. Roy. Soc. Queensland 50: 85 (1938).
Queensland. 50 QLD.

Actephila merrilliana Chun, Sunyatsenia 3: 26 (1935).
Hainan, China (Guangdong). 36 CHH CHS. Nanophan.
Actephila inopinata Croizat, J. Arnold Arbor. 21: 490 (1940).

Actephila nitidula Gagnep., Bull. Soc. Bot. France 71: 566 (1924).
Vietnam, Cambodia. 41 CBD VIE.

Actephila ovalis (Ridl.) Gage, Rec. Bot. Surv. India 9: 219 (1922).
S. Thailand, Pen. Malaysia. 41 THA 42 MLY. Nanophan. or phan.
* *Dimorphocalyx ovalis* Ridl., J. Straits Branch Roy. Asiat. Soc. 59: 178 (1911).

Actephila pierrei Gagnep., Bull. Soc. Bot. France 71: 567 (1924).
Vietnam. 41 VIE.

Actephila platysepala Gagnep., Bull. Soc. Bot. France 71: 567 (1924).
Laos. 41 LAO.

Actephila sessilifolia Benth., Fl. Austral. 6: 90 (1873).
Queensland. 50 QLD. Nanophan.

Actephila subsessilis Gagnep., Bull. Soc. Bot. France 71: 569 (1924).
China (S. Yunnan), Vietnam. 36 CHC 41 VIE. Phan.

Actephila trichogyna Airy Shaw, Kew Bull. 25: 499 (1971).
Papua New Guinea (Wissi Kussa River). 42 NWG. Phan.

Synonyms:
Actephila africana Pax === **Pentabrachion reticulatum** Müll.Arg.
Actephila bantamensis Miq. === **Actephila excelsa** (Dalzell) Müll.Arg. var. **excelsa**
Actephila dispersa (Elmer) Merr. === **Actephila excelsa** var. **javanica** (Miq.) Pax & K.Hoffm.
Actephila dolichantha Croizat === **Actephila excelsa** (Dalzell) Müll.Arg. var. **excelsa**
Actephila excelsa var. *genuinum* Pax & K.Hoffm. === **Actephila excelsa** (Dalzell) Müll.Arg.
Actephila excelsa var. *thomsonii* (Müll.Arg.) Pax & K.Hoffm. === **Actephila excelsa** (Dalzell) Müll.Arg. var. **excelsa**
Actephila excelsa var. *zeylanica* (Baill.) Pax & K.Hoffm. === **Actephila excelsa** (Dalzell) Müll.Arg. var. **excelsa**
Actephila flaviflora K.Schum. & Lauterb. === **Phyllanthus flaviflorus** (Lauterb. & K.Schum.) Airy Shaw
Actephila gigantifolia Koord. === **Actephila excelsa** var. **javanica** (Miq.) Pax & K.Hoffm.
Actephila gitingensis Elmer === **Actephila excelsa** var. **javanica** (Miq.) Pax & K.Hoffm.
Actephila grandifolia (Müll.Arg.) Baill. === **Actephila lindleyi** (Steud.) Airy Shaw
Actephila inopinata Croizat === **Actephila merrilliana** Chun
Actephila javanica Miq. === **Actephila excelsa** var. **javanica** (Miq.) Pax & K.Hoffm.
Actephila magnifolia Elmer === **Actephila excelsa** var. **javanica** (Miq.) Pax & K.Hoffm.
Actephila major Müll.Arg. === **Actephila excelsa** var. **javanica** (Miq.) Pax & K.Hoffm.
Actephila megistophylla Quisumb. & Merr. === **Cleidion megistophyllum** (Quisumb. & Merr.) Airy Shaw
Actephila minahassae Koord. === **Actephila excelsa** var. **javanica** (Miq.) Pax & K.Hoffm.
Actephila mooriana Baill. === **Actephila lindleyi** (Steud.) Airy Shaw
Actephila neilgherrensis Wight === **Actephila excelsa** (Dalzell) Müll.Arg. var. **excelsa**
Actephila nitida (Müll.Arg.) Benth. & Hook. ex Drake === **Actephila lindleyi** (Steud.) Airy Shaw
Actephila petiolaris Benth. === **Actephila lindleyi** (Steud.) Airy Shaw
Actephila puberula Kurz === **Actephila excelsa** var. **puberula** (Kurz) Pax & K.Hoffm.
Actephila rectinervis Kurz === **Excoecaria rectinervis** (Kurz) Kurz
Actephila reticulata (Müll.Arg.) Pax === **Pentabrachion reticulatum** Müll.Arg.
Actephila siamensis Pierre ex Gagnep. === **Actephila collinsiae** Hunter ex Craib
Actephila thomsonii Müll.Arg. === **Actephila excelsa** (Dalzell) Müll.Arg. var. **excelsa**
Actephila zeylanica (Baill.) Müll.Arg. === **Actephila excelsa** (Dalzell) Müll.Arg. var. **excelsa**

Actephilopsis

Synonyms:
Actephilopsis Ridl. === **Trigonostemon** Blume
Actephilopsis malayana Ridl. === **Trigonostemon aurantiacus** (Kurz ex Teijsm. & Binn.) Boerl. var. **aurantiacus**

Actinostema

An orthographic variant of *Actinostemon.*

Synonyms:
Actinostema Lindl. === **Actinostemon** Mart. ex Klotzsch

Actinostemon

30 species, Middle and S. America, south to northernmost Argentina (Misiones); most diverse in Brazil where several species are fairly local. This genus of shrubs and small trees was revised firstly by Pax (1912) and then by Jablonski (1969). It is very close to *Gymnanthes* and Webster (in Webster & Huft 1988; see **Americas**) united them; however, Esser (1994;

see **Euphorbiaceae** (**except Euphorbieae**)) considers it distinct. He supports Jablonski who indicated as distinctive 'the strobilaceous tegmentum enclosing the undeveloped inflorescence is very conspicuous and a good character of the genus' but notes that this structure persists into mature inflorescences. A modern revision may considerably reduce the number of species, as Esser hints (1994, p. 33). He also suggests that the earlier *Gussonia* Spreng., here included with *Sebastiania*, may actually be part of *Actinostemon*. (Euphorbioideae (except Euphorbieae))

Pax, F. (with K. Hoffmann) (1912). *Actinostemon*. In A. Engler (ed.), Das Pflanzenreich, IV 147 V (Euphorbiaceae-Hippomaneae): 57-81. Berlin. (Heft 52.) La/Ge. — c. 33 species (4 doubtful), in 2 subgenera each with 2 sections.

Jablonski, E. (1967). *Actinostemon*. Euphorbiaceae, Guayana Highland (Mem. New York Bot. Gard. 17(1)): 177-178. New York. En. — 2 species, neither new. The writer noted here that *Gymnanthes* and *Actinostemon* were very closely related.

Jablonski, E. (1969). Notes on neotropical Euphorbiaceae, 4. Monograph of the genus *Actinostemon*. Phytologia 18: 213-240, illus., map. En. — Introduction to and description of genus along with key to species; synopsis (13 species) with synonymy, typifications, indication of distribution, localities with exsiccatae, and commentary; separate index to names.

Actinostemon Mart. ex Klotzsch, Arch. Naturgesch. 7: 184 (1841).
S. America. 80 81 82 84 85.
Dactylostemon Klotzsch, Arch. Naturgesch. 7: 181 (1841).
Actinostema Lindl., Veg. Kingd.: 281 (1846).

Actinostemon amazonicus Pax & K.Hoffm. in H.G.A.Engler, Pflanzenr., IV, 147, V: 63 (1912).
Brazil (Acre, Amazonas). 84 BZN. Phan.

Actinostemon appendiculatus Jabl., Phytologia 18: 229 (1969).
Brazil (Pernambuco, Bahia, Espírito Santo, Rio de Janeiro). 84 BZE BZL. Nanophan. or phan.

Actinostemon brachypodus (Griseb.) Urb., Repert. Spec. Nov. Regni Veg. 28: 231 (1930).
Cuba (Pinar del Río, La Habana). 81 CUB. Nanophan.
* *Excoecaria brachypoda* Griseb., Nachr. Königl. Ges. Wiss. Georg-Augusts-Univ. 1: 178 (1865). *Sebastiania brachypoda* (Griseb.) C.Wright, Anales Acad. Ci. Méd. Habana 7: 156 (1870). *Gymnanthes brachypoda* (Griseb.) Pax & K.Hoffm. in H.G.A.Engler, Pflanzenr., IV, 147, V: 87 (1912).
Excoecaria brachyandra Müll.Arg. in A.P.de Candolle, Prodr. 15(2): 1224 (1866).

Actinostemon caribaeus Griseb., Abh. Königl. Ges. Wiss. Göttingen 7: 168 (1857).
Excoecaria caribaea (Griseb.) Griseb., Fl. Brit. W. I.: 51 (1859). *Actinostemon concolor* var. *caribaeus* (Griseb.) Müll.Arg. in A.P.de Candolle, Prodr. 15(2): 1193 (1866).
C. America, Caribbean, N. Venezuela. 80 COS NIC 81 LEE TRT WIN 82 VEN. Nanophan. or phan.
Actinostemon furcatus Klotzsch ex Baill., Étude Euphorb.: 532 (1858).

Actinostemon concepcionis (Chodat & Hassl.) Hochr., Bull. New York Bot. Gard. 6: 278 (1910).
Brazil (São Paulo), Paraguay. 84 BZL 85 PAR. Nanophan. or phan.
* *Dactylostemon klotzschii* var. *concepcionis* Chodat & Hassl., Bull. Herb. Boissier, II, 5: 678 (1905).

Actinostemon concolor (Spreng.) Müll.Arg. in A.P.de Candolle, Prodr. 15(2): 1193 (1866).
E. Brazil, Uruguay, NE. Argentina, Paraguay. 84 BZE BZL 85 AGE PAR URU. Nanophan. or phan.

* *Gussonia concolor* Spreng., Neue Entd. 2: 119 (1821). *Excoecaria concolor* (Spreng.) Spreng., Syst. Veg. 3: 24 (1826). *Gymnanthes concolor* (Spreng.) Müll.Arg., Linnaea 32: 103 (1863). *Stillingia concolor* (Spreng.) Baill., Adansonia 5: 327 (1865). *Actinostemon concolor* var. *genuinus* Müll.Arg. in A.P.de Candolle, Prodr. 15(2): 1194 (1866), nom. inval

Actinostemon sessilifolius Klotzsch, Index Seminum (B) 1851, App.: 13 (1851).

Actinostemon acuminatus Baill., Étude Euphorb.: 532 (1858). *Actinostemon concolor* var. *acuminatus* (Baill.) Müll.Arg. in A.P.de Candolle, Prodr. 15(2): 1194 (1866).

Actinostemon grandifolius Baill., Étude Euphorb.: 532 (1858). *Actinostemon polymorphus* var. *grandifolius* (Baill.) Müll.Arg., Linnaea 32: 110 (1863). *Actinostemon concolor* var. *grandifolius* (Baill.) Müll.Arg. in A.P.de Candolle, Prodr. 15(2): 1190 (1866).

Actinostemon marginatus Klotzsch ex Baill., Étude Euphorb.: 532 (1858).

Actinostemon angustifolius Klotzsch ex Regel, Gartenflora 1859: 363 (1859). *Actinostemon polymorphus* f. *angustifolius* (Klotzsch ex Regel) Müll.Arg., Linnaea 32: 109 (1863). *Actinostemon concolor* var. *angustifolius* (Klotzsch ex Regel) Müll.Arg. in A.P.de Candolle, Prodr. 15(2): 1193 (1866).

Actinostemon polymorphus Müll.Arg., Linnaea 32: 108 (1863).

Actinostemon polymorphus f. *battenuatus* Müll.Arg., Linnaea 32: 109 (1863).

Actinostemon polymorphus f. *latifolius* Müll.Arg., Linnaea 32: 169 (1863). *Actinostemon concolor* f. *latifolius* (Müll.Arg.) Müll.Arg. in A.P.de Candolle, Prodr. 15(2): 1194 (1866).

Actinostemon polymorphus f. *microphyllus* Müll.Arg., Linnaea 32: 109 (1863). *Actinostemon concolor* f. *microphyllus* (Müll.Arg.) Müll.Arg. in C.F.P.von Martius, Fl. Bras. 11(2): 109 (1874).

Actinostemon polymorphus f. *minor* Müll.Arg., Linnaea 32: 108 (1863).

Actinostemon polymorphus f. *platyphyllos* Müll.Arg., Linnaea 32: 108 (1863).

Actinostemon polymorphus var. *acutissimus* Müll.Arg., Linnaea 32: 110 (1863). *Actinostemon concolor* var. *acutissimus* (Müll.Arg.) Müll.Arg. in C.F.P.von Martius, Fl. Bras. 11(2): 597 (1874).

Actinostemon polymorphus var. *angustatus* Müll.Arg., Linnaea 32: 109 (1863).

Actinostemon polymorphus var. *ellipticus* Müll.Arg., Linnaea 32: 169 (1863). *Actinostemon concolor* var. *ellipticus* (Müll.Arg.) Müll.Arg. in A.P.de Candolle, Prodr. 15(2): 1193 (1866).

Actinostemon polymorphus var. *gardneri* Müll.Arg., Linnaea 32: 169 (1863). *Actinostemon concolor* var. *gardneri* (Müll.Arg.) Müll.Arg. in C.F.P.von Martius, Fl. Bras. 11(2): 594 (1874).

Actinostemon polymorphus var. *intermedius* Müll.Arg., Linnaea 32: 109 (1863). *Actinostemon concolor* var. *intermedius* (Müll.Arg.) Müll.Arg. in A.P.de Candolle, Prodr. 15(2): 1193 (1866).

Actinostemon polymorphus var. *longifolius* Müll.Arg., Linnaea 32: 110 (1863). *Actinostemon concolor* var. *longifolius* (Müll.Arg.) Müll.Arg. in A.P.de Candolle, Prodr. 15(2): 1194 (1866).

Actinostemon polymorphus var. *mucronatus* Müll.Arg., Linnaea 32: 110 (1863). *Actinostemon concolor* var. *mucronatus* (Müll.Arg.) Müll.Arg. in A.P.de Candolle, Prodr. 15(2): 1194 (1866).

Actinostemon polymorphus var. *obovatus* Müll.Arg., Linnaea 32: 108 (1863). *Actinostemon concolor* var. *obovatus* (Müll.Arg.) Müll.Arg. in A.P.de Candolle, Prodr. 15(2): 1193 (1866).

Actinostemon polymorphus var. *sellowii* Müll.Arg., Linnaea 32: 103 (1863). *Actinostemon concolor* var. *sellowii* (Müll.Arg.) Müll.Arg. in A.P.de Candolle, Prodr. 15(2): 1194 (1866).

Actinostemon polymorphus var. *variifolius* Müll.Arg., Linnaea 32: 109 (1863). *Actinostemon concolor* var. *variifolius* (Müll.Arg.) Müll.Arg. in A.P.de Candolle, Prodr. 15(2): 1194 (1866).

Actinostemon concolor f. *sessilifolius* (Klotzsch) Müll.Arg. in A.P.de Candolle, Prodr. 15(2): 1193 (1866).

Actinostemon concolor var. *bicolor* Müll.Arg. in C.F.P.von Martius, Fl. Bras. 11(2): 597 (1874).

Actinostemon concolor var. *riedelii* Müll.Arg. in C.F.P.von Martius, Fl. Bras. 11(2): 595 (1874).

Actinostemon cantagallensis Glaz., Bull. Soc. Bot. France 59(3): 634 (1912 publ. 1913), nom. nud.
Actinostemon tortuosus Glaz., Bull. Soc. Bot. France 59(3): 634 (1912 publ. 1913), nom. nud.
Actinostemon jamaicensis Britton, Bull. Torrey Bot. Club 39: 7 (1912).

Actinostemon desertorum (Müll.Arg.) Pax in H.G.A.Engler, Pflanzenr., IV, 147, V: 70 (1912).
SE. Brazil. 84 BZL. Nanophan.
* *Dactylostemon desertorum* Müll.Arg. in C.F.P.von Martius, Fl. Bras. 11(2): 604 (1874).

Actinostemon echinatus Müll.Arg., Linnaea 32: 107 (1863).
Brazil (Pernambuco, Rio de Janeiro). 84 BZE BZL. Nanophan. or phan. – Provisionally accepted.
Actinostemon echinatus var. *major* Müll.Arg., Linnaea 32: 108 (1863).
Actinostemon echinatus var. *minor* Müll.Arg., Linnaea 32: 108 (1863).
Actinostemon echinatus var. *spathularis* Müll.Arg., Linnaea 32: 108 (1863).
Actinostemon trachycarpus Müll.Arg. in C.F.P.von Martius, Fl. Bras. 11(2): 591 (1873).
Actinostemon echinatus var. *obovatus* Müll.Arg. in C.F.P.von Martius, Fl. Bras. 11(2): 592 (1874).

Actinostemon estrellensis (Müll.Arg.) Pax in H.G.A.Engler, Pflanzenr., IV, 147, V: 71 (1912).
SE. Brazil. 84 BZL. Nanophan.
* *Dactylostemon estrellensis* Müll.Arg. in C.F.P.von Martius, Fl. Bras. 11(2): 609 (1874).
Actinostemon estrellensis var. *genuinus* Pax in H.G.A.Engler, Pflanzenr., IV, 147, V: 72 (1912), nom. inval.
Actinostemon estrellensis var. *latifolius* Pax in H.G.A.Engler, Pflanzenr., IV, 147, V: 72 (1912).

Actinostemon gardneri (Müll.Arg.) Pax in H.G.A.Engler, Pflanzenr., IV, 147, V: 63 (1912).
SE. Brazil. 84 BZL. Nanophan.
* *Dactylostemon gardneri* Müll.Arg. in C.F.P.von Martius, Fl. Bras. 11(2): 602 (1874).

Actinostemon glabrescens Pax & K.Hoffm. in H.G.A.Engler, Pflanzenr., IV, 147, V: 64 (1912).
Brazil (Rio de Janeiro). 84 BZL. Nanophan. or phan.
Dactylostemon angustifolius Müll.Arg. in C.F.P.von Martius, Fl. Bras. 11(2): 604 (1874).
Actinostemon glabrescens var. *angustifolius* (Müll.Arg.) Pax in H.G.A.Engler, Pflanzenr., IV, 147, V: 65 (1912). *Actinostemon angustifolius* (Müll.Arg.) Pax in H.G.A.Engler, Pflanzenr., IV, 147, V: 64 (1912), nom. illeg.
Dactylostemon klotzschii var. *acuminatus* Müll.Arg. in C.F.P.von Martius, Fl. Bras. 11(2): 606 (1874). *Actinostemon glabrescens* var. *acuminatus* (Müll.Arg.) Pax in H.G.A.Engler, Pflanzenr., IV, 147, V: 65 (1912).
Dactylostemon klotzschii var. *tenuifolius* Müll.Arg. in C.F.P.von Martius, Fl. Bras. 11(2): 607 (1874). *Actinostemon glabrescens* var. *tenuifolius* (Müll.Arg.) Pax in H.G.A.Engler, Pflanzenr., IV, 147, V: 65 (1912).
Actinostemon glabrescens var. *macrophyllus* Pax & K.Hoffm. in H.G.A.Engler, Pflanzenr., IV, 147, V: 65 (1912).

Actinostemon glaziovii Pax & K.Hoffm. in H.G.A.Engler, Pflanzenr., IV, 147, V: 74 (1912).
Brazil (Rio de Janeiro, Minas Gerais). 84 BZL. Nanophan. or phan.

Actinostemon grandifolius (Müll.Arg.) Pax in H.G.A.Engler, Pflanzenr., IV, 147, V: 61 (1912).
SE. Brazil. 84 BZL. Nanophan.
* *Dactylostemon grandifolius* Müll.Arg., Linnaea 32: 111 (1863).
Actinostemon klotzschianus Baill., Adansonia 5: 334 (1865).

Actinostemon guyanensis Pax in H.G.A.Engler, Pflanzenr., IV, 147, V: 80 (1912).
Guyana. 82 GUY. Nanophan.

Actinostemon imbricatus Müll.Arg., Linnaea 34: 216 (1865).
NW. Brazil. 84 BZN. Nanophan.

Actinostemon klotzschii (Didr.) Pax in H.G.A.Engler, Pflanzenr., IV, 147, V: 69 (1912).
Brazil (Maranhão, Bahia, Minas Gerais, São Paulo, Rio de Janeiro). 84 BZE BZL. Nanophan. or phan.
* *Dactylostemon klotzschii* Didr., Vidensk. Meddel. Dansk Naturhist. Foren. Kjøbenhavn 1857: 127 (1857). *Excoecaria klotzschii* (Didr.) Baill., Hist. Pl. 5: 135 (1874).
Actinostemon sprengelii Baill., Étude Euphorb.: 333 (1859).
Actinostemon communis f. *glabrescens* Müll.Arg., Linnaea 32: 112 (1863).
Actinostemon communis var. *cordatus* Müll.Arg., Linnaea 32: 113 (1863).
Actinostemon communis var. *obtusatus* Müll.Arg., Linnaea 32: 112 (1863).
Actinostemon communis var. *weddellianus* Müll.Arg., Linnaea 32: 113 (1863).
Actinostemon cuneatus var. *angustifolius* Müll.Arg., Linnaea 32: 114 (1863).
Actinostemon cuneatus var. *latifolius* Müll.Arg., Linnaea 32: 114 (1863).
Dactylostemon communis Müll.Arg., Linnaea 32: 112 (1863). *Actinostemon communis* (Müll.Arg.) Pax in H.G.A.Engler, Pflanzenr., IV, 147, V: 65 (1912).
Dactylostemon cuneatus Müll.Arg., Linnaea 32: 114 (1863). *Actinostemon cuneatus* (Müll.Arg.) Baill., Adansonia 5: 535 (1865).
Actinostemon communis var. *grandifolius* Müll.Arg. in C.F.P.von Martius, Fl. Bras. 11(2): 604 (1874).
Actinostemon communis var. *heterophyllus* Müll.Arg. in C.F.P.von Martius, Fl. Bras. 11(2): 605 (1874).
Actinostemon communis var. *intermedius* Müll.Arg. in C.F.P.von Martius, Fl. Bras. 11(2): 605 (1874).
Actinostemon communis var. *obovatus* Müll.Arg. in C.F.P.von Martius, Fl. Bras. 11(2): 606 (1874).
Actinostemon communis var. *spathularis* Müll.Arg. in C.F.P.von Martius, Fl. Bras. 11(2): 606 (1874).
Dactylostemon australis Müll.Arg. in C.F.P.von Martius, Fl. Bras. 11(2): 608 (1874). *Actinostemon australis* (Müll.Arg.) Pax in H.G.A.Engler, Pflanzenr., IV, 147, V: 69 (1912).

Actinostemon lagoensis (Müll.Arg.) Pax in H.G.A.Engler, Pflanzenr., IV, 147, V: 62 (1912).
SE. Brazil. 84 BZL. Nanophan.
* *Dactylostemon lagoensis* Müll.Arg. in C.F.P.von Martius, Fl. Bras. 11(2): 602 (1874).

Actinostemon lanceolatus Saldanha ex Baill., Adansonia 8: 263 (1868).
Brazil (Rio de Janeiro). 84 BZL. Nanophan.

Actinostemon lasiocarpus (Müll.Arg.) Baill., Adansonia 5: 334 (1865).
E. Brazil. 84 BZE BZL. Nanophan.
* *Dactylostemon lasiocarpus* Müll.Arg., Linnaea 32: 111 (1863).

Actinostemon leptopus (Müll.Arg.) Pax in H.G.A.Engler, Pflanzenr., IV, 147, V: 69 (1912).
SE. Brazil. 84 BZL. Nanophan.
* *Dactylostemon leptopus* Müll.Arg. in C.F.P.von Martius, Fl. Bras. 11(2): 607 (1874).

Actinostemon lundianus (Didr.) Pax in H.G.A.Engler, Pflanzenr., IV, 147, V: 70 (1912).
Brazil (Amapá, Bahia, Pernambuco, Minas Gerais, Rio de Janeiro). 84 BZE BZL BZN. Nanophan. or phan.
* *Dactylostemon lundianus* Didr., Vidensk. Meddel. Dansk Naturhist. Foren. Kjøbenhavn 1857: 126 (1857).
Dactylostemon lasiocarpoides Müll.Arg., Linnaea 32: 114 (1863). *Actinostemon lasiocarpoides* (Müll.Arg.) Baill., Adansonia 5: 334 (1865).

Actinostemon luquense Morong, Ann. New York Acad. Sci. 7: 228 (1893).
Paraguay. 85 PAR. Nanophan.

Actinostemon macrocarpus Müll.Arg. in C.F.P.von Martius, Fl. Bras. 11(2): 597 (1873).
SE. Brazil. 84 BZL. Nanophan.

Actinostemon mandiocanus (Müll.Arg.) Pax in H.G.A.Engler, Pflanzenr., IV, 147, V: 61 (1912).
Brazil (Rio de Janeiro). 84 BZL. Nanophan.
* *Dactylostemon mandiocanus* Müll.Arg. in C.F.P.von Martius, Fl. Bras. 11(2): 601 (1874).

Actinostemon multiflorus Müll.Arg., Linnaea 32: 111 (1863).
Brazil (Rio de Janeiro). 84 BZL. Nanophan.

Actinostemon oligandrus (Müll.Arg.) Baill., Adansonia 5: 335 (1865).
SE. Brazil. 84 BZL. Nanophan.
* *Dactylostemon oligandrus* Müll.Arg., Linnaea 32: 115 (1863).

Actinostemon schomburgkii (Klotzsch) Hochr., Bull. New York Bot. Gard. 6: 278 (1910).
Guiana, Guyana, Brazil (Roraima), Venezuela (Bolívar). 82 FRG GUY SUR? VEN 84 BZN. Nanophan. or phan.
* *Dactylostemon schomburgkii* Klotzsch, Hooker's J. Bot. Kew Gard. Misc. 2: 45 (1843).
Gymnanthes schomburgkii (Klotzsch) G.L.Webster, Ann. Missouri Bot. Gard. 81: 122 (1994).
Actinostemon depauperatus Pax & K.Hoffm. in H.G.A.Engler, Pflanzenr., IV, 147, XIV: 58 (1919).
Actinostemon parvifolius Pittier, Bol. Soc. Venez. Ci. Nat. 5: 306 (1939).

Actinostemon sparsifolius (Müll.Arg.) Pax in H.G.A.Engler, Pflanzenr., IV, 147, V: 72 (1912).
SE. Brazil. 84 BZL. Nanophan.
* *Dactylostemon sparsifolius* Müll.Arg. in C.F.P.von Martius, Fl. Bras. 11(2): 610 (1874).

Actinostemon unciformis Jabl., Phytologia 18: 223 (1969).
Brazil (Bahia). 84 BZE. Nanophan. or phan.

Actinostemon verticillatus (Klotzsch) Baill., Adansonia 5: 334 (1865).
Brazil (Bahia, Minas Gerais, Rio de Janeiro). 84 BZE BZL. Nanophan. or phan.
* *Dactylostemon verticillatus* Klotzsch, Index Seminum (B) 1851, App.: 13 (1851).
Actinostemon verticillatus f. *subinermis* Müll.Arg. in C.F.P.von Martius, Fl. Bras. 11(2): 603 (1874).

Synonyms:
Actinostemon acuminatus Baill. === **Actinostemon concolor** (Spreng.) Müll.Arg.
Actinostemon angustifolius Klotzsch ex Regel === **Actinostemon concolor** (Spreng.) Müll.Arg.
Actinostemon angustifolius (Müll.Arg.) Pax === **Actinostemon glabrescens** Pax & K.Hoffm.
Actinostemon anisandrus (Griseb.) Pax === **Sebastiania sp.**
Actinostemon australis (Müll.Arg.) Pax === **Actinostemon klotzschii** (Didr.) Pax
Actinostemon brasiliensis (Spreng.) Pax === **Sebastiania brasiliensis** Spreng.
Actinostemon cantagallensis Glaz. === **Actinostemon concolor** (Spreng.) Müll.Arg.
Actinostemon communis (Müll.Arg.) Pax === **Actinostemon klotzschii** (Didr.) Pax
Actinostemon communis var. *cordatus* Müll.Arg. === **Actinostemon klotzschii** (Didr.) Pax
Actinostemon communis f. *glabrescens* Müll.Arg. === **Actinostemon klotzschii** (Didr.) Pax
Actinostemon communis var. *grandifolius* Müll.Arg. === **Actinostemon klotzschii** (Didr.) Pax
Actinostemon communis var. *heterophyllus* Müll.Arg. === **Actinostemon klotzschii** (Didr.) Pax
Actinostemon communis var. *intermedius* Müll.Arg. === **Actinostemon klotzschii** (Didr.) Pax
Actinostemon communis var. *obovatus* Müll.Arg. === **Actinostemon klotzschii** (Didr.) Pax
Actinostemon communis var. *obtusatus* Müll.Arg. === **Actinostemon klotzschii** (Didr.) Pax

Actinostemon communis var. *spathularis* Müll.Arg. === **Actinostemon klotzschii** (Didr.) Pax
Actinostemon communis var. *weddellianus* Müll.Arg. === **Actinostemon klotzschii** (Didr.) Pax
Actinostemon concolor var. *acuminatus* (Baill.) Müll.Arg. === **Actinostemon concolor** (Spreng.) Müll.Arg.
Actinostemon concolor var. *acutissimus* (Müll.Arg.) Müll.Arg. === **Actinostemon concolor** (Spreng.) Müll.Arg.
Actinostemon concolor var. *angustifolius* (Klotzsch ex Regel) Müll.Arg. === **Actinostemon concolor** (Spreng.) Müll.Arg.
Actinostemon concolor var. *bicolor* Müll.Arg. === **Actinostemon concolor** (Spreng.) Müll.Arg.
Actinostemon concolor var. *caribaeus* (Griseb.) Müll.Arg. === **Actinostemon caribaeus** Griseb.
Actinostemon concolor var. *ellipticus* (Müll.Arg.) Müll.Arg. === **Actinostemon concolor** (Spreng.) Müll.Arg.
Actinostemon concolor var. *gardneri* (Müll.Arg.) Müll.Arg. === **Actinostemon concolor** (Spreng.) Müll.Arg.
Actinostemon concolor var. *genuinus* Müll.Arg. === **Actinostemon concolor** (Spreng.) Müll.Arg.
Actinostemon concolor var. *grandifolius* (Baill.) Müll.Arg. === **Actinostemon concolor** (Spreng.) Müll.Arg.
Actinostemon concolor var. *intermedius* (Müll.Arg.) Müll.Arg. === **Actinostemon concolor** (Spreng.) Müll.Arg.
Actinostemon concolor f. *latifolius* (Müll.Arg.) Müll.Arg. === **Actinostemon concolor** (Spreng.) Müll.Arg.
Actinostemon concolor var. *longifolius* (Müll.Arg.) Müll.Arg. === **Actinostemon concolor** (Spreng.) Müll.Arg.
Actinostemon concolor f. *microphyllus* (Müll.Arg.) Müll.Arg. === **Actinostemon concolor** (Spreng.) Müll.Arg.
Actinostemon concolor var. *mucronatus* (Müll.Arg.) Müll.Arg. === **Actinostemon concolor** (Spreng.) Müll.Arg.
Actinostemon concolor var. *obovatus* (Müll.Arg.) Müll.Arg. === **Actinostemon concolor** (Spreng.) Müll.Arg.
Actinostemon concolor var. *riedelii* Müll.Arg. === **Actinostemon concolor** (Spreng.) Müll.Arg.
Actinostemon concolor var. *sellowii* (Müll.Arg.) Müll.Arg. === **Actinostemon concolor** (Spreng.) Müll.Arg.
Actinostemon concolor f. *sessilifolius* (Klotzsch) Müll.Arg. === **Actinostemon concolor** (Spreng.) Müll.Arg.
Actinostemon concolor var. *variifolius* (Müll.Arg.) Müll.Arg. === **Actinostemon concolor** (Spreng.) Müll.Arg.
Actinostemon cuneatus (Müll.Arg.) Baill. === **Actinostemon klotzschii** (Didr.) Pax
Actinostemon cuneatus var. *angustifolius* Müll.Arg. === **Actinostemon klotzschii** (Didr.) Pax
Actinostemon cuneatus var. *latifolius* Müll.Arg. === **Actinostemon klotzschii** (Didr.) Pax
Actinostemon depauperatus Pax & K.Hoffm. === **Actinostemon schomburgkii** (Klotzsch) Hochr.
Actinostemon echinatus var. *major* Müll.Arg. === **Actinostemon echinatus** Müll.Arg.
Actinostemon echinatus var. *minor* Müll.Arg. === **Actinostemon echinatus** Müll.Arg.
Actinostemon echinatus var. *obovatus* Müll.Arg. === **Actinostemon echinatus** Müll.Arg.
Actinostemon echinatus var. *spathularis* Müll.Arg. === **Actinostemon echinatus** Müll.Arg.
Actinostemon estrellensis var. *genuinus* Pax === **Actinostemon estrellensis** (Müll.Arg.) Pax
Actinostemon estrellensis var. *latifolius* Pax === **Actinostemon estrellensis** (Müll.Arg.) Pax
Actinostemon furcatus Klotzsch ex Baill. === **Actinostemon caribaeus** Griseb.
Actinostemon glabrescens var. *acuminatus* (Müll.Arg.) Pax === **Actinostemon glabrescens** Pax & K.Hoffm.
Actinostemon glabrescens var. *angustifolius* (Müll.Arg.) Pax === **Actinostemon glabrescens** Pax & K.Hoffm.
Actinostemon glabrescens var. *macrophyllus* Pax & K.Hoffm. === **Actinostemon glabrescens** Pax & K.Hoffm.
Actinostemon glabrescens var. *tenuifolius* (Müll.Arg.) Pax === **Actinostemon glabrescens** Pax & K.Hoffm.

Actinostemon grandifolius Baill. === **Actinostemon concolor** (Spreng.) Müll.Arg.
Actinostemon jamaicensis Britton === **Actinostemon concolor** (Spreng.) Müll.Arg.
Actinostemon klotzschianus Baill. === **Actinostemon grandifolius** (Müll.Arg.) Pax
Actinostemon lasiocarpoides (Müll.Arg.) Baill. === **Actinostemon lundianus** (Didr.) Pax
Actinostemon marginatus Klotzsch ex Baill. === **Actinostemon concolor** (Spreng.) Müll.Arg.
Actinostemon parvifolius Pittier === **Actinostemon schomburgkii** (Klotzsch) Hochr.
Actinostemon polymorphus Müll.Arg. === **Actinostemon concolor** (Spreng.) Müll.Arg.
Actinostemon polymorphus var. *acutissimus* Müll.Arg. === **Actinostemon concolor** (Spreng.) Müll.Arg.
Actinostemon polymorphus var. *angustatus* Müll.Arg. === **Actinostemon concolor** (Spreng.) Müll.Arg.
Actinostemon polymorphus f. *angustifolius* (Klotzsch ex Regel) Müll.Arg. === **Actinostemon concolor** (Spreng.) Müll.Arg.
Actinostemon polymorphus f. *battenuatus* Müll.Arg. === **Actinostemon concolor** (Spreng.) Müll.Arg.
Actinostemon polymorphus var. *ellipticus* Müll.Arg. === **Actinostemon concolor** (Spreng.) Müll.Arg.
Actinostemon polymorphus var. *gardneri* Müll.Arg. === **Actinostemon concolor** (Spreng.) Müll.Arg.
Actinostemon polymorphus var. *grandifolius* (Baill.) Müll.Arg. === **Actinostemon concolor** (Spreng.) Müll.Arg.
Actinostemon polymorphus var. *intermedius* Müll.Arg. === **Actinostemon concolor** (Spreng.) Müll.Arg.
Actinostemon polymorphus f. *latifolius* Müll.Arg. === **Actinostemon concolor** (Spreng.) Müll.Arg.
Actinostemon polymorphus var. *longifolius* Müll.Arg. === **Actinostemon concolor** (Spreng.) Müll.Arg.
Actinostemon polymorphus f. *microphyllus* Müll.Arg. === **Actinostemon concolor** (Spreng.) Müll.Arg.
Actinostemon polymorphus f. *minor* Müll.Arg. === **Actinostemon concolor** (Spreng.) Müll.Arg.
Actinostemon polymorphus var. *mucronatus* Müll.Arg. === **Actinostemon concolor** (Spreng.) Müll.Arg.
Actinostemon polymorphus var. *obovatus* Müll.Arg. === **Actinostemon concolor** (Spreng.) Müll.Arg.
Actinostemon polymorphus f. *platyphyllos* Müll.Arg. === **Actinostemon concolor** (Spreng.) Müll.Arg.
Actinostemon polymorphus var. *sellowii* Müll.Arg. === **Actinostemon concolor** (Spreng.) Müll.Arg.
Actinostemon polymorphus var. *variifolius* Müll.Arg. === **Actinostemon concolor** (Spreng.) Müll.Arg.
Actinostemon sessilifolius Klotzsch === **Actinostemon concolor** (Spreng.) Müll.Arg.
Actinostemon sprengelii Baill. === **Actinostemon klotzschii** (Didr.) Pax
Actinostemon tortuosus Glaz. === **Actinostemon concolor** (Spreng.) Müll.Arg.
Actinostemon trachycarpus Müll.Arg. === **Actinostemon echinatus** Müll.Arg.
Actinostemon verrucosum Glaz. === ? (Celastraceae)
Actinostemon verticillatus f. *subinermis* Müll.Arg. === **Actinostemon verticillatus** (Klotzsch) Baill.

Adelia

12 species, Americas (southern Texas and Mexico to Panama, West Indies, and central Brazil to Paraguay and Argentina); disjunct with three centres of distribution in mainly seasonal areas. The branchlets of these shrubs or small trees are often spinescent and the leaves are alternate or tufted. No revision of this genus has appeared since 1914. Its closest ally

appears to be *Crotonogynopsis* in central Africa. Some authors have combined the genus with *Bernardia* but according to Webster (Synopsis, 1994; see **General**) the pollen morphology differs. The name is conserved against *Adelia* P.Br. (=*Bernardia* Mill.) (Acalyphoideae).

Pax, F. (with K. Hoffmann) (1914). *Adelia*. In A. Engler (ed.), Das Pflanzenreich, IV 147 VII (Euphorbiaceae-Acalypheae-Mercurialinae): 64-71. Berlin. (Heft 63.) La/Ge. — 11 species, Americas.

Adelia L., Syst. Nat. ed. 10: 1298 (1759), nom. & typ. cons.
Texas, Mexico, Trop. America. 77 79 80 81 83 94 95.
Ricinella Müll.Arg., Linnaea 34: 153 (1865).

Adelia barbinervis Cham. & Schltdl., Linnaea 6: 362 (1831).
Mexico to Nicaragua. 79 MXE MXE MXS 80 GUA NIC. Nanophan. or phan.

Adelia haemiolandra (Griseb.) Pax & K.Hoffm. in H.G.A.Engler, Pflanzenr., IV, 147, VII: 67 (1914).
Jamaica. 81 JAM. Nanophan. or phan.
* *Ditaxis haemiolandra* Griseb., Fl. Brit. W. I.: 44 (1859).

Adelia membranifolia (Müll.Arg.) Chodat & Hassl., Bull. Herb. Boissier, II, 5: 604 (1905).
Brazil (Bahia). 84 BZE. Nanophan. or phan.
* *Ricinella membranifolia* Müll.Arg. in C.F.P.von Martius, Fl. Bras. 11(2): 306 (1874).

Adelia oaxacana (Müll.Arg.) Hemsl., Biol. Cent.-Amer., Bot. 3: 129 (1883).
Mexico (Oaxaca). 79 MXS. Nanophan. or phan.
* *Ricinella oaxacana* Müll.Arg., Linnaea 34: 154 (1865).

Adelia obovata Wiggins & Rollins, Contr. Dudley Herb. 3: 271 (1943).
N. Mexico. 79 MXE MXN. Nanophan.

Adelia panamensis Pax & K.Hoffm. in H.G.A.Engler, Pflanzenr., IV, 147, VII: 67 (1914).
Panama. 80 PAN. Nanophan. or phan.

Adelia peduncularis (Kuntze) Pax & K.Hoffm. in H.G.A.Engler, Pflanzenr., IV, 147, VII: 65 (1914).
Brazil (Mato Grosso). 84 BZC. Nanophan. or phan.
* *Ricinella peduncularis* Kuntze, Rev. Gen. 3: 290 (1898).

Adelia ricinella L., Syst. Nat. ed. 10: 1298 (1759).
Caribbean, Colombia, Venezuela. 81 CUB DOM HAI JAM LEE NLA PUE TRT WIN 82 VEN 83 CLM. Nanophan. or phan.
Adelia pedunculosa A.Rich. in R.de la Sagra, Hist. Fis. Cuba, Bot. 2: 210 (1850). *Ricinella pedunculosa* (A.Rich.) Müll.Arg., Linnaea 34: 154 (1865).
Adelia sylvestris Griseb., Nachr. Königl. Ges. Wiss. Georg-Augusts-Univ. 1: 174 (1865).
Croton sylvestris Poepp. ex Griseb., Nachr. Königl. Ges. Wiss. Georg-Augusts-Univ. 1: 174 (1865).
Adelia macrophylla Urb., Repert. Spec. Nov. Regni Veg. 15: 409 (1919).

Adelia spinosa (Chodat & Hassl.) Pax & K.Hoffm. in H.G.A.Engler, Pflanzenr., IV, 147, VII: 66 (1914).
Paraguay, N. Argentina. 85 AGE PAR. Nanophan. or phan.
Adelia membranifolia f. *hirsuta* Chodat & Hassl., Bull. Herb. Boissier, II, 5: 604 (1905). *Adelia spinosa* var. *hirsuta* (Chodat & Hassl.) Pax & K.Hoffm. in H.G.A.Engler, Pflanzenr., IV, 147, VII: 66 (1914).

Adelia membranifolia var. *spinosa* Chodat & Hassl., Bull. Herb. Boissier, II, 5: 604 (1905).
Adelia spinosa var. *hassleri* Pax & K.Hoffm. in H.G.A.Engler, Pflanzenr., IV, 147, VII: 66 (1914).

Adelia tenuifolia (Urb.) Moscoso, Cat. Fl. Doming. Pt. 1: 303 (1943).
Hispaniola. 81 DOM HAI. Nanophan.
* *Bernardia tenuifolia* Urb., Symb. Antill. 7: 260 (1912).

Adelia triloba (Müll.Arg.) Hemsl., Biol. Cent.-Amer., Bot. 3: 130 (1883).
C. America. 80 COS NIC PAN. Nanophan. or phan.
* *Ricinella triloba* Müll.Arg., Linnaea 34: 153 (1865).

Adelia vaseyi (Coult.) Pax & K.Hoffm. in H.G.A.Engler, Pflanzenr., IV, 147, VII: 69 (1914).
S. Texas. 77 TEX. Nanophan.
* *Euphorbia vaseyi* Coult., Contr. U. S. Natl. Herb. 1: 48 (1890).

Synonyms:
Adelia acidoton Blanco === **Doryxylon spinosum** Zoll.
Adelia acidoton L. === **Flueggea acidoton** (L.) G.L.Webster
Adelia anomala Juss. ex Poir. === **Erythrococca anomala** (Juss. ex Poir.) Prain
Adelia barbata Blanco === **Mallotus mollissimus** (Geiseler) Airy Shaw
Adelia bernardia L. === **Bernardia dichotoma** (Willd.) Müll.Arg. var. **dichotoma**
Adelia bernardia Blanco === **Mallotus mollissimus** (Geiseler) Airy Shaw
Adelia caperoniifolia Baill. === **Bernardia caperoniifolia** (Baill.) Müll.Arg.
Adelia castanocarpa Roxb. === **Chaetocarpus castanocarpus** (Roxb.) Thwaites
Adelia celastrinea Baill. === **Bernardia celastrinea** (Baill.) Müll.Arg.
Adelia cordifolia Roxb. === **Macaranga cordifolia** (Roxb.) Müll.Arg.
Adelia cuneata Wall. === **Homonoia retusa** (Graham ex Wight) Müll.Arg.
Adelia dodecandra Sessé ex Cav. === **Bernardia dodecandra** (Sessé ex Cav.) Govaerts
Adelia ferruginea Poit. ex Baill. === **Leucocroton leprosus** (Willd.) Pax & K.Hoffm.
Adelia glandulosa Blanco === **Alchornea rugosa** (Lour.) Müll.Arg.
Adelia gracilis Salisb. === **Flueggea tinctoria** (L.) G.L.Webster
Adelia hirsutissima Baill. === **Bernardia hirsutissima** (Baill.) Müll.Arg.
Adelia houlletiana Baill. === **Bernardia axillaris** subsp. **houlletiana** (Baill.) Müll.Arg.
Adelia javanica Miq. === **Spathiostemon javensis** Blume
Adelia leprosa (Willd.) Moscoso === **Leucocroton leprosus** (Willd.) Pax & K.Hoffm.
Adelia macrophylla Urb. === **Adelia ricinella** L.
Adelia martii Spreng. === **Bernardia axillaris** (Spreng.) Müll.Arg. subsp. **axillaris**
Adelia membranifolia f. *hirsuta* Chodat & Hassl. === **Adelia spinosa** (Chodat & Hassl.) Pax & K.Hoffm.
Adelia membranifolia var. *spinosa* Chodat & Hassl. === **Adelia spinosa** (Chodat & Hassl.) Pax & K.Hoffm.
Adelia microphylla A.Rich. === **Leucocroton microphyllus** (A.Rich.) Pax & K.Hoffm.
Adelia monoica Blanco === **Aleurites moluccana** (L.) Willd.
Adelia myrtifolia Vent. ex Spreng. === **Adelia** sp.
Adelia neriifolia B.Heyne ex Roth === **Homonoia riparia** Lour.
Adelia papillaris Blanco === **Mallotus tiliifolius** (Blume) Müll.Arg.
Adelia patens Baill. === **Adelia** sp.
Adelia pedunculosa A.Rich. === **Adelia ricinella** L.
Adelia pulchella Baill. === **Bernardia pulchella** (Baill.) Müll.Arg.
Adelia resinosa Blanco === **Mallotus resinosus** (Blanco) Merr.
Adelia retusa Graham ex Wight === **Homonoia retusa** (Graham ex Wight) Müll.Arg.
Adelia scabrida Baill. === **Bernardia axillaris** subsp. **scabrida** (Baill.) Pax & K.Hoffm.

Adelia scandens Span. === ?
Adelia spartioides (Baill.) Baill. === **Bernardia spartioides** (Baill.) Müll.Arg.
Adelia spinosa var. *hassleri* Pax & K.Hoffm. === **Adelia spinosa** (Chodat & Hassl.) Pax & K.Hoffm.
Adelia spinosa var. *hirsuta* (Chodat & Hassl.) Pax & K.Hoffm. === **Adelia spinosa** (Chodat & Hassl.) Pax & K.Hoffm.
Adelia sylvestris Griseb. === **Adelia ricinella** L.
Adelia tamanduana Baill. === **Bernardia tamanduana** (Baill.) Müll.Arg.
Adelia timoriana Span. === ?
Adelia virgata Poir. === **Flueggea tinctoria** (L.) G.L.Webster

Ademo

of Post & Kuntze = *Euphorbia* L.

Adenoceras

Synonyms:
Adenoceras Rchb. & Zoll. ex Baill. === **Macaranga** Thouars

Adenochlaena

2 species, Sri Lanka, Madagascar, Comores; related to *Cephalocroton*. The two species, both monoecious shrubs or subshrubs, included by Pax (1910) are still those accepted. [The Sri Lankan species, however, was for the family account in *Revised Handbook to the Flora of Ceylon* 11 (1997) retained in *Cephalocroton* following Radcliffe-Smith (1973). It has not apparently been collected since 1890. The Malagasy species, with highly dissected stipules and sepals, is illustrated by Pax.] (Acalyphoideae)

Pax, F. (1910). *Adenochlaena*. In A. Engler (ed.), Das Pflanzenreich, IV 147 II (Euphorbiaceae-Adrianae): 12-14. Berlin. (Heft 44.) La/Ge. — 2 species, 1 in Sri Lanka, 1 in Madagascar. [A third reported species, *A. calycina*, belongs to *Koilodepas*.]
Radcliffe-Smith, A. (1973). An account of the genus *Cephalocroton* Hochst. (Euphorbiaceae). Kew Bull. 28: 123-132. En. — Synoptic treatment with key, synonymy, references and citations, localities with exsiccatae, and commentary. [*Adenochlaena* on pp. 125-126.]

Adenochlaena Boivin ex Baill., Étude Euphorb.: 472 (1858).
Comoros, Madagascar, Sri Lanka. 29 40.
Centrostylis Baill., Étude Euphorb.: 469 (1858).
Niedenzua Pax, Bot. Jahrb. Syst. 19: 106 (1894).

Adenochlaena leucocephala Baill., Étude Euphorb.: 473 (1858). *Cephalocroton leucocephalus* (Baill.) Baill., Adansonia 5: 148 (1865).
Madagascar, Comoros. 29 COM MDG. Nanophan.
Croton acuminatus A.Rich. ex Baill., Adansonia 1: 276 (1861), nom. illeg.
Cephalocroton cordifolius Baker, J. Linn. Soc., Bot. 21: 520 (1887).

Adenochlaena zeylanica (Baill.) Thwaites, Enum. Pl. Zeyl.: 270 (1861).
Sri Lanka. 40 SRL. Nanophan.
* *Centrostylis zeylanica* Baill., Étude Euphorb.: 470 (1858). *Cephalocroton zeylanicus* (Baill.) Baill., Adansonia 5: 148 (1865).

Synonyms:
Adenochlaena calycina Bedd. === **Koilodepas calycinum** Bedd.
Adenochlaena indica Bedd. ex Hook.f. === **Epiprinus mallotiformis** (Müll.Arg.) Croizat
Adenochlaena siamensis Ridl. === **Cladogynos orientalis** Zipp. ex Span.
Adenochlaena siletensis (Baill.) Benth. === **Epiprinus siletianus** (Baill.) Croizat

Adenoclina

An orthographic variant of *Adenocline.*

Adenocline

3 species, S. tropical and southern Africa; annual or perennial herbs, sometimes woody at the base. The narrow species limits proposed by Prain (1913) have not found favour; indeed, some of the varieties of *A. pauciflora*, documented by Pax (1914), may themselves not warrant recognition pending a new study of the genus. Only *A. acuta* extends much beyond the Cape and KwaZulu-Natal. (Crotononoideae)

Prain, D. (1913(1914)). Mercurialineae and Adenoclineae of South Africa. Ann. Bot. 27: 371-410. En. — Includes (pp. 402-409) a synopsis of *Adenocline* (8 species, in former Cape Province, South Africa) with key, synonymy, references and citations, localities with exsiccatae, and brief comments; biogeographical review at end of paper.
Pax, F. (with K. Hoffmann) (1914). *Adenocline.* In A. Engler (ed.), Das Pflanzenreich, IV 147 VII [Euphorbiaceae-Additamentum V]: 409-413, illus. Berlin. (Heft 63.) La/Ge. — 3 species, S Africa, one with eight non-nominate varieties.

Adenocline Turcz., Bull. Soc. Imp. Naturalistes Moscou 16: 59 (1843).
SC. Trop. & S. Africa. 26 27.
Diplostylis H.Karst. & Triana, Linnaea 28: 433 (1856).
Paradenocline Müll.Arg. in A.P.de Candolle, Prodr. 15(2): 1141 (1866).

Adenocline acuta (Thunb.) Baill., Étude Euphorb.: 457 (1858).
S. Africa, Malawi, Zimbabwe. 26 MLW ZIM 27 CPP. Ther. or cham.
* *Acalypha acuta* Thunb., Fl. Cap., ed. 2: 546 (1823).
Adenocline mercurialis Turcz., Bull. Soc. Imp. Naturalistes Moscou 16: 61 (1843).
Mercurialis caffra Meisn., Hooker's J. Bot. Kew Gard. Misc. 2: 558 (1843).
Mercurialis dregeana Buchinger ex Krauss, Flora 28: 84 (1845).
Mercurialis subcordata Buchinger ex Krauss, Flora 28: 84 (1845).

Adenocline pauciflora Turcz., Bull. Soc. Imp. Naturalistes Moscou 16: 61 (1843).
Cape Prov., KwaZulu-Natal. 27 CPP NAT. Ther. or cham.

var. **bupleuroides** (Meisn.) Müll.Arg. in A.P.de Candolle, Prodr. 15(2): 1140 (1866).
KwaZulu-Natal. 27 NAT. Cham.
* *Mercurialis bupleuroides* Meisn., Hooker's J. Bot. Kew Gard. Misc. 2: 557 (1843).
Adenocline bupleuroides (Meisn.) Prain, Ann. Bot. (Usteri) 27: 407 (1913 publ. 1914).
Adenocline bupleuroides var. *peglerae* Prain, Ann. Bot. (Usteri) 27: 408 (1913 publ. 1914).

var. **cuneata** Pax & K.Hoffm. in H.G.A.Engler, Pflanzenr., IV, 147, VII: 412 (1914).
KwaZulu-Natal. 27 NAT. Cham.

var. **ovalifolia** (Turcz.) Müll.Arg. in A.P.de Candolle, Prodr. 15(2): 1139 (1866).
KwaZulu-Natal. 27 NAT. Cham.
* *Adenocline ovalifolia* Turcz., Bull. Soc. Imp. Naturalistes Moscou 16: 60 (1843).

Mercurialis serrata Meisn., Hooker's J. Bot. Kew Gard. Misc. 2: 557 (1843). *Adenocline serrata* (Meisn.) Turcz., Bull. Soc. Imp. Naturalistes Moscou 25(2): 180 (1852). *Adenocline pauciflora* var. *serrata* (Meisn.) Müll.Arg. in A.P.de Candolle, Prodr. 15(2): 1139 (1866).

var. **pauciflora**
Cape Prov. 27 CPP. Ther. or cham.
Adenocline humilis Turcz., Bull. Soc. Imp. Naturalistes Moscou 16: 61 (1843). *Adenocline pauciflora* var. *humilis* (Turcz.) Müll.Arg. in A.P.de Candolle, Prodr. 15(2): 1140 (1866).
Mercurialis pauciflora Baill., Adansonia 3: 159 (1863).

var. **rotundifolia** Müll.Arg. in A.P.de Candolle, Prodr. 15(2): 1139 (1866). *Adenocline ovalifolia* var. *rotundifolia* (Müll.Arg.) Prain, Ann. Bot. (Usteri) 27: 406 (1913 publ. 1914).
Cape Prov. 27 CPP. Hemicr. or cham.

var. **sessilifolia** (Turcz.) Müll.Arg. in A.P.de Candolle, Prodr. 15(2): 1140 (1866).
Cape Prov. 27 CPP. Hemicr.
* *Adenocline sessilifolia* Turcz., Bull. Soc. Imp. Naturalistes Moscou 16: 61 (1843).
Adenocline sessiliflora Baill., Étude Euphorb.: 457 (1858).

var. **stricta** (Prain) Pax & K.Hoffm. in H.G.A.Engler, Pflanzenr., IV, 147, VII: 412 (1914).
Cape Prov. 27 CPP. Cham.
* *Adenocline stricta* Prain, Bull. Misc. Inform. Kew 1912: 338 (1912).

var. **tenella** (Meisn.) Müll.Arg. in A.P.de Candolle, Prodr. 15(2): 1140 (1866).
Cape Prov. 27 CPP.
* *Mercurialis tenella* Meisn., Hooker's J. Bot. Kew Gard. Misc. 2: 556 (1843).

var. **transiens** Müll.Arg. in A.P.de Candolle, Prodr. 15(2): 1140 (1866).
Cape Prov. 27 CPP. Ther. or cham.
Mercurialis zeyheri Kunze, Linnaea 20: 54 (1847). *Adenocline pauciflora* var. *zeyheri* (Kunze) Pax & K.Hoffm. in H.G.A.Engler, Pflanzenr., IV, 147, VII: 412 (1914), nom. illeg. *Adenocline zeyheri* (Kunze) Prain, Ann. Bot. (Usteri) 27: 408 (1913 publ. 1914).

Adenocline violifolia (Kunze) Prain, Ann. Bot. (Usteri) 27: 403 (1913 publ. 1914).
Cape Prov. 27 CPP. Ther.
Acalypha obtusata Spreng. ex Steud., Nomencl. Bot., ed. 2, 1: 10 (1840), nom. nud.
* *Mercurialis violifolia* Kunze, Index Seminum (LZ) 1846: (1846).
Mercurialis tricocca E.Mey. ex Sond., Linnaea 23: 111 (1850). *Adenocline mercurialis* Baill., Étude Euphorb.: 457 (1858).
Adenocline procumbens Benth. ex Pax in H.G.A.Engler & K.A.E.Prantl, Nat. Pflanzenfam. 3(5): 49 (1890), no basionym.

Synonyms:
Adenocline bupleuroides (Meisn.) Prain === **Adenocline pauciflora** var. **bupleuroides** (Meisn.) Müll.Arg.
Adenocline bupleuroides var. *peglerae* Prain === **Adenocline pauciflora** var. **bupleuroides** (Meisn.) Müll.Arg.
Adenocline humilis Turcz. === **Adenocline pauciflora** Turcz. var. **pauciflora**
Adenocline mercurialis Turcz. === **Adenocline acuta** (Thunb.) Baill.
Adenocline mercurialis Baill. === **Adenocline violifolia** (Kunze) Prain
Adenocline ovalifolia Turcz. === **Adenocline pauciflora** var. **ovalifolia** (Turcz.) Müll.Arg.
Adenocline ovalifolia var. *rotundifolia* (Müll.Arg.) Prain === **Adenocline pauciflora** var. **rotundifolia** Müll.Arg.
Adenocline pauciflora var. *humilis* (Turcz.) Müll.Arg. === **Adenocline pauciflora** Turcz. var. **pauciflora**
Adenocline pauciflora var. *serrata* (Meisn.) Müll.Arg. === **Adenocline pauciflora** var. **ovalifolia** (Turcz.) Müll.Arg.
Adenocline pauciflora var. *zeyheri* (Kunze) Pax & K.Hoffm. === **Adenocline pauciflora** var. **transiens** Müll.Arg.

Adenocline procumbens Benth. ex Pax === **Adenocline violifolia** (Kunze) Prain
Adenocline procumbens (L.) Druce === **Leidesia procumbens** (L.) Prain
Adenocline serrata (Meisn.) Turcz. === **Adenocline pauciflora** var. **ovalifolia** (Turcz.) Müll.Arg.
Adenocline sessiliflora Baill. === **Adenocline pauciflora** var. **sessilifolia** (Turcz.) Müll.Arg.
Adenocline sessilifolia Turcz. === **Adenocline pauciflora** var. **sessilifolia** (Turcz.) Müll.Arg.
Adenocline stricta Prain === **Adenocline pauciflora** var. **stricta** (Prain) Pax & K.Hoffm.
Adenocline zeyheri (Kunze) Prain === **Adenocline pauciflora** var. **transiens** Müll.Arg.

Adenocrepis

Synonyms:
Adenocrepis Blume === **Baccaurea** Lour.
Adenocrepis javanica Blume === **Baccaurea javanica** (Blume) Müll.Arg.
Adenocrepis lanceolata (Miq.) Müll.Arg. === **Baccaurea lanceolata** (Miq.) Müll.Arg.
Adenocrepis tetrandra Baill. === **Baccaurea tetrandra** (Baill.) Müll.Arg.

Adenogyne

Synonyms:
Adenogyne Klotzsch === **Sebastiania** Spreng.
Adenogyne discolor Klotzsch === **Sebastiania klotzschiana** (Müll.Arg.) Müll.Arg.
Adenogyne marginata Klotzsch === **Sebastiania klotzschiana** (Müll.Arg.) Müll.Arg.
Adenogyne pachystachya Klotzsch === **Sebastiania pachystachys** (Klotzsch) Müll.Arg.

Adenogynum

Synonyms:
Adenogynum Rchb.f. & Zoll. === **Cladogynos** Zipp. ex Span.

Adenopeltis

1 species, central Chile; a bushy glabrous shrub to 1 m or so with erect branches and coarsely denticulate leaves (illustrated by Pax 1912). The genus is related to *Colliguaja*, more widely distributed in South America. *A. serrata* was originally described from plants grown at Kew from an introduction by Archibald Menzies. The supposed Peruvian record is based on a Dombey collection; as he was also active in Chile it was probably wrongly labelled. (Euphorbioideae (except Euphorbieae))

Pax, F. (with K. Hoffmann) (1912). *Adenopeltis*. In A. Engler (ed.), Das Pflanzenreich, IV 147 V (Euphorbiaceae-Hippomaneae): 264-265, illus. Berlin. (Heft 52.) La/Ge. — 1 species, Chile and (?) Peru).

Adenopeltis Bertero ex A.Juss., Ann. Sc. Nat. (Paris) 25: 24 (1832).
Chile. 85.

Adenopeltis serrata (Aiton) I.M.Johnst., Contr. Gray Herb. 68: 84 (1923).
C. Chile. 85 CLC. Nanophan.
* *Excoecaria serrata* W.T.Aiton, Hortus Kew. 5: 418 (1813).
Adenopeltis colliguaya Bertero ex A.Juss., Ann. Sc. Nat. (Paris) 25: 24 (1832). *Excoecaria colliguaya* (Bertero ex A.Juss.) Baill., Hist. Pl. 5: 134 (1874).
Excoecaria marginata Kunze ex Baill., Étude Euphorb.: 532 (1858).
Stillingia glandulosa Dombey ex Baill., Étude Euphorb., Atlas: 17 (1858).

Synonyms:
Adenopeltis colliguaya Bertero ex A.Juss. === **Adenopeltis serrata** (Aiton) I.M.Johnst.

Adenopetalum

A Klotzsch segregate from *Euphorbia*. Being a later homonym of *Adenopetalum* Turcz. (= *Vitis*, Vitaceae), Millspaugh (1916; in Publ. Field Mus. Nat. Hist. 2: 412-415) substituted *Eumecanthus* Klotzsch & Garcke (q.v.).

Synonyms:
Adenopetalum Klotzsch & Garcke === **Euphorbia** L.
Adenopetalum barnesii Millsp. === **Euphorbia ocymoidea** L.
Adenopetalum boerhaviifolium Klotzsch & Garcke === **Euphorbia boerhaviifolia** (Klotzsch & Garcke) Boiss.
Adenopetalum bracteatum Klotzsch & Garcke === **Euphorbia triphylla** (Klotzsch & Garcke) Oudejans
Adenopetalum discolor Klotzsch & Garcke === **Euphorbia graminea** Jacq. var. **graminea**
Adenopetalum hoffmannii Klotzsch & Garcke === **Euphorbia graminea** Jacq. var. **graminea**
Adenopetalum pubescens Klotzsch & Garcke === **Euphorbia graminea** Jacq. var. **graminea**
Adenopetalum subsinuatum Klotzsch & Garcke === **Euphorbia graminea** Jacq. var. **graminea**

Adenophaedra

3 species, Middle and South America, from Costa Rica to Brazil and Peru; shrubs or small trees with moderately large, more or less tufted leaves, most closely related to *Bernardia* (Punt 1962; see **Special**). *A. grandifolia* is by far the most widely distributed species; the remainder are Brazilian. No recent treatments are available apart from floras and then mostly for *A. grandifolia*. (Acalyphoideae)

Pax, F. (with K. Hoffmann) (1914). *Adenophaedra*. In A. Engler (ed.), Das Pflanzenreich, IV 147 VII (Euphorbiaceae-Acalypheae-Mercurialinae): 261-263. Berlin. (Heft 63.) La/Ge. — 2 species, S America.

Jablonski, E. (1967). *Adenophaedra*. Euphorbiaceae, Guayana Highland (Mem. New York Bot. Gard. 17(1)): 140. New York. En. — 1 species, *A. grandifolia*.

Adenophaedra (Müll.Arg.) Müll.Arg. in C.F.P.von Martius, Fl. Bras. 11(2): 385 (1874).
Costa Rica to Brazil. 80 82 93 94.

Adenophaedra grandifolia (Klotzsch) Müll.Arg. in C.F.P.von Martius, Fl. Bras. 11(2): 386 (1874).
Costa Rica, Panama, Guyana (Mt. Roraima), Venezuela (Bolivar), Guiana, Brazil (Venezuela), Peru. 80 COS PAN 82 FRG GUY VEN 83 PER 84 BZN. Nanophan. or phan.
* *Tragia grandifolia* Klotzsch, Hooker's J. Bot. Kew Gard. Misc. 2: 46 (1843). *Bernardia grandifolia* (Klotzsch) Müll.Arg., Linnaea 34: 173 (1865).
Cleidion denticulatum Standl., Publ. Field Mus. Nat. Hist., Bot. Ser. 4: 218 (1929). *Bernardia denticulata* (Standl.) G.L.Webster, Ann. Missouri Bot. Gard. 54: 200 (1967).

Adenophaedra megalophylla (Müll.Arg.) Müll.Arg. in C.F.P.von Martius, Fl. Bras. 11(2): 386 (1874).
Brazil (Bahia). 84 BZE. Nanophan. or phan.
* *Bernardia megalophylla* Müll.Arg., Linnaea 34: 173 (1865).

Adenophaedra prealta (Croizat) Croizat, Trop. Woods 88: 31 (1946).
Brazil (Amazonas). 84 BZN. Nanophan.
* *Cleidion prealtum* Croizat, J. Arnold Arbor. 24: 167 (1943).

Synonyms:
Adenophaedra minor Ducke === **Tetrorchidium duckei** Radcl.-Sm. & Govaerts
Adenophaedra woodsoniana (Croizat) Croizat === **Cleidion membranaceum** Pax & K.Hoffm.

Adenorhopium

Synonyms:
Adenorhopium Rchb. === **Jatropha** L.

Adenorima

Synonyms:
Adenorima Raf. === **Euphorbia** L.

Adenoropium

Synonyms:
Adenoropium Pohl === **Jatropha** L.
Adenoropium divergens Pohl === **Jatropha mollissima** (Pohl) Baill. var. **mollissima**
Adenoropium elegans Pohl === **Jatropha gossypiifolia** var. **elegans** (Pohl) Müll.Arg.
Adenoropium ellipticum Pohl === **Jatropha elliptica** (Pohl) Oken
Adenoropium luxurians Pohl === **Jatropha mollissima** (Pohl) Baill. var. **mollissima**
Adenoropium martiusii Pohl === **Jatropha martiusii** (Pohl) Baill.
Adenoropium mollissimum Pohl === **Jatropha mollissima** (Pohl) Baill.
Adenoropium mutabile Pohl === **Jatropha mutabilis** (Pohl) Baill.
Adenoropium ribifolium Pohl === **Jatropha ribifolia** (Pohl) Baill.
Adenoropium villosum Pohl === **Jatropha mollissima** (Pohl) Baill. var. **mollissima**

Adisa

Synonyms:
Adisa Steud. === **Mallotus** Lour.

Adisca

Synonyms:
Adisca Blume === **Mallotus** Lour.
Adisca acuminata Blume === **Mallotus peltatus** (Geiseler) Müll.Arg.
Adisca albicans Blume === **Sumbaviopsis albicans** (Blume) J.J.Sm.
Adisca floribunda Blume === **Mallotus floribundus** (Blume) Müll.Arg.
Adisca subpeltata Blume === **Mallotus subpeltatus** (Blume) Müll.Arg.

Adriana

2 species, Australia; small to medium-sized dioecious shrubs with unlobed to 3-5-lobed leaves. Recently revised by Gross & Whalen (1996). *A. tomentosa* var. *hookeri* is widely distributed in interior dry grassland and desert areas. The southern *A. quadripartita* has been used for landscaping in South Australia. In the Webster system it is considered related to *Ricinus*. A recent revision is that by Gross & Whalen (1996). (Acalyphoideae)

Pax, F. (1910). *Adriana*. In A. Engler (ed.), Das Pflanzenreich, IV 147 II (Euphorbiaceae-Adrianae): 17-21. Berlin. (Heft 44.) La/Ge. — 5 species in 2 sections, Australia. [Now superseded.]

* Gross, C. L. & M. A. Whalen (1996). A revision of *Adriana*. Austral. Syst. Bot. 9: 749-771, illus. En. — Introduction; character review and general notes on distribution, habitat, conservation and ethnobiology; revision of 2 species (one with 1 additional variety) including key, synonymy, types, descriptions, phenology, illustration references, representative exsiccatae, overall distribution, and commentary; literature list at end but no separate index. [Succeeds the treatments of Pax (1910) and Airy-Shaw (1980; see **Australasia**).]

Adriana Gaudich., Ann. Sc. Nat. (Paris) 5: 223 (1825).
Australia. 50.
Meialisa Raf., Sylva Tellur.: 63 (1838).
Trachycaryon Klotzsch in J.G.C.Lehmann, Pl. Preiss. 1: 175 (1845).

Adriana quadripartita (Labill.) Gaudich., Voy. Uranie 12: 489 (1830).
S. Australia. 50 SOA VIC WAU. Nanophan.
* *Croton quadripartitus* Labill., Nov. Holl. Pl. 2: 73 (1806).
Trachycaryon klotzschii F.Muell., Trans. Philos. Soc. Victoria 1: 15 (1855). *Adriana klotzschii* (F.Muell.) Müll.Arg. in A.P.de Candolle, Prodr. 15(2): 892 (1866).
Adriana billardieri Baill., Étude Euphorb., Atlas: 6 (1858).
Adriana bloudowskyana Müll.Arg. in H.G.A.Engler, Pflanzenr., IV, 147, II: 21 (1910).

Adriana tomentosa (Thunb.) Gaudich., Ann. Sc. Nat. (Paris) 5: 223 (1825). – FIGURE, p. 132.
Australia. 50 NSW NTA QLD SOA VIC WAU. Nanophan.
* *Ricinus tomentosus* Thunb., Ricin.: 6 (1815). *Adriana gaudichaudii* Baill., Adansonia 6: 312 (1866).

var. **hookeri** (F.Muell.) C.L.Gross & M.A.Whalen, Austr. Syst. Bot. 9: 767 (1996).
Australia. 50 NSW NTA QLD SOA VIC WAU. Nanophan.
* *Trachycaryon hookeri* F.Muell., Trans. Philos. Soc. Victoria 1: 16 (1855). *Adriana hookeri* (F.Muell.) Müll.Arg. in A.P.de Candolle, Prodr. 15(2): 891 (1866).
Trachycaryon hookeri var. *glabriusculum* F.Muell., Hooker's J. Bot. Kew Gard. Misc. 8: 210 (1856). *Adriana hookeri* var. *glabriuscula* (F.Muell.) Müll.Arg. in A.P.de Candolle, Prodr. 15(2): 891 (1866).
Trachycaryon hookeri var. *velutinum* F.Muell., Hooker's J. Bot. Kew Gard. Misc. 8: 210 (1856).

var. **tomentosa**
Australia. 50 NSW NTA SOA VIC WAU. Nanophan.
Adriana glabrata Gaudich., Ann. Sc. Nat. (Paris) 5: 223 (1825). *Adriana acerifolia* var. *glabrata* (Gaudich.) Benth., Fl. Austral. 6: 134 (1873).
Croton urticoides A.Cunn. ex Steud., Nomencl. Bot., ed. 2, 1: 447 (1840).
Adriana acerifolia Hook. in T.L.Mitchell, J. Exped. Trop. Australia: 371 (1848). *Adriana acerifolia* var. *genuina* Müll.Arg. in A.P.de Candolle, Prodr. 15(2): 890 (1866), nom. inval. *Adriana glabrata* var. *acerifolia* (Hook.) Pax in H.G.A.Engler, Pflanzenr., IV, 147, II: 18 (1910).
Adriana heterophylla Hook. in T.L.Mitchell, J. Exped. Trop. Australia: 371 (1848). *Adriana glabrata* var. *heterophylla* (Hook.) Müll.Arg. in A.P.de Candolle, Prodr. 15(2): 891 (1866).
Croton acerifolius A.Cunn. ex Hook. in T.L.Mitchell, J. Exped. Trop. Australia: 371 (1848).
Trachycaryon cunninghamii F.Muell., Trans. Philos. Soc. Victoria 1: 15 (1855). *Adriana glabrata* var. *cunninghamii* (F.Muell.) Müll.Arg. in A.P.de Candolle, Prodr. 15(2): 890 (1866).
Trachycaryon cunninghamii var. *glabrum* F.Muell., Trans. Philos. Soc. Victoria 1: 15 (1855).
Trachycaryon cunninghamii var. *tomentosum* F.Muell., Trans. Philos. Soc. Victoria 1: 15 (1855).

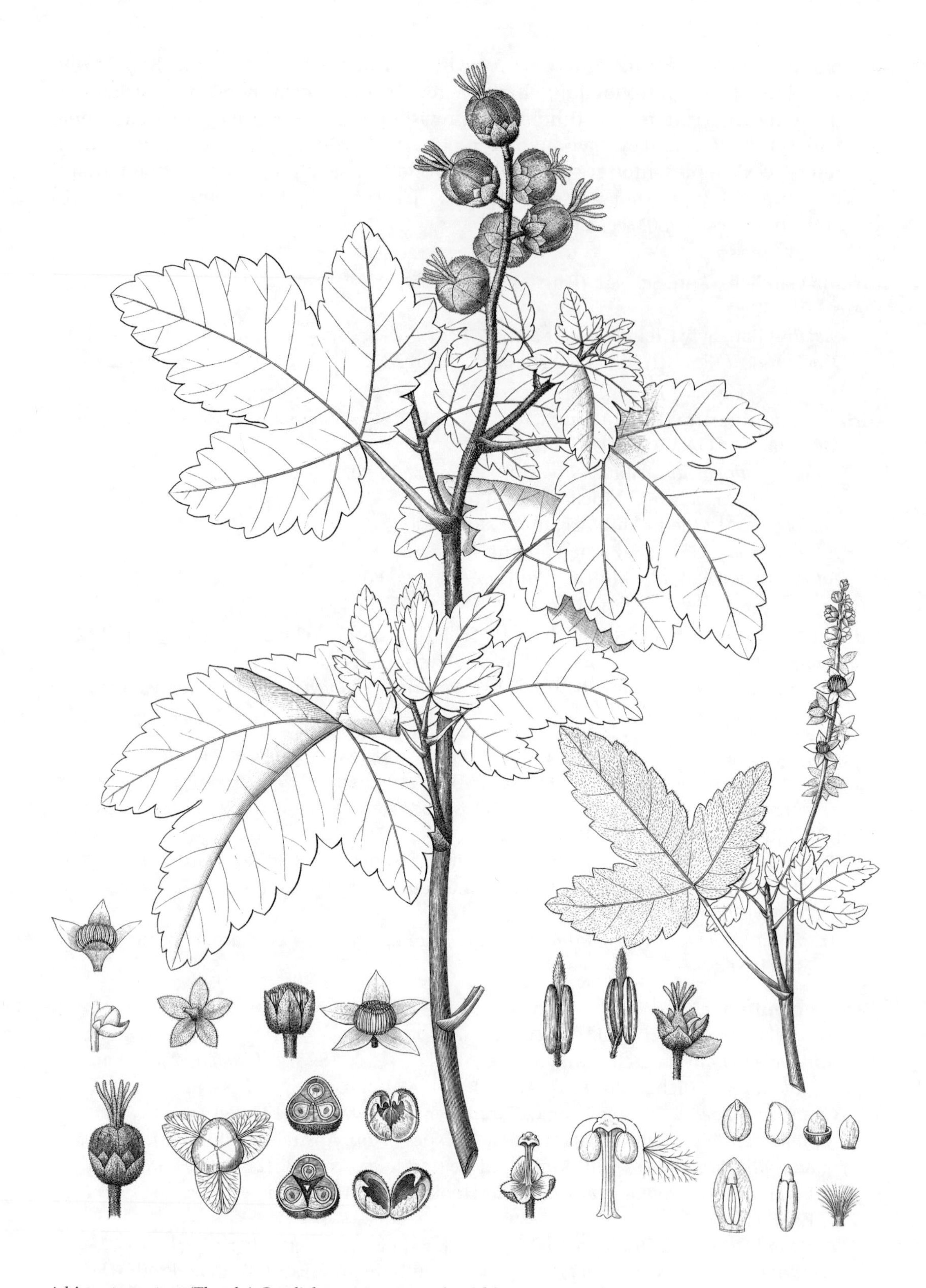

Adriana tomentosa (Thunb.) Gaudich. var. *tomentosa* (as *Adriana tomentosa*)
Artist: A. Poiret fils
Gaudichaud, Voy. Uranie, Atlas, pl. 116 (1830)
KEW ILLUSTRATIONS COLLECTION

Adriana acerifolia var. *lessonii* Baill., Adansonia 6: 313 (1866).
Adriana acerifolia var. *puberula* Müll.Arg. in A.P.de Candolle, Prodr. 15(2): 891 (1866).
Adriana gaudichaudii var. *subglabra* Baill., Adansonia 6: 312 (1866). *Adriana glabrata* var. *subglabra* (Baill.) Airy Shaw, Kew Bull. 35: 591 (1980).

Synonyms:
Adriana acerifolia Hook. === **Adriana tomentosa** (Thunb.) Gaudich. var. **tomentosa**
Adriana acerifolia var. *genuina* Müll.Arg. === **Adriana tomentosa** (Thunb.) Gaudich. var. **tomentosa**
Adriana acerifolia var. *glabrata* (Gaudich.) Benth. === **Adriana tomentosa** (Thunb.) Gaudich. var. **tomentosa**
Adriana acerifolia var. *lessonii* Baill. === **Adriana tomentosa** (Thunb.) Gaudich. var. **tomentosa**
Adriana acerifolia var. *puberula* Müll.Arg. === **Adriana tomentosa** (Thunb.) Gaudich. var. **tomentosa**
Adriana billardieri Baill. === **Adriana quadripartita** (Labill.) Gaudich.
Adriana bloudowskyana Müll.Arg. === **Adriana quadripartita** (Labill.) Gaudich.
Adriana gaudichaudii Baill. === **Adriana tomentosa** (Thunb.) Gaudich.
Adriana gaudichaudii var. *subglabra* Baill. === **Adriana tomentosa** (Thunb.) Gaudich. var. **tomentosa**
Adriana gaudichaudii var. *thomasiifolia* Baill. === ?
Adriana glabrata Gaudich. === **Adriana tomentosa** (Thunb.) Gaudich. var. **tomentosa**
Adriana glabrata var. *acerifolia* (Hook.) Pax === **Adriana tomentosa** (Thunb.) Gaudich. var. **tomentosa**
Adriana glabrata var. *cunninghamii* (F.Muell.) Müll.Arg. === **Adriana tomentosa** (Thunb.) Gaudich. var. **tomentosa**
Adriana glabrata var. *heterophylla* (Hook.) Müll.Arg. === **Adriana tomentosa** (Thunb.) Gaudich. var. **tomentosa**
Adriana glabrata var. *subglabra* (Baill.) Airy Shaw === **Adriana tomentosa** (Thunb.) Gaudich. var. **tomentosa**
Adriana heterophylla Hook. === **Adriana tomentosa** (Thunb.) Gaudich. var. **tomentosa**
Adriana hookeri (F.Muell.) Müll.Arg. === **Adriana tomentosa** var. **hookeri** (F.Muell.) C.L.Gross & M.A.Whalen
Adriana hookeri var. *glabriuscula* (F.Muell.) Müll.Arg. === **Adriana tomentosa** var. **hookeri** (F.Muell.) C.L.Gross & M.A.Whalen
Adriana klotzschii (F.Muell.) Müll.Arg. === **Adriana quadripartita** (Labill.) Gaudich.

Aegopicron

An orthographic variant of *Aegopricum.*

Aegopricon

An orthographic variant of *Aegopricum.*

Aegopricum

Synonyms:
Aegopricum L. === **Maprounea** Aubl.

Aerisilvaea

Reduced in 1997 to *Lingelsheimia.*

Synonyms:
Aerisilvaea Radcl.-Sm. === **Lingelsheimia** Pax
Aerisilvaea serrata Radcl.-Sm. === **Maytenus undata** (Thunb.) Blakelock (Celastraceae)
Aerisilvaea sylvestris Radcl.-Sm. === **Lingelsheimia sylvestris** (Radcl.-Sm.) Radcl.-Sm.

Afrotrewia

1 species, Africa (southern Cameroon); small trees with stellate indumentum. Listed in Lebrun and Stork, *Énumeration des plantes à fleurs d'Afrique tropicale* 1 (1991) and accepted by Brummitt in *Vascular plant families and genera* (1992) and in USDA but missing from Webster (Synopsis, 1994; see **General**). All original material apparently destroyed in Berlin and no new collections yet known. The genus was allied by Pax with *Necepsia* and *Neopalissya* (= *Necepsia*), also in Africa. (Acalyphoideae)

Pax, F. (with K. Hoffmann) (1914). *Afrotrewia*. In A. Engler (ed.), Das Pflanzenreich, IV 147 VII (Euphorbiaceae-Acalypheae-Mercurialinae): 14. Berlin. (Heft 63.) La/Ge. — 1 species, Africa (Cameroon).

Afrotrewia Pax & K.Hoffm. in H.G.A.Engler, Pflanzenr., IV, 147, VII: 14 (1914).
WC. Trop. Africa. 23.

Afrotrewia kamerunica Pax & K.Hoffm. in H.G.A.Engler, Pflanzenr., IV, 147, VII: 14 (1914).
Cameroon. 23 CMN. Phan.

Agaloma

Usually treated as a subgenus of *Euphorbia*. Most synonyms have been omitted here.

Synonyms:
Agaloma Raf. === **Euphorbia** L.
Agaloma purpurea Raf. === **Euphorbia purpurea** (Raf.) Fernald
Agaloma marginata (Pursh) A.Löve & D.Löve === **Euphorbia marginata** Pursh

Agelandra

An orthographic variant of *Angelandra*.

Agirta

Synonyms:
Agirta Baill. === **Tragia** Plum. ex L.

Agrostistachys

8 species, S Asia and Malesia (Sri Lanka and India to New Guinea), shrubs or small to medium trees. *AA. borneensis* and *indica* are the most widely distributed species. *A. sessilifolia* is monocaulous, a representative of the 'Corner Model' with narrow spathulate-oblanceolate leaves to 50 × 10 cm. Last revised in 1912 but as of writing (1998) has been under study by Sofia Sevilla as part of the collective revision of the family led by the Rijksherbarium/Hortus Botanicus (Leiden). Webster (Synopsis, 1994) has reported that some species are hard to distinguish. The two sections recognised by Pax appear to be without value, with reductions of taxa having taken place from one to the other. (Acalyphoideae)

Pax, F. (with K. Hoffmann) (1912). *Agrostistachys*. In A. Engler (ed.), Das Pflanzenreich, IV 147 VI (Euphorbiaceae-Acalypheae-Chrozophorinae): 98-105. Berlin. (Heft 57.) La/Ge. — 11 species in 2 subgenera, S Asia and Malesia.

Airy-Shaw, H. K. (1960). Notes on Malaysian Euphorbiaceae, XVII. The identity of the genus *Heterocalyx* Gagnep. Kew Bull. 14: 472. En. —*Heterocalyx* a synonym of *Agrostistachys*, the single species being part of *A. indica*.

Airy-Shaw, H. K. (1966). Notes on Malaysian and other Asiatic Euphorbiaceae, LVII. *Agrostistachys sessiliflora* in Sumatra and Borneo. Kew Bull. 20: 26. En. — Range extensions.

Airy-Shaw, H. K. (1974). Notes on Malesian and other Asiatic Euphorbiaceae, CLXXVIII. Range extensions for *Agrostistachys indica* Dalz. Kew Bull. 29: 312-313. En. — Includes complete reduction of *A. maesoana* Vidal (Philippines); discussion of its distribution.

Agrostistachys Dalzell, Hooker's J. Bot. Kew Gard. Misc. 2: 41 (1850).
Trop. Asia. 40 41 42.
Sarcoclinium Wight, Icon. Pl. Ind. Orient. 5(2): 24 (1852).
Heterocalyx Gagnep., Notul. Syst. (Paris) 14: 33 (1950).

Agrostistachys borneensis Becc., Nelle Foreste di Borneo: 331 (1904).
India, Sri Lanka, Malesia. 40 IND SRL 42 BOR MLY NWG PHI. Phan.
Sarcoclinium longifolium Wight, Icon. Pl. Ind. Orient. 5: t. 1887-1888 (1852). *Agrostistachys longifolia* (Wight) Trimen, Syst. Cat. Fl. Pl. Ceylon: 81 (1885), nom. illeg.
Agrostistachys longifolia var. *latifolia* Hook.f., Fl. Brit. India 5: 407 (1887). *Agrostistachys latifolia* (Hook.f.) Pax & K.Hoffm. in H.G.A.Engler, Pflanzenr., IV, 147, VI: 100 (1912).
Agrostistachys leptostachya Pax & K.Hoffm. in H.G.A.Engler, Pflanzenr., IV, 147, VI: 102 (1912). *Agrostistachys longifolia* var. *leptostachya* (Pax & K.Hoffm.) Whitmore, Gard. Bull. Singapore 26: 52 (1972).

Agrostistachys coriacea Alston in H.Trimen, Handb. Fl. Ceylon 6(Suppl.): 265 (1931).
Sri Lanka. 40 SRL.

Agrostistachys gaudichaudii (Baill.) Müll.Arg., Linnaea 34: 144 (1865).
S. Thailand, Pen. Malaysia. 41 THA 42 MLY. Nanophan.
* *Sarcoclinium gaudichaudii* Baill., Étude Euphorb.: 310 (1858).
Agrostistachys longifolia Kurz, Forest Fl. Burma 2: 377 (1877).
Agrostistachys maingayi Hook.f., Fl. Brit. India 5: 406 (1887).
Agrostistachys filipendula Hook.f., Fl. Brit. India 5: 487 (1888).

Agrostistachys hookeri (Thwaites) Benth. & Hook.f., Gen. Pl. 3: 303 (1880).
Sri Lanka. 40 SRL. Phan.
* *Sarcoclinium hookeri* Thwaites, Enum. Pl. Zeyl.: 279 (1861).

Agrostistachys indica Dalzell, Hooker's J. Bot. Kew Gard. Misc. 2: 41 (1850). *Agrostistachys indica* var. *genuina* Müll.Arg. in A.P.de Candolle, Prodr. 15(2): 726 (1866), nom. inval. *Agrostistachys indica* subsp. *genuina* (Müll.Arg.) Pax & K.Hoffm.in H.G.A.Engler, Pflanzenr., IV, 147, VI: 104 (1912), nom. inval. – FIGURE, p. 136.
SW. India, Sri Lanka, Burma, Thailand, Vietnam, Borneo (Banguey Is.), Philippines, New Guinea. 40 IND SRL 41 BMA THA VIE 42 BOR NWG PHI. Nanophan. or phan.
Agrostistachys indica var. *longifolia* Müll.Arg. in A.P.de Candolle, Prodr. 15(2): 726 (1866). *Agrostistachys longifolia* (Müll.Arg.) Kurz, Prelim. Rep. Forest Pegu, App. A: 91 (1875). *Agrostistachys indica* subsp. *longifolia* (Müll.Arg.) Pax & K.Hoffm. in H.G.A.Engler, Pflanzenr., IV, 147, VI: 105 (1912).
Agrostistachys maesoana Vidal, Revis. Pl. Vasc. Filip.: 342 (1886). *Agrostistachys indica* var. *maesoana* (Vidal) Pax & K.Hoffm. in H.G.A.Engler, Pflanzenr., IV, 147, VI: 105 (1912).
Agrostistachys indica var. *subintegra* Pax & K.Hoffm. in H.G.A.Engler, Pflanzenr., IV, 147, VI: 105 (1912).

Agrostistachys intramarginalis Philcox, Kew Bull. 50: 119 (1995).
Sri Lanka. 40 SRL. Nanophan. or phan.

Agrostistachys meeboldii Pax & K.Hoffm. in H.G.A.Engler, Pflanzenr., IV, 147, VI: 100 (1912).
India (Kerala). 40 IND. Nanophan.

Agrostistachys sessilifolia (Kurz) Pax & K.Hoffm. in H.G.A.Engler, Pflanzenr., IV, 147, VI: 102 (1912).

Agrostistachys indica Dalzell
Artist: W. Stocks
Unpubl.
KEW ILLUSTRATIONS COLLECTION

Pen. Malaysia (incl. Singapore), Sumatera, Borneo. 42 BOR MLY SUM. Nanophan. or phan.
* *Sarcoclinium sessilifolium* Kurz, Flora 58: 31 (1875).

var. **graciliflora** Airy Shaw, Kew Bull., Addit. Ser. 4: 27 (1975).
Borneo (Sabah, E. Kalimantan). 42 BOR. Nanophan. or phan.

var. **sessilifolia**
Pen. Malaysia (incl. Singapore), Sumatera. 42 MLY SUM. Nanophan. or phan.

Synonyms:
Agrostistachys africana Müll.Arg. === **Pseudagrostistachys africana** (Müll.Arg.) Pax & K.Hoffm.
Agrostistachys comorensis Pax === **Tannodia cordifolia** (Baill.) Baill.
Agrostistachys filipendula Hook.f. === **Agrostistachys gaudichaudii** (Baill.) Müll.Arg.
Agrostistachys indica subsp. *genuina* (Müll.Arg.) Pax & K.Hoffm. === **Agrostistachys indica** Dalzell
Agrostistachys indica var. *genuina* Müll.Arg. === **Agrostistachys indica** Dalzell
Agrostistachys indica var. *longifolia* Müll.Arg. === **Agrostistachys indica** Dalzell
Agrostistachys indica subsp. *longifolia* (Müll.Arg.) Pax & K.Hoffm. === **Agrostistachys indica** Dalzell
Agrostistachys indica var. *maesoana* (Vidal) Pax & K.Hoffm. === **Agrostistachys indica** Dalzell
Agrostistachys indica var. *subintegra* Pax & K.Hoffm. === **Agrostistachys indica** Dalzell
Agrostistachys latifolia (Hook.f.) Pax & K.Hoffm. === **Agrostistachys borneensis** Becc.
Agrostistachys leptostachya Pax & K.Hoffm. === **Agrostistachys borneensis** Becc.
Agrostistachys longifolia Kurz === **Agrostistachys gaudichaudii** (Baill.) Müll.Arg.
Agrostistachys longifolia (Wight) Trimen === **Agrostistachys borneensis** Becc.
Agrostistachys longifolia (Müll.Arg.) Kurz === **Agrostistachys indica** Dalzell
Agrostistachys longifolia var. *latifolia* Hook.f. === **Agrostistachys borneensis** Becc.
Agrostistachys longifolia var. *leptostachya* (Pax & K.Hoffm.) Whitmore === **Agrostistachys borneensis** Becc.
Agrostistachys maesoana Vidal === **Agrostistachys indica** Dalzell
Agrostistachys maingayi Hook.f. === **Agrostistachys gaudichaudii** (Baill.) Müll.Arg.
Agrostistachys pubescens Merr. === **Wetria insignis** (Steud.) Airy Shaw
Agrostistachys ugandensis Hutch. === **Pseudagrostistachys ugandensis** (Hutch.) Pax & K.Hoffm.

Agyneia

Rejected against *Glochidion.*

Synonyms:
Agyneia L. === **Glochidion** J.R.Forst. & G.Forst.
Agyneia affinis Kurz ex Teijsm. & Binn. === **Sauropus bacciformis** (L.) Airy Shaw
Agyneia bacciformis (L.) A.Juss. === **Sauropus bacciformis** (L.) Airy Shaw
Agyneia bacciformis var. *angustifolia* Müll.Arg. === **Sauropus bacciformis** (L.) Airy Shaw
Agyneia bacciformis var. *genuina* Müll.Arg. === **Sauropus bacciformis** (L.) Airy Shaw
Agyneia bacciformis var. *oblongifolia* Müll.Arg. === **Sauropus bacciformis** (L.) Airy Shaw
Agyneia berteri Spreng. === **Phyllanthus grandifolius** L.
Agyneia ciliata Wall. === **Trigonostemon semperflorens** (Roxb.) Müll.Arg.
Agyneia coccinea Buch.-Ham. === **Glochidion coccineum** (Buch.-Ham.) Müll.Arg.
Agyneia flexuosa B.Heyne ex Wall. === **Glochidion zeylanicum** (Gaertn.) A.Juss. var. **zeylanicum**
Agyneia glomerulata Miq. === **Glochidion glomerulatum** (Miq.) Boerl.
Agyneia gonioclada (Merr. & Chun) H.Keng === **Sauropus bacciformis** (L.) Airy Shaw
Agyneia hirsuta Miq. === **Glochidion zeylanicum** var. **talbotii** (Hook.f.) Haines
Agyneia impuber (Roxb.) Miq. === **Glochidion impuber** (Roxb.) Govaerts
Agyneia impuber Wall. === **Glochidion littorale** Blume

Agyneia impubes L. === **Glochidion puberum** (L.) Hutch.
Agyneia lanceolata F.Dietr. === ?
Agyneia latifolia Moon === **Aporusa latifolia** Thwaites
Agyneia multiflora Hassk. === **Phyllanthus hasskarlianus** Müll.Arg.
Agyneia multilocularis Rottler ex Willd. === **Glochidion multiloculare** (Rottler ex Willd.) Voigt
Agyneia multilocularis Moon === **Aporusa lanceolata** (Tul.) Thwaites
Agyneia obliqua Willd. === **Glochidion zeylanicum** (Gaertn.) A.Juss. var. **zeylanicum**
Agyneia ovata Poir. === **Sauropus androgynus** (L.) Merr.
Agyneia phyllanthoides Spreng. === **Sauropus bacciformis** (L.) Airy Shaw
Agyneia pinnata Miq. === **Glochidion puberum** (L.) Hutch.
Agyneia pubera L. === **Glochidion puberum** (L.) Hutch.
Agyneia pubera Wall. === **Glochidion multiloculare** (Rottler ex Willd.) Voigt var. **multiloculare**
Agyneia sinica Miq. === **Glochidion puberum** (L.) Hutch.
Agyneia taiwaniana H.Keng === **Sauropus bacciformis** (L.) Airy Shaw
Agyneia tenera Zoll. & Moritzi ex Miq. === **Andrachne australis** Zoll. & Moritzi
Agyneia tetrandra Wall. === **Trigonostemon semperflorens** (Roxb.) Müll.Arg.
Agyneia tetrandra Buch.-Ham. === **Phyllanthus sikkimensis** Müll.Arg.

Agyneja

An orthographic variant of *Agyneia.*

Agyneja

of Ventanat = *Sauropus*

Aklema

A segregate of *Euphorbia* within subgen. *Agaloma;* an earlier name for *Alectroctonum* revived by Millspaugh (1916; in Publ. Field Mus. Nat. Hist. 2: 415-417) with numerous new combinations (not, however, listed here).

Synonyms:
Aklema Raf. === **Euphorbia** L.

Alchornea

62 species, tropics (W., C. & S. Africa, Madagascar (3), India to E. Asia and through Malesia (3) to Australia, Mexico to N. Argentina); includes *Coelobogyne* (see Airy-Shaw 1980b under **Australasia**), a genus retained by Webster (Synopsis, 1994). Shrubs or small to large trees (the widely distributed *A. triplinervia* may reach 30 m), the leaves alternate or spirally arranged and inflorescences axillary. Of the 3 sections formed by Pax (1914) only sect. *Alchornea*, with stellate as opposed to simple hairs and 2-locular ovary, is in the Americas. Nguyen Nghia Thin has proposed elevation of at least this section to subgeneric rank. *A. rugosa* is widely distributed in the eastern tropics, *A. cordifolia* similarly in Africa, and (as already noted) *A. triplinervia* in the Americas; the majority, however, appear to be local or simply remain poorly known or documented. There has been apart from flora accounts almost no revisionary work in more than 80 years. According to Webster (Synopsis, 1994), the relationships among these and segregate genera require clarification. (Acalyphoideae)

Pax, F. (with K. Hoffmann) (1914). *Alchornea.* In A. Engler (ed.), Das Pflanzenreich, IV 147 VII (Euphorbiaceae-Acalypheae-Mercurialinae): 220-253. Berlin. (Heft 63.) La/Ge. — 46 species, tropics; the 3 sections have differing foci but only sect. *Alchornea* is in the Americas. [Additions in ibid, XIV (Additamentum VI): 20-21 (1919).]

Pax, F. (with K. Hoffmann) (1914). *Caelebogyne*. In A. Engler (ed.), Das Pflanzenreich, IV 147 VII (Euphorbiaceae-Acalypheae-Mercurialinae): 255-257. Berlin. (Heft 63.) La/Ge. — 2 species, E Australia (one imperfectly known). [Genus merged with *Alchornea* at Kew and in USDA (following Airy-Shaw 1980) but not by Webster, who spells it *Coelebogyne*.

Jablonski, E. (1967). *Alchornea*. Euphorbiaceae, Guayana Highland (Mem. New York Bot. Gard. 17(1)): 137-139. New York. En. — 5 species, none new.

Airy-Shaw, H. K. (1980). Notes on Euphorbiaceae from Indomalesia, Australia and the Pacific, CCXXXIX. *Alchornea* Sw. Kew Bull. 35: 395-396. En. — Description of *A. petalostyla* from Nueva Vizcaya province, from old material.

Alchornea Sw., Prodr.: 98 (1788).
Trop. & Subtrop. 22 23 24 25 26 27 29 36 38 40 41 42 50 79 80 81 82 83 84 85.
Caturus Lour., Fl. Cochinch.: 612 (1790).
Cladodes Lour., Fl. Cochinch.: 574 (1790).
Hermesia Humb. & Bonpl. ex Willd., Sp. Pl. 4: 809 (1806).
Schousboea Schumach. & Thonn. in C.F.Schumacher, Beskr. Guin. Pl.: 449 (1827).
Coelebogyne Js.Sm., Proc. Linn. Soc. Lond. 1: 41 (1839).
Stipellaria Benth., Hooker's J. Bot. Kew Gard. Misc. 6: 2 (1854).
Lepidoturus Bojer ex Baill., Étude Euphorb.: 448 (1858).
Bleekeria Miq., Fl. Ned. Ind. 1(2): 407 (1859), nom. illeg.

Alchornea acroneura Pax & K.Hoffm. in H.G.A.Engler, Pflanzenr., IV, 147, VII: 229 (1914).
Peru (Loreto). 83 PER. Nanophan.

Alchornea acutifolia Müll.Arg., Linnaea 34: 171 (1865).
Peru (San Martin). 83 PER. Cl. nanophan.

Alchornea adenophila Pax & K.Hoffm. in H.G.A.Engler, Pflanzenr., IV, 147, VII: 251 (1914).
Pen. Malaysia, S. Sumatera. 42 MLY SUM. Nanophan.
Alchornea villosa var. *glabrata* Hook.f., Fl. Brit. India 5: 421 (1887).

Alchornea alnifolia (Bojer ex Baill.) Pax & K.Hoffm. in H.G.A.Engler, Pflanzenr., IV, 147, VII: 250 (1914).
Madagascar, Comoros (Mayotte). 29 COM MDG. Nanophan. or phan.
* *Lepidoturus alnifolius* Bojer ex Baill., Étude Euphorb.: 449 (1858).

Alchornea androgyna Croizat, J. Arnold Arbor. 23: 47 (1942).
N. Vietnam. 41 VIE.

Alchornea annamica Gagnep., Bull. Soc. Bot. France 71: 137 (1924).
Vietnam. 41 VIE.

Alchornea aquifolia (Js.Sm.) Domin, Biblioth. Bot. 89: 332 (1927).
SE. Queensland, New South Wales. 50 NSW QLD. Nanophan. or phan.
* *Sapium aquifolium* Js.Sm., Proc. Linn. Soc. Lond. 1: 41 (1839). *Coelebogyne aquifolium* (Js.Sm.) Domin, Biblioth. Bot. 89: 333 (1927).
Coelebogyne ilicifolia Js.Sm., Trans. Linn. Soc. London 18: 512 (1841). *Alchornea ilicifolia* (Js.Sm.) Müll.Arg., Linnaea 34: 170 (1865).
Cladodes thozetiana Baill., Adansonia 6: 321 (1866). *Alchornea thozetiana* (Baill.) Benth., Fl. Austral. 6: 137 (1873). *Coelebogyne thozetiana* (Baill.) Pax & K.Hoffm. in H.G.A.Engler, Pflanzenr., IV, 147, VII: 257 (1914).
Sapium berberifolium Meisn. in A.P.de Candolle, Prodr. 15(2): 907 (1866).
Alchornea thozetiana var. *longifolia* Benth., Fl. Austral. 6: 137 (1873).

Alchornea bogotensis Pax & K.Hoffm. in H.G.A.Engler, Pflanzenr., IV, 147, VII: 235 (1914).
Colombia. 83 CLM. Nanophan.

Alchornea borneensis Pax & K.Hoffm., Mitt. Inst. Allg. Bot. Hamburg 7: 227 (1931).
Borneo (SW. Kalimantan). 42 BOR. Phan. – Close to A. villosa.

Alchornea brevistyla Pax & K.Hoffm. in H.G.A.Engler, Pflanzenr., IV, 147, VII: 227 (1914).
Peru (Loreto). 83 PER. Nanophan.

Alchornea castaneifolia (Humb. & Bonpl. ex Willd.) A.Juss., Euphorb. Gen.: 42 (1824).
Colombia, S. Venezuela, Brazil, Paraguay, Peru. 82 VEN 83 CLM PER 84 BZE BZL BZN 85 PAR. Nanophan.
* *Hermesia castaneifolia* Humb. & Bonpl. ex Willd., Sp. Pl. 4: 809 (1806). *Alchornea castaneifolia* var. *genuina* Müll.Arg. in A.P.de Candolle, Prodr. 15(2): 912 (1866), nom. inval.
Hermesia salicifolia Baill., Étude Euphorb.: 447 (1858). *Alchornea castaneifolia* var. *salicifolia* (Baill.) Baill., Adansonia 5: 238 (1865).
Alchornea castaneifolia var. *puberula* Müll.Arg. in C.F.P.von Martius, Fl. Bras. 11(2): 283 (1873).
Alchornea passargei Pax & K.Hoffm. in H.G.A.Engler, Pflanzenr., IV, 147, VII: 237 (1914).

Alchornea cerifera Croizat, Caldasia 2: 128 (1944).
Colombia. 83 CLM.

Alchornea chiapasana Miranda, Ceiba 4: 131 (1954).
Mexico. 79 MXT.

Alchornea coelophylla Pax & K.Hoffm. in H.G.A.Engler, Pflanzenr., IV, 147, VII: 226 (1914).
Colombia (Cauca). 83 CLM. Phan.

Alchornea columnularis Müll.Arg. in C.F.P.von Martius, Fl. Bras. 11(2): 378 (1874).
Brazil (Amazonas). 84 BZN. Phan.

Alchornea cordifolia (Schumach. & Thonn.) Müll.Arg., Linnaea 34: 170 (1865).
– FIGURE, p. 141.
Trop. Africa. 22 BEN GAM GHA GNB GUI IVO LBR MLI NGA SEN SIE TOG 23 CMN CON EQG GAB GGI RWA ZAI 24 SUD 25 KEN TAN UGA 26 ANG. Nanophan. or phan.
* *Schousboea cordifolia* Schumach. & Thonn. in C.F.Schumacher, Beskr. Guin. Pl.: 449 (1827).
Alchornea cordata Benth. in W.J.Hooker, Niger Fl.: 507 (1849).

Alchornea costaricensis Pax & K.Hoffm. in H.G.A.Engler, Pflanzenr., IV, 147, VII: 235 (1914).
Honduras to Colombia. 80 COS HON PAN 83 CLM. Phan.
Alchornea costaricensis f. *longispicata* Pax & K.Hoffm. in H.G.A.Engler, Pflanzenr., IV, 147, XIV: 20 (1919).

Alchornea davidii Franch., Pl. David. 1: 264 (1884).
S. China. 36 CHS. Nanophan.
Acalypha giraldii Pamp., Nuovo Giorn. Bot. Ital. 15: 438 (1908), nom. illeg.
Acalypha silvestri Pamp., Nuovo Giorn. Bot. Ital. 17: 409 (1910).

Alchornea floribunda Müll.Arg., J. Bot. 1: 336 (1864).
Sierra Leone to Uganda. 22 GHA IVO LBR MLI NGA SIE 23 CMN CON EQG GAB GGI ZAI 25 UGA. Nanophan. or phan.

Alchornea fluviatilis Secco, Bol. Mus. Paraense Emilio Gouldi, N. S., Bot. 9: 60 (1993).
S. Venezuela, Guiana, N. Brazil. 82 FRG VEN 84 BZN. Nanophan. or phan.

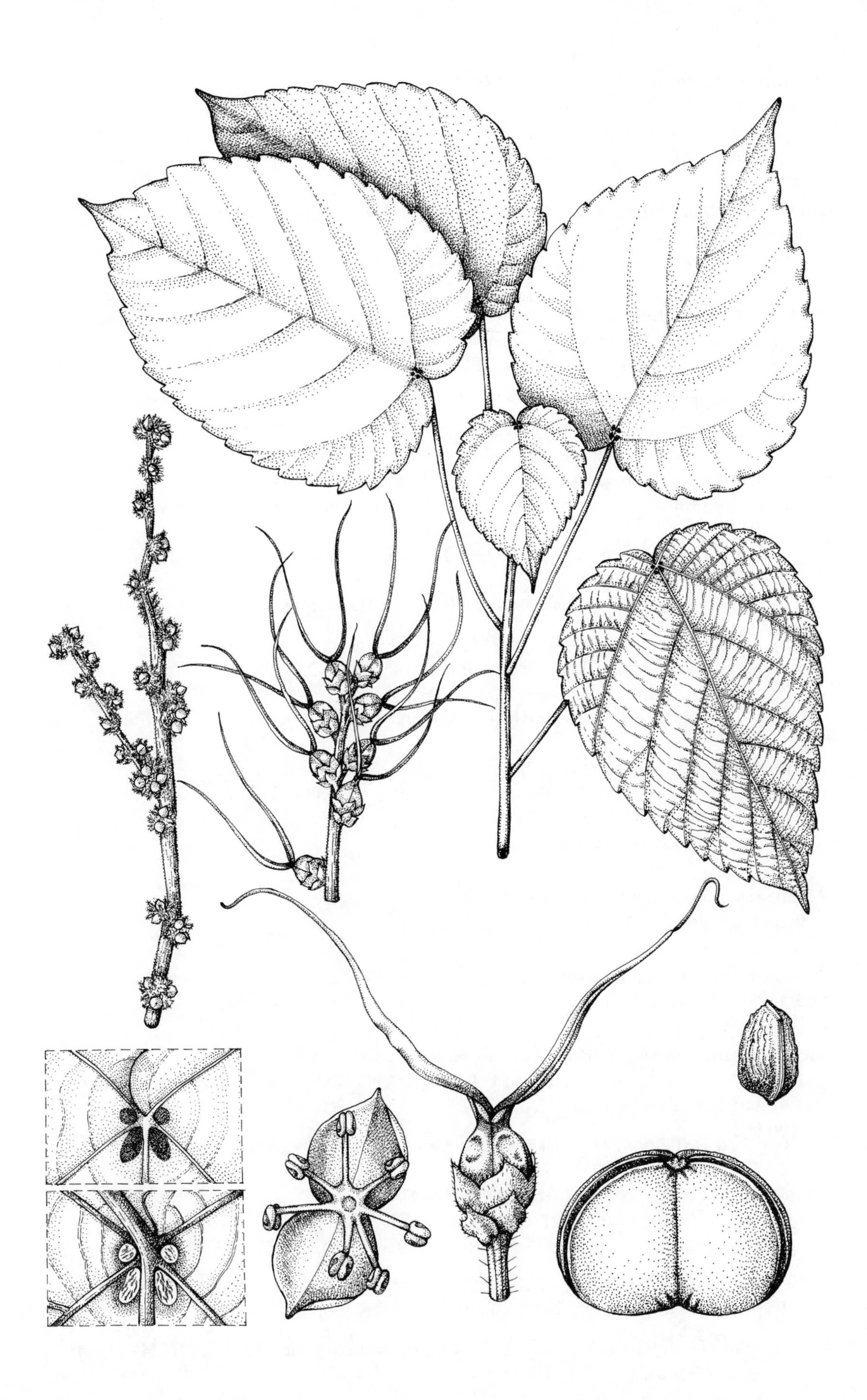

Alchornea cordifolia (Schumach. & Thonn.) Müll. Arg.
Artist: J.C. Dunkley
Fl. Trop. East Africa, Euphorbiaceae 1: 253, fig. 50 (1987)
KEW ILLUSTRATIONS COLLECTION

Alchornea gardneri Müll.Arg., Flora 47: 435 (1864).
Brazil (Pernambuco: Rio Preto). 84 BZE. Phan.
Alchornea schomburgkii Benth., Hooker's J. Bot. Kew Gard. Misc. 6: 330 (1854), nom. illeg.

Alchornea glabra (Merr.) Hurus., J. Fac. Sci. Univ. Tokyo, Sect. 3, Bot. 6: 304 (1954).
China (Anhui). 36 CHS.
* *Discocleidion glabrum* Merr., J. Arnold Arbor. 8: 8 (1927).

Alchornea glandulosa Poepp. & Endl., Nov. Gen. Sp. Pl. 3: 18 (1841). *Alchornea glandulosa* var. *genuina* Müll.Arg. in A.P.de Candolle, Prodr. 15(2): 911 (1866), nom. inval.
C. & S. Trop. America. 80 COS PAN 82 VEN 83 CLM ECU PER 84 BZN. Phan.

var. **glandulosa**
C. & S. Trop. America. 80 COS PAN 82 VEN 83 CLM ECU PER 84 BZN. Phan.
Alchornea subrotunda Baill., Étude Euphorb.: 447 (1858).
Alchornea nemoralis var. *glandulosa* Baill., Adansonia 5: 240 (1865).
Alchornea glandulosa var. *pavoniana* Müll.Arg. in A.P.de Candolle, Prodr. 15(2): 911 (1866).
Alchornea glandulosa var. *hispida* Pax & K.Hoffm. in H.G.A.Engler, Pflanzenr., IV, 147, VII: 234 (1914).
Alchornea glandulosa var. *poeppingii* Müll.Arg. in H.G.A.Engler, Pflanzenr., IV, 147, VII: 234 (1914).

var. **pittieri** (Pax) Pax in H.G.A.Engler, Pflanzenr., IV, 63: 235 (1914).
C. America, Colombia (Choco). 80 COS PAN 83 CLM.
* *Alchornea pittieri* Pax, Bot. Jahrb. Syst. 33: 291 (1903).

Alchornea grandiflora Müll.Arg., Linnaea 34: 170 (1865).
Costa Rica to Bolivia. 80 COS PAN 82 VEN 83 BOL CLM PER. Phan.
Alchornea hederifolia H.Karst. ex Pax & K.Hoffm. in H.G.A.Engler, Pflanzenr., IV, 147, VII: 226 (1914), pro syn.

Alchornea grandis Benth., Bot. Voy. Sulphur: 164 (1846).
Panama, Colombia (Nariño). 80 PAN 83 CLM. Phan.

Alchornea hilariana Baill., Adansonia 5: 240 (1865).
Brazil (Goiás). 84 BZC. Phan.

Alchornea hirtella Benth. in W.J.Hooker, Niger Fl.: 507 (1849).
Trop. & S. Africa. 22 GHA GNB GUI IVO LBR NGA SEN SIE 23 BUR RWA CMN GAB GGI ZAI 25 KEN TAN UGA 26 ANG MOZ ZAM 27 NAT TVL. Nanophan. or phan.

f. **comoensis** (Beille) Pax & K.Hoffm. in H.G.A.Engler, Pflanzenr., IV, 147, VII: 241 (1914).
W. & WC. Trop. Africa. 22 GHA IVO NGA 23 CMN GAB GGI. Nanophan. or phan.
Alchornea duparquetiana Baill., Adansonia 11: 175 (1874).
* *Alchornea comoensis* Beille, Bull. Soc. Bot. France 57(8): 124 (1910).

f. **cuneata** Pax & K.Hoffm. in H.G.A.Engler, Pflanzenr., IV, 147, VII: 241 (1914).
Cameroon, Zaire, Angola. 23 CMN ZAI 26 ANG. Nanophan. or phan.

f. **glabrata** (Prain) Pax & K.Hoffm. in H.G.A.Engler, Pflanzenr., IV, 147, VII: 242 (1914).
Trop. & S. Africa. 22 GHA GNB GUI IVO LBR NGA SEN SIE 23 BUR RWA CMN ZAI 25 KEN TAN UGA 26 ANG MOZ ZAM 27 NAT TVL. Nanophan. or phan.
* *Alchornea floribunda* var. *glabrata* Müll.Arg., J. Bot. 1: 336 (1864). *Alchornea glabrata* (Müll.Arg.) Prain, Bull. Misc. Inform. Kew 1910: 342 (1910).

f. **hirtella**
Trop. Africa. 22 GHA GNB GUI IVO LBR NGA SEN SIE 23 BUR RWA CMN ZAI 25 TAN 26 ZAM. Nanophan. or phan.

Alchornea humbertii Leandri, Notul. Syst. (Paris) 9: 181 (1941).
Madagascar. 29 MDG.

Alchornea hunanensis H.S.Kiu, Acta Phytotax. Sin. 26: 458 (1988).
China (Hunan). 36 CHS.

Alchornea integrifolia Pax & K.Hoffm. in H.G.A.Engler, Pflanzenr., IV, 147, VII: 237 (1914).
Guatemala (Coban). 80 GUA. Phan.

Alchornea iricurana Casar., Nov. Stirp. Bras.: 24 (1842). *Alchornea iricurana* f. *genuina* Pax & K.Hoffm. in H.G.A.Engler, Pflanzenr., IV, 147, VII: 232 (1914), nom. inval.
Brazil, Paraguay, Bolivia. 83 BOL 84 BZC BZL 85 PAR. Phan.
Alchornea erythrosperma Klotzsch ex Benth., Hooker's J. Bot. Kew Gard. Misc. 6: 331 (1854).
Alchornea puberula Klotzsch ex Benth., Hooker's J. Bot. Kew Gard. Misc. 6: 331 (1854).
Conceveiba pubescens Britton, Bull. Torrey Bot. Club 28: 306 (1901). *Alchornea iricurana* f. *pubescens* (Britton) Pax & K.Hoffm. in H.G.A.Engler, Pflanzenr., IV, 147, VII: 233 (1914).
Alchornea iricurana f. *villosula* Pax & K.Hoffm. in H.G.A.Engler, Pflanzenr., IV, 147, VII: 233 (1914).

Alchornea kelungensis Hayata, Icon. Pl. Formosan. 9: 102 (1920).
Taiwan. 38 TAI.

Alchornea latifolia Sw., Prodr.: 98 (1788).
Mexico, Trop. America. 79 MXT 80 COS GUA PAN 81 CUB DOM HAI JAM PUE 82 VEN 83 PER. Phan.
Manettia serrata Spreng. ex Schult. & Schult.f. in J.J.Roemer & J.A.Schultes, Mant. 3: 147 (1827).
Alchornea glandulosa Poit. ex Baill., Étude Euphorb.: 446 (1858).
Alchornea platyphylla Müll.Arg., Linnaea 34: 171 (1865).
Alchornea haitiensis Urb., Repert. Spec. Nov. Regni Veg. 18: 188 (1922).
Alchornea cyclophylla Croizat, J. Arnold Arbor. 24: 166 (1943).
Alchornea latifolia var. *islaensis* Kitan., Fitologiya 11: 47 (1979).

Alchornea laxiflora (Benth.) Pax & K.Hoffm. in H.G.A.Engler, Pflanzenr., IV, 147, VII: 245 (1914).
Trop. & S. Africa. 22 NGA 23 CMN CON GGI ZAI 24 ETH SUD 25 KEN TAN UGA 26 MLW MOZ ZAM ZIM 27 SWZ TVL. (Cl.) nanophan. or phan.
* *Lepidoturus laxiflorus* Benth., Hooker's Icon. Pl. 13: 76, t. 1297 (1879).
Alchornea engleri Pax, Bot. Jahrb. Syst. 43: 80 (1909).
Alchornea schlechteri Pax, Bot. Jahrb. Syst. 43: 221 (1909).
Macaranga thonneri De Wild., Etudes Fl. Bangala & Ubangi: 227 (1911).

Alchornea leptogyna Diels, Biblioth. Bot. 29(116): 103 (1937).
Ecuador. 83 ECU.

Alchornea megalophylla Müll.Arg., Flora 47: 434 (1864).
Panama, Colombia (Antioquia). 80 PAN 83 CLM. Nanophan. or phan.

Alchornea megalostylis Rusby, Phytologia 1: 63 (1934).
Bolivia. 83 BOL.

Alchornea mildbraedii Pax & K.Hoffm., Bot. Jahrb. Syst. 58(130): 39 (1923).
Cameroon. 23 CMN.

Alchornea mollis (Benth.) Müll.Arg., Linnaea 34: 168 (1865).
SC. Himalaya to Assam. 40 ASS EHM NEP. Phan.
Sapium cordifolium Roxb., Fl. Ind. ed. 1832, 3: 693 (1832).
* *Stipellaria mollis* Benth., Hooker's J. Bot. Kew Gard. Misc. 6: 3 (1854).

Alchornea obovata Pax & K.Hoffm. in H.G.A.Engler, Pflanzenr., IV, 147, VII: 223 (1914).
Colombia (Cauca). 83 CLM. Nanophan. or phan.

Alchornea occidentalis (Müll.Arg.) Pax & K.Hoffm. in H.G.A.Engler, Pflanzenr., IV, 147, VII: 245 (1914).
Angola, Zaire, Cabinda, Zambia. 23 CAB ZAI 26 ANG ZAM. Nanophan. or phan.
* *Lepidoturus occidentalis* Müll.Arg., J. Bot. 2: 332 (1864).

Alchornea pearcei Britton, Bull. Torrey Bot. Club 28: 305 (1901).
W. Bolivia, Peru. 83 BOL PER. Nanophan. or phan.
Alchornea coriacea Ule, Verh. Bot. Vereins Prov. Brandenburg 50: 76 (1908 publ. 1909). *Alchornea pearcei* var. *coriacea* (Ule) Pax in H.G.A.Engler, Pflanzenr., IV, 147, VII: 225 (1914).
Alchornea sclerophylla Pax, Repert. Spec. Nov. Regni Veg. 7: 242 (1909).
Alchornea pearcei var. *sclerophylla* Pax in H.G.A.Engler, Pflanzenr., IV, 147, VII: 225 (1914).

Alchornea perrieri Leandri, Notul. Syst. (Paris) 9: 182 (1941).
Madagascar. 29 MDG.

Alchornea petalostyla Airy Shaw, Kew Bull. 35: 395 (1980).
Philippines. 42 PHI.

Alchornea polyantha Pax & K.Hoffm. in H.G.A.Engler, Pflanzenr., IV, 147, VII: 225 (1914).
Colombia (Cauca). 83 CLM. Phan.

Alchornea pubescens Merr., Philipp. J. Sci. 20: 399 (1922).
Philippines (Luzon). 42 PHI.

Alchornea rhodophylla Pax & K.Hoffm. in H.G.A.Engler, Pflanzenr., IV, 147, VII: 249 (1914).
Pen. Malaysia. 42 MLY. Nanophan.
Alchornea discolor Hook.f., Fl. Brit. India 5: 420 (1887), nom. illeg.

Alchornea rugosa (Lour.) Müll.Arg., Linnaea 34: 170 (1865).
Hainan, Trop. Asia, N. Queensland. 36 CHH 41 BMA NCB THA 42 BIS BOR JAW LSI MLY MOL NWG PHI SUL SUM 50 QLD. Nanophan. or phan.
* *Cladodes rugosa* Lour., Fl. Cochinch.: 574 (1790).
Croton apetalum Blume, Catalogus: 104 (1823).
Conceveiba javanensis Blume, Bijdr.: 614 (1826). *Aparisthmium javanense* (Blume) Hassk., Cat. Hort. Bot. Bogor.: 235 (1844). *Alchornea javanensis* (Blume) Müll.Arg., Linnaea 34: 170 (1865). *Alchornea javanensis* (Blume) Backer & Bakh.f., Fl. Java 1: 485 (1963), nom. illeg.
Adelia glandulosa Blanco, Fl. Filip.: 814 (1837).
Conceveiba latifolia Zipp. ex Span., Linnaea 15: 349 (1841), pro syn.
Tragia innocua Blanco, Fl. Filip., ed. 2: 479 (1845).
Aparisthmium javanicum Baill., Étude Euphorb.: 468 (1858).
Alchornea hainanensis Pax & K.Hoffm. in H.G.A.Engler, Pflanzenr., IV, 147, VII: 242 (1914).

Alchornea hainanensis var. *glabrescens* Pax & K.Hoffm. in H.G.A.Engler, Pflanzenr., IV, 147, VII: 242 (1914).
Alchornea hainanensis var. *pubescens* Pax & K.Hoffm. in H.G.A.Engler, Pflanzenr., IV, 147, VII: 243 (1914).
Alchornea rugosa var. *macrocarpa* Airy Shaw, Kew Bull. 26: 211 (1972).

Alchornea scandens (Lour.) Müll.Arg., Linnaea 34: 170 (1865).
Vietnam. 41 VIE. Phan.
* *Caturus scandens* Lour., Fl. Cochinch.: 612 (1790).

Alchornea schomburgkii Klotzsch, Hooker's J. Bot. Kew Gard. Misc. 2: 46 (1843).
Colombia, Guyana, Venezuela (Amazonas), Surinam, Guiana, Brazil (Amazonas, Pará, Goiás). 82 FRG GUY SUR VEN 83 CLM 84 BZC BZN. Phan.
Alchornea discolor Poepp. & Endl., Nov. Gen. Sp. Pl. 3: 19 (1845).
Alchornea glaziovii Pax & K.Hoffm. in H.G.A.Engler, Pflanzenr., IV, 147, VII: 238 (1914).
Alchornea brachygyne Pax & K.Hoffm. in H.G.A.Engler, Pflanzenr., IV, 147, XIV: 20 (1919).

Alchornea sicca (Blanco) Merr., Philipp. J. Sci., C 5: 192 (1910).
Philippines. 42 PHI. Nanophan.
* *Excoecaria sicca* Blanco, Fl. Filip.: 787 (1837). *Homalanthus populneus* var. *siccus* (Blanco) Pax in H.G.A.Engler, Pflanzenr., IV, 147, V: 46 (1912).
Stipellaria parviflora Benth., Hooker's J. Bot. Kew Gard. Misc. 6: 4 (1854). *Alchornea parviflora* (Benth.) Müll.Arg., Linnaea 34: 168 (1865).
Alchornea philippinensis Pax & K.Hoffm. in H.G.A.Engler, Pflanzenr., IV, 147, VII: 249 (1914).

Alchornea sidifolia Müll.Arg., Linnaea 34: 169 (1865). *Alchornea sidifolia* f. *eusidifolia* Pax & K.Hoffm. in H.G.A.Engler, Pflanzenr., IV, 147, VII: 231 (1914), nom. inval.
SE. Brazil. 84 BZL. Phan.
Alchornea sidifolia Baill., Étude Euphorb.: 447 (1858), nom. nud.
Alchornea pycnogyne Müll.Arg. in C.F.P.von Martius, Fl. Bras. 11(2): 378 (1874). *Alchornea sidifolia* f. *pycnogyne* (Müll.Arg.) Pax & K.Hoffm. in H.G.A.Engler, Pflanzenr., IV, 147, VII: 233 (1914).
Alchornea sidifolia f. *intermedia* Pax & K.Hoffm. in H.G.A.Engler, Pflanzenr., IV, 147, VII: 233 (1914).

Alchornea similis Müll.Arg., Flora 47: 434 (1864).
Mexico (Sierra de San Pedro Nolasco). 79 MXS. Nanophan. or phan.

Alchornea sodiroi Pax & K.Hoffm. in H.G.A.Engler, Pflanzenr., IV, 147, VII: 234 (1914).
Ecuador. 83 ECU. Nanophan.

Alchornea tiliifolia (Benth.) Müll.Arg., Linnaea 34: 168 (1865).
Assam, Burma, Thailand, Pen. Malaysia, Andaman Is., S. China, N. Vietnam. 36 CHC 40 ASS 41 AND BMA MLY THA VIE. Nanophan. or phan.
Croton chiamala Wall., Numer. List: 7775 (1847), nom. inval.
* *Stipellaria tiliifolia* Benth., Hooker's J. Bot. Kew Gard. Misc. 6: 4 (1854).

Alchornea trewioides (Benth.) Müll.Arg., Linnaea 34: 168 (1865).
S. China (incl. Hong Kong), Nansei-shoto, Taiwan, Vietnam, Laos, Cambodia, N. Thailand. 36 CHS 38 NNS TAI 41 CBD LAO THA VIE. Nanophan.
* *Stipellaria trewioides* Benth., Hooker's J. Bot. Kew Gard. Misc. 6: 3 (1854). *Alchornea trewioides* var. *genuina* Pax & K.Hoffm. in H.G.A.Engler, Pflanzenr., IV, 147, VII: 248 (1914), nom. inval.

var. **sinica** H.S.Kiu, Acta Phytotax. Sin. 26: 460 (1988).
S. China. 36 CHS. Nanophan.

var. **trewioides**
S. China (incl. Hong Kong), Nansei-shoto, Taiwan, Vietnam, Laos, Cambodia, N. Thailand. 36 CHS 38 NNS TAI 41 CBD LAO THA VIE. Nanophan.
Alchornea liukiuensis Hayata, J. Coll. Sci. Imp. Univ. Tokyo 30: 268 (1911).
Alchornea formosae Müll.Arg. in H.G.A.Engler, Pflanzenr., IV, 147, VII: 248 (1914), pro syn.
Alchornea trewioides var. *formosae* Pax & K.Hoffm. in H.G.A.Engler, Pflanzenr., IV, 147, VII: 248 (1914).
Alchornea loochooensis Hayata, Icon. Pl. Formosan. 9: 103 (1920).
Alchornea coudercii Gagnep., Bull. Soc. Bot. France 71: 138 (1924).

Alchornea triplinervia (Spreng.) Müll.Arg. in A.P.de Candolle, Prodr. 15(2): 909 (1866).
Trop. America. 80 COS GUA PAN 81 TRT 82 FRG GUY SUR VEN 83 BOL CLM PER 84 BZC BZE BZL BZN 85 PAR. Nanophan. or phan.
* *Antidesma triplinervium* Spreng., Neue Entd. 2: 116 (1821). *Alchornea triplinervia* var. *genuina* Müll.Arg. in C.F.P.von Martius, Fl. Bras. 11(2): 380 (1874), nom. inval.

var. **janeirensis** (Casar.) Müll.Arg. in A.P.de Candolle, Prodr. 15(2): 909 (1866).
Brazil, Paraguay, Bolivia. 83 BOL 84 BZC BZL 85 PAR. Nanophan. or phan.
* *Alchornea janeirensis* Casar., Atti Riunione Sci. Ital. 3: 515 (1841). *Alchornea nemoralis* var. *janeirensis* (Casar.) Baill., Adansonia 5: 239 (1865).
Alchornea parvifolia Klotzsch ex Benth., Hooker's J. Bot. Kew Gard. Misc. 6: 331 (1854), nom. illeg. *Alchornea nemoralis* var. *parvifolia* Baill., Adansonia 5: 239 (1865).
Alchornea nemoralis var. *intermedia* Baill., Adansonia 5: 239 (1865). *Alchornea triplinervia* f. *intermedia* (Baill.) Müll.Arg. in A.P.de Candolle, Prodr. 15(2): 909 (1866).

var. **triplinervia**
Trop. America. 80 COS GUA PAN 81 TRT 82 FRG GUY SUR VEN 83 BOL CLM PER 84 BZE BZL BZN 85 PAR. Nanophan. or phan.
Alchornea nemoralis Mart., Herb. Fl. Bras.: 271 (1841). *Alchornea triplinervia* var. *nemoralis* (Mart.) Pax & K.Hoffm. in H.G.A.Engler, Pflanzenr., IV, 147, VII: 228 (1914).
Alchornea parvifolia Miq., Linnaea 22: 797 (1849). *Alchornea triplinervia* var. *parvifolia* (Miq.) Müll.Arg. in A.P.de Candolle, Prodr. 15(2): 910 (1866). *Alchornea glandulosa* var. *parvifolia* (Miq.) Benth., Hooker's J. Bot. Kew Gard. Misc. 6: 381 (1874).
Alchornea intermedia Klotzsch ex Benth., Hooker's J. Bot. Kew Gard. Misc. 6: 331 (1854).
Alchornea psilorhachis Klotzsch ex Benth., Hooker's J. Bot. Kew Gard. Misc. 6: 331 (1854).
Alchornea rotundifolia Moric. ex Baill., Étude Euphorb.: 447 (1858).
Alchornea nemoralis var. *psilorhachis* Baill., Adansonia 5: 239 (1865).
Alchornea nemoralis var. *rotundifolia* Baill., Adansonia 5: 239 (1865).
Alchornea nemoralis var. *lanceolata* Baill., Adansonia 7: 910 (1866). *Alchornea triplinervia* var. *lanceolata* (Baill.) Müll.Arg. in A.P.de Candolle, Prodr. 15(2): 910 (1866).
Alchornea triplinervia f. *psilorhachis* Müll.Arg. in A.P.de Candolle, Prodr. 15(2): 909 (1866).
Alchornea triplinervia var. *crassifolia* Müll.Arg. in A.P.de Candolle, Prodr. 15(2): 909 (1866).
Alchornea triplinervia var. *laevigata* Müll.Arg. in A.P.de Candolle, Prodr. 15(2): 910 (1866).
Alchornea triplinervia var. *meridensis* Müll.Arg. in A.P.de Candolle, Prodr. 15(2): 910 (1866).
Alchornea triplinervia var. *tomentella* Müll.Arg. in C.F.P.von Martius, Fl. Bras. 11(2): 380 (1874).
Alchornea triplinervia var. *iricuranoides* Chodat & Hassl., Bull. Herb. Boissier, II, 5: 603 (1905).
Alchornea nemoralis var. *major* Müll.Arg. ex Pax & K.Hoffm. in H.G.A.Engler, Pflanzenr., IV, 147, VII: 229 (1914), pro syn.
Alchornea triplinervia var. *boliviana* Pax & K.Hoffm. in H.G.A.Engler, Pflanzenr., IV, 147, VII: 228 (1914).
Alchornea guatemalensis Lundell, Wrightia 6: 10 (1978).

Alchornea ulmifolia (Müll.Arg.) Hurus., J. Fac. Sci. Univ. Tokyo, Sect. 3, Bot. 6: 304 (1954).
Nansei-shoto. 38 NNS. Nanophan. or phan. - Of dubious origin.
* *Cleidion ulmifolium* Müll.Arg., Flora 47: 481 (1864). *Discocleidion ulmifolium* (Müll.Arg.) Pax & K.Hoffm. in H.G.A.Engler, Pflanzenr., IV, 147, VII: 46 (1914).

Alchornea umboensis Croizat, Caldasia 2: 357 (1944).
Colombia. 83 CLM.

Alchornea vaniotii H.Lév., Cat. Pl. Yun-Nan: 95 (1916).
China (Yunnan). 36 CHC.

Alchornea villosa (Benth.) Müll.Arg., Linnaea 34: 168 (1865).
Pen. Malaysia, Sumatera, Jawa. 42 JAW MLY SUM. Nanophan.
* *Stipellaria villosa* Benth., Hooker's J. Bot. Kew Gard. Misc. 6: 4 (1854). *Alchornea villosa* var. *genuina* Müll.Arg. in A.P.de Candolle, Prodr. 15(2): 902 (1866), nom. inval.
Alchornea zollingeri Hassk., Retzia: 156 (1855). *Bleekeria zollingeri* (Hassk.) Miq., Fl. Ned. Ind. 1(2): 407 (1859).
Aparisthmium sumatranum Rchb. & Zoll., Linnaea 28: 331 (1856).
Alchornea villosa var. *lanceolata* Müll.Arg. in A.P.de Candolle, Prodr. 15(2): 902 (1866).
Alchornea villosa var. *latisepala* Hook.f., Fl. Brit. India 5: 421 (1887).

Alchornea yambuyaensis De Wild., Ann. Mus. Congo Belge, Bot., V, 2: 280 (1908).
Zaire, Angola, W. Tanzania, Zambia, Mozambique (Tete). 23 ZAI 25 TAN 26 ANG MOZ ZAM. Nanophan.
Alchornea verrucosa Pax, Bot. Jahrb. Syst. 43: 321 (1909).
Alchornea bangweolensis R.E.Fr., Wiss. Erg. Schwed. Rhod.-Kongo Exped. 1: 123 (1911-1912 publ. 1914).

Synonyms:
Alchornea amentiflora Airy Shaw === **Sampantaea amentiflora** (Airy Shaw) Airy Shaw
Alchornea arborea Elmer === **Neoscortechinia nicobarica** (Hook.f.) Pax & K.Hoffm.
Alchornea bangweolensis R.E.Fr. === **Alchornea yambuyaensis** De Wild.
Alchornea blumeana Müll.Arg. === **Wetria insignis** (Steud.) Airy Shaw
Alchornea brachygyne Pax & K.Hoffm. === **Alchornea schomburgkii** Klotzsch
Alchornea caloneura Pax === **Discoglypremna caloneura** (Pax) Prain
Alchornea castaneifolia (Baill.) Müll.Arg. === **Necepsia castaneifolia** (Baill.) Bouchat & J.Léonard
Alchornea castaneifolia var. *genuina* Müll.Arg. === **Alchornea castaneifolia** (Humb. & Bonpl. ex Willd.) A.Juss.
Alchornea castaneifolia var. *puberula* Müll.Arg. === **Alchornea castaneifolia** (Humb. & Bonpl. ex Willd.) A.Juss.
Alchornea castaneifolia var. *salicifolia* (Baill.) Baill. === **Alchornea castaneifolia** (Humb. & Bonpl. ex Willd.) A.Juss.
Alchornea comoensis Beille === **Alchornea hirtella** f. **comoensis** (Beille) Pax & K.Hoffm.
Alchornea cordata (A.Juss.) Müll.Arg. === **Aparisthmium cordatum** (A.Juss.) Baill.
Alchornea cordata Benth. === **Alchornea cordifolia** (Schumach. & Thonn.) Müll.Arg.
Alchornea coriacea Ule === **Alchornea pearcei** Britton
Alchornea coriacea (Baill.) Müll.Arg. === **Orfilea coriacea** Baill.
Alchornea costaricensis f. *longispicata* Pax & K.Hoffm. === **Alchornea costaricensis** Pax & K.Hoffm.
Alchornea coudercii Gagnep. === **Alchornea trewioides** (Benth.) Müll.Arg. var. **trewioides**
Alchornea cuneata Miq. === **Trigonostemon heteranthus** Wight
Alchornea cuneifolia (Miq.) Müll.Arg. === **Trigonostemon heteranthus** Wight
Alchornea cyclophylla Croizat === **Alchornea latifolia** Sw.
Alchornea discolor Poepp. & Endl. === **Alchornea schomburgkii** Klotzsch
Alchornea discolor Hook.f. === **Alchornea rhodophylla** Pax & K.Hoffm.

Alchornea duparquetiana Baill. === **Alchornea hirtella** f. **comoensis** (Beille) Pax & K.Hoffm.
Alchornea engleri Pax === **Alchornea laxiflora** (Benth.) Pax & K.Hoffm.
Alchornea erythrosperma Klotzsch ex Benth. === **Alchornea iricurana** Casar.
Alchornea floribunda var. *glabrata* Müll.Arg. === **Alchornea hirtella** f. **glabrata** (Prain) Pax & K.Hoffm.
Alchornea formosae Müll.Arg. === **Alchornea trewioides** (Benth.) Müll.Arg. var. **trewioides**
Alchornea glabrata (Müll.Arg.) Prain === **Alchornea hirtella** f. **glabrata** (Prain) Pax & K.Hoffm.
Alchornea glandulosa Poit. ex Baill. === **Alchornea latifolia** Sw.
Alchornea glandulosa var. *floribunda* Benth. === **Alchorneopsis floribunda** (Benth.) Müll.Arg.
Alchornea glandulosa var. *genuina* Müll.Arg. === **Alchornea glandulosa** Poepp. & Endl.
Alchornea glandulosa var. *hispida* Pax & K.Hoffm. === **Alchornea glandulosa** Poepp. & Endl. var. **glandulosa**
Alchornea glandulosa var. *parvifolia* (Miq.) Benth. === **Alchornea triplinervia** (Spreng.) Müll.Arg. var. **triplinervia**
Alchornea glandulosa var. *pavoniana* Müll.Arg. === **Alchornea glandulosa** Poepp. & Endl. var. **glandulosa**
Alchornea glandulosa var. *poeppingii* Müll.Arg. === **Alchornea glandulosa** Poepp. & Endl. var. **glandulosa**
Alchornea glaziovii Pax & K.Hoffm. === **Alchornea schomburgkii** Klotzsch
Alchornea guatemalensis Lundell === **Alchornea triplinervia** (Spreng.) Müll.Arg. var. **triplinervia**
Alchornea hainanensis Pax & K.Hoffm. === **Alchornea rugosa** (Lour.) Müll.Arg.
Alchornea hainanensis var. *glabrescens* Pax & K.Hoffm. === **Alchornea rugosa** (Lour.) Müll.Arg.
Alchornea hainanensis var. *pubescens* Pax & K.Hoffm. === **Alchornea rugosa** (Lour.) Müll.Arg.
Alchornea haitiensis Urb. === **Alchornea latifolia** Sw.
Alchornea hederifolia H.Karst. ex Pax & K.Hoffm. === **Alchornea grandiflora** Müll.Arg.
Alchornea ilicifolia (Js.Sm.) Müll.Arg. === **Alchornea aquifolia** (Js.Sm.) Domin
Alchornea intermedia Klotzsch ex Benth. === **Alchornea triplinervia** (Spreng.) Müll.Arg. var. **triplinervia**
Alchornea iricurana f. *genuina* Pax & K.Hoffm. === **Alchornea iricurana** Casar.
Alchornea iricurana f. *pubescens* (Britton) Pax & K.Hoffm. === **Alchornea iricurana** Casar.
Alchornea iricurana f. *villosula* Pax & K.Hoffm. === **Alchornea iricurana** Casar.
Alchornea janeirensis Casar. === **Alchornea triplinervia** var. **janeirensis** (Casar.) Müll.Arg.
Alchornea javanensis (Blume) Müll.Arg. === **Alchornea rugosa** (Lour.) Müll.Arg.
Alchornea javanensis (Blume) Backer & Bakh.f. === **Alchornea rugosa** (Lour.) Müll.Arg.
Alchornea latifolia Klotzsch === **Aparisthmium cordatum** (A.Juss.) Baill.
Alchornea latifolia var. *islaensis* Kitan. === **Alchornea latifolia** Sw.
Alchornea liukiuensis Hayata === **Alchornea trewioides** (Benth.) Müll.Arg. var. **trewioides**
Alchornea loochooensis Hayata === **Alchornea trewioides** (Benth.) Müll.Arg. var. **trewioides**
Alchornea macrophylla Mart. === **Aparisthmium cordatum** (A.Juss.) Baill.
Alchornea madagascariensis Müll.Arg. === **Necepsia castaneifolia** (Baill.) Bouchat & J.Léonard subsp. **castaneifolia**
Alchornea mairei H.Lév. === **Cnesmone mairei** (H.Lév.) Croizat
Alchornea mappa (L.) Oken === **Macaranga mappa** (L.) Müll.Arg.
Alchornea martiana (Baill.) Müll.Arg. === **Conceveiba martiana** Baill.
Alchornea multispicata (Baill.) Müll.Arg. === **Orfilea multispicata** (Baill.) G.L.Webster
Alchornea nemoralis Mart. === **Alchornea triplinervia** (Spreng.) Müll.Arg. var. **triplinervia**
Alchornea nemoralis var. *floribunda* (Benth.) Baill. === **Alchorneopsis floribunda** (Benth.) Müll.Arg.
Alchornea nemoralis var. *glandulosa* Baill. === **Alchornea glandulosa** Poepp. & Endl. var. **glandulosa**
Alchornea nemoralis var. *intermedia* Baill. === **Alchornea triplinervia** var. **janeirensis** (Casar.) Müll.Arg.

Alchornea nemoralis var. *janeirensis* (Casar.) Baill. === **Alchornea triplinervia** var. **janeirensis** (Casar.) Müll.Arg.
Alchornea nemoralis var. *lanceolata* Baill. === **Alchornea triplinervia** (Spreng.) Müll.Arg. var. **triplinervia**
Alchornea nemoralis var. *major* Müll.Arg. ex Pax & K.Hoffm. === **Alchornea triplinervia** (Spreng.) Müll.Arg. var. **triplinervia**
Alchornea nemoralis var. *parvifolia* Baill. === **Alchornea triplinervia** var. **janeirensis** (Casar.) Müll.Arg.
Alchornea nemoralis var. *psilorhachis* Baill. === **Alchornea triplinervia** (Spreng.) Müll.Arg. var. **triplinervia**
Alchornea nemoralis var. *rotundifolia* Baill. === **Alchornea triplinervia** (Spreng.) Müll.Arg. var. **triplinervia**
Alchornea oblongifolia Standl. === **Cleidion castaneifolium** Müll.Arg.
Alchornea orinocensis Croizat === **Aparisthmium cordatum** (A.Juss.) Baill.
Alchornea parviflora (Benth.) Müll.Arg. === **Alchornea sicca** (Blanco) Merr.
Alchornea parvifolia Miq. === **Alchornea triplinervia** (Spreng.) Müll.Arg. var. **triplinervia**
Alchornea parvifolia Klotzsch ex Benth. === **Alchornea triplinervia** var. **janeirensis** (Casar.) Müll.Arg.
Alchornea passargei Pax & K.Hoffm. === **Alchornea castaneifolia** (Humb. & Bonpl. ex Willd.) A.Juss.
Alchornea pearcei var. *coriacea* (Ule) Pax === **Alchornea pearcei** Britton
Alchornea pearcei var. *sclerophylla* Pax === **Alchornea pearcei** Britton
Alchornea philippinensis Pax & K.Hoffm. === **Alchornea sicca** (Blanco) Merr.
Alchornea pittieri Pax === **Alchornea glandulosa** var. **pittieri** (Pax) Pax
Alchornea platyphylla Müll.Arg. === **Alchornea latifolia** Sw.
Alchornea psilorhachis Klotzsch ex Benth. === **Alchornea triplinervia** (Spreng.) Müll.Arg. var. **triplinervia**
Alchornea puberula Klotzsch ex Benth. === **Alchornea iricurana** Casar.
Alchornea pycnogyne Müll.Arg. === **Alchornea sidifolia** Müll.Arg.
Alchornea rotundifolia Moric. ex Baill. === **Alchornea triplinervia** (Spreng.) Müll.Arg. var. **triplinervia**
Alchornea rufescens Franch. === **Discocleidion rufescens** (Franch.) Pax & K.Hoffm.
Alchornea rugosa var. *macrocarpa* Airy Shaw === **Alchornea rugosa** (Lour.) Müll.Arg.
Alchornea schlechteri Pax === **Alchornea laxiflora** (Benth.) Pax & K.Hoffm.
Alchornea schomburgkii Benth. === **Alchornea gardneri** Müll.Arg.
Alchornea sclerophylla Pax === **Alchornea pearcei** Britton
Alchornea sidifolia Baill. === **Alchornea sidifolia** Müll.Arg.
Alchornea sidifolia f. *eusidifolia* Pax & K.Hoffm. === **Alchornea sidifolia** Müll.Arg.
Alchornea sidifolia f. *intermedia* Pax & K.Hoffm. === **Alchornea sidifolia** Müll.Arg.
Alchornea sidifolia f. *pycnogyne* (Müll.Arg.) Pax & K.Hoffm. === **Alchornea sidifolia** Müll.Arg.
Alchornea subrotunda Baill. === **Alchornea glandulosa** Poepp. & Endl. var. **glandulosa**
Alchornea thozetiana (Baill.) Benth. === **Alchornea aquifolia** (Js.Sm.) Domin
Alchornea thozetiana var. *longifolia* Benth. === **Alchornea aquifolia** (Js.Sm.) Domin
Alchornea trewioides var. *formosae* Pax & K.Hoffm. === **Alchornea trewioides** (Benth.) Müll.Arg. var. **trewioides**
Alchornea trewioides var. *genuina* Pax & K.Hoffm. === **Alchornea trewioides** (Benth.) Müll.Arg.
Alchornea triplinervia var. *boliviana* Pax & K.Hoffm. === **Alchornea triplinervia** (Spreng.) Müll.Arg. var. **triplinervia**
Alchornea triplinervia var. *crassifolia* Müll.Arg. === **Alchornea triplinervia** (Spreng.) Müll.Arg. var. **triplinervia**
Alchornea triplinervia var. *genuina* Müll.Arg. === **Alchornea triplinervia** (Spreng.) Müll.Arg.
Alchornea triplinervia f. *intermedia* (Baill.) Müll.Arg. === **Alchornea triplinervia** var. **janeirensis** (Casar.) Müll.Arg.

Alchornea triplinervia var. *iricuranoides* Chodat & Hassl. === **Alchornea triplinervia** (Spreng.) Müll.Arg. var. **triplinervia**
Alchornea triplinervia var. *laevigata* Müll.Arg. === **Alchornea triplinervia** (Spreng.) Müll.Arg. var. **triplinervia**
Alchornea triplinervia var. *lanceolata* (Baill.) Müll.Arg. === **Alchornea triplinervia** (Spreng.) Müll.Arg. var. **triplinervia**
Alchornea triplinervia var. *meridensis* Müll.Arg. === **Alchornea triplinervia** (Spreng.) Müll.Arg. var. **triplinervia**
Alchornea triplinervia var. *nemoralis* (Mart.) Pax & K.Hoffm. === **Alchornea triplinervia** (Spreng.) Müll.Arg. var. **triplinervia**
Alchornea triplinervia var. *parvifolia* (Miq.) Müll.Arg. === **Alchornea triplinervia** (Spreng.) Müll.Arg. var. **triplinervia**
Alchornea triplinervia f. *psilorhachis* Müll.Arg. === **Alchornea triplinervia** (Spreng.) Müll.Arg. var. **triplinervia**
Alchornea triplinervia var. *tomentella* Müll.Arg. === **Alchornea triplinervia** (Spreng.) Müll.Arg. var. **triplinervia**
Alchornea verrucosa Pax === **Alchornea yambuyaensis** De Wild.
Alchornea villosa var. *genuina* Müll.Arg. === **Alchornea villosa** (Benth.) Müll.Arg.
Alchornea villosa var. *glabrata* Hook.f. === **Alchornea adenophila** Pax & K.Hoffm.
Alchornea villosa var. *lanceolata* Müll.Arg. === **Alchornea villosa** (Benth.) Müll.Arg.
Alchornea villosa var. *latisepala* Hook.f. === **Alchornea villosa** (Benth.) Müll.Arg.
Alchornea zollingeri Hassk. === **Alchornea villosa** (Benth.) Müll.Arg.

Alchorneopsis

2 species, Central and South America (Costa Rica and Panama, Venezuela, the Guianas, northern Brazil and Peru), West Indies (Puerto Rico); forest trees to 20 m tall. Of the three closely related species noted by Webster (Synopsis, 1994), *A. trimera* has been reduced to the widely distributed *A. floribunda*. A vicariant African genus is *Discoglypremna*; Webster (Synopsis, 1994) believes they could as well be a single genus with two subgenera. This could be said to lend support to what appears to be a 'relict' distribution in *Alchorneopsis*. (Acalyphoideae)

Pax, F. (with K. Hoffmann) (1914). *Alchorneopsis*. In A. Engler (ed.), Das Pflanzenreich, IV 147 VII (Euphorbiaceae-Acalypheae-Mercurialinae): 267-268. Berlin. (Heft 63.) La/Ge. — 2 species, S America, West Indies (in the latter only in Puerto Rico).

Jablonski, E. (1967). *Alchorneopsis*. Euphorbiaceae, Guayana Highland (Mem. New York Bot. Gard. 17(1)): 139-140. New York. En. — 1 species, *A. floribunda*.

Alchorneopsis Müll.Arg., Linnaea 34: 156 (1865).
Trop. America. 80 81 82 83 84.

Alchorneopsis floribunda (Benth.) Müll.Arg., Linnaea 34: 156 (1865).
Trop. America. 80 COS PAN 81 DOM PUE 82 FRG GUY SUR VEN 83 PER 84 BZN. Phan.
* *Alchornea glandulosa* var. *floribunda* Benth., Hooker's J. Bot. Kew Gard. Misc. 6: 331 (1854). *Alchorneopsis floribunda* var. *genuina* Müll.Arg., Linnaea 34: 156 (1865), nom. inval. *Alchornea nemoralis* var. *floribunda* (Benth.) Baill., Adansonia 5: 239 (1865).
Alchorneopsis floribunda var. *sessiliflora* Müll.Arg., Linnaea 34: 156 (1865).
Alchorneopsis trimera Lanj., Euphorb. Surinam: 23 (1931).

Alchorneopsis portoricensis Urb., Symb. Antill. 1: 337 (1899).
Puerto Rico. 81 PUE. Phan.

Synonyms:
Alchorneopsis floribunda var. *genuina* Müll.Arg. === **Alchorneopsis floribunda** (Benth.) Müll.Arg.

Alchorneopsis floribunda var. *sessiliflora* Müll.Arg. === **Alchorneopsis floribunda** (Benth.) Müll.Arg.
Alchorneopsis trimera Lanj. === **Alchorneopsis floribunda** (Benth.) Müll.Arg.

Alcinaeanthus

Synonyms:
Alcinaeanthus Merr. === **Neoscortechinia** Pax
Alcinaeanthus parvifolius Merr. === **Neoscortechinia philippinensis** (Merr.) Belzen
Alcinaeanthus philippinensis Merr. === **Neoscortechinia philippinensis** (Merr.) Belzen

Alcoceria

Synonyms:
Alcoceria Fernald === **Dalembertia** Baill.

Aldinia

Synonyms:
Aldinia Raf. === **Croton** L.

Alectoroctonum

A Klotzsch & Garcke segregate of *Euphorbia*.

Synonyms:
Alectoroctonum Schltdl. === **Euphorbia** L.
Alectoroctonum caracasanum Klotzsch & Garcke === **Euphorbia caracasana** (Klotzsch & Garcke) Boiss.
Alectoroctonum yavalquahuitl Schltdl. === **Euphorbia cotinifolia** L.
Alectoroctonum riedelianum Klotzsch & Garcke === **Euphorbia cotinifolia** L.
Alectoroctonum viride Klotzsch & Garcke === **Euphorbia viridis** (Klotzsch & Garcke) Boiss.

Aleurites

2 species, Malesia and adjacent areas (original range now somewhat obscured); medium to large trees. *A. moluccana*, the candlenut tree or kemiri, has been since antiquity introduced beyond its natural range, in the first instance for the oil in the seeds. *A. remyi*, described from Hawaii, is synonymous with *A. moluccana* and represents a Polynesian introduction; until the arrival of paraffin (kerosene) in the 1860s this oil was used for lighting. Other species, including those yielding tung oil, are now in *Vernicia* and *Reutealis*. [All these as well as *Deutzianthus* and *Tapoïdes* are perhaps living relics of a formerly larger group.] (Crotonoideae)

Langeron, M. (1902). Le genre *Aleurites* Forst. (Euphorbiacées): systématique — anatomie — pharmacologie. 160 pp., illus., map. Paris. Fr. — The first part comprises a taxonomic review with historical account, descriptions, synonymy, lists of specimens seen, and commentary (in an idiosyncratic arrangement) along with a distribution map (p. 22). Three species in three sections are accepted. The remainder of the work comprises an anatomical and morphological study and a review of economic aspects. [The individual sections are now accorded generic rank corresponding to *Aleurites* s.s., *Reutealis* and *Vernicia*.]

Pax, F. (1910). *Aleurites*. In A. Engler (ed.), Das Pflanzenreich, IV 147 [I] (Euphorbiaceae-Jatropheae): 128-133. Berlin. (Heft 42.) La/Ge. — 4 species, of which 1 in sect. *Camirium* (= sect. *Aleurites*).

Airy-Shaw, H. K. (1966). Notes on Malaysian and other Asiatic Euphorbiaceae, LVIII. A New Guinea variety of *Aleurites moluccana*. Kew Bull. 20: 26-27. En. — Description of *A. moluccana* var. *floccosa*. [Not accepted as distinct in the *Checklist*.]

Airy-Shaw, H. K. (1966). Notes on Malaysian and other Asiatic Euphorbiaceae, LXXII. Generic segregation in the affinity of *Aleurites* J. R. et G. Forst. Kew Bull. 20: 393-395. En. —*Vernicia* and *Rutealis* separated from *Aleurites*; the last-named left with 2 spp., *A. moluccana* and *A. remyi* (the latter since reduced to the former). Necessary new combinations are made but no key is presented. [A precursor to this was improved knowledge of *Deutzianthus*.]

Forster, P. I. (1996). A taxonomic revision of *Aleurites* J. R. Forst. & G. Forst. (Euphorbiaceae) in Australia and New Guinea. Muelleria 9: 5-13, illus. En. — Regional revision (2 species); includes key, descriptions, notes on distribution, habitat, vegetation, uses and properties, representative localities with exsiccatae, and commentary. [*A. moluccana* var. *rockinghamensis* is here elevated to species rank. Both species are known in New Guinea as well as Australia, *A. rockinghamensis* in New Guinea mostly in the southeastern peninsula.]

Stuppy, W. et al. (1999). Revision of the genera *Aleurites*, *Reutealis* and *Vernicia* (Euphorbiaceae). Blumea 44: 73-98, illus. En. — *Aleurites*, pp. 79-85; treatment of 2 species with key, descriptions, synonymy, references, types, indication of distribution and habitat, and commentary; all general references, identification list and index to botanical names at end of paper. *A. erratica* O.Deg., I.Deg. & K.Hummel, based on drift seeds collected at Canton Atoll in the Phoenix Islands (Pacific Ocean), was found to be undeterminable to genus; it was referable not to *Aleurites* but merely to a uniovulate but non-aleuritoid species of Euphorbiaceae.

Aleurites J.R.Forst. & G.Forst., Char. Gen. Pl.: 56 (1775).
Trop. & Subtrop. Asia, NE. Australia. 36 38 40 41 42 50 (51) (60) (63).
Ambinux Comm. ex A.Juss., Gen. Pl.: 389 (1789).
Camirium Gaertn., Fruct. Sem. Pl. 2: 194 (1791).
Telopea Sol. ex Baill., Étude Euphorb.: 345 (1858).

Aleurites moluccana (L.) Willd., Sp. Pl. 4: 590 (1805). – FIGURE, p. 153.
Trop. & Subtrop. Asia, NE. Queensland. 36 CHC CHH CHS 38 TAI 40 ASS IND SRL 41 BMA CBD THA VIE 42 BOR JAW MLY NWG PHI 50 QLD (51) ker (60) fij nwc (63) haw (84) bze. Phan. – Oil obtained from the seeds.
* *Jatropha moluccana* L., Sp. Pl.: 1006 (1753). *Manihot moluccana* (L.) Crantz in ?, . *Mallotus moluccanus* var. *genuinus* Müll.Arg., Linnaea 34: 185 (1865), nom. inval. *Mallotus moluccanus* (L.) Müll.Arg., Linnaea 34: 185 (1865).
Aleurites triloba J.R.Forst. & G.Forst., Char. Gen. Pl.: 56 (1775).
Juglans camirium Lour., Fl. Cochinch. 1: 573 (1790). *Camirium cordifolium* Gaertn., Fruct. Sem. Pl. 2: 194 (1791). *Aleurites cordifolia* (Gaertn.) Steud., Nomencl. Bot., ed. 2, 1: 49 (1840).
Aleurites ambinux Pers., Syn. Pl. 2: 579 (1807).
Aleurites commutata Geiseler, Croton. Monogr.: 82 (1807).
Aleurites lanceolata Blanco, Fl. Filip.: 757 (1837).
Aleurites lobata Blanco, Fl. Filip.: 756 (1837).
Adelia monoica Blanco, Fl. Filip., ed. 2: 561 (1845).
Aleurites pentaphylla Wall., Numer. List: 7959 (1847), nom. inval.
Aleurites angustifolia Vieill., Ann. Sci. Nat., Bot., IV, 16: 60 (1862).
Aleurites integrifolia Vieill., Ann. Sci. Nat., Bot., IV, 16: 59 (1862), in syn.
Aleurites javanica Gand., Bull. Soc. Bot. France 60: 27 (1913).
Aleurites remyi Sherff, Publ. Field Mus. Nat. Hist., Bot. Ser. 17: 558 (1939). *Aleurites moluccana* var. *remyi* (Sherff) Stone, Pacific Sci. 21: 553 (1967).
Aleurites moluccana var. *serotina* O.Deg. & Sherff, Amer. J. Bot. 38: 57 (1951).
Aleurites moluccana var. *floccosa* Airy Shaw, Kew Bull. 20: 26, 393 (1966).
Aleurites moluccana var. *aulanii* O.Deg. & I.Deg., Phytologia 21: 316 (1971).
Aleurites moluccana var. *katoi* O.Deg., I.Deg. & Stone, Phytologia 21: 315 (1971).

Aleurites moluccana (L.) Willd.
Artist: P. Halliday
Fl. Trop. East Africa, Euphorbiaceae 1: 177, fig. 34 (1987)
KEW ILLUSTRATIONS COLLECTION

Aleurites rockinghamensis (Baill.) P.I.Forst., Muelleria 9: 8 (1996).
Papua New Guinea, NE. Queensland. 42 NWG 50 QLD. Phan.
* *Aleurites moluccana* var. *rockinghamensis* Baill., Adansonia 6: 297 (1866).

Synonyms:
Aleurites ambinux Pers. === **Aleurites moluccana** (L.) Willd.
Aleurites angustifolia Vieill. === **Aleurites moluccana** (L.) Willd.
Aleurites commutata Geiseler === **Aleurites moluccana** (L.) Willd.
Aleurites cordata (Thunb.) R.Br. ex Steud. === **Vernicia cordata** (Thunb.) Airy Shaw
Aleurites cordifolia (Gaertn.) Steud. === **Aleurites moluccana** (L.) Willd.
Aleurites erratica O.Deg., I.Deg. & K.Hummel === ? (Stuppy et al., 1999: 95)
Aleurites fordii Hemsl. === **Vernicia fordii** (Hemsl.) Airy Shaw
Aleurites integrifolia Vieill. === **Aleurites moluccana** (L.) Willd.
Aleurites japonica Blume === **Vernicia cordata** (Thunb.) Airy Shaw
Aleurites javanica Gand. === ?
Aleurites laccifera (L.) Willd. === **Croton laccifer** L.
Aleurites lanceolata Blanco === **Aleurites moluccana** (L.) Willd.
Aleurites lobata Blanco === **Aleurites moluccana** (L.) Willd.
Aleurites moluccana var. *alaunii* O.Deg. & I.Deg. === **Aleurites moluccana** (L.) Willd.
Aleurites moluccana var. *floccosa* Airy Shaw === **Aleurites moluccana** (L.) Willd.
Aleurites moluccana var. *katoi* O.Deg., I.Deg. & Stone === **Aleurites moluccana** (L.) Willd.
Aleurites moluccana var. *remyi* (Sherff) Stone === **Aleurites moluccana** (L.) Willd.
Aleurites moluccana var. *rockinghamensis* Baill. === **Aleurites rockinghamensis** (Baill.) P.I.Forst.
Aleurites moluccana var. *serotina* O.Deg. & Sherff === **Aleurites moluccana** (L.) Willd.
Aleurites montana (Lour.) E.H.Wilson === **Vernicia montana** Lour.
Aleurites peltata Geiseler === **Mallotus peltatus** (Geiseler) Müll.Arg.
Aleurites pentaphylla Wall. === **Aleurites moluccana** (L.) Willd.
Aleurites remyi Sherff === **Aleurites moluccana** (L.) Willd.
Aleurites saponaria Blanco === **Reutealis trisperma** (Blanco) Airy Shaw
Aleurites triloba J.R.Forst. & G.Forst. === **Aleurites moluccana** (L.) Willd.
Aleurites trisperma Blanco === **Reutealis trisperma** (Blanco) Airy Shaw
Aleurites vernicia (Corrêa) Hassk. === **Vernicia montana** Lour.
Aleurites verniciflua Baill. === **Vernicia cordata** (Thunb.) Airy Shaw

Alevia

Synonyms:
Alevia Baill. === **Bernardia** Houst. ex Mill.

Algernonia

6 species, S. America (E. Brazil); glabrous, laticiferous shrubs or small trees, presumably of the *mata atlantica*, closely related to *Tetraplandra* with some arguing for their merger. Both are in turn related to *Pachystroma* and possibly also *Ophthalmoblapton*. The leaves and branching in *A. brasiliensis* are reminiscent of some species of *Drypetes*. (Euphorbioideae (except Euphorbieae))

Pax, F. (with K. Hoffmann) (1912). *Algernonia*. In A. Engler (ed.), Das Pflanzenreich, IV 147 V (Euphorbiaceae-Hippomaneae): 276-278. Berlin. (Heft 52.) La/Ge. — 2 species, E Brazil (described from around Rio de Janeiro). [See also Croizat (1943) under *Cnidoscolus*.]

• Emmerich, M. (1981). Revisão taxinômica dos gêneros *Algernonia* Baill. e *Tetraplandra* Baill. (Euphorbiaceae-Hippomaneae). Arq. Mus. Nac. Rio de Janeiro 56: 91-110, 11 pls. in text. Pt. — *Algernonia*, pp. 92-94; revision (6 species, 2 new, one transferred from *Tetraplandra*) with key, descriptions, synonymy, references and citations, types, indication of distribution and habitat, and localities with exsiccatae; phenological and ecological data, literature and well-drawn plates at end of paper.

Algernonia Baill., Ann. Sci. Nat., Bot., IV, 9: 198 (1858).
Brazil. 84.
Dendrobryon Klotzsch ex Pax in H.G.A.Engler, Pflanzenr., IV, 147, V: 275, 277 (1912).

Algernonia brasiliensis Baill., Ann. Sci. Nat., Bot., IV, 9: 198 (1858).
Brazil (Rio de Janeiro). 84 BZL. Phan.
Algernonia brasiliensis var. *cuneata* Müll.Arg. in A.P.de Candolle, Prodr. 15(2): 1231 (1866).

Algernonia gibbosa (Pax & K.Hoffm.) Emmerich, Arq. Mus. Nac. Rio de Janeiro 56: 93 (1981).
Brazil (Rio de Janeiro). 84 BZL. Phan.
* *Tetraplandra gibbosa* Pax & K.Hoffm. in H.G.A.Engler, Pflanzenr., IV, 147, V: 276 (1912).

Algernonia glazioui Emmerich, Arq. Mus. Nac. Rio de Janeiro 56: 94 (1981).
Brazil. 84 BZL.

Algernonia obovata (Müll.Arg.) Müll.Arg. in C.F.P.von Martius, Fl. Bras. 11(2): 536 (1874).
Brazil (Rio de Janeiro). 84 BZL. Phan.
* *Algernonia brasiliensis* var. *obovata* Müll.Arg. in A.P.de Candolle, Prodr. 15(2): 1231 (1866).

Algernonia pardina Croizat, Trop. Woods 76: 14 (1943).
Brazil (Bahia). 84 BZE. Phan.

Algernonia paulae Emmerich, Bradea 3: 148 (1981).
Brazil. 84 BZL.

Synonyms:
Algernonia brasiliensis var. *cuneata* Müll.Arg. === **Algernonia brasiliensis** Baill.
Algernonia brasiliensis var. *obovata* Müll.Arg. === **Algernonia obovata** (Müll.Arg.) Müll.Arg.

Allenia

Synonyms:
Allenia Ewart === **Micrantheum** Desf.
Allenia blackiana Ewart, Jean White & Rees === **Micrantheum demissum** F.Muell.
Allenia blackiana var. *microphylla* Ewart, Jean White & Rees === **Micrantheum demissum** F.Muell.

Allobia

Synonyms:
Allobia Raf. === **Euphorbia** L.

Allosandra

Synonyms:
Allosandra Raf. === **Tragia** Plum. ex L.

Alphandia

3 species, New Guinea, Vanuatu (Aneityum I.) and New Caledonia; shrubs or small trees with distinctive hard, shining leaves. The leaves exude a copious resinous sap which coats the upper surfaces and petioles upon drying (Airy-Shaw, 1966). *A. furfuracea* of New Caledonia, a maquis dweller, is by some considered to encompass Vanuatu. The Papuasian

species was first collected near Jayapura and may thus be associated with the Cyclops geological terrane. Webster (Synopsis, 1994) has included the genus together with *Beyeria* and *Ricinocarpos* in his subtribe *Ricinocarpinae*, in our opinion a somewhat odd move and ultimately largely related to pollen form. (Crotonoideae)

Pax, F. (with K. Hoffmann) (1911). *Alphandia*. In A. Engler (ed.), Das Pflanzenreich, IV 147 III (Euphorbiaceae-Cluytieae): 22. Berlin. (Heft 47.) La/Ge. — 2 species, New Caledonia.

Airy-Shaw, H. K. (1966). Notes on Malaysian and other Asiatic Euphorbiaceae, LXXIII. *Alphandia* Baill. in W. New Guinea. Kew Bull. 20: 395-398. En. — New species, *A. verniciflua*, in New Guinea; extensive discussion of genus.

McPherson, G. & C. Tirel (1987). *Alphandia*. Fl. Nouvelle-Calédonie, 14 (Euphorbiacées, I): 86-90. Paris. Fr. — Flora treatment (2 species, one only in Île Art north of Grande-Terre); key.

Alphandia Baill., Adansonia 11: 85 (1873).
New Guinea, Vanuatu, New Caledonia. 42 60.

Alphandia furfuracea Baill., Adansonia 11: 86 (1873).
New Caledonia, Vanuatu (Aneityum I.). 60 NWC VAN. Nanophan. or phan.

Alphandia resinosa Baill., Adansonia 11: 86 (1873).
New Caledonia (I. Art). 60 NWC. Nanophan. or phan.

Alphandia verniciflua Airy Shaw, Kew Bull. 20: 395 (1966).
Irian Jaya. 42 NWG. Phan.

Altora

Synonyms:
Altora Adans. === **Clutia** Boerh. ex L.

Amanoa

16 species, Mesoamerica, Caribbean and tropical S. America; W. & WC. tropical Africa (2 species). Monoecious forest trees to as high as 35 m (*A. congesta*) with alternate, pinnately veined leaves. *A. guianensis* is widely distributed from Belize to northern Brazil. The African species are limited to the wet forest zones of Sierra Leone to the Ivory Coast and from Cameroon to Congo (Zaïre). Of the 3 sections, that in Africa is endemic; the other two are American. A precursor to a modern revision for the Americas was published by Hayden (1990) who showed that Aublet's material of *A. guianensis* included discordant elements, necessitating lectotypification and description of a novelty, *A. neglecta* (only collected three times in more than 200 years!). (Phyllanthoideae)

Pax, F. & K. Hoffmann (1922). *Amanoa*. In A. Engler (ed.), Das Pflanzenreich, IV 147 XV (Euphorbiaceae-Phyllanthoideae-Phyllantheae): 191-201. Berlin. (Heft 81.) La/Ge. — 9 species in 3 sections; 3 species in Africa, the rest in the Americas. One additional species (*A. muricata*, Bolivia) dubious.

Jablonski, E. (1967). *Amanoa*. Euphorbiaceae, Guayana Highland (Mem. New York Bot. Gard. 17(1)): 82-84. New York. En. — 4 species, none new; key. [Now partly superseded.]

Hayden, W.J. (1990). Notes on neotropical *Amanoa* (Euphorbiaceae). Brittonia 42: 260-270, illus. En. — Novelties, notes and reductions; key to all species (pp. 269-270) with indication of synonymy. Reliance was placed on systematic anatomical data in support of delineation of species limits.

Amanoa Aubl., Hist. Pl. Guiane 1: 256 (1775).
Trop. America, W. & WC. Trop. Africa. 22 23 80 81 82 83 84. Nanophan. or phan.

Amanoa almerindae Leal, Arch. Jard. Bot. Rio de Janeiro 11: 68 (1951).
Brazil (Amazonas), Venezuela (Amazonas). 82 VEN 84 BZN.
Amanoa pubescens Steyerm., Fieldiana, Bot. 28: 304 (1952).

Amanoa anomala Little, Phytologia 18: 413 (1969).
Ecuador. 83 ECU.

Amanoa bracteosa Planch., Hooker's Icon. Pl.8: t. 797 (1848).
W. Trop. Africa. 22 IVO LBR SIE. Phan.
Amanoa strobilantha Planch. in W.J.Hooker, Niger Fl.: 47 (1849).

Amanoa caribaea Krug & Urb., Notizbl. Bot. Gart. Berlin-Dahlem 1: 326 (1897).
Leeward Is., Windward Is. 81 LEE WIN. Phan.

Amanoa congesta W.J.Hayden, Brittonia 42: 261 (1990).
Guiana, Brazil (Amapá, Pará). 82 FRG 84 BZN. Phan.

Amanoa cupatensis Huber, Bol. Mus. Goeldi Paraense Hist. Nat. Ethnogr. 7: 296 (1910).
Brazil (NW. Amazonas), Venezuela (Amazonas). 82 VEN 84 BZN. Phan.

Amanoa glaucophylla Müll.Arg. in C.F.P.von Martius, Fl. Bras. 11(2): 11 (1873).
Brazil (Piauí, Goiás). 84 BZC BZE. Nanophan. or phan.

Amanoa gracillima W.J.Hayden, Brittonia 42: 262 (1990).
Brazil (Amazonas). 84 BZN. Phan.

Amanoa guianensis Aubl., Hist. Pl. Guiane 1: 256 (1775). *Amanoa guianensis* var. *genuina* Müll.Arg. in A.P.de Candolle, Prodr. 15(2): 219 (1866), nom. inval.
Trop. America. 80 BLZ COS GUA HON NIC PAN 81 TRT 82 FRG GUY SUR VEN 83 CLM 84 BZN. Nanophan. or phan.
Amanoa bracteata Rich. ex Baill., Étude Euphorb.: 581 (1858).
Amanoa guianensis var. *oblonga* Benth. ex Baill., Adansonia 5: 345 (1865).
Amanoa guianensis var. *grandiflora* Müll.Arg. in A.P.de Candolle, Prodr. 15(2): 219 (1866). *Amanoa grandiflora* (Müll.Arg.) Müll.Arg., Flora 55: 2 (1872).
Amanoa guianensis var. *poeppigii* Müll.Arg. in A.P.de Candolle, Prodr. 15(2): 220 (1866).
Amanoa potamophila Croizat, Amer. Midl. Naturalist 29: 475 (1943).
Amanoa macrocarpa Cuatrec., Brittonia 11: 164 (1959).

Amanoa muricata Rusby, Bull. New York Bot. Gard. 8: 100 (1912).
Bolivia. 83 BOL. Nanophan.

Amanoa nanayensis W.J.Hayden, Brittonia 42: 265 (1990).
Peru (Loreto). 83 PER. Phan.

Amanoa neglecta W.J.Hayden, Brittonia 42: 267 (1990).
Guiana, Surinam. 82 FRG SUR. Nanophan. or phan.

Amanoa oblongifolia Müll.Arg., Linnaea 32: 77 (1863). *Amanoa guianensis* var. *oblonga* Benth. ex Baill., Adansonia 5: 345 (1865).
Venezuela (Amazonas), Brazil (Amazonas), Peru (Loreto). 82 VEN 83 PER 84 BZN. Phan.

Amanoa sinuosa W.J.Hayden, Brittonia 42: 268 (1990).
Brazil (Amazonas), Peru (Loreto). 83 PER 84 BZN. Phan.
* *Amanoa robusta* Leal, Arch. Jard. Bot. Rio de Janeiro 11: 69 (1951), nom. illeg.

Amanoa steyermarkii Jabl., Acta Bot. Venez. 2: 237 (1967).
Venezuela (Bolívar). 82 VEN. Phan.

Amanoa strobilacea Müll.Arg., Flora 47: 515 (1864).
W. Trop. Africa to Cameroon. 22 GHA LBR 23 CMN. Phan.

Synonyms:
Amanoa acuminata Thwaites === **Cleistanthus acuminatus** (Thwaites) Müll.Arg.
Amanoa boiviniana Baill. === **Cleistanthus boivinianus** (Baill.) Müll.Arg.
Amanoa bracteata Rich. ex Baill. === **Amanoa guianensis** Aubl.
Amanoa brasiliensis Baill. === **Gonatogyne brasiliensis** (Baill.) Müll.Arg.
Amanoa chartacea Baill. ex Müll.Arg. === **Cleistanthus oblongifolius** (Roxb.) Müll.Arg.
Amanoa collina (Roxb.) Baill. === **Cleistanthus collinus** (Roxb.) Benth.
Amanoa cunninghamii (Müll.Arg.) Baill. === **Cleistanthus cunninghamii** (Müll.Arg.) Müll.Arg.
Amanoa dallachyana Baill. === **Cleistanthus dallachyanus** (Baill.) Benth.
Amanoa divaricata Poepp. & Endl. === **Richeria grandis** Vahl var. **grandis**
Amanoa faginea Baill. === **Bridelia leichardtii** Baill. ex Müll.Arg.
Amanoa ferruginea Thwaites === **Cleistanthus ferrugineus** (Thwaites) Müll.Arg.
Amanoa grandiflora (Müll.Arg.) Müll.Arg. === **Amanoa guianensis** Aubl.
Amanoa guianensis var. *genuina* Müll.Arg. === **Amanoa guianensis** Aubl.
Amanoa guianensis var. *grandiflora* Müll.Arg. === **Amanoa guianensis** Aubl.
Amanoa guianensis var. *oblonga* Benth. ex Baill. === **Amanoa guianensis** Aubl.
Amanoa guianensis var. *poeppigii* Müll.Arg. === **Amanoa guianensis** Aubl.
Amanoa indica Wight === **Cleistanthus patulus** (Roxb.) Müll.Arg.
Amanoa indica Thwaites === **Cleistanthus robustus** (Thwaites) Müll.Arg.
Amanoa indica f. *minor* Thwaites === **Cleistanthus patulus** (Roxb.) Müll.Arg.
Amanoa laurifolia Pax === **Pentabrachion reticulatum** Müll.Arg.
Amanoa leichardtii Baill. === **Bridelia leichardtii** Baill. ex Müll.Arg.
Amanoa macrocarpa Cuatrec. === **Amanoa guianensis** Aubl.
Amanoa ovata Baill. === **Bridelia exaltata** F.Muell.
Amanoa pallida Thwaites === **Cleistanthus pallidus** (Thwaites) Müll.Arg.
Amanoa patula (Roxb.) Thwaites === **Cleistanthus patulus** (Roxb.) Müll.Arg.
Amanoa potamophila Croizat === **Amanoa guianensis** Aubl.
Amanoa pubescens Steyerm. === **Amanoa almerindae** Leal
Amanoa purpurascens Poepp. ex Baill. === **Richeria grandis** Vahl var. **grandis**
Amanoa racemosa Poepp. & Endl. === **Richeria grandis** Vahl var. **grandis**
Amanoa ramiflora Poepp. & Endl. === **Richeria grandis** Vahl var. **grandis**
Amanoa robusta Thwaites === **Cleistanthus robustus** (Thwaites) Müll.Arg.
Amanoa robusta Leal === **Amanoa sinuosa** W.J.Hayden
Amanoa schweinfurthii Baker & Hutch. === **Erythroxylum fischeri** Engl. (Erythroxylaceae)
Amanoa stenonia Baill. === **Cleistanthus stenonia** (Baill.) Jabl.
Amanoa strobilantha Planch. === **Amanoa bracteosa** Planch.
Amanoa tomentosa Baill. === **Bridelia tomentosa** Blume

Ambinux

Synonyms:
Ambinux Comm. ex A.Juss. === **Aleurites** J.R.Forst. & G.Forst.

Amperea

8 species, Australia (mainly in southwestern Western Australia; only *A. xiphoclada* is eastern). Low microphyllous perennials, the short-lived stems arising continuously (or perhaps periodically) from a woody, fire-resistant stock; *A. volubilis* is a twining perennial. The narrow cotyledons resulted it being placed in a subdivision known as 'Stenolobae' by Mueller and by Pax and Hoffmann; however, like other Australian genera of that group it appears instead to be specialised, in this case within the Acalyphoideae (Webster 1994). The most recent account is by Henderson (1992) who believes the genus most closely related to *Monotaxis* with one species, *Amperea spicata*, apparently transitional. (Acalyphoideae)

Gruening, G. (1913). *Amperea*. In A. Engler (ed.), Das Pflanzenreich, IV 147 [Stenolobieae] (Euphorbiaceae-Porantheroideae et Ricinocarpoideae): 86-92. Berlin. (Heft 58.) La/Ge. — 6 species in 2 sections, Australia (5 in W Australia).

- Henderson, R. J. F. (1992). Studies in Euphorbiaceae A. L. Juss. sens. lat., I: A revision of *Amperea* Adr. Juss. (Acalyphoideae Ascherson, Ampereae Muell.-Arg.). Austral. Syst. Bot. 5: 1-27, illus., maps. En. — Full treatment (8 species) with key, descriptions, synonymy, references, types, localities with exsiccatae, indication of distribution and habitat, and commentary; references but no separate index. An extensive general part covers the history of the genus, its relationships, character assessment and conservation status. [The sections of Gruening have not been maintained.]

Amperea A.Juss., Euphorb. Gen.: 35 (1824).
Australia. 50.

Amperea conferta Benth., Fl. Austral. 6: 83 (1873).
SE. Western Australia. 50 WAU. Cham.
Amperea podperae Domin, Mem. Soc. Boheme 2: 59 (1921-1922 publ. 1923).

Amperea ericoides A.Juss., Euphorb. Gen.: 112 (1824).
SW. Western Australia. 50 WAU. Cham.
Amperea rosmarinifolia Klotzsch in J.G.C.Lehmann, Pl. Preiss. 1: 176 (1845).
Amperea ericoides var. *linearis* Grüning in H.G.A.Engler, Pflanzenr., IV, 147: 89 (1913).

Amperea micrantha Benth., Fl. Austral. 6: 83 (1873).
SW. Western Australia. 50 WAU. Cham.

Amperea protensa Nees in J.G.C.Lehmann, Pl. Preiss. 2: 229 (1848). *Amperea protensa* var. *genuina* Müll.Arg. in A.P.de Candolle, Prodr. 15(2): 214 (1866), nom. inval.
S. Western Australia. 50 WAU. Cham.
Amperea protensa var. *tenuiramea* Müll.Arg. in A.P.de Candolle, Prodr. 15(2): 213 (1866).

Amperea simulans R.J.F.Hend., Austral. Syst. Bot. 5: 18 (1992).
SW. Western Australia. 50 WAU.
Amperea ericoides var. *planifolia* Grüning in H.G.A.Engler, Pflanzenr., IV, 147: 90 (1913).

Amperea spicata Airy Shaw, Kew Bull. 37: 380 (1982).
Northern Territory (Central Australia South). 50 NTA. Cham.

Amperea volubilis F Muell. ex Benth., Fl. Austral. 6: 82 (1873).
SW. Western Australia. 50 WAU. Cl. cham.

Amperea xiphoclada (Sieber ex Spreng.) Druce, Bot. Soc. Exch. Club Brit. Isles 4: 604 (1916 publ. 1917). – FIGURE, p. 160.
E. & SE. Australia. 50 NSW QLD SOA TAS VIC. Cham.
* *Leptomeria xiphoclada* Sieber ex Spreng., Syst. Veg. 4(2): 109 (1827).

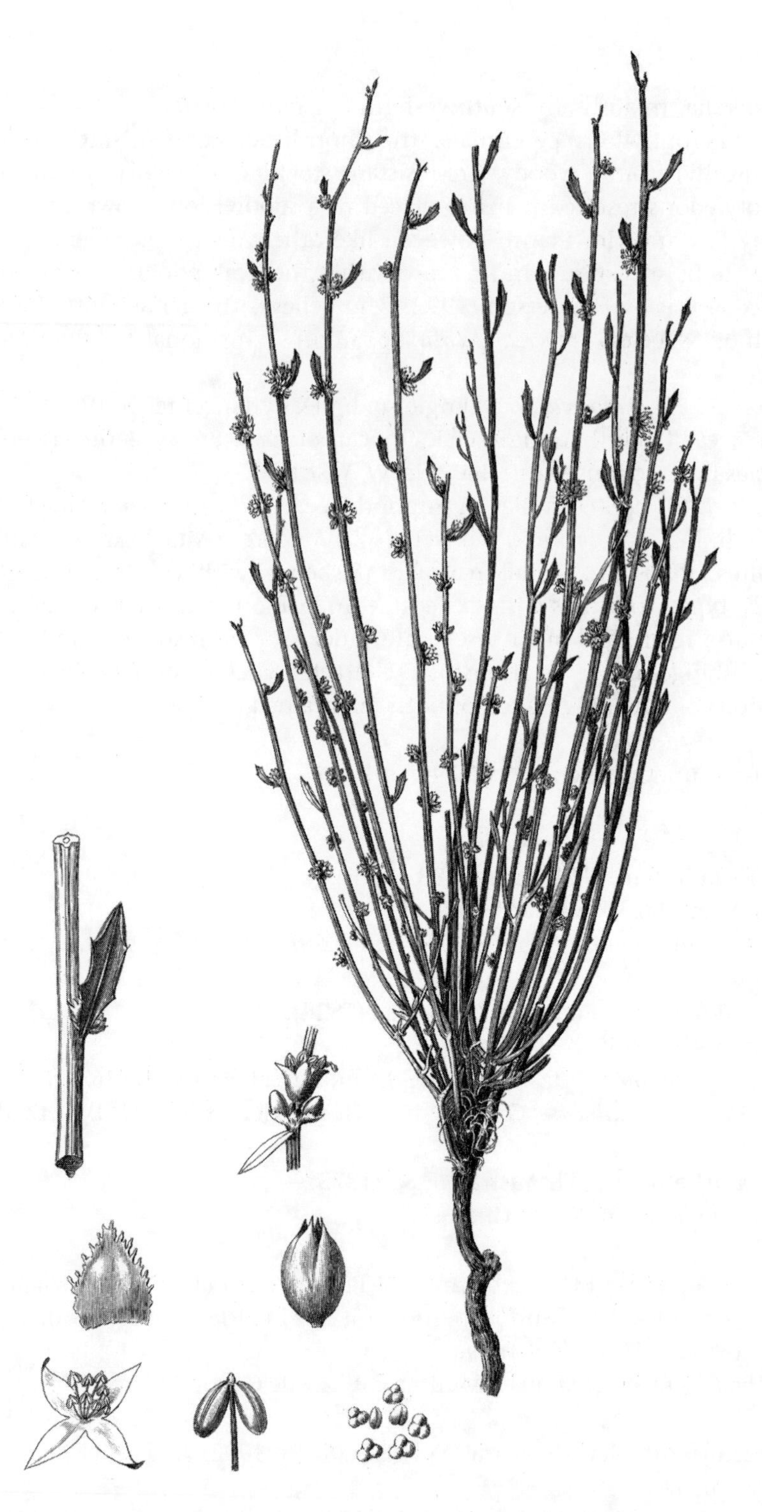

Amperea xiphoclada (Seiber ex Spreng.) Druce var. *xiphoclada* (as *Amperea spartioides* Brongn.)
Artist: Bessa
Duperrey, Voy. Monde, pl. 49, left [1834]
KEW ILLUSTRATIONS COLLECTION

var. **papillata** R.J.F.Hend., Austral. Syst. Bot. 5: 24 (1992).
EC. New South Wales. 50 NSW. Cham.

var. **pedicellata** R.J.F.Hend., Austral. Syst. Bot. 5: 24 (1992).
New South Wales (Sydney). 50 NSW. Cham.

var. **xiphoclada**
E. to SE. Australia. 50 NSW QLD SOA TAS VIC. Cham.

Amperea spartioides Brongn. in L.I.Duperrey, Voy. Monde: 226 (1829).
Amperea cuneiformis F.Muell. ex Baill., Étude Euphorb.: 455 (1858), nom. inval.

Synonyms:
Amperea cuneiformis F.Muell. ex Baill. === **Amperea xiphoclada** (Sieber ex Spreng.) Druce var. **xiphoclada**
Amperea ericoides var. *linearis* Grüning === **Amperea ericoides** A.Juss.
Amperea ericoides var. *planifolia* Grüning === **Amperea simulans** R.J.F.Hend.
Amperea podperae Domin === **Amperea conferta** Benth.
Amperea protensa var. *genuina* Müll.Arg. === **Amperea protensa** Nees
Amperea protensa var. *tenuiramea* Müll.Arg. === **Amperea protensa** Nees
Amperea rosmarinifolia Klotzsch === **Amperea ericoides** A.Juss.
Amperea spartioides Brongn. === **Amperea xiphoclada** (Sieber ex Spreng.) Druce var. **xiphoclada**
Amperea subnuda Nees === **Gyrostemon subnudus** (Nees) Baill. (Gyrostemonaceae)

Amyrea

10 species, Madagascar, with *A. humbertii* also in the Comoros (Mayotte); shrubs or small trees to 10 m of submontane and montane forest with alternate, sometimes toothed leaves, simple, flexuous, 'agrostiform' axillary inflorescences, and apetalous flowers. The bracts are moreover glume-like, with the whole recalling *Agrostistachys* and *Pseudagrostistachys*. Webster (Synopsis, 1994; see **General**) considered it allied to *Claoxylon*, while Radcliffe-Smith (1998) opted for an affinity with *Necepsia* (Bernardieae). The genus also shows a superficial similarity to *Tannodia* (Crotonoideae). (Acalyphoideae)

Leandri, J. (1941). Contributions à l'étude des Euphorbiacées de Madagascar, IV: Acalyphées. Notul. Syst. (Paris) 9: 156-188, illus. Fr. — Includes (pp. 168-170) protologue of genus and description of *AA. sambiranensis* and *humbertii*; figure of *A. sambiranensis* in pl. 1 (p. 160).

- Radcliffe-Smith, A. (1998). A synopsis of the genus *Amyrea* Leandri (Euphorbiaceae-Acalyphoideae). Kew Bull. 53: 437-451, illus. En. — Synoptic revision (11 species, 9 of them new) with key, descriptions, Latin diagnoses of novelties, synonymy, vernacular names (where recorded), localities with exsiccatae (though rather few since 1970), and indication of distribution and habitat, little taxonomic discussion. Distinctive features of the genus, along with its putative relationships, are discussed in the introduction.

Amyrea Leandri, Notul. Syst. (Paris) 9: 168 (1941).
Comoros, Madagascar. 29.

Amyrea celastroides Radcl.-Sm., Kew Bull. 53: 440 (1998).
Madagascar. 29 MDG. Phan.

Amyrea eucleoides Radcl.-Sm., Kew Bull. 53: 440 (1998).
Madagascar. 29 MDG. Nanophan.

Amyrea gracillima Radcl.-Sm., Kew Bull. 53: 442 (1998).
Madagascar. 29 MDG. Nanophan.

Amyrea humbertii Leandri, Notul. Syst. (Paris) 9: 169 (1941).
Comoros (Mayotte), Madagascar. 29 MDG. Nanophan. or phan.

Amyrea lancifolia Radcl.-Sm., Kew Bull. 53: 445 (1998).
Madagascar. 29 MDG. Phan.

Amyrea maprouneifolia Radcl.-Sm., Kew Bull. 53: 447 (1998).
Madagascar. 29 MDG. Nanophan.

Amyrea myrtifolia Radcl.-Sm., Kew Bull. 53: 447 (1998).
Madagascar. 29 MDG. Nanophan.

Amyrea remotiflora Radcl.-Sm., Kew Bull. 53: 449 (1998).
Madagascar. 29 MDG. Nanophan.

Amyrea sambiranensis Leandri, Notul. Syst. (Paris) 9: 169 (1941).
Madagascar. 29 MDG. Nanophan.

Amyrea stenocarpa Radcl.-Sm., Kew Bull. 53: 451 (1998).
Madagascar. 29 MDG. Phan.

Anabaena

Rejected against *Anabaena* Bory ex Bornet & Flahault (Cyanophyceae, Cyanobacteria).

Synonyms:
Anabaena A.Juss. === **Romanoa** Trevis.
Anabaena tamnoides A.Juss. === **Romanoa tamnoides** (A.Juss.) Radcl.-Sm.

Anabaenella

Synonyms:
Anabaenella Pax & K.Hoffm. === **Romanoa** Trevis.
Anabaenella tamnoides (A.Juss.) Pax & K.Hoffm. === **Romanoa tamnoides** (A.Juss.) Radcl.-Sm.
Anabaenella tamnoides var. *sinuata* (Ule) Pax & K.Hoffm. === **Romanoa tamnoides** var. **sinuata** (Ule) Radcl.-Sm.

Anaua

Synonyms:
Anaua Miq. === **Drypetes** Vahl
Anaua sumatrana Miq. === **Drypetes sumatrana** (Miq.) Pax & K.Hoffm.

Anda

Synonyms:
Anda A.Juss. === **Joannesia** Vell.

Andicus

Synonyms:
Andicus Vell. === **Joannesia** Vell.

Andiscus

Synonyms:
Andiscus Vell. === **Joannesia** Vell.

Andrachne

43 species, tropics and Mediterranean (in wider sense), south to southern Africa, south to Peru and north from the Caribbean (Cuba) to the southern and southwestern United States in the Americas, and east through Middle and South Asia to China, Malesia and Australia; often many-stemmed shrubs or perennials with a woody base. As interpreted here (following Pax & Hoffmann, 1922, and Pojarkova 1940) inclusive of *Leptopus*, an opinion contested by some (including Pojarkova 1960 and Webster) who point to a different (anatropous) arrangement of the ovule in that taxon. The German authors recognised 4 sections in 1922 and again in 1931; Pojarkova in 1960 accepted 3 sections in *Andrachne* and 2 in *Leptopus*. Hoffmann (1994) added 3 species from *Savia* sect. *Maschalanthus*, reviving Müller's sects. *Phyllanthopsis* and *Pseudophyllanthus*. *AA. aspera* and *telephoides* are together very widespread from the Cape Verdes across northern Africa and the Mediterranean to Tadzhikistan and Socotra. Some species are very similar to *Phyllanthus* save for the presence here of petals; any possible affinity is as yet unproven. The most recent overall study is by Pojarkova (1960). The genus has also made its appearance in many floras and partial studies. Further studies on this and related genera including *Savia* have been carried out by Petra Hoffmann; it is hoped these will be continued. No testable geographical or phylogenetic system has been attempted; Pax & Hoffmann (1922) believed it to be 'eine phylogenetisch alte Gattung' and its modern distribution a remnant of something larger. (Phyllanthoideae)

Pax, F. & K. Hoffmann (1922). *Andrachne*. In A. Engler (ed.), Das Pflanzenreich, IV 147 XV (Euphorbiaceae-Phyllanthoideae-Phyllantheae): 169-179. Berlin. (Heft 81.) La/Ge. — 16 species, tropics and Mediterranean. [In wider and currently accepted sense; divided by Pojarkova (1960) with re-segregation of *Leptopus*.]

Pojarkova, A. I. (1940). K sistematike kavkazskikh i vostochno-sredizemnomorskikh predstavitelej *Andrachne* s.l./Contribution à la systématique des représentatives du genre *Andrachne* s.l. habitant le Caucase et la partie de la région méditerranéenne. Bot. Zhurn. SSSR 25: 341-348. Ru/Fr. — *Andrachne* and *Leptopus* here combined following Pax.

• Pojarkova, A. I. (1960). Materialy k monografii rodov *Andrachne* L. i *Leptopus* Decne./ Incrementa ad monographiam generum *Andrachne* L. et *Leptopus* Decne. Notul. Syst. (Leningrad) 20: 251-274, map. Ru. — *Leptopus* segregated from *Andrachne* in a wider sense; synopses of both genera (including some novelties) with 22 *Andrachne* in 3 sections (chiefly Afro-SW Asian but scattered elsewhere) and 20 *Leptopus* in 2 sections (mainly in monsoonal Asia through to Australia). The synopses include synonymy, references and citations, typifications, descriptions of novelties, indication of distribution, and commentary but no keys. Several new infrageneric taxa are described

Warnock, B. H. & M. C. Johnston (1960). The genus *Savia* (Euphorbiaceae) in extreme western Texas. Southwestern Nat. 5: 1-6. En. — Includes key to 2 species (one new) along with descriptions, synonymy, and indication of distribution, ecology and phenology; no map. [Both species now referable to *Andrachne*.]

Airy-Shaw, H. K. (1965). Notes on Malaysian and other Asiatic Euphorbiaceae, XLIX. On the identity of *Thelypetalum* Gagnep. Kew Bull. 19: 299-300. En. — Genus reduced to *Andrachne*; discussion.

Airy-Shaw, H. K. (1971). Notes on Malesian and other Asiatic Euphorbiaceae, CXXVI. A new *Leptopus* from Indochina. Kew Bull. 25: 495-496. En. — *L. robinsonii* described. [Recombined in *Andrachne* for the *World Checklist*.]

Airy-Shaw, H. K. (1978). Notes on Malesian and other Asiatic Euphorbiaceae, CXCI. The genus *Leptopus* Decne. in New Guinea. Kew Bull. 32: 379. En. — Range extension of *L. decaisnei* (=*Andrachne decaisnei*) as var. *orbicularis*, originally known from northern Australia. [The relationship and extent of this and the nominate variety require further study.]

Li Ping-t'ao [P.T. Li] (1983). A synopsis of the genus *Leptopus* Decne. (Euphorbiaceae) in China. Notes Roy. Bot. Gard. Edinburgh 40: 467-474, illus., map. En. — Synoptic revision (8 species, 2 new) with key, descriptions of novelties, synonymy, references and citations, localities with exsiccatae, indication of distribution, and commentary.

Gilbert, M. G. & M. Thulin (1988). *Andrachne fragilis* (Euphorbiaceae), a new species from Somalia. Nordic J. Bot. 8: 159-160. En. — Description, with indication of habitat; no key.

Hoffmann, P. (1994). A contribution to the systematics of *Andrachne* sect. *Phyllanthopsis* and sect. *Pseudophyllanthus* compared with *Savia* s.l. (Euphorbiaceae) with special reference to floral morphology. Bot. Jahrb. Syst. 116: 321-331, illus. En. — Analysis of the generic position of three *Andrachne* species, two in sect. *Phyllanthopsis* and one in sect. *Pseudophyllanthus*, with synonymy, references and types indicated. Floral morphology was in particular studied, and some chromosome counts are also given. The first group was found to accord well with *Leptopus*, while the second combined features of both it and *Andrachne*. [The study formed part of a larger programme covering *Savia* s.l. Both sections studied were included in that genus by Pax & Hoffmann (1922) and assigned to its sect. *Maschalanthus*.]

Andrachne L., Sp. Pl.: 1014 (1753).
EC. & SE. U.S.A., Mexico, Caribbean, Medit. to W. Himalaya. 11 12 13 20 21 22 23 24 26 27 32 33 34 35 36 40 41 42 50 74 77 78 79 81. Hemicr., cham. or nanophan.
Eraclissa Forssk., Fl. Aegypt.-Arab.: 208 (1775).
Arachne Neck., Elem. Bot. 2: 348 (1790).
Telephioides Tourn. ex Moench, Suppl. Meth.: 310 (1802).
Lepidanthus Nutt., Trans. Amer. Philos. Soc., n.s. 5: 175 (1837).
Maschalanthus Nutt., Trans. Amer. Philos. Soc., n.s. 5: 175 (1837).
Leptopus Decne. in V.Jacquemont, Voy. Inde 4: 155 (1844).
Phyllanthidea Didr., Vidensk. Meddel. Dansk Naturhist. Foren. Kjøbenhavn 1857: 150 (1857).
Thelypetalum Gagnep., Bull. Soc. Bot. France 71: 876 (1924 publ. 1925).

Andrachne afghanica Pojark., Bot. Mater. Gerb. Bot. Inst. Komarova Akad. Nauk S.S.S.R. 20: 260 (1960).
Afghanistan. 34 AFG. Hemicr. or cham.

Andrachne arida (Warnock & M.C.Johnst.) G.L.Webster, J. Arnold Arbor. 48: 328 (1967).
Texas, Mexico (Coahuila). 77 TEX 79 MXE. Nanophan.
* *Savia arida* Warnock & M.C.Johnst., SouthW. Naturalist 5: 3 (1960).

Andrachne aspera Spreng., Syst. Veg. 3: 884 (1826).
Cape Verde Is., N. Africa to Pakistan, Sudan, Ethiopia, Somalia, Cameroon, Socotra (Abd-al-Kuri). 20 EGY MOR 21 CVI 22 MTN 23 CMN 24 ETH SOC SOM SUD 25 KEN 34 IRN PAL SAU SIN TUR 35 YEM 40 PAK. Hemicr. or cham.
Andrachne aspera var. *glandulosa* Hochst. ex A.Rich., Tent. Fl. Abyss. 2: 254 (1850).
Andrachne aspera var. *maritima* N.Terracc., Annuario Reale Ist. Bot. Roma 5: 98 (1894).

Andrachne australis Zoll. & Moritzi, Natuur.-Geneesk. Arch. Ned.-Indië 2: 17 (1845).
Leptopus australis (Zoll. & Moritzi) Pojark., Bot. Mater. Gerb. Bot. Inst. Komarova Akad. Nauk S.S.S.R. 20: 270 (1960).
SW. Hainan, Indo-China to Philippines. 36 CHH 41 THA VIE 42 JAW LSI MLY PHI. Cham. or nanophan.
Agyneia tenera Zoll. & Moritzi ex Miq., Fl. Ned. Ind. 1(2): 365 (1859).
Andrachne tenera Miq., Fl. Ned. Ind. 1(2): 365 (1859).
Andrachne australis var. *lanceolata* Müll.Arg. in A.P.de Candolle, Prodr. 15(2): 235 (1866).
Andrachne fruticosa Hook.f., Fl. Brit. India 5: 284 (1887), nom. illeg.
Andrachne lanceolata Pierre ex Beille in H.Lecomte, Fl. Indo-Chine 5: 539 (1927).
Leptopus lanceolatus (Pierre ex Beille) Pojark., Bot. Mater. Gerb. Bot. Inst. Komarova Akad. Nauk S.S.S.R. 20: 271 (1960).
Leptopus philippinensis Pojark., Bot. Mater. Gerb. Bot. Inst. Komarova Akad. Nauk S.S.S.R. 20: 270 (1960).

Andrachne brittonii Urb., Symb. Antill. 7: 245 (1912).
E. Cuba, Haiti. 81 CUB HAI. Cham.
Securinega abeggii Urb. & Ekman, Ark. Bot. 20A(15): 46 (1926).

Andrachne buschiana Pojark., Bot. Zhurn. S.S.S.R. 25: 342 (1940).
S. Transcaucasus. 33 TCS. Cham.
Andrachne fruticulosa Schischk., Bull. Mus. Geogr. 1: 21 (1922), nom. illeg.

Andrachne calcarea Ridl., Bull. Misc. Inform. Kew 1923: 361 (1923). *Leptopus calcareus* (Ridl.) Pojark., Bot. Mater. Gerb. Bot. Inst. Komarova Akad. Nauk S.S.S.R. 20: 271 (1960).
S. Thailand, Pen. Malaysia (incl. Lankawi I.). 41 THA 42 MLY. Nanophan.

Andrachne chinensis Bunge, Enum. Pl. China Bor.: 59 (1833). *Leptopus chinensis* (Bunge) Pojark., Bot. Mater. Gerb. Bot. Inst. Komarova Akad. Nauk S.S.S.R. 20: 274 (1960).
China, Tibet. 36 CHC CHI CHM CHN CHQ CHS CHT. Nanophan.
Flueggea capillipes Pax, Bot. Jahrb. Syst. 29: 427 (1900). *Andrachne capillipes* (Pax) Hutch. in C.S.Sargent, Pl. Wilson. 2: 516 (1916). *Leptopus capillipes* (Pax) Pojark., Bot. Mater. Gerb. Bot. Inst. Komarova Akad. Nauk S.S.S.R. 20: 273 (1960).
Andrachne bodinieri H.Lév., Repert. Spec. Nov. Regni Veg. 12: 187 (1913).
Andrachne cavaleriei H.Lév., Repert. Spec. Nov. Regni Veg. 12: 187 (1913).
Andrachne capillipes var. *pubescens* Hutch. in C.S.Sargent, Pl. Wilson. 2: 516 (1916). *Leptopus chinensis* var. *pubescens* (Hutch.) S.B.Ho, Fl. Tsinlingensis 1(3): 168 (1981).
Andrachne hirsuta Hutch. in C.S.Sargent, Pl. Wilson. 2: 516 (1916). *Leptopus hirsutus* (Hutch.) Pojark., Bot. Mater. Gerb. Bot. Inst. Komarova Akad. Nauk S.S.S.R. 20: 271 (1960). *Leptopus chinensis* var. *hirsutus* (Hutch.) P.T.Li, Notes Roy. Bot. Gard. Edinburgh 40: 474 (1983).
Andrachne montana Hutch. in C.S.Sargent, Pl. Wilson. 2: 517 (1916). *Leptopus montanus* (Hutch.) Pojark., Bot. Mater. Gerb. Bot. Inst. Komarova Akad. Nauk S.S.S.R. 20: 274 (1960).

Andrachne clarkei Hook.f., Fl. Brit. India 5: 285 (1887). *Leptopus clarkei* (Hook.f.) Pojark., Bot. Mater. Gerb. Bot. Inst. Komarova Akad. Nauk S.S.S.R. 20: 272 (1960).
E. Assam, China (Yunnan). 36 CHC 40 ASS. Nanophan.

Andrachne colchica Fisch. & C.A.Mey. ex Boiss., Fl. Orient. 4: 1137 (1879). *Leptopus colchicus* (Fisch. & C.A.Mey. ex Boiss.) Pojark., Bot. Mater. Gerb. Bot. Inst. Komarova Akad. Nauk S.S.S.R. 20: 274 (1960).
W. Caucasus. 33 TCS. Nanophan.
Andrachne cadishan Roxb. ex Hook.f., Fl. Brit. India 5: 275 (1887), pro syn.

Andrachne cordifolia (Wall. ex Decne.) Müll.Arg. in A.P.de Candolle, Prodr. 15(2): 234 (1866).
Pakistan, Kashmir, N. India, Nepal, SW. China. 36 CHC 40 IND NEP PAK WHM. Nanophan.
* *Phyllanthus cordifolius* Wall. ex Decne. in V.Jacquemont, Voy. Inde 4: 155 (1844). *Leptopus cordifolius* Decne. in V.Jacquemont, Voy. Inde 4: 155 (1844).
Andrachne decaisneana Baill., Étude Euphorb.: 577 (1858).

Andrachne decaisnei Benth., Fl. Austral. 6: 88 (1873). *Leptopus decaisnei* (Benth.) Pojark., Bot. Mater. Gerb. Bot. Inst. Komarova Akad. Nauk S.S.S.R. 20: 271 (1960).
E. Jawa to N. Australia. 42 JAW LSI NWG 50 NTA QLD WAU. Cham.
* *Andrachne fruticosa* Decne. ex Müll.Arg. in A.P.de Candolle, Prodr. 15(2): 235 (1866), nom. illeg.

var. **decaisnei**
E. Jawa, Lesser Sunda Is., N. Australia. 42 JAW LSI 50 NTA QLD WAU. Cham.

var. **orbicularis** Benth., Fl. Austral. 6: 88 (1973). *Andrachne orbicularis* (Benth.) Domin, Biblioth. Bot. 89: 315 (1927). *Leptopus orbicularis* (Benth.) Pojark., Bot. Mater. Gerb. Bot. Inst. Komarova Akad. Nauk S.S.S.R. 20: 272 (1960). *Leptopus decaisnei* var. *orbicularis* (Benth.) Airy Shaw, Kew Bull. 32: 379 (1978).
Papua New Guinea, N. Australia. 42 NWG 50 NTA QLD WAU. Cham.
Leptopus dominianus Pojark., Bot. Mater. Gerb. Bot. Inst. Komarova Akad. Nauk S.S.S.R. 20: 511 (1960), nom. nud.

Andrachne emicans Dunn, Bull. Misc. Inform. Kew 1920: 210 (1920). *Leptopus emicans* (Dunn) Pojark., Bot. Mater. Gerb. Bot. Inst. Komarova Akad. Nauk S.S.S.R. 20: 272 (1960).
Arunachal Pradesh (Abor Hills). 40 EHM. Nanophan.

Andrachne ephemera M.G.Gilbert, Kew Bull. 42: 351 (1987).
Ethiopia. 24 ETH.

Andrachne esquirolii H.Lév., Repert. Spec. Nov. Regni Veg. 9: 327 (1911). *Leptopus esquirolii* (H.Lév.) P.T.Li, Notes Roy. Bot. Gard. Edinburgh 40: 471 (1983).
China (Guangxi, Sichuan, Yunnan, Guizhou). 36 CHC CHS. Nanophan.
Andrachne hypoglauca H.Lév., Repert. Spec. Nov. Regni Veg. 12: 187 (1913).
Andrachne persicariifolia H.Lév., Repert. Spec. Nov. Regni Veg. 12: 187 (1913).
Andrachne attenuata Hand.-Mazz., Anz. Akad. Wiss. Wien, Math.-Naturwiss. Kl. 58: 178 (1921). *Leptopus attenuatus* (Hand.-Mazz.) Pojark., Bot. Mater. Gerb. Bot. Inst. Komarova Akad. Nauk S.S.S.R. 20: 271 (1960).
Andrachne attenuata var. *microcalyx* Hand.-Mazz., Symb. Sin. 7: 220 (1931). *Andrachne esquirolii* var. *microcalyx* (Hand.-Mazz.) Rehder, J. Arnold Arbor. 18: 257 (1937).
Leptopus kwangsiensis Pojark., Bot. Mater. Gerb. Bot. Inst. Komarova Akad. Nauk S.S.S.R. 20: 272 (1960).
Leptopus esquirolii var. *villosus* P.T.Li, Acta Phytotax. Sin. 26: 61 (1988).

Andrachne fedtschenkoi Kossinsky, Bot. Mater. Gerb. Glavn. Bot. Sada RSFSR 2: 89 (1921).
Turkmenistan, Tadzhikistan (Pamir Mts.). 32 TKM TZK. Cham.
Andrachne hapladena Pazij, Bot. Mater. Gerb. Inst. Bot. Zool. Akad. Nauk Uzbeksk. S.S.R. 11: 20 (1948).
Andrachne rupestris Pazij, Bot. Mater. Gerb. Inst. Bot. Zool. Akad. Nauk Uzbeksk. S.S.R. 11: 21 (1948).

Andrachne filiformis Pojark., Bot. Zhurn. S.S.S.R. 25: 344 (1940).
S. Transcaucasus. 33 TCS. Cham.

Andrachne fragilis M.G.Gilbert & Thulin, Nordic J. Bot. 8: 159 (1988).
Somalia. 24 SOM.

Andrachne fruticulosa Boiss., Diagn. Pl. Orient. 7: 86 (1846).
SW. Iran. 34 IRN. Cham.

Andrachne gruvelii Daveau, Actes Soc. Linn. Bordeaux 60: 13 (1905).
Morocco. 20 MOR. Cham.

Andrachne hainanensis Merr. & Chun, Sunyatsenia 5: 102 (1940). *Leptopus hainanensis* (Merr. & Chun) Pojark., Bot. Mater. Gerb. Bot. Inst. Komarova Akad. Nauk S.S.S.R. 20: 271 (1960).
SW. Hainan. 36 CHH. Nanophan.
Andrachne hainanensis var. *nummulariifolia* Merr. & Chun, Sunyatsenia 5: 102 (1940).

Andrachne hirta Ridl., Bull. Misc. Inform. Kew 1923: 362 (1923). *Leptopus hirtus* (Ridl.) Pojark., Bot. Mater. Gerb. Bot. Inst. Komarova Akad. Nauk S.S.S.R. 20: 271 (1960).
Pen. Malaysia. 42 MLY. Nanophan.

Andrachne lolonum Hand.-Mazz., Anz. Akad. Wiss. Wien, Math.-Naturwiss. Kl. 58: 178 (1921). *Leptopus lolonus* (Hand.-Mazz.) Pojark., Bot. Mater. Gerb. Bot. Inst. Komarova Akad. Nauk S.S.S.R. 20: 274 (1960).
China (Guizhou, Yunnan, Sichuan). 36 CHC. Cham. or nanophan.

Andrachne maroccana Ball, J. Bot. 13: 205 (1875).
Morocco. 20 MOR. Nanophan.

Andrachne merxmuelleri Rech.f., Mitt. Bot. Staatssamml. München, Beih. 16: 38 (1980).
Iran. 34 IRN.

Andrachne microphylla (Lam.) Baill., Étude Euphorb.: 577 (1858).
Mexico (C. & S. Baja California, Sonora). Peru. 79 MXN 83 PER. Ther.
* *Croton microphyllus* Lam., Encycl. 2: 212 (1786). *Phyllanthidea microphylla* (Lam.) Didr., Vidensk. Meddel. Dansk Naturhist. Foren. Kjøbenhavn 1857: 151 (1857).
Phyllanthus ciliatoglandulosus Millsp., Proc. Calif. Acad. Sci., II, 2: 219 (1889). *Andrachne ciliatoglandulosa* (Millsp.) Croizat, J. Wash. Acad. Sci. 33: 11 (1943).

Andrachne minutifolia Pojark., Bot. Mater. Gerb. Bot. Inst. Komarova Akad. Nauk S.S.S.R. 20: 264 (1960).
N. Iran. 34 IRN. Cham.

Andrachne nana (P.T.Li) Govaerts in R.Govaerts, D.G.Frodin & A.Radcliffe-Smith, World Checklist Bibliogr. Euphorbiaceae: 167 (2000).
China (Hebei). 36 CHN. Cham. or nanophan.
* *Leptopus nanus* P.T.Li, Notes Roy. Bot. Gard. Edinburgh 40: 474 (1983).

Andrachne ovalis (E.Mey. ex Sond.) Müll.Arg., Linnaea 32: 78 (1863).
S. Africa, Zimbabwe. 26 ZIM 27 CPP NAT SWZ TVL. Nanophan.
* *Phyllanthus ovalis* E.Mey. ex Sond., Linnaea 23: 135 (1850). *Clutia ovalis* (E.Mey. ex Sond.) Scheele, Linnaea 25: 583 (1852). *Savia ovalis* (E.Mey. ex Sond.) Pax & K.Hoffm. in H.G.A.Engler, Pflanzenr., IV, 147, XV: 186 (1922).
Phyllanthus dregeanus Scheele, Linnaea 25: 585 (1852). *Andrachne dregeana* (Scheele) Baill., Adansonia 3: 164 (1863).
Andrachne capensis Baill., Adansonia 3: 163 (1863).

Andrachne pachyphylla (X.X.Chen) Govaerts in R.Govaerts, D.G.Frodin & A.Radcliffe-Smith, World Checklist Bibliogr. Euphorbiaceae: 167 (2000).
China (NW. Guangxi). 36 CHS. Nanophan.
* *Leptopus pachyphyllus* X.X.Chen, Guihaia 8: 233 (1988).

Andrachne phyllanthoides (Nutt.) Müll.Arg. in A.P.de Candolle, Prodr. 15(2): 435 (1866).
EC. & SE. U.S.A. 74 MSO OKL 75 ALA ARK 77 TEX. Nanophan.
* *Lepidanthus phyllanthoides* Nutt., Trans. Amer. Philos. Soc., n.s. 5: 175 (1837). *Savia phyllanthoides* (Nutt.) Pax & K.Hoffm. in H.G.A.Engler, Pflanzenr., IV, 147, XV: 184 (1922). *Leptopus phyllanthoides* (Nutt.) G.L.Webster, Ann. Missouri Bot. Gard. 81: 40 (1994).
Maschalanthus obovatus Nutt., Trans. Amer. Philos. Soc., n.s. 5: 175 (1837).
Maschalanthus polygonoides Nutt., Trans. Amer. Philos. Soc., n.s. 5: 175 (1837).
Phyllanthus roemerianus Scheele, Linnaea 25: 583 (1852). *Andrachne roemeriana* (Scheele) Müll.Arg. in A.P.de Candolle, Prodr. 15(2): 234 (1866). *Savia phyllanthoides* var. *roemeriana* (Scheele) Pax & K.Hoffm. in H.G.A.Engler, Pflanzenr., IV, 147, XV: 184 (1922).
Andrachne reverchonii Coult., Contr. U. S. Natl. Herb. 2: 396 (1894). *Savia phyllanthoides* var. *reverchonii* (Coult.) Pax & K.Hoffm. in H.G.A.Engler, Pflanzenr., IV, 147, XV: 185 (1922).

Andrachne polypetala Kuntze, Revis. Gen. Pl. 2: 592 (1891). *Leptopus polypetalus* (Kuntze) Pojark., Bot. Mater. Gerb. Bot. Inst. Komarova Akad. Nauk S.S.S.R. 20: 271 (1960).
Vietnam. 41 VIE. Nanophan.

Andrachne pulvinata Pojark., Bot. Mater. Gerb. Bot. Inst. Komarova Akad. Nauk S.S.S.R. 20: 267 (1960).
N. Iran. 34 IRN. Cham.

Andrachne pusilla Pojark. in V.L.Komarov, Fl. URSS 14: 733 (1949).
Tadzhikistan (Pamir Mts.). 32 TZK. Cham.

Andrachne pygmaea Kossinsky, Bot. Mater. Gerb. Glavn. Bot. Sada RSFSR 2: 88 (1921).
Tadzhikistan (Alai). 32 TZK. Cham.

Andrachne ramosa Pojark., Bot. Mater. Gerb. Bot. Inst. Komarova Akad. Nauk S.S.S.R. 20: 262 (1960).
Iran. 34 IRN. Cham.

Andrachne reflexa Stapf, Denkschr. Kaiserl. Akad. Wiss., Wien. Math.-Naturwiss. Kl. 51: 315 (1886).
Iran. 34 IRN. Cham.

Andrachne robinsonii (Airy Shaw) Govaerts in R.Govaerts, D.G.Frodin & A.Radcliffe-Smith, World Checklist Bibliogr. Euphorbiaceae: 168 (2000).
Vietnam. 41 VIE. Nanophan.
* *Leptopus robinsonii* Airy Shaw, Kew Bull. 25: 495 (1971).

Andrachne schweinfurthii (Balf.f.) Radcl.-Sm., Hooker's Icon. Pl. 37: t. 3697 (1971).
Somalia, Socotra. 24 SOC SOM. Cham.
* *Securinega schweinfurthii* Balf.f., Proc. Roy. Soc. Edinburgh 12: 411 (1884).

var. **papillosa** Radcl.-Sm., Kew Bull. 44: 447 (1989).
Socotra (Khlohat). 24 SOC. Cham.

var. **schweinfurthii**
Somalia, Socotra. 24 SOC SOM. Cham.
Andrachne somalensis Pax, Bot. Jahrb. Syst. 15: 322 (1893).

Andrachne stenophylla Kossinsky, Bot. Mater. Gerb. Glavn. Bot. Sada RSFSR 2: 91 (1921).
S. Turkmenistan, N. Iran. 32 TKM 34 IRN. Cham.

Andrachne telephioides L., Sp. Pl.: 1014 (1753). – FIGURE, p. 169.
Cape Verde Is., Medit. to Pakistan, Tadzhikistan, Ethiopia, Somalia, Socotra (Abd-al-Kuri). 12 SPA 13 BUL GRC ITA KRI SIC YUG 14 KRY 20 ALG EGY LBY MOR TUN 21 CVI 22 MTN 24 ETH SOC SOM 32 TZK 33 TCS 34 AFG CYP IRN IRQ LBS PAL SIN TUR 35 SAU YEM 40 PAK. Hemicr. or cham.
Eraclissa hexagyna Forssk., Fl. Aegypt.-Arab.: 208 (1775).
Telephioides procumbens Moench, Suppl. Meth.: 310 (1802).
Andrachne rotundifolia Eichw. ex C.A.Mey., Verz. Pfl. Casp. Meer.: 18 (1831). *Andrachne telephioides* var. *rotundifolia* (Eichw. ex C.A.Mey.) Müll.Arg. in A.P.de Candolle, Prodr. 15(2): 236 (1866).
Andrachne telephioides var. *brevifolia* Müll.Arg. in A.P.de Candolle, Prodr. 15(2): 236 (1866).
Andrachne nummulariifolia Stapf, Denkschr. Kaiserl. Akad. Wiss., Wien. Math.-Naturwiss. Kl. 60: 314 (1886).
Andrachne virescens Stapf, Denkschr. Kaiserl. Akad. Wiss., Wien. Math.-Naturwiss. Kl. 51: 314 (1887).
Andrachne asperula Nevski, Trudy Bot. Inst. Armjansk. Fil. Akad. Nauk SSSR 4: 263 (1937).

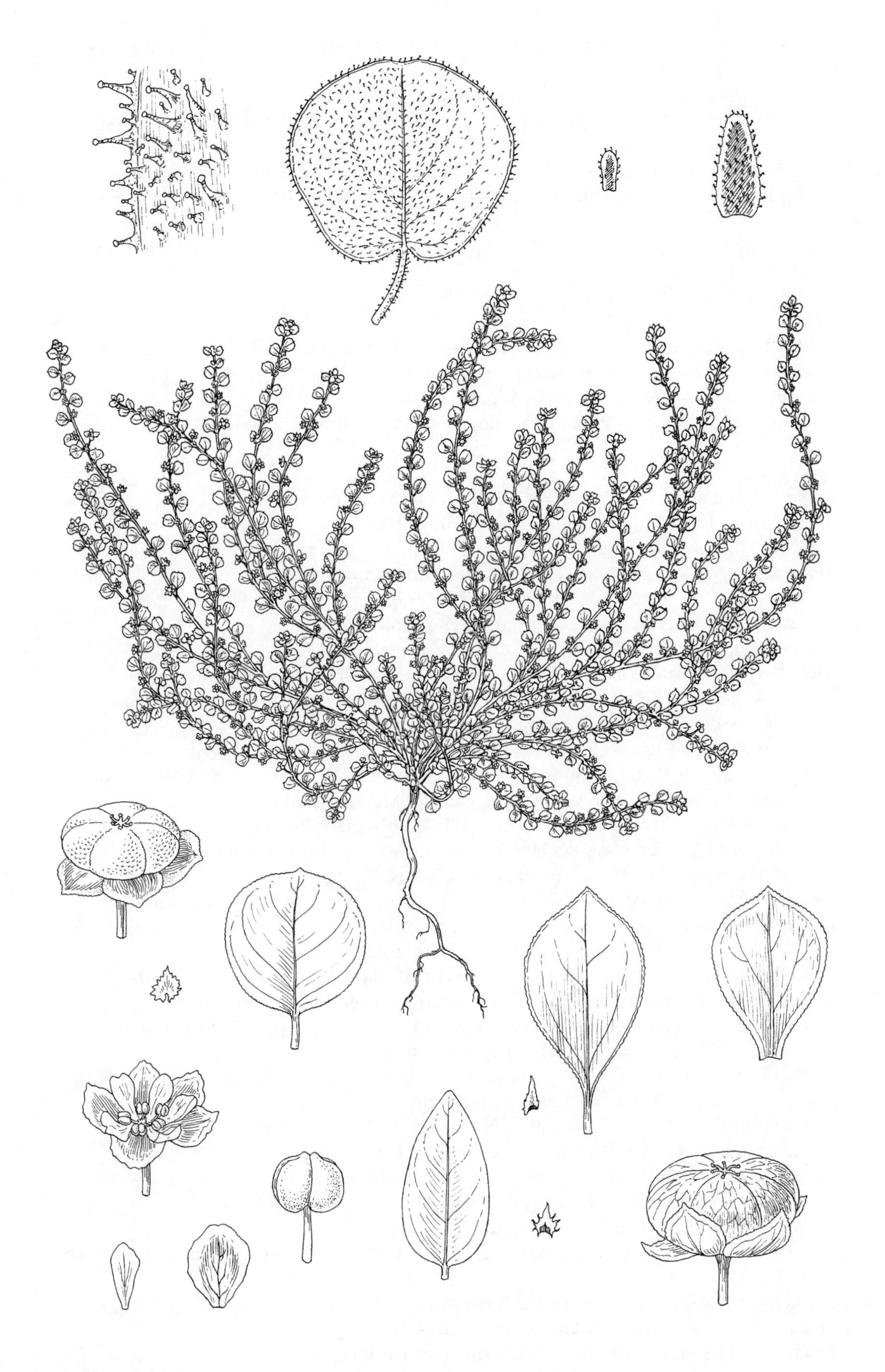

Andrachne telephioides L.
Artist: D. Erasmus
Fl. Iraq 4: 313, pl. 57 (1980)
KEW ILLUSTRATIONS COLLECTION

Andrachne asperula Nevski, Trudy Bot. Inst. Akad. Nauk S.S.S.R., Ser. 1, Fl. Sist. Vyssh. Rast. 4: 263 (1937).
Andrachne cretica Pojark., Bot. Zhurn. S.S.S.R. 25: 364 (1940).
Andrachne vvedenskyi Pazij, Bot. Mater. Gerb. Inst. Bot. Zool. Akad. Nauk Uzbeksk. S.S.R. 11: 22 (1948).
Andrachne pojarkoviae Kovatsch., Novosti Sist. Vyssh. Rast. 1965: 168 (1965).
Andrachne telephioides subsp. *oreocretensis* Aldén, Willdenowia 12: 211 (1982).

Andrachne virgatenuis Nevski, Trudy Bot. Inst. Akad. Nauk S.S.S.R., Ser. 1, Fl. Sist. Vyssh. Rast. 4: 264 (1937).
Tadzhikistan (S. Pamir Mts.). 32 TZK. Cham.

Andrachne yunnanensis (P.T.Li) Govaerts in R.Govaerts, D.G.Frodin & A.Radcliffe-Smith, World Checklist Bibliogr. Euphorbiaceae: 170 (2000).
China (Yunnan). 36 CHC. Cham. or nanophan.
* *Leptopus yunnanensis* P.T.Li, Notes Roy. Bot. Gard. Edinburgh 40: 469 (1983).

Synonyms:
Andrachne apetala Roxb. ex Wall. === **Bischofia javanica** Blume
Andrachne arborea Mill. === **Phyllanthus grandifolius** L.
Andrachne aspera var. *glandulosa* Hochst. ex A.Rich. === **Andrachne aspera** Spreng.
Andrachne aspera var. *maritima* N.Terracc. === **Andrachne aspera** Spreng.
Andrachne asperula Nevski === **Andrachne telephioides** L.
Andrachne asperula Nevski === **Andrachne telephioides** L.
Andrachne attenuata Hand.-Mazz. === **Andrachne esquirolii** H.Lév.
Andrachne attenuata var. *microcalyx* Hand.-Mazz. === **Andrachne esquirolii** H.Lév.
Andrachne australis var. *lanceolata* Müll.Arg. === **Andrachne australis** Zoll. & Moritzi
Andrachne bodinieri H.Lév. === **Andrachne chinensis** Bunge
Andrachne cadishaco Roxb. ex Wall. === **Cleistanthus collinus** (Roxb.) Benth.
Andrachne cadishan Roxb. ex Hook.f. === **Andrachne colchica** Fisch. & C.A.Mey. ex Boiss.
Andrachne capensis Baill. === **Andrachne ovalis** (E.Mey. ex Sond.) Müll.Arg.
Andrachne capillipes (Pax) Hutch. === **Andrachne chinensis** Bunge
Andrachne capillipes var. *pubescens* Hutch. === **Andrachne chinensis** Bunge
Andrachne cavaleriei H.Lév. === **Andrachne chinensis** Bunge
Andrachne ciliatoglandulosa (Millsp.) Croizat === **Andrachne microphylla** (Lam.) Baill.
Andrachne cretica Pojark. === **Andrachne telephioides** L.
Andrachne cuneifolia Britton === **Phyllanthus orbicularis** Kunth
Andrachne decaisneana Baill. === **Andrachne cordifolia** (Wall. ex Decne.) Müll.Arg.
Andrachne doonkyboisca B.Heyne ex Wall. === **Bridelia retusa** (L.) A.Juss.
Andrachne dregeana (Scheele) Baill. === **Andrachne ovalis** (E.Mey. ex Sond.) Müll.Arg.
Andrachne elliptica Roth === **Bridelia montana** (Roxb.) Willd.
Andrachne esquirolii var. *microcalyx* (Hand.-Mazz.) Rehder === **Andrachne esquirolii** H.Lév.
Andrachne frutescens Ehret === **Arbutus andrachne** L. (Ericaceae)
Andrachne fruticosa Hook.f. === **Andrachne australis** Zoll. & Moritzi
Andrachne fruticosa L. === **Breynia fruticosa** (L.) Hook.f.
Andrachne fruticosa B.Heyne ex Hook.f. === **Meineckia parvifolia** (Wight) G.L.Webster
Andrachne fruticosa Decne. ex Müll.Arg. === **Andrachne decaisnei** Benth.
Andrachne fruticulosa Schischk. === **Andrachne buschiana** Pojark.
Andrachne hainanensis var. *nummulariifolia* Merr. & Chun === **Andrachne hainanensis** Merr. & Chun
Andrachne hapladena Pazij === **Andrachne fedtschenkoi** Kossinsky
Andrachne hirsuta Hutch. === **Andrachne chinensis** Bunge
Andrachne hypoglauca H.Lév. === **Andrachne esquirolii** H.Lév.
Andrachne lanceolata Pierre ex Beille === **Andrachne australis** Zoll. & Moritzi
Andrachne montana Hutch. === **Andrachne chinensis** Bunge
Andrachne nummulariifolia Stapf === **Andrachne telephioides** L.

Andrachne orbicularis (Benth.) Domin === **Andrachne decaisnei** var. **orbicularis** Benth.
Andrachne orbiculata Roth === **Cleistanthus collinus** (Roxb.) Benth.
Andrachne ovata Lam. ex Poir. === **Sauropus androgynus** (L.) Merr.
Andrachne persicariifolia H.Lév. === **Andrachne esquirolii** H.Lév.
Andrachne pojarkoviae Kovatsch. === **Andrachne telephioides** L.
Andrachne pumila (Raf.) Rydb. === **Phyllanthus caroliniensis** Walter subsp. **caroliniensis**
Andrachne reverchonii Coult. === **Andrachne phyllanthoides** (Nutt.) Müll.Arg.
Andrachne roemeriana (Scheele) Müll.Arg. === **Andrachne phyllanthoides** (Nutt.) Müll.Arg.
Andrachne rotundifolia Eichw. ex C.A.Mey. === **Andrachne telephioides** L.
Andrachne rupestris Pazij === **Andrachne fedtschenkoi** Kossinsky
Andrachne schaffneriana Pax & K.Hoffm. === **Phyllanthus caroliniensis** Walter subsp. **caroliniensis**
Andrachne somalensis Pax === **Andrachne schweinfurthii** (Balf.f.) Radcl.-Sm. var. **schweinfurthii**
Andrachne telephioides var. *brevifolia* Müll.Arg. === **Andrachne telephioides** L.
Andrachne telephioides subsp. *oreocretensis* Aldén === **Andrachne telephioides** L.
Andrachne telephioides var. *rotundifolia* (Eichw. ex C.A.Mey.) Müll.Arg. === **Andrachne telephioides** L.
Andrachne tenera Miq. === **Andrachne australis** Zoll. & Moritzi
Andrachne trifoliata Roxb. === **Bischofia javanica** Blume
Andrachne virescens Stapf === **Andrachne telephioides** L.
Andrachne vvedenskyi Pazij === **Andrachne telephioides** L.

Androphoranthus

Synonyms:
Androphoranthus H.Karst. === **Caperonia** A.St.-Hil.

Androstachys

1 species, Zimbabwe to South Africa and in Madagascar. *A. johnsonii*, a tree with decussate foliage reaching as much as 36 m and yielding a valuable hardwood (Lebombo ironwood), is dominant in dry forests in parts of Swaziland where some cloud cover prevails (Alvin 1987). Other Malagasy species formerly here were transferred by Radcliffe-Smith to *Stachyandra*. The name is conserved against *Androstachys* Grand'Eury (a fossil genus). [Airy-Shaw (1965; see General) proposed for the genus a separate family Androstachydaceae but this generally has not found favour.] (Oldfieldioideae)

Pax, F. & K. Hoffmann (1922). *Androstachys*. In A. Engler (ed.), Das Pflanzenreich, IV 147 XV (Euphorbiaceae-Phyllanthoideae-Phyllantheae): 287. Berlin. (Heft 81.) La/Ge. — 1 species; Africa (S Africa).

Leandri, J. (1958). *Androstachys*. Fl. Madag. Comores 111 (Euphorbiacées), I: 197-199. Paris. Fr. — Flora treatment (1 species, shared with Africa).

Alvin, K. L. (1987). Leaf anatomy of *Androstachys johnsonii* Prain and its functional significance. Ann. Bot., n.s., 59: 579-591, illus. En. — Anatomical investigations related to a review of the family status of the species. The unusual features leading to proposals for family status were found probably to be related to prevailing environmental factors. [Samples came from Swaziland; although the range of this tree in Africa is reviewed nothing is said about its presence in Madagascar and its environmental circumstances therein.]

Androstachys Prain, Bull. Misc. Inform. Kew 1908: 438 (1908).
SE. Trop. & S. Africa, Madagascar. 26 27 29. Phan.

Androstachys johnsonii Prain, Bull. Misc. Inform. Kew 1908: 439 (1908).
Mozambique, Transvaal, Swaziland, Madagascar. 26 MOZ 27 SWZ TVL 29 MDG. Phan.
Weihea subpeltata Sim, Forest Fl. Port. E. Afr.: 66 (1909). *Androstachys subpeltata* (Sim) E.Phillips, Mem. Bot. Surv. S. Africa 25: 460 (1951).

Synonyms:
Androstachys imberbis Airy Shaw === **Stachyandra imberbis** (Airy Shaw) Radcl.-Sm.
Androstachys merana Airy Shaw === **Stachyandra merana** (Airy Shaw) J.-F.Leroy ex Radcl.-Sm.
Androstachys rufibarbis Airy Shaw === **Stachyandra rufibarbis** (Airy Shaw) Radcl.-Sm.
Androstachys subpeltata (Sim) E.Phillips === **Androstachys johnsonii** Prain
Androstachys viticifolia Airy Shaw === **Stachyandra viticifolia** (Airy Shaw) Radcl.-Sm.

Angelandra

Synonyms:
Angelandra Endl. === **Croton** L.

Angostyles

2 species, N. South America (south to N. Brazil); slender shrubs or small trees with tufted, narrowly oblanceolate, basally hastate foliage. Pax & Hoffmann (1919) introduced the spelling *Angostylis* which has become widely used; a proposal for conservation of this variant has been made (Radcliffe-Smith 1996). (Acalyphoideae)

Pax, F. & K. Hoffmann (1919). *Angostylis*. In A. Engler (ed.), Das Pflanzenreich, IV 147 IX (Euphorbiaceae-Acalypheae-Plukenetiinae): 29-30. Berlin. (Heft 68.) La/Ge. — 1 species.
Radcliffe-Smith, A. (1996). Proposal to conserve the name *Angostylis* (Euphorbiaceae) with a conserved spelling. Taxon 45: 705. En. — Nomenclatural. [A decision on this was pending in 1998.]

Angostylis Benth., Hooker's J. Bot. Kew Gard. Misc. 6: 328 (1854).
N. South America, N. Brazil. 82 84.

Angostylis longifolia Benth., Hooker's J. Bot. Kew Gard. Misc. 6: 328 (1845).
Brazil (Amazonas). 84 BZN. Nanophan. or phan.

Angostylis tabulamontana Croizat, Bull. Torrey Bot. Club 75: 403 (1948).
Surinam. 82 SUR. Nanophan. or phan.

Angostylidium

Synonyms:
Angostylidium Pax & K.Hoffm. === **Tetracarpidium** Pax

Angostylis

An orthographic variant of *Angostyles* (currently proposed for conservation).

Anisepta

Synonyms:
Anisepta Raf. === **Croton** L.

Anisonema

Synonyms:

Anisonema A.Juss. === **Phyllanthus** L.
Anisonema dubium Blume === **Phyllanthus reticulatus** Poir. var. **reticulatus**
Anisonema eglandulosum Decne. === **Margaritaria anomala** (Baill.) Fosberg
Anisonema glaucinum Miq. === **Phyllanthus glaucinus** Müll.Arg.
Anisonema hypoleucum Miq. === **Glochidion lutescens** Blume
Anisonema intermedium Decne. === **Phyllanthus reticulatus** Poir. var. **reticulatus**
Anisonema jamaicense (Griseb.) Griseb. === **Phyllanthus reticulatus** Poir. var. **reticulatus**
Anisonema multiflorum (Poir.) Wight === **Phyllanthus multiflorus** Poir.
Anisonema puberulum (Miq. ex Baill.) Baill. === **Phyllanthus reticulatus** Poir. var. **reticulatus**
Anisonema reticulatum (Poir.) A.Juss. === **Phyllanthus reticulatus** Poir.
Anisonema wrightianum Baill. === **Phyllanthus reticulatus** Poir. var. **reticulatus**
Anisonema zollingeri (Müll.Arg.) Miq. === **Phyllanthus pulcher** Wall. ex Müll.Arg.

Anisophyllum

A segregate of *Euphorbia* proposed by Haworth and adopted by Klotzsch; congruent with the later *Chamaesyce* S. F. Gray but itself a homonym of *Anisophyllum* Jacq. (Phanerogamae, incertae sedis).

Synonyms:

Anisophyllum Haw. === **Euphorbia** L.
Anisophyllum acutifolium Boivin ex Baill. === **Croton muricatus** Vahl
Anisophyllum aegyptiacum (Boiss.) Schweinf. === **Euphorbia forskalii** J.Gay
Anisophyllum affine Klotzsch & Garcke === **Euphorbia balbisii** Boiss.
Anisophyllum amoenum Klotzsch & Garcke === **Euphorbia adenoptera** Bertol. subsp. **adenoptera**
Anisophyllum arabicum (Hochst. & Steud. ex Anderson) Schweinf. === **Euphorbia arabica** Hochst. & Steud. ex Anderson
Anisophyllum arenarium Klotzsch & Garcke === **Euphorbia missurica** Raf.
Anisophyllum atoto (G.Forst.) Klotzsch & Garcke === **Euphorbia atoto** G.Forst.
Anisophyllum bahiense Klotzsch & Garcke === **Euphorbia bahiensis** (Klotzsch & Garcke) Boiss.
Anisophyllum berteroanum (Balb. ex Spreng.) Klotzsch & Garcke === **Euphorbia berteroana** Balb. ex Spreng.
Anisophyllum besseri Klotzsch & Garcke === **Euphorbia besseri** (Klotzsch & Garcke) Boiss.
Anisophyllum burmannianum (J.Gay) Klotzsch & Garcke === **Euphorbia forskalii** J.Gay
Anisophyllum caecorum Klotzsch & Garcke === **Euphorbia potentilloides** Boiss.
Anisophyllum californicum Klotzsch & Garcke === **Euphorbia deppeana** Boiss.
Anisophyllum centunculoides (Kunth) Klotzsch & Garcke === **Euphorbia centunculoides** Kunth
Anisophyllum chamaesyce (L.) Haw. === **Euphorbia chamaesyce** L.
Anisophyllum chamissonis Klotzsch & Garcke ex Klotzsch === **Euphorbia chamissonis** (Klotzsch & Garcke ex Klotzsch) Boiss.
Anisophyllum cheirolepis (Fisch. & C.A.Mey. ex Karelin) Klotzsch & Garcke === **Euphorbia cheirolepis** Fisch. & C.A.Mey. ex Karelin
Anisophyllum congenerum (Blume) Klotzsch & Garcke === **Euphorbia reniformis** Blume
Anisophyllum convolvuloides (Hochst. ex Benth.) Klotzsch & Garcke === **Euphorbia convolvuloides** Hochst. ex Benth.
Anisophyllum cordatum Klotzsch & Garcke === **Euphorbia degeneri** Sherff
Anisophyllum crassipes Klotzsch === **Euphorbia macropus** (Klotzsch) Boiss.
Anisophyllum densiflorum Klotzsch === **Euphorbia densiflora** (Klotzsch) Klotzsch
Anisophyllum dentatum (Michx.) Haw. === **Euphorbia dentata** Michx.
Anisophyllum dioecium (Kunth) Klotzsch & Garcke === **Euphorbia dioeca** Kunth
Anisophyllum emarginatum Klotzsch & Garcke === **Euphorbia serpens** Kunth

Anisophyllum fendleri (Torr. & A.Gray) Klotzsch & Garcke === **Euphorbia fendleri** Torr. & A.Gray
Anisophyllum flexuosum (Kunth) Klotzsch & Garcke === **Euphorbia mesembryanthemifolia** Jacq.
Anisophyllum forskalii (J.Gay) Klotzsch & Garcke === **Euphorbia forskalii** J.Gay
Anisophyllum geyeri (Engelm. & A.Gray) Klotzsch & Garcke === **Euphorbia geyeri** Engelm. & A.Gray
Anisophyllum glabratum (Sw.) Klotzsch & Garcke === **Euphorbia mesembryanthemifolia** Jacq.
Anisophyllum glaucophyllum (Poir.) Klotzsch & Garcke === **Euphorbia trinervia** Schumach. & Thonn.
Anisophyllum golianum (Comm. ex Lam.) Klotzsch & Garcke === **Euphorbia goliana** Comm. ex Lam.
Anisophyllum granulatum (Forssk.) Schweinf. === **Euphorbia granulata** Forssk.
Anisophyllum humboldtii (Donn) Haw. === ?
Anisophyllum humifusum (Willd.) Klotzsch & Garcke === **Euphorbia humifusa** Willd.
Anisophyllum humistratum (Engelm. ex A.Gray) Klotzsch & Garcke === **Euphorbia humistrata** Engelm. ex A.Gray
Anisophyllum hypericifolium (L.) Haw. === **Euphorbia hypericifolia** L.
Anisophyllum hyssopifolium (L.) Haw. === **Euphorbia hyssopifolia** L.
Anisophyllum inaequale Klotzsch & Garcke === **Euphorbia adenoptera** Bertol. subsp. **adenoptera**
Anisophyllum inaequilaterum (Sond.) Klotzsch & Garcke === **Euphorbia inaequilatera** Sond.
Anisophyllum indicum (Lam.) Schweinf. === **Euphorbia indica** Lam.
Anisophyllum ipecacuanha (L.) Haw. === **Euphorbia ipecacuanhae** L.
Anisophyllum laevigatum (Vahl) Klotzsch & Garcke === **Euphorbia atoto** G.Forst.
Anisophyllum lasiocarpum (Klotzsch) Klotzsch & Garcke === **Euphorbia lasiocarpa** Klotzsch
Anisophyllum leiospermum Klotzsch & Garcke === **Euphorbia obliqua** F.A.Bauer ex Endl.
Anisophyllum leucanthum Klotzsch & Garcke === **Euphorbia mendezii** Boiss.
Anisophyllum lindenianum (A.Rich.) Klotzsch & Garcke === **Euphorbia adenoptera** Bertol. subsp. **adenoptera**
Anisophyllum lineare (Retz.) Klotzsch & Garcke === **Euphorbia articulata** Burm.
Anisophyllum macropus Klotzsch === **Euphorbia macropus** (Klotzsch) Boiss.
Anisophyllum maculatum (L.) Haw. === **Euphorbia maculata** L.
Anisophyllum melanadenium (Torr.) Klotzsch & Garcke === **Euphorbia melanadenia** Torr.
Anisophyllum mexicanum Klotzsch & Garcke === **Euphorbia graminea** Jacq. var. **graminea**
Anisophyllum meyenianum (Klotzsch) Klotzsch & Garcke === **Euphorbia meyeniana** Klotzsch
Anisophyllum mossambicense Klotzsch & Garcke === **Euphorbia mossambicensis** (Klotzsch & Garcke) Boiss.
Anisophyllum multiforme (Gaudich. ex Hook. & Arn.) Klotzsch & Garcke === **Euphorbia multiformis** Gaudich. ex Hook. & Arn.
Anisophyllum mundtii Klotzsch & Garcke === **Euphorbia inaequilatera** Sond. var. **inaequilatera**
Anisophyllum nagleri Klotzsch & Garcke === **Euphorbia nagleri** (Klotzsch & Garcke) Boiss.
Anisophyllum nanum Klotzsch & Garcke === **Euphorbia chamaerrhodos** Boiss.
Anisophyllum nodosum Klotzsch & Garcke === **Euphorbia clusiifolia** Hook. & Arn.
Anisophyllum novomexicanum Klotzsch & Garcke === **Euphorbia serpyllifolia** Pers. var. **serpyllifolia**
Anisophyllum ocymoideum (L.) Haw. === **Euphorbia ocymoidea** L.
Anisophyllum orbiculatum (Kunth) Klotzsch & Garcke === **Euphorbia orbiculata** Kunth
Anisophyllum ovalifolium Engelm. ex Klotzsch === **Euphorbia klotzschii** Oudejans
Anisophyllum peplis (L.) Haw. === **Euphorbia peplis** L.
Anisophyllum piluliferum (L.) Haw. === **Euphorbia hirta** L.

Anisophyllum polycnemoides (Hochst. ex Boiss.) Klotzsch & Garcke === **Euphorbia polycnemoides** Hochst. ex Boiss.
Anisophyllum polygonifolium (L.) Haw. === **Euphorbia polygonifolia** L.
Anisophyllum prostratum (Aiton) Haw. === **Euphorbia prostrata** Aiton
Anisophyllum ramosissimum Klotzsch & Garcke === **Euphorbia garkeana** Boiss.
Anisophyllum rhytispermum Klotzsch & Garcke === **Euphorbia rhytisperma** (Klotzsch & Garcke) Boiss.
Anisophyllum roseum (Retz.) Haw. === **Euphorbia rosea** Retz.
Anisophyllum ruizianum Klotzsch & Garcke === **Euphorbia ruiziana** (Klotzsch & Garcke) Boiss.
Anisophyllum scordiifolium (Jacq.) Klotzsch & Garcke === **Euphorbia scordiifolia** Jacq.
Anisophyllum scutelligerum Boivin ex Baill. === **Croton adenophorus** Baill.
Anisophyllum selloi Klotzsch & Garcke === **Euphorbia selloi** (Klotzsch & Garcke) Boiss.
Anisophyllum selloi Klotzsch & Garcke === **Euphorbia selloi** (Klotzsch & Garcke) Boiss.
Anisophyllum senile Klotzsch & Garcke === **Euphorbia stictospora** Engelm.
Anisophyllum serpens (Kunth) Klotzsch & Garcke === **Euphorbia serpens** Kunth
Anisophyllum setigerum Klotzsch & Garcke === **Euphorbia inaequilatera** Sond. var. **inaequilatera**
Anisophyllum tenuiflorum Klotzsch & Garcke === **Euphorbia velligera** Schauer
Anisophyllum tettense Klotzsch & Garcke === **Euphorbia tettensis** Klotzsch
Anisophyllum thymifolium (L.) Haw. === **Euphorbia thymifolia** L.
Anisophyllum trapezoidale Baill. === **Anisophyllea disticha** (Jack) Baill. (Anisophylleaceae)
Anisophyllum vahlii Willd. ex Klotzsch & Garcke === **Euphorbia articulata** Burm.
Anisophyllum vaticanum Gand. === **Euphorbia chamaesyce** L.
Anisophyllum velleriflorum Klotzsch & Garcke === **Euphorbia velleriflora** (Klotzsch & Garcke) Boiss.
Anisophyllum virgatum Klotzsch & Garcke === **Euphorbia arnottiana** Endl.

Anisophyllum

A later homonym of *Anisophyllum* Jacq. and *Anisophyllum* Haw.

Synonyms:
Anisophyllum Boivin ex Baill. === **Croton** L.

Annesijoa

1 species, New Guinea; small to medium trees up to 25 m with palmately compound leaves having the appearance of *Hevea* rubber trees but actually related to *Joannesia* (South America) and *Leeuwenbergia* (Africa). Not 'rediscovered' for more than 40 years after being first collected prior to World War I. (Crotonoideae)

Pax, F. & K. Hoffmann (1919). *Annesijoa*. In A. Engler (ed.), Das Pflanzenreich, IV 147 XIV (Euphorbiaceae-Additamentum VI): 9. Berlin. (Heft 68.) La/Ge. — 1 species, New Guinea.
Airy-Shaw, H. K. (1963). Notes on Malaysian and other Asiatic Euphorbiaceae, XXVII. Rediscovery of *Annesijoa* Pax and K. Hoffm. in New Guinea. Kew Bull. 16: 345. En. — New record.
Airy-Shaw, H. K. (1974). *Annesijoa novoguineensis*. Ic. Pl. 38: pl. 3713. En. — Plant portrait with description and commentary.

Annesijoa Pax & K.Hoffm. in H.G.A.Engler, Pflanzenr., IV, 147, XIV: 9 (1919).
New Guinea. 42.

Annesijoa novoguineensis Pax & K.Hoffm. in H.G.A.Engler, Pflanzenr., IV, 147, XIV: 9 (1919).
New Guinea. 42 NWG. Phan. – FIGURE, p. 176.

Annesijoa novoguineensis Pax & K. Hoffm. in H.G.A. Engler
Artist: Mary Grierson
Ic. Pl. 38: pl. 3713 (1974)
KEW ILLUSTRATIONS COLLECTION

Anomalocalyx

1 species, S America (Amazonian Brazil). *A. uleanus* is a small to large (up to 40 m) tree of riverbanks and swampy places in the central and eastern Amazon. In Secco (1990) the genus is keyed out next to *Dodecastigma*, but Webster (Synopsis, 1994) regards it as of uncertain systematic position. (Crotonoideae)

Ducke, A. (1933). *Anomalocalyx*. Arq. Jard. Bot. Rio de Janeiro 6 (in Plantes nouvelles .. V): 60-61, illus. Fr/La. — Protologue and description of 1 species, *A. uleanus* (also new; illustrated in plate 6) from the vicinity of Manaus; segregated from *Cunuria* (or *Micrandra*). [The genus was, however, actually first described in 1932; this account does not have priority.]

Secco, R. de S. (1990). Revisão dos gêneros *Anomalocalyx* Ducke, *Dodecastigma* Ducke, *Pausandra* Radlk., *Pogonophora* Miers ex Benth. e *Sagotia* Baill. (Euphorbiaceae-Crotonoideae) para a América do Sul. 133 pp., illus., maps. Belém: Museu Paraense 'Emilio Goeldi'. Pt. — Includes (pp. 38-42) revision of *Anomalocalyx* with description, exsiccatae and commentary; map, p. 46.

Anomalocalyx Ducke, Notizbl. Bot. Gart. Berlin-Dahlem 11: 334 (1932).
Brazil (Amapá, Amazonas). 84.

Anomalocalyx uleanus (Pax & K.Hoffm.) Ducke, Notizbl. Bot. Gart. Berlin-Dahlem 11: 344 (1932).
Brazil (Amapá, Amazonas). 84 BZN. Phan. - Most often seen near Manaus.
* *Cunuria uleana* Pax & K.Hoffm. in H.G.A.Engler, Pflanzenr., IV, 147, XIV: 51 (1919).

Anomospermum

Synonyms:
Anomospermum Dalzell === **Actephila** Blume
Anomospermum excelsum Dalzell === **Actephila excelsa** (Dalzell) Müll.Arg.

Anomostachys

1 species, Madagascar; long referred to *Excoecaria* following Müller (1866; see **General**) and Pax & Hoffmann (1912) but first raised to generic rank by Hurusawa in 1954 and now restored following Kruijt (1996). Laticiferous shrubs or small trees to 15 m, the distichously arranged leaves with strongly contrasting surfaces and many fine lateral veins, the inflorescences axillary and the capsules very hard and apparently indehiscent. Long known only from the type, the species is now better understood thanks to several collections made since about 1920. Related to *Duvigneaudia* (Kruijt, 1996: 5) and certainly similar in general appearance. (Euphorbioideae (except Euphorbieae))

Pax, F. (with K. Hoffmann) (1912). *Excoecaria*. In A. Engler (ed.), Das Pflanzenreich, IV 147 V (Euphorbiaceae-Hippomaneae): 157-174. Berlin. (Heft 52.) La/Ge. — No. 1 (p. 159), *E. lastellei*, comprises sect. *Anomostachys* (raised by Hurusawa to generic rank); only the type collection then available.

• Kruijt, R. C. (1996). A taxonomic monograph of *Sapium* Jacq., *Anomostachys* (Baill.) Hurus., *Duvigneaudia* J. Léonard and *Sclerocroton* Hochst. (Euphorbiaceae tribe Hippomaneae). 109 pp., illus. (Biblioth. Bot. 146). Stuttgart. En. — *Anomostachys*, pp. 8-12; treatment of 1 species, *A. lastellei*, with description, synonymy, distribution and habitat, localities with exsiccatae, and notes and illustrations.

Anomostachys (Baill.) Hurus., J. Fac. Sci. Univ. Tokyo, Sect. 3, Bot. 6: 311 (1954).
Madagascar. 29.

Anomostachys lastellei (Müll.Arg.) Kruijt, Biblioth. Bot. 146: 11 (1996).
Madagascar. 29 MDG. Nanophan. or phan.
Stillingia lastellei Baill., Étude Euphorb.: 525 (1858), nom. inval.
* *Excoecaria lastellei* Müll.Arg. in A.P.de Candolle, Prodr. 15(2): 1218 (1866).
Sapium gymnogyne Leandri, Bull. Soc. Bot. France 85: 532 (1938 publ. 1939).
Sapium loziense Leandri, Bull. Soc. Bot. France 85: 532 (1938 publ. 1939).
Sapium perrieri Leandri, Bull. Soc. Bot. France 85: 531 (1938 publ. 1939).

Anthacantha

Synonyms:
Anthacantha Lem. === **Euphorbia** L.

Anthostema

3 species, west and west-central Africa (from Gambia to Angola) and in Madagascar (1 or perhaps 2); laticiferous, straight-boled trees to 25 m with foliage resembling *Excoecaria*. Along with *Dichostemma* the cyathia are tetramerous and somewhat bilaterally symmetrical. *A. aubryanum*, a swamp-dweller, exhibits a form of sympodial growth with 'whorls' of lateral branches bearing distichously arranged leaves. For a key to known species, see Denis (1921: 10-11). No separate revision is available and modern treatments must be sought in floras. (Euphorbioideae (Euphorbieae))

Denis, M. (1921). Les Euphorbiées des îles australes d'Afrique. 153 pp., illus. Nemours. (Thèse, Nemours, 1921. Reissued in Rev. Gén. Bot. 34 (1922): 1-64, 96-123, 170-177, 214-236, 287-299, 346-366.) Fr. — Includes (pp. 10-11) key to the three then (and currently) known species. [For fuller description of the work, see under *Euphorbia*.]
Pax, F. & K. Hoffmann (1931). *Anthostema*. In A. Engler (ed.), Die natürlichen Pflanzenfamilien, 2. Aufl., 19c: 207. Leipzig. Ge. — Synopsis with description of genus; 2 spp. in W Africa, 1 in Madagascar.

Anthostema A.Juss., Euphorb. Gen.: 56 (1824).
W. Trop. Africa to Angola, Madagascar. 22 23 26 29.

Anthostema aubryanum Baill., Adansonia 5: 366 (1865).
W. & WC. Trop. Africa. 22 GHA IVO 23 CAB GAB GGI. Phan.

Anthostema madagascariensis Baill., Étude Euphorb.: 60, 544 (1858).
Madagascar. 29 MDG. Phan.

Anthostema senegalensis A.Juss., Euphorb. Gen.: 117 (1824).
W. Trop. Africa. 22 GAM GUI IVO LBR MLI SEN SIE. Phan.

Antidesma

154 species, mostly in Asia, Malesia, Australia and the Pacific with but 10 to the west (7 in tropical and southern Africa, 3 in Madagascar, Comoros and the Mascarene Islands). Shrubs or small to medium trees; most species are in forest but *AA. ghaesembilla* and *venosum* are generally found in more or less open woodlands. Taxonomically, the species everywhere are mutually so intimately related that internal subdivision of the genus is impossible, contrary to Pax & Hoffmann (1922) who recognised 8 sections. It is moreover certain that too many species have been described; many African names were, for example, reduced by Léonard (1988). Some species are cultivated for fruit, especially *A. bunius*. Of particular note is the inclusion of the genus, sometimes with the American *Hieronyma* (and *Celianella*), *Thecacoris* and other members of subtribe Antidesminae (Webster, Synopsis 1994), by some authors –

beginning with C. A. Agardh – in a distinct family Stilaginaceae. H. K. Airy-Shaw and A. D. J. Meeuse have been among more recent advocates of this view; they call attention to the unilocular ovary and a fruit and seed structure apparently discordant with Euphorbiaceae in general and Phyllanthoideae in particular. Airy-Shaw in particular drew a likeness with *Rhyticaryum* (Icacinaceae). Webster has, on the other hand, advocated retention in Euphorbiaceae on the basis of pollen and foliar anatomical evidence. The fruit structure represents an autapomorphic specialisation in a line also including other Antidesminae. — Petra Hoffmann has recently completed a revision of Thai and West Malesian species, recognising some 70 taxa, and has also carried out studies on the Papuasian and Malagasy representatives (with only one species now considered to be represented in Madagascar). Publication of her Thai and West Malesian work is expected shortly. (Phyllanthoideae)

- Pax, F. & K. Hoffmann (1922). *Antidesma*. In A. Engler (ed.), Das Pflanzenreich, IV 147 XV (Euphorbiaceae-Phyllanthoideae-Phyllantheae): 107-168, illus. Berlin. (Heft 81.) La/Ge. — 146 species in 8 sections, Africa, Asia, Malesia, the Pacific and Australia. See also Additamentum VII: 179.

Leandri, J. (1958). *Antidesma*. Fl. Madag. Comores 111 (Euphorbiacées), I: 14-21, illus. Paris. Fr. — Flora treatment (2 species); key.

Airy-Shaw, H. K. (1969). New or noteworthy Asiatic species of *Antidesma* (Stilaginaceae). Kew Bull. 23: 277-290. En. — Novelties and notes including range extensions; sections indicated.

Airy-Shaw, H. K. (1972). New or noteworthy Asiatic species of *Antidesma* (Stilaginaceae), II. Kew Bull. 26: 457-468. En. — Novelties and notes including range extensions; sections indicated.

Airy-Shaw, H. K. (1973). New or noteworthy Asiatic species of *Antidesma* (Stilaginaceae), III. Kew Bull. 28: 269-281. En. — Descriptions of four new species along with other notes.

Airy-Shaw, H. K. (1978). New or noteworthy Asiatic species of *Antidesma* (Stilaginaceae), IV. Kew Bull. 33: 15-18. En. — Descriptions of three novelties along with other notes; sections indicated.

Airy-Shaw, H. K. (1979). New or noteworthy Asiatic species of *Antidesma* (Stilaginaceae), V. Kew Bull. 33: 423-427. En. — Descriptions of four new species from New Guinea.

Airy-Shaw, H. K. (1981). New species of *Antidesma* (Stilaginaceae) from Malesia and Australia. Kew Bull. 36: 635-637. En. — Descriptions of three new species, respectively from Borneo, Sulawesi and Australia; note on *A. erostre*.

Léonard, J. (1988). Révision du genre *Antidesma* L. (Euphorbiaceae) en Afrique centrale. Bull. Jard. Bot. Natl. Belg. 58: 3-46, illus., maps. Fr. — Detailed descriptive treatment (5 species and some additional infraspecific taxa) with key, synonymy, references and citations, types, vernacular names, observations, localities with exsiccatae, and indication of distribution (with maps), uses, etc.; index (p. 46). [There has here been a considerable reduction in names.]

Hoffmann, P. (1999, in press). The genus *Antidesma* (Euphorbiaceae) in Madagascar and the Comoro Islands. Kew Bull.

Hoffmann, P. (1999, in press). New taxa and new combinations in Asian *Antidesma*. Kew Bull.

Antidesma L., Sp. Pl.: 1027 (1753).
Trop. & Subtrop. Old World. 22 23 24 25 26 27 29 36 38 40 41 42 50 60 61 63.
Bestram Adans., Fam. Pl. 2: 354 (1763).
Stilago L., Mant. Pl. 1: 16 (1767).
Rhytis Lour., Fl. Cochinch.: 660 (1790).
Rubina Noronha, Verh. Batav. Genootsch. Kunsten 5(4): 3 (1790).
Coulejia Dennst., Schlüssel Hortus Malab.: 31 (1818).
Minutalia Fenzl, Flora 27: 312 (1844), nom. nud.

Antidesma acidum Retz., Observ. Bot. 5: 30 (1788). – FIGURE, p. 180.
S. China, Trop. Asia. 36 CHC 40 ASS EHM IND NEP PAK WHM 41 BMA THA VIE 42 JAW.
Nanophan. or phan.

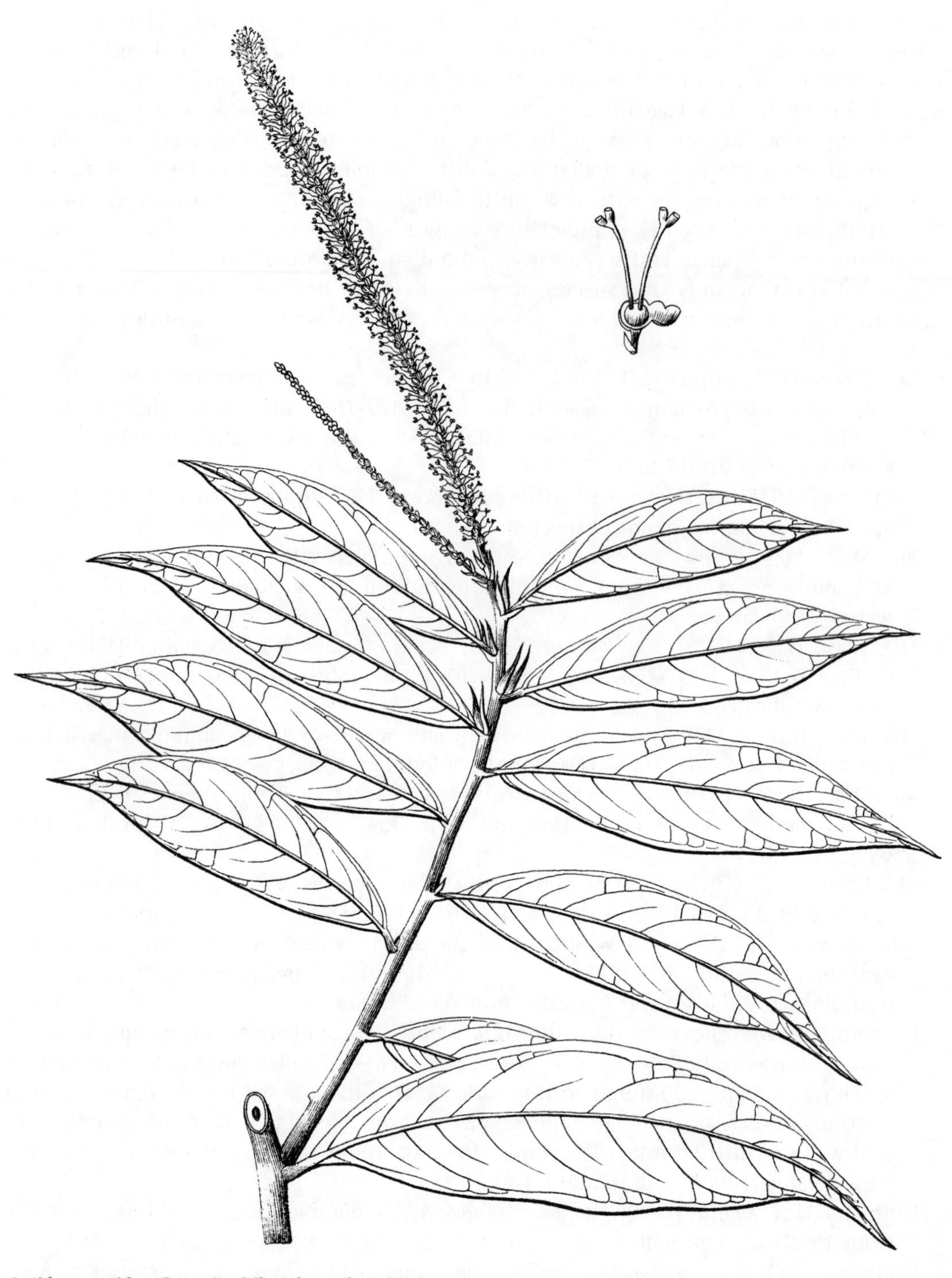

Antidesma acidum Retz. (as *Stilago lanceolaria* Roxb.)
Artist: Anonymous, India
Icones Roxburghianae, pl. 766 (apparently drawn from col. pl. 2554); unpubl.
KEW ILLUSTRATIONS COLLECTION

Stilago diandra Roxb., Pl. Coromandel 2: t. 166 (1802). *Antidesma diandrum* (Roxb.) Roth, Nov. Pl. Sp.: 369 (1821). *Antidesma diandrum* (Roxb.) Spreng., Syst. Veg. 1: 826 (1824). *Antidesma diandrum* var. *genuinum* Müll.Arg. in A.P.de Candolle, Prodr. 15(2): 267 (1866), nom. inval.

Antidesma sylvestre Wall., Numer. List: 7281 (1832), nom. inval.

Stilago lanceolaria Roxb., Fl. Ind. ed. 1832, 3: 760 (1832). *Antidesma lanceolarium* (Roxb.) Wight, Icon. Pl. Ind. Orient. 3(1): 4 (1843).

Antidesma diandrum var. *lanceolatum* Tul., Ann. Sci. Nat., Bot., III, 15: 199 (1851).

Antidesma diandrum var. *ovatum* Tul., Ann. Sci. Nat., Bot., III, 15: 189 (1851).

Antidesma diandrum var. *parvifolium* Tul., Ann. Sci. Nat., Bot., III, 15: 199 (1851).
Antidesma wallichianum C.Presl, Epimel. Bot.: 235 (1851).
Antidesma diandrum f. *javanicum* J.J.Sm. in S.H.Koorders & T.Valeton, Bijdr. Boomsoort. Java 12: 275 (1910). *Antidesma diandrum* var. *javanicum* (J.J.Sm.) Pax & K.Hoffm. in H.G.A.Engler, Pflanzenr., IV, 147, XV: 144 (1922).
Antidesma henryi Pax & K.Hoffm. in H.G.A.Engler, Pflanzenr., IV, 147, XV: 132 (1922), nom. illeg. *Antidesma paxii* F.P.Metcalf, Lingnan Sci. J. 10: 485 (1931).
Antidesma parviflorum Ham. ex Pax & K.Hoffm. in H.G.A.Engler, Pflanzenr., IV, 147, XV: 143 (1922).

Antidesma acuminatissimum Quisumb. & Merr., Philipp. J. Sci. 37: 159 (1928).
Philippines (Luzon). 42 PHI.

Antidesma acuminatum Wight, Icon. Pl. Ind. Orient. 6: t. 1991 (1853).
Sikkim to Burma. 40 ASS EHM 41 BMA. Nanophan.
Antidesma macrophyllum Wall. ex Voigt, Hort. Suburb. Calcutt.: 295 (1845), nom. illeg.
Antidesma refractum Müll.Arg. in A.P.de Candolle, Prodr. 15(2): 257 (1866).
Antidesma simile Müll.Arg. in A.P.de Candolle, Prodr. 15(2): 259 (1866).

Antidesma agusanense Elmer, Leafl. Philipp. Bot. 7: 2632 (1915).
Philippines (Mindanao). 42 PHI.

Antidesma alexiteria L., Sp. Pl.: 1027 (1753).
S. India, Sri Lanka. 40 IND SRL. Nanophan.
Antidesma zeylanicum Lam., Encycl. 1: 207 (1783).
Antidesma alexiteria Gaertn., Fruct. Sem. Pl. 1: 188 (1788), sphalm.
Antidesma zeylanicum C.Presl, Epimel. Bot.: 234 (1851), nom. illeg.

Antidesma ambiguum Pax & K.Hoffm. in H.G.A.Engler, Pflanzenr., IV, 147, XV: 127 (1922).
China (Yunnan), Vietnam. 36 CHC 41 VIE. Nanophan.

Antidesma andamanicum Hook.f., Fl. Brit. India 5: 364 (1887).
Andaman Is. 41 AND. Nanophan.

Antidesma angustifolium (Merr.) Pax & K.Hoffm. in H.G.A.Engler, Pflanzenr., IV, 147, XV: 165 (1922).
Philippines. 42 PHI.
* *Antidesma pentandrum* var. *angustifolium* Merr., Philipp. J. Sci., C 9: 464 (1914 publ. 1915).

Antidesma annamense Gagnep., Bull. Soc. Bot. France 70: 117 (1923).
Vietnam. 41 VIE.

Antidesma aruanum Pax & K.Hoffm. in H.G.A.Engler, Pflanzenr., IV, 147, XV: 149 (1922).
New Guinea (Aru Is.). 42 NWG. Nanophan. or phan. – No material now known.

Antidesma baccatum Airy Shaw, Kew Bull. 23: 287 (1969).
Irian Jaya. 42 NWG. Nanophan. or phan.

Antidesma boridiense Airy Shaw, Kew Bull. 33: 16 (1978).
Papua New Guinea. 42 NWG. Phan.

Antidesma brachybotrys Airy Shaw, Kew Bull. 26: 457 (1972).
Pen. Malaysia, Borneo (Sarawak, Kalimantan). 42 BOR MLY. Nanophan.

Antidesma bunius (L.) Spreng., Syst. Veg. 1: 826 (1824).
China, Trop. Asia, Queensland. 36 CHC CHH CHS CHT 40 ASS IND SRL 41 BMA LAO MLY THA VIE 42 BOR JAW MOL NWG PHI SUL SUM 50 QLD. Phan.

* *Stilago bunius* L., Mant. Pl. 1: 122 (1767). *Antidesma stilago* Poir. in J.B.A.M.de Lamarck, Encycl., Suppl. 1: 403 (1811). *Antidesma bunius* var. *genuinum* Müll.Arg. in A.P.de Candolle, Prodr. 15(2): 262 (1866), nom. inval.
Antidesma sylvestre Lam., Encycl. 1: 207 (1783).
Antidesma glabellum K.D.Koenig ex Benn., Pl. Jav. Rar.: 192 (1840), nom. nud.
Antidesma retusum Zipp. ex Span., Linnaea 15: 350 (1841), nom. nud.
Antidesma ciliatum C.Presl, Epimel. Bot.: 234 (1851).
Antidesma cordifolium C.Presl, Epimel. Bot.: 235 (1851). *Antidesma bunius* var. *cordifolium* (C.Presl) Müll.Arg. in A.P.de Candolle, Prodr. 15(2): 262 (1866).
Antidesma floribundum Tul., Ann. Sci. Nat., Bot., III, 15: 189 (1851). *Antidesma bunius* var. *floribundum* (Tul.) Müll.Arg. in A.P.de Candolle, Prodr. 15(2): 262 (1866).
Antidesma glabrum Tul., Ann. Sci. Nat., Bot., III, 15: 188 (1851).
Antidesma rumphii Tul., Ann. Sci. Nat., Bot., III, 15: 238 (1851).
Antidesma bunius var. *sylvestre* Müll.Arg. in A.P.de Candolle, Prodr. 15(2): 262 (1866).
Antidesma bunius var. *wallichii* Müll.Arg. in A.P.de Candolle, Prodr. 15(2): 262 (1866).
Antidesma dallachyanum Baill., Adansonia 6: 337 (1866).
Sapium crassifolium Elmer, Leafl. Philipp. Bot. 2: 485 (1908). *Antidesma crassifolium* (Elmer) Merr., Philipp. J. Sci., C 7: 383 (1912 publ. 1913).
Antidesma collettii Craib, Bull. Misc. Inform. Kew 1911: 461 (1911).
Antidesma thorelianum Gagnep., Bull. Soc. Bot. France 70: 124 (1923).

Antidesma catanduanense Merr., Philipp. J. Sci. 16: 549 (1920).
Philippines (Catanduanes). 42 PHI.

Antidesma celebicum Miq., Ann. Mus. Bot. Lugduno-Batavi 1: 218 (1864).
Sulawesi. 42 SUL.

Antidesma chalaranthum Airy Shaw, Kew Bull. 33: 424 (1979).
Papua New Guinea. 42 NWG. Phan.

Antidesma chonmon Gagnep., Bull. Soc. Bot. France 70: 119 (1923).
China (Yunnan), Vietnam. 36 CHC 41 VIE. Phan.

Antidesma cinnamomifolium Pax & K.Hoffm. in H.G.A.Engler, Pflanzenr., IV, 147, XV: 154 (1922).
NE. Papua New Guinea. 42 NWG. Nanophan. or phan.

Antidesma cochinchinense Gagnep., Bull. Soc. Bot. France 70: 119 (1923).
S. Vietnam. 41 VIE.

Antidesma comptum Tul., Ann. Sci. Nat., Bot., III, 15: 190 (1851).
E. India. 40 IND. Nanophan.

Antidesma concinnum Airy Shaw, Kew Bull. 33: 425 (1979).
Papua New Guinea. 42 NWG. Nanophan. or phan.

Antidesma contractum J.J.Sm., Nova Guinea 8: 229 (1910).
New Guinea. 42 NWG. Nanophan. or phan.

Antidesma cordatum Airy Shaw, Kew Bull. 26: 465 (1972).
Pen. Malaysia. 42 BOR MLY. Phan.

Antidesma coriaceum Tul., Ann. Sci. Nat., Bot., III, 15: 204 (1851).
Pen. Malaysia, Borneo, Sumatera (incl. Bangka). 42 BOR MLY SUM. Phan.
Antidesma minus Wall., Numer. List: 7288 (1832), nom. inval.
Antidesma fallax Müll.Arg., Linnaea 34: 68 (1865).

Aporusa griffithii Hook.f., Fl. Brit. India 5: 353 (1887).
Antidesma pachyphyllum Merr., Philipp. J. Sci., C 11: 58 (1916).
Antidesma nitens Pax & K.Hoffm. in H.G.A.Engler, Pflanzenr., IV, 147, XV: 136 (1922).

Antidesma coriifolium Pax & K.Hoffm. in H.G.A.Engler, Pflanzenr., IV, 147, XV: 154 (1922).
NE. Papua New Guinea. 42 NWG. Phan.

Antidesma costulatum Pax & K.Hoffm. in H.G.A.Engler, Pflanzenr., IV, 147, XV: 129 (1922).
China (Sichuan, Yunnan). 36 CHC. Phan.

Antidesma cruciforme Gage, Rec. Bot. Surv. India 9: 226 (1922).
Pen. Malaysia. 42 MLY.

Antidesma curranii Merr., Philipp. J. Sci., C 9: 466 (1914 publ. 1915).
Philippines (Luzon). 42 PHI. Phan.

Antidesma cuspidatum Müll.Arg., Linnaea 34: 67 (1865).
Burma, W. Malesia. 41 BMA 42 BOR MLY. Nanophan. or phan.

var. **borneense** Airy Shaw, Kew Bull., Addit. Ser. 4: 210 (1975).
Borneo (Sarawak). 42 BOR. Phan.

var. **cuspidatum**
Burma, Pen. Malaysia (incl. Singapore). 41 BMA 42 MLY. Nanophan. or phan.
Antidesma rotatum Müll.Arg. in A.P.de Candolle, Prodr. 15(2): 256 (1866).

Antidesma digitaliforme Tul., Ann. Sci. Nat., Bot., III, 15: 191 (1851).
Philippines. 42 PHI.
Antidesma lucidum Merr., Philipp. J. Sci. 1(Suppl.): 78 (1906).

Antidesma discolor Airy Shaw, Kew Bull., Addit. Ser. 8: 212 (1980).
Irian Jaya. 42 NWG. – No material known.
* *Antidesma bicolor* Pax & K.Hoffm. in H.G.A.Engler, Pflanzenr., IV, 147, XV: 126 (1922).

Antidesma eberhardtii Gagnep., Bull. Soc. Bot. France 70: 120 (1923).
Laos, N. Vietnam. 41 LAO VIE.

Antidesma elassophyllum A.C.Sm., J. Arnold Arbor. 33: 371 (1952).
Fiji. 60 FIJ.

Antidesma erostre F.Muell. in G.Bentham, Fl. Austral. 6: 87 (1873).
Papua New Guinea, Queensland. 42 NWG 50 QLD. Phan.

Antidesma excavatum Miq., Ann. Mus. Bot. Lugduno-Batavi 1: 218 (1864).
Sri Lanka to Philippines (Palawan). 40 SRL 41 AND BMA THA VIE 42 BOR PHI SUL SUM. Phan.
Antidesma thwaitesianum Müll.Arg. in A.P.de Candolle, Prodr. 15(2): 263 (1866).
Antidesma bunius var. *thwaitesianum* (Müll.Arg.) Trimen, Syst. Cat. Fl. Pl. Ceylon: 81 (1885).

Antidesma ferrugineum Airy Shaw, Kew Bull. 26: 462 (1972).
NE. Papua New Guinea. 42 NWG. Phan.

Antidesma fleuryi Gagnep., Bull. Soc. Bot. France 70: 121 (1923).
Laos, Vietnam. 41 LAO VIE.

Antidesma forbesii Pax & K.Hoffm. in H.G.A.Engler, Pflanzenr., IV, 147, XV: 153 (1922).
Sumatera. 42 SUM. Nanophan. or phan.
Antidesma salicifolium Miq., Fl. Ned. Ind., Eerste Bijv.: 467 (1861), nom. illeg.

Antidesma fordii Hemsl., J. Linn. Soc., Bot. 26: 430 (1894).
China, Indo-China. 36 CHC CHH CHS 41 LAO VIE. Phan.
Antidesma yunnanense Pax & K.Hoffm. in H.G.A.Engler, Pflanzenr., IV, 147, XV: 157 (1922).

Antidesma fruticosum (Lour.) Müll.Arg. in A.P.de Candolle, Prodr. 15(2): 259 (1866).
Vietnam. 41 VIE. Nanophan.
* *Rhytis fruticosa* Lour., Fl. Cochinch.: 660 (1790).

Antidesma fruticulosum Kurz, J. Asiat. Soc. Bengal, Pt. 2, Nat. Hist. 42(2): 237 (1873).
Burma. 41 BMA. Nanophan.

Antidesma fusicarpum Elmer, Leafl. Philipp. Bot. 8: 3081 (1919).
Philippines (Luzon). 42 PHI.

Antidesma ghaesembilla Gaertn., Fruct. Sem. Pl. 1: 189 (1788). *Antidesma ghaesembilla* var. *genuinum* Müll.Arg. in A.P.de Candolle, Prodr. 15(2): 251 (1866), nom. inval.
S. China, Trop. Asia, N. Australia. 36 CHC CHH CHS 40 ASS IND SRL 41 BMA MLY THA VIE 42 BIS BOR JAW NWG PHI SUM SUL 50 NTA QLD WAU. Nanophan. or phan.
Antidesma pubescens Roxb., Pl. Coromandel 2: 35 (1802).
Antidesma frutescens Jack, Malayan Misc. 2: 91 (1822).
Antidesma paniculatum Blume, Bijdr.: 1128 (1827). *Antidesma ghaesembilla* var. *paniculatum* (Blume) Müll.Arg. in A.P.de Candolle, Prodr. 15(2): 251 (1866).
Antidesma rhamnoides Brongn. ex Tul., Ann. Sci. Nat., Bot., III, 15: 217 (1851).
Antidesma vestitum C.Presl, Epimel. Bot.: 232 (1851). *Antidesma ghaesembilla* var. *vestitum* (C.Presl) Müll.Arg. in A.P.de Candolle, Prodr. 15(2): 251 (1866).
Antidesma schultzii Benth., Fl. Austral. 6: 86 (1873).

Antidesma gillespieanum A.C.Sm., J. Arnold Arbor. 33: 370 (1952).
Fiji. 60 FIJ.

Antidesma globuligerum Airy Shaw, Kew Bull. 36: 635 (1981).
Sulawesi. 42 SUL.

Antidesma hainanense Merr., Philipp. J. Sci. 21: 347 (1922).
China, Indo-China. 36 CHC CHH CHS 41 LAO VIE. Nanophan. or phan.

Antidesma helferi Hook.f., Fl. Brit. India 5: 357 (1887).
Burma, Thailand, Pen. Malaysia. 41 BMA THA 42 MLY. Nanophan.
Antidesma pachystemon Airy Shaw, Kew Bull. 23: 279 (1969).

Antidesma heterophyllum Blume, Bijdr.: 1123 (1827).
Jawa, Lesser Sunda Is. 42 JAW LSI. Nanophan.
Antidesma ovalifolium Zipp. ex Span., Linnaea 15: 350 (1841), nom. nud.

Antidesma hildebrandtii Pax & K.Hoffm. in H.G.A.Engler, Pflanzenr., IV, 147, XV: 122 (1922).
Madagascar. 29 MDG. Phan. – Probably to be reduced to *A. madagascariense*.

Antidesma hosei Pax & K.Hoffm. in H.G.A.Engler, Pflanzenr., IV, 147, XV: 138 (1922).
Pen. Malaysia, Sumatera, Borneo (Brunei, Kalimantan), Sulawesi. 42 BOR MLY SUL SUM. Phan.

var. **hosei**

Pen. Malaysia, Borneo (Brunei, Kalimantan), Sulawesi. 42 BOR MLY SUL. Phan.

Antidesma plumbeum Pax & K.Hoffm. in H.G.A.Engler, Pflanzenr., IV, 147, XV: 133 (1922).

Antidesma hosei var. *microcarpum* Airy Shaw, Kew Bull. 28: 271 (1973).

Antidesma neurocarpum var. *angustatum* Airy Shaw, Kew Bull. 28: 270 (1973). *Antidesma hosei* var. *angustatum* (Airy Shaw) Airy Shaw, Kew Bull., Addit. Ser. 4: 212 (1975).

var. **oxyurum** Airy Shaw, Kew Bull. 36: 362 (1981).

Sumatera. 42 SUM.

Antidesma hylandii Airy Shaw, Kew Bull. 36: 636 (1981).

Queensland. 50 QLD.

Antidesma ilocanum Merr., Philipp. J. Sci. 16: 549 (1920).

Philippines (Luzon). 42 PHI.

Antidesma impressinerve Merr., Philipp. J. Sci. 16: 548 (1920).

Philippines (Panay). 42 PHI.

Antidesma japonicum Siebold & Zucc., Abh. Math.-Phys. Cl. Königl. Bayer. Akad. Wiss. 4(2): 212 (1846).

China, S. Japan, Taiwan, Vietnam, Cambodia, Thailand, Pen. Malaysia. 36 CHC CHH CHQ CHS CHT 38 JAP TAI 41 CBD THA VIE 42 MLY. Nanophan. or phan.

var. **japonicum**

China, S. Japan, Taiwan, Vietnam, Cambodia, Thailand, Pen. Malaysia. 36 CHC CHH CHQ CHS CHT 38 JAP TAI 41 CBD THA VIE 42 MLY. Nanophan. or phan.

Antidesma gracile Hemsl., J. Linn. Soc., Bot. 26: 431 (1894).

Antidesma kuroiwai Makino, Bot. Mag. (Tokyo) 20: 6 (1906).

Antidesma delicatulum Hutch. in C.S.Sargent, Pl. Wilson. 2: 522 (1916).

Antidesma acutisepalum Hayata, Icon. Pl. Formosan. 9: 97 (1920).

Antidesma hiiranense Hayata, Icon. Pl. Formosan. 9: 98 (1920).

Antidesma gracillimum Gage, Rec. Bot. Surv. India 9: 227 (1922).

Antidesma cambodianum Gagnep., Bull. Soc. Bot. France 70: 118 (1923).

Antidesma filipes Hand.-Mazz., Symb. Sin. 7: 218 (1931).

var. **robustium** Airy Shaw, Kew Bull. 26: 355 (1972).

Thailand, Vietnam. 41 THA VIE. Nanophan. or phan.

Antidesma jucundum Airy Shaw, Kew Bull. 33: 426 (1979).

Papua New Guinea. 42 NWG. Nanophan. or phan.

Antidesma × kapuae Rock, Indig. Trees Haw. Isl.: 249 (1913). A. platyphyllum × A. pulvinatum.

Hawaiian Is. (Hawaii: South Kona). 63 HAW. Phan.

Antidesma katikii Airy Shaw, Kew Bull. 28: 278 (1973).

NE. Papua New Guinea. 42 NWG. Phan.

Antidesma kerrii Craib, Bull. Misc. Inform. Kew 1911: 462 (1911).

Thailand. 41 THA. Nanophan. or phan.

Antidesma khasianum Hook.f., Fl. Brit. India 5: 362 (1887).

Assam. 40 ASS.

Antidesma lanceolatum Hook.f. & Thomson ex Hook.f., Fl. Brit. India 5: 362 (1887), nom. inval.

Antidesma kunstleri Gage, Rec. Bot. Surv. India 9: 225 (1922).
Pen. Malaysia. 42 MLY. Phan.

Antidesma kusaiense Kaneh., Bot. Mag. (Tokyo) 46: 456 (1932).
Caroline Is. (Kusai I.). 62 CRL.

Antidesma laciniatum Müll.Arg., Flora 47: 520 (1864). *Antidesma laciniatum* var. *genuinum* Pax & K.Hoffm. in H.G.A.Engler, Pflanzenr., IV, 147, XV: 145 (1922), nom. inval.
W. Trop. Africa to Uganda. 22 GHA GUI IVO LBR NGA SIE TOG 23 CMN EQG GAB GGI ZAI 24 SUD 25 UGA. Phan.

var. **laciniatum**
W. Trop. Africa to Uganda. 22 GUI IVO NGA 23 CMN EQG GAB GGI ZAI 24 SUD 25 UGA. Phan.
Antidesma chevalieri Beille, Bull. Soc. Bot. France 55(8): 65 (1908).

var. **membranaceum** Müll.Arg., Flora 47: 520 (1864). *Antidesma laciniatum* subsp. *membranaceum* (Müll.Arg.) J.Léonard, Bull. Jard. Bot. Belg. 58: 20 (1988).
W. & WC. Trop. Africa. 22 GHA GUI IVO LBR SIE TOG 23 CMN EQG ZAI. Phan.
Antidesma pseudolaciniatum Beille, Bull. Soc. Bot. France 57(8): 122 (1910).

Antidesma laurifolium Airy Shaw, Kew Bull. 26: 356, 458 (1972).
Thailand, Pen. Malaysia. 41 THA 42 MLY. Nanophan. or phan.

Antidesma ledermannianum Pax & K.Hoffm. in H.G.A.Engler, Pflanzenr., IV, 147, XV: 133 (1922).
NE. Papua New Guinea. 42 NWG. Nanophan.

Antidesma leptodictyum Airy Shaw, Kew Bull. 36: 635 (1981).
Borneo (Sabah). 42 BOR.

Antidesma leucocladon Hook.f., Fl. Brit. India 5: 358 (1887).
S. Thailand, Pen. Malaysia, Sumatera. 41 THA 42 MLY SUM. Nanophan. or phan.

Antidesma leucopodum Miq., Fl. Ned. Ind., Eerste Bijv.: 465 (1861).
S. Thailand, Pen. Malaysia, Sumatera (incl. Anamba Is.), Borneo, Philippines. 41 THA 42 BOR MLY PHI SUM. Nanophan. or phan.

var. **kinabaluense** Airy Shaw, Kew Bull. 28: 273 (1973).
Borneo (Sabah). 42 BOR. Nanophan. or phan.

var. **leucopodum**
S. Thailand, Pen. Malaysia, Sumatera (incl. Anamba Is.), Borneo, Philippines. 41 THA 42 BOR MLY PHI SUM. Nanophan. or phan.
Antidesma clementis Merr., Philipp. J. Sci., C 9: 465 (1914 publ. 1915).
Antidesma cauliflorum W.W.Sm., Notes Roy. Bot. Gard. Edinburgh 8: 316 (1915).
Antidesma cauliflorum Merr., J. Straits Branch Roy. Asiat. Soc. 76: 89 (1917), nom. illeg. *Antidesma trunciflorum* Merr., J. Straits Branch Roy. Asiat. Soc., Spec. Nr.: 333 (1921).
Antidesma hirtellum Ridl., Bull. Misc. Inform. Kew 1923: 366 (1923).
Antidesma caudatum Pax & K.Hoffm., Mitt. Inst. Allg. Bot. Hamburg 7: 223 (1931).

var. **platyphyllum** Airy Shaw, Kew Bull. 28: 273 (1973).
Borneo (Sarawak, Brunei). 42 BOR. Nanophan. or phan.

Antidesma linearifolium Pax & K.Hoffm. in H.G.A.Engler, Pflanzenr., IV, 147, XV: 130 (1922).
Borneo (Sarawak, Sabah). 42 BOR. Nanophan.

Antidesma luzonicum Merr., Philipp. J. Sci., C 9: 464 (1914 publ. 1915).
Philippines (Luzon). 42 PHI.

Antidesma macgregorii C.B.Rob., Philipp. J. Sci., C 6: 207 (1911).
Taiwan, Philippines (Polillo, Luzon). 38 TAI 42 PHI. Phan.

Antidesma maclurei Merr., Philipp. J. Sci. 23: 248 (1923).
Hainan, Vietnam. 36 CHH 41 VIE. Phan.

Antidesma madagascariense Lam., Encycl. 1: 206 (1783).
Madagascar, Comoros, Mascarenes. 29 COM MAU MDG REU. Nanophan. or phan.
Antidesma lancifolium Bojer, Hortus Maurit.: 289 (1837).
Antidesma longifolium Bojer, Hortus Maurit.: 289 (1837).
Antidesma rotundifolium Bojer, Hortus Maurit.: 289 (1837).
Antidesma tulasneanum Baill., Étude Euphorb.: 602 (1858), nom. nud.
Antidesma madagascariense f. *aporosum* Müll.Arg. in A.P.de Candolle, Prodr. 15(2): 265 (1866).
Antidesma madagascariense f. *trichophorum* Müll.Arg. in A.P.de Candolle, Prodr. 15(2): 265 (1866).
Antidesma boutonii Baker, Fl. Mauritius: 306 (1877).
Antidesma comorense Vatke & Pax ex Pax, Bot. Jahrb. Syst. 15: 529 (1893).
Antidesma hildebrandtii var. *comorense* (Pax & Vatke) Leandri, Notul. Syst. (Paris) 6: 27 (1937).

Antidesma martabanicum C.Presl, Epimel. Bot.: 232 (1851).
S. Burma, S. Thailand. 41 BMA THA. Phan.
Antidesma oblongifolium var. *wallichii* Tul., Ann. Sci. Nat., Bot., III, 15: 221 (1851).
Antidesma menasu Kurz, Forest Fl. Burma 2: 360 (1877).
Antidesma oblongum Wall. ex Hook.f., Fl. Brit. India 5: 364 (1887), pro syn.

Antidesma megalophyllum Merr., Philipp. J. Sci. 16: 551 (1920).
Philippines (Babuyan). 42 PHI.

Antidesma membranaceum Müll.Arg., Linnaea 34: 68 (1865).
Trop. & S. Africa. 22 GHA GNB GUI IVO NGA SEN SIE TOG 23 CMN ZAI 24 SUD 25 KEN TAN UGA 26 ANG MOZ ZIM 27 TVL. Nanophan. or phan.
Antidesma membranaceum var. *molle* Müll.Arg., Linnaea 34: 68 (1865).
Antidesma membranaceum var. *tenuifolium* Müll.Arg. in A.P.de Candolle, Prodr. 15(2): 261 (1866).
Antidesma meiocarpum J.Léonard, Bull. Jard. Bot. État 17: 260 (1945).

Antidesma menasu (Tul.) Müll.Arg. in A.P.de Candolle, Prodr. 15(2): 257 (1866).
India. 40 IND. Nanophan. or phan.
* *Antidesma pubescens* var. *menasu* Tul., Ann. Sci. Nat., Bot., III, 15: 215 (1851).
Antidesma lanceolatum Dalzell & Gibson, Bombay Fl.: 237 (1861), nom. illeg.
Antidesma menasu var. *linearifolium* Hook.f., Fl. Brit. India 5: 364 (1887).

Antidesma messianianum Guillaumin, Arch. Bot. Mém. 2(3): 26 (1929).
C. New Caledonia. 60 NWC. Phan.

Antidesma microcarpum Elmer, Leafl. Philipp. Bot. 2: 487 (1908).
Philippines (Mindanao). 42 PHI.
Antidesma maesoides Pax & K.Hoffm. in H.G.A.Engler, Pflanzenr., IV, 147, XV: 164 (1922).
Antidesma frutiferum Elmer, Leafl. Philipp. Bot. 10: 3731 (1939), no latin descr.

Antidesma minus Blume, Bijdr.: 1123 (1827).
W. Jawa. 42 JAW. Nanophan. or phan.
Antidesma lanceolarium Moritzi, Syst. Verz.: 73 (1846), nom. illeg.
Antidesma lanceolatum Tul., Ann. Sci. Nat., Bot., III, 15: 195 (1851), nom. illeg.
Antidesma zollingeri Müll.Arg. ex Pax & K.Hoffm. in H.G.A.Engler, Pflanzenr., IV, 147, XV: 132 (1922).

Antidesma moluccanum Airy Shaw, Kew Bull. 23: 284 (1969).
Maluku, New Guinea (incl. Admiralty Is., D'Entrecasteaux Is.) to Solomon Is. 42 BIS MOL NWG 60 SOL. Phan.
Antidesma moluccanum var. *indutum* Airy Shaw, Kew Bull. 33: 16 (1978).

Antidesma montanum Blume, Bijdr.: 1124 (1827).
China, Trop. Asia. 36 CHC CHH CHS CHT 41 BMA CBD LAO THA VIE 42 BOR JAW LSI MLY PHI SUL SUM. Phan.

var. **microcarpum** Airy Shaw, Kew Bull. 36: 363 (1981).
Sumatera. 42 SUM. Phan.

var. **montanum**
China, Trop. Asia. 36 CHC CHH CHS CHT 41 BMA CBD LAO THA VIE 42 BOR JAW LSI MLY PHI SUL SUM. Phan.
Antidesma oblongifolium Blume, Bijdr.: 1125 (1827).
Antidesma pubescens Moritzi, Syst. Verz.: 73 (1846), nom. illeg.
Antidesma alexiterium C.Presl, Epimel. Bot.: 234 (1851).
Antidesma heterophyllum C.Presl, Epimel. Bot.: 234 (1851), nom. illeg.
Antidesma leptocladum Tul., Ann. Sci. Nat., Bot., III, 15: 199 (1851). *Antidesma leptocladum* var. *genuinum* Müll.Arg. in A.P.de Candolle, Prodr. 15(2): 253 (1866), nom. inval.
Antidesma nitidum Tul., Ann. Sci. Nat., Bot., III, 15: 193 (1851). *Antidesma leptocladum* var. *nitidum* (Tul.) Müll.Arg. in A.P.de Candolle, Prodr. 15(2): 253 (1866).
Antidesma pubescens var. *moritzii* Tul., Ann. Sci. Nat., Bot., III, 15: 215 (1851). *Antidesma moritzii* (Tul.) Müll.Arg., Linnaea 34: 67 (1865).
Antidesma diversifolium Miq., Fl. Ned. Ind., Eerste Bijv.: 468 (1861).
Antidesma palembanicum Miq., Fl. Ned. Ind., Eerste Bijv.: 465 (1861).
Antidesma erythrocarpum Müll.Arg. in A.P.de Candolle, Prodr. 15(2): 258 (1866).
Antidesma leptocladum var. *glabrum* Müll.Arg. in A.P.de Candolle, Prodr. 15(2): 253 (1866).
Antidesma apiculatum Hemsl., J. Linn. Soc., Bot. 26: 430 (1894).
Antidesma henryi Hemsl., J. Linn. Soc., Bot. 26: 431 (1894).
Antidesma oblongifolium Boerl. & Koord. in A.Koorders-Schumacher, Syst. Verz. 2: 28 (1910), nom. illeg.
Antidesma mindanaense Merr., Philipp. J. Sci., C 7: 383 (1912 publ. 1913).
Antidesma palawanense Merr., Philipp. J. Sci., C 9: 467 (1914 publ. 1915).
Antidesma phanerophlebium Merr., Philipp. J. Sci., C 11: 59 (1916).
Antidesma calvescens Pax & K.Hoffm. in H.G.A.Engler, Pflanzenr., IV, 147, XV: 118 (1922).
Antidesma pseudomontanum Pax & K.Hoffm. in H.G.A.Engler, Pflanzenr., IV, 147, XV: 163 (1922).
Antidesma leptocladum var. *schmutzii* Airy Shaw, Kew Bull. 37: 5 (1982).

Antidesma montis-silam Airy Shaw, Kew Bull. 28: 269 (1973).
Borneo (Sabah). 42 BOR. Phan.

Antidesma mucronatum Boerl. & Koord. in A.Koorders-Schumacher, Syst. Verz. 2: 27 (1910).
C. Sumatera. 42 SUM. Phan.

Antidesma myriocarpum Airy Shaw, Kew Bull. 26: 467 (1972).
New Guinea. 42 NWG. Nanophan. or phan.
Antidesma myriocarpum var. *puberulum* Airy Shaw, Kew Bull. 33: 17 (1978).

Antidesma neurocarpum Miq., Fl. Ned. Ind., Eerste Bijv.: 466 (1861).
Thailand, Pen. Malaysia, Sumatera (incl. Bangka), Borneo. 41 THA 42 BOR MLY SUM. Phan.
Antidesma microcarpum Miq., Fl. Ned. Ind., Eerste Bijv.: 184 (1861), sphalm.
Antidesma alatum Hook.f., Fl. Brit. India 5: 358 (1887).
Antidesma hallieri Merr., Philipp. J. Sci., C 11: 57 (1916).
Antidesma rubiginosum Merr., Philipp. J. Sci., C 11: 61 (1916).
Antidesma inflatum Merr., J. Straits Branch Roy. Asiat. Soc. 76: 91 (1917).
Antidesma urophyllum Pax & K.Hoffm., Mitt. Inst. Allg. Bot. Hamburg 7: 224 (1931).

Antidesma nienkui Merr. & Chun, Sunyatsenia 2: 263 (1935).
N. Thailand, China (Guangdong, Hainan). 36 CHH CHS 41 THA. Phan.

Antidesma nigricans Tul., Ann. Sci. Nat., Bot., III, 15: 225 (1851).
Assam. 40 ASS.
Antidesma alexiteria Willd., Sp. Pl. 4: 762 (1806), nom. illeg.
Antidesma bunius Wall., Numer. List: 7282 (1832), nom. inval.
Antidesma flexuosum Tul., Ann. Sci. Nat., Bot., III, 15: 225 (1851).

Antidesma novoguineense Pax & K.Hoffm. in H.G.A.Engler, Pflanzenr., IV, 147, XV: 153 (1922).
NE. Papua New Guinea. 42 NWG. Phan.

Antidesma obliquinervium Merr., Philipp. J. Sci., C 9: 466 (1914 publ. 1915).
Philippines (Palawan). 42 PHI.

Antidesma oblongatum Müll.Arg. in A.P.de Candolle, Prodr. 15(2): 254 (1866).
Assam. 40 ASS.

Antidesma oblongum (Hutch.) Keay, Bull. Jard. Bot. État 26: 184 (1956).
Liberia, Ivory Coast. 22 IVO LBR. Nanophan. or phan.
* *Maesobotrya oblonga* Hutch. in D.Oliver, Fl. Trop. Afr. 6(1): 670 (1912).

Antidesma oligoneurum Lauterb. in K.M.Schumann & C.A.G.Lauterbach, Fl. Schutzgeb. Südsee, Nachtr.: 294 (1905).
NE. Papua New Guinea. 42 NWG. Nanophan.

Antidesma olivaceum K.Schum. in K.M.Schumann & U.M.Hollrung, Fl. Kais. Wilh. Land: 76 (1889).
New Guinea, Bismarck Archip. 42 BIS NWG. Nanophan. or phan.

Antidesma orarium Airy Shaw, Kew Bull. 33: 17 (1978).
NE. Papua New Guinea. 42 NWG. Phan.

Antidesma orthogyne (Hook.f.) Airy Shaw, Kew Bull. 26: 359 (1972).
S. Thailand, Pen. Malaysia. 41 THA 42 MLY. Nanophan.
* *Antidesma velutinosum* var. *orthogyne* Hook.f., Fl. Brit. India 5: 357 (1887).

Antidesma pachybotryum Pax & K.Hoffm., Bot. Jahrb. Syst. 45: 236 (1910).
E. Cameroon. 23 CMN. Nanophan. or phan.

Antidesma pachystachys Hook.f., Fl. Brit. India 5: 355 (1887).
Pen. Malaysia. 42 MLY.
Antidesma pachystachys var. *palustre* Airy Shaw, Kew Bull. 28: 269 (1973).

Antidesma pacificum Müll.Arg. in A.P.de Candolle, Prodr. 15(2): 254 (1866).
Fiji. 60 FIJ. Nanophan.

Antidesma pahangense Airy Shaw, Kew Bull. 23: 277 (1969).
Pen. Malaysia. 42 MLY.

Antidesma parvifolium F.Muell., Fragm. 4: 86 (1864).
N. Australia. 50 NTA QLD. Nanophan.

Antidesma pedicellare Pax & K.Hoffm. in H.G.A.Engler, Pflanzenr., IV, 147, XV: 162 (1922).
NE. Papua New Guinea (Sattelberg). 42 NWG. Nanophan. – No material now known.

Antidesma pendulum Hook.f., Fl. Brit. India 5: 355 (1887).
Pen. Malaysia, Borneo, Sumatera. 42 BOR MLY SUM. Phan.
Antidesma batuense J.J.Sm., Icon. Bogor. 4: t. 251 (1914).
Antidesma sumatranum Pax & K.Hoffm. in H.G.A.Engler, Pflanzenr., IV, 147, XV: 120 (1922).

Antidesma pentandrum (Blanco) Merr., Philipp. J. Sci., C 9: 462 (1914 publ. 1915).
SE. Taiwan, Philippines, Jawa. 38 TAI 42 JAW PHI.
* *Cansjera pentandra* Blanco, Fl. Filip.: 71 (1837).
Antidesma barbatum C.Presl, Epimel. Bot.: 233 (1851). *Antidesma rostratum* var. *barbatum* (C.Presl) Müll.Arg. in A.P.de Candolle, Prodr. 15(2): 257 (1866). *Antidesma pentandrum* var. *barbatum* (C.Presl) Merr., Philipp. J. Sci., C 9: 463 (1914 publ. 1915).
Antidesma rostratum Tul., Ann. Sci. Nat., Bot., III, 15: 218 (1851). *Antidesma rostratum* var. *genuinum* Müll.Arg. in A.P.de Candolle, Prodr. 15(2): 257 (1866), nom. inval. *Antidesma pentandrum* var. *genuinum* (Müll.Arg.) Pax & K.Hoffm.in H.G.A.Engler, Pflanzenr., IV, 147, XV: 125 (1922), nom. inval.
Antidesma rostratum var. *lobbianum* Tul., Ann. Sci. Nat., Bot., III, 15: 219 (1851). *Antidesma pentandrum* var. *lobbianum* (Tul.) Merr., Philipp. J. Sci., C 9: 463 (1914 publ. 1915).
Antidesma salicifolium C.Presl, Epimel. Bot.: 233 (1851).
Antidesma lobbianum Müll.Arg. in A.P.de Candolle, Prodr. 15(2): 254 (1866).
Antidesma kotoense Kaneh., Formos. Trees: 472 (1917), no latin descr.
Antidesma rotundisepalum Hayata, Icon. Pl. Formosan. 9: 98 (1920).

Antidesma perakense Pax & K.Hoffm. in H.G.A.Engler, Pflanzenr., IV, 147, XV: 117 (1922).
Pen. Malaysia. 42 MLY. Nanophan. or phan.

Antidesma petiolare Tul., Ann. Sci. Nat., Bot., III, 15: 207 (1851).
Madagascar. 29 MDG. Phan. – Probably to be reduced to *A. madagascariense*.
Antidesma erythroxyloides Tul., Ann. Sci. Nat., Bot., III, 15: 208 (1851).
Antidesma alnifolium Baker, J. Linn. Soc., Bot. 22: 519 (1887), nom. illeg.
Antidesma arbutifolium Baker, J. Linn. Soc., Bot. 22: 519 (1887).
Antidesma brachyscyphum Baker, J. Linn. Soc., Bot. 22: 519 (1887). *Antidesma petiolare* var. *brachyscyphum* (Baker) Leandri, Notul. Syst. (Paris) 6: 27 (1937).
Antidesma petiolare var. *perrieri* Leandri, Notul. Syst. (Paris) 6: 28 (1937).
Antidesma petiolare f. *elliotii* Leandri, in Fl. Madag. 111: 20 (1958).
Antidesma petiolare f. *humbertii* Leandri, in Fl. Madag. 111: 20 (1958).

Antidesma petiolatum Airy Shaw, Kew Bull. 33: 16 (1978).
New Guinea. 42 NWG. Nanophan. or phan.
Antidesma petiolare Airy Shaw, Kew Bull. 26: 466 (1972), nom. illeg.

Antidesma phanrangense Gagnep., Bull. Soc. Bot. France 72: 459 (1925).
Vietnam. 41 VIE.

Antidesma plagiorrhynchum Airy Shaw, Kew Bull. 28: 272 (1973).
Sumatera. 42 SUM. Phan.

Antidesma platyphyllum H.Mann, Proc. Amer. Acad. Arts 7: 202 (1867). *Antidesma platyphyllum* var. *genuinum* Pax & K.Hoffm. in H.G.A.Engler, Pflanzenr., IV, 147, XV: 162 (1922), nom. inval.
Hawaiian Is. 63 HAW. Phan.
Antidesma platyphyllum f. *rubrum* O.Deg. & Sherff in ?, .
Antidesma platyphyllum var. *hamakuaense* Fosberg in ?, .
Antidesma platyphyllum var. *subamplexicaule* Sherff in ?, .
Antidesma pulvinatum var. *contractum* O.Deg. & Sherff in ?, .
Antidesma pulvinatum var. *leiogonum* Sherff in ?, .
Antidesma platyphyllum var. *hillebrandii* Pax & K.Hoffm. in H.G.A.Engler, Pflanzenr., IV, 147, XV: 162 (1922).

Antidesma pleuricum Tul., Ann. Sci. Nat., Bot., III, 15: 213 (1851).
Taiwan, Philippines. 38 TAI 40 SRL? 42 PHI. Nanophan.
Antidesma obliquicarpum Elmer, Leafl. Philipp. Bot. 7: 2633 (1915).
Antidesma tenuifolium Pax & K.Hoffm. in H.G.A.Engler, Pflanzenr., IV, 147, XV: 137 (1922).
Antidesma hontaushanense C.E.Chang, Forest. J. For. Ass. Taiwan Prov. Inst. Agric. Pintung 6: 2 (1964).

Antidesma poilanei Gagnep., Bull. Soc. Bot. France 70: 122 (1923).
S. Vietnam. 41 VIE.

Antidesma polyanthum K.Schum. & Lauterb., Fl. Schutzgeb. Südsee: 392 (1900).
New Guinea. 42 NWG. Nanophan. or phan.
Antidesma warburgii K.Schum. in K.M.Schumann & C.A.G.Lauterbach, Fl. Schutzgeb. Südsee, Nachtr.: 293 (1905).

Antidesma polystylum Airy Shaw, Kew Bull. 26: 460 (1972).
Borneo (Sabah, Kalimantan ?). 42 BOR. Phan.

Antidesma pseudomicrophyllum Croizat, J. Arnold Arbor. 21: 496 (1940).
SE. China, Hainan. 36 CHH CHS. Phan.

Antidesma pseudopentandrum Hurus., Iconogr. Pl. Asiae Orient. 4: 351 (1941).
Nansei-shoto (Iriomote). 38 NNS.

Antidesma pseudopetiolatum Airy Shaw, Kew Bull. 33: 423 (1979).
Irian Jaya. 42 NWG. Nanophan. or phan.

Antidesma pulvinatum Hillebr., Fl. Hawaiian Isl.: 403 (1888).
Hawaiian Is. 63 HAW. Phan.
Antidesma wawraeanum Beck ex Wawra, Itin. Princ. S. Coburgi 2: 77 (1888).
Antidesma barbigerum Hildebr. ex Pax & K.Hoffm. in H.G.A.Engler, Pflanzenr., IV, 147, XV: 165 (1922).

Antidesma puncticulatum Miq., Fl. Ned. Ind., Eerste Bijv.: 468 (1861).
Sumatera. 42 SUM. Phan.

Antidesma pyrifolium Müll.Arg., Linnaea 34: 68 (1865).
Sri Lanka. 40 SRL. Phan.
Antidesma montanum Thwaites, Enum. Pl. Zeyl.: 289 (1861), nom. illeg.
Antidesma brunneum Hook.f., Fl. Brit. India 5: 359 (1887).

Antidesma ramosii Merr., Philipp. J. Sci., C 9: 468 (1914 publ. 1915). *Antidesma pentandrum* var. *ramosii* (Merr.) Pax & K.Hoffm. in H.G.A.Engler, Pflanzenr., IV, 147, XV: 126 (1922).
Philippines (Luzon). 42 PHI.

Antidesma rhynchophyllum K.Schum. in K.M.Schumann & C.A.G.Lauterbach, Fl. Schutzgeb. Südsee, Nachtr.: 294 (1905).
New Guinea. 42 NWG. Nanophan. or phan.
Antidesma obovatum J.J.Sm., Nova Guinea 8: 230 (1910).
Antidesma densiflorum Pax & K.Hoffm. in H.G.A.Engler, Pflanzenr., IV, 147, XV: 121 (1922).

Antidesma riparium Airy Shaw, Kew Bull. 23: 282 (1969).
Borneo (Sarawak). 42 BOR.

Antidesma roxburghii Wall. ex Tul., Ann. Sci. Nat., Bot., III, 15: 234 (1851).
NE. India, Assam. 40 ASS IND. Nanophan.
* *Stilago tomentosa* Roxb., Fl. Ind. ed. 1832, 3: 757 (1832). *Antidesma tomentosum* (Roxb.) Voigt, Hort. Suburb. Calcutt.: 294 (1845), nom. illeg.

Antidesma rufescens Tul., Ann. Sci. Nat., Bot., III, 15: 231 (1851).
Trop. & S. Africa. 22 BEN GHA GNB GUI IVO LBR NGA SEN SIE TOG 23 CAF CMN CON ZAI 25 TAN 26 ANG MOZ ZAM ZIM 27 BOT NAM.
Antidesma membranaceum var. *glabrescens* Müll.Arg. in A.P.de Candolle, Prodr. 15(2): 261 (1866).
Antidesma sassandrae Beille, Bull. Soc. Bot. France 57(8): 123 (1910).

Antidesma salicinum Ridl., Fl. Malay Penins. 3: 228 (1924).
S. Thailand, Pen. Malaysia, Sumatera, Borneo (Kalimantan). 41 THA 42 BOR MLY SUM. Nanophan. – Close to *A. montanum*.

Antidesma samarense Merr., Philipp. J. Sci., C 9: 469 (1914 publ. 1915).
Philippines (Samar). 42 PHI. – Close to *A. tomentosum*.

Antidesma santosii Merr., Philipp. J. Sci. 16: 550 (1920).
Philippines (Luzon). 42 PHI.

Antidesma sarcocarpum Airy Shaw, Kew Bull. 24: 288 (1969).
New Guinea (incl. Woodlark I., Louisiade Archip.), Philippines. 42 NWG PHI. Nanophan. or phan.

Antidesma sinuatum Benth., Fl. Austral. 6: 87 (1873).
Queensland (North Kennedy). 50 QLD. Nanophan. – Perhaps a hybrid between *A. bunius* & *A. erostre*.

Antidesma sootepense Craib, Bull. Misc. Inform. Kew 1911: 463 (1911).
China (S. Yunnan), Indo-China. 36 CHC 41 BMA LAO THA. Nanophan. or phan.

Antidesma spaniothix Airy Shaw, Kew Bull. 33: 15 (1978).
Thailand. 41 THA.

Antidesma spatulifolium Airy Shaw, Kew Bull. 23: 283 (1969).
Papua New Guinea. 42 NWG. Nanophan. or phan.

Antidesma sphaerocarpum Müll.Arg. in A.P.de Candolle, Prodr. 15(2): 255 (1866).
New Guinea, Bismarck Archip., Samoa. 42 BIS NWG 60 SAM. Phan.

Antidesma spicatum Blanco, Fl. Filip.: 794 (1837).
Philippines. 42 PHI.
Antidesma edule Merr., Philipp. Gov. Lab. Bur. Bull. 17: 26 (1904).

Antidesma stenocarpum Airy Shaw, Kew Bull. 23: 281 (1969).
Borneo (Sarawak). 42 BOR. Nanophan.

Antidesma stenophyllum Merr., Philipp. J. Sci., C 11: 62 (1916).
Borneo (Sarawak). 42 BOR. Nanophan.

Antidesma stipulare Blume, Bijdr.: 1125 (1827).
Pen. Malaysia, Sumatera, Borneo, Maluku, Jawa, Philippines. 42 BOR JAW MLY MOL PHI SUM. Nanophan.
Antidesma diepenhorstii Miq., Fl. Ned. Ind., Eerste Bijv.: 467 (1861).
Antidesma amboinense Miq., Ann. Mus. Bot. Lugduno-Batavi 1: 218 (1864). *Antidesma stipulare* f. *amboinense* (Miq.) J.J.Sm. in ?, : 262 (1910).
Antidesma cordatostipulaceum Merr., Philipp. J. Sci., C 4: 275 (1909).
Antidesma grandistipulum Merr., Philipp. J. Sci., C 11: 56 (1916).
Antidesma sarawakense Merr., Philipp. J. Sci., C 11: 57 (1916).
Antidesma tsenophyllum Gage, Rec. Bot. Surv. India 9: 225 (1922).
Antidesma cordatostipulaceum var. *lanceifolium* Merr. in ?, : 413 (1923).

Antidesma subbicolor Gagnep., Bull. Soc. Bot. France 70: 123 (1923).
S. Vietnam. 41 VIE.

Antidesma subcordatum Merr., Philipp. J. Sci., C 4: 275 (1909). *Antidesma subcordatum* var. *genuinum* Pax & K.Hoffm. in H.G.A.Engler, Pflanzenr., IV, 147, XV: 156 (1922), nom. inval.
Philippines. 42 PHI. Phan.
Antidesma subcordatum var. *glabrescens* Pax & K.Hoffm. in H.G.A.Engler, Pflanzenr., IV, 147, XV: 157 (1922).

Antidesma tagulae Airy Shaw, Kew Bull. 23: 289 (1969).
New Guinea (Louisiade Archip.). 42 NWG. Phan.

Antidesma tetrandrum Blume, Bijdr.: 1124 (1827).
Jawa, Sumatera. 42 JAW SUM. Nanophan. or phan.
Antidesma salaccense Zoll. & Moritzi in A.Moritzi, Syst. Verz.: 74 (1846).
Antidesma auritum Tul., Ann. Sci. Nat., Bot., III, 15: 203 (1851).
Antidesma blumei Tul., Ann. Sci. Nat., Bot., III, 15: 211 (1851).

Antidesma teysmannianum Pax & K.Hoffm. in H.G.A.Engler, Pflanzenr., IV, 147, XV: 144 (1922).
Jawa. 42 JAW. Nanophan. or phan.

Antidesma tomentosum Blume, Bijdr.: 1126 (1827). *Antidesma tomentosum* var. *genuinum* Pax & K.Hoffm. in H.G.A.Engler, Pflanzenr., IV, 147, XV: 116 (1922), nom. inval.
Nicobar Is., W. & C. Malesia. 41 NCB 42 BOR JAW MLY PHI SUL SUM. Nanophan. or phan.

var. **bangueyense** (Merr.) Airy Shaw, Kew Bull., Addit. Ser. 4: 218 (1975).
Borneo (Banguey). 42 BOR.
Antidesma bangueyense Merr., Philipp. J. Sci. 24: 114 (1924).

var. **tomentosum**

Nicobar Is., W. & C. Malesia. 41 NCB 42 BOR JAW MLY PHI SUL SUM. Nanophan. or phan.

Antidesma cumingii Müll.Arg. in A.P.de Candolle, Prodr. 15(2): 249 (1866).
Antidesma persimile Kurz, J. Bot. 13: 330 (1875).
Antidesma kingii Hook.f., Fl. Brit. India 5: 356 (1887).
Antidesma longipes Hook.f., Fl. Brit. India 5: 355 (1887).
Antidesma perserrula Hook.f., Fl. Brit. India 5: 365 (1887), orth. var.
Antidesma membranifolium Elmer, Leafl. Philipp. Bot. 1: 313 (1908).
Antidesma subolivaceum Elmer, Leafl. Philipp. Bot. 4: 1272 (1911).
Antidesma gibbsiae Hutch., J. Linn. Soc., Bot. 42: 134 (1914).
Antidesma urdanetense Elmer, Leafl. Philipp. Bot. 7: 2635 (1915).
Antidesma foxworthyi Merr., Philipp. J. Sci., C 11: 55 (1916).
Antidesma rivulare Merr., Philipp. J. Sci., C 11: 60 (1916). *Antidesma tomentosum* var. *rivulare* (Merr.) Pax & K.Hoffm. in H.G.A.Engler, Pflanzenr., IV, 147, XV: 117 (1922).
Antidesma clementis Merr., J. Straits Branch Roy. Asiat. Soc. 76: 90 (1917), nom. illeg.
Antidesma tomentosum var. *giganteum* Pax & K.Hoffm. in ?, : 224 (1931).

Antidesma tonkinense Gagnep., Bull. Soc. Bot. France 70: 124 (1923).
N. Vietnam. 41 VIE.

Antidesma trichophyllum A.C.Sm., J. Arnold Arbor. 33: 373 (1952).
Fiji. 60 FIJ.

Antidesma vaccinioides Airy Shaw, Kew Bull. 28: 280 (1973).
NE. New Guinea. 42 NWG. Phan.

Antidesma velutinosum Blume, Bijdr.: 112 (1825).
Burma, Pen. Malaysia, Thailand, Vietnam, Sumatera (incl. Bangka, Anambas), Jawa. 41 BMA THA VIE 42 JAW MLY SUM. Nanophan. or phan.

Antidesma attenuatum Wall. ex Tul., Ann. Sci. Nat., Bot., III, 15: 235 (1851).
Antidesma tomentosum Miq., Fl. Ned. Ind. 1(2): 427 (1859).
Antidesma molle Wall. ex Müll.Arg., Linnaea 34: 67 (1865).
Antidesma velutinosum var. *lancifolium* Hook.f., Fl. Brit. India 5: 357 (1887).

Antidesma velutinum Tul., Ann. Sci. Nat., Bot., III, 15: 223 (1851).
Burma, Thailand. 41 BMA THA. Nanophan. or phan.

Antidesma velutinum f. *polystachya* Müll.Arg. in A.P.de Candolle, Prodr. 15(2): 259 (1866).
Antidesma gymnogyne Pax & K.Hoffm. in H.G.A.Engler, Pflanzenr., IV, 147, XV: 135 (1922).

Antidesma venenosum J.J.Sm., Icon. Bogor.: t. 313 (1910). – FIGURE, p. 195.
Borneo. 42 BOR. Nanophan. or phan.

Antidesma venosum E.Mey. ex Tul., Ann. Sci. Nat., Bot., III, 15: 232 (1851).
Trop. & S. Africa, Madagascar, China, Indo-China. 22 BEN BKN GAM GHA GUI IVO MLI NGA NGR SEN SIE TOG 23 BUR CAF CMN GAB RWA ZAI 24 CHA ETH SOM SUD 25 KEN TAN UGA 26 ANG MLW MOZ ZAM ZIM 27 BOT CPP NAM NAT SWZ TVL 29 MDG 36 CHC CHH CHS 41 LAO THA VIE. Nanophan. or phan.

Antidesma bifrons Tul., Ann. Sci. Nat., Bot., III, 15: 229 (1851).
Antidesma venosum var. *thouarsianum* Tul., Ann. Sci. Nat., Bot., III, 15: 234 (1851).
Antidesma boivinianum Baill., Adansonia 2: 45 (1861).
Antidesma natalense Harv., Thes. Cap. 2: 45 (1863).
Antidesma microphyllum Hemsl., J. Linn. Soc., Bot. 26: 432 (1894).
Antidesma fuscocinereum Beille, Bull. Soc. Bot. France 55(8): 64 (1908).

Antidesma venenosum J.J. Sm.
Artist: R. Natadipoera
Ic. Bogor. 4: pl. 313 (1910)
KEW ILLUSTRATIONS COLLECTION

Antidesma nervosum De Wild., Études Fl. Bas- Moyen-Congo 2: 270 (1908), sphalm.
Antidesma seguinii H.Lév., Repert. Spec. Nov. Regni Veg. 9: 460 (1911).
Antidesma neriifolium Pax & K.Hoffm. in H.G.A.Engler, Pflanzenr., IV, 147, XV: 130 (1922).

Antidesma vogelianum Müll.Arg., Flora 47: 529 (1864).
Trop. Africa. 22 NGA 23 CAF CMN CON GAB GGI ZAI 24 SUD 25 KEN TAN UGA 26 ANG MLW ZAM ZIM. Nanophan. or phan.
Antidesma staudtii Pax, Bot. Jahrb. Syst. 26: 327 (1899).
Antidesma venosum f. *glabrescens* De Wild., Contr. Fl. Katanga: 79 (1908).
Antidesma membranaceum var. *crassifolium* Pax & K.Hoffm. in H.G.A.Engler, Pflanzenr., IV, 147, XV: 141 (1922).

Antidesma walkeri (Tul.) Pax & K.Hoffm. in H.G.A.Engler, Pflanzenr., IV, 147, XV: 118 (1922).
Sri Lanka. 40 SRL. Nanophan.
* *Antidesma lanceolatum* var. *walkeri* Tul., Ann. Sci. Nat., Bot., III, 15: 196 (1851).

Antidesma wattii Hook.f., Fl. Brit. India 5: 366 (1887).
Assam (Manipur). 40 ASS. Phan.

Antidesma zippelii Airy Shaw, Kew Bull. 37: 6 (1982).
Lesser Sunda Is. 42 LSI.

Synonyms:
Antidesma acutisepalum Hayata === **Antidesma japonicum** Siebold & Zucc. var. **japonicum**
Antidesma alatum Hook.f. === **Antidesma neurocarpum** Miq.
Antidesma alexiteria Gaertn. === **Antidesma alexiteria** L.
Antidesma alexiteria Willd. === **Antidesma nigricans** Tul.
Antidesma alexiterium C.Presl === **Antidesma montanum** Blume var. **montanum**
Antidesma alnifolium Baker === **Antidesma petiolare** Tul.
Antidesma alnifolium Hook. === **Trimeria grandifolia** (Hochst.) Warb. (Flacourtiaceae)
Antidesma amboinense Miq. === **Antidesma stipulare** Blume
Antidesma apiculatum Hemsl. === **Antidesma montanum** Blume var. **montanum**
Antidesma arbutifolium Baker === **Antidesma petiolare** Tul.
Antidesma attenuatum Wall. ex Tul. === **Antidesma velutinosum** Blume
Antidesma auritum Tul. === **Antidesma tetrandrum** Blume
Antidesma bangueyense Merr. === **Antidesma tomentosum** var. **bangueyense** (Merr.) Airy Shaw
Antidesma barbatum C.Presl === **Antidesma pentandrum** (Blanco) Merr.
Antidesma barbigerum Hildebr. ex Pax & K.Hoffm. === **Antidesma pulvinatum** Hillebr.
Antidesma batuense J.J.Sm. === **Antidesma pendulum** Hook.f.
Antidesma bicolor Hassk. === **Excoecaria cochinchinensis** Lour.
Antidesma bicolor Pax & K.Hoffm. === **Antidesma discolor** Airy Shaw
Antidesma bifrons Tul. === **Antidesma venosum** E.Mey. ex Tul.
Antidesma blumei Tul. === **Antidesma tetrandrum** Blume
Antidesma boivinianum Baill. === **Antidesma venosum** E.Mey. ex Tul.
Antidesma boutonii Baker === **Antidesma madagascariense** Lam.
Antidesma brachyscyphum Baker === **Antidesma petiolare** Tul.
Antidesma brunneum Hook.f. === **Antidesma pyrifolium** Müll.Arg.
Antidesma bunius Wall. === **Antidesma nigricans** Tul.
Antidesma bunius var. *cordifolium* (C.Presl) Müll.Arg. === **Antidesma bunius** (L.) Spreng.
Antidesma bunius var. *floribundum* (Tul.) Müll.Arg. === **Antidesma bunius** (L.) Spreng.
Antidesma bunius var. *genuinum* Müll.Arg. === **Antidesma bunius** (L.) Spreng.
Antidesma bunius var. *sylvestre* Müll.Arg. === **Antidesma bunius** (L.) Spreng.
Antidesma bunius var. *thwaitesianum* (Müll.Arg.) Trimen === **Antidesma excavatum** Miq.
Antidesma bunius var. *wallichii* Müll.Arg. === **Antidesma bunius** (L.) Spreng.
Antidesma calvescens Pax & K.Hoffm. === **Antidesma montanum** Blume var. **montanum**
Antidesma cambodianum Gagnep. === **Antidesma japonicum** Siebold & Zucc. var. **japonicum**

Antidesma caudatum Pax & K.Hoffm. === **Antidesma leucopodum** Miq. var. **leucopodum**
Antidesma cauliflorum Merr. === **Antidesma leucopodum** Miq. var. **leucopodum**
Antidesma cauliflorum W.W.Sm. === **Antidesma leucopodum** Miq. var. **leucopodum**
Antidesma chevalieri Beille === **Antidesma laciniatum** Müll.Arg. var. **laciniatum**
Antidesma ciliatum C.Presl === **Antidesma bunius** (L.) Spreng.
Antidesma clementis Merr. === **Antidesma leucopodum** Miq. var. **leucopodum**
Antidesma clementis Merr. === **Antidesma tomentosum** Blume var. **tomentosum**
Antidesma collettii Craib === **Antidesma bunius** (L.) Spreng.
Antidesma comoense Beille === **Thecacoris stenopetala** (Müll.Arg.) Müll.Arg.
Antidesma comorense Vatke & Pax ex Pax === **Antidesma madagascariense** Lam.
Antidesma cordatostipulaceum Merr. === **Antidesma stipulare** Blume
Antidesma cordatostipulaceum var. *lanceifolium* Merr. === **Antidesma stipulare** Blume
Antidesma cordifolium C.Presl === **Antidesma bunius** (L.) Spreng.
Antidesma crassifolium (Elmer) Merr. === **Antidesma bunius** (L.) Spreng.
Antidesma crenatum H.St.John === **Xylosma crenatum** (H.St.John) H.St.John (Flacourtiaceae)
Antidesma cumingii Müll.Arg. === **Antidesma tomentosum** Blume var. **tomentosum**
Antidesma dallachyanum Baill. === **Antidesma bunius** (L.) Spreng.
Antidesma delicatulum Hutch. === **Antidesma japonicum** Siebold & Zucc. var. **japonicum**
Antidesma densiflorum Pax & K.Hoffm. === **Antidesma rhynchophyllum** K.Schum.
Antidesma diandrum (Roxb.) Spreng. === **Antidesma acidum** Retz.
Antidesma diandrum (Roxb.) Roth === **Antidesma acidum** Retz.
Antidesma diandrum var. *genuinum* Müll.Arg. === **Antidesma acidum** Retz.
Antidesma diandrum var. *javanicum* (J.J.Sm.) Pax & K.Hoffm. === **Antidesma acidum** Retz.
Antidesma diandrum f. *javanicum* J.J.Sm. === **Antidesma acidum** Retz.
Antidesma diandrum var. *lanceolatum* Tul. === **Antidesma acidum** Retz.
Antidesma diandrum var. *ovatum* Tul. === **Antidesma acidum** Retz.
Antidesma diandrum var. *parvifolium* Tul. === **Antidesma acidum** Retz.
Antidesma diepenhorstii Miq. === **Antidesma stipulare** Blume
Antidesma diversifolium Miq. === **Antidesma montanum** Blume var. **montanum**
Antidesma edule Merr. === **Antidesma spicatum** Blanco
Antidesma erythrocarpum Müll.Arg. === **Antidesma montanum** Blume var. **montanum**
Antidesma erythroxyloides Tul. === **Antidesma petiolare** Tul.
Antidesma fallax Müll.Arg. === **Antidesma coriaceum** Tul.
Antidesma filiforme Blume === **Galearia filiformis** (Blume) Boerl. (Pandaceae)
Antidesma filipes Hand.-Mazz. === **Antidesma japonicum** Siebold & Zucc. var. **japonicum**
Antidesma flexuosum Tul. === **Antidesma nigricans** Tul.
Antidesma floribundum Tul. === **Antidesma bunius** (L.) Spreng.
Antidesma foxworthyi Merr. === **Antidesma tomentosum** Blume var. **tomentosum**
Antidesma frutescens Jack === **Antidesma ghaesembilla** Gaertn.
Antidesma frutiferum Elmer === **Antidesma microcarpum** Elmer
Antidesma fuscocinereum Beille === **Antidesma venosum** E.Mey. ex Tul.
Antidesma ghaesembilla var. *genuinum* Müll.Arg. === **Antidesma ghaesembilla** Gaertn.
Antidesma ghaesembilla var. *paniculatum* (Blume) Müll.Arg. === **Antidesma ghaesembilla** Gaertn.
Antidesma ghaesembilla var. *vestitum* (C.Presl) Müll.Arg. === **Antidesma ghaesembilla** Gaertn.
Antidesma gibbsiae Hutch. === **Antidesma tomentosum** Blume var. **tomentosum**
Antidesma glabellum K.D.Koenig ex Benn. === **Antidesma bunius** (L.) Spreng.
Antidesma glabrum Tul. === **Antidesma bunius** (L.) Spreng.
Antidesma gracile Hemsl. === **Antidesma japonicum** Siebold & Zucc. var. **japonicum**
Antidesma gracillimum Gage === **Antidesma japonicum** Siebold & Zucc. var. **japonicum**
Antidesma grandistipulum Merr. === **Antidesma stipulare** Blume
Antidesma grossularia Raeusch. === **Embelia ribes** Burm.f. (Myrsinaceae)
Antidesma guineensis G.Don ex Hook. === **Uapaca guineensis** Müll.Arg. var. **guineensis**
Antidesma gymnogyne Pax & K.Hoffm. === **Antidesma velutinum** Tul.
Antidesma hallieri Merr. === **Antidesma neurocarpum** Miq.

Antidesma henryi Hemsl. === **Antidesma montanum** Blume var. **montanum**
Antidesma henryi Pax & K.Hoffm. === **Antidesma acidum** Retz.
Antidesma heterophyllum C.Presl === **Antidesma montanum** Blume var. **montanum**
Antidesma hiiranense Hayata === **Antidesma japonicum** Siebold & Zucc. var. **japonicum**
Antidesma hildebrandtii var. *comorense* (Pax & Vatke) Leandri === **Antidesma madagascariense** Lam.
Antidesma hirtellum Ridl. === **Antidesma leucopodum** Miq. var. **leucopodum**
Antidesma hontaushanense C.E.Chang === **Antidesma pleuricum** Tul.
Antidesma hosei var. *angustatum* (Airy Shaw) Airy Shaw === **Antidesma hosei** Pax & K.Hoffm. var. **hosei**
Antidesma hosei var. *microcarpum* Airy Shaw === **Antidesma hosei** Pax & K.Hoffm. var. **hosei**
Antidesma inflatum Merr. === **Antidesma neurocarpum** Miq.
Antidesma kingii Hook.f. === **Antidesma tomentosum** Blume var. **tomentosum**
Antidesma kotoense Kaneh. === **Antidesma pentandrum** (Blanco) Merr.
Antidesma kuroiwai Makino === **Antidesma japonicum** Siebold & Zucc. var. **japonicum**
Antidesma laciniatum var. *genuinum* Pax & K.Hoffm. === **Antidesma laciniatum** Müll.Arg.
Antidesma laciniatum subsp. *membranaceum* (Müll.Arg.) J.Léonard === **Antidesma laciniatum** var. **membranaceum** Müll.Arg.
Antidesma lanceolarium Moritzi === **Antidesma minus** Blume
Antidesma lanceolarium (Roxb.) Wight === **Antidesma acidum** Retz.
Antidesma lanceolatum Tul. === **Antidesma minus** Blume
Antidesma lanceolatum Hook.f. & Thomson ex Hook.f. === **Antidesma khasianum** Hook.f.
Antidesma lanceolatum Dalzell & Gibson === **Antidesma menasu** (Tul.) Müll.Arg.
Antidesma lanceolatum var. *walkeri* Tul. === **Antidesma walkeri** (Tul.) Pax & K.Hoffm.
Antidesma lancifolium Bojer === **Antidesma madagascariense** Lam.
Antidesma leptobotryum Müll.Arg. === **Thecacoris leptobotrya** (Müll.Arg.) Brenan
Antidesma leptocladum Tul. === **Antidesma montanum** Blume var. **montanum**
Antidesma leptocladum var. *genuinum* Müll.Arg. === **Antidesma montanum** Blume var. **montanum**
Antidesma leptocladum var. *glabrum* Müll.Arg. === **Antidesma montanum** Blume var. **montanum**
Antidesma leptocladum var. *nitidum* (Tul.) Müll.Arg. === **Antidesma montanum** Blume var. **montanum**
Antidesma leptocladum var. *schmutzii* Airy Shaw === **Antidesma montanum** Blume var. **montanum**
Antidesma litorale Blume === **Polyosma integrifolia** Blume (Escalloniaceae)
Antidesma lobbianum Müll.Arg. === **Antidesma pentandrum** (Blanco) Merr.
Antidesma longifolium Decne. ex Baill. === **Richeria grandis** Vahl var. **grandis**
Antidesma longifolium Bojer === **Antidesma madagascariense** Lam.
Antidesma longipes Pax === **Maesobotrya longipes** (Pax) Hutch.
Antidesma longipes Hook.f. === **Antidesma tomentosum** Blume var. **tomentosum**
Antidesma lucidum Merr. === **Antidesma digitaliforme** Tul.
Antidesma lunatum Miq. === **Aporusa lunata** (Miq.) Kurz
Antidesma macrophyllum Wall. ex Voigt === **Antidesma acuminatum** Wight
Antidesma madagascariense f. *aporosum* Müll.Arg. === **Antidesma madagascariense** Lam.
Antidesma madagascariense f. *trichophorum* Müll.Arg. === **Antidesma madagascariense** Lam.
Antidesma maesoides Pax & K.Hoffm. === **Antidesma microcarpum** Elmer
Antidesma mannianum Müll.Arg. === **Thecacoris manniana** (Müll.Arg.) Müll.Arg.
Antidesma maximowiczii Conw. === **Fossil**
Antidesma megalocarpum S.Moore === **Rhyticaryum longifolium** K.Schum. & Lauterb. (Icacinaceae)
Antidesma meiocarpum J.Léonard === **Antidesma membranaceum** Müll.Arg.
Antidesma membranaceum var. *crassifolium* Pax & K.Hoffm. === **Antidesma vogelianum** Müll.Arg.
Antidesma membranaceum var. *glabrescens* Müll.Arg. === **Antidesma rufescens** Tul.
Antidesma membranaceum var. *molle* Müll.Arg. === **Antidesma membranaceum** Müll.Arg.

Antidesma membranaceum var. *tenuifolium* Müll.Arg. === **Antidesma membranaceum** Müll.Arg.
Antidesma membranifolium Elmer === **Antidesma tomentosum** Blume var. **tomentosum**
Antidesma menasu Kurz === **Antidesma martabanicum** C.Presl
Antidesma menasu var. *linearifolium* Hook.f. === **Antidesma menasu** (Tul.) Müll.Arg.
Antidesma microcarpum Miq. === **Antidesma neurocarpum** Miq.
Antidesma microphyllum Hemsl. === **Antidesma venosum** E.Mey. ex Tul.
Antidesma mindanaense Merr. === **Antidesma montanum** Blume var. **montanum**
Antidesma minus Wall. === **Antidesma coriaceum** Tul.
Antidesma molle Wall. ex Müll.Arg. === **Antidesma velutinosum** Blume
Antidesma moluccanum var. *indutum* Airy Shaw === **Antidesma moluccanum** Airy Shaw
Antidesma montanum Thwaites === **Antidesma pyrifolium** Müll.Arg.
Antidesma moritzii (Tul.) Müll.Arg. === **Antidesma montanum** Blume var. **montanum**
Antidesma myriocarpum var. *puberulum* Airy Shaw === **Antidesma myriocarpum** Airy Shaw
Antidesma natalense Harv. === **Antidesma venosum** E.Mey. ex Tul.
Antidesma neriifolium Pax & K.Hoffm. === **Antidesma venosum** E.Mey. ex Tul.
Antidesma nervosum Wall. === **Gironniera nervosa** Planch. (Ulmaceae)
Antidesma nervosum De Wild. === **Antidesma venosum** E.Mey. ex Tul.
Antidesma neurocarpum var. *angustatum* Airy Shaw === **Antidesma hosei** Pax & K.Hoffm. var. **hosei**
Antidesma nitens Pax & K.Hoffm. === **Antidesma coriaceum** Tul.
Antidesma nitidum Tul. === **Antidesma montanum** Blume var. **montanum**
Antidesma obliquicarpum Elmer === **Antidesma pleuricum** Tul.
Antidesma oblongifolium Boerl. & Koord. === **Antidesma montanum** Blume var. **montanum**
Antidesma oblongifolium Blume === **Antidesma montanum** Blume var. **montanum**
Antidesma oblongifolium var. *wallichii* Tul. === **Antidesma martabanicum** C.Presl
Antidesma oblongum Wall. ex Hook.f. === **Antidesma martabanicum** C.Presl
Antidesma obovatum J.J.Sm. === **Antidesma rhynchophyllum** K.Schum.
Antidesma ovalifolium Zipp. ex Span. === **Antidesma heterophyllum** Blume
Antidesma pachyphyllum Merr. === **Antidesma coriaceum** Tul.
Antidesma pachystachys var. *palustre* Airy Shaw === **Antidesma pachystachys** Hook.f.
Antidesma pachystemon Airy Shaw === **Antidesma helferi** Hook.f.
Antidesma palawanense Merr. === **Antidesma montanum** Blume var. **montanum**
Antidesma palembanicum Miq. === **Antidesma montanum** Blume var. **montanum**
Antidesma paniculatum Blume === **Antidesma ghaesembilla** Gaertn.
Antidesma parasitica Dillwyn === **Scleropyrum pentandrum** (Dennst.) D.J.Mabb. (Santalaceae)
Antidesma parviflorum Ham. ex Pax & K.Hoffm. === **Antidesma acidum** Retz.
Antidesma paxii F.P.Metcalf === **Antidesma acidum** Retz.
Antidesma pentandrum var. *angustifolium* Merr. === **Antidesma angustifolium** (Merr.) Pax & K.Hoffm.
Antidesma pentandrum var. *barbatum* (C.Presl) Merr. === **Antidesma pentandrum** (Blanco) Merr.
Antidesma pentandrum var. *genuinum* (Müll.Arg.) Pax & K.Hoffm. === **Antidesma pentandrum** (Blanco) Merr.
Antidesma pentandrum var. *lobbianum* (Tul.) Merr. === **Antidesma pentandrum** (Blanco) Merr.
Antidesma pentandrum var. *ramosii* (Merr.) Pax & K.Hoffm. === **Antidesma ramosii** Merr.
Antidesma perserrula Hook.f. === **Antidesma tomentosum** Blume var. **tomentosum**
Antidesma persimile Kurz === **Antidesma tomentosum** Blume var. **tomentosum**
Antidesma petiolare Airy Shaw === **Antidesma petiolatum** Airy Shaw
Antidesma petiolare var. *brachyscyphum* (Baker) Leandri === **Antidesma petiolare** Tul.
Antidesma petiolare f. *elliotii* Leandri === **Antidesma petiolare** Tul.
Antidesma petiolare f. *humbertii* Leandri === **Antidesma petiolare** Tul.
Antidesma petiolare var. *perrieri* Leandri === **Antidesma petiolare** Tul.
Antidesma phanerophlebium Merr. === **Antidesma montanum** Blume var. **montanum**
Antidesma platyphyllum var. *genuinum* Pax & K.Hoffm. === **Antidesma platyphyllum** H.Mann

Antidesma platyphyllum var. *hamakuaense* Fosberg === **Antidesma platyphyllum** H.Mann
Antidesma platyphyllum var. *hillebrandii* Pax & K.Hoffm. === **Antidesma platyphyllum** H.Mann
Antidesma platyphyllum f. *rubrum* O.Deg. & Sherff === **Antidesma platyphyllum** H.Mann
Antidesma platyphyllum var. *subamplexicaule* Sherff === **Antidesma platyphyllum** H.Mann
Antidesma plumbeum Pax & K.Hoffm. === **Antidesma hosei** Pax & K.Hoffm. var. **hosei**
Antidesma ponapense Kaneh. === ?
Antidesma praegrandifolium S.Moore === **Aporusa sp.**
Antidesma pseudolaciniatum Beille === **Antidesma laciniatum** var. **membranaceum** Müll.Arg.
Antidesma pseudomontanum Pax & K.Hoffm. === **Antidesma montanum** Blume var. **montanum**
Antidesma puberum Zipp. ex Span. === ?
Antidesma pubescens Moritzi === **Antidesma montanum** Blume var. **montanum**
Antidesma pubescens Roxb. === **Antidesma ghaesembilla** Gaertn.
Antidesma pubescens var. *menasu* Tul. === **Antidesma menasu** (Tul.) Müll.Arg.
Antidesma pubescens var. *moritzii* Tul. === **Antidesma montanum** Blume var. **montanum**
Antidesma pulvinatum var. *contractum* O.Deg. & Sherff === **Antidesma platyphyllum** H.Mann
Antidesma pulvinatum var. *leiogonum* Sherff === **Antidesma platyphyllum** H.Mann
Antidesma refractum Müll.Arg. === **Antidesma acuminatum** Wight
Antidesma reticulata Planch. === **Hieronyma reticulata** (Planch.) Britton ex Rusby
Antidesma retusum Zipp. ex Span. === **Antidesma bunius** (L.) Spreng.
Antidesma rhamnoides Brongn. ex Tul. === **Antidesma ghaesembilla** Gaertn.
Antidesma ribes (Burm.f.) Raeusch. === **Embelia ribes** Burm.f. (Myrsinaceae)
Antidesma rivulare Merr. === **Antidesma tomentosum** Blume var. **tomentosum**
Antidesma rosaurianum M.Gómez === **Hieronyma clusioides** (Tul.) Griseb.
Antidesma rostratum Tul. === **Antidesma pentandrum** (Blanco) Merr.
Antidesma rostratum var. *barbatum* (C.Presl) Müll.Arg. === **Antidesma pentandrum** (Blanco) Merr.
Antidesma rostratum var. *genuinum* Müll.Arg. === **Antidesma pentandrum** (Blanco) Merr.
Antidesma rostratum var. *lobbianum* Tul. === **Antidesma pentandrum** (Blanco) Merr.
Antidesma rotatum Müll.Arg. === **Antidesma cuspidatum** Müll.Arg. var. **cuspidatum**
Antidesma rotundifolium Bojer === **Antidesma madagascariense** Lam.
Antidesma rotundisepalum Hayata === **Antidesma pentandrum** (Blanco) Merr.
Antidesma rubiginosum Merr. === **Antidesma neurocarpum** Miq.
Antidesma rugosum Wall. ex Voigt === ?
Antidesma rumphii Tul. === **Antidesma bunius** (L.) Spreng.
Antidesma salaccense Zoll. & Moritzi === **Antidesma tetrandrum** Blume
Antidesma salicifolium C.Presl === **Antidesma pentandrum** (Blanco) Merr.
Antidesma salicifolium Miq. === **Antidesma forbesii** Pax & K.Hoffm.
Antidesma sarawakense Merr. === **Antidesma stipulare** Blume
Antidesma sassandrae Beille === **Antidesma rufescens** Tul.
Antidesma scandens Lour. === **Humulus scandens** (Lour.) Merr. (Cannabaceae)
Antidesma schultzii Benth. === **Antidesma ghaesembilla** Gaertn.
Antidesma schweinfurthii Pax === **Maesobotrya floribunda** Benth.
Antidesma seguinii H.Lév. === **Antidesma venosum** E.Mey. ex Tul.
Antidesma simile Müll.Arg. === **Antidesma acuminatum** Wight
Antidesma staudtii Pax === **Antidesma vogelianum** Müll.Arg.
Antidesma stenopetalum Müll.Arg. === **Thecacoris stenopetala** (Müll.Arg.) Müll.Arg.
Antidesma stilago Poir. === **Antidesma bunius** (L.) Spreng.
Antidesma stipulare f. *amboinense* (Miq.) J.J.Sm. === **Antidesma stipulare** Blume
Antidesma subcordatum var. *genuinum* Pax & K.Hoffm. === **Antidesma subcordatum** Merr.
Antidesma subcordatum var. *glabrescens* Pax & K.Hoffm. === **Antidesma subcordatum** Merr.
Antidesma subolivaceum Elmer === **Antidesma tomentosum** Blume var. **tomentosum**
Antidesma sumatranum Pax & K.Hoffm. === **Antidesma pendulum** Hook.f.

Antidesma sylvestre Lam. === **Antidesma bunius** (L.) Spreng.
Antidesma sylvestre Wall. === **Antidesma acidum** Retz.
Antidesma tenuifolium Pax & K.Hoffm. === **Antidesma pleuricum** Tul.
Antidesma thorelianum Gagnep. === **Antidesma bunius** (L.) Spreng.
Antidesma thwaitesianum Müll.Arg. === **Antidesma excavatum** Miq.
Antidesma tomentosum Miq. === **Antidesma velutinosum** Blume
Antidesma tomentosum (Roxb.) Voigt === **Antidesma roxburghii** Wall. ex Tul.
Antidesma tomentosum var. *genuinum* Pax & K.Hoffm. === **Antidesma tomentosum** Blume
Antidesma tomentosum var. *giganteum* Pax & K.Hoffm. === **Antidesma tomentosum** Blume var. **tomentosum**
Antidesma tomentosum var. *rivulare* (Merr.) Pax & K.Hoffm. === **Antidesma tomentosum** Blume var. **tomentosum**
Antidesma triplinervium Spreng. === **Alchornea triplinervia** (Spreng.) Müll.Arg.
Antidesma trunciflorum Merr. === **Antidesma leucopodum** Miq. var. **leucopodum**
Antidesma tsenophyllum Gage === **Antidesma stipulare** Blume
Antidesma tulasneanum Baill. === **Antidesma madagascariense** Lam.
Antidesma urdanetense Elmer === **Antidesma tomentosum** Blume var. **tomentosum**
Antidesma urophyllum Pax & K.Hoffm. === **Antidesma neurocarpum** Miq.
Antidesma velutinosum var. *lancifolium* Hook.f. === **Antidesma velutinosum** Blume
Antidesma velutinosum var. *orthogyne* Hook.f. === **Antidesma orthogyne** (Hook.f.) Airy Shaw
Antidesma velutinum f. *polystachya* Müll.Arg. === **Antidesma velutinum** Tul.
Antidesma venosum f. *glabrescens* De Wild. === **Antidesma vogelianum** Müll.Arg.
Antidesma venosum var. *thouarsianum* Tul. === **Antidesma venosum** E.Mey. ex Tul.
Antidesma vestitum C.Presl === **Antidesma ghaesembilla** Gaertn.
Antidesma wallichianum C.Presl === **Antidesma acidum** Retz.
Antidesma warburgii K.Schum. === **Antidesma polyanthum** K.Schum. & Lauterb.
Antidesma wawraeanum Beck ex Wawra === **Antidesma pulvinatum** Hillebr.
Antidesma yunnanense Pax & K.Hoffm. === **Antidesma fordii** Hemsl.
Antidesma zeylanicum C.Presl === **Antidesma alexiteria** L.
Antidesma zeylanicum Lam. === **Antidesma alexiteria** L.
Antidesma zollingeri Müll.Arg. ex Pax & K.Hoffm. === **Antidesma minus** Blume

Aonikena

Synonyms:
Aonikena Speg. === **Chiropetalum** A.Juss.
Aonikena patagonica Speg. === **Chiropetalum patagonicum** (Speg.) O'Donell & Lourteig

Aparisthmium

1 species, widely distributed in tropical South America; shrubs or small trees with large cordate leaves in secondary forest at generally low elevations (illustration in Pax 1914). The present generic name requires conservation; technically it is a synonym of *Conceveiba* (Webster, Synopsis 1994: 82). (Acalyphoideae)

Pax, F. (with K. Hoffmann) (1914). *Aparisthmium.* In A. Engler (ed.), Das Pflanzenreich, IV 147 VII (Euphorbiaceae-Acalypheae-Mercurialinae): 257-259, illus. Berlin. (Heft 63.) La/Ge. — Monotypic, S America (widely distributed).
Jablonski, E. (1967). *Aparisthmium.* Euphorbiaceae, Guayana Highland (Mem. New York Bot. Gard. 17(1)): 135-137. New York. En. — 1 species, *A. cordatum*; many collections cited.

Aparisthmium Endl., Gen. Pl.: 1112 (1840).
C. & S. Trop. America. 80 82 83 84 85?

Aparisthmium cordatum (A.Juss.) Baill., Adansonia 5: 307 (1865).
C & S. Trop. America. 80 COS 82 FRG GUY SUR VEN 83 BOL CLM ECU PER 84 BZC BZL BZN BZS 85? Nanophan. or phan.
* *Conceveiba cordata* A.Juss., Euphorb. Gen.: 43 (1824). *Alchornea cordata* (A.Juss.) Müll.Arg. in A.P.de Candolle, Prodr. 15(2): 901 (1866), nom. illeg.
Alchornea macrophylla Mart., Herb. Fl. Bras.: 271 (1841). *Conceveiba macropylla* (Mart.) Klotzsch ex Benth., Hooker's J. Bot. Kew Gard. Misc. 6: 333 (1854). *Aparisthmium macrophyllum* (Mart.) Baill., Étude Euphorb.: 468 (1858).
Alchornea latifolia Klotzsch, Hooker's J. Bot. Kew Gard. Misc. 2: 46 (1843).
Aparisthmium spruceanum Baill., Adansonia 5: 307 (1865).
Conceveiba poeppigiana Klotzsch ex Pax & K.Hoffm. in H.G.A.Engler, Pflanzenr., IV, 147, VII: 258 (1914), nom. inval.
Alchornea orinocensis Croizat, J. Arnold Arbor. 26: 191 (1945).

Synonyms:
Aparisthmium javanense (Blume) Hassk. === **Alchornea rugosa** (Lour.) Müll.Arg.
Aparisthmium javanicum Baill. === **Alchornea rugosa** (Lour.) Müll.Arg.
Aparisthmium macrophyllum (Mart.) Baill. === **Aparisthmium cordatum** (A.Juss.) Baill.
Aparisthmium spruceanum Baill. === **Aparisthmium cordatum** (A.Juss.) Baill.
Aparisthmium sumatranum Rchb. & Zoll === **Alchornea villosa** (Benth.) Müll.Arg.

Aphora

Synonyms:
Aphora Nutt. === **Ditaxis** Vahl ex A.Juss.
Aphora blodgettii Torr. ex Chapm. === **Ditaxis argothamnoides** (Bertol. ex Spreng.) Radcl.-Sm. & Govaerts
Aphora catamarcensis Griseb. === **Ditaxis catamarcensis** (Griseb.) Pax
Aphora humilis Engelm. & A.Gray === **Ditaxis humilis** (Engelm. & A.Gray) Pax
Aphora laevis A.Gray ex Torr. === **Ditaxis laevis** (A.Gray ex Torr.) A.Heller
Aphora mercurialina Nutt. === **Ditaxis mercurialina** (Nutt.) Coult.
Aphora serrata Torr. === **Ditaxis serrata** (Torr.) A.Heller

Aplarina

Synonyms:
Aplarina Raf. === **Euphorbia** L.

Apodandra

Retained in Brummitt (1992) following Pax & Hoffmann but reduced to *Plukenetia* by Webster and by Gillespie (1993; see that genus).

Synonyms:
Apodandra Pax & K.Hoffm. === **Plukenetia** L.
Apodandra brachybotrya (Müll.Arg.) J.F.Macbr. === **Plukenetia brachybotrya** Müll.Arg.
Apodandra buchtienii (Pax) Pax === **Plukenetia brachybotrya** Müll.Arg.
Apodandra loretensis (Ule) Pax & K.Hoffm. === **Plukenetia loretensis** Ule

Apodiscus

1 species, W Africa (Guinea, Liberia and possibly Sierra Leone); small trees. Said by Pax & Hoffmann (1922) to be related to *Maesobotrya*. [Illustration in Ic. Pl. 31: pl. 3032 (1915).] (Phyllanthoideae)

Pax, F. & K. Hoffmann (1922). *Apodiscus*. In A. Engler (ed.), Das Pflanzenreich, IV 147 XV (Euphorbiaceae-Phyllanthoideae-Phyllantheae): 45. Berlin. (Heft 81.) La/Ge. — 1 species, Africa (W Africa; described from the present Guinea Republic).

Apodiscus Hutch., Bull. Soc. Bot. France 58(8): 205 (1911 publ. 1912).
W. Trop. Africa. 22. Phan.

Apodiscus chevalieri Hutch., Bull. Soc. Bot. France 58(8): 206 (1911 publ. 1912).
Guinea, Liberia, Sierra Leone ? 22 GUI LBR SIE? Phan.

Aporosa

An orthographic variant of *Aporusa* which has continued here and there to be taken up.

Aporosella

Synonyms:
Aporosella Chodat === **Phyllanthus** L.
Aporosella chacoensis Speg. === **Phyllanthus chacoensis** Morong
Aporosella chacorensis Pax & K.Hoffm. === **Phyllanthus chacoensis** Morong
Aporosella hassleriana Chodat === **Phyllanthus chacoensis** Morong

Aporusa

90 species, Asia, Malesia, Australia and SW Pacific (India and China to Queensland and Solomon Islands); shrubs or small to medium trees, generally in forest. The leaves of many species have a distinctive reticulum which is useful in field recognition. Related to *Maesobotrya* (Africa) and, less closely, to *Richeria* (Americas). *A. octandra* occurs over most of the range of the genus; some formal varieties have been listed here. The genus is most speciose in West Malesia and wet western parts of SE Asia, with secondary centres in Sri Lanka/S India, Philippines and New Guinea. 2 sections, 1 with three subsections, were accepted in the last overall revision by Pax & Hoffmann (1922) who also furnished a geographical survey. Anne M. Schot (Leiden) was by 1998 far advanced with a revision within the framework of the Rijksherbarium/Hortus Botanicus programme on Malesian Euphorbiaceae; she has to date contributed a preliminary paper (Schot 1995) as well as a biogeographical survey in *Biogeography and geological evolution of Southeast Asia* (ed. R. Hall and J. D. Holloway; Leiden, Backhuys, 1998). (Phyllanthoideae)

- Pax, F. & K. Hoffmann (1922). *Aporosa*. In A. Engler (ed.), Das Pflanzenreich, IV 147 XV (Euphorbiaceae-Phyllanthoideae-Phyllantheae): 80-105, illus. Berlin. (Heft 81.) La/Ge. — 62 species, Asia and Malesia.

Croizat, L. (1942). Notes on the Euphorbiaceae, III. Bull. Bot. Gard. Buitenzorg, III, 17: 209-219. En. — Includes (among other topics) a discussion of the correct orthography of *Aporusa*.

Airy-Shaw, H. K. (1966). Notes on Malaysian and other Asiatic Euphorbiaceae, LVI. A new *Aporosa* from New Guinea. Kew Bull. 20: 25-26. En. — Description of *A. squarrosa*.

Airy-Shaw, H. K. (1966). Notes on Malaysian and other Asiatic Euphorbiaceae, LXVII. New species of *Aporusa* Bl. Kew Bull. 20: 379-383. En. — 3 species and 1 variety described.

Airy-Shaw, H. K. (1968). Notes on Malesian and other Asiatic Euphorbiaceae, LXXXIV. A new *Aporusa* from Borneo. Kew Bull. 21: 355-357. En. — Description of *A. lagenocarpa*; several exsiccatae.

Airy-Shaw, H. K. (1969). Notes on Malesian and other Asiatic Euphorbiaceae, XCVII. New or noteworthy species of *Aporusa* Bl. Kew Bull. 23: 2-6. En. — Notes on 4 species with one new; some reductions.

Airy-Shaw, H. K. (1971). Notes on Malesian and other Asiatic Euphorbiaceae, CXX. New or noteworthy species of *Aporusa* Bl. Kew Bull. 25: 474-481. En. — Descriptions of novelties; description of male inflorescence and fruit of *A. selangorica*.

Airy-Shaw, H. K. (1974). Notes on Malesian and other Asiatic Euphorbiaceae, CLXX. New or noteworthy species of *Aporusa* Bl. Kew Bull. 29: 281-287. En. — Novelties and notes; removal of *A. incisa* from the Euphorbiaceae to *Prunus* (Rosaceae).

Airy-Shaw, H. K. (1978). Notes on Malesian and other Asiatic Euphorbiaceae, CCVII. *Aporusa* Bl. Kew Bull. 33: 25-27. En. — Two novelties (Malaya, New Guinea).

Airy-Shaw, H. K. (1978). Notes on Malesian and other Asiatic Euphorbiaceae, CLXXXVII. New or noteworthy species of *Aporusa* Bl. Kew Bull. 32: 361-365. En. — Treatment of 5 species (some new), all from W. Malesia; sections indicated.

Airy-Shaw, H. K. (1980). Notes on Euphorbiaceae from Indomalesia, Australia and the Pacific, CCXXX. *Aporusa* Bl. Kew Bull. 35: 384-385. En. — New variety for *A. wallichii* from northern Thailand.

Schot, A. M. (1995). A synopsis of taxonomic changes in *Aporosa* Blume (Euphorbiaceae). Blumea 40: 449-460. En. — Includes new taxa along with notes on others; no keys, illustrations or maps. [The paper is preliminary to a full revision.]

Schot, A (1997(1998)). Systematics of Aporosa (Euphorbiaceae). In J. Dransfield, M. J. E. Coode & D. A. Simpson (eds), *Plant diversity in Malesia, III: proceedings of the third international Flora Malesiana symposium* (Kew, 1995); 265-284, illus., maps. Kew: Royal Botanic Gardens. En. — Character review (with some illustrations, in particular figures of inflorescences); species concepts; phylogenetic analyses and reconstructions, with discussion (final consensus, p. 283); no formal synopsis or key (but the recommendations for sections - one of them paraphyletic - on p. 282).

Schot, A. (1998). Biogeography of *Aporosa* (Euphorbiaceae): testing a phylogenetic hypothesis using geology and distribution patterns. In R. Hall and J. D. Holloway (eds), *Biogeography and geological evolution of SE Asia*: 279-290, illus., maps. Leiden: Backhuys. En. — Analysis of phylogenies within the genus, with discussion; tectonic background; comparison of phylogenies, distribution and geological patterns (with the significant findings that the genus may have arrived in New Guinea by two main routes (map, p. 286) and that some species there are the result of hybridisation between the lines in association with the accretion and unification of the various terranes).

Schot, A. (in preparation). Taxonomy, phylogeny, and biogeography of *Aporosa* Blume (Euphorbiaceae). Blumea, Suppl.) Leiden. En. (Ph.D. dissertation, Rijksherbarium/ Hortus Botanicus, Univ. of Leiden.) — Includes a full revision of the genus (82 species with another 7 imperfectly known; 8 varieties and 6 forms have also been recognised).

Aporusa Blume, Bijdr.: 514 (1826).
Malesia to Solomon Is. 36 40 41 42 60. Nanophan. or phan.
Leiocarpus Blume, Bijdr.: 581 (1826).
Lepidostachys Wall. ex Lindl., Intr. Nat. Syst. Bot., ed. 2: 441 (1836).
Scepa Lindl., Intr. Nat. Syst. Bot., ed. 2: 441 (1836).
Tetractinostigma Hassk., Flora 40: 533 (1857).

Aporusa acuminata Thwaites, Enum. Pl. Zeyl.: 288 (1861).
S. India, Sri Lanka. 40 IND SRL.

Aporusa alia Schot, Blumea 40: 453 (1995).
Borneo. 42 BOR. Phan.

Aporusa annulata Schot, Blumea 40: 454 (1995).
Irian Jaya. 42 NWG. Phan.

Aporusa antennifera (Airy Shaw) Airy Shaw, Kew Bull., Addit. Ser. 4: 32 (1975).
Pen. Malaysia, Sumatera (Bangka), Borneo. 42 BOR MLY SUM. Phan.
* *Aporusa nigropunctata* var. *antennifera* Airy Shaw, Kew Bull. 20: 382 (1966).

Aporusa arborea (Blume) Müll.Arg. in A.P.de Candolle, Prodr. 15(2): 470 (1866).
S. Thailand, Pen. Malaysia, Sumatera, Jawa, N. Borneo. 41 THA 42 BOR JAW MLY SUM. Nanophan. or phan.

* *Leiocarpus arboreus* Blume, Bijdr.: 582 (1826).
  Leiocarpus arborescens Hassk., Hort. Bogor. Desc.: 59 (1858). *Aporusa arborescens* (Hassk.) Müll.Arg. in A.P.de Candolle, Prodr. 15(2): 470 (1866).

Aporusa aurea Hook.f., Fl. Brit. India 5: 351 (1887).
S. Thailand, Pen. Malaysia, Borneo (Sarawak, Sabah), N. Sumatera. 41 THA 42 BOR MLY SUM. Phan.
Excoecaria integrifolia Roxb., Fl. Ind. ed. 1832, 3: 757 (1832), nom. rejic. prop.

Aporusa basilanensis Merr., J. Sci. Bot. 9: 471 (1914 publ. 1915).
Philippines (Basilan). 42 PHI.

Aporusa benthamiana Hook.f., Hooker's Icon. Pl. 16: t. 1583 (1887).
Pen. Malaysia, Borneo, Philippines. 42 BOR MLY PHI. Phan.
Aporusa euphlebia Merr., Philipp. J. Sci., C 11: 62 (1916).
Aporusa stipulosa Merr., Philipp. J. Sci. 16: 547 (1920). *Aporusa lunata* var. *stipulosa* (Merr.) Merr. in ?, : 410 (1923), nom. illeg.
Aporusa lunata var. *philippinensis* Pax & K.Hoffm. in H.G.A.Engler, Pflanzenr., IV, 147, XV: 82 (1922).
Aporusa grandifolia Merr., Univ. Calif. Publ. Bot. 15: 143 (1929).

Aporusa bourdillonii Stapf, Hooker's Icon. Pl. 23: t. 2204 (1892).
India. 40 IND.

Aporusa bracteosa Pax & K.Hoffm. in H.G.A.Engler, Pflanzenr., IV, 147, XV: 95 (1922).
Pen. Malaysia, Borneo (Sarawak), Sumatera. 42 BOR MLY SUM. Phan. – Very close to A. subcaudata.

Aporusa brassii Mansf., J. Arnold Arbor. 10: 77 (1929).
Papua New Guinea, Bismarck Archip. 42 BIS NWG. Phan. – Perhaps conspecific with A. reticulata.

Aporusa brevicaudata Pax & K.Hoffm. in H.G.A.Engler, Pflanzenr., IV, 147, XV: 98 (1922).
Papua New Guinea. 42 NWG. Nanophan. or phan.

Aporusa bullatissima Airy Shaw, Kew Bull. 20: 379 (1966).
Borneo. 42 BOR. Phan.

Aporusa caloneura Airy Shaw, Kew Bull. 23: 5 (1969).
Borneo (Sabah, Sarawak). 42 BOR. Phan.

Aporusa cardiosperma (Gaertn.) Merr., J. Arnold Arbor. 35: 139 (1954).
S. India, Sri Lanka. 40 IND SRL. Phan.
* *Croton cardiospermus* Gaertn., Fruct. Sem. Pl. 2: 120 (1790).
  Scepa lindleyana Wight, Icon. Pl. Ind. Orient. 2: 5, t. 361 (1840). *Aporusa lindleyana* (Wight) Baill., Étude Euphorb.: 645 (1858).
  Aporusa affinis Baill., Étude Euphorb.: 645 (1858), nom. nud.
  Aporusa sphaerocarpa Müll.Arg., Flora 47: 51 (1864).

Aporusa carrii Schot, Blumea 40: 454 (1995).
New Guinea. 42 NWG. Phan.

Aporusa chondroneura (Airy Shaw) Schot, Blumea 40: 451 (1995).
Borneo. 42 BOR. Phan.
* *Aporusa prainiana* var. *chondroneura* Airy Shaw, Kew Bull. 25: 476 (1971). *Aporusa symplocoides* var. *chondroneura* (Airy Shaw) Airy Shaw, Kew Bull., Addit. Ser. 4: 42 (1975).

Aporusa clellandii Hook.f., Fl. Brit. India 5: 348 (1887).
S. Burma (Pegu). 41 BMA.

Aporusa confusa Gage, Rec. Bot. Surv. India 9: 229 (1922).
W. Malesia. 42 BOR MLY SUM. Phan.
Aporusa mollis Merr., Univ. Calif. Publ. Bot. 15: 144 (1929).

Aporusa decipiens Pax & K.Hoffm. in H.G.A.Engler, Pflanzenr., IV, 147, XV: 83 (1922).
Papua New Guinea. 42 NWG. Phan.

Aporusa dendroidea Schot, Blumea 40: 455 (1995).
Maluku. 42 MOL. Phan.

Aporusa duthieana King ex Pax & K.Hoffm. in H.G.A.Engler, Pflanzenr., IV, 147, XV: 99 (1922).
Burma. 41 BMA.

Aporusa egregia Airy Shaw, Kew Bull. 29: 285 (1974).
Irian Jaya. 42 NWG. Phan.

Aporusa elliptifolia Merr., Philipp. J. Sci., C 9: 472 (1914 publ. 1915).
Philippines (Palawan). 42 PHI. Phan.

Aporusa elmeri Merr., Univ. Calif. Publ. Bot. 15: 142 (1929).
Borneo (Sabah, E. Kalimantan). 42 BOR. Phan.

Aporusa falcifera Hook.f., Fl. Brit. India 5: 352 (1887).
S. Thailand, Pen. Malaysia, Borneo, Sumatera. 41 THA 42 BOR MLY SUM. Phan.
Aporusa hosei Merr., Philipp. J. Sci., C 11: 63 (1916).
Aporusa acuminatissima Merr., Univ. Calif. Publ. Bot. 15: 142 (1929), nom. illeg. *Aporusa merrilliana* Govaerts & Radcl.-Sm., Kew Bull. 51: 175 (1996).

Aporusa ficifolia Baill., Adansonia 11: 177 (1874).
Indo-China, Pen. Malaysia. 41 BMA THA VIE 42 MLY. Phan.

Aporusa flexuosa Pax & K.Hoffm. in H.G.A.Engler, Pflanzenr., IV, 147, XV: 91 (1922).
NE. Papua New Guinea. 42 NWG. Nanophan.

Aporusa frutescens Blume, Bijdr.: 514 (1826).
S. Burma, Thailand, Pen. Malaysia, Sumatera, Jawa, Borneo, Maluku, Philippines. 41 BMA THA 42 BOR JAW MLY MOL PHI SUM. Nanophan. or phan.
Leiocarpus fruticosus Blume, Bijdr.: 582 (1826). *Aporusa fruticosa* (Blume) Müll.Arg. in A.P.de Candolle, Prodr. 15(2): 475 (1866).
Baccaurea banahaensis Elmer, Leafl. Philipp. Bot. 4: 1475 (1912). *Aporusa banahaensis* (Elmer) Merr., Enum. Philipp. Fl. Pl. 2: 410 (1923).
Aporusa similis Merr., Philipp. J. Sci., C 9: 472 (1914 publ. 1915).
Aporusa agusanensis Elmer, Leafl. Philipp. Bot. 7: 2636 (1915).

Aporusa fulvovittata Schott, Blumea 40: 455 (1995).
N. Borneo. 42 BOR. Phan.

Aporusa fusiformis Thwaites, Enum. Pl. Zeyl.: 288 (1861).
Sri Lanka. 40 SRL.
Aporusa thwaitesii Baill., Étude Euphorb.: 645 (1858), nom. nud.
Lepidostachys grandifolia Planch. ex Müll.Arg. in A.P.de Candolle, Prodr. 15(2): 471 (1866).

Aporusa globifera Hook.f., Fl. Brit. India 5: 347 (1887).
Pen. Malaysia. 42 MLY.

Aporusa grandistipula Merr., Philipp. J. Sci. 21: 521 (1922).
Borneo, Sulawesi. 42 BOR SUL. Phan.

Aporusa granularis Airy Shaw, Kew Bull. 29: 283 (1974).
Borneo (Sarawak, E. Kalimantan). 42 BOR. Nanophan. or phan.

Aporusa hermaphrodita Airy Shaw, Kew Bull. 25: 478 (1971).
Papua New Guinea. 42 NWG. Phan.

Aporusa heterodoxa Airy Shaw, Kew Bull. 25: 479 (1971).
Solomon Is. (Bougainville I.). 60 SOL. Phan.

Aporusa illustris Airy Shaw, Kew Bull. 29: 281 (1974).
Borneo (Sarawak, Brunei). 42 BOR. Phan.

Aporusa isabellina Airy Shaw, Kew Bull. 25: 475 (1971).
Pen. Malaysia. 42 MLY.

Aporusa lagenocarpa Airy Shaw, Kew Bull. 21: 355 (1968). – FIGURE, p. 208.
Borneo. 42 BOR. Phan.

Aporusa lamellata Airy Shaw, Kew Bull. 33: 26 (1978).
Irian Jaya. 42 NWG.

Aporusa lanceolata (Tul.) Thwaites, Enum. Pl. Zeyl.: 288 (1861).
Sri Lanka. 40 SRL. Phan.
Agyneia multilocularis Moon, Cat. Pl. Ceylon: 65 (1824), nom. nud.
* *Lepidostachys lanceolata* Tul., Ann. Sci. Nat., Bot., III, 15: 253 (1851).

Aporusa latifolia Thwaites, Enum. Pl. Zeyl.: 288 (1861).
Sri Lanka. 40 SRL.
* *Agyneia latifolia* Moon, Numer. List: 65 (1824), nom. nud.

Aporusa laxiflora Pax & K.Hoffm. in H.G.A.Engler, Pflanzenr., IV, 147, XV: 31 (1922).
New Guinea, Bismarck Archip. 42 BIS NWG. Phan.

Aporusa ledermanniana Pax & K.Hoffm. in H.G.A.Engler, Pflanzenr., IV, 147, XV: 84 (1922).
New Guinea (incl. Louisiade Archip.), Bismarck Archip. 42 BIS NWG. Phan.

Aporusa leptochrysandra Airy Shaw, Kew Bull. 20: 381 (1966).
Papua New Guinea. 42 NWG. Phan.

Aporusa leytensis Merr., Philipp. J. Sci., C 9: 368 (1914 publ. 1915).
Philippines (Luzon, Leyte). 42 PHI. Phan.
Aporusa alvarezii Merr., Philipp. J. Sci., C 9: 470 (1914 publ. 1915).

Aporusa longicaudata Kaneh. & Hatus. ex Schot, Blumea 40: 456 (1995).
New Guinea. 42 NWG. Phan.

Aporusa lophodonta Airy Shaw, Kew Bull. 33: 25 (1978).
Pen. Malaysia. 42 MLY.

Aporusa lucida (Miq.) Airy Shaw, Kew Bull., Addit. Ser. 4: 38 (1975).
W. & C. Malesia. 42 BOR JAW MLY MOL SUM. Phan.
* *Tetractinostigma lucidum* Miq., Fl. Ned. Ind., Eerste Bijv.: 471 (1861).

var. **ellipsoidea** Airy Shaw, Kew Bull. 32: 362 (1978).
Sumatera. 42 SUM. Phan. – Provisionally accepted.

var. **lucida**
Pen. Malaysia, Sumatera (incl. Bangka), Borneo, Jawa. 42 BOR JAW MLY SUM. Phan.
Aporusa miqueliana Müll.Arg. in A.P.de Candolle, Prodr. 15(2): 474 (1866).
Aporusa microsphaera Hook.f., Fl. Brit. India 5: 350 (1887).
Aporusa borneensis Pax & K.Hoffm. in H.G.A.Engler, Pflanzenr., IV, 147, XV: 87 (1922).

var. **pubescens** Schot, Blumea 40: 457 (1995).
W. & C. Malesia. 42 BOR JAW MOL SUM. Phan.

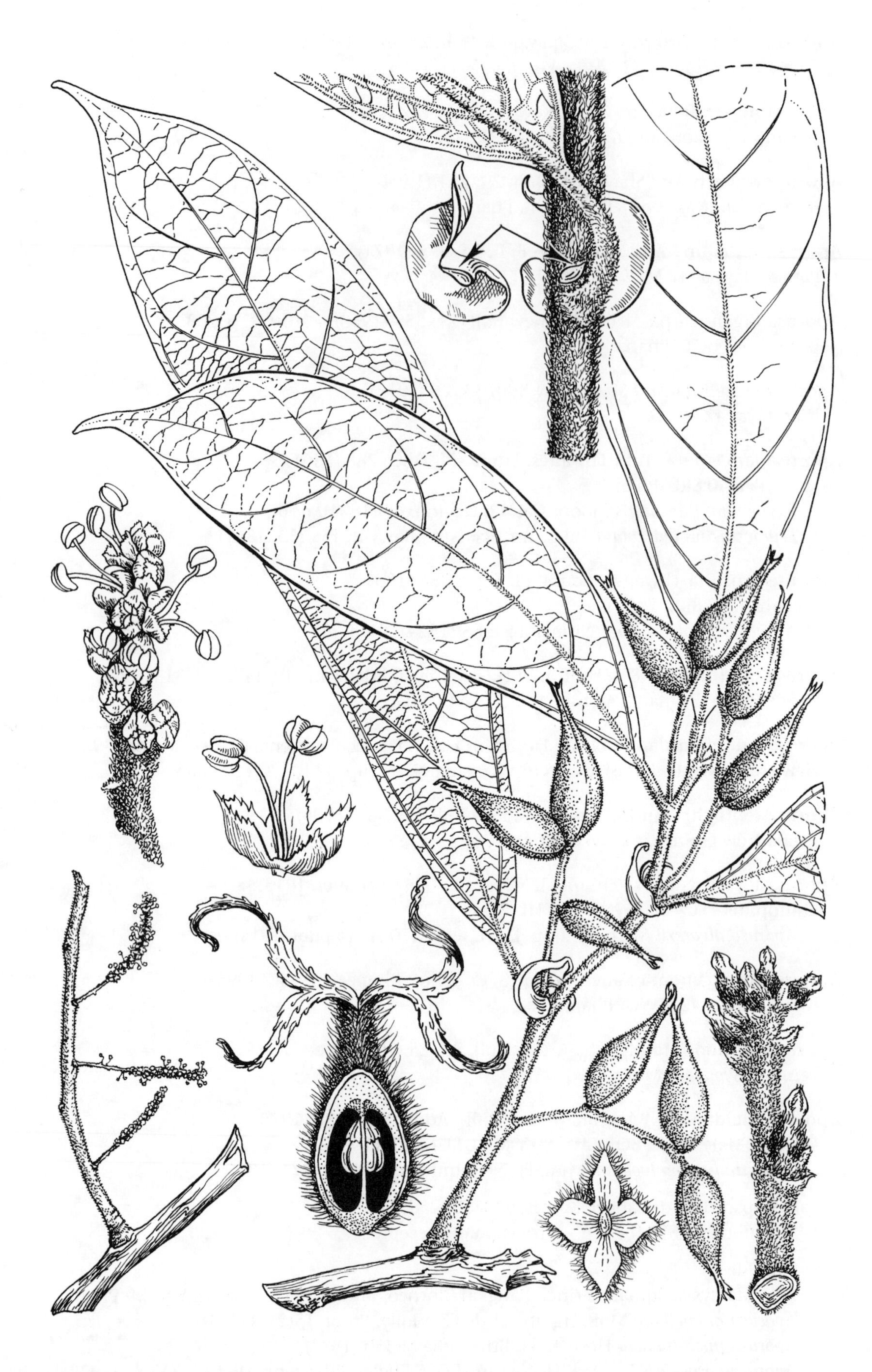

Aporusa lagenocarpa Airy Shaw
Artist: M. Grierson
Ic. Pl. 38(1), pl . 3701 (1974)
KEW ILLUSTRATIONS COLLECTION

var. **trilocularis** Schot, Blumea 40: 457 (1995).
Borneo. 42 BOR. Phan.

Aporusa lunata (Miq.) Kurz, J. Asiat. Soc. Bengal, Pt. 2, Nat. Hist. 42(2): 239 (1873).
S. Thailand, Pen. Malaysia, Sumatera, Jawa, Borneo. 41 THA 42 BOR JAW MLY SUM. Phan.
* *Antidesma lunatum* Miq., Fl. Ned. Ind., Eerste Bijv.: 467 (1861).
Bridelia rugosa Miq., Fl. Ned. Ind., Eerste Bijv.: 445 (1861).

Aporusa macrophylla (Tul.) Müll.Arg. in A.P.de Candolle, Prodr. 15(2): 470 (1866).
Burma. 41 BMA.
* *Lepidostachys macrophylla* Tul., Ann. Sci. Nat., Bot., III, 15: 253 (1851).

Aporusa microstachya (Tul.) Müll.Arg. in A.P.de Candolle, Prodr. 15(2): 474 (1866).
Burma, Thailand, Pen. Malaysia (incl. Singapore). 41 BMA THA 42 MLY. Nanophan. or phan.
* *Scepa microstachya* Tul., Ann. Sci. Nat., Bot., III, 15: 255 (1851). *Aporusa tulasneana* Baill., Étude Euphorb.: 645 (1858), nom. illeg.
Aporusa maingayi Hook.f., Fl. Brit. India 5: 348 (1887).
Lepidostachys griffithii Planch. ex Pax & K.Hoffm. in H.G.A.Engler, Pflanzenr., IV, 147, XV: 102 (1922).

Aporusa minahassae Koord., Meded. Lands Plantentuin 19: 625 (1898).
Sulawesi. 42 SUL.

Aporusa misimana Airy Shaw ex Schot, Blumea 40: 457 (1995).
Papua New Guinea (Is.). 42 NWG. Phan.

Aporusa nervosa Hook.f., Fl. Brit. India 5: 350 (1887).
Pen. Malaysia, Sumatera (incl. Bangka), Borneo, Philippines ? 42 BOR MLY PHI? SUM. Phan.

Aporusa nigricans Hook.f., Fl. Brit. India 5: 347 (1887).
S. Thailand, Pen. Malaysia, Borneo, Sumatera. 41 THA 42 BOR MLY SUM. Phan.

Aporusa nigropunctata Pax & K.Hoffm. in H.G.A.Engler, Pflanzenr., IV, 147, XV: 98 (1922).
Papua New Guinea. 42 NWG. Phan.

Aporusa nitida Merr., Univ. Calif. Publ. Bot. 15: 143 (1929).
Borneo. 42 BOR. Phan.

Aporusa oblonga Müll.Arg., Linnaea 32: 78 (1863).
Burma. 41 BMA.
Lepidostachys oblonga Wall., Numer. List: 7299 (1832), nom. nud.

Aporusa obovata Pax & K.Hoffm. in H.G.A.Engler, Pflanzenr., IV, 147, XV: 100 (1922).
S. Burma (Mergui Archip.). 41 BMA.

Aporusa octandra (Buch.-Ham. ex D.Don) Vickery, Enum. Fl. Pl. Nepal 3: 193 (1982).
S. China, Trop. Asia, Queensland. 36 CHC CHS 40 ASS EHM NEP 41 AND BMA MLY THA VIE 42 BOR JAW MLY NWG PHI SUM 50 QLD. Phan.
* *Myrica octandra* Buch.-Ham. ex D.Don, Prodr. Fl. Nepal.: 56 (1825).

var. **chinensis** (Champ. ex Benth.) Schot, Blumea 40: 452 (1995).
SE. China. 36 CHS. Phan.
* *Scepa chinensis* Champ. ex Benth., Hooker's J. Bot. Kew Gard. Misc. 6: 72 (1854).
Aporusa microcalyx var. *chinensis* (Champ. ex Benth.) Müll.Arg. in A.P.de Candolle, Prodr. 15(2): 472 (1866). *Aporusa chinensis* (Champ. ex Benth.) Merr., Lingnan Sci. J. 13: 34 (1934).
Aporusa leptostachya Benth., Fl. Hongk.: 317 (1861).

var. **malesiana** Schot, Blumea 40: 452 (1995).
Malesia, Queensland. 42 BOR JAW MLY NWG PHI SUM 50 QLD. Phan.
* *Scepa aurita* Tul., Ann. Sci. Nat., Bot., III, 15: 254 (1851). *Aporusa aurita* (Tul.) Miq., Fl. Ned. Ind. 1(1): 431 (1855).
Tetractinostigma microcalyx Hassk., Flora 40: 533 (1857). *Aporusa microcalyx* (Hassk.) Hassk., Bull. Soc. Bot. France 6: 714 (1859).
Aporusa cumingiana Baill., Étude Euphorb.: 645 (1858), nom. nud.

var. **octandra**
Trop. Asia. 40 ASS EHM NEP 41 AND BMA MLY THA VIE 42 BOR JAW PHI SUM. Phan.
Alnus dioica Roxb., Fl. Ind. ed. 1832, 3: 580 (1832). *Aporusa dioica* (Roxb.) Müll.Arg. in A.P.de Candolle, Prodr. 15(2): 472 (1866).
Lepidostachys roxburghii Wall. ex Lindl., Intr. Nat. Syst. Bot., ed. 2: 441 (1836). *Aporusa roxburghii* (Wall. ex Lindl.) Baill., Étude Euphorb.: 645 (1858).
Scepa stipulacea Lindl., Intr. Nat. Syst. Bot., ed. 2: 441 (1836).
Alnus integrifolia Roxb. ex Steud., Nomencl. Bot., ed. 2, 1: 56 (1840).
Leiocarpus serratus Hassk., Hort. Bogor. Desc.: 55 (1858).
Aporusa frutescens Benth., Fl. Hongk.: 317 (1861).
Aporusa villosula Kurz, J. Asiat. Soc. Bengal, Pt. 2, Nat. Hist. 42(2): 23 (1873).
Aporusa microcalyx var. *intermedia* Pax & K.Hoffm. in H.G.A.Engler, Pflanzenr., IV, 147, XV: 102 (1922).
Leiocarpus tinctorius Blume ex Pax & K.Hoffm. in H.G.A.Engler, Pflanzenr., IV, 147, XV: 101 (1922).

Aporusa papuana Pax & K.Hoffm. in H.G.A.Engler, Pflanzenr., IV, 147, XV: 83 (1922).
New Guinea (incl. Admiralty Is.), Solomon Is. 42 BIS NWG 60 SOL. Phan.

Aporusa parvula Schot, Blumea 40: 457 (1995).
Irian Jaya. 42 NWG. Phan.

Aporusa penangensis (Ridl.) Airy Shaw, Kew Bull. 23: 3 (1969).
Pen. Malaysia, Borneo (Brunei), Sumatera. 42 BOR MLY SUM. Phan.
* *Aporusa maingayi* var. *penangensis* Ridl., Fl. Malay Penins. 3: 242 (1924).

Aporusa petiolaris Airy Shaw, Kew Bull. 25: 474 (1971).
Papua New Guinea. 42 NWG. Nanophan. or phan.

Aporusa planchoniana Baill., Étude Euphorb.: 645 (1858).
China (Yunnan, Guangxi, Hainan), Indo-China. 36 CHC CHH CHS 41 BMA CBD THA VIE. Nanophan. or phan.
Lepidostachys parviflora Planch. ex Baill., Étude Euphorb.: 645 (1858), nom. nud.
Aporusa lanceolata Müll.Arg. in A.P.de Candolle, Prodr. 15(2): 475 (1866).
Aporusa lanceolata var. *murtonii* F.N.Williams, Bull. Herb. Boissier, II, 5: 30 (1905).

Aporusa prainiana King ex Gage, Rec. Bot. Surv. India 9: 228 (1922).
Pen. Malaysia, Sumatera, Borneo. 42 BOR MLY SUM. Phan.

Aporusa pseudoficifolia Pax & K.Hoffm. in H.G.A.Engler, Pflanzenr., IV, 147, XV: 94 (1922).
N. Thailand, Pen. Malaysia. 41 THA 42 MLY. Phan.

Aporusa quadrilocularis (Miq.) Müll.Arg. in A.P.de Candolle, Prodr. 15(2): 476 (1866).
Pen. Malaysia, Sumatera. 42 MLY SUM. Phan.
* *Leiocarpus quadrilocularis* Miq., Fl. Ned. Ind., Eerste Bijv.: 443 (1861).
Aporusa claviflora Airy Shaw, Kew Bull. 32: 364 (1978).

Aporusa reticulata Pax & K.Hoffm. in H.G.A.Engler, Pflanzenr., IV, 147, XV: 100 (1922).
Papua New Guinea. 42 NWG. Phan.

Aporusa rhacostyla Airy Shaw, Kew Bull. 29: 282 (1974).
Borneo (Sarawak). 42 BOR. Phan.

Aporusa sarawakensis Schot, Blumea 40: 458 (1995).
Borneo. 42 BOR. Phan.

Aporusa sclerophylla Pax & K.Hoffm. in H.G.A.Engler, Pflanzenr., IV, 147, XV: 98 (1922).
NE. Papua New Guinea. 42 NWG. Phan.

Aporusa selangorica Pax & K.Hoffm. in H.G.A.Engler, Pflanzenr., IV, 147, XV: 105 (1922).
Pen. Malaysia. 42 MLY.

Aporusa serrata Gagnep., Bull. Soc. Bot. France 70: 233 (1923).
Laos. 41 LAO.

Aporusa sphaeridiophora Merr., Philipp. J. Sci. 1(Suppl.): 76 (1906).
Philippines. 42 PHI. Phan.
Aporusa campanulata J.J.Sm., Icon. Bogor.: t. 229 (1907).
Aporusa acuminatissima Merr., Philipp. J. Sci. 16: 546 (1920). *Aporusa merrilliana* Govaerts & Radcl.-Sm., Kew Bull. 51: 175 (1996).
Aporusa bulusanensis Elmer ex Merr., Enum. Philipp. Fl. Pl. 2: 410 (1923).
Aporusa rotundifolia Elmer ex Merr., Enum. Philipp. Fl. Pl. 2: 410 (1923).
Aporusa sorsogonensis Elmer ex Merr., Enum. Philipp. Fl. Pl. 2: 410 (1923).

Aporusa sphaerosperma Gagnep., Bull. Soc. Bot. France 70: 234 (1923).
Vietnam. 41 VIE.

Aporusa squarrosa Airy Shaw & Gage, Kew Bull. 20: 25 (1966).
Irian Jaya. 42 NWG. Nanophan.

Aporusa stellifera Hook.f., Fl. Brit. India 5: 352 (1887).
Pen. Malaysia. 42 MLY. Phan.

Aporusa stenostachys Airy Shaw, Kew Bull. 29: 285 (1974).
Borneo (Sarawak: Bintulu). 42 BOR. Phan.

Aporusa subcaudata Merr., Philipp. J. Sci., C 11: 64 (1916).
Pen. Malaysia, Borneo. 42 BOR MLY. Phan.

Aporusa sylvestri Airy Shaw, Kew Bull. 32: 365 (1978).
Borneo (Sarawak). 42 BOR.

Aporusa symplocifolia Merr., Philipp. J. Sci. 1(Suppl.): 77 (1906).
Philippines. 42 PHI.
Aporusa symplocosifolia Merr. in ?, , nom. illeg.

Aporusa symplocoides (Hook.f.) Gage, Rec. Bot. Surv. India 9: 229 (1922).
S. Thailand, Pen. Malaysia, Sumatera (incl. Bangka), Borneo (Sarawak, Sabah). 41 THA 42 BOR MLY SUM. Phan.
* *Baccaurea symplocoides* Hook.f., Fl. Brit. India 5: 376 (1887).

var. **chalarocarpa** (Airy Shaw) Schot, Blumea 40: 453 (1995).
Borneo. 42 BOR. Phan.
* *Aporusa chalarocarpa* Airy Shaw, Kew Bull. 20: 380 (1966).

var. **symplocoides**
S. Thailand, Pen. Malaysia, Sumatera (incl. Bangka), Borneo (Sarawak, Sabah). 41 THA 42 BOR MLY SUM. Phan.

Aporusa tetrapleura Hance, J. Bot. 14: 260 (1876).
Cambodia. 41 CBD.

Aporusa vagans Schot, Blumea 40: 459 (1995).
New Guinea. 42 NWG. Phan.

Aporusa villosa (Lindl.) Baill., Étude Euphorb.: 645 (1858).
S. China, Hainan, Indo-China. 36 CHC CHH CHS 41 BMA NCB THA VIE. Nanophan. or phan.
Lepidostachys villosa Wall., Numer. List: 7298 (1832), nom. nud.
* *Scepa villosa* Lindl., Intr. Nat. Syst. Bot., ed. 2: 441 (1836).
Aporusa glabrifolia Kurz, J. Bot. 13: 330 (1875).
Aporusa microcalyx var. *yunnanensis* Pax & K.Hoffm. in H.G.A.Engler, Pflanzenr., IV, 147, XV: 102 (1922). *Aporusa dioica* var. *yunnanensis* (Pax & K.Hoffm.) H.S.Kiu, Guihaia 11: 17 (1991). *Aporusa octandra* var. *yunnanensis* (Pax & K.Hoffm.) Schot, Blumea 40: 452 (1995).

Aporusa wallichii Hook.f., Fl. Brit. India 5: 350 (1887). *Aporusa wallichii* var. *genuina* Pax & K.Hoffm. in H.G.A.Engler, Pflanzenr., IV, 147, XV: 88 (1922), nom. inval.
Assam, Burma, Thailand. 40 ASS 41 BMA THA. Phan.

var. **ambigua** Airy Shaw, Kew Bull. 35: 384 (1980).
Thailand. 41 THA. Phan.

var. **wallichii**
Assam, Burma, Thailand. 40 ASS 41 BMA THA. Phan.
Lepidostachys roxburghii Hook., Fl. Brit. India 5: 350 (1887), pro syn.

Aporusa whitmorei Airy Shaw, Kew Bull. 32: 361 (1978).
Pen. Malaysia. 42 MLY.

Aporusa yunnanensis (Pax & K.Hoffm.) F.P.Metcalf, Lingnan Sci. J. 10: 486 (1931).
S. China, Hainan, Vietnam, Thailand, Burma, Assam. 36 CHC CHH CHS 40 ASS 41 BMA THA VIE. Nanophan. or phan.
Aporusa wallichii f. *yunnanensis* Pax & K.Hoffm. in H.G.A.Engler, Pflanzenr., IV, 147, XV: 90 (1922).

Synonyms:
Aporusa aberrans Gagnep. === **Antidesma sp.**
Aporusa acuminatissima Merr. === **Aporusa sphaeridiophora** Merr.
Aporusa acuminatissima Merr. === **Aporusa falcifera** Hook.f.
Aporusa affinis Baill. === **Aporusa cardiosperma** (Gaertn.) Merr.
Aporusa agusanensis Elmer === **Aporusa frutescens** Blume
Aporusa alvarezii Merr. === **Aporusa leytensis** Merr.
Aporusa arborescens (Hassk.) Müll.Arg. === **Aporusa arborea** (Blume) Müll.Arg.
Aporusa aurita (Tul.) Miq. === **Aporusa octandra** var. **malesiana** Schot
Aporusa australiana F.Muell. === ?
Aporusa banahaensis (Elmer) Merr. === **Aporusa frutescens** Blume
Aporusa bilitonensis Pax & K.Hoffm. === **Baccaurea minor** Hook.f.
Aporusa borneensis Pax & K.Hoffm. === **Aporusa lucida** (Miq.) Airy Shaw var. **lucida**
Aporusa bulusanensis Elmer ex Merr. === **Aporusa sphaeridiophora** Merr.
Aporusa calocarpa Kurz ex Hook.f. === **Drypetes longifolia** (Blume) Pax & K.Hoffm.
Aporusa campanulata J.J.Sm. === **Aporusa sphaeridiophora** Merr.
Aporusa chalarocarpa Airy Shaw === **Aporusa symplocoides** var. **chalarocarpa** (Airy Shaw) Schot
Aporusa chinensis (Champ. ex Benth.) Merr. === **Aporusa octandra** var. **chinensis** (Champ. ex Benth.) Schot
Aporusa claviflora Airy Shaw === **Aporusa quadrilocularis** (Miq.) Müll.Arg.

Aporusa cumingiana Baill. === **Aporusa octandra** var. **malesiana** Schot
Aporusa dioica (Roxb.) Müll.Arg. === **Aporusa octandra** (Buch.-Ham. ex D.Don) Vickery var. **octandra**
Aporusa dioica var. *yunnanensis* (Pax & K.Hoffm.) H.S.Kiu === **Aporusa villosa** (Lindl.) Baill.
Aporusa dolichocarpa Pax & K.Hoffm. === **Baccaurea minutiflora** Müll.Arg.
Aporusa euphlebia Merr. === **Aporusa benthamiana** Hook.f.
Aporusa frutescens Benth. === **Aporusa octandra** (Buch.-Ham. ex D.Don) Vickery var. **octandra**
Aporusa fruticosa (Blume) Müll.Arg. === **Aporusa frutescens** Blume
Aporusa glabrifolia Kurz === **Aporusa villosa** (Lindl.) Baill.
Aporusa grandifolia Merr. === **Aporusa benthamiana** Hook.f.
Aporusa griffithii Hook.f. === **Antidesma coriaceum** Tul.
Aporusa hosei Merr. === **Aporusa falcifera** Hook.f.
Aporusa inaequalis Pax & K.Hoffm. === **Drypetes leonensis** Pax
Aporusa incisa Airy Shaw === **Prunus arborea** var. **montana** (Hook.f.) Kalkman (Rosaceae)
Aporusa lanceolata Müll.Arg. === **Aporusa planchoniana** Baill.
Aporusa lanceolata Hance === ?
Aporusa lanceolata var. *murtonii* F.N.Williams === **Aporusa planchoniana** Baill.
Aporusa leptostachya Benth. === **Aporusa octandra** var. **chinensis** (Champ. ex Benth.) Schot
Aporusa lindleyana (Wight) Baill. === **Aporusa cardiosperma** (Gaertn.) Merr.
Aporusa lunata var. *philippinensis* Pax & K.Hoffm. === **Aporusa benthamiana** Hook.f.
Aporusa lunata var. *stipulosa* (Merr.) Merr. === **Aporusa benthamiana** Hook.f.
Aporusa maingayi Hook.f. === **Aporusa microstachya** (Tul.) Müll.Arg.
Aporusa maingayi var. *penangensis* Ridl. === **Aporusa penangensis** (Ridl.) Airy Shaw
Aporusa merrilliana Govaerts & Radcl.-Sm. === **Aporusa falcifera** Hook.f.
Aporusa microcalyx (Hassk.) Hassk. === **Aporusa octandra** var. **malesiana** Schot
Aporusa microcalyx var. *chinensis* (Champ. ex Benth.) Müll.Arg. === **Aporusa octandra** var. **chinensis** (Champ. ex Benth.) Schot
Aporusa microcalyx var. *intermedia* Pax & K.Hoffm. === **Aporusa octandra** (Buch.-Ham. ex D.Don) Vickery var. **octandra**
Aporusa microcalyx var. *yunnanensis* Pax & K.Hoffm. === **Aporusa villosa** (Lindl.) Baill.
Aporusa microsphaera Hook.f. === **Aporusa lucida** (Miq.) Airy Shaw var. **lucida**
Aporusa miqueliana Müll.Arg. === **Aporusa lucida** (Miq.) Airy Shaw var. **lucida**
Aporusa mollis Merr. === **Aporusa confusa** Gage
Aporusa nigropunctata var. *antennifera* Airy Shaw === **Aporusa antennifera** (Airy Shaw) Airy Shaw
Aporusa octandra var. *yunnanensis* (Pax & K.Hoffm.) Schot === **Aporusa villosa** (Lindl.) Baill.
Aporusa prainiana var. *chondroneura* Airy Shaw === **Aporusa chondroneura** (Airy Shaw) Schot
Aporusa rotundifolia Elmer ex Merr. === **Aporusa sphaeridiophora** Merr.
Aporusa roxburghii (Wall. ex Lindl.) Baill. === **Aporusa octandra** (Buch.-Ham. ex D.Don) Vickery var. **octandra**
Aporusa similis Merr. === **Aporusa frutescens** Blume
Aporusa somalensis Mattei === **Shirakiopsis elliptica** (Hochst.) Esser
Aporusa sorsogonensis Elmer ex Merr. === **Aporusa sphaeridiophora** Merr.
Aporusa sphaerocarpa Müll.Arg. === **Aporusa cardiosperma** (Gaertn.) Merr.
Aporusa stipulosa Merr. === **Aporusa benthamiana** Hook.f.
Aporusa symplocoides var. *chondroneura* (Airy Shaw) Airy Shaw === **Aporusa chondroneura** (Airy Shaw) Schot
Aporusa symplocosifolia Merr. === **Aporusa symplocifolia** Merr.
Aporusa thwaitesii Baill. === **Aporusa fusiformis** Thwaites
Aporusa tulasneana Baill. === **Aporusa microstachya** (Tul.) Müll.Arg.
Aporusa villosula Kurz === **Aporusa octandra** (Buch.-Ham. ex D.Don) Vickery var. **octandra**
Aporusa wallichii var. *genuina* Pax & K.Hoffm. === **Aporusa wallichii** Hook.f.
Aporusa wallichii f. *yunnanensis* Pax & K.Hoffm. === **Aporusa yunnanensis** (Pax & K.Hoffm.) F.P.Metcalf

Arachne

Synonyms:

Arachne Neck. === **Andrachne** L.

Arachnodes

Synonyms:

Arachnodes Gagnep. === **Phyllanthus** L.
Arachnodes chevalieri Gagnep. === **Phyllanthus arachnodes** Govaerts & Radcl.-Sm.

Archileptopus

1 species, China (Guangxi). Shrubs to 1.5 m in hill forest or scrub on limestone; distinguished from *Andrachne* by extrorse anthers and a 4-5-celled ovary. [Not yet represented in Kew Herbarium.] (Phyllanthoideae)

Li, P.-T. (1991). A new genus of Euphorbiaceae and some new nomenclatural combination[s] of the Asclepiadaceous plants. J. South China Agric. Univ. 12(3): 38-42. La/Ch. — Pp. 38-39 feature the protologue of *Archileptopus* and description of *A. fangdingianus* from Guangxi.

Archileptopus P.T.Li, J. S. China Agric. Univ. 12(3): 38 (1991).
SE. China. 36.

Archileptopus fangdingianus P.T.Li, J. S. China Agric. Univ. 12(3): 39 (1991).
China (W. Guangxi). 36 CHS. Nanophan.

Ardinghalia

Synonyms:

Ardinghalia Comm. ex A.Juss. === **Phyllanthus** L.

Argithamnia

An orthographic variant of *Argythamnia*

Argomuellera

11 species, tropical Africa, Comoros and Madagascar; shrubs or small understorey trees. Present limits were established by Léonard (1959) who made several transfers from *Pycnocoma*. *A. macrophylla* is distributed throughout most of tropical Africa in light forest or in edge habitats; other species are relatively local (e.g. *A. pierlotiana*) or imperfectly known. (Acalyphoideae)

Pax, F. (with K. Hoffmann) (1914). *Wetriaria*. In A. Engler (ed.), Das Pflanzenreich, IV 147 VII (Euphorbiaceae-Acalypheae-Mercurialinae): 49-52. Berlin. (Heft 63.) La/Ge. — 7 species, 3 in Africa, 4 in Madagascar; all mutually closely related. [Now, for nomenclatural reasons, *Argomuellera*.]

Léonard, J. (1959). Observations sur les genres *Pycnocoma* et *Argomuellera* (Euphorbiacées africaines). Bull. Soc. Roy. Bot. Belg. 91: 267-281. Fr. — General discussion of the two genera with evaluation of characters and key; synopsis of *Argomuellera* (pp. 274-276) with synonymy but no key or indication of distribution. 9 species are accepted.

- Léonard, J. (1996). Le genre *Argomuellera* Pax en Afrique centrale (Zaïre, Rwanda, Burundi)(Euphorbiaceae). Bull. Jard. Bot. Natl. Belg. 65: 23-35, illus., maps. Fr. — Regional revision of *Argomuellera* (2 species) with key, synonymy, references and citations, types, localities with exsiccatae, indication of habitat and chorology, and notes on uses and biology; list of known species without key (p. 26); index at end.

Argomuellera Pax, Bot. Jahrb. Syst. 19: 90 (1894).
Trop. Africa, Madagascar. 22 23 24 25 26 29.
Wetriaria (Müll.Arg.) Kuntze in T.E.von Post & C.E.O.Kuntze, Lex. Gen. Phan.: 592 (1903).
Neopycnocoma Pax, Bot. Jahrb. Syst. 43: 222 (1909).

Argomuellera basicordata Peter ex Radcl.-Sm., Kew Bull. 30: 677 (1975 publ. 1976).
Tanzania (Lushoto). 25 TAN. Nanophan.

Argomuellera calcicola (Leandri) J.Léonard, Bull. Soc. Roy. Bot. Belgique 91: 275 (1959).
Madagascar. 29 MDG.
* *Pycnocoma gigantea* var. *calcicola* Leandri, Notul. Syst. (Paris) 9: 166 (1941).

Argomuellera danguyana (Leandri) J.Léonard, Bull. Soc. Roy. Bot. Belgique 91: 275 (1959).
Madagascar. 29 MDG.
* *Pycnocoma danguyana* Leandri, Notul. Syst. (Paris) 9: 166 (1941).

Argomuellera decaryana (Leandri) J.Léonard, Bull. Soc. Roy. Bot. Belgique 91: 275 (1959).
Madagascar. 29 MDG.
* *Pycnocoma decaryana* Leandri, Notul. Syst. (Paris) 9: 167 (1941).

Argomuellera gigantea (Baill.) Pax & K.Hoffm. in H.G.A.Engler, Nat. Pflanzenfam. ed. 2, 19C: 108 (1931).
Madagascar (Nosi Bé I.). 29 MDG. Nanophan.
* *Pycnocoma gigantea* Baill., Étude Euphorb.: 411 (1858).

Argomuellera lancifolia (Pax) Pax in H.G.A.Engler, Nat. Pflanzenfam. ed. 2, 19C: 108 (1931).
Equatorial Guinea. 23 EQG. Nanophan. or phan.
* *Neopycnocoma lancifolia* Pax, Bot. Jahrb. Syst. 43: 223 (1909).

Argomuellera macrophylla Pax, Bot. Jahrb. Syst. 19: 90 (1894). – FIGURE, p. 216.
Trop. Africa. 22 GHA GUI IVO LBR NGA SIE 23 BUR CMN CON EQG GAB RWA ZAI 24 ETH SUD 25 KEN TAN UGA 26 ANG MLW MOZ ZAM ZIM. Nanophan. or phan.
Pycnocoma laurentii De Wild., Ann. Mus. Congo Belge, Bot., V, 2: 285 (1908).
Argomuellera macrophylla var. *laurentii* (De Wild.) Prain in D.Oliver, Fl. Trop. Afr. 6(1): 926 (1912). *Argomuellera macrophylla* f. *laurentii* (De Wild.) Prain ex De Wild., Ann. Mus. Congo Belge, Bot., V, 3: 427 (1912).
Pycnocoma sapinii De Wild., Ann. Mus. Congo Belge, Bot., V, 2: 285 (1908).
Pycnocoma hirsuta Prain, Bull. Misc. Inform. Kew 1909: 51 (1909).
Pycnocoma parviflora Pax, Bot. Jahrb. Syst. 43: 81 (1909).
Pycnocoma hutchinsonii Beille, Bull. Soc. Bot. France 61(8): 295 (1914 publ. 1917).
Pycnocoma sassandrae Beille, Bull. Soc. Bot. France 61(8): 296 (1914 publ. 1917).

Argomuellera perrieri (Leandri) J.Léonard, Bull. Soc. Roy. Bot. Belgique 91: 276 (1959).
Madagascar. 29 MDG.
* *Pycnocoma perrieri* Leandri, Notul. Syst. (Paris) 9: 167 (1941).

Argomuellera pierlotiana J.Léonard, Bull. Jard. Bot. Belg. 65: 31 (1996).
Zaire. 23 ZAI.

Argomuellera sessilifolia Prain, Bull. Misc. Inform. Kew 1912: 191 (1912).
Congo. 23 CON. Nanophan. or phan.

Argomuellera trewioides (Baill.) Pax & K.Hoffm. in H.G.A.Engler, Nat. Pflanzenfam. ed. 2, 19c: 108 (1931).
Comoros. 29 COM.
* *Pycnocoma trewioides* Baill., Étude Euphorb.: 411 (1858).

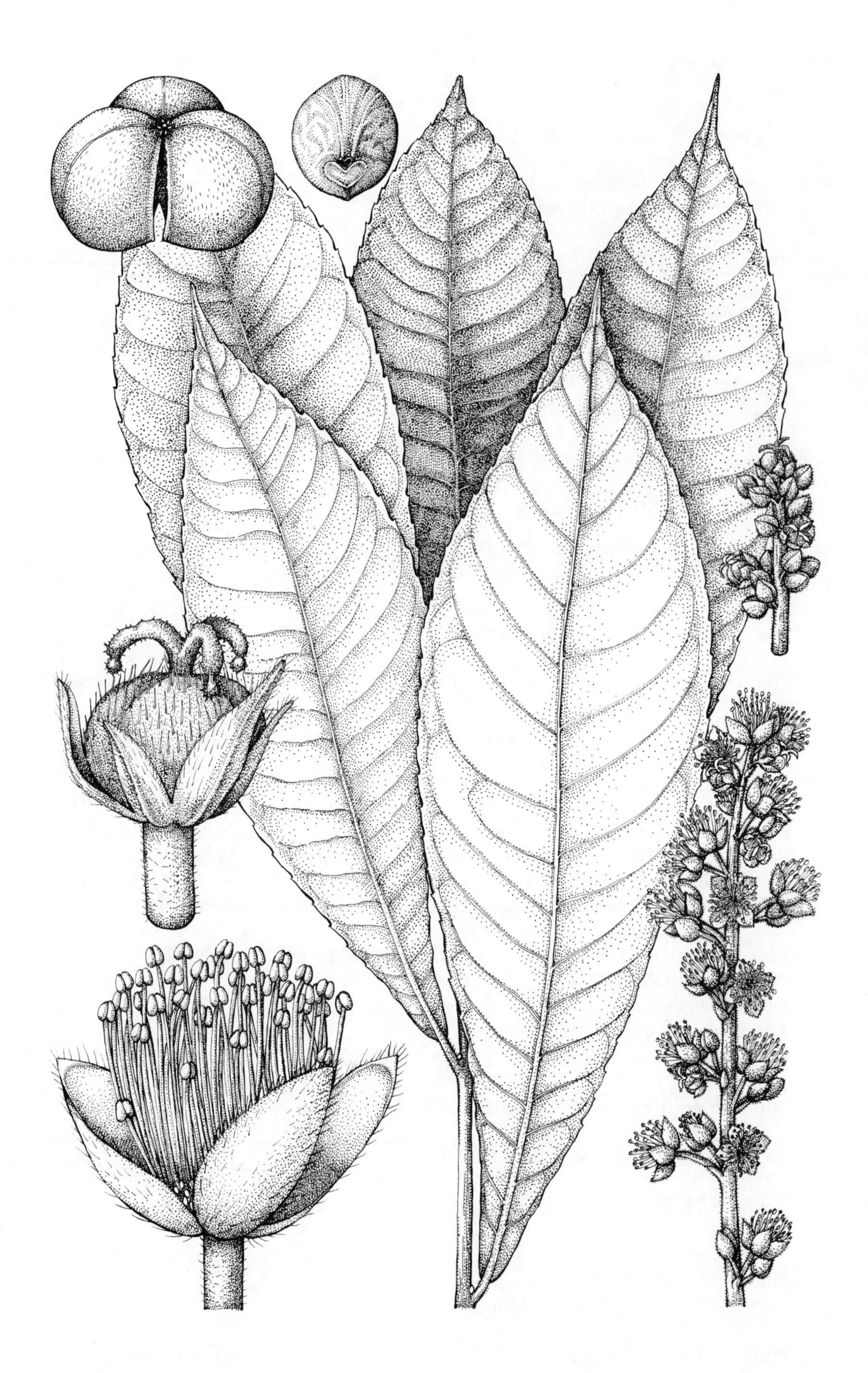

Argomuellera macrophylla Pax
Artist: J.C. Dunkley
Fl. Trop. East Africa, Euphorbiaceae 1: 226, fig. 45 (1987)
KEW ILLUSTRATIONS COLLECTION

Synonyms:
Argomuellera macrophylla var. *laurentii* (De Wild.) Prain === **Argomuellera macrophylla** Pax
Argomuellera macrophylla f. *laurentii* (De Wild.) Prain ex De Wild. === **Argomuellera macrophylla** Pax
Argomuellera reticulata (Baill.) Pax & K.Hoffm. === **Droceloncia sp.**
Argomuellera rigidifolia (Baill.) Pax & K.Hoffm. === **Droceloncia rigidifolia** (Baill.) J.Léonard

Argothamnia

An orthographic variant of *Argythamnia.*

Argyra

Synonyms:
Argyra Noronha ex Baill. === **Croton** L.

Argyrodendron

Synonyms:
Argyrodendron Klotzsch === **Croton** L.
Argyrodendron bicolor Klotzsch === **Croton menyharthii** Pax
Argyrodendron ovatum Boivin ex Baill. === **Croton bernierus** Baill.
Argyrodendron petersii Klotzsch === **Combretum imberbe** Wawra (Combretaceae)

Argyrothamnia

An orthographic variant of *Argythamnia.*

Argytamnia

An orthographic variant of *Argythamnia.*

Argythamnia

23 species, SW. N America (Texas southwards) through Central America to NW. South America (Colombia) and in the West Indies (where well-represented); shrubs, subshrubs with herbaceous stems, or perennial herbs with a woody stock (*A. candicans* may rarely be a small tree). Arguments remain over its union with *Chiropetalum*; such was advocated by Croizat (1945), Ingram (1979) and Mabberley (in *The Plant-Book*) but not by Pax, Brummitt, USDA-GRIN or Webster. The latter opinion is followed here. Croizat and Ingram have further argued for reduction of *Ditaxis* but, on the grounds of pollen morphology, this was not accepted by Webster (nor has it been adopted here). The most recent revision, by Ingram (1967), recognised 18 species. (Acalyphoideae)

Pax, F. (with K. Hoffmann) (1912). *Argithamnia*. In A. Engler (ed.), Das Pflanzenreich, IV 147 VI (Euphorbiaceae-Acalypheae-Chrozophorinae): 78-86. Berlin. (Heft 57.) La/Ge. — 7 species, 6 in West Indies, 1 in C America. [Superseded by Ingram 1967, 1980.]

- Ingram, J. (1967). A revisional study of *Argythamnia*, subgenus *Argythamnia* (Euphorbiaceae). Gentes Herb. 10(1): 1-38, illus., maps. En. — Descriptive revision (18 species, some new) with key, synonymy, references, types, indication of distribution and localities with exsiccatae; list of references but no separate index. The general part includes a historical review and character survey, but little on relationships or evolutionary trends.

• Ingram, J. (1980). The generic limits of *Argythamnia* (Euphorbiaceae) defined. Gentes Herb. 11(7): 426-436, illus., maps. En. — Biogeography (with postulated centre of origin and dispersal tracks); morphology and taxonomic relationships; revised generic description along with a key to and distribution maps of subgenera (*Argythamnia*, *Ditaxis* and *Chiropetalum*). [The reduction of *Ditaxis* by Croizat (1945) is accepted, along with the reduction (here, for the first time) of *Chiropetalum*.]

Argythamnia P.Browne, Civ. Nat. Hist. Jamaica: 338 (1756).
Caribbean, Mexico, C. America, NE. South America, Brazil ? 79 80 81 82 84?
Odotalon Raf., Sylva Tellur.: 66 (1838).

Argythamnia acutangula Croizat, Ciencia (Mexico) 6: 353 (1946).
Colombia. 83 CLM.

Argythamnia argentea Millsp., Publ. Field Columbian Mus., Bot. Ser. 2: 154 (1906). *Ditaxis argentea* (Millsp.) Pax & K.Hoffm. in H.G.A.Engler, Pflanzenr., IV, 147, VII: 426 (1914).
Bahamas (South Caicos). 81 BAH. Nanophan.
Argythamnia sericea var. *lanceolata* Müll.Arg. in A.P.de Candolle, Prodr. 15(2): 742 (1866). *Argythamnia lanceolata* (Müll.Arg.) Pax & K.Hoffm. in H.G.A.Engler, Pflanzenr., IV, 147, VI: 79 (1912), nom. illeg.

Argythamnia argyaea Cory, Madroño 8: 92 (1945).
Texas. 77 TEX.

Argythamnia bicolor M.E.Jones, Contr. W. Bot. 15: 127 (1929).
Mexico. 79 MXS.

Argythamnia candicans Sw., Prodr.: 39 (1788).
Caribbean. 81 BAH CUB DOM HAI JAM LEE PUE. Nanophan. or phan.
Argythamnia candicans var. *serratifolia* Urb., Repert. Spec. Nov. Regni Veg. 28: 221 (1930). *Argythamnia candicans* subsp. *serratifolia* (Urb.) Borhidi & O.Muñiz, Bot. Közlem. 62: 25 (1975).

Argythamnia coatepensis (Brandegee) Croizat, J. Arnold Arbor. 26: 191 (1945).
Mexico (Puebla: near San Luis Tultitlanapa). 79 MXC. Nanophan.
* *Croton coatepensis* Brandegee, Zoe 5: 249 (1908).

Argythamnia cubensis Britton & P.Wilson, Mem. Torrey Bot. Club 16: 75 (1920).
SE. Cuba. 81 CUB. Hemicr. or cham.

Argythamnia discolor Brandegee, Zoe 5: 242 (1908).
Mexico (Sinaloa). 79 MXN.

Argythamnia ecdyomena J.W.Ingram, Gentes Herb. 10: 16 (1967).
Guatemala (Alta Verapaz). 80 GUA. Nanophan.

Argythamnia haplostigma Pax & K.Hoffm. in H.G.A.Engler, Pflanzenr., IV, 147, VI: 81 (1912).
Rotan I. 81 SWC. Nanophan.

Argythamnia heteropilosa J.W.Ingram, Gentes Herb. 10: 32 (1967).
E. Cuba. 81 CUB. Cham.

Argythamnia lottiae J.W.Ingram, Phytologia 55: 229 (1984).
Mexico (Jalisco). 79 MXS. Nanophan.

Argythamnia lucayana Millsp., Publ. Field Columbian Mus., Bot. Ser. 2: 145 (1906).
Bahamas. 81 BAH. Cham. or nanophan.

Argythamnia lundellii J.W.Ingram, Gentes Herb. 10: 16 (1967).
Mexico (Yucatán). 79 MXT. Nanophan.

Argythamnia microphylla Pax in H.G.A.Engler, Pflanzenr., IV, 147, VI: 82 (1912).
Cuba (N. Camagüey). 81 CUB. Hemicr.

Argythamnia moorei J.W.Ingram, Gentes Herb. 10: 11 (1967).
Mexico (Guerrero). 79 MXS. Nanophan.

Argythamnia oblongifolia Urb., Symb. Antill. 5: 386 (1908).
Haiti, Dominican Rep. 81 DOM HAI. Hemicr. or cham.

Argythamnia proctorii J.W.Ingram, Gentes Herb. 10: 25 (1967).
Cayman Is. (Grand Cayman). 81 CAY. Nanophan.

Argythamnia sericea Griseb., Fl. Brit. W. I.: 44 (1859). *Argythamnia sericea* var. *genuina* Müll.Arg. in A.P.de Candolle, Prodr. 15(2): 742 (1866), nom. inval.
S. Bahamas. 81 BAH. Nanophan.

Argythamnia sitiens (Brandegee) J.W.Ingram, Gentes Herb. 10: 11 (1967).
Mexico (Veracruz). 79 MXG. Nanophan.
* *Croton sitiens* Brandegee, Univ. Calif. Publ. Bot. 10: 185 (1922).

Argythamnia stahlii Urb., Symb. Antill. 1: 336 (1899).
S. Puerto Rico, Leeward Is. (Anegada). 81 LEE PUE. Cham.

Argythamnia tinctoria Millsp., Fieldiana, Bot. 1: 302 (1896). *Ditaxis tinctoria* (Millsp.) Pax & K.Hoffm. in H.G.A.Engler, Pflanzenr., IV, 147, VI: 59 (1912).
Mexico (Yucatán). 79 MXT. Nanophan.

Argythamnia wheeleri J.W.Ingram, Brittonia 16: 274 (1964).
Mexico (Yucatán: South Kancabzonot). 79 MXT. Nanophan.

Synonyms:
Argythamnia acalyphifolia (Griseb.) Kuntze === **Byttneria sp.** (Byttneriaceae)
Argythamnia acaulis (Herter ex Arechav.) J.W.Ingram === **Ditaxis acaulis** Herter ex Arechav.
Argythamnia aculeolata (Müll.Arg.) Kuntze === **Caperonia aculeolata** Müll.Arg.
Argythamnia adenophora A.Gray === **Ditaxis adenophora** (A.Gray) Pax & K.Hoffm.
Argythamnia angustissima (Klotzsch) Kuntze === **Caperonia angustissima** Klotzsch
Argythamnia anisotricha Müll.Arg. === **Chiropetalum anisotrichum** (Müll.Arg.) Pax & K.Hoffm.
Argythamnia aphoroides Müll.Arg. === **Ditaxis aphoroides** (Müll.Arg.) Pax
Argythamnia argentinense (Skottsb.) Allem & Irgang === ?
Argythamnia argothamnoides (Bertol. ex Spreng.) J.W.Ingram === **Ditaxis argothamnoides** (Bertol. ex Spreng.) Radcl.-Sm. & Govaerts
Argythamnia arlynniana J.W.Ingram === **Ditaxis arlynniana** (J.W.Ingram) Radcl.-Sm. & Govaerts
Argythamnia astroplethos J.W.Ingram === **Chiropetalum astroplethos** (J.W.Ingram) Radcl.-Sm. & Govaerts
Argythamnia bahiensis Kuntze === **Caperonia bahiensis** Müll.Arg.
Argythamnia berteriana (Schltdl.) Müll.Arg. === **Chiropetalum berterianum** Schltdl.
Argythamnia berteriana var. *psiladenia* (Skottsb.) J.W.Ingram === **Chiropetalum berterianum** var. **psiladenium** Skottsb.
Argythamnia blodgettii (Torr. ex Chapm.) Chapm. === **Ditaxis argothamnoides** (Bertol. ex Spreng.) Radcl.-Sm. & Govaerts
Argythamnia boliviensis Müll.Arg. === **Chiropetalum boliviense** (Müll.Arg.) Pax & K.Hoffm.

Argythamnia brandegeei Millsp. === **Ditaxis brandegeei** (Millsp.) Rose & Standl.
Argythamnia brasiliensis (Klotzsch) Müll.Arg. === **Philyra brasiliensis** Klotzsch
Argythamnia breviramea Müll.Arg. === **Ditaxis breviramea** (Müll.Arg.) Pax & K.Hoffm.
Argythamnia buettneriacea (Müll.Arg.) Kuntze === **Caperonia buettneriacea** Müll.Arg.
Argythamnia californica Brandegee === **Ditaxis serrata** var. **californica** (Brandegee) Steinm. & Felger
Argythamnia calycina Müll.Arg. === **Ditaxis calycina** (Müll.Arg.) Pax & K.Hoffm.
Argythamnia candicans subsp. *serratifolia* (Urb.) Borhidi & O.Muñiz === **Argythamnia candicans** Sw.
Argythamnia candicans var. *serratifolia* Urb. === **Argythamnia candicans** Sw.
Argythamnia canescens (Phil.) F.Phil. === **Chiropetalum canescens** Phil.
Argythamnia cantonensis Hance === **Speranskia cantonensis** (Hance) Pax & K.Hoffm.
Argythamnia castaneifolia (L.) Kuntze === **Caperonia castaneifolia** (L.) A.St.-Hil.
Argythamnia catamarcensis (Griseb.) Hieron. === **Ditaxis catamarcensis** (Griseb.) Pax
Argythamnia clariana Jeps. === **Ditaxis clariana** (Jeps.) G.L.Webster
Argythamnia corchorodes (Müll.Arg.) Kuntze === **Caperonia corchoroides** Müll.Arg.
Argythamnia cordata (A.St.-Hil.) Kuntze === **Caperonia cordata** A.St.-Hil.
Argythamnia cremnophila (I.M.Johnst.) J.W.Ingram === **Chiropetalum cremnophilum** I.M.Johnst.
Argythamnia cuneifolia (Pax & K.Hoffm.) J.W.Ingram === **Ditaxis cuneifolia** Pax & K.Hoffm.
Argythamnia cyanophylla (Wooton & Standl.) J.W.Ingram === **Ditaxis cyanophylla** Wooton & Standl.
Argythamnia depressa (Greenm.) J.W.Ingram === **Ditaxis depressa** (Greenm.) Pax & K.Hoffm.
Argythamnia desertorum Müll.Arg. === **Ditaxis desertorum** (Müll.Arg.) Pax & K.Hoffm.
Argythamnia dioica (Kunth) Müll.Arg. === **Ditaxis dioica** Kunth
Argythamnia dressleriana J.W.Ingram === **Ditaxis dressleriana** (J.W.Ingram) Radcl.-Sm. & Govaerts
Argythamnia erubescens J.R.Johnst. === **Ditaxis erubescens** (Johnst.) Pax & K.Hoffm.
Argythamnia fasciculata (Vahl ex A.Juss.) Müll.Arg. === **Ditaxis fasciculata** Vahl ex A.Juss.
Argythamnia fendleri Müll.Arg. === **Ditaxis argothamnoides** (Bertol. ex Spreng.) Radcl.-Sm. & Govaerts
Argythamnia foliosa Müll.Arg. === **Chiropetalum foliosum** (Müll.Arg.) Pax & K.Hoffm.
Argythamnia gardneri Müll.Arg. === **Ditaxis gardneri** (Müll.Arg.) Pax & K.Hoffm.
Argythamnia gentryi J.W.Ingram === **Ditaxis manzanilloana** (Rose) Pax & K.Hoffm.
Argythamnia gracilis Brandegee === **Ditaxis neomexicana** (Müll.Arg.) A.Heller
Argythamnia grisea (Griseb.) Allem & Irgang === **Chiropetalum griseum** Griseb.
Argythamnia guatemalensis Müll.Arg. === **Ditaxis guatemalensis** (Müll.Arg.) Pax & K.Hoffm.
Argythamnia guatemalensis var. *barrancana* McVaugh === **Ditaxis guatemalensis** var. **barrancana** (McVaugh) Radcl.-Sm. & Govaerts
Argythamnia gymnadenia Müll.Arg. === **Chiropetalum gymnadenium** (Müll.Arg.) Pax & K.Hoffm.
Argythamnia haitiensis (Urb.) J.W.Ingram === **Ditaxis haitiensis** Urb.
Argythamnia herbacea Spreng. === **Croton monanthogynus** Michx.
Argythamnia heterantha (Zucc.) Müll.Arg. === **Ditaxis heterantha** Zucc.
Argythamnia heteropetala (Didr.) Kuntze === **Caperonia heteropetala** Didr.
Argythamnia heteropetalodes (Müll.Arg.) Kuntze === **Caperonia heteropetala** Didr.
Argythamnia hochstetteri Kuntze === **Caperonia serrata** (Turcz.) C.Presl
Argythamnia humilis (Engelm. & A.Gray) Müll.Arg. === **Ditaxis humilis** (Engelm. & A.Gray) Pax
Argythamnia humilis var. *leiosperma* Waterf. === **Ditaxis humilis** var. **leiosperma** (Waterf.) Radcl.-Sm. & Govaerts
Argythamnia illimaniensis (Baill.) Müll.Arg. === **Ditaxis illimaniensis** Baill.
Argythamnia intermedia (Pax & K.Hoffm.) Allem & Irgang === **Chiropetalum intermedium** Pax & K.Hoffm.

Argythamnia jablonszkyana (Pax & K.Hoffm.) J.W.Ingram === **Ditaxis jablonszkyana** Pax & K.Hoffm.
Argythamnia katharinae (Pax) Croizat === **Ditaxis katharinae** Pax
Argythamnia laevis (A.Gray ex Torr.) Müll.Arg. === **Ditaxis laevis** (A.Gray ex Torr.) A.Heller
Argythamnia lanceolata (Benth.) Müll.Arg. === **Ditaxis lanceolata** (Benth.) Pax & K.Hoffm.
Argythamnia lanceolata (Müll.Arg.) Pax & K.Hoffm. === **Argythamnia argentea** Millsp.
Argythamnia lancifolia (Schltdl.) Müll.Arg. === **Ditaxis polygama** (Jacq.) Wheeler
Argythamnia langsdorffii (Müll.Arg.) Kuntze === **Caperonia langsdorffii** Müll.Arg.
Argythamnia linearifolia (A.St.-Hil.) Kuntze === **Caperonia linearifolia** A.St.-Hil.
Argythamnia lineata (Klotzsch) Baill. === **Chiropetalum tricoccum** (Vell.) Chodat & Hassl.
Argythamnia macrantha (Pax & K.Hoffm.) Croizat === **Ditaxis macrantha** Pax & K.Hoffm.
Argythamnia macrobotrys (Pax & K.Hoffm.) J.W.Ingram === **Ditaxis macrobotrys** Pax & K.Hoffm.
Argythamnia malpighiacea Ule === **Ditaxis malpighiacea** (Ule) Pax & K.Hoffm.
Argythamnia malpighiphila (Hicken) J.W.Ingram === **Ditaxis malpighipila** (Hicken) Wheeler
Argythamnia manzanilloana Rose === **Ditaxis manzanilloana** (Rose) Pax & K.Hoffm.
Argythamnia melochiiflora Müll.Arg. === **Ditaxis simoniana** Casar.
Argythamnia mercurialina (Nutt.) Müll.Arg. === **Ditaxis mercurialina** (Nutt.) Coult.
Argythamnia micrandra Croizat === **Ditaxis micrandra** (Croizat) Radcl.-Sm. & Govaerts
Argythamnia mollis Baill. === **Chiropetalum molle** (Baill.) Pax & K.Hoffm.
Argythamnia montevidensis (Didr.) Müll.Arg. === **Ditaxis montevidensis** (Didr.) Pax
Argythamnia muellerargoviana Kuntze === **Caperonia gardneri** Müll.Arg.
Argythamnia multicostata (Müll.Arg.) Kuntze === **Caperonia multicostata** Müll.Arg.
Argythamnia neomexicana Müll.Arg. === **Ditaxis neomexicana** (Müll.Arg.) A.Heller
Argythamnia neomexicana var. *depressa* Greenm. === **Ditaxis depressa** (Greenm.) Pax & K.Hoffm.
Argythamnia palmeri S.Watson === **Ditaxis lanceolata** (Benth.) Pax & K.Hoffm.
Argythamnia paludosa (Klotzsch) Kuntze === **Caperonia castaneifolia** (L.) A.St.-Hil.
Argythamnia palustris (L.) Kuntze === **Caperonia palustris** (L.) A.St.-Hil.
Argythamnia pauciflora Mirb. ex Steud. === ?
Argythamnia pavoniana Müll.Arg. === **Chiropetalum pavonianum** (Müll.Arg.) Pax
Argythamnia phalacradenia J.W.Ingram === **Chiropetalum phalacradenium** (J.W.Ingram) L.B.Sm. & Downs
Argythamnia pilosissima (Benth.) Müll.Arg. === **Ditaxis pilosissima** (Benth.) A.Heller
Argythamnia pilosistyla Allem & Irgang === **Chiropetalum pilosistylum** (Allem & Irgang) Radcl.-Sm. & Govaerts
Argythamnia polygama (Jacq.) Kuntze === **Ditaxis polygama** (Jacq.) Wheeler
Argythamnia pringlei Greenm. === **Ditaxis pringlei** (Greenm.) Pax & K.Hoffm.
Argythamnia purpurascens S.Moore === **Ditaxis purpurascens** (S.Moore) Pax & K.Hoffm.
Argythamnia quinquecuspidata (A.Juss.) Müll.Arg. === **Chiropetalum quinquecuspidatum** (A.Juss.) Pax & K.Hoffm.
Argythamnia ramboi Allem & Irgang === **Chiropetalum ramboi** (Allem & Irgang) Radcl.-Sm. & Govaerts
Argythamnia regnellii (Müll.Arg.) Kuntze === **Caperonia regnellii** Müll.Arg.
Argythamnia rhizantha (Pax & K.Hoffm.) J.W.Ingram === **Ditaxis rhizantha** Pax & K.Hoffm.
Argythamnia rosularis (Pax & K.Hoffm.) J.W.Ingram === **Ditaxis rosularis** Pax & K.Hoffm.
Argythamnia rubricaulis (Pax & K.Hoffm.) Croizat === **Ditaxis rubricaulis** Pax & K.Hoffm.
Argythamnia ruiziana Müll.Arg. === **Chiropetalum ruizianum** (Müll.Arg.) Pax & K.Hoffm.
Argythamnia rutenbergii (Müll.Arg.) Kuntze === **Caperonia rutenbergii** Müll.Arg.
Argythamnia salina (Pax & K.Hoffm.) J.W.Ingram === **Ditaxis salina** Pax & K.Hoffm.
Argythamnia savanillensis Kuntze === **Ditaxis argothamnoides** (Bertol. ex Spreng.) Radcl.-Sm. & Govaerts
Argythamnia schiedeana Müll.Arg. === **Chiropetalum schiedeanum** (Müll.Arg.) Pax
Argythamnia schiedeana var. *major* Müll.Arg. === **Chiropetalum schiedeanum** (Müll.Arg.) Pax
Argythamnia schiedeana var. *minor* Müll.Arg. === **Chiropetalum astroplethos** (J.W.Ingram) Radcl.-Sm. & Govaerts

Argythamnia sellowiana (Pax & K.Hoffm.) J.W.Ingram === **Ditaxis sellowiana** Pax & K.Hoffm.
Argythamnia senegalensis (Müll.Arg.) Kuntze === **Caperonia serrata** (Turcz.) C.Presl
Argythamnia sericea var. *genuina* Müll.Arg. === **Argythamnia sericea** Griseb.
Argythamnia sericea var. *lanceolata* Müll.Arg. === **Argythamnia argentea** Millsp.
Argythamnia sericophylla A.Gray === **Ditaxis lanceolata** (Benth.) Pax & K.Hoffm.
Argythamnia serrata Brandegee === **Ditaxis serrata** (Torr.) A.Heller
Argythamnia serrata (Torr.) Müll.Arg. === **Ditaxis serrata** (Torr.) A.Heller
Argythamnia simoniana (Casar.) Müll.Arg. === **Ditaxis simoniana** Casar.
Argythamnia simulans J.W.Ingram === **Ditaxis simulans** (J.W.Ingram) Radcl.-Sm. & Govaerts
Argythamnia sponiella Müll.Arg. === **Chiropetalum canescens** Phil.
Argythamnia stenophylla (Müll.Arg.) Kuntze === **Caperonia stenophylla** Müll.Arg.
Argythamnia triandra (Griseb.) Allem & Irgang === **Chiropetalum boliviense** (Müll.Arg.) Pax & K.Hoffm.
Argythamnia tricocca (Vell.) Müll.Arg. === **Chiropetalum tricoccum** (Vell.) Chodat & Hassl.
Argythamnia tricuspidata (Lam.) Müll.Arg. === **Chiropetalum tricuspidatum** (Lam.) A.Juss.
Argythamnia tricuspidata var. *genuina* Müll.Arg. === **Chiropetalum tricuspidatum** (Lam.) A.Juss.
Argythamnia tricuspidata var. *lanceolata* (Cav.) Müll.Arg. === **Chiropetalum tricuspidatum** (Lam.) A.Juss.
Argythamnia tuberculata (Bunge) Müll.Arg. === **Speranskia tuberculata** (Bunge) Baill.

Aristogeitonia

6 species, Africa (3) and Madagascar (3); includes the formerly separate *Paragelonium*. Shrubs or small to medium trees to 20 m with alternate or spirally arranged simple (actually unifoliolate) or (in *A. limoniifolia*) unequally trifoliolate leaves. The inflorescences are ramiflorous, a feature unusual in the family (but cf. *Baccaurea*). The leaf arrangement distinguishes the genus from others in Mueller's subtribe Mischodontinae. *AA. monophylla* and *lophirifolia* are associated with limestone. No modern revision has been published. (Oldfieldioideae)

Pax, F. & K. Hoffmann (1922). *Aristogeitonia*. In A. Engler (ed.), Das Pflanzenreich, IV 147 XV (Euphorbiaceae-Phyllanthoideae-Phyllantheae): 296. Berlin. (Heft 81.) La/Ge. — 1 species; Africa.
Airy-Shaw, H. K. (1972). A second species of the genus *Aristogeitonia* Prain (Euphorbiaceae), from East Africa. Kew Bull. 26: 495-498. En. — Description; discussion of related genera.
Radcliffe-Smith, A. (1988). Notes on Madagascan Euphorbiaceae, I. On the identity of *Paragelonium* and on the affinities of *Benoistia* and *Claoxylopsis*. Kew Bull. 43: 625-647, illus., map. En. — Pp. 625-630 comprise reduction of the Malagasy genus *Paragelonium* and addition of two species (one new) to the African *Aristogeitonia*; distribution map but no key.
Radcliffe-Smith, A. (1996). A second species of *Aristogeitonia* (Euphorbiaceae) for Tanzania. Kew Bull. 51: 799-801, illus. En. — A new large-leaved shrub, *A. magnistipulata*, from coastal forest formations.
Radcliffe-Smith, A. (1998). A third species of *Aristogeitonia* (Euphorbiaceae-Oldenfieldioideae) for Madagascar. Kew Bull. 53: 977-980, illus. — Description of *A. uapacifolia*.

Aristogeitonia Prain, Bull. Misc. Inform. Kew 1908: 438 (1908).
E. & SC. Trop. Africa, Madagascar. 25 26 29.
Paragelonium Leandri, Bull. Soc. Bot. France 85: 531 (1938 publ. 1939).

Aristogeitonia limoniifolia Prain, Bull. Misc. Inform. Kew 1908: 438 (1908).
Angola. 26 ANG. Phan.

Aristogeitonia lophirifolia Radcl.-Sm., Kew Bull. 43: 629 (1988). – FIGURE, p. 223.
N. & W. Madagascar. 29 MDG. Phan.

Aristogeitonia lophirifolia Radcl.-Sm.
Artist: Christine Grey-Wilson
Kew Bull. 43: 628 (1988)
KEW ILLUSTRATIONS COLLECTION

Aristogeitonia magnistipula Radcl.-Sm., Kew Bull. 51: 801 (1996).
Tanzania. 25 TAN. Nanophan.

Aristogeitonia monophylla Airy Shaw, Kew Bull. 26: 495 (1972).
SE. Kenya, E. Tanzania. 25 KEN TAN. Phan.

Aristogeitonia perrieri (Leandri) Radcl.-Sm., Kew Bull. 43: 627 (1988).
N. & W. Madagascar. 29 MDG. Phan.
* *Paragelonium perrieri* Leandri, Bull. Soc. Bot. France 85: 531 (1938 publ. 1939).

Aristogeitonia uapacifolia Radcl.-Sm., Kew Bull. 53: 978 (1998).
Madagascar. 29 MDG. Phan.

Aroton

Synonyms:
Aroton Neck. === **Croton** L.

Arthrothamnus

A Klotzsch segregate from *Euphorbia*; part of the *Tirucalli* group.

Synonyms:
Arthrothamnus Klotzsch & Garcke === **Euphorbia** L.
Arthrothamnus brachiatus Klotzsch & Garcke === **Euphorbia brachiata** (Klotzsch & Garcke) E.Mey. ex Boiss.
Arthrothamnus burmannii Klotzsch & Garcke === **Euphorbia burmannii** (Klotzsch & Garcke) E.Mey. ex Boiss.
Arthrothamnus scopiformis Klotzsch & Garcke === **Euphorbia arceuthobioides** Boiss.

Ashtonia

2 species, Thailand and W. Malesia (Peninsular Malaysia, Borneo). *A. excelsa* is a big forest tree to 33 m in transitions between mixed dipterocarp (on red/yellow soils) and kerangas (on white sand) forest. Its male inflorescences resemble those of *Richeria* but it appears closest to *Aporusa*. (Phyllanthoideae)

Airy-Shaw, H. K. (1968). Notes on Malesian and other Asiatic Euphorbiaceae, LXXXV. A new genus from Borneo. Kew Bull. 21: 357-360. En. — Protologue of *Ashtonia* and description of *A. excelsa*, a tree to 33 m in kerangas/mixed dipterocarp forest transitions; discussion. [For illustration, see Ic. Pl. 38: pl. 3702 (1974).]
Airy-Shaw, H. K. (1972). Notes on Malesian and other Asiatic Euphorbiaceae, CL: A new *Ashtonia* from Malaya. Kew Bull. 27: 4-5. En. —*A. praeterita* in the Main Range; genus previously known only from Borneo.

Ashtonia Airy Shaw, Kew Bull. 21: 357 (1968).
Thailand, Pen. Malaysia, Borneo. 41 42.

Ashtonia excelsa Airy Shaw, Kew Bull. 21: 357 (1968).
Borneo. 42 BOR. Phan.

Ashtonia praeterita Airy Shaw, Kew Bull. 27: 4 (1972). – FIGURE, p. 225.
Thailand, Pen. Malaysia. 41 THA 42 MLY. Phan.

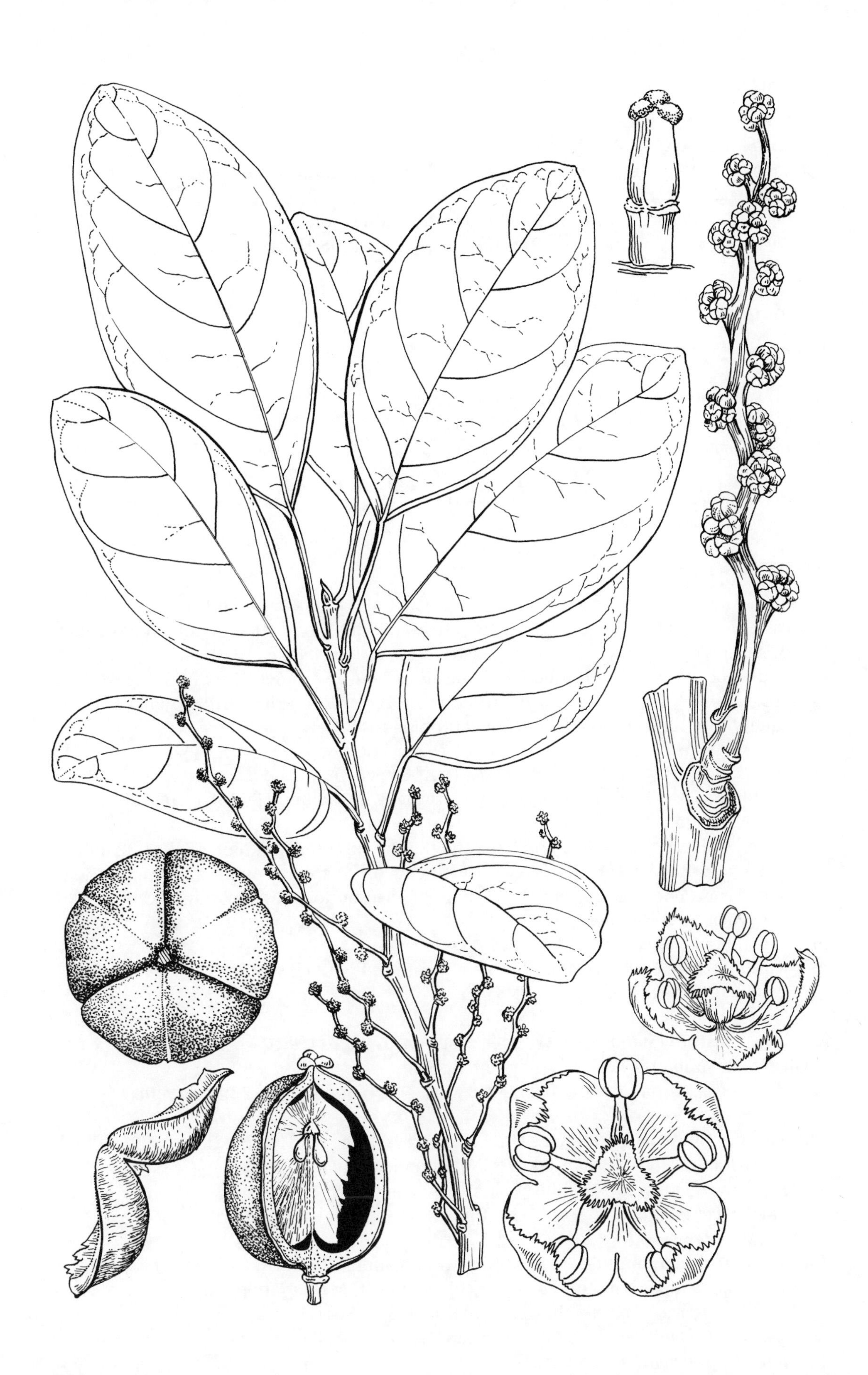

Ashtonia praeterita Airy Shaw
Artist: Mary Grierson
Ic. Pl. 38(1), pl. 3702 (1974)
KEW ILLUSTRATIONS COLLECTION

Asterandra

Synonyms:
Asterandra Klotzsch === **Phyllanthus** L.
Asterandra cornifolia (Kunth) Klotzsch === **Phyllanthus juglandifolius** subsp. **cornifolius** (Kunth) G.L.Webster
Asterandra grandifolia (L.) Britton === **Phyllanthus grandifolius** L.
Asterandra sellowiana Klotzsch === **Phyllanthus sellowianus** (Klotzsch) Müll.Arg.

Astraea

Synonyms:
Astraea Klotzsch === **Croton** L.
Astraea klotzschii Didr. === **Croton glandulosus** L.
Astraea manihot Klotzsch === **Croton lobatus** L.
Astraea palmata Klotzsch === **Croton lobatus** L.
Astraea paulina Didr. === **Croton paulinus** (Didr.) Müll.Arg.
Astraea seemannii Klotzsch === **Croton lobatus** L.

Astrocasia

5 species, Middle and South America (Mexico, Central America and West Indies (4); Bolivia and eastern Brazil (1)); shrubs or small trees to 10 m with long-petiolate, sometimes peltate leaves. No species is particularly widely distributed; *A. jacobinensis* is known only from two widely disjunct parts of South America. Some at least are associated with limestone. Webster (Synopsis, 1994) has assigned this to a monogeneric subtribe within Phyllantheae, consequent to a full revision (1992). (Phyllanthoideae)

Pax, F. & K. Hoffmann (1922). *Astrocasia*. In A. Engler (ed.), Das Pflanzenreich, IV 147 XV (Euphorbiaceae-Phyllanthoideae-Phyllantheae): 189. Berlin. (Heft 81.) La/Ge. — 1 species, Middle America (Mexico).

• Webster, G. L. (1992). Revision of *Astrocasia* (Euphorbiaceae). Syst. Bot. 17: 311-323, illus., maps. En. — Character review; treatment of 5 species with key, descriptions, synonymy, references, types, indication of distribution and habitat, localities with exsiccatae, and commentary.

Astrocasia B.L.Rob. & Millsp., Bot. Jahrb. Syst. 36(80): 19 (1905).
Mexico to Bolivia & E. Brazil. 79 80 83 84.

Astrocasia austinii (Standl.) G.L.Webster, Syst. Bot. 17: 322 (1992).
Guatemala (Izabal). 80 GUA. Nanophan. or phan.
* *Phyllanthus austinii* Standl., Publ. Field Mus. Nat. Hist., Bot. Ser. 22: 38 (1940).

Astrocasia jacobinensis (Müll.Arg.) G.L.Webster, Syst. Bot. 17: 316 (1992).
Brazil (Bahia), Bolivia. 83 BOL 84 BZE. Nanophan. or phan.
* *Phyllanthus jacobinensis* Müll.Arg., Linnaea 32: 6 (1863).
Phyllanthus inaequalis Rusby, Mem. Torrey Bot. Club 6: 118 (1910).

Astrocasia neurocarpa (Müll.Arg.) I.M.Johnst. ex Standl., Contr. Dudley Herb. 1: 74 (1927).
Mexico (Tamaulipas, San Luis Potosí, Puebla). 79 MXC MXE. Nanophan.
* *Phyllanthus neurocarpus* Müll.Arg., Linnaea 34: 69 (1865).
Astrocasia populifolia I.M.Johnst., Contr. Gray Herb. 68: 84 (1923).
Jatropha cercidiphylla Standl., Contr. U. S. Natl. Herb. 23: 639 (1923).

Astrocasia peltata Standl., Contr. Dudley Herb. 1: 74 (1927).
Mexico (Sinaloa, Nayarit, Jalisco). 79 MXN MXS. Nanophan. or phan.

Astrocasia tremula (Griseb.) G.L.Webster, J. Arnold Arbor. 39: 208 (1958).
S. Mexico, C. America, E. Cuba to Venezuela. 79 MXS MXT 80 BLZ COS GUA PAN 81 CAY CUB JAM 82 VEN 83 CLM? 84 BZE? BZN? Nanophan. or phan.
* *Phyllanthus tremulus* Griseb., Fl. Brit. W. I.: 34 (1859).
Astrocasia phyllanthoides B.L.Rob. & Millsp., Bot. Jahrb. Syst. 36(80): 20 (1905).

Synonyms:
Astrocasia phyllanthoides B.L.Rob. & Millsp. === **Astrocasia tremula** (Griseb.) G.L.Webster
Astrocasia populifolia I.M.Johnst. === **Astrocasia neurocarpa** (Müll.Arg.) I.M.Johnst. ex Standl.

Astrococcus

1 species, S America (Amazonian Brazil, extending into Venezuela); trees. Most closely related to *Haematostemon*. (Acalyphoideae)

Pax, F. & K. Hoffmann (1919). *Astrococcus*. In A. Engler (ed.), Das Pflanzenreich, IV 147 IX (Euphorbiaceae-Acalypheae-Plukenetiinae): 30-31. Berlin. (Heft 68.) La/Ge. — 1 species, Amazonian Brazil.

Astrococcus Benth., Hooker's J. Bot. Kew Gard. Misc. 6: 327 (1845).
S. Venezuela, N. Brazil. 82 84.

Astrococcus cornutus Benth., Hooker's J. Bot. Kew Gard. Misc. 6: 327 (1854).
Venezuela (Amazonas), Brazil (Amazonas). 82 VEN 84 BZN. Phan.

Synonyms:
Astrococcus coriaceus Baill. === **Haematostemon coriaceus** (Baill.) Pax & K.Hoffm.

Astrogyne

Synonyms:
Astrogyne Benth. === **Croton** L.

Astylis

Synonyms:
Astylis Wight === **Drypetes** Vahl
Astylis venusta Wight === **Drypetes venusta** (Wight) Pax & K.Hoffm.

Ateramnus

Synonyms:
Ateramnus P.Browne === **Gymnanthes** Sw.

Athroandra

Synonyms:
Athroandra Pax & K.Hoffm. === **Erythrococca** Benth.
Athroandra africana (Baill.) Pax & K.Hoffm. === **Erythrococca africana** (Baill.) Prain
Athroandra angolensis (Müll.Arg.) Pax & K.Hoffm. === **Erythrococca angolensis** (Müll.Arg.) Prain
Athroandra atrovirens (Pax) Pax & K.Hoffm. === **Erythrococca atrovirens** (Pax) Prain
Athroandra atrovirens var. *flaccida* (Pax) Pax & K.Hoffm. === **Erythrococca atrovirens** var. **flaccida** (Pax) Radcl.-Sm.

Athroandra atrovirens var. *schweinfurthii* (Pax) Pax & K.Hoffm. === **Erythrococca atrovirens** (Pax) Prain var. **atrovirens**
Athroandra chevalieri (Beille) Pax & K.Hoffm. === **Erythrococca chevalieri** (Beille) Prain
Athroandra columnaris (Müll.Arg.) Pax & K.Hoffm. === **Erythrococca columnaris** (Müll.Arg.) Prain
Athroandra dewevrei (Pax) Pax & K.Hoffm. === **Erythrococca dewevrei** (Pax) Prain
Athroandra hispida (Pax) Pax & K.Hoffm. === **Erythrococca hispida** (Pax) Prain
Athroandra inopinata (Prain) Pax & K.Hoffm. === **Erythrococca dewevrei** (Pax) Prain
Athroandra macrophylla Pax & K.Hoffm. === **Erythrococca macrophylla** Prain
Athroandra mannii (Hook.f.) Pax & K.Hoffm. === **Erythrococca mannii** (Hook.f.) Prain
Athroandra membranacea (Müll.Arg.) Pax & K.Hoffm. === **Erythrococca membranacea** (Müll.Arg.) Prain
Athroandra molleri (Pax) Pax & K.Hoffm. === **Erythrococca molleri** (Pax) Prain
Athroandra pallidifolia Pax & K.Hoffm. === **Erythrococca pallidifolia** (Pax & K.Hoffm.) Keay
Athroandra patula (Prain) Pax & K.Hoffm. === **Erythrococca patula** (Prain) Prain
Athroandra poggei (Prain) Pax & K.Hoffm. === **Erythrococca poggei** (Prain) Prain
Athroandra rivularis (Müll.Arg.) Pax & K.Hoffm. === **Erythrococca rivularis** (Müll.Arg.) Prain
Athroandra welwitschiana (Müll.Arg.) Pax & K.Hoffm. === **Erythrococca welwitschiana** (Müll.Arg.) Prain

Athroisma

Synonyms:
Athroisma Griff. === **Trigonostemon** Blume

Athymalus

Synonyms:
Athymalus Neck. === **Euphorbia** L.

Aubertia

Synonyms:
Aubertia Chapel. ex Baill. === **Croton** L.

Austrobuxus

20 species, Malesia, NE and E Australia, and Pacific Islands (New Caledonia and Fiji); most strongly represented in New Caledonia. Shrubs or small to large trees to as much as 40 m with opposite or alternate leaves, possibly light-demanding (*A. petiolaris* was recorded from old garden sites). *Longetia*, though by Brummitt (1992) included in *Austrobuxus*, was maintained as distinct by Webster (Synopsis, 1994), a position supported by A. Radcliffe-Smith (personal communication). Differences of opinion remain on the limits of *Austrobuxus* and in New Caledonia some species have been segregated into *Canaca* (a position supported by Webster (Synopsis, 1994) but not by Brummitt or McPherson and Tirel and not here) and *Scagea* (accepted by Brummitt and Webster and upheld here). The Australian species were revised by Forster (1997) who indicates that the genus is most closely related to *Dissiliaria* (for key, see Forster (1995) under *Sankowskya*). *A. nitidus* is an important timber tree in West Malesia. (Oldfieldioideae)

Pax, F. & K. Hoffmann (1922). *Longetia*. In A. Engler (ed.), Das Pflanzenreich, IV 147 XV (Euphorbiaceae-Phyllanthoideae-Phyllantheae): 289-291. Berlin. (Heft 81.) La/Ge. — 6 species; New Caledonia and W Malesia. [*L. depauperata* was transferred to *Scagea* in 1985, and the Malesian and other New Caledonian species are in *Austrobuxus*.]

Steenis, C. G. G. J. van (1964). *Austrobuxus*. Blumea 12: 362. En. — Nomenclatural; recognition of genus as euphorbiaceous, referable to *Longetia* as delimited by Pax and Hoffmann.

Airy-Shaw, H. K. (1971). Notes on Malesian and other Asiatic Euphorbiaceae, CXXXI. New combinations and new taxa in *Austrobuxus* Miq. Kew Bull. 25: 506-510. En. — Novelties and notes; no key. *Canaca* of New Caledonia, misplaced in Monimiaceae, is not distinct from *Austrobuxus*.

Airy-Shaw, H. K. (1974). Notes on Malesian and other Asiatic Euphorbiaceae, CLXXV. New species of *Austrobuxus* Miq. with a key to the whole genus. Kew Bull. 29: 303-309. En. — Five novelties from New Guinea and New Caledonia; key to 16 species (throughout range). *A. petiolaris* (New Guinea) is, however, *Ryparosa javanica* (Flacourtiaceae).

Airy-Shaw, H. K. (1978). Notes on Malesian and other Asiatic Euphorbiaceae, CCXII. *Austrobuxus* Miq. Kew Bull. 33: 39-43. En. — Further additions to this genus, with revision to part of the key.

Airy-Shaw, H. K. (1979). Notes on Malesian and other Asiatic Euphorbiaceae, CCXXIV. *Austrobuxus* Miq. Kew Bull. 33: 531-532. En. — New combination.

Airy-Shaw, H. K. (1980). *Austrobuxus*. Kew Bull. 35: 597-598. (Euphorbiaceae-Platylobeae of Australia.) En. — Treatment of 2 species, one endemic (*A. swainii*). [The latter is reminiscent of the Atlantic-Brazilian species of *Paradrypetes*. Its position in *Austrobuxus* has been questioned on account of its dentate leaves.]

• McPherson, G. & C. Tirel (1987). *Austrobuxus*. Fl. Nouvelle-Calédonie, 14 (Euphorbiacées, I): 193-222. Paris. Fr. — Flora treatment (15 species); key. [Nos. 9-15, including *A. vieillardii*, would seem to belong in *Canaca* if arguments for its segregation from *Austrobuxus* were accepted. Nos. 1-8 are in *Austrobuxus* s.s.]

Forster, P. I. (1997). A taxonomic revision of *Austrobuxus* Miq. (Euphorbiaceae: Dissiliariinae) in Australia. Austrobaileya 4: 619-626, illus., map. En. — Descriptive treatment (2 species) with key, synonymy, references, types, localities with (sometimes selected) exsiccatae, indication of distribution and habitat, commentary, notes on conservation status, and critical figures; references but no separate index.

Austrobuxus Miq., Fl. Ned. Ind., Eerste Bijv.: 444 (1861).
Malesia to Fiji. 42 50 60.
Choriophyllum Benth., Hooker's Icon. Pl. 13: t. 1280 (1829).
Buraeavia Baill., Adansonia 11: 83 (1873).
Canaca Guillaumin, Arch. Bot. Bull. Mens. 1: 74 (1927).

Austrobuxus alticola McPherson, Bull. Mus. Natl. Hist. Nat., B, Adansonia 6: 461 (1984 publ. 1985).
NE. New Caledonia. 60 NWC. Nanophan.

Austrobuxus brevipes Airy Shaw, Kew Bull. 33: 41 (1978).
E. New Caledonia. 60 NWC. Phan.

Austrobuxus carunculatus (Baill.) Airy Shaw, Kew Bull. 25: 507 (1971).
New Caledonia. 60 NWC. Nanophan. or phan.
* *Baloghia carunculata* Baill., Adansonia 2: 215 (1862). *Codiaeum carunculatum* (Baill.) Müll.Arg. in A.P.de Candolle, Prodr. 15(2): 1117 (1866). *Buraeavia carunculata* (Baill.) Baill., Adansonia 11: 84 (1873). *Longetia carunculata* (Baill.) Pax & K.Hoffm. in H.G.A.Engler, Pflanzenr., IV, 147, XV: 290 (1922).
Austrobuxus angustus Airy Shaw, Kew Bull. 37: 377 (1982).

Austrobuxus clusiaceus (Baill.) Airy Shaw, Kew Bull. 25: 507 (1971).
New Caledonia (near Païta). 60 NWC. Nanophan. or phan.
* *Buraeavia clusiaceus* Baill., Adansonia 11: 84 (1873). *Longetia clusiacea* (Baill.) Pax & K.Hoffm. in H.G.A.Engler, Pflanzenr., IV, 147, XV: 290 (1922).

Austrobuxus cracens McPherson, in Fl. N. Caled. & Depend. 14: 216 (1987).
EC. New Caledonia. 60 NWC. Phan.

Austrobuxus cuneatus (Airy Shaw) Airy Shaw, Kew Bull. 33: 40 (1978).
S. New Caledonia. 60 NWC. Phan.
* *Austrobuxus clusiaceus* var. *cuneatus* Airy Shaw, Kew Bull. 25: 507 (1971).

Austrobuxus ellipticus McPherson, Bull. Mus. Natl. Hist. Nat., B, Adansonia 6: 462 (1984 publ. 1985).
C. & EC. New Caledonia. 60 NWC. Phan.

Austrobuxus eugeniifolius (Guillaumin) Airy Shaw, Kew Bull. 29: 308 (1974).
New Caledonia. 60 NWC. Phan.
* *Longetia eugeniifolia* Guillaumin, Mém. Mus. Natl. Hist. Nat., B, Bot. 8: 251 (1962).

Austrobuxus horneanus (A.C.Sm.) Airy Shaw, Kew Bull. 23: 342 (1969).
Fiji. 60 FIJ.
* *Buraeavia horneana* A.C.Sm., J. Arnold Arbor. 33: 374 (1969).

Austrobuxus huerlimannii Airy Shaw, Kew Bull. 29: 305 (1974).
SE. New Caledonia. 60 NWC. Phan.

Austrobuxus mandjelicus McPherson, Bull. Mus. Natl. Hist. Nat., B, Adansonia 6: 462 (1984 publ. 1985).
N. New Caledonia. 60 NWC. Phan.

Austrobuxus megacarpus P.I.Forst., Austrobaileya 4: 622 (1997).
Queensland (SE. Cook Distr.). 50 QLD. Phan.

Austrobuxus montis-do Airy Shaw, Kew Bull. 33: 41 (1978).
New Caledonia (Mt. Do). 60 NWC. Nanophan.

Austrobuxus nitidus Miq., Fl. Ned. Ind., Eerste Bijv.: 444 (1861). *Longetia nitida* (Miq.) Steenis, Blumea 12: 362 (1964). – FIGURE, p. 231.
Pen. Malaysia, Sumatera, Borneo, Queensland (Cook Distr.). 42 BOR MLY SUM. Phan. – Important timber tree.

var. **macrocarpus** Airy Shaw, Kew Bull. 25: 506 (1971).
Pen. Malaysia (Selangor). 42 MLY. Phan.

var. **montanus** (Ridl.) Whitmore, Gard. Bull. Singapore 26: 51 (1972).
Pen. Malaysia (Pehang). 42 MLY. Phan.
* *Choriophyllum montanum* Ridl., J. Linn. Soc., Bot. 38: 322 (1908). *Longetia montana* (Ridl.) Pax & K.Hoffm. in H.G.A.Engler, Pflanzenr., IV, 147, XV: 291 (1922). *Austrobuxus montanus* (Ridl.) Airy Shaw, Kew Bull. 25: 507 (1971).

var. **nitidus**
Pen. Malaysia, Sumatera, Borneo, Queensland (Cook Distr.). 42 BOR MLY SUM. Phan. – Important timber tree.
Choriophyllum malayanum Benth., Hooker's Icon. Pl. 13: t. 1280 (1879). *Longetia malayana* (Benth.) Pax & K.Hoffm. in H.G.A.Engler, Pflanzenr., IV, 147, XV: 291 (1922).

Austrobuxus ovalis Airy Shaw, Kew Bull. 29: 307 (1974).
NW. & C. New Caledonia. 60 NWC. Nanophan. or phan.
Austrobuxus lugubris Airy Shaw, Kew Bull. 33: 43 (1978).

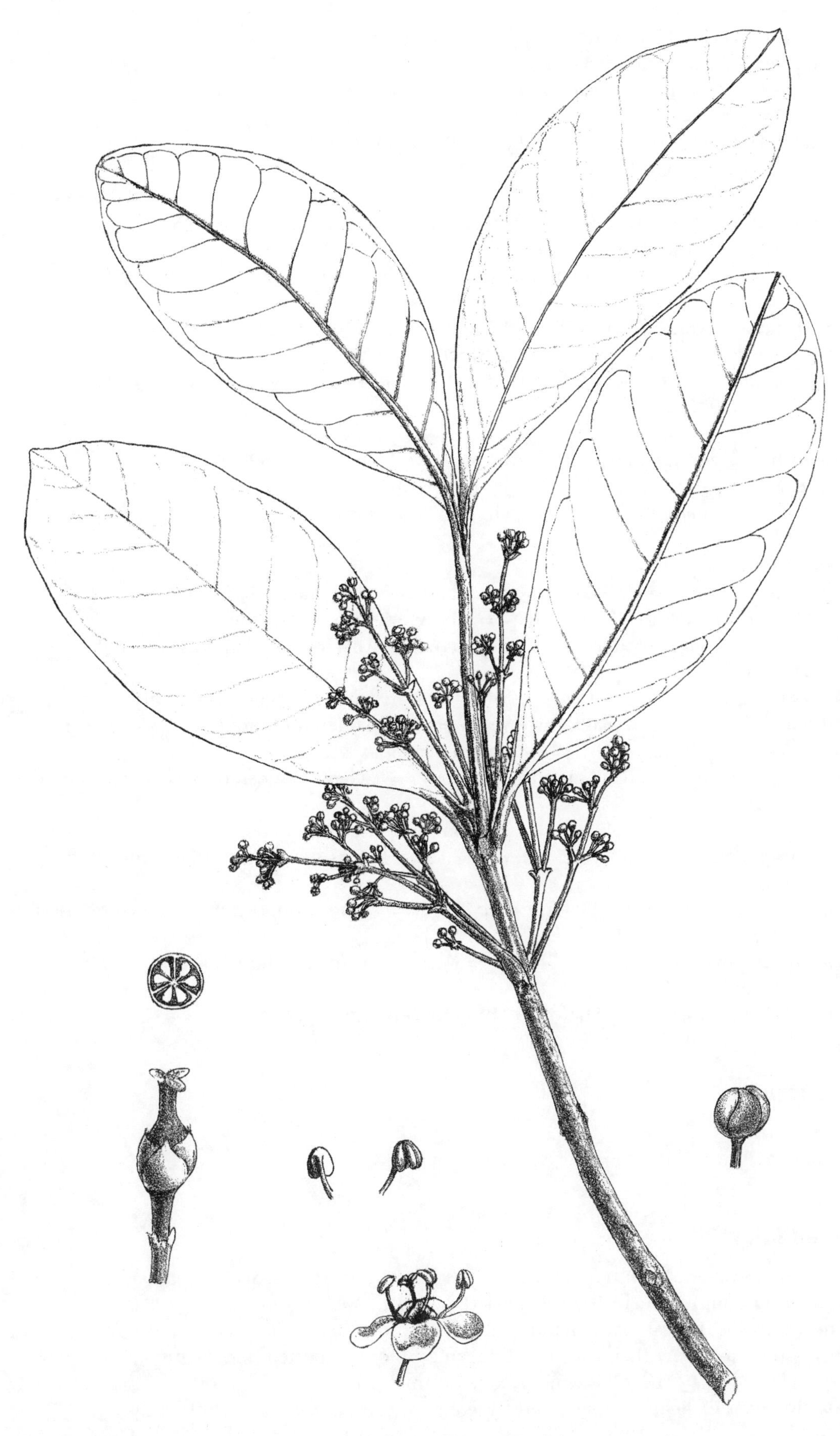

Austrobuxus nitidus Miq. var *nitidus* (as *Choriophyllum malayanum* Benth.)
Artist: 'A.M.C.'
Ic. Pl. 13: pl. 1280 (1879)
KEW ILLUSTRATIONS COLLECTION

Austrobuxus pauciflorus Airy Shaw, Kew Bull. 29: 306 (1974).
SE. New Caledonia. 60 NWC. Phan.
Austrobuxus clusiaceus var. *opacus* Airy Shaw, Kew Bull. 37: 378 (1982).

Austrobuxus petiolaris Airy Shaw, Kew Bull. 29: 304 (1974).
Papua New Guinea. 42 NWG. Phan.

Austrobuxus rubiginosus (Guillaumin) Airy Shaw, Kew Bull. 33: 39 (1978).
E. New Caledonia. 60 NWC. Nanophan. or phan.
* *Buraeavia rubiginosa* Guillaumin, Mém. Mus. Natl. Hist. Nat., B, Bot. 4: 69 (1953).

Austrobuxus swainii (Beuzev. & C.T.White) Airy Shaw, Kew Bull. 25: 508 (1971).
SE. Queensland, NE. New South Wales. 50 NSW. Phan.
* *Longetia swainii* Beuzev. & C.T.White, Proc. Linn. Soc. New South Wales, II, 71: 236 (1947).

Austrobuxus vieillardii (Guillaumin) Airy Shaw, Kew Bull. 25: 508 (1971).
NE. New Caledonia. 60 NWC. Nanophan. or phan.
* *Canaca vieillardii* Guillaumin, Arch. Bot. Bull. Mens. 1: 74 (1927).

Synonyms:
Austrobuxus angustus Airy Shaw === **Austrobuxus carunculatus** (Baill.) Airy Shaw
Austrobuxus buxoides (Baill.) Airy Shaw === **Longetia buxoides** Baill.
Austrobuxus clusiaceus var. *cuneatus* Airy Shaw === **Austrobuxus cuneatus** (Airy Shaw) Airy Shaw
Austrobuxus clusiaceus var. *opacus* Airy Shaw === **Austrobuxus pauciflorus** Airy Shaw
Austrobuxus depauperatus (Baill.) Airy Shaw === **Scagea depauperata** (Baill.) McPherson
Austrobuxus gracilis Airy Shaw === **Longetia buxoides** Baill.
Austrobuxus gynotrichus (Guillaumin) Airy Shaw === **Scagea oligostemon** (Guillaumin) McPherson
Austrobuxus lugubris Airy Shaw === **Austrobuxus ovalis** Airy Shaw
Austrobuxus montanus (Ridl.) Airy Shaw === **Austrobuxus nitidus** var. **montanus** (Ridl.) Whitmore
Austrobuxus oligostemon (Guillaumin) Airy Shaw === **Scagea oligostemon** (Guillaumin) McPherson
Austrobuxus phyllanthoides Airy Shaw === **Kairothamnus phyllanthoides** (Airy Shaw) Airy Shaw
Austrobuxus pisocarpus Airy Shaw === **Longetia buxoides** Baill.

Autrandra

Synonyms:
Autrandra Pierre ex Prain === **Erythrococca** Benth.

Avellanita

1 species, S America (central Chile); nondescript shrubs to 1 m with apetalous flowers and branching reminiscent of *Actinostemon* (Pax 1910). Webster (Synopsis, 1994) lists this as 'incertae sedis' but an examination of pollen and floral characters supports inclusion in Acalyphoideae rather than Crotonoideae, an opinion supported here. Bentham in 1880 had assigned it to his Crotoneae while Pax & Hoffmann in 1931 opted for their Acalypheae. It may well be a Chilean analogue to the Mediterranean *Chrozophora*. (Crotonoideae)

Pax, F. (1910). *Avellanita*. In A. Engler (ed.), Das Pflanzenreich, IV 147 [I] (Euphorbiaceae-Jatropheae): 15-16. Berlin. (Heft 42.) La/Ge. — 1 species, Chile.

Avellanita Phil., Linnaea 33: 237 (1864).
Chile. 85.

Avellanita bustillosii Phil., Linnaea 33: 238 (1864).
C. Chile. 85 CLC. Nanophan.

Axenfeldia

Synonyms:
Axenfeldia Baill. === **Mallotus** Lour.
Axenfeldia intermedia Baill. === **Mallotus resinosus** (Blanco) Merr.

Baccaurea

56 species, Asia, Malesia and the Pacific; most strongly developed in W Malesia. This genus of trees, which may reach 25 m or more but frequently are smaller, typically features tufted, obovate leaves and often hanging clusters of fruit. These latter are often cauli- and ramiflorous as well as being borne in leaf axils. Some species when well-grown have fluted trunks, among them *B. motleyana* (rambai). Five sections, some originally proposed as distinct genera, were recognised by Pax and Hoffmann (1922); since then, no full revision has been published. Sect. *Calyptröon* was elevated to subgeneric rank by Nguyen Nghia Thin in 1986. The four Philippines species were revised by Fernando (1979 publ. 1980) and a partial revision covering 28 species was made by Chakrabarty & Gangopadhyay (1994). The genus is of considerable value in the East for its fruit and as timber and pulpwood trees (Soejarto 1965); four widely cultivated species are therein extensively described including *BB. motleyana, racemosa, dulcis* and *sapida*. Where the fruit is eaten it is in most cases the arilloid which is comestible; in some, however, the pericarp is also consumed. Raoul Haegens (Leiden) is as of writing engaged in a revision of Malesian species and has made an examination of potential new characters. Vegetative features have proved useful in species identification, and male inflorescences and fruits were also found helpful. Good characters are also furnished by wood anatomy, as reported by Janssonius (references in Soejarto 1965). The genus is closely related to the African *Maesobotrya* and according to Haegens (unpubl.) phylogenetic analyses suggest a partial overlap. (Phyllanthoideae)

- Pax, F. & K. Hoffmann (1922). *Baccaurea*. In A. Engler (ed.), Das Pflanzenreich, IV 147 XV (Euphorbiaceae-Phyllanthoideae-Phyllantheae): 45-72. Berlin. (Heft 81.) La/Ge. — 61 species in 5 sections, Asia, Malesia and the Pacific; most strongly developed in W. Malesia.

Airy-Shaw, H. K. (1960). Notes on Malaysian Euphorbiaceae, II. The identity of the genus *Gatnaia* Gagnep. Kew Bull. 14: 353-354. En. —*Gatnaia* is a synonym of *Baccaurea*, the single species being part of *B. oxycarpa* Gagnep.

Airy-Shaw, H. K. (1963). Notes on Malaysian and other Asiatic Euphorbiaceae, XX. *Baccaurea minor* in Billiton and Borneo. Kew Bull. 16: 342. En. — Range extension.

- Soejarto, D.D. (1965). *Baccaurea* and its uses. Bot. Mus. Leafl. 21: 65-104, illus. En. — Introduction, with description of genus and list of 'economic' species (amounting to some two-fifths of those known); systematic treatment of the four main cultivated species with descriptions, synonymy, vernacular names, commentary and illustrations; discussions of origin, distribution, propagation and cultivation, pathology, marketing of the fruit, and nutritional value; summary treatment of timber- and pulp-yielding species (20 in all, including some also in the main treatment); list of references.

Airy-Shaw, H. K. (1968). Notes on Malesian and other Asiatic Euphorbiaceae, LXXXIII. A new *Baccaurea* from Lower Burma. Kew Bull. 21: 354-355. En. — Description of *B. ptychopyxis* from Tenasserim.

Airy-Shaw, H. K. (1979). Notes on Malesian and other Asiatic Euphorbiaceae, CCXXII. *Baccaurea* Lour. Kew Bull. 33: 529-530. En. — Description of *B. nesophila* from New Guinea (Louisiades), with one variety.

Fernando, E.S. (1979(1980)). *Baccaurea* (Euphorbiaceae) in the Philippines. Kalikasan 8: 301-312, illus. En. — Area revision with key, descriptions, synonymy, references and citations, localities with exsiccatae, indication of distribution and habitat, commentary and illustrations of all species; list of references.

Airy-Shaw, H. K. (1980). Notes on Euphorbiaceae from Indomalesia, Australia and the Pacific, CCXXIX. *Baccaurea* Lour. Kew Bull. 35: 383-384. En. — Description of *B. sancta-crucis* from Vanikoro; possibly also in Vanuatu.

Chakrabarty, T. & M. Gangopadhyay (1994). Notes on the genus *Baccaurea* (Euphorbiaceae). J. Econ. Taxon. Bot. 18: 419-426, illus. En. — 28 Malesian species in all accounted for where types are present in herb. CAL; three here newly described (from Peninsular Malaysia and Sumatra), all from nineteenth-century material. No key is given. [Some of the species have been lectotypified.]

Kiew, R., S. Madhavan & H. Selamat (1997(1998)). *Baccaurea scortechinii* distinct from *B. parviflora* (Euphorbiaceae). Gard. Bull. Singapore 49: 37-47, illus.

Baccaurea Lour., Fl. Cochinch.: 651 (1790).
S. China, Trop. Asia, SW. Pacific. 36 40 41 42 43 60.
Coccomelia Reinw. ex Blume, Catalogus: 110 (1823).
Hedycarpus Jack, Trans. Linn. Soc. London 14: 118 (1823).
Adenocrepis Blume, Bijdr.: 579 (1826).
Pierardia Roxb., Fl. Ind. ed. 1832, 2: 254 (1832).
Calyptroon Miq., Fl. Ned. Ind., Eerste Bijv.: 471 (1861).
Microsepala Miq., Fl. Ned. Ind., Eerste Bijv.: 444 (1861).
Everettiodendron Merr., Philipp. J. Sci., C 4: 279 (1909).
Gatnaia Gagnep., Bull. Soc. Bot. France 71: 870 (1924 publ. 1925).

Baccaurea angulata Merr., Univ. Calif. Publ. Bot. 15: 148 (1929).
Borneo. 42 BOR.

Baccaurea annamensis Gagnep., Bull. Soc. Bot. France 70: 235 (1923).
Vietnam. 41 VIE.

Baccaurea bakeri Elmer ex Merr., Enum. Philipp. Fl. Pl. 2: 412 (1923).
Philippines. 42 PHI.

Baccaurea bracteata Müll.Arg. in A.P.de Candolle, Prodr. 15(2): 466 (1866).
Pen. Malaysia (incl. Singapore), S. Thailand, Sumatera, Borneo. 41 THA 42 BOR MLY SUM. Phan.

var. **bracteata**
Pen. Malaysia (incl. Singapore), S. Thailand, Sumatera, Borneo. 41 THA 42 BOR MLY SUM. Phan.
Sapium sterculiaceum Wall., Numer. List: 7974 (1847), nom. nud.

var. **crassifolia** (J.J.Sm.) Airy Shaw, Kew Bull., Addit. Ser. 4: 47 (1975).
Borneo, Sumatera. 42 BOR SUM. Phan.
* *Baccaurea crassifolia* J.J.Sm., Bull. Jard. Bot. Buitenzorg, III, 1: 394 (1920).

Baccaurea brevipes Hook.f., Fl. Brit. India 5: 372 (1887).
Pen. Malaysia, Sumatera, Borneo (Sarawak). 42 BOR MLY SUM. Phan.

Baccaurea celebica Pax & K.Hoffm. in H.G.A.Engler, Pflanzenr., IV, 147, XV: 69 (1922).
Sulawesi. 42 SUL. Phan.

Baccaurea costulata (Miq.) Müll.Arg. in A.P.de Candolle, Prodr. 15(2): 464 (1866).
Sumatera (incl. Bangka), Borneo (Sabah). 42 BOR SUM. Phan.
* *Mappa costulata* Miq., Fl. Ned. Ind., Eerste Bijv.: 459 (1861). *Pierardia costulata* (Miq.) Müll.Arg., Flora 47: 469 (1864).
Baccaurea deflexa Müll.Arg. in A.P.de Candolle, Prodr. 15(2): 462 (1866).

Baccaurea courtallensis (Wight) Müll.Arg. in A.P.de Candolle, Prodr. 15(2): 459 (1866).
W. & S. India. 40 IND. Phan.
* *Pierardia courtallensis* Wight, Icon. Pl. Ind. Orient. 5: t. 1912 (1852).

Baccaurea dasystachya (Miq.) Müll.Arg. in A.P.de Candolle, Prodr. 15(2): 458 (1866).
Sumatera. 42 SUM.
* *Pierardia dasystachya* Miq., Fl. Ned. Ind. 1(2): 358 (1859).

Baccaurea dulcis (Jack) Müll.Arg. in A.P.de Candolle, Prodr. 15(2): 460 (1866).
S. Sumatera. 42 SUM. Phan. – Fruit edible.
* *Pierardia dulcis* Jack, Trans. Linn. Soc. London 14: 120 (1823).

Baccaurea edulis Merr., Univ. Calif. Publ. Bot. 15: 149 (1929).
Borneo. 42 BOR. Phan. – Fruit edible.

Baccaurea esquirolii H.Lév., Fl. Kouy-Tchéou: 159 (1914).
China (Guizhou). 36 CHC.

Baccaurea flaccida Müll.Arg. in A.P.de Candolle, Prodr. 15(2): 459 (1866).
Burma. 41 BMA.
Pierardia flaccida Wall., Numer. List: 8074 (1847), nom. nud.

Baccaurea forbesii Pax & K.Hoffm. in H.G.A.Engler, Pflanzenr., IV, 147, XV: 56 (1922).
Sumatera. 42 SUM. Nanophan. or phan.

Baccaurea harmandii Gagnep., Bull. Soc. Bot. France 70: 235 (1923).
Laos. 41 LAO.

Baccaurea henii Thin, J. Biol. (Vietnam) 9: 37 (1987).
Vietnam. 41 VIE.

Baccaurea hookeri Gage, Rec. Bot. Surv. India 9: 232 (1922).
Pen. Malaysia, Borneo. 42 BOR MLY. Phan.
Baccaurea dolichobotrys Merr., Univ. Calif. Publ. Bot. 15: 147 (1929).

Baccaurea javanica (Blume) Müll.Arg. in A.P.de Candolle, Prodr. 15(2): 465 (1866).
Borneo, Jawa, Sulawesi, Sumatera. 42 BOR JAW SUL SUM. Phan.
* *Adenocrepis javanica* Blume, Bijdr.: 579 (1826).
Microsepala acuminata Miq., Fl. Ned. Ind., Eerste Bijv.: 444 (1861). *Baccaurea acuminata* (Miq.) Müll.Arg. in A.P.de Candolle, Prodr. 15(2): 463 (1866).
Baccaurea minahassae Koord., Meded. Lands Plantentuin 19: 625 (1898).
Baccaurea leucodermis Hook.f. ex Ridl., Fl. Malay Penins. 3: 244 (1924).

Baccaurea kunstleri King ex Gage, Rec. Bot. Surv. India 9: 230 (1922).
S. Thailand, Pen. Malaysia, Sumatera (incl. Bangka), Borneo. 41 THA 42 BOR MLY SUM. Phan.
Baccaurea cordata Merr., Univ. Calif. Publ. Bot. 15: 147 (1929).

Baccaurea lanceolata (Miq.) Müll.Arg. in A.P.de Candolle, Prodr. 15(2): 457 (1866).
S. Thailand, Pen. Malaysia, Sumatera, Borneo, Philippines (Palawan). 41 THA 42 BOR MLY PHI SUM. Phan.
* *Hedycarpus lanceolatus* Miq., Fl. Ned. Ind. 1(2): 359 (1859). *Adenocrepis lanceolata* (Miq.) Müll.Arg., Linnaea 32: 82 (1863).
Baccaurea glabriflora Pax & K.Hoffm. in H.G.A.Engler, Pflanzenr., IV, 147, XV: 59 (1922).

Baccaurea macrocarpa (Miq.) Müll.Arg. in A.P.de Candolle, Prodr. 15(2): 459 (1866).
Pen. Malaysia, S. & W. Sumatera (incl. Bangka), Borneo. 42 BOR MLY SUM. Phan.
* *Pierardia macrocarpa* Miq., Fl. Ned. Ind., Eerste Bijv.: 441 (1861).
Baccaurea borneensis Müll.Arg. in A.P.de Candolle, Prodr. 15(2): 460 (1866).
Baccaurea griffithii Hook.f., Fl. Brit. India 5: 371 (1887).

Baccaurea macrophylla (Müll.Arg.) Müll.Arg. in A.P.de Candolle, Prodr. 15(2): 46O (1866).
S. Thailand, Pen. Malaysia, Sumatera, Borneo. 41 THA 42 BOR MLY SUM. Phan.
* *Pierardia macrophylla* Müll.Arg., Flora 47: 516 (1864).
Baccaurea beccariana Pax & K.Hoffm. in H.G.A.Engler, Pflanzenr., IV, 147, XV: 62 (1922).

Baccaurea maingayi Hook.f., Fl. Brit. India 5: 370 (1887).
Pen. Malaysia (incl. Singapore), Borneo (Sarawak). 42 BOR MLY. Phan.

Baccaurea malayana (Jack) King ex Hook.f., Fl. Brit. India 5: 374 (1887).
Pen. Malaysia, Sumatera. 42 MLY SUM. Phan.
* *Hedycarpus malayanus* Jack, Trans. Linn. Soc. London 14: 118 (1823).

Baccaurea minor Hook.f., Fl. Brit. India 5: 370 (1887).
Pen. Malaysia, Sumatera (Bangka, Billiton), Borneo. 42 BOR MLY SUM. Phan.
Aporusa bilitonensis Pax & K.Hoffm. in H.G.A.Engler, Pflanzenr., IV, 147, XV: 97 (1922).
Baccaurea pendula Merr., Univ. Calif. Publ. Bot. 15: 152 (1929).

Baccaurea minutiflora Müll.Arg. in A.P.de Candolle, Prodr. 15(2): 463 (1866).
Sumatera. 42 SUM. Phan.
Baccaurea sanguinea J.J.Sm., Icon. Bogor.: t. 319 (1910).
Aporusa dolichocarpa Pax & K.Hoffm. in H.G.A.Engler, Pflanzenr., IV, 147, XV: 98 (1922).

Baccaurea montana Pax & K.Hoffm. in H.G.A.Engler, Pflanzenr., IV, 147, XV: 55 (1922).
Papua New Guinea. 42 NWG. Nanophan. or phan.

Baccaurea motleyana (Müll.Arg.) Müll.Arg. in A.P.de Candolle, Prodr. 15(2): 461 (1866).
Sumatera, Borneo. (36) chc 42 BOR jaw mly SUM. Phan. – Fruit edible.
* *Pierardia motleyana* Müll.Arg., Flora 47: 516 (1864).

Baccaurea multiflora Burck ex J.J.Sm., Icon. Bogor. 4: t. 37 (1910). – FIGURE, p. 237.
Sumatera (Bangka). 42 SUM. Phan.

Baccaurea nanihua Merr., Interpr. Herb. Amboin.: 315 (1917).
Maluku, Borneo. 42 BOR MOL. Phan.

var. **nanihua**
Maluku. 42 MOL. Phan.

var. **oblongata** J.J.Sm., Bull. Jard. Bot. Buitenzorg, III, 6: 94 (1924).
Borneo (Kalimantan). 42 BOR. Phan.

Baccaurea nesophila Airy Shaw, Kew Bull. 33: 529 (1979).
Papua New Guinea (incl. Louisiade Archip.). 42 NWG. Phan.
Baccaurea nesophila var. *microcarpa* Airy Shaw, Kew Bull. 33: 530 (1979).

Baccaurea papuana F.M.Bailey, Proc. Roy. Soc. Queensland 18: 3 (1904).
Papua New Guinea. 42 NWG. Phan.
Baccaurea plurilocularis J.J.Sm., Nova Guinea 8: 228 (1910).

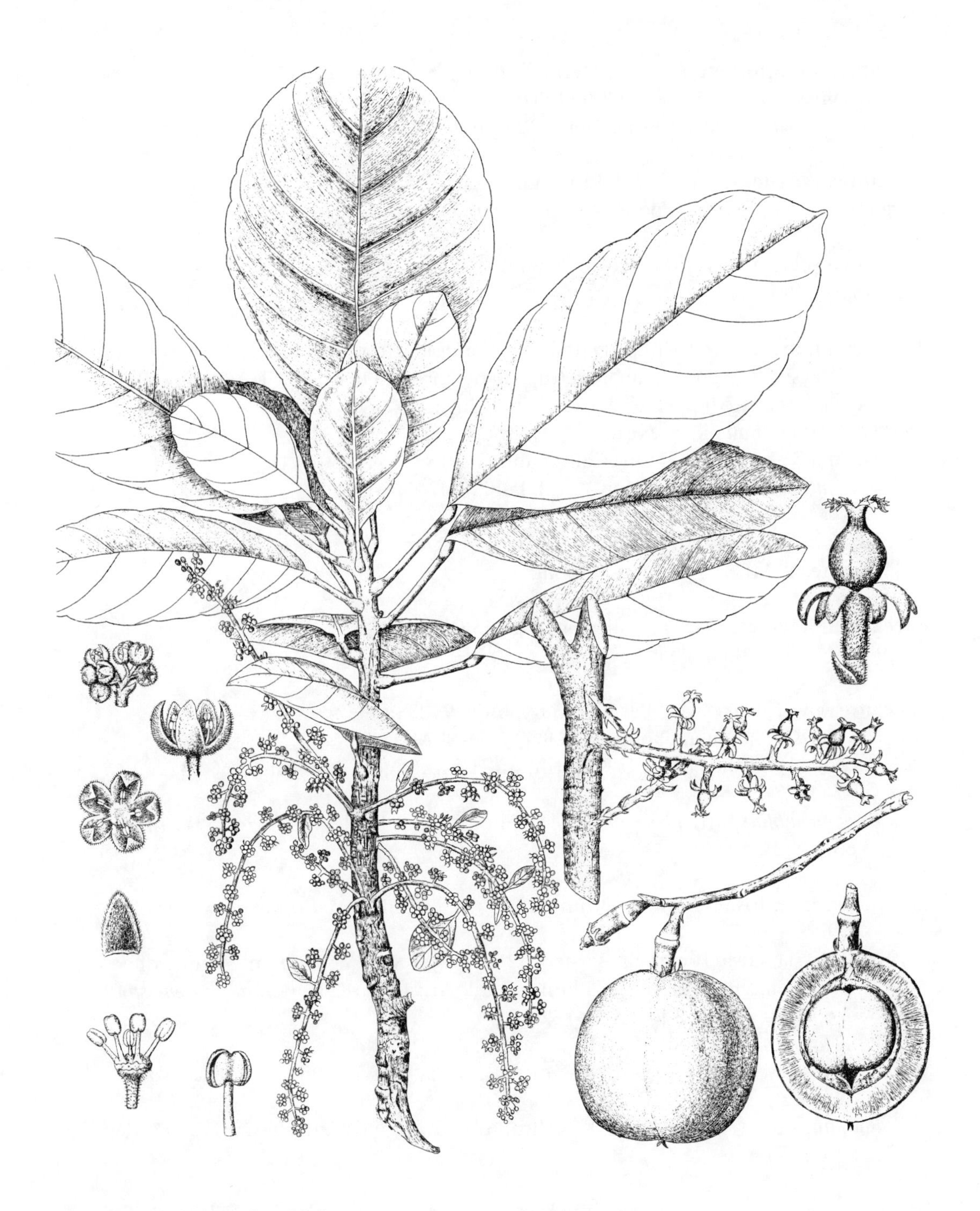

Baccaurea multiflora Burck ex J.J. Sm.
Artist: Natadipoera
Icon. Bogor. 4: pl. 312 (1910)
KEW ILLUSTRATIONS COLLECTION

Baccaurea parviflora (Müll.Arg.) Müll.Arg. in A.P.de Candolle, Prodr. 15(2): 462 (1866).
Burma, Thailand, Pen. Malaysia (incl. Singapore), Sumatera, Borneo, Philippines (Palawan). 41 BMA THA 42 BOR MLY PHI SUM. Phan.
* *Pierardia parviflora* Müll.Arg., Linnaea 32: 82 (1863).
Baccaurea affinis Müll.Arg. in A.P.de Candolle, Prodr. 15(2): 459 (1866).
Baccaurea odoratissima Elmer, Leafl. Philipp. Bot. 4: 1276 (1911).
Baccaurea singaporica Pax & K.Hoffm. in H.G.A.Engler, Pflanzenr., IV, 147, XV: 54 (1922).
Baccaurea rostrata Merr., Univ. Calif. Publ. Bot. 15: 150 (1929).

Baccaurea philippinensis (Merr.) Merr., Philipp. J. Sci., C 10: 275 (1915).
Philippines, Borneo (Sabah). 42 BOR PHI. Phan.
* *Everettiodendron philippinense* Merr., Philipp. J. Sci., C 4: 279 (1909).

Baccaurea polyneura Hook.f., Fl. Brit. India 5: 369 (1887).
Pen. Malaysia (incl. Singapore). 42 MLY. Phan.

Baccaurea ptychopyxis Airy Shaw, Kew Bull. 22: 354 (1968).
Burma. 41 BMA.

Baccaurea pubera (Miq.) Müll.Arg. in A.P.de Candolle, Prodr. 15(2): 458 (1866).
Pen. Malaysia, Borneo, E. Sumatera (incl. Bangka). 42 BOR MLY SUM. Phan.
* *Pierardia pubera* Miq., Fl. Ned. Ind., Eerste Bijv.: 177, 442 (1861).
Baccaurea latifolia King ex Hook.f., Fl. Brit. India 5: 373 (1887).
Baccaurea elmeri Merr., Univ. Calif. Publ. Bot. 15: 146 (1929).
Baccaurea puberula Merr., Univ. Calif. Publ. Bot. 15: 145 (1929).

Baccaurea pubescens Pax & K.Hoffm. in H.G.A.Engler, Pflanzenr., IV, 147, XV: 61 (1922).
Singapore. 42 MLY. Nanophan. or phan.

Baccaurea pulvinata A.C.Sm., Allertonia 1: 386 (1978).
Fiji (Viti Levu). 60 FIJ. Phan.

Baccaurea pyriformis Gage, Rec. Bot. Surv. India 9: 233 (1922).
Pen. Malaysia, Sumatera, Borneo. 42 BOR MLY SUM. Phan.
Baccaurea platyphylla Pax & K.Hoffm. in H.G.A.Engler, Pflanzenr., IV, 147, XV: 67 (1922).
Baccaurea platyphylloides Pax & K.Hoffm. in H.G.A.Engler, Pflanzenr., IV, 147, XV: 68 (1922).

Baccaurea racemosa (Reinw. ex Blume) Müll.Arg. in A.P.de Candolle, Prodr. 15(2): 461 (1866).
Pen. Malaysia, Jawa, Sumatera, Borneo. 42 BOR JAW MLY SUM. Phan. – Fruit edible.
* *Coccomelia racemosa* Reinw. ex Blume, Catalogus: 110 (1823). *Pierardia racemosa* (Reinw. ex Blume) Blume, Bijdr.: 579 (1826).
Baccaurea wallichii Hook.f., Fl. Brit. India 5: 375 (1887).
Baccaurea membranacea Pax & K.Hoffm. in H.G.A.Engler, Pflanzenr., IV, 147, XV: 49 (1922).
Baccaurea sarawakensis Pax & K.Hoffm. in H.G.A.Engler, Pflanzenr., IV, 147, XV: 54 (1922).

Baccaurea ramiflora Lour., Fl. Cochinch.: 661 (1790).
NE. India, Assam, Andaman Is., Burma, Thailand, Pen. Malaysia, Vietnam, SW. China. 36 CHC 40 ASS IND 41 AND BMA THA VIE 42 MLY. Phan. – Fruit edible.
Baccaurea cauliflora Lour., Fl. Cochinch.: 661 (1790).
Pierardia sapida Roxb., Fl. Ind. ed. 1832, 2: 254 (1832). *Baccaurea sapida* (Roxb.) Müll.Arg. in A.P.de Candolle, Prodr. 15(2): 459 (1866).
Pierardia macrostachya Wight & Arn. in R.Wight, Icon. Pl. Ind. Orient. 5: t. 1912 (1852). *Baccaurea macrostachya* (Wight & Arn.) Hook.f., Fl. Brit. India 5: 371 (1887).
Baccaurea propinqua Müll.Arg. in A.P.de Candolle, Prodr. 15(2): 463 (1866).
Baccaurea wrayi King ex Hook.f., Fl. Brit. India 5: 374 (1887).
Baccaurea oxycarpa Gagnep., Bull. Soc. Bot. France 70: 431 (1923).

Baccaurea reniformis Chakrab. & M.G.Gangop., J. Econ. Taxon. Bot. 19: 449 (1995).
Pen. Malaysia. 42 MLY. Phan.

Baccaurea reticulata Hook.f., Fl. Brit. India 5: 373 (1887).
Pen. Malaysia, Sumatera, Borneo (Sarawak, Brunei). 42 BOR MLY SUM. Phan.

Baccaurea sanctae-crucis Airy Shaw, Kew Bull. 35: 383 (1980).
Santa Cruz Is., Vanuatu. 60 SCZ VAN. Phan.

Baccaurea seemannii (Müll.Arg.) Müll.Arg. in A.P.de Candolle, Prodr. 15(2): 462 (1866).
Fiji (Viti Levu, Ovalau), Niue. 60 FIJ NUE. Phan.
* *Pierardia seemannii* Müll.Arg., Flora 47: 469 (1864).
Baccaurea wilkesiana Müll.Arg. in A.P.de Candolle, Prodr. 15(2): 461 (1866).

Baccaurea scortechinii Hook.f., Fl. Brit. India 5: 368 (1887).
Pen. Malaysia. 42 MLY.

Baccaurea stipulata J.J.Sm., Icon. Bogor.: t. 311 (1910).
Borneo. 42 BOR. Phan.
Baccaurea brevipedicellata Pax & K.Hoffm. in H.G.A.Engler, Pflanzenr., IV, 147, XV: 55 (1922).

Baccaurea stylaris Müll.Arg. in A.P.de Candolle, Prodr. 15(2): 465 (1866).
Fiji. 60 FIJ. Phan.
Baccaurea obtusa A.C.Sm., Bernice P. Bishop Mus. Bull. 141: 84 (1936).

Baccaurea sumatrana (Miq.) Müll.Arg. in A.P.de Candolle, Prodr. 15(2): 466 (1866).
Pen. Malaysia, Sumatera (incl. Bangka), Borneo. 42 BOR MLY SUM. Phan.
* *Calyptroon sumatranum* Miq., Fl. Ned. Ind., Eerste Bijv.: 472 (1861).
Baccaurea kingii Gage, Rec. Bot. Surv. India 9: 231 (1922).
Baccaurea bivalvis Merr., Univ. Calif. Publ. Bot. 15: 148 (1929).

Baccaurea sylvestris Lour., Fl. Cochinch.: 662 (1790).
Vietnam. 41 VIE. Phan.

Baccaurea taitensis Müll.Arg. in A.P.de Candolle, Prodr. 15(2): 463 (1866).
Western Samoa, American Samoa. 60 SAM. Phan.
Baccaurea seemannii var. *samoensis* Christoph., Bernice P. Bishop Mus. Bull. 128: 117 (1935).

Baccaurea tetrandra (Baill.) Müll.Arg. in A.P.de Candolle, Prodr. 15(2): 465 (1866).
Philippines. 42 PHI. Nanophan.
* *Adenocrepis tetrandra* Baill., Étude Euphorb.: 601 (1858).
Baccaurea terminaliifolia Elmer, Leafl. Philipp. Bot. 4: 1277 (1911).

Baccaurea trigonocarpa Merr., Univ. Calif. Publ. Bot. 15: 152 (1929).
Borneo. 42 BOR. Phan.

Baccaurea tristis Pax & K.Hoffm. in H.G.A.Engler, Pflanzenr., IV, 147, XV: 69 (1922).
Papua New Guinea. 42 NWG. Nanophan. or phan.

Baccaurea trunciflora Merr., Univ. Calif. Publ. Bot. 15: 151 (1929).
Borneo. 42 BOR. Phan.

Baccaurea velutina (Ridl.) Ridl., J. Bot. 62: 300 (1924).
S. Thailand, Pen. Malaysia. 41 THA 42 MLY. Phan.
* *Baccaurea reticulata* var. *velutina* Ridl., Fl. Malay Penins. 3: 246 (1924).

Synonyms:
Baccaurea acuminata (Miq.) Müll.Arg. === **Baccaurea javanica** (Blume) Müll.Arg.
Baccaurea affinis Müll.Arg. === **Baccaurea parviflora** (Müll.Arg.) Müll.Arg.

Baccaurea banahaensis Elmer === **Aporusa frutescens** Blume
Baccaurea barteri (Baill.) Müll.Arg. === **Maesobotrya barteri** (Baill.) Hutch.
Baccaurea beccariana Pax & K.Hoffm. === **Baccaurea macrophylla** (Müll.Arg.) Müll.Arg.
Baccaurea bipindensis Pax === **Maesobotrya bipindensis** (Pax) Hutch.
Baccaurea bivalvis Merr. === **Baccaurea sumatrana** (Miq.) Müll.Arg.
Baccaurea bonnetii Beille === **Maesobotrya barteri** var. **sparsiflora** (Scott-Elliot) Keay
Baccaurea borneensis Müll.Arg. === **Baccaurea macrocarpa** (Miq.) Müll.Arg.
Baccaurea brevipedicellata Pax & K.Hoffm. === **Baccaurea stipulata** J.J.Sm.
Baccaurea caillei Beille === **Maesobotrya barteri** var. **sparsiflora** (Scott-Elliot) Keay
Baccaurea capensis Spreng. ex Pax & K.Hoffm. === ? (Flacourtiaceae)
Baccaurea cauliflora Lour. === **Baccaurea ramiflora** Lour.
Baccaurea cavaleriei H.Lév. === **Cleidiocarpon cavaleriei** (H.Lév.) Airy Shaw
Baccaurea cavalliensis Beille === **Maesobotrya barteri** var. **sparsiflora** (Scott-Elliot) Keay
Baccaurea cordata Merr. === **Baccaurea kunstleri** King ex Gage
Baccaurea crassifolia J.J.Sm. === **Baccaurea bracteata** var. **crassifolia** (J.J.Sm.) Airy Shaw
Baccaurea deflexa Müll.Arg. === **Baccaurea costulata** (Miq.) Müll.Arg.
Baccaurea dolichobotrys Merr. === **Baccaurea hookeri** Gage
Baccaurea elmeri Merr. === **Baccaurea pubera** (Miq.) Müll.Arg.
Baccaurea gagnepainii Beille === **Maesobotrya barteri** var. **sparsiflora** (Scott-Elliot) Keay
Baccaurea glabriflora Pax & K.Hoffm. === **Baccaurea lanceolata** (Miq.) Müll.Arg.
Baccaurea glaziovii Beille === **Maesobotrya barteri** var. **sparsiflora** (Scott-Elliot) Keay
Baccaurea gracilis Merr. === **Richeriella gracilis** (Merr.) Pax & K.Hoffm.
Baccaurea griffithii Hook.f. === **Baccaurea macrocarpa** (Miq.) Müll.Arg.
Baccaurea griffoniana (Baill.) Müll.Arg. === **Maesobotrya griffoniana** (Baill.) Pierre ex Hutch.
Baccaurea kingii Gage === **Baccaurea sumatrana** (Miq.) Müll.Arg.
Baccaurea latifolia King ex Hook.f. === **Baccaurea pubera** (Miq.) Müll.Arg.
Baccaurea leucodermis Hook.f. ex Ridl. === **Baccaurea javanica** (Blume) Müll.Arg.
Baccaurea longispicata Beille === **Maesobotrya barteri** var. **sparsiflora** (Scott-Elliot) Keay
Baccaurea macrophylla Pax === **Protomegabaria macrophylla** (Pax) Hutch.
Baccaurea macrostachya (Wight & Arn.) Hook.f. === **Baccaurea ramiflora** Lour.
Baccaurea membranacea Pax & K.Hoffm. === **Baccaurea racemosa** (Reinw. ex Blume) Müll.Arg.
Baccaurea minahassae Koord. === **Baccaurea javanica** (Blume) Müll.Arg.
Baccaurea nesophila var. *microcarpa* Airy Shaw === **Baccaurea nesophila** Airy Shaw
Baccaurea obtusa A.C.Sm. === **Baccaurea stylaris** Müll.Arg.
Baccaurea odoratissima Elmer === **Baccaurea parviflora** (Müll.Arg.) Müll.Arg.
Baccaurea oxycarpa Gagnep. === **Baccaurea ramiflora** Lour.
Baccaurea pendula Merr. === **Baccaurea minor** Hook.f.
Baccaurea platyphylla Pax & K.Hoffm. === **Baccaurea pyriformis** Gage
Baccaurea platyphylloides Pax & K.Hoffm. === **Baccaurea pyriformis** Gage
Baccaurea plurilocularis J.J.Sm. === **Baccaurea papuana** F.M.Bailey
Baccaurea poissonii Beille === **Maesobotrya barteri** var. **sparsiflora** (Scott-Elliot) Keay
Baccaurea propinqua Müll.Arg. === **Baccaurea ramiflora** Lour.
Baccaurea puberula Merr. === **Baccaurea pubera** (Miq.) Müll.Arg.
Baccaurea pynaertii De Wild. === **Maesobotrya pynaertii** (De Wild.) Pax
Baccaurea pyrrhodasya (Miq.) Müll.Arg. === **Terminalia sp.** (Combretaceae)
Baccaurea reticulata var. *velutina* Ridl. === **Baccaurea velutina** (Ridl.) Ridl.
Baccaurea rostrata Merr. === **Baccaurea parviflora** (Müll.Arg.) Müll.Arg.
Baccaurea sanguinea J.J.Sm. === **Baccaurea minutiflora** Müll.Arg.
Baccaurea sapida (Roxb.) Müll.Arg. === **Baccaurea ramiflora** Lour.
Baccaurea sarawakensis Pax & K.Hoffm. === **Baccaurea racemosa** (Reinw. ex Blume) Müll.Arg.
Baccaurea seemannii var. *samoensis* Christoph. === **Baccaurea taitensis** Müll.Arg.
Baccaurea singaporica Pax & K.Hoffm. === **Baccaurea parviflora** (Müll.Arg.) Müll.Arg.
Baccaurea sparsiflora Scott-Elliot === **Maesobotrya barteri** var. **sparsiflora** (Scott-Elliot) Keay
Baccaurea staudtii Pax === **Maesobotrya staudtii** (Pax) Hutch.
Baccaurea symplocoides Hook.f. === **Aporusa symplocoides** (Hook.f.) Gage

Baccaurea terminaliifolia Elmer === **Baccaurea tetrandra** (Baill.) Müll.Arg.
Baccaurea vermeulenii De Wild. === **Maesobotrya vermeulenii** (De Wild.) J.Léonard
Baccaurea wallichii Hook.f. === **Baccaurea racemosa** (Reinw. ex Blume) Müll.Arg.
Baccaurea wilkesiana Müll.Arg. === **Baccaurea seemannii** (Müll.Arg.) Müll.Arg.
Baccaurea wrayi King ex Hook.f. === **Baccaurea ramiflora** Lour.

Baccaureopsis

Synonyms:
Baccaureopsis Pax === **Thecacoris** A.Juss.
Baccaureopsis lucida Pax === **Thecacoris lucida** (Pax) Hutch.

Balakata

2 species, China, S. & SE. Asia, and Malesia; light-demanding and fast-growing glabrous trees to 36 m with distichously arranged, penniveined leaves and terminal or (in upper axils) lateral inflorescences, these sometimes branching. The genus, based on *Sapium* sect. *Pleurostachya*, has only recently been segregated; it was 'Unnamed Genus 1' in Esser et al. (1997; see **Euphorbioideae** (**except Euphorbieae**)). A notable distinguishing feature, unusual in the Hippomaneae, is marked elongation of pedicels after anthesis; this causes the mature fruit to be exposed beyond the leaves. (Euphorbioideae (except Euphorbieae))

Pax, F. (with K. Hoffmann) (1912). *Sapium*. In A. Engler (ed.), Das Pflanzenreich, IV 147 V (Euphorbiaceae-Hippomaneae): 199-258. Berlin. (Heft 52.) La/Ge. — Nos. 73 (p. 240), *S. baccatum*, and 77 (p. 243), *S. merrillianum* (*S. lateriflorum*), are referable to *Balakata* (the latter as a new combination based on *Myrica luzonica* Vidal or *Sapium luzonicum* (Vidal) Merr.). [The latter species was also the sole member of the authors' sect. *Pleurostachya*.]

Esser, H.-J. (1999). A partial revision of the Hippomaneae (Euphorbiaceae) in Malesia. Blumea 44: 149-215, illus., maps. En. — *Balakata*, pp. 152-159; protologue of genus and treatment of 2 species with key, descriptions, synonymy, references, types, indication of distribution and habitat, illustration (*B. luzonica*), map and commentary; all general references, identification list and index to botanical names at end of paper. [*B. baccata* used for timber and as a wayside tree.]

Balakata Esser, Blumea 44: 154 (1999).
S. China, Trop. Asia. 36 40 41 42.

Balakata baccata (Roxb.) Esser, Blumea 44: 155 (1999).
Nepal, Bhutan, Assam, Indo-China, S. China, Andaman Is., Sumatera, Borneo (W. Kalimantan). 36 CHS 40 ASS EHM NEP 41 AND BMA MLY THA VIE 42 BOR SUM. Phan.
* *Sapium baccatum* Roxb., Fl. Ind. ed. 1832, 3: 694 (1832). *Stillingia baccata* (Roxb.) Baill., Étude Euphorb.: 513 (1858). *Excoecaria baccata* (Roxb.) Müll.Arg. in A.P.de Candolle, Prodr. 15(2): 1211 (1866).
Sapium dacdece Buch.-Ham. ex Wall., Numer. List: 7965 (1847), nom. nud.
Sapium hexandrum Wall., Numer. List: 7965 (1847), nom. nud.
Excoecaria affinis Griff., Not. Pl. Asiat. 4: 486 (1851), nom. illeg. *Sapium populifolium* Wall. ex Wight, Icon. Pl. Ind. Orient. 6: t. 1940 (1853).
Stillingia paniculata Miq., Fl. Ned. Ind., Eerste Bijv.: 461 (1861).

Balakata luzonica (Vidal) Esser, Blumea 44: 157 (1999).
Philippines to New Guinea. 42 PHI SUL NWG. Phan.
* *Myrica luzonica* Vidal, Sin. Gen. Pl. Leãos. Filip.: 40 (1883). *Sapium luzonicum* (Vidal) Merr., Philipp. J. Sci. 16: 577 (1920).
Sapium lateriflorum Merr., Philipp. J. Sci. 1(Suppl.): 83 (1906), nom. illeg. *Sapium merrillianum* Pax & K.Hoffm. in H.G.A.Engler, Pflanzenr., IV, 147, V: 243 (1912).

Baliospermum

5 species, Asia and W Malesia; mostly shrubs or geoxylic herbs. Two sections are now recognised; according to Chakraborty & Balakrishnan (1990 publ. 1992) each is composed of one, highly variable species with numerous formal varieties accepted for *B. calycinum*. The leaves of *B. montanum* are reminiscent of *Hibiscus rosa-sinensis* and its relatives. (Crotonoideae)

Pax, F. (with K. Hoffmann) (1912). *Baliospermum*. In A. Engler (ed.), Das Pflanzenreich, IV 147 IV (Euphorbiaceae-Gelonieae): 24-29. Berlin. (Heft 52.) La/Ge. — 9 species, 3 more or less doubtful; Asia, W Malesia. [One of the doubtful species collected in Honolulu in a garden. Another, *B. reidioides*, now referred to *Trigonostemon*.]

Pax, F. (with K. Hoffmann) (1914). *Baliospermum*. In A. Engler (ed.), Das Pflanzenreich, IV 147 VII [Euphorbiaceae-Additamentum V]: 414-415. Berlin. (Heft 63.) La/Ge. — 2 species, SE Asia.

• Chakrabarty, T. & N. P. Balakrishnan (1990(1992)). A revision of the genus *Baliospermum* Bl. (Euphorbiaceae) for the Indian subcontinent. Bull. Bot. Surv. India 32: 1-27, illus. En. — Introduction with historical review, general features and highlights; well-illustrated revision of 2 species (one with many varieties) with key, descriptions, synonymy, references and citations, vernacular names, localities with exsiccatae, indication of phenology, distribution, habitat, uses and properties, and taxonomic notes; list of intergrading specimens in *B. calycinum*; no separate index. [Features heavy reduction of the taxa accepted by Pax and Hoffmann and later workers.]

Baliospermum Blume, Bijdr.: 603 (1826).
Himalaya, Trop. Asia. 36 40 41 42.

Baliospermum angustifolium Y.T.Chang, Acta Phytotax. Sin. 27: 148 (1989).
Tibet (Medong). 36 CHT. Nanophan. or phan.

Baliospermum bilobatum T.L.Chin, Acta Phytotax. Sin. 18: 252 (1980).
Tibet (Medong). 36 CHT. Nanophan. or phan.

Baliospermum calycinum Müll.Arg., Flora 47: 470 (1864).
E. Himalaya, India to SC. China. 36 CHC 40 ASS EHM IND NEP 41 BMA THA VIE. Nanophan.

var. **balansae** (Gagnep.) Chakrab. & N.P.Balakr., Bull. Bot. Surv. India 32: 27 (1990 publ. 1992).
N. Vietnam. 41 VIE. Nanophan.
* *Baliospermum balansae* Gagnep., Bull. Soc. Bot. France 72: 460 (1925).

var. **bracteatum** Chakrab. & N.P.Balakr., J. Econ. Taxon. Bot. 7: 359 (1985 publ. 1986).
Arunachal Pradesh. 40 EHM. Nanophan.

var. **calycinum**
E. Himalaya, Assam. 40 ASS EHM. Nanophan.

var. **corymbiferum** (Hook.f.) Chakrab. & N.P.Balakr., Bull. Bot. Surv. India 32: 13 (1990 publ. 1992).
E. Nepal to Burma. 40 EHM NEP 41 BMA. Nanophan.
* *Baliospermum corymbiferum* Hook.f., Fl. Brit. India 5: 463 (1888).

var. **densiflorum** (D.G.Long) Chakrab. & N.P.Balakr., Bull. Bot. Surv. India 32: 15 (1990 publ. 1992).
Bhutan, China (Yunnan), Burma. 36 CHC 40 EHM 41 BMA. Cham. or nanophan.
* *Baliospermum densiflorum* D.G.Long, Notes Roy. Bot. Gard. Edinburgh 44: 171 (1986).

var. **effusum** (Pax & K.Hoffm.) Chakrab. & N.P.Balakr., Bull. Bot. Surv. India 32: 16 (1990 publ. 1992).
China (SW. Yunnan), Burma, NW. Thailand. 36 CHC 41 BMA THA. Nanophan.
* *Baliospermum effusum* Pax & K.Hoffm. in H.G.A.Engler, Pflanzenr., IV, 147, IV: 27 (1912).

var. **micranthum** (Müll.Arg.) Chakrab. & N.P.Balakr., Bull. Bot. Surv. India 32: 16 (1990 publ. 1992).
China (S. Yunnan), Assam to NW. Thailand. 36 CHC 40 ASS 41 BMA THA. Nanophan.
* *Baliospermum micranthum* Müll.Arg., Linnaea 34: 215 (1865).
Baliospermum meeboldii Pax & K.Hoffm. in H.G.A.Engler, Pflanzenr., IV, 147, VII: 414 (1914).
Baliospermum suffruticosum Pax & K.Hoffm. in H.G.A.Engler, Pflanzenr., IV, 147, VII: 414 (1914).

var. **nepalense** (Hurus. & Yu.Tanaka) Chakrab. & N.P.Balakr., Bull. Bot. Surv. India 32: 20 (1990 publ. 1992).
E. Nepal to Bhutan. 40 EHM NEP. Nanophan.
* *Baliospermum nepalense* Hurus. & Yu.Tanaka, Fl. E. Himal.: 174 (1966).

var. **racemiferum** Bhujel & Yonzone, J. Econ. Taxon. Bot. 18: 613 (1994 publ. 1995).
India (West Bengal). 40 IND. Nanophan.

var. **siamense** (Craib) Chakrab. & N.P.Balakr., Bull. Bot. Surv. India 32: 22 (1990 publ. 1992).
Burma, N. & C. Thailand. 41 BMA THA. Nanophan.
* *Baliospermum siamense* Craib, Bull. Misc. Inform. Kew 1911: 467 (1911).

var. **sinuatum** (Müll.Arg.) Chakrab. & N.P.Balakr., Bull. Bot. Surv. India 32: 24 (1990 publ. 1992).
Arunachal Pradesh. 40 EHM. Nanophan.
* *Baliospermum sinuatum* Müll.Arg., Flora 47: 470 (1864).

Baliospermum montanum (Willd.) Müll.Arg. in A.P.de Candolle, Prodr. 15(2): 1125 (1866).
China (SW. Yunnan), Trop. Asia. 36 CHC 40 BAN IND NEP WHM 41 BMA THA VIE 42 JAW LSI MLY SUM. Cham. or nanophan.
* *Jatropha montana* Willd., Sp. Pl. 4: 563 (1805). *Ricinus montanus* (Willd.) Wall., Numer. List: 7727 (1847).
Croton solanifolius Geiseler, Croton. Monogr.: 74 (1807). *Baliospermum solanifolium* (Geiseler) Suresh in D.H.Nicolson & al., Interpret. Rheede's Hort. Malab.: 106 (1988).
Baliospermum axillare Blume, Bijdr.: 604 (1826).
Croton polyandrus Roxb., Fl. Ind. ed. 1832, 3: 682 (1832). *Baliospermum polyandrum* (Roxb.) Wight, Icon. Pl. Ind. Orient. 5: t. 1885 (1852).
Baliospermum indicum Decne. in V.Jacquemont, Voy. Inde 4: 154 (1841).
Baliospermum angulare Decne. ex Baill., Étude Euphorb.: 395 (1858).
Baliospermum moritzianum Baill., Étude Euphorb.: 395 (1858).
Baliospermum pendulinum Pax in H.G.A.Engler, Pflanzenr., IV, 147, IV: 28 (1912).
Baliospermum raziana Keshaw, Murthy & Yogan., Curr. Sci. 56: 486 (1987).

Baliospermum yui Y.T.Chang, Acta Bot. Yunnan. 11: 413 (1989).
China (SW. Yunnan), Burma. 36 CHC 41 BMA. Nanophan.

Synonyms:
Baliospermum analayanum Hook.f. ex B.D.Jacks. === **Cheilosa montana** Blume
Baliospermum angulare Decne. ex Baill. === **Baliospermum montanum** (Willd.) Müll.Arg.
Baliospermum axillare Blume === **Baliospermum montanum** (Willd.) Müll.Arg.
Baliospermum balansae Gagnep. === **Baliospermum calycinum** var. **balansae** (Gagnep.) Chakrab. & N.P.Balakr.
Baliospermum corymbiferum Hook.f. === **Baliospermum calycinum** var. **corymbiferum** (Hook.f.) Chakrab. & N.P.Balakr.

Baliospermum densiflorum D.G.Long === **Baliospermum calycinum** var. **densiflorum** (D.G.Long) Chakrab. & N.P.Balakr.
Baliospermum effusum Pax & K.Hoffm. === **Baliospermum calycinum** var. **effusum** (Pax & K.Hoffm.) Chakrab. & N.P.Balakr.
Baliospermum indicum Decne. === **Baliospermum montanum** (Willd.) Müll.Arg.
Baliospermum malayanum Hook.f. === **Cheilosa montana** Blume
Baliospermum meeboldii Pax & K.Hoffm. === **Baliospermum calycinum** var. **micranthum** (Müll.Arg.) Chakrab. & N.P.Balakr.
Baliospermum micranthum Müll.Arg. === **Baliospermum calycinum** var. **micranthum** (Müll.Arg.) Chakrab. & N.P.Balakr.
Baliospermum moritzianum Baill. === **Baliospermum montanum** (Willd.) Müll.Arg.
Baliospermum nepalense Hurus. & Yu.Tanaka === **Baliospermum calycinum** var. **nepalense** (Hurus. & Yu.Tanaka) Chakrab. & N.P.Balakr.
Baliospermum pendulinum Pax === **Baliospermum montanum** (Willd.) Müll.Arg.
Baliospermum polyandrum (Roxb.) Wight === **Baliospermum montanum** (Willd.) Müll.Arg.
Baliospermum raziana Keshaw, Murthy & Yogan. === **Baliospermum montanum** (Willd.) Müll.Arg.
Baliospermum reidioides Kurz === **Trigonostemon reidioides** (Kurz) Craib
Baliospermum siamense Craib === **Baliospermum calycinum** var. **siamense** (Craib) Chakrab. & N.P.Balakr.
Baliospermum sinuatum Müll.Arg. === **Baliospermum calycinum** var. **sinuatum** (Müll.Arg.) Chakrab. & N.P.Balakr.
Baliospermum solanifolium (Geiseler) Suresh === **Baliospermum montanum** (Willd.) Müll.Arg.
Baliospermum suffruticosum Pax & K.Hoffm. === **Baliospermum calycinum** var. **micranthum** (Müll.Arg.) Chakrab. & N.P.Balakr.

Baloghia

15 species, Vanuatu, New Caledonia, Lord Howe and Norfolk Islands and eastern Australia; best developed in New Caledonia. *B. inophylla* (formerly *B. lucida*) is a shrub or tree to 15 m with opposite, finely-veined leaves, occuring over most of the range of the genus (illustrated by Pax & Hoffmann 1911, fig. 3). Other species have alternate or spirally arranged leaves, in some cases tuft-like (*B. bureavii*, New Caledonia). The genus has in the past been variously interpreted; with the type species, *B. lucida*, somewhat distinct from the rest there may be a case for revival of *Steigeria* for the remainder. Its type species, *B. montana* of New Caledonia and Vanuatu, is described as an 'arbor pulchra' by Pax & Hoffmann. Some of the other New Caledonian species are serpentine maquis shrubs. (Crotonoideae)

Pax, F. (with K. Hoffmann) (1911). *Baloghia*. In A. Engler (ed.), Das Pflanzenreich, IV 147 III (Euphorbiaceae-Cluytieae): 12-16. Berlin. (Heft 47.) La/Ge. — 9 species, New Caledonia (one, *B. lucida*, also in Australia).

Airy-Shaw, H. K. (1981). Notes on Asiatic, Malesian and Melanesian Euphorbiaceae, CCL. *Baloghia* Baill. Kew Bull. 36: 610. En. — New variety of *B. montana* from Vanuatu.

Green, P. S. (1986). New combinations in *Baloghia* and *Codiaeum* (Euphorbiaceae). Kew Bull. 41: 1026. En. — *B. lucida* is renamed *B. inophylla*, based on *Croton inophyllum* G. Forst. Collected in New Caledonia in Cook's second voyage, this was long wrongly used for *Chrozophora peltata* (= Codiaeum peltatum & (Labill.) P.S. Green).

• McPherson, G. & C. Tirel (1987). *Baloghia*. Fl. Nouvelle-Calédonie, 14 (Euphorbiacées, I): 43-72. Paris. Fr. — Illustrated flora treatment, with key. [Accounts for 13 species (of which 12 endemic) out of 15 known for the genus.]

Baloghia Endl., Prodr. Fl. Norfolk.: 84 (1833).
Australia to New Caledonia. 50 60.
Steigeria Müll.Arg., Linnaea 34: 215 (1865).

Baloghia alternifolia Baill., Adansonia 2: 216 (1862). *Codiaeum alternifolium* (Baill.) Müll.Arg. in A.P.de Candolle, Prodr. 15(2): 1117 (1866).
New Caledonia. 60 NWC. Nanophan. or phan.

Baloghia anisomera Guillaumin, Bull. Mus. Natl. Hist. Nat. 28: 105 (1922).
C. & SE. New Caledonia. 60 NWC. Phan.

Baloghia balansae (Baill.) Pax in H.G.A.Engler, Pflanzenr., IV, 147, III: 14 (1911).
C. New Caledonia. 60 NWC. Nanophan. or phan.
* *Codiaeum balansae* Baill., Adansonia 11: 77 (1873).

Baloghia brongniartii (Baill.) Pax in H.G.A.Engler, Pflanzenr., IV, 147, III: 16 (1911).
EC. New Caledonia. 60 NWC. Nanophan. or phan.
* *Codiaeum brongniartii* Baill., Adansonia 11: 76 (1873).

Baloghia buchholzii Guillaumin, Bull. Mus. Natl. Hist. Nat., II, 21: 263 (1949).
New Caledonia. 60 NWC. Nanophan. or phan.
Baloghia mackeeana Guillaumin, Mém. Mus. Natl. Hist. Nat., B, Bot. 8: 180 (1959).

Baloghia bureavii (Baill.) Schltr., Bot. Jahrb. Syst. 39: 152 (1906).
C. & SE. New Caledonia. 60 NWC. Nanophan. or phan.
* *Codiaeum bureavii* Baill., Adansonia 11: 74 (1873).

Baloghia deplanchei (Baill.) Pax in H.G.A.Engler, Pflanzenr., IV, 147, III: 16 (1911).
SE. New Caledonia. 60 NWC. Nanophan.
* *Codiaeum deplanchei* Baill., Adansonia 11: 75 (1873).

Baloghia drimiflora (Baill.) Schltr., Bot. Jahrb. Syst. 39: 152 (1906).
NW. & NWC. New Caledonia. 60 NWC. Nanophan. or phan.
* *Codiaeum drimiflorum* Baill., Adansonia 11: 75 (1873).

Baloghia inophylla (G.Forst.) P.S.Green, Kew Bull. 41: 1026 (1986). – FIGURE, p. 246.
Queensland, New South Wales, Norfolk I., Lord Howe I., New Caledonia (incl. Loyalty Is.). 50 LHN QLD NSW 60 NWC. Nanophan. or phan.
* *Croton inophyllus* G.Forst., Fl. Ins. Austr.: 67 (1786). *Trewia inophyllum* (G.Forst.) Spreng., Syst. Veg. 3: 906 (1826). *Codiaeum inophyllum* (G.Forst.) Müll.Arg. in A.P.de Candolle, Prodr. 15(2): 1120 (1866).
Baloghia lucida Endl., Prodr. Fl. Norfolk.: 84 (1833). *Codiaeum lucidum* (Endl.) Müll.Arg. in A.P.de Candolle, Prodr. 15(2): 1116 (1866).

Baloghia marmorata C.T.White, Proc. Roy. Soc. Queensland 53: 226 (1942).
Queensland (Moreton), New South Wales (North Coast). 50 NSW QLD. Phan.

Baloghia montana (Müll.Arg.) Pax in H.G.A.Engler, Pflanzenr., IV, 147, III: 15 (1911).
New Caledonia, Vanuatu. 60 NWC VAN. Phan.
* *Steigeria montana* Müll.Arg., Linnaea 34: 215 (1865). *Codiaeum montanum* (Müll.Arg.) Baill., Adansonia 11: 74 (1873).
Baloghia montana var. *neohebridensis* Airy Shaw, Kew Bull. 36: 610 (1981).

Baloghia neocaledonica (S.Moore) McPherson, in Fl. N. Caléd. Depend. 14: 54 (1987).
SE. New Caledonia. 60 NWC. Nanophan.
* *Ricinocarpos neocaledonicus* S.Moore, J. Linn. Soc., Bot. 45: 394 (1921).

Baloghia parviflora C.T.White, Proc. Roy. Soc. Queensland 53: 227 (1942).
Queensland (Altherton Tableland). 50 QLD. Phan.

Baloghia inophylla (G. Forst.) P.S. Green (as *B. lucida* Endl.)
Artist: Margaret Flockton
Maiden, Forest Fl. New South Wales 1: pl. 28 (1904)
KEW ILLUSTRATIONS COLLECTION

Baloghia pininsularis Guillaumin, Mém. Mus. Natl. Hist. Nat., B, Bot. 8: 260 (1962).
New Caledonia (Ile des Pins). 60 NWC. Nanophan. or phan.

Baloghia pulchella Schltr. ex Pax in H.G.A.Engler, Pflanzenr., IV, 147, III: 14 (1911).
New Caledonia (Mt. Dzumac). 60 NWC. Nanophan. or phan.

Synonyms:
Baloghia carunculata Baill. === **Austrobuxus carunculatus** (Baill.) Airy Shaw
Baloghia lucida Endl. === **Baloghia inophylla** (G.Forst.) P.S.Green
Baloghia mackeeana Guillaumin === **Baloghia buchholzii** Guillaumin

Baloghia montana var. *neohebridensis* Airy Shaw === **Baloghia montana** (Müll.Arg.) Pax
Baloghia oligostemon Guillaumin === **Scagea oligostemon** (Guillaumin) McPherson
Baloghia pancheri Baill. === **Fontainea pancheri** (Baill.) Heckel

Banalia

Synonyms:
Banalia Raf. === **Croton** L.
Banalia muricata Raf. === **Croton** sp.

Baprea

Synonyms:
Baprea Pierre ex Pax & K.Hoffm. === **Cladogynos** Zipp. ex Span.
Baprea bicolor Pierre ex Pax & K.Hoffm. === **Cladogynos orientalis** Zipp. ex Span.

Barhamia

An orthographic variant of *Berhamia*

Barrettia

Synonyms:
Barrettia Sim === **Ricinodendron** Müll.Arg.
Barrettia umbrosa Sim === **Ricinodendron heudelotii** (Baill.) Heckel subsp. **heudelotii**

Bellevalia

of Roemer & Schultes, a later homonym of *Bellevalia* Lapeyr. (Hyacinthaceae); = *Richeria* Vahl

Beltrania

Replaced for nomenclatural reasons by *Enriquebeltrania*.

Synonyms:
Beltrania Miranda === **Enriquebeltrania** Rzed.
Beltrania crenatifolia Miranda === **Enriquebeltrania crenatifolia** (Miranda) Rzed.

Benoistia

3 species, Madagascar. Considered to be the sole member of tribe Benoistieae (Radcliffe-Smith, 1988) but included in subtribe Neoboutonieae of Aleuritidae by Webster (Synopsis, 1994). *B. orientalis* is a tree to 16 m with loosely spirally arranged leaves (illustration in Radcliffe-Smith 1988: 638). (Acalyphoideae)

Leandri, J. (1939). Euphorbiacées malgaches nouvelles récoltées par M. H. Perrier de la Bâthie. Bull. Soc. Bot. France 85: 523-533. Fr. — Description of *Benoistia*, p. 528.

Radcliffe-Smith, A. (1988). Notes on Madagascan Euphorbiaceae, I. On the identity of *Paragelonium* and on the affinities of *Benoistia* and *Claoxylopsis*. Kew Bull. 43: 625-647, illus., maps. En. — Includes (pp. 632-640, 642) a phenetic analysis of *Benoistia* inclusive of putative relatives; position isolated and a new tribe in Crotonoideae proposed. 2 new species also described (for a total of 3); no key.

Benoistia H.Perrier & Leandri, Bull. Soc. Bot. France 85: 528 (1938 publ. 1939).
Madagascar. 29.

Benoistia orientalis Radcl.-Sm., Kew Bull. 43: 633 (1988).
N. & E. Madagascar. 29 MDG. Phan.

Benoistia perrieri H.Perrier & Leandri, Bull. Soc. Bot. France 85: 528 (1938 publ. 1939).
Madagascar. 29 MDG. Phan.

Benoistia sambiranensis H.Perrier & Leandri, Bull. Soc. Bot. France 85: 528 (1938 publ. 1939).
N. Madagascar. 29 MDG. Phan.

Berhamia

A Klotzsch segregate from *Croton.*

Synonyms:
Berhamia Klotzsch === **Croton** L.

Bernarda

An orthographic variant of *Bernardia*, proposed by Adanson.

Bernardia

68 species, Americas (as far north as the SW US); the 7 sections recognized by Pax and Hoffmann are reportedly diverse. The largest concentration of these shrubs or subshrubs (rarely annuals; sect. *Traganthus*) is in southern Brazil, Paraguay and northern Argentina; there is another center in the West Indies. 7 sections were recognised by Pax & Hoffmann (1914), divided in the first instance on trichome morphology. In sect. *Phyllopassaea* of south-central South America the branching is from a single-stemmed woody base. *B. pulchella* reaches 4 m. Sect. *Traganthus* (*B. sidoides*) may deserve generic rank according to Webster, Synopsis (1994: 75). [If *Bernardia* is combined with *Adelia* following Baillon (1858) then it must be rejected in favour of the latter. However, the pollen morphology of the two is mutually quite different, one reason that the two genera are now placed in different tribes (Webster, Synopsis 1994).] (Acalyphoideae)

Pax, F. (with K. Hoffmann) (1914). *Bernardia.* In A. Engler (ed.), Das Pflanzenreich, IV 147 VII (Euphorbiaceae-Acalypheae-Mercurialinae): 21-45. Berlin. (Heft 63.) La/Ge. — 35 species, Americas (most in tropics); 7 sections recognized.

Büchheim, G. (1960). Nomenklatorische und systematische Bemerkungen über die Gattung *Bernardia* (Euphorbiaceae). Willdenowia 2: 291-318. Ge. — A literature-based survey with respect to the history of *Bernardia* and the inclusion of otherwise of *Adelia*; numerous literature citations included in the synonymy. No key or formal synopsis of the genus is presented. [The author includes *Adelia* in *Bernardia* following Baillon (1858; see **General**), an opinion not now generally accepted.]

Büchheim, G. (1962). Über die Typusart der Gattung *Bernardia* (Euphorbiaceae). Willdenowia 3: 217-220. Ge. — Corrections to 1960 paper; selection of a different type species.

Jablonski, E. (1967). *Bernardia.* Euphorbiaceae, Guayana Highland (Mem. New York Bot. Gard. 17(1)): 130. New York. En. — 1 species, *B. amazonica*.

• McVaugh, R. (1995). Euphorbiacearum sertum Novo-Galicianarum revisarum. Contr. Univ. Michigan Herb. 20: 173-215, illus. En. — Revision precursory to treatment in *Flora Novo-Galiciana*; includes (pp. 191-201) a treatment of *Bernardia* (9 species, one unnamed) with key.

Bernardia Houst. ex Mill., Gard. Dict. Abr. ed. 4 (1754).
Trop. & Subtrop. America. 76 77 79 80 81 82 83 84 85.
Bivonia Spreng., Neue Entd. 2: 116 (1821).
Traganthus Klotzsch, Arch. Naturgesch. 7: 188 (1841).
Phaedra Klotzsch in S.L.Endlicher, Gen. Pl., Suppl. 4(3): 88 (1850).
Polyboea Klotzsch in S.L.Endlicher, Gen. Pl., Suppl. 4(3): 88 (1850).
Tyria Klotzsch in S.L.Endlicher, Gen. Pl., Suppl. 4(3): 88 (1850).
Alevia Baill., Étude Euphorb.: 508 (1858).
Passaea Baill., Étude Euphorb.: 507 (1858).

Bernardia alarici Allem & Irgang, Bol. Soc. Argent. Bot. 17: 301 (1976).
Brazil (Rio Grande do Sul). 84 BZS. Hemicr.

Bernardia albida Lundell, Wrightia 5: 245 (1976).
Mexico (San Luis Potosí, Hidalgo). 79 MXE. Nanophan.

Bernardia amazonica Croizat, J. Arnold Arbor. 24: 166 (1943).
S. Venezuela. 82 VEN.

Bernardia ambigua Pax & K.Hoffm. in H.G.A.Engler, Pflanzenr., IV, 147, VII: 30 (1914).
Brazil (Rio Grande do Sul). 84 BZS. Nanophan.

Bernardia argentinensis Lourteig & O'Donell, Ark. Bot., n.s., 3: 75 (1955).
N. Argentina. 85 AGE.

Bernardia asplundii Lourteig, Ark. Bot., n.s., 3: 78 (1955).
N. Argentina. 85 AGE.

Bernardia aurantiaca Lundell, Wrightia 1: 57 (1945).
Belize, Guatemala. 80 BLZ GUA. Phan.

Bernardia axillaris (Spreng.) Müll.Arg., Linnaea 34: 174 (1865).
SE. Brazil. 84 BZL. Nanophan.
* *Bivonia axillaris* Spreng., Neue Entd. 2: 116 (1821). *Bernardia axillaris* var. *genuina* Müll.Arg., Linnaea 34: 175 (1865), nom. inval.

subsp. **axillaris**
Brazil (Minas Gerais, Rio de Janeiro). 84 BZL. Nanophan.
Adelia martii Spreng., Syst. Veg. 3: 147 (1826).

subsp. **houlletiana** (Baill.) Müll.Arg. in A.P.de Candolle, Prodr. 15(2): 921 (1866).
Brazil. 84 BZL. Nanophan.
* *Adelia houlletiana* Baill., Adansonia 4: 373 (1864).
Bernardia axillaris var. *trichoclada* Müll.Arg., Linnaea 34: 175 (1865).

subsp. **scabrida** (Baill.) Pax & K.Hoffm. in H.G.A.Engler, Pflanzenr., IV, 147, VII: 34 (1914).
SE. Brazil. 84 BZL. Nanophan.
* *Adelia scabrida* Baill., Adansonia 4: 373 (1864).
Bernardia axillaris var. *obovata* Müll.Arg., Linnaea 34: 175 (1865).
Bernardia axillaris var. *spathulata* Müll.Arg., Linnaea 34: 175 (1865).

Bernardia brevipes Müll.Arg., Linnaea 34: 176 (1865).
Brazil (Rio de Janeiro). 84 BZL. Nanophan.

Bernardia caperoniifolia (Baill.) Müll.Arg. in A.P.de Candolle, Prodr. 15(2): 920 (1866).
Uruguay, Paraguay. 85 PAR URU. Nanophan.
* *Adelia caperoniifolia* Baill., Adansonia 4: 376 (1864).
Bernardia guaranitica Chodat & Hassl., Bull. Herb. Boissier, II, 5: 505 (1905).

Bernardia celastrinea (Baill.) Müll.Arg. in A.P.de Candolle, Prodr. 15(2): 921 (1866).
Brazil (Rio de Janeiro, São Paulo). 84 BZL. Nanophan.
* *Adelia celastrinea* Baill., Adansonia 4: 375 (1864). *Bernardia celastrinea* var. *genuina* Müll.Arg. in C.F.P.von Martius, Fl. Bras. 11(2): 397 (1874), nom. inval.
Bernardia capitellata Müll.Arg., Linnaea 34: 176 (1865). *Bernardia celastrinea* var. *capitellata* (Müll.Arg.) Müll.Arg. in C.F.P.von Martius, Fl. Bras. 11(2): 396 (1874).
Bernardia celastrinea var. *intermedia* Müll.Arg. in C.F.P.von Martius, Fl. Bras. 11(2): 397 (1874).
Bernardia celastrinea var. *obscura* Müll.Arg. in C.F.P.von Martius, Fl. Bras. 11(2): 397 (1874).
Bernardia celastrinea var. *serratifolia* Müll.Arg. in C.F.P.von Martius, Fl. Bras. 11(2): 397 (1874).

Bernardia chiapensis Lundell, Wrightia 5: 246 (1976).
Mexico (Chiapas). 79 MXT. Nanophan. or phan.

Bernardia cinerea Wiggins & Rollins, Contr. Dudley Herb. 3: 273 (1943).
Mexico (WC. Sonora). 79 MXN. Nanophan.

Bernardia colombiana Croizat, Ciencia (Mexico) 6: 353 (1946).
Colombia. 83 CLM.

Bernardia confertifolia Müll.Arg., Linnaea 34: 175 (1865). *Bernardia confertifolia* var. *latifolia* Müll.Arg., Linnaea 34: 175 (1865), nom. illeg.
S. Brazil (?). 84 BZS. Nanophan.
Bernardia confertifolia var. *lanceolata* Müll.Arg., Linnaea 34: 175 (1865).

Bernardia corensis (Jacq.) Müll.Arg., Linnaea 34: 173 (1865).
Caribbean, Venezuela. 81 CUB HAI JAM LEE NLA WIN 82 VEN. Nanophan.
* *Acalypha corensis* Jacq., Enum. Syst. Pl.: 32 (1760).
Acalypha capensis D.Dietr., Syn. Pl. 5: 378 (1852), sphalm.

Bernardia crassifolia Müll.Arg. in C.F.P.von Martius, Fl. Bras. 11(2): 394 (1874).
Brazil (Minas Gerais). 84 BZL. Nanophan.

Bernardia dichotoma (Willd.) Müll.Arg., Linnaea 34: 172 (1865).
Caribbean. 81 BAH CAY CUB DOM HAI JAM LEE. Nanophan. or phan.
* *Croton dichotomus* Willd., Sp. Pl. 4: 537 (1805). *Bernardia dichotoma* var. *genuina* Müll.Arg., Linnaea 34: 172 (1865), nom. inval.

var. **dichotoma**
Caribbean. 81 BAH CAY CUB DOM HAI JAM LEE. Nanophan. or phan.
Adelia bernardia L., Syst. Nat. ed. 10: 1289 (1759). *Bernardia bernardia* (L.) Millsp., Publ. Field Columbian Mus., Bot. Ser. 2(1): 58 (1900), nom. inval.
Bernardia carpinifolia Griseb., Fl. Brit. W. I.: 45 (1859). *Bernardia dichotoma* var. *carpinifolia* (Griseb.) Pax & K.Hoffm. in H.G.A.Engler, Pflanzenr., IV, 147, VII: 23 (1914), nom. illeg.
Bernardia dichotoma var. *macrocarpa* Pax & K.Hoffm. in H.G.A.Engler, Pflanzenr., IV, 147, VII: 23 (1914).

var. **venosa** (Griseb.) Müll.Arg., Linnaea 34: 172 (1865).
Cuba. 81 CUB. Nanophan. or phan.
Bernardia intermedia Griseb., Mem. Amer. Acad. Arts, n.s., 8: 160 (1861).
* *Bernardia venosa* Griseb., Mem. Amer. Acad. Arts, n.s., 8: 59 (1861).

Bernardia dodecandra (Sessé ex Cav.) Govaerts in R.Govaerts, D.G.Frodin & A.Radcliffe-Smith, World Checklist Bibliogr. Euphorbiaceae: 250 (2000).
Mexico. 79 MXC MCE MXG. Nanophan. or phan.

Adelia dodecandra Sessé ex Cav., Anales Ci. Nat. 5: 254 (1802), nom. nud.
Acalypha interrupta Schltdl., Linnaea 7: 386 (1832). *Bernardia interrupta* (Schltdl.) Müll.Arg., Linnaea 34: 171 (1865).

Bernardia flexuosa Pax & K.Hoffm. in H.G.A.Engler, Pflanzenr., IV, 147, XVII: 181 (1924).
Brazil (Rio Grande do Sul). 84 BZS.

Bernardia fruticulosa Alain, Phytologia 22: 164 (1971).
Dominican Rep. 81 DOM. Cham.

Bernardia gambosa Müll.Arg. in C.F.P.von Martius, Fl. Bras. 11(2): 391 (1874).
E. Brazil. 84 BZE BZL. Nanophan.
Bernardia brasiliensis var. *major* Müll.Arg., Linnaea 34: 174 (1865). *Bernardia tamanduana* var. *major* (Müll.Arg.) Müll.Arg. in A.P.de Candolle, Prodr. 15(2): 920 (1866). *Bernardia pulchella* var. *major* (Müll.Arg.) Müll.Arg. in C.F.P.von Martius, Fl. Bras. 11(2): 392 (1874). *Bernardia major* (Müll.Arg.) Pax & K.Hoffm. in H.G.A.Engler, Pflanzenr., IV, 147, VII: 30 (1914).

Bernardia gardneri Müll.Arg. in C.F.P.von Martius, Fl. Bras. 11(2): 394 (1874).
Brazil (Piauí, Goiás). 84 BZC BZE. Nanophan.

Bernardia geniculata Allem & J.L.Wächt., Revista Brasil. Biol. 37: 88 (1977).
Brazil (Rio Grande do Sul). 84 BZS. Cham.

Bernardia gentryana Croizat, J. Arnold Arbor. 24: 165 (1943).
W. Mexico. 79 MXN MXS. Nanophan. or phan.

Bernardia hagelundii Allem & Irgang, Bol. Soc. Argent. Bot. 17: 302 (1976).
Brazil (Rio Grande do Sul). 84 BZS. Hemicr.

Bernardia hassleriana Chodat, Bull. Herb. Boissier, II, 1: 397 (1901).
Paraguay. 85 PAR. Cham. or nanophan.
Bernardia hassleriana var. *tobatyensis* Chodat & Hassl., Bull. Herb. Boissier, II, 5: 503 (1905).

Bernardia heteropilosa McVaugh, Brittonia 13: 155 (1961).
Mexico (Nayarit). 79 MXS. Phan.

Bernardia hirsutissima (Baill.) Müll.Arg. in A.P.de Candolle, Prodr. 15(2): 922 (1866).
C. & SE. Brazil. 84 BZC BZE. Cham.
Adelia hirsutissima Baill., Adansonia 4: 372 (1864). *Bernardia peduncularis* var. *hirsutissima* (Baill.) Müll.Arg. in C.F.P.von Martius, Fl. Bras. 11(2): 399 (1874). *Bernardia hirsutissima* var. *genuina* Pax & K.Hoffm. in H.G.A.Engler, Pflanzenr., IV, 147, VII: 38 (1914), nom. inval.
Bernardia peduncularis Müll.Arg., Linnaea 34: 176 (1865).
Bernardia peduncularis var. *pubescens* Müll.Arg. in C.F.P.von Martius, Fl. Bras. 11(2): 400 (1874). *Bernardia hirsutissima* var. *pubescens* (Müll.Arg.) Pax in H.G.A.Engler, Pflanzenr., IV, 147, VII: 28 (1914).

Bernardia jacquiniana Müll.Arg., Linnaea 34: 173 (1865).
Venezuela, Colombia, Ecuador. 82 VEN 83 CLM ECU. Nanophan. or phan.

Bernardia kochii McVaugh, Contr. Univ. Michigan Herb. 20: 193 (1995).
Mexico (Jalisco). 79 MXS. Nanophan.

Bernardia lagunensis (M.E.Jones) Wheeler, Contr. U. S. Natl. Herb. 29: 106 (1945).
Mexico (Baja California). 79 MXN.
Croton lagunensis M.E.Jones, Contr. W. Bot. 18: 55 (1933).

Bernardia lanceifolia (Lundell) Lundell, Phytologia 57: 367 (1985).
Mexico (Chiapas). 79 MXT. Nanophan.
* *Bernardia mollis* var. *lanceifolia* Lundell, Contr. Univ. Michigan Herb. 4: 13 (1940).

Bernardia laurentii R.A.Howard, Phytologia 61: 2 (1986).
St. Lucia. 81 WIN. Nanophan.

Bernardia leptostachys Chodat & Hassl., Bull. Herb. Boissier, II, 5: 506 (1905).
Paraguay. 85 PAR. Cham.

Bernardia longipedunculata (Chodat & Hassl.) Pax & K.Hoffm. in H.G.A.Engler, Pflanzenr., IV, 147, VII: 38 (1914).
Paraguay. 85 PAR. Cham.
Bernardia peduncularis f. *spiciflora* Chodat & Hassl., Bull. Herb. Boissier, II, 5: 505 (1905).
Bernardia peduncularis f. *subcapituliflora* Chodat & Hassl., Bull. Herb. Boissier, II, 5: 505 (1905).
* *Bernardia peduncularis* var. *longipedunculata* Chodat & Hassl., Bull. Herb. Boissier, II, 5: 505 (1905).

Bernardia lorentzii Müll.Arg., J. Bot. 12: 229 (1874).
Paraguay, N. Argentina. 85 AGE PAR. Cham. or nanophan.
Bernardia apaensis Chodat & Hassl., Bull. Herb. Boissier, II, 5: 504 (1905). *Bernardia lorentzii* var. *apaensis* (Chodat & Hassl.) Pax & K.Hoffm. in H.G.A.Engler, Pflanzenr., IV, 147, VII: 41 (1914).
Bernardia apaensis var. *subintegra* Chodat & Hassl., Bull. Herb. Boissier, II, 5: 505 (1905). *Bernardia lorentzii* var. *subintegra* (Chodat & Hassl.) Pax & K.Hoffm. in H.G.A.Engler, Pflanzenr., IV, 147, VII: 41 (1914).
Bernardia lorentzii var. *fistulosa* Pax & K.Hoffm. in H.G.A.Engler, Pflanzenr., IV, 147, VII: 41 (1914).
Bernardia lorentzii var. *obovata* Pax & K.Hoffm. in H.G.A.Engler, Pflanzenr., IV, 147, VII: 41 (1914).

Bernardia macrophylla Standl., J. Wash. Acad. Sci. 15: 103 (1925).
Panama. 80 PAN.

Bernardia mayana Lundell, Wrightia 5: 247 (1976).
Guatemala. 80 GUA. Nanophan.

Bernardia mazatlana M.E.Jones, Contr. W. Bot. 18: 49 (1933).
Mexico (Sinaloa). 79 MXN.

Bernardia mexicana (Hook. & Arn.) Müll.Arg., Linnaea 34: 172 (1865).
Mexico, C. America, Venezuela. 79 MXE MXS MXT 80 GUA NIC 82 VEN. Nanophan.
* *Hermesia mexicana* Hook. & Arn., Bot. Beechey Voy.: 309 (1838). *Bernardia mexicana* var. *genuina* Müll.Arg., Linnaea 34: 171 (1865), nom. inval.
Bernardia mexicana var. *cinerascens* Müll.Arg., Linnaea 34: 172 (1865).
Bernardia mexicana var. *subbiflora* Müll.Arg., Linnaea 34: 172 (1865).
Bernardia brandegei Millsp., Proc. Calif. Acad. Sci., II, 3: 172 (1891).
Bernardia aspera Pax & K.Hoffm. in H.G.A.Engler, Pflanzenr., IV, 147, VII: 24 (1914).
Bernardia mexicana var. *albida* Pax & K.Hoffm. in H.G.A.Engler, Pflanzenr., IV, 147, VII: 24 (1914).

Bernardia micrantha Pax & K.Hoffm. in H.G.A.Engler, Pflanzenr., IV, 147, VII: 30 (1914).
Brazil (Rio de Janeiro). 84 BZL. Cham.

Bernardia mollis Lundell, Contr. Univ. Michigan Herb. 4 : 12 (1940).
Mexico (Chiapas). 79 MXT. Phan.

Bernardia multicaulis Müll.Arg., Linnaea 34: 177 (1865).
Brazil (Minas Gerais). 84 BZL. Cham.

Bernardia myricifolia (Scheele) S.Watson, Bot. California 2: 70 (1880).
S. California to W. Texas, Mexico (Sonora, Coahuila). 76 ARI CAL 77 NWM TEX 79 MXE MXN. Nanophan.
* *Tyria myricifolia* Scheele, Linnaea 25: 581 (1852).
Bernardia incana J.K.Morton, J. Wash. Acad. Sci. 29: 376 (1939).
Bernardia myricifolia var. *incanoides* M.C.Johnst., Phytologia 46: 281 (1980).

Bernardia nicaraguensis Standl. & L.O.Williams, Ceiba 1: 85 (1950).
Nicaragua, Honduras. 80 HON NIC.

Bernardia oblanceolata Lundell, Contr. Univ. Michigan Herb. 4: 13 (1940).
Mexico (Chiapas). 79 MXT. Nanophan.

Bernardia obovata L.M.Johnst., J. Arnold Arbor. 21: 261 (1940).
Texas, Mexico (Chihuahua). 77 TEX 79 MXE.

Bernardia odonellii Villa, Lilloa 30: 137 (1960).
Argentina (Misiones). 85 AGE. Hemicr. or cham.

Bernardia ovalifolia Lundell, Wrightia 5: 247 (1976).
Mexico (Durango). 79 MXE. Nanophan.

Bernardia ovata (Chodat & Hassl.) Pax & K.Hoffm. in H.G.A.Engler, Pflanzenr., IV, 147, VII: 41 (1914).
Paraguay. 85 PAR. Cham.
* *Bernardia hassleriana* var. *ovata* Chodat & Hassl., Bull. Herb. Boissier, II, 5: 503 (1905).

Bernardia paraguariensis Chodat & Hassl., Bull. Herb. Boissier, II, 5: 503 (1905).
Bolivia, Paraguay. 83 BOL 85 PAR. Nanophan.
Bernardia paraguariensis f. *orbiculata* Chodat & Hassl., Bull. Herb. Boissier, II, 5: 504 (1905).
Bernardia paraguariensis f. *rhombifolia* Chodat & Hassl., Bull. Herb. Boissier, II, 5: 504 (1905).
Bernardia paraguariensis var. *fruticosa* Chodat & Hassl., Bull. Herb. Boissier, II, 5: 504 (1905).
Bernardia paraguariensis var. *parvifolia* Chodat & Hassl., Bull. Herb. Boissier, II, 5: 504 (1905).
Bernardia rotundifolia Herzog, Repert. Spec. Nov. Regni Veg. 7: 59 (1909).

Bernardia polymorpha Chodat & Hassl., Bull. Herb. Boissier, II, 5: 505 (1905).
Paraguay. 85 PAR. Cham.

var. **curuguatensis** Chodat & Hassl., Bull. Herb. Boissier, II, 5: 506 (1905).
Paraguay. 85 PAR. Cham.
Bernardia polymorpha f. *latifolia* Chodat & Hassl., Bull. Herb. Boissier, II, 5: 506 (1905).
Bernardia polymorpha f. *salicina* Chodat & Hassl., Bull. Herb. Boissier, II, 5: 506 (1905).

var. **polymorpha**
Paraguay. 85 PAR. Cham.

var. **setosa** Chodat & Hassl., Bull. Herb. Boissier, II, 5: 506 (1905).
Paraguay. 85 PAR. Cham.
Bernardia polymorpha f. *angustifolia* Pax & K.Hoffm. in H.G.A.Engler, Pflanzenr., IV, 147, VII: 39 (1914).
Bernardia polymorpha f. *elliptica* Pax & K.Hoffm. in H.G.A.Engler, Pflanzenr., IV, 147, VII: 39 (1914).

Bernardia pooleae Lundell, Phytologia 57: 367 (1985).
Honduras. 80 HON.

Bernardia pulchella (Baill.) Müll.Arg. in C.F.P.von Martius, Fl. Bras. 11(2): 392 (1874).
SE. & S. Brasil, Paraguay, N. Argentina. 84 BZL BZS 85 AGE PAR. Nanophan. or phan.
* *Adelia pulchella* Baill., Adansonia 4: 374 (1864). *Bernardia tamanduana* var. *pulchella* (Baill.) Müll.Arg. in A.P.de Candolle, Prodr. 15(2): 920 (1866). *Bernardia pulchella* f. *genuina* Müll.Arg. in C.F.P.von Martius, Fl. Bras. 11(2): 393 (1874), nom. inval. *Bernardia pulchella* var. *genuina* Chodat & Hassl., Bull. Herb. Boissier, II, 5: 503 (1905), nom. inval.
Bernardia brasiliensis var. *lanceolata* Müll.Arg., Linnaea 34: 174 (1865). *Bernardia pulchella* f. *lanceolata* (Müll.Arg.) Müll.Arg. in C.F.P.von Martius, Fl. Bras. 11(2): 392 (1874).
Bernardia brasiliensis var. *longifolia* Müll.Arg., Linnaea 34: 174 (1865).
Bernardia brasiliensis var. *parvifolia* Müll.Arg., Linnaea 34: 174 (1865). *Bernardia tamanduana* var. *parvifolia* (Müll.Arg.) Müll.Arg. in A.P.de Candolle, Prodr. 15(2): 920 (1866). *Bernardia pulchella* f. *parvifolia* (Müll.Arg.) Müll.Arg. in C.F.P.von Martius, Fl. Bras. 11(2): 393 (1874).
Bernardia pulchella f. *latifolia* Müll.Arg. in C.F.P.von Martius, Fl. Bras. 11(2): 392 (1874).
Bernardia pulchella f. *acuminata* Chodat & Hassl., Bull. Herb. Boissier, II, 5: 503 (1905).
Bernardia pulchella f. *acutidens* Chodat & Hassl., Bull. Herb. Boissier, II, 5: 503 (1905).
Bernardia pulchella f. *breviserrata* Chodat & Hassl., Bull. Herb. Boissier, II, 5: 503 (1905).

Bernardia santanae McVaugh, Contr. Univ. Michigan Herb. 20: 194 (1995).
Mexico (Jalisco). 79 MXS. Nanophan.

Bernardia scabra Müll.Arg. in C.F.P.von Martius, Fl. Bras. 11(2): 396 (1874).
Brazil (Rio de Janeiro). 84 BZL. Nanophan.
Bernardia scabra var. *brevipila* Pax & K.Hoffm. in H.G.A.Engler, Pflanzenr., IV, 147, VII: 34 (1914).
Bernardia scabra var. *longipila* Pax & K.Hoffm. in H.G.A.Engler, Pflanzenr., IV, 147, VII: 34 (1914).

Bernardia sellowii Müll.Arg., Linnaea 34: 177 (1865).
S. Brazil, Uruguay. 84 BZS 85 URU. Cham.

Bernardia sidoides (Klotzsch) Müll.Arg., Linnaea 34: 177 (1865).
Guyana, Venezuela (Guarico), N. Brazil. 82 GUY VEN 84 BZE BZN. Ther.
* *Traganthus sidoides* Klotzsch, Arch. Naturgesch. 7: 188 (1841).

Bernardia similis Pax & K.Hoffm. in H.G.A.Engler, Pflanzenr., IV, 147, VII: 35 (1914).
Brazil (Rio de Janeiro). 84 BZL. Nanophan.

Bernardia simplex Chodat & Hassl., Bull. Herb. Boissier, II, 5: 505 (1905).
Paraguay. 85 PAR. Cham.

Bernardia spartioides (Baill.) Müll.Arg., Linnaea 34: 177 (1865).
Brazil (São Paulo, Goiás). 84 BZC BZL. Cham.
* *Passaea spartioides* Baill., Étude Euphorb.: 508 (1858). *Adelia spartioides* (Baill.) Baill., Adansonia 5: 319 (1865). *Bernardia spartioides* var. *pubescens* Müll.Arg. in C.F.P.von Martius, Fl. Bras. 11(2): 401 (1874), nom. illeg.
Bernardia spartioides var. *glabrata* Müll.Arg. in C.F.P.von Martius, Fl. Bras. 11(2): 402 (1874).

Bernardia spongiosa McVaugh, Brittonia 13: 157 (1961).
Mexico (Colima). 79 MXS. Nanophan. or phan.

Bernardia tamanduana (Baill.) Müll.Arg. in A.P.de Candolle, Prodr. 15(2): 920 (1866).
Brazil (Bahia, Rio de Janeiro). 84 BZE BZL. Nanophan.
* *Adelia tamanduana* Baill., Adansonia 4: 374 (1864). *Bernardia tamanduana* var. *genuina* Müll.Arg. in A.P.de Candolle, Prodr. 15(2): 920 (1866), nom. inval.
Bernardia oligandra Müll.Arg., Linnaea 34: 173 (1865).

Bernardia venezuelana Steyerm., Fieldiana, Bot. 28: 305 (1952).
Venezuela. 82 VEN.

Bernardia viridis Millsp., Proc. Calif. Acad. Sci., II, 2: 223 (1889).
Mexico (Baja California Sur, Sonora, Chihuahua). 79 MXE MXN. Nanophan. or phan.
Croton crenulatus M.E.Jones, Contr. W. Bot. 18: 48 (1933).

Bernardia wilburii McVaugh, Brittonia 8: 157 (1961).
Mexico (Jalisco). 79 MXS. Nanophan. or phan.

Bernardia yucatanensis Lundell, Contr. Univ. Michigan Herb. 4: 14 (1940).
Mexico (Campeche), Guatemala. 79 MXT 80 GUA. Nanophan.

Synonyms:
Bernardia apaensis Chodat & Hassl. === **Bernardia lorentzii** Müll.Arg.
Bernardia apaensis var. *subintegra* Chodat & Hassl. === **Bernardia lorentzii** Müll.Arg.
Bernardia aspera Pax & K.Hoffm. === **Bernardia mexicana** (Hook. & Arn.) Müll.Arg.
Bernardia axillaris var. *genuina* Müll.Arg. === **Bernardia axillaris** (Spreng.) Müll.Arg.
Bernardia axillaris var. *obovata* Müll.Arg. === **Bernardia axillaris** subsp. **scabrida** (Baill.) Pax & K.Hoffm.
Bernardia axillaris var. *spathulata* Müll.Arg. === **Bernardia axillaris** subsp. **scabrida** (Baill.) Pax & K.Hoffm.
Bernardia axillaris var. *trichoclada* Müll.Arg. === **Bernardia axillaris** subsp. **houlletiana** (Baill.) Müll.Arg.
Bernardia bernardia (L.) Millsp. === **Bernardia dichotoma** (Willd.) Müll.Arg. var. **dichotoma**
Bernardia brandegei Millsp. === **Bernardia mexicana** (Hook. & Arn.) Müll.Arg.
Bernardia brasiliensis var. *lanceolata* Müll.Arg. === **Bernardia pulchella** (Baill.) Müll.Arg.
Bernardia brasiliensis var. *longifolia* Müll.Arg. === **Bernardia pulchella** (Baill.) Müll.Arg.
Bernardia brasiliensis var. *major* Müll.Arg. === **Bernardia gambosa** Müll.Arg.
Bernardia brasiliensis var. *parvifolia* Müll.Arg. === **Bernardia pulchella** (Baill.) Müll.Arg.
Bernardia capitellata Müll.Arg. === **Bernardia celastrinea** (Baill.) Müll.Arg.
Bernardia carpinifolia Griseb. === **Bernardia dichotoma** (Willd.) Müll.Arg. var. **dichotoma**
Bernardia celastrinea var. *capitellata* (Müll.Arg.) Müll.Arg. === **Bernardia celastrinea** (Baill.) Müll.Arg.
Bernardia celastrinea var. *genuina* Müll.Arg. === **Bernardia celastrinea** (Baill.) Müll.Arg.
Bernardia celastrinea var. *intermedia* Müll.Arg. === **Bernardia celastrinea** (Baill.) Müll.Arg.
Bernardia celastrinea var. *obscura* Müll.Arg. === **Bernardia celastrinea** (Baill.) Müll.Arg.
Bernardia celastrinea var. *serratifolia* Müll.Arg. === **Bernardia celastrinea** (Baill.) Müll.Arg.
Bernardia confertifolia var. *lanceolata* Müll.Arg. === **Bernardia confertifolia** Müll.Arg.
Bernardia confertifolia var. *latifolia* Müll.Arg. === **Bernardia confertifolia** Müll.Arg.
Bernardia denticulata (Standl.) G.L.Webster === **Adenophaedra grandifolia** (Klotzsch) Müll.Arg.
Bernardia dichotoma var. *carpinifolia* (Griseb.) Pax & K.Hoffm. === **Bernardia dichotoma** (Willd.) Müll.Arg. var. **dichotoma**
Bernardia dichotoma var. *genuina* Müll.Arg. === **Bernardia dichotoma** (Willd.) Müll.Arg.
Bernardia dichotoma var. *macrocarpa* Pax & K.Hoffm. === **Bernardia dichotoma** (Willd.) Müll.Arg. var. **dichotoma**
Bernardia fasciculata S.Watson === **Tetracoccus fasciculatus** (S.Watson) Croizat
Bernardia grandifolia (Klotzsch) Müll.Arg. === **Adenophaedra grandifolia** (Klotzsch) Müll.Arg.

Bernardia guaranitica Chodat & Hassl. === **Bernardia caperoniifolia** (Baill.) Müll.Arg.
Bernardia hassleriana var. *ovata* Chodat & Hassl. === **Bernardia ovata** (Chodat & Hassl.) Pax & K.Hoffm.
Bernardia hassleriana var. *tobatyensis* Chodat & Hassl. === **Bernardia hassleriana** Chodat
Bernardia hirsutissima var. *genuina* Pax & K.Hoffm. === **Bernardia hirsutissima** (Baill.) Müll.Arg.
Bernardia hirsutissima var. *pubescens* (Müll.Arg.) Pax === **Bernardia hirsutissima** (Baill.) Müll.Arg.
Bernardia incana J.K.Morton === **Bernardia myricifolia** (Scheele) S.Watson
Bernardia intermedia Griseb. === **Bernardia dichotoma** var. **venosa** (Griseb.) Müll.Arg.
Bernardia interrupta (Schltdl.) Müll.Arg. === **Bernardia dodecandra** (Sessé ex Cav.) Govaerts
Bernardia leprosa (Willd.) Müll.Arg. === **Leucocroton leprosus** (Willd.) Pax & K.Hoffm.
Bernardia lorentzii var. *apaensis* (Chodat & Hassl.) Pax & K.Hoffm. === **Bernardia lorentzii** Müll.Arg.
Bernardia lorentzii var. *fistulosa* Pax & K.Hoffm. === **Bernardia lorentzii** Müll.Arg.
Bernardia lorentzii var. *obovata* Pax & K.Hoffm. === **Bernardia lorentzii** Müll.Arg.
Bernardia lorentzii var. *subintegra* (Chodat & Hassl.) Pax & K.Hoffm. === **Bernardia lorentzii** Müll.Arg.
Bernardia lycioides Müll.Arg. ex Pax & K.Hoffm. === **Leucocroton microphyllus** (A.Rich.) Pax & K.Hoffm.
Bernardia major (Müll.Arg.) Pax & K.Hoffm. === **Bernardia gambosa** Müll.Arg.
Bernardia megalophylla Müll.Arg. === **Adenophaedra megalophylla** (Müll.Arg.) Müll.Arg.
Bernardia mexicana var. *albida* Pax & K.Hoffm. === **Bernardia mexicana** (Hook. & Arn.) Müll.Arg.
Bernardia mexicana var. *cinerascens* Müll.Arg. === **Bernardia mexicana** (Hook. & Arn.) Müll.Arg.
Bernardia mexicana var. *genuina* Müll.Arg. === **Bernardia mexicana** (Hook. & Arn.) Müll.Arg.
Bernardia mexicana var. *subbiflora* Müll.Arg. === **Bernardia mexicana** (Hook. & Arn.) Müll.Arg.
Bernardia microphylla (A.Rich.) Müll.Arg. === **Leucocroton microphyllus** (A.Rich.) Pax & K.Hoffm.
Bernardia mollis var. *lanceifolia* Lundell === **Bernardia lanceifolia** (Lundell) Lundell
Bernardia myricifolia var. *incanoides* M.C.Johnst. === **Bernardia myricifolia** (Scheele) S.Watson
Bernardia oligandra Müll.Arg. === **Bernardia tamanduana** (Baill.) Müll.Arg.
Bernardia paraguariensis var. *fruticosa* Chodat & Hassl. === **Bernardia paraguariensis** Chodat & Hassl.
Bernardia paraguariensis f. *orbiculata* Chodat & Hassl. === **Bernardia paraguariensis** Chodat & Hassl.
Bernardia paraguariensis var. *parvifolia* Chodat & Hassl. === **Bernardia paraguariensis** Chodat & Hassl.
Bernardia paraguariensis f. *rhombifolia* Chodat & Hassl. === **Bernardia paraguariensis** Chodat & Hassl.
Bernardia peduncularis Müll.Arg. === **Bernardia hirsutissima** (Baill.) Müll.Arg.
Bernardia peduncularis var. *hirsutissima* (Baill.) Müll.Arg. === **Bernardia hirsutissima** (Baill.) Müll.Arg.
Bernardia peduncularis var. *longipedunculata* Chodat & Hassl. === **Bernardia longipedunculata** (Chodat & Hassl.) Pax & K.Hoffm.
Bernardia peduncularis var. *pubescens* Müll.Arg. === **Bernardia hirsutissima** (Baill.) Müll.Arg.
Bernardia peduncularis f. *spiciflora* Chodat & Hassl. === **Bernardia longipedunculata** (Chodat & Hassl.) Pax & K.Hoffm.
Bernardia peduncularis f. *subcapituliflora* Chodat & Hassl. === **Bernardia longipedunculata** (Chodat & Hassl.) Pax & K.Hoffm.
Bernardia polymorpha f. *angustifolia* Pax & K.Hoffm. === **Bernardia polymorpha** var. **setosa** Chodat & Hassl.

Bernardia polymorpha f. *elliptica* Pax & K.Hoffm. === **Bernardia polymorpha** var. **setosa** Chodat & Hassl.
Bernardia polymorpha f. *latifolia* Chodat & Hassl. === **Bernardia polymorpha** var. **curuguatensis** Chodat & Hassl.
Bernardia polymorpha f. *salicina* Chodat & Hassl. === **Bernardia polymorpha** var. **curuguatensis** Chodat & Hassl.
Bernardia pulchella f. *acuminata* Chodat & Hassl. === **Bernardia pulchella** (Baill.) Müll.Arg.
Bernardia pulchella f. *acutidens* Chodat & Hassl. === **Bernardia pulchella** (Baill.) Müll.Arg.
Bernardia pulchella f. *breviserrata* Chodat & Hassl. === **Bernardia pulchella** (Baill.) Müll.Arg.
Bernardia pulchella f. *genuina* Müll.Arg. === **Bernardia pulchella** (Baill.) Müll.Arg.
Bernardia pulchella var. *genuina* Chodat & Hassl. === **Bernardia pulchella** (Baill.) Müll.Arg.
Bernardia pulchella f. *lanceolata* (Müll.Arg.) Müll.Arg. === **Bernardia pulchella** (Baill.) Müll.Arg.
Bernardia pulchella f. *latifolia* Müll.Arg. === **Bernardia pulchella** (Baill.) Müll.Arg.
Bernardia pulchella var. *major* (Müll.Arg.) Müll.Arg. === **Bernardia gambosa** Müll.Arg.
Bernardia pulchella f. *parvifolia* (Müll.Arg.) Müll.Arg. === **Bernardia pulchella** (Baill.) Müll.Arg.
Bernardia rotundifolia Herzog === **Bernardia paraguariensis** Chodat & Hassl.
Bernardia scabra var. *brevipila* Pax & K.Hoffm. === **Bernardia scabra** Müll.Arg.
Bernardia scabra var. *longipila* Pax & K.Hoffm. === **Bernardia scabra** Müll.Arg.
Bernardia spartioides var. *glabrata* Müll.Arg. === **Bernardia spartioides** (Baill.) Müll.Arg.
Bernardia spartioides var. *pubescens* Müll.Arg. === **Bernardia spartioides** (Baill.) Müll.Arg.
Bernardia tamanduana var. *genuina* Müll.Arg. === **Bernardia tamanduana** (Baill.) Müll.Arg.
Bernardia tamanduana var. *major* (Müll.Arg.) Müll.Arg. === **Bernardia gambosa** Müll.Arg.
Bernardia tamanduana var. *parvifolia* (Müll.Arg.) Müll.Arg. === **Bernardia pulchella** (Baill.) Müll.Arg.
Bernardia tamanduana var. *pulchella* (Baill.) Müll.Arg. === **Bernardia pulchella** (Baill.) Müll.Arg.
Bernardia tenuifolia Urb. === **Adelia tenuifolia** (Urb.) Moscoso
Bernardia venosa Griseb. === **Bernardia dichotoma** var. **venosa** (Griseb.) Müll.Arg.

Bernhardia

Proposed by Post & Kuntze as an orthographic variant of *Bernardia*, but in fact a later homonym of *Bernhardia* Willd. ex Bernh. (Psilotaceae).

Bertya

25 species, Australia. The narrow cotyledons in these shrubs, many of them microphyllous, resulted it being placed in a subdivision known as 'Stenolobae' by Mueller and by Pax and Hoffmann; however, like other Australian genera of that group it appears instead to be specialised; in this case it is related to Crotoneae (Webster 1994). Known species were keyed out by Guymer (1988) who indicated that a complete new revision was required for a better understanding of the genus. (Crotonoideae)

- Gruening, G. (1913). *Bertya*. In A. Engler (ed.), Das Pflanzenreich, IV 147 [Stenolobieae] (Euphorbiaceae-Porantheroideae et Ricinocarpoideae): 49-63. Berlin. (Heft 58.) La/Ge. — 19 species in 2 sections. [Additions in ibid., XIV (Additamentum VI): 63 (1919).]

  Airy-Shaw, H. K. (1971). New or noteworthy Euphorbiaceae-Ricinocarpoideae from Western Australia. Kew Bull. 26: 67-71. En. — Additions to Gruening (1913), the greater part in *Bertya*.

  Guymer, G. (1988). Notes on *Bertya* Planchon (Euphorbiaceae). Austrobaileya 2: 427-431, illus. En. — Notes, one novelty (*B. sharpeana*) and key to known species (now 25).

Bertya Planch., London J. Bot. 4: 472 (1845).
Australia. 50.

Bertya brownii S.Moore, J. Bot. 43: 147 (1905).
E. New South Wales. 50 NSW. Nanophan.
Bertya astrotricha Blakeley, Contr. New South Wales Natl. Herb. 1: 120 (1941).

Bertya cunninghamii Planch., London J. Bot. 4: 473 (1845).
SE. Queensland, New South Wales, Victoria. 50 NSW QLD VIC. Nanophan.

Bertya cupressoidea (Grüning) Airy Shaw, Kew Bull. 26: 67 (1971).
Western Australia (Coolgardie). 50 WAU. Nanophan.
* *Bertya dimerostigma* var. *cupressoidea* Grüning in H.G.A.Engler, Pflanzenr., IV, 147: 62 (1913).

Bertya dimerostigma F.Muell., S. Sci. Rec. 2: 98 (1882). *Bertya dimerostigma* var. *genuina* Grüning in H.G.A.Engler, Pflanzenr., IV, 147: 62 (1913), nom. inval.
Western Australia (Austin, Coolgardie). 50 WAU. Nanophan.

Bertya findlayi F.Muell., Fragm. 8: 141 (1874).
SE. New South Wales, Victoria. 50 NSW VIC. Nanophan.

Bertya glabrescens (C.T.White) Guymer, Austrobaileya 2: 429 (1988).
Queensland (Burnett, Leichhardt). 50 QLD. Nanophan.
* *Bertya oleifolia* var. *glabrescens* C.T.White, Proc. Roy. Soc. Queensland 50: 86 (1939).

Bertya glandulosa Grüning in H.G.A.Engler, Pflanzenr., IV, 147: 59 (1913).
N. New South Wales. 50 NSW. Nanophan. – Provisionally accepted.

Bertya gummifera Planch., London J. Bot. 4: 473 (1845). *Bertya gummifera* var. *genuina* Müll.Arg., Flora 47: 471 (1864), nom. inval.
EC. New South Wales. 50 NSW. Nanophan.
Croton gummiferus A.Cunn. ex Planch., Hooker's J. Bot. Kew Gard. Misc. 4: 473 (1845).
Bertya polymorpha var. *mitchelliana* Baill., Adansonia 6: 299 (1866).

Bertya ingramii T.A.James, Telopea 3: 285 (1988).
New South Wales (Northern Tablelands). 50 NSW. Nanophan.

Bertya mitchellii (Sond.) Müll.Arg., Linnaea 34: 63 (1865).
Southern Australia, Victoria, New South Wales. 50 NSW SOA VIC. Nanophan.
* *Ricinocarpos mitchellii* Sond., Linnaea 28: 563 (1856). *Bertya mitchellii* var. *genuina* Grüning in H.G.A.Engler, Pflanzenr., IV, 147: 61 (1913), nom. inval.
Bertya mitchellii var. *vestita* Grüning in H.G.A.Engler, Pflanzenr., IV, 147: 61 (1913).

Bertya mollissima Blakeley, Contr. New South Wales Natl. Herb. 1: 120 (1941).
New South Wales. 50 NSW. Nanophan.

Bertya oblonga Blakeley, Proc. Linn. Soc. New South Wales, II, 54: 682 (1929).
New South Wales (C.-W. Slopes, N.-W. Plains). 50 NSW. Nanophan.

Bertya oblongifolia Müll.Arg., Flora 47: 471 (1864).
SE. New South Wales. 50 NSW. Nanophan.

Bertya oleifolia Planch., London J. Bot. 4: 473 (1845). *Bertya polymorpha* Baill., Adansonia 6: 298 (1866), nom. illeg. *Bertya polymorpha* var. *genuina* Baill., Adansonia 6: 299 (1966), nom. inval. – FIGURE, p. 259.
Queensland, New South Wales. 50 NSW QLD. Nanophan.

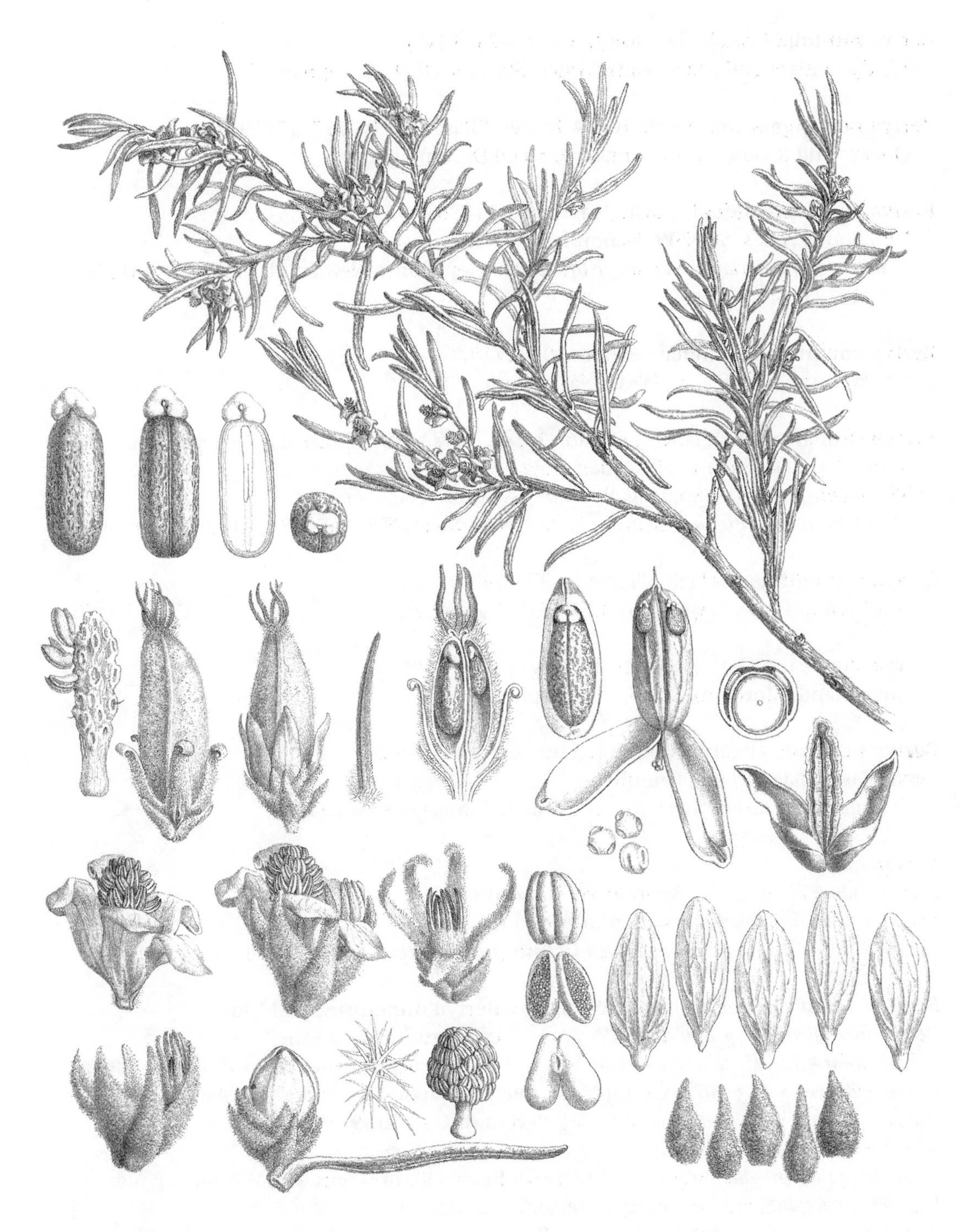

Bertya oleifolia Planch. (as *B. oleaefolia*)
Artist: F. Schoenfeld
F. Mueller, Pl. Victoria, [2]: pl. 20 (1864-65)
KEW ILLUSTRATIONS COLLECTION

Bertya opponens (F.Muell. ex Benth.) Guymer, Austrobaileya 2: 147 (1985).
Queensland (Leichhardt, Maranoa). 50 QLD. Nanophan.
* *Croton opponens* F.Muell. ex Benth., Fl. Austral. 6: 125 (1873).

Bertya oppositifolia F.Muell. & O'Shanesy ex F.Muell., S. Sci. Rec. 2: 98 (1882).
Queensland (Expedition Range). 50 QLD. Nanophan.

Bertya pedicellata F.Muell., Fragm. 4: 143 (1864).
Queensland (Leichhardt). 50 QLD. Nanophan.

Bertya pinifolia Planch., London J. Bot. 4: 473 (1845).
SE. Queensland, NE. New South Wales. 50 NSW QLD. Nanophan.

Bertya polystigma Grüning in H.G.A.Engler, Pflanzenr., IV, 147: 57 (1913).
Queensland (Cook, North Kennedy). 50 QLD. Nanophan.

Bertya pomaderrioides F.Muell., Fragm. 4: 34 (1863).
New South Wales. 50 NSW. Nanophan.
Bertya pomaderrioides var. *angustifolia* Blakeley, Contr. New South Wales Natl. Herb. 1: 121 (1941).

Bertya quadrisepala F.Muell., Fragm. 10: 52 (1876).
SW. Western Australia. 50 WAU. Nanophan.

Bertya rosmarinifolia Planch., London J. Bot. 4: 473 (1845). *Bertya polymorpha* var. *rosmarinifolia* (Planch.) Baill., Adansonia 6: 300 (1866).
SE. Queensland, New South Wales. 50 NSW QLD. Nanophan.
* *Croton rosmarinifolius* A.Cunn., Field Geogr. Mem. N.S.Wales: 355 (1825), nom. illeg.

Bertya rotundifolia F.Muell., Fragm. 4: 34 (1863).
Southern Australia (Kangaroo I.). 50 SOA. Nanophan.

Bertya sharpeana Guymer, Austrobaileya 2: 427 (1988).
Queensland (Moreton). 50 QLD. Nanophan.

Bertya tasmanica (Sond.) Müll.Arg., Linnaea 34: 63 (1865).
NE. Tasmania. 50 TAS. Nanophan.
* *Ricinocarpos tasmanicus* Sond. & F.Muell., Linnaea 28: 562 (1856).

Synonyms:
Bertya andrewsii Fitzg. === **Ricinocarpos stylosus** Diels
Bertya astrotricha Blakeley === **Bertya brownii** S.Moore
Bertya dimerostigma var. *cupressoidea* Grüning === **Bertya cupressoidea** (Grüning) Airy Shaw
Bertya dimerostigma var. *genuina* Grüning === **Bertya dimerostigma** F.Muell.
Bertya gummifera var. *genuina* Müll.Arg. === **Bertya gummifera** Planch.
Bertya gummifera var. *psiloclada* Müll.Arg. === **Ricinocarpos psilocladus** (Müll.Arg.) Benth.
Bertya mitchellii var. *genuina* Grüning === **Bertya mitchellii** (Sond.) Müll.Arg.
Bertya mitchellii var. *vestita* Grüning === **Bertya mitchellii** (Sond.) Müll.Arg.
Bertya neglecta Dummer === ?
Bertya oleifolia var. *glabrescens* C.T.White === **Bertya glabrescens** (C.T.White) Guymer
Bertya polymorpha Baill. === **Bertya oleifolia** Planch.
Bertya polymorpha var. *genuina* Baill. === **Bertya oleifolia** Planch.
Bertya polymorpha var. *mitchelliana* Baill. === **Bertya gummifera** Planch.
Bertya polymorpha var. *rosmarinifolia* (Planch.) Baill. === **Bertya rosmarinifolia** Planch.
Bertya pomaderrioides var. *angustifolia* Blakeley === **Bertya pomaderrioides** F.Muell.
Bertya psiloclada Müll.Arg. ex B.D.Jacks. === **Ricinocarpos psilocladus** (Müll.Arg.) Benth.

Bessera

Synonyms:
Bessera Spreng. === **Flueggea** Willd.
Bessera inermis Spreng. === **Flueggea virosa** (Roxb. ex Willd.) Voigt
Bessera spinosa Spreng. === **Drypetes alba** Poit.

Bestram

Synonyms:
Bestram Adans. === **Antidesma** L.

Beyeria

17 species, Australia. The narrow cotyledons in these shrubs or undershrubs resulted it being placed in a subdivision known as 'Stenolobae' by Mueller and by Pax and Hoffmann; however, like other Australian genera of that group it appears instead to be specialised, in this case related to *Ricinocarpos* and closest to Crotoneae (Webster 1994). (Crotonoideae)

- Gruening, G. (1913). *Beyeria.* In A. Engler (ed.), Das Pflanzenreich, IV 147 [Stenolobieae] (Euphorbiaceae-Porantheroideae et Ricinocarpoideae): 63-75. Berlin. (Heft 58.) La/Ge. — 12 species in 2 sections.

  Airy-Shaw, H. K. (1972). *Beyeria* — a correction. Kew Bull. 27: 566. En.

  Barker, W. R.; del. G. R. M. Dashorst (1984). Plant portraits, 10: *Beyeria.* J. Adelaide Bot. Gard. 7: 137-139, illus. En. — Line drawing of *B. subtecta* along with a description; includes key to 2 species. [This species is endemic to Kangaroo Island.]

Beyeria Miq., Ann. Sci. Nat., Bot., III, 1: 350 (1844).
Australia. 50.
Calyptrostigma Klotzsch in J.G.C.Lehmann, Pl. Preiss. 1: 175 (1845).
Beyeriopsis Müll.Arg., Linnaea 34: 56 (1865).
Clavipodium Desv. ex Grüning in H.G.A.Engler, Pflanzenr., IV, 147: 67 (1913).

Beyeria bickertonensis Specht, Rec. Amer.-Austral. Sci. Exped. Arnhem Land 3: 249 (1958).
Northern Territory (Darwin). 50 NTA. Nanophan.

Beyeria brevifolia (Müll.Arg.) Baill., Adansonia 6: 309 (1866).
S. Western Australia. 50 WAU. Nanophan.
* *Beyeriopsis brevifolia* Müll.Arg., Linnaea 34: 58 (1865).
Beyeria brevifolia var. *brevipes* Airy Shaw, Kew Bull. 27: 566 (1971).
Beyeria brevifolia var. *robustior* Airy Shaw, Kew Bull. 26: 69 (1971).
Beyeria brevifolia var. *truncata* Airy Shaw, Kew Bull. 26: 69 (1971).

Beyeria calycina Airy Shaw, Kew Bull. 26: 70 (1971).
Western Australia (Avon, Roe). 50 WAU. Nanophan.
Beyeria calycina var. *minor* Airy Shaw, Kew Bull. 26: 70 (1971).

Beyeria cinerea (Müll.Arg.) Baill., Adansonia 6: 309 (1866).
W. Western Australia (Carnarvon, Irwin, Drummond). 50WAU. Nanophan.
* *Beyeriopsis cinerea* Müll.Arg., Linnaea 34: 57 (1865).

Beyeria cyanescens (Müll.Arg.) Benth., Fl. Austral. 6: 66 (1873).
Western Australia (Dorre I., Lyndon). 50 WAU. Nanophan.
* *Beyeriopsis cyanescens* Müll.Arg. in A.P.de Candolle, Prodr. 15(2): 200 (1866).

Beyeria cygnorum (Müll.Arg.) Benth., Fl. Austral. 6: 66 (1873).
Western Australia (Drummond). 50 WAU. Nanophan.
* *Beyeriopsis cygnorum* Müll.Arg., Linnaea 34: 56 (1865).

Beyeria gardneri Airy Shaw, Kew Bull. 26: 68 (1971).
Western Australia (Irwin). 50 WAU. Nanophan.

Beyeria lasiocarpa (F.Muell.) Müll.Arg., Linnaea 34: 59 (1865).
Queensland, New South Wales, Victoria. 50 NSW QLD VIC. Nanophan. or phan.
* *Beyeria viscosa* var. *lasiocarpa* F.Muell., Fragm. 2: 182 (1861).
Beyeria lasiocarpa f. *denudata* Baill., Adansonia 6: 307 (1866).

Beyeria latifolia Baill., Adansonia 6: 304 (1866).
SW. Western Australia. 50 WAU. Nanophan.

Beyeria lepidopetala F.Muell., Fragm. 1: 230 (1859).
Western Australia (Murchison river). 50 WAU. Nanophan.

Beyeria leschenaultii (DC.) Baill., Adansonia 6: 307 (1866).
Western Australia, Southern Australia, Victoria, New South Wales, Tasmania. 50 NSW SOA TAS VIC WAU. Nanophan.
* *Hemistemma leschenaultii* DC., Syst. Nat. 1: 414 (1817). *Beyeria leschenaultii* f. *genuina* Baill., Adansonia 6: 307 (1866), nom. inval. *Beyeria leschenaultii* var. *genuina* (Baill.) Grüningin H.G.A.Engler, Pflanzenr., IV, 147: 70 (1913), nom. inval.
Calyptrostigma ledifolium Klotzsch in J.G.C.Lehmann, Pl. Preiss. 1: 176 (1845). *Beyeria ledifolia* (Klotzsch) Sond., Linnaea 28: 565 (1856). *Beyeria ledifolia* var. *genuina* Müll.Arg. in A.P.de Candolle, Prodr. 15(2): 203 (1866), nom. inval. *Beyeria leschenaultii* var. *ledifolia* (Klotzsch) Grüning in H.G.A.Engler, Pflanzenr., IV, 147: 70 (1913).
Beyeria backhousii Hook.f., Fl. Tasman. 1: 339 (1857). *Beyeria ledifolia* var. *backhousii* (Hook.f.) Müll.Arg. in A.P.de Candolle, Prodr. 15(2): 203 (1866). *Beyeria leschenaultii* var. *backhousii* (Hook.f.) Grüning in H.G.A.Engler, Pflanzenr., IV, 147: 70 (1913).
Beyeria drummondii F.Muell., Linnaea 34: 58 (1865). *Beyeria leschenaultii* var. *drummondii* (F.Muell.) Grüning in H.G.A.Engler, Pflanzenr., IV, 147: 70 (1913).
Beyeria ledifolia var. *angustifolia* Müll.Arg. in A.P.de Candolle, Prodr. 15(2): 203 (1866).
Beyeria leschenaultii f. *eloeagnoides* Baill., Adansonia 6: 308 (1866).
Beyeria leschenaultii f. *myrtoides* Baill., Adansonia 6: 308 (1866).
Beyeria leschenaultii f. *pernettioides* Baill., Adansonia 6: 307 (1866).
Beyeria leschenaultii f. *rosmarinoides* Baill., Adansonia 6: 308 (1866).
Beyeria leschenaultii f. *salsoloides* Baill., Adansonia 6: 308 (1866).
Beyeria leschenaultii f. *vaccinioides* Baill., Adansonia 6: 308 (1866).
Beyeria leschenaultii var. *latifolia* Grüning in H.G.A.Engler, Pflanzenr., IV, 147: 71 (1913).
Beyeria leschenaultii var. *rosmarinoides* (Baill.) Grüning in H.G.A.Engler, Pflanzenr., IV, 147: 70 (1913).
Beyeria leschenaultii var. *latifolia* Ewart, Fl. Victoria: 726 (1931), nom. illeg.

Beyeria opaca F.Muell., Trans. Philos. Soc. Victoria 1: 16 (1855). *Beyeria opaca* var. *typica* Grüning in H.G.A.Engler, Pflanzenr., IV, 147: 69 (1913), nom. inval.
Southern Australia, Victoria, New South Wales. 50 NSW SOA VIC. Nanophan.
Beyeria opaca var. *linearis* Benth., Fl. Austral. 6: 65 (1873).
Beyeria opaca var. *longifolia* Grüning in H.G.A.Engler, Pflanzenr., IV, 147: 69 (1913).
Beyeria opaca var. *latifolia* J.M.Black, Fl. S. Austral. 2: 357 (1924).

Beyeria similis (Müll.Arg.) Baill., Adansonia 6: 309 (1865).
Western Australia. 50 WAU. Nanophan.
* *Beyeriopsis similis* Müll.Arg., Linnaea 34: 58 (1865).

Beyeria subtecta J.M.Black, Fl. S. Austral. 2: 357 (1924).
Southern Australia (Kangaroo I.). 50 SOA. Nanophan.

Beyeria tristigma F.Muell., Fragm. 6: 181 (1868).
Northern Territory (Darwin), Queensland (Cook, North Kennedy). 50 NTA QLD. Nanophan.

Beyeria virgata Ewart, Proc. Roy. Soc. Victoria 33: 226 (1921).
Western Australia (near Lefroy). 50 WAU. Nanophan. – Provisionally accepted.

Beyeria viscosa (Labill.) Miq., Ann. Sci. Nat., Bot., III, 1: 350 (1844).
Western Australia, New South Wales, Victoria, Tasmania, Queensland. 50 NSW QLD TAS WAU. Nanophan. or phan.
* *Croton viscosus* Labill., Nov. Holl. Pl. 2: 72 (1806). *Beyeria viscosa* var. *genuina* Müll.Arg. in A.P.de Candolle, Prodr. 15(2): 202 (1866), nom. inval.
Calyptrostigma oblongifolium Klotzsch in J.G.C.Lehmann, Pl. Preiss. 1: 176 (1845). *Beyeria oblongifolia* (Klotzsch) Sond., Linnaea 28: 564 (1856). *Beyeria viscosa* var. *oblongifolia* (Klotzsch) Müll.Arg. in A.P.de Candolle, Prodr. 15(2): 202 (1866).
Beyeria viscosa var. *preissii* Sond., Linnaea 28: 564 (1856).
Beyeria viscosa var. *amoena* Müll.Arg. in A.P.de Candolle, Prodr. 15(2): 202 (1866).
Beyeria viscosa var. *minor* Müll.Arg. in A.P.de Candolle, Prodr. 15(2): 202 (1866).
Beyeria viscosa var. *latifolia* Benth., Fl. Austral. 6: 65 (1873).
Beyeria viscosa var. *obovata* C.T.White, Proc. Roy. Soc. Queensland 50: 86 (1939).

Synonyms:
Beyeria backhousii Hook.f. === **Beyeria leschenaultii** (DC.) Baill.
Beyeria brevifolia var. *brevipes* Airy Shaw === **Beyeria brevifolia** (Müll.Arg.) Baill.
Beyeria brevifolia var. *robustior* Airy Shaw === **Beyeria brevifolia** (Müll.Arg.) Baill.
Beyeria brevifolia var. *truncata* Airy Shaw === **Beyeria brevifolia** (Müll.Arg.) Baill.
Beyeria calycina var. *minor* Airy Shaw === **Beyeria calycina** Airy Shaw
Beyeria drummondii F.Muell. === **Beyeria leschenaultii** (DC.) Baill.
Beyeria lasiocarpa f. *denudata* Baill. === **Beyeria lasiocarpa** (F.Muell.) Müll.Arg.
Beyeria ledifolia (Klotzsch) Sond. === **Beyeria leschenaultii** (DC.) Baill.
Beyeria ledifolia var. *angustifolia* Müll.Arg. === **Beyeria leschenaultii** (DC.) Baill.
Beyeria ledifolia var. *backhousii* (Hook.f.) Müll.Arg. === **Beyeria leschenaultii** (DC.) Baill.
Beyeria ledifolia var. *genuina* Müll.Arg. === **Beyeria leschenaultii** (DC.) Baill.
Beyeria leschenaultii var. *backhousii* (Hook.f.) Grüning === **Beyeria leschenaultii** (DC.) Baill.
Beyeria leschenaultii var. *drummondii* (F.Muell.) Grüning === **Beyeria leschenaultii** (DC.) Baill.
Beyeria leschenaultii f. *eloeagnoides* Baill. === **Beyeria leschenaultii** (DC.) Baill.
Beyeria leschenaultii f. *genuina* Baill. === **Beyeria leschenaultii** (DC.) Baill.
Beyeria leschenaultii var. *genuina* (Baill.) Grüning === **Beyeria leschenaultii** (DC.) Baill.
Beyeria leschenaultii var. *latifolia* Grüning === **Beyeria leschenaultii** (DC.) Baill.
Beyeria leschenaultii var. *latifolia* Ewart === **Beyeria leschenaultii** (DC.) Baill.
Beyeria leschenaultii var. *ledifolia* (Klotzsch) Grüning === **Beyeria leschenaultii** (DC.) Baill.
Beyeria leschenaultii f. *myrtoides* Baill. === **Beyeria leschenaultii** (DC.) Baill.
Beyeria leschenaultii f. *pernettioides* Baill. === **Beyeria leschenaultii** (DC.) Baill.
Beyeria leschenaultii f. *rosmarinoides* Baill. === **Beyeria leschenaultii** (DC.) Baill.
Beyeria leschenaultii var. *rosmarinoides* (Baill.) Grüning === **Beyeria leschenaultii** (DC.) Baill.
Beyeria leschenaultii f. *salsoloides* Baill. === **Beyeria leschenaultii** (DC.) Baill.
Beyeria leschenaultii f. *vaccinioides* Baill. === **Beyeria leschenaultii** (DC.) Baill.
Beyeria loranthoides Baill. === **Drimys sp.** (Winteraceae)
Beyeria oblongifolia (Klotzsch) Sond. === **Beyeria viscosa** (Labill.) Miq.
Beyeria opaca var. *latifolia* J.M.Black === **Beyeria opaca** F.Muell.
Beyeria opaca var. *linearis* Benth. === **Beyeria opaca** F.Muell.
Beyeria opaca var. *longifolia* Grüning === **Beyeria opaca** F.Muell.
Beyeria opaca var. *typica* Grüning === **Beyeria opaca** F.Muell.
Beyeria uncinata Baill. === **Cryptandra uncinata** (Baill.) Grünig (Rhamnaceae)
Beyeria viscosa var. *amoena* Müll.Arg. === **Beyeria viscosa** (Labill.) Miq.
Beyeria viscosa var. *genuina* Müll.Arg. === **Beyeria viscosa** (Labill.) Miq.
Beyeria viscosa var. *lasiocarpa* F.Muell. === **Beyeria lasiocarpa** (F.Muell.) Müll.Arg.
Beyeria viscosa var. *latifolia* Benth. === **Beyeria viscosa** (Labill.) Miq.

Beyeria viscosa var. *minor* Müll.Arg. === **Beyeria viscosa** (Labill.) Miq.
Beyeria viscosa var. *oblongifolia* (Klotzsch) Müll.Arg. === **Beyeria viscosa** (Labill.) Miq.
Beyeria viscosa var. *obovata* C.T.White === **Beyeria viscosa** (Labill.) Miq.
Beyeria viscosa var. *preissii* Sond. === **Beyeria viscosa** (Labill.) Miq.

Beyeriopsis

Synonyms:
Beyeriopsis Müll.Arg. === **Beyeria** Miq.
Beyeriopsis brevifolia Müll.Arg. === **Beyeria brevifolia** (Müll.Arg.) Baill.
Beyeriopsis cinerea Müll.Arg. === **Beyeria cinerea** (Müll.Arg.) Baill.
Beyeriopsis cyanescens Müll.Arg. === **Beyeria cyanescens** (Müll.Arg.) Benth.
Beyeriopsis cygnorum Müll.Arg. === **Beyeria cygnorum** (Müll.Arg.) Benth.
Beyeriopsis similis Müll.Arg. === **Beyeria similis** (Müll.Arg.) Baill.

Bia

Synonyms:
Bia Klotzsch === **Tragia** Plum. ex L.
Bia alienata Didr. === **Tragia alienata** (Didr.) Múlgura & M.M.Gut.
Bia lessertiana Baill. === **Tragia lessertiana** (Baill.) Müll.Arg.
Bia sellowiana Klotzsch === **Tragia alienata** (Didr.) Múlgura & M.M.Gut.

Bischofia

2 species, Asia (India to China), Malesia, Pacific Islands, Australia; trees to 40m. Much information was summarised by Airy-Shaw (1967) following the earlier treatment of Pax & Hoffmann (1922). A more recent account will be found in PROSEA series 5 on timber trees. *B. javanica* is of very wide distribution in the East. The several varieties of Müller listed (but not really accepted) by Pax and Hoffmann are not accounted for here. There has long been controversy over whether or not the genus belongs in Euphorbiaceae; this was renewed in the 1960s after a long period of quiescence, when Airy-Shaw gave it family status and allied it with Staphyleaceae. Webster (Synopsis, 1994; see **General**), however, argues that embryology and leaf anatomy call for its retention in the family although with its trifoliolate leaves (which reportedly may have pinnate ancestry) it sits oddly, to put it mildly, in Phyllanthoideae. Pollen morphology also appears to be discordant. (Phyllanthoideae)

Pax, F. & K. Hoffmann (1922). *Bischofia*. In A. Engler (ed.), Das Pflanzenreich, IV 147 XV (Euphorbiaceae-Phyllanthoideae-Phyllantheae): 312-315, illus. Berlin. (Heft 81.) La/Ge. — 1 species; Asia (China to India), Malesia, Pacific Islands, Australia. [2 species now recognized, the second, *B. polycarpa*, in China.]

Airy-Shaw, H. K. (1967). Notes on the genus *Bischofia* Bl. (Bischofiaceae). Kew Bull. 21: 327-329. En. — Additional records; commentary (including the author's reasons for excluding the genus from Euphorbiaceae and establishing a separate family). [Among other points it was argued that the leaves fundamentally were pinnately compound. One suggested affinity was the Staphyleaceae.]

Bischofia Blume, Bijdr.: 1168 (1827).
Trop. & Subtrop. Asia, Trop. Australia, Pacific. 36 38 40 41 42 50 60 61 63?
Stylodiscus Benn., Pl. Jav. Rar.: 133 (1838).
Microelus Wight & Arn., Edinburgh New Philos. J. 14: 298 (Apr. 1833).

Bischofia javanica Blume, Bijdr.: 1168 (1827). *Bischofia javanica* var. *genuina* Müll.Arg. in A.P.de Candolle, Prodr. 15(2): 478 (1866), nom. inval. – FIGURE, p. 265.
Trop. & Subtrop. Asia to Cook Is. 36 CHC CHH CHS 38 TAI NNS 40 IND ASS 41 BMA CBD LAO THA VIE 42 BOR JAW LSI PHI SUL SUM NWG 50 QLD 60 NWC SAM 61 COO. Phan. – Known in timber trade as 'Java Cedar'.

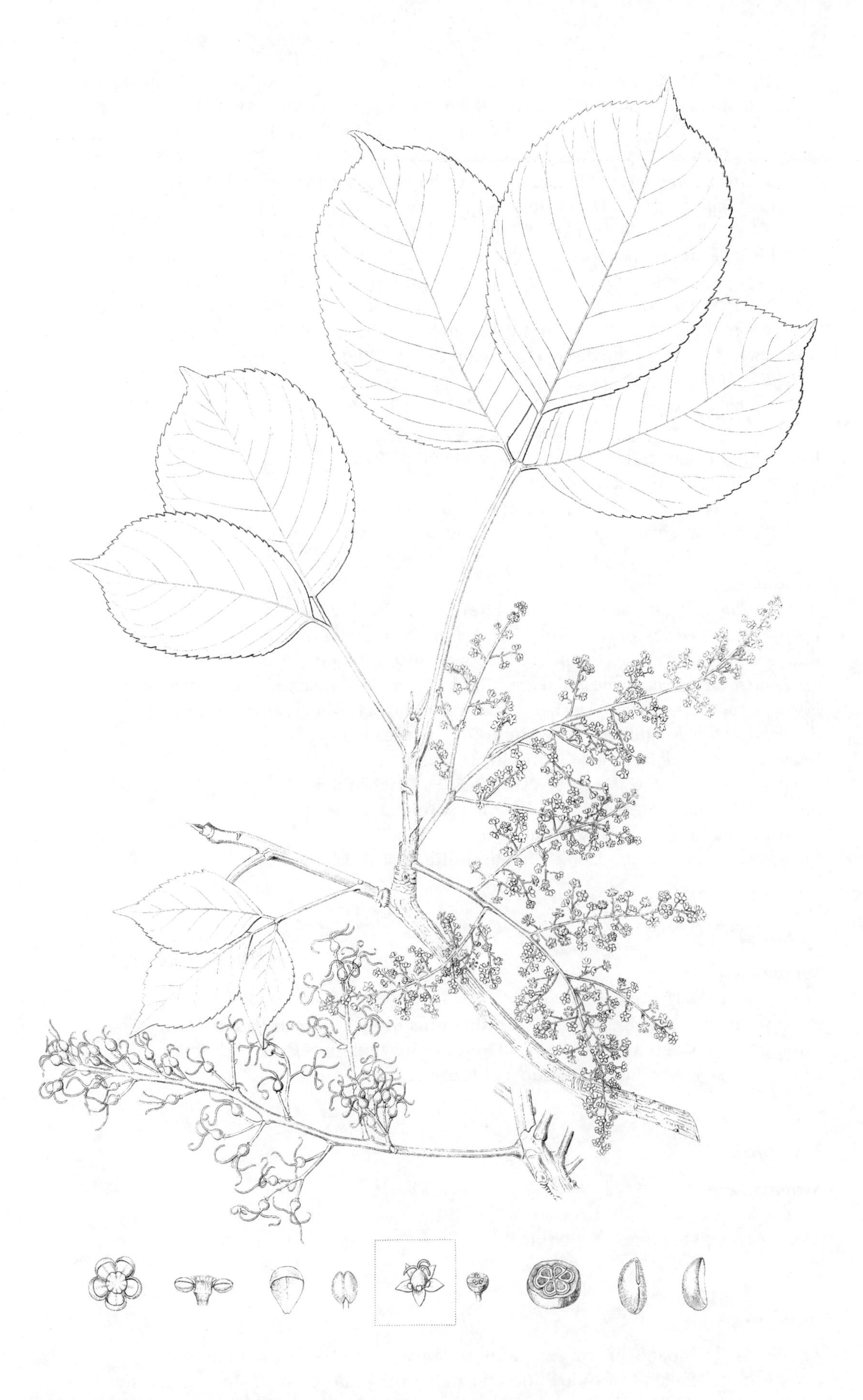

Bischofia javanica Blume (as *Stylodiscus trifoliatus* (Roxb.) Benn.)
Artist: J. & C. Curtis
Horsfield, Bennett & Brown, Pl. Jav. Rar.: pl. 29 (1840), uncoloured version
KEW ILLUSTRATIONS COLLECTION

Andrachne trifoliata Roxb., Fl. Ind. ed. 1832, 3: 728 (1832). *Stylodiscus trifoliatus* (Roxb.) Benn., Pl. Jav. Rar.: 133 (1838). *Bischofia trifoliata* (Roxb.) Hook., Hooker's Icon. Pl. 9: t. 844 (1851). *Bischofia javanica* var. *lanceolata* Müll.Arg. in A.P.de Candolle, Prodr. 15(2): 479 (1866).
Bischofia cummingiana Decne. in V.Jacquemont, Voy. Inde 4: 153 (1844).
Bischofia oblongifolia Decne. in V.Jacquemont, Voy. Inde 4: 152 (1844). *Bischofia javanica* var. *oblongifolia* (Decne.) Müll.Arg. in A.P.de Candolle, Prodr. 15(2): 479 (1866).
Bischofia roeperiana Decne. in V.Jacquemont, Voy. Inde 4: 153 (1844). *Microelus roeperianus* (Decne.) Wight & Arn., Edinburgh New Philos. J. 14: 298 (Apr. 1833).
Bischofia toui Decne. in V.Jacquemont, Voy. Inde 4: 153 (1844). *Bischofia javanica* var. *toui* (Decne.) Müll.Arg. in A.P.de Candolle, Prodr. 15(2): 478 (1866).
Andrachne apetala Roxb. ex Wall., Numer. List: 7956 (1847), nom. inval.
Phyllanthus gymnanthus Baill., Adansonia 2: 240 (1862).
Bischofia leptopoda Müll.Arg. in A.P.de Candolle, Prodr. 15(2): 479 (1866).

Bischofia polycarpa (H.Lév.) Airy Shaw, Kew Bull. 27: 271 (1972).
C. & S. China. 36 CHC CHS. Phan.
* *Celtis polycarpa* H.Lév., Repert. Spec. Nov. Regni Veg. 11: 296 (1912).
Bischofia racemosa W.C.Cheng & W.M.Chu, Sci. Silvae Sin. 8: 13 (1963).

Synonyms:
Bischofia cummingiana Decne. === **Bischofia javanica** Blume
Bischofia javanica var. *genuina* Müll.Arg. === **Bischofia javanica** Blume
Bischofia javanica var. *lanceolata* Müll.Arg. === **Bischofia javanica** Blume
Bischofia javanica var. *oblongifolia* (Decne.) Müll.Arg. === **Bischofia javanica** Blume
Bischofia javanica var. *toui* (Decne.) Müll.Arg. === **Bischofia javanica** Blume
Bischofia leptopoda Müll.Arg. === **Bischofia javanica** Blume
Bischofia oblongifolia Decne. === **Bischofia javanica** Blume
Bischofia racemosa W.C.Cheng & W.M.Chu === **Bischofia polycarpa** (H.Lév.) Airy Shaw
Bischofia roeperiana Decne. === **Bischofia javanica** Blume
Bischofia toui Decne. === **Bischofia javanica** Blume
Bischofia trifoliata (Roxb.) Hook. === **Bischofia javanica** Blume

Bivonea

Synonyms:
Bivonea Raf. === **Cnidoscolus** Pohl
Bivonea stimulosa (Michx.) Raf. === **Cnidoscolus urens** var. **stimulosus** (Michx.)
Bivonea texana (Müll.Arg.) House === **Cnidoscolus texanus** (Müll.Arg.) Small
Bivonea urens (L.) Arthur === **Cnidoscolus urens** (L.) Arthur

Bivonia

Synonyms:
Bivonia Spreng. === **Bernardia** Houst. ex Mill.
Bivonia axillaris Spreng. === **Bernardia axillaris** (Spreng.) Müll.Arg.

Blachia

11 species, S China, S and SE Asia, and Andaman Is. to the Philippines and Sulawesi (most diverse from Thailand to Indochina and Hainan); shrubs or small trees generally found in evergreen forest. *B. andamanica* is the only species extending beyond mainland Asia and nearby east Asian islands (notably Hainan). No full revision has appeared since that of Pax (1911), though Indian species were further revised in 1989. (Crotonoideae)

Pax, F. (with K. Hoffmann) (1911). *Blachia*. In A. Engler (ed.), Das Pflanzenreich, IV 147 III (Euphorbiaceae-Cluytieae): 36-39. Berlin. (Heft 47.) La/Ge. — 7 species, S and SE Asia and Andaman Is.

Airy-Shaw, H. K. (1969). Notes on Malesian and other Asiatic Euphorbiaceae, CXIV. *Blachia* Baill. in the Malay Peninsula and Celebes. Kew Bull. 23: 121-122. En. — Additional records, with extension of range of *B. andamanica* west to Assam and east to Sulawesi.

Balakrishnan, N. P. & T. Chakrabarty (1989). Genus *Blachia* Baill. (Euphorbiaceae) in India. Proc. Indian Acad. Sci. (Pl. Sci.) 99: 567-578. En. — General notes including history of the genus; national revision (3 species) with key, lengthy descriptions, synonymy, references and citations, types, indication of phenology, distribution and habitat, localities with exsiccatae, and commentary; list of literature at end.

Blachia Baill., Étude Euphorb.: 388 (1858).
S. China, Trop. Asia. 36 40 41 42.
Deonia Pierre ex Pax in H.G.A.Engler, Pflanzenr., IV, 147, III: 39 (1911).

Blachia andamanica (Kurz) Hook.f., Fl. Brit. India 5: 403 (1887).
W. & S. India, Andaman Is., Assam, Burma, Thailand, Pen. Malaysia, Philippines, Sulawesi. 40 ASS IND 41 AND BMA THA 42 MLY PHI SUL. Nanophan. or phan.
* *Codiaeum andamanicum* Kurz, J. Asiat. Soc. Bengal, Pt. 2, Nat. Hist. 42(2): 246 (1873).
Dimorphocalyx andamanicus (Kurz) Benth. in G.Bentham & J.D.Hooker, Gen. Pl. 3: 302 (1880).

subsp. **andamanica**
W. & S. India, Andaman Is., Assam, Burma, Thailand, Pen. Malaysia, Philippines, Sulawesi. 40 ASS IND 41 AND BMA THA 42 MLY PHI SUL. Nanophan. or phan.
Blachia philippinensis Merr., Philipp. J. Sci., C 4: 277 (1909).

subsp. **denudata** (Benth.) N.P.Balakr. & Chakrab., Proc. Indian Acad. Sci., Pl. Sci. 99: 571 (1989).
W. India. 40 IND. Nanophan. or phan.
Croton umbellatus Dalzell & Gibson, Bombay Fl.: 231 (1861), nom. illeg.
* *Blachia denudata* Benth., J. Linn. Soc., Bot. 17: 226 (1880).

Blachia calycina Benth., J. Linn. Soc., Bot. 17: 226 (1880).
India. 40 IND. Nanophan. or phan.

Blachia cotoneaster Gagnep., Bull. Soc. Bot. France 71: 619 (1924).
Laos. 41 LAO.

Blachia jatrophifolia Pax & K.Hoffm. in H.G.A.Engler, Pflanzenr., IV, 147, III: 39 (1911).
Vietnam. 41 VIE. Nanophan.

Blachia longzhouensis X.X.Chen, Acta Phytotax. Sin. 26: 76 (1988).
China (Guangxi). 36 CHS. Nanophan.

Blachia pentzii (Müll.Arg.) Benth., J. Linn. Soc., Bot. 17: 226 (1880).
China (S. Guangdong, Hainan), Vietnam. 36 CHH CHS 41 VIE. Nanophan. or phan.
* *Codiaeum pentzii* Müll.Arg. in A.P.de Candolle, Prodr. 15(2): 1118 (1866).

Blachia poilanei Gagnep., Bull. Soc. Bot. France 71: 619 (1924).
Vietnam. 41 VIE.

Blachia siamensis Gagnep., Bull. Soc. Bot. France 71: 620 (1924).
S. Thailand, Hainan. 36 CHH 41 THA. (Cl.) nanophan. or phan.
Blachia jatrophifolia var. *siamensis* Craib, Bull. Misc. Inform. Kew 1924: 98 (1924).

Blachia thorelii Gagnep., Bull. Soc. Bot. France 72: 461 (1925).
Laos. 41 LAO.

Blachia umbellata (Willd.) Baill., Étude Euphorb.: 387 (1858).
India, Sri Lanka. 40 IND SRL. Nanophan. or phan.
* *Croton umbellatus* Willd., Sp. Pl. 4: 545 (1805). *Codiaeum umbellatum* (Willd.) Müll.Arg. in A.P.de Candolle, Prodr. 15(2): 1118 (1866).
Blachia reflexa Benth., J. Linn. Soc., Bot. 17: 226 (1880).

Blachia yaihsienensis W.Xing & Z.X.Li, Bull. Bot. Res., Harbin 11: 57 (1991).
Hainan. 36 CHH.

Synonyms:
Blachia chunii Y.T.Chang & P.T.Li === ?
Blachia denudata Benth. === **Blachia andamanica** subsp. **denudata** (Benth.) N.P.Balakr. & Chakrab.
Blachia glandulosa Pierre ex Pax === **Strophioblachia fimbricalyx** Boerl.
Blachia jatrophifolia var. *siamensis* Craib === **Blachia siamensis** Gagnep.
Blachia philippinensis Merr. === **Blachia andamanica** (Kurz) Hook.f. subsp. **andamanica**
Blachia reflexa Benth. === **Blachia umbellata** (Willd.) Baill.
Blachia viridissima (Kurz) King === **Trigonostemon viridissimus** (Kurz) Airy Shaw

Bleekeria

Synonyms:
Bleekeria Miq. === **Alchornea** Sw.
Bleekeria zollingeri (Hassk.) Miq. === **Alchornea villosa** (Benth.) Müll.Arg.

Blotia

5 species, Madagascar; closely related to *Petalodiscus* and possibly not distinct (Webster, Synopsis 1994). Shrubs or small trees to 10 m with small, closely spaced, distichously arranged leaves, sometimes in limestone country or in coastal sands; flowers on long, very slender pedicels. Formerly included with *Savia* in its sect. *Maschalanthus* (Pax & Hoffmann 1922). The genus has had two revisions since that of Pax and Hoffmann: a flora treatment by Leandri (1958) and a full study by Hoffmann and McPherson (1998). (Phyllanthoideae)

Pax, F. & K. Hoffmann (1922). *Savia*. In A. Engler (ed.), Das Pflanzenreich, IV 147 XV (Euphorbiaceae-Phyllanthoideae-Phyllantheae): 181-188. Berlin. (Heft 81.) La/Ge. — Includes 2 species (in sect. *Maschalanthus*) now referable to *Blotia* (nos. 14 and 17).

Leandri, J. (1958). *Blotia*. Fl. Madag. Comores 111 (Euphorbiacées), I: 126-135. Paris. Fr. — Flora treatment, with key; 5 species accepted.

- Hoffmann, P. & G. McPherson (1998). Revision of the genus *Blotia* (Euphorbiaceae-Phyllanthoideae). Adansonia, III, 20: 247-261, illus., maps. — Brief introduction; treatment of 5 species with key, descriptions, synonymy, references, literature citations, vernacular names, indication of distribution and habitat, references to figures, commentary, and localities with exsiccatae; short bibliography at end. [One new combination.]

Blotia Leandri, Mém. Inst. Sci. Madagascar, Sér. B, Biol. Vég. 8: 240 (1957).
Madagascar. 29.
Charidia Baill., Étude Euphorb.: 572 (1858).

Blotia bemarensis (Leandri) Leandri, Mém. Inst. Sci. Madagascar, Sér. B, Biol. Vég. 8: 242 (1957).
W. & C. Madagascar. 29 MDG. Nanophan. or phan.
* *Savia bemarensis* Leandri, Bull. Soc. Bot. France 81: 588 (1934).
Blotia ankaranae Leandri, Mém. Inst. Sci. Madagascar, Sér. B, Biol. Vég. 8: 242 (1957).
Blotia ankaranae var. *sambiranensis* Leandri, Mém. Inst. Sci. Madagascar, Sér. B, Biol. Vég. 8: 242 (1957).

Blotia leandriana Petra Hoffm. & McPherson, Novon 7: 249 (1997).
E. Madagascar. 29 MDG. Nanophan. or phan.

Blotia mimosoides (Baill.) Petra Hoffm. & McPherson, Adansonia, III, 20: 253 (1998). *Savia mimosoides Baill., Adansonia, 2: 34 (1861). Petalodiscus mimosoides (Baill.) Pax (1890).
C. & E. Madagascar. 29 MDG. Nanophan. or phan.
Savia hildebrandtii Baill. in A.Grandidier, Hist. Phys. Madagascar, Atlas: 209 (1892). *Blotia hildebrandtii* (Baill.) Leandri, Mém. Inst. Sci. Madagascar, Sér. B, Biol. Vég. 8: 242 (1957).
Savia maroando Danguy, Notul. Syst. (Paris) 5: 1 (1935).

Blotia oblongifolia (Baill.) Leandri, Mém. Inst. Sci. Madagascar, Sér. B, Biol. Vég. 8: 240 (1957).
C. & E. Madagascar. 29 MDG. Nanophan. or phan.
* *Wielandia oblongifolia* Baill., Étude Euphorb.: 569 (1858). *Savia oblongifolia* (Baill.) Baill., Adansonia 2: 35 (1861). *Petalodiscus oblongifolius* (Baill.) Pax in H.G.A.Engler & K.A.E.Prantl, Nat. Pflanzenfam. 3(5): 15 (1890).
Savia decaryi Leandri, Bull. Soc. Bot. France 81: 589 (1934).

Blotia tanalorum Leandri, Mém. Inst. Sci. Madagascar, Sér. B, Biol. Vég. 8: 243 (1957).
C. Madagascar. 29 MDG. Phan.

Synonyms:
Blotia hildebrandtii (Baill.) Leandrii === **Blotia mimosoides** (Baill.) Petra Hoffmn. & McPherson
Blotia oblongifolia var. *louvelii* Leandri === **Savia danguyana** Leandri

Blumeodendron

5 species, Andaman Islands, SE Asia and Malesia (most diverse in Borneo). *BB. kurzii* and *tokbrai* are widely distributed, serious forest trees with long-petioled entire leaves; the latter is up to 40 m in height. The genus has been under study by Balu Perumal as part of the Rijksherbarium/Hortus Botanicus (Leiden) project on Malesian Euphorbiaceae. When published in *Flora Malesiana* and elsewhere the results will succeed the revision of Pax (1914; additions in 1919), now well out of date. (Acalyphoideae)

Pax, F. (with K. Hoffmann) (1914). *Blumeodendron*. In A. Engler (ed.), Das Pflanzenreich, IV 147 VII (Euphorbiaceae-Acalypheae-Mercurialinae): 47-49. Berlin. (Heft 63.) La/Ge. — 3 species, Malesia (*B. subrotundifolium* uncertain). Additions in ibid., XIV (Additamentum VI): 14 (1919).

Airy-Shaw, H. K. (1963). Notes on Malaysian and other Asiatic Euphorbiaceae, XXX. An unexpected synonym in *Blumeodendron* (Muell. Arg.) Kurz. Kew Bull. 16: 348-349. En. — Two species, formerly misplaced in Flacourtiaceae, reduced to *Blumeodendron papuanum*.

Airy-Shaw, H. K. (1965). Notes on Malaysian and other Asiatic Euphorbiaceae, LII. New species of *Blumeodendron* (Muell. Arg.) Kurz. Kew Bull. 19: 309-311. En. — Two new species described.

Airy-Shaw, H. K. (1971). Notes on Malesian and other Asiatic Euphorbiaceae, CXXXV. The male flower of *Blumeodendron calophyllum*. Kew Bull. 25: 518-519. En. — Amplification.

Airy-Shaw, H. K. (1972). Notes on Malesian and other Asiatic Euphorbiaceae, CLXI: Note on *Blumeodendron bullatum*. Kew Bull. 27: 86. En. — An indication of a frequent problem: a single replicate is not always enough to convey an adequate impression of a given collection. The leaves in the isotype cited were almost 50% larger than those in the holotype.

Blumeodendron (Müll.Arg.) Kurz, J. Asiat. Soc. Bengal, Pt. 2, Nat. Hist. 42(2): 245 (1873).
Indo-China, Malesia. 41 42.

Blumeodendron bullatum Airy Shaw, Kew Bull. 19: 310 (1965).
Borneo (Sarawak). 42 BOR. Phan.

Blumeodendron calophyllum Airy Shaw, Kew Bull. 19: 309 (1965).
Pen. Malaysia, Borneo (Brunei, Sarawak). 42 BOR MLY. Phan.

Blumeodendron concolor Gage, Rec. Bot. Surv. India 9: 244 (1922).
Pen. Malaysia (Perak, Lankawi Is.), Borneo (Sabah, N. Kalimantan). 42 BOR MLY. Phan.

Blumeodendron kurzii (Hook.f.) J.J.Sm. ex Koord. & Valeton, Meded. Dept. Landb. Ned.-Indië 10: 463 (1910).
Andaman Is., Burma, S. Thailand, Pen. Malaysia, Sumatera, Jawa, Borneo, Philippines, Irian Jaya. 41 AND BMA THA 42 BOR JAW MLY NWG PHI SUM. Nanophan. or phan.
* *Mallotus kurzii* Hook.f., Fl. Brit. India 5: 427 (1888).
Sapium subrotundifolium Elmer, Leafl. Philipp. Bot. 3: 930 (1910). *Blumeodendron subrotundifolium* (Elmer) Merr., Philipp. J. Sci., C 7: 384 (1912 publ. 1913).
Blumeodendron philippinense Merr. & Rolfe, Philipp. J. Sci. 16: 555 (1920).
Blumeodendron subcaudatum Merr., Philipp. J. Sci. 16: 557 (1920).
Blumeodendron verticillatum Merr., Philipp. J. Sci. 16: 557 (1920).
Blumeodendron cuneatum S.Moore, J. Bot. 63(Suppl.): 103 (1925).
Blumeodendron sumatranum S.Moore, J. Bot. 63(Suppl.): 102 (1925).

Blumeodendron tokbrai (Blume) Kurz, J. Asiat. Soc. Bengal, Pt. 2, Nat. Hist. 42(2): 245 (1873).
Malesia. 42 BIS BOR JAW MLY MOL NWG PHI SUL SUM. Phan.
* *Elateriospermum tokbrai* Blume, Bijdr.: 621 (1826). *Mallotus tokbrai* (Blume) Müll.Arg. in A.P.de Candolle, Prodr. 15(2): 956 (1866). *Blumeodendron elateriospermum* J.J.Sm., Bull. Jard. Bot. Buitenzorg, II, 8: 56 (1912).

var. **borneense** (Pax & K.Hoffm.) J.J.Sm. ex Airy Shaw, Kew Bull. 36: 269 (1981).
Sumatera, Borneo (Sarawak, Sabah). 42 BOR SUM. Phan.
* *Blumeodendron borneense* Pax & K.Hoffm. in H.G.A.Engler, Pflanzenr., IV, 147, XIV: 14 (1919).

var. **tokbrai**
Pen. Malaysia, Jawa, Borneo, Philippines, Maluku, New Guinea, Bismarck Archip. 42 BIS BOR JAW MLY MOL NWG PHI SUL. Phan.
Mallotus vernicosus Hook.f., Fl. Brit. India 5: 443 (1887). *Blumeodendron vernicosum* (Hook.f.) Gage, Rec. Bot. Surv. India 9: 244 (1922).
Elateriospermum paucinervium Elmer, Leafl. Philipp. Bot. 2: 484 (1908). *Blumeodendron paucinervium* (Elmer) Merr., Philipp. J. Sci. 16: 555 (1920).
Blumeodendron papuanum Pax & K.Hoffm. in H.G.A.Engler, Pflanzenr., IV, 147, XIV: 14 (1919).

Synonyms:

Blumeodendron borneense Pax & K.Hoffm. === **Blumeodendron tokbrai** var. **borneense** (Pax & K.Hoffm.) J.J.Sm. ex Airy Shaw
Blumeodendron cuneatum S.Moore === **Blumeodendron kurzii** (Hook.f.) J.J.Sm. ex Koord. & Valeton

Blumeodendron elateriospermum J.J.Sm. === **Blumeodendron tokbrai** (Blume) Kurz
Blumeodendron muelleri Kurz === **Paracroton pendulus** (Hassk.) Miq. subsp. **pendulus**
Blumeodendron papuanum Pax & K.Hoffm. === **Blumeodendron tokbrai** (Blume) Kurz var. **tokbrai**
Blumeodendron paucinervium (Elmer) Merr. === **Blumeodendron tokbrai** (Blume) Kurz var. **tokbrai**
Blumeodendron philippinense Merr. & Rolfe === **Blumeodendron kurzii** (Hook.f.) J.J.Sm. ex Koord. & Valeton
Blumeodendron subcaudatum Merr. === **Blumeodendron kurzii** (Hook.f.) J.J.Sm. ex Koord. & Valeton
Blumeodendron subrotundifolium (Elmer) Merr. === **Blumeodendron kurzii** (Hook.f.) J.J.Sm. ex Koord. & Valeton
Blumeodendron sumatranum S.Moore === **Blumeodendron kurzii** (Hook.f.) J.J.Sm. ex Koord. & Valeton
Blumeodendron vernicosum (Hook.f.) Gage === **Blumeodendron tokbrai** (Blume) Kurz var. **tokbrai**
Blumeodendron verticillatum Merr. === **Blumeodendron kurzii** (Hook.f.) J.J.Sm. ex Koord. & Valeton

Bocquillonia

14 species, New Caledonia; shrubs or small trees to 15 m, sometimes few-branched, in forest or maquis. Fully revised by McPherson & Tirel (1987). Related to *Alchornea*. (Acalyphoideae)

Pax, F. (with K. Hoffmann) (1914). *Bocquillonia*. In A. Engler (ed.), Das Pflanzenreich, IV 147 VII (Euphorbiaceae-Acalypheae-Mercurialinae): 260-261. Berlin. (Heft 63.) La/Ge. — 5 species, New Caledonia. [Genus now also includes *Ramelia*.]

Pax, F. & K. Hoffmann (1919). *Ramelia*. In A. Engler (ed.), Das Pflanzenreich, IV 147 IX (Euphorbiaceae-Acalypheae-Plukenetiinae): 107-108. Berlin. (Heft 68.) La/Ge. — 1 species, New Caledonia. [Now reduced to *Bocquillonia*.]

Airy-Shaw, H. K. (1968). Notes on Malesian and other Asiatic Euphorbiaceae, XCII. Note on the affinity of *Ramelia* Baill. Kew Bull. 21: 401-403. En. — Extensive discussion of *R. codonostylis* (fortunately now with more collections!); closest to *Bocquillonia* (to which it is now reduced). [Fuller treatment in McPherson and Tirel (1987) under *Bocquillonia*.]

Airy-Shaw, H. K. (1972). Notes on Malesian and other Asiatic Euphorbiaceae, CLXIV: A misplaced species of *Bocquillonia*. Kew Bull. 27: 88-89. En. — Includes revised key; *B. rhomboidea* formerly in *Excoecaria*.

Airy-Shaw, H. K. (1974). Notes on Malesian and other Asiatic Euphorbiaceae, CLXXXI. New or noteworthy species of *Bocquillonia* Baill., with a revised key to the genus. Kew Bull. 29: 315-322. En. — Novelties, new combinations and notes; revised key covering 11 species.

Airy-Shaw, H. K. (1978). Notes on Malesian and other Asiatic Euphorbiaceae, CCI. A new variety in *Bocquillonia* Baill. Kew Bull. 32: 408. En. — Description of var. *trichogyne* (eastern New Caledonia).

Airy-Shaw, H. K. (1978). Notes on Malesian and other Asiatic Euphorbiaceae, CCXVII. *Bocquillonia* Baill. Kew Bull. 33: 66-67. En. — Description of *B. schistophila* from New Caledonia. [Now combined with *B. nervosa*; see McPherson and Tirel (1987).]

Airy-Shaw, H. K. (1980). Notes on Euphorbiaceae from Indomalesia, Australia and the Pacific, CCXL. *Bocquillonia* Baill. Kew Bull. 35: 396-398. En. — Descriptions of *BB. spinuligera* and *arborea* from New Caledonia, the latter from the Isle of Pines southeast of the mainland. The latter is a tree to 15 m.

Airy-Shaw, H. K. (1981). Notes on Asiatic, Malesian and Melanesian Euphorbiaceae, CCXLVII. *Bocquillonia* Baill. Kew Bull. 36: 607-608. En. — Description of *B. goniorrhachis* from New Caledonia.

• McPherson, G. & C. Tirel (1987). *Bocquillonia*. Fl. Nouvelle-Calédonie, 14 (Euphorbiacées, I): 114-143. Paris. Fr. — Flora treatment (14 species) with key.

Bocquillonia Baill., Adansonia 2: 225 (1862).
New Caledonia. 60.
Ramelia Baill., Adansonia 11: 132 (1874).

Bocquillonia arborea Airy Shaw, Kew Bull. 35: 397 (1980).
New Caledonia (Ile des Pins). 60 NWC. Phan.

Bocquillonia brachypoda Baill., Adansonia 11: 127 (1874).
New Caledonia (incl. Ile Art, Ile des Pins). 60 NWC. Nanophan.
Bocquillonia brachypoda var. *spathulata* Airy Shaw, Kew Bull. 29: 320 (1974).

Bocquillonia brevipes Müll.Arg., Linnaea 34: 166 (1865).
NW. & NC. New Caledonia. 60 NWC. Nanophan.
Bocquillonia brevipes var. *trichogyne* Airy Shaw, Kew Bull. 32: 408 (1978).

Bocquillonia castaneifolia Guillaumin, Ann. Soc. Bot. Lyon 38: 36 (1913 publ. 1914).
NW. New Caledonia (incl. Ile Art). 60 NWC. Nanophan.
Bocquillonia spinuligera Airy Shaw, Kew Bull. 35: 396 (1980).

Bocquillonia codonostylis (Baill.) Airy Shaw, Kew Bull. 29: 319 (1974).
NW. New Caledonia (Mt. Panié Reg.). 60 NWC. Nanophan. or phan.
* *Ramelia codonostylis* Baill., Adansonia 11: 132 (1874).
Cleidion platystigma Schltr., Bot. Jahrb. Syst. 39: 150 (1906).

Bocquillonia goniorrhachis Airy Shaw, Kew Bull. 36: 607 (1981).
NC. New Caledonia. 60 NWC. Nanophan.

Bocquillonia grandidens Baill., Adansonia 11: 128 (1874).
C. & SC. New Caledonia. 60 NWC. Nanophan. or phan.

Bocquillonia longipes McPherson, in Fl. N. Caléd. Depend. 14: 128 (1987).
NW. & WC. New Caledonia. 60 NWC. Nanophan. or phan.

Bocquillonia lucidula Airy Shaw, Kew Bull. 29: 315 (1974).
C. New Caledonia. 60 NWC. Nanophan. or phan.

Bocquillonia nervosa Airy Shaw, Kew Bull. 29: 317 (1974).
NW. New Caledonia. 60 NWC. Nanophan. or phan.
Bocquillonia schistophila Airy Shaw, Kew Bull. 33: 66 (1978).

Bocquillonia phenacostigma Airy Shaw, Kew Bull. 29: 318 (1974).
C. New Caledonia (Mt. Aoupinié). 60 NWC. Nanophan.

Bocquillonia rhomboidea (Schltr.) Airy Shaw, Kew Bull. 27: 88 (1972).
SE. New Caledonia. 60 NWC. Nanophan.
* *Excoecaria rhomboidea* Schltr., Bot. Jahrb. Syst. 39: 153 (1906).
Excoecaria unequidentata Baill. ex Guillaumin, Ann. Inst. Bot.-Géol. Colon. Marseille, II, 9: 229 (1911), nom. illeg.
Bocquillonia rhomboidea var. *bullata* Airy Shaw, Kew Bull. 29: 320 (1974).

Bocquillonia sessiliflora Baill., Adansonia 2: 226 (1862).
New Caledonia (incl. Loyalty Is.). 60 NWC. Nanophan. or phan.

Bocquillonia spicata Baill., Adansonia 2: 227 (1862).
NW. & SE. New Caledonia. 60 NWC. Nanophan. or phan.

Synonyms:
Bocquillonia brachypoda var. *spathulata* Airy Shaw === **Bocquillonia brachypoda** Baill.
Bocquillonia brevipes var. *trichogyne* Airy Shaw === **Bocquillonia brevipes** Müll.Arg.
Bocquillonia rhomboidea var. *bullata* Airy Shaw === **Bocquillonia rhomboidea** (Schltr.) Airy Shaw
Bocquillonia schistophila Airy Shaw === **Bocquillonia nervosa** Airy Shaw
Bocquillonia spinuligera Airy Shaw === **Bocquillonia castaneifolia** Guillaumin

Bojeria

of Rafinesque; a later homonym of *Bojeria* DC. (Compositae). = *Euphorbia* L.

Bonania

7 species, Caribbean (Bahamas, Cuba, Hispaniola); includes *Hypocoton*. Shrubs with a box-like habit; most nearly related to *Grimmeodendron* but featuring axillary rather than terminal inflorescences. (Euphorbioideae (except Euphorbieae))

Pax, F. (with K. Hoffmann) (1912). *Bonania*. In A. Engler (ed.), Das Pflanzenreich, IV 147 V (Euphorbiaceae-Hippomaneae): 259-261. Berlin. (Heft 52.) La/Ge. — 3 species, Cuba.
Pax, F. (with K. Hoffmann) (1914). *Hypocoton*. In A. Engler (ed.), Das Pflanzenreich, IV 147 VII [Euphorbiaceae-Additamentum V]: 423-424. Berlin. (Heft 63.) La/Ge. — 1 species, *H. domingensis* (since placed in *Bonania*).
Borhidi, A. & O. Muñiz (1976(1977)). Plantas nuevas en Cuba, V. Acta Bot. Acad. Sci. Hungar. 22: 295-320. Sp. — Includes novelties and notes in *Bonania*. [Not seen.]
Borhidi, A. (1983). New names and new species in the flora of Cuba and Antilles, III. Acta Bot. Acad. Sci. Hungar. 29: 181-215, illus. En. —*Bonania*, pp. 183-184; reductions of some species to infraspecific rank.

Bonania A.Rich. in R.de la Sagra, Hist. Fis. Cuba, Bot. 11: 201 (1850).
Caribbean. 81.
Hypocoton Urb., Symb. Antill. 7: 263 (1912).

Bonania cubana A.Rich. in R.de la Sagra, Hist. Fis. Cuba, Bot. 11: 201 (1850). *Stillingia cubana* (A.Rich.) Baill., Étude Euphorb.: 515 (1858). *Excoecaria cubensis* (A.Rich.) Müll.Arg., Linnaea 32: 122 (1863).
Bahamas, Cuba. 81 BAH CUB. Nanophan.

subsp. **acunae** (Borhidi) Borhidi, Acta Bot. Hung. 29: 183 (1983).
Cuba. 81 CUB. Nanophan.
* *Bonania acunae* Borhidi, Acta Bot. Acad. Sci. Hung. 22: 306 (1976 publ. 1977).

subsp. **cubana**
Bahamas, Cuba. 81 BAH CUB. Nanophan.
Bonania cubensis A.Rich. in R.de la Sagra, Hist. Fis. Cuba, Bot. 11: 201 (1850).

subsp. **microphylla** (Urb.) Borhidi, Acta Bot. Hung. 29: 183 (1983).
E. Cuba. 81 CUB. Nanophan.
* *Bonania microphylla* Urb., Symb. Antill. 3: 311 (1902).

Bonania domingensis (Urb.) Urb., Symb. Antill. 9: 215 (1924).
Hispaniola. 81 DOM HAI. Nanophan.
* *Hypocoton domingensis* Urb., Symb. Antill. 7: 264 (1912).

Bonania elliptica Urb., Symb. Antill. 9: 214 (1924).
S. & E. Cuba. 81 CUB. Nanophan.

var. **elliptica**
S. & E. Cuba. 81 CUB. Nanophan.

var. **spinosa** (Urb.) Borhidi, Acta Bot. Hung. 29: 183 (1983).
E. Cuba. 81 CUB. Nanophan.
* *Bonania spinosa* Urb., Symb. Antill. 9: 215 (1924).

Bonania emarginata C.Wright ex Griseb., Nachr. Ges. Wiss. Göttingen Jahresber. 1865: 78 (1865). *Excoecaria emarginata* (C.Wright ex Griseb.) Müll.Arg. in A.P.de Candolle, Prodr. 15(2): 1212 (1866).
Cuba. 81 CUB. Nanophan.

subsp. **emarginata**
W. Cuba. 81 CUB. Nanophan.

subsp. **nipensis** (Urb. & Ekman) Borhidi, Acta Bot. Hung. 29: 183 (1983).
E. Cuba. 81 CUB. Nanophan.
* *Bonania nipensis* Urb. & Ekman, Repert. Spec. Nov. Regni Veg. 28: 231 (1930).

subsp. **suborbiculata** (Borhidi & Urbino) Borhidi, Acta Bot. Hung. 29: 183 (1983).
E. Cuba. 81 CUB. Nanophan.
* *Bonania suborbiculata* Borhidi & Urbino, Acta Bot. Acad. Sci. Hung. 22: 305 (1976 publ. 1977).

Bonania erythrosperma (Griseb.) Benth. & Hook.f. ex B.D.Jacks., Index Kew. 1: 321 (1895).
Cuba. 81 CUB.
* *Excoecaria erythrosperma* Griseb., Mem. Amer. Acad. Arts, n.s. 8: 161 (1861). *Sapium erythrospermum* (Griseb.) Müll.Arg., Linnaea 32: 119 (1863).

Bonania linearifolia Urb. & Ekman, Ark. Bot. 22A(8): 63 (1929).
Haiti. 81 HAI. Nanophan.

Bonania myricifolia (Griseb.) Benth. & Hook.f., Gen. Pl. 3: 335 (1880).
Cuba (Guantánamo). 81 CUB. Nanophan.
* *Excoecaria myricifolia* Griseb., Nachr. Königl. Ges. Wiss. Georg-Augusts-Univ. 1: 178 (1865). *Sebastiania myricifolia* (Griseb.) C.Wright, Anales Acad. Ci. Méd. Habana 7: 156 (1870).
Sapium myricifolium C.Wright ex Griseb., Nachr. Königl. Ges. Wiss. Georg-Augusts-Univ. 1: 178 (1865).

Synonyms:
Bonania acunae Borhidi === **Bonania cubana** subsp. **acunae** (Borhidi) Borhidi
Bonania adenodon (Griseb.) Benth. & Hook.f. === **Sapium adenodon** Griseb.
Bonania cubensis A.Rich. === **Bonania cubana** A.Rich. subsp. **cubana**
Bonania microphylla Urb. === **Bonania cubana** subsp. **microphylla** (Urb.) Borhidi
Bonania nipensis Urb. & Ekman === **Bonania emarginata** subsp. **nipensis** (Urb. & Ekman) Borhidi
Bonania spinosa Urb. === **Bonania elliptica** var. **spinosa** (Urb.) Borhidi
Bonania suborbiculata Borhidi & Urbino === **Bonania emarginata** subsp. **suborbiculata** (Borhidi & Urbino) Borhidi

Borneodendron

1 species, Malesia (Borneo); trees to 30 m with verticillate leaves in threes, at least partly in areas of ultramafic rock. Related to *Cocconerion*. A revised treatment of the genus, for *Flora Malesiana*, is in the hands of Jamili Nais (Sabah, Malaysia). (Crotonoideae)

Airy-Shaw, H. K. (1963). Notes on Malaysian and other Asiatic Euphorbiaceae, XL. *Borneodendron* Airy Shaw, a remarkable new genus and species from North Borneo. Kew Bull. 16: 358-362. En. — Protologue of genus and description of *B. aenigmaticum*; considered related to *Baloghia*, though remotely. [A comparison with *Cocconerion* was later made by him in *Euphorbiaceae of Borneo* (see **Malesia**).]

Borneodendron Airy Shaw, Kew Bull. 16: 359 (1963).
Borneo. 42.

Borneodendron aenigmaticum Airy Shaw, Kew Bull. 16: 359 (1963). – FIGURE, p. 276.
Borneo (Sabah). 42 BOR. Phan.

Bossera

1 species, Madagascar; shrubs of dry forest with baobabs (*Adansonia*). Some workers combine this with *Alchornea* from which it is only weakly distinguished (in ovary and stamen number). It is also close to *Orfilea*, likewise Malagasy. The cristate ovary is the main distinctive feature. (Acalyphoideae)

Leandri, J. (1962). Notes sur les Euphorbiacées malgaches. Adansonia, II, 2: 216-223, illus. Fr. — Part 1 (pp. 216-220), on *Bossera*, comprises the protologue of the genus and a description of *B. cristatocarpa*, also new; extensive discussion.

Bossera Leandri, Adansonia, n.s., 2: 216 (1962).
Madagascar. 29.

Bossera cristatocarpa Leandri, Adansonia, n.s. 2: 218 (1962).
Madagascar. 29 MDG.

Botryanthe

Synonyms:
Botryanthe Klotzsch === **Plukenetia** L.
Botryanthe concolor Klotzsch === **Plukenetia serrata** (Vell.) L.J.Gillespie
Botryanthe discolor Klotzsch === **Plukenetia serrata** (Vell.) L.J.Gillespie

Botryophora

1 species, SE Asia and W Malesia (Burma to Jawa and Borneo); a large shrub or small to medium forest tree with pendulous, much-branched male inflorescences related to *Blumeodendron*. The name has been conserved against *Botryophora* Bomgard (Chlorophycophyta). [Not revised in Pflanzenreich; by Pax and Hoffmann (1931:228) treated as 'uncertain'. A full treatment appears in Airy-Shaw (1962). Our current knowledge of this species again is based on assembly of a jigsaw-puzzle with only scattered pieces! Further studies have been undertaken by Balu Perumal towards a treatment for *Flora Malesiana*.] (Acalyphoideae)

Airy-Shaw, H. K. (1949). [Notes on Malaysian Euphorbiaceae, I.] A note on the euphorbiaceous genus *Botryophora*. Kew Bull. 3: 484. En. — Notes with literature citations and localities with exsiccatae along with a new combination.
Airy-Shaw, H. K. (1960). Notes on Malaysian Euphorbiaceae, IX. The female inflorescence, and affinities, of *Botryophora* Hook. fil. Kew Bull. 14: 374-378. En. — Includes the new

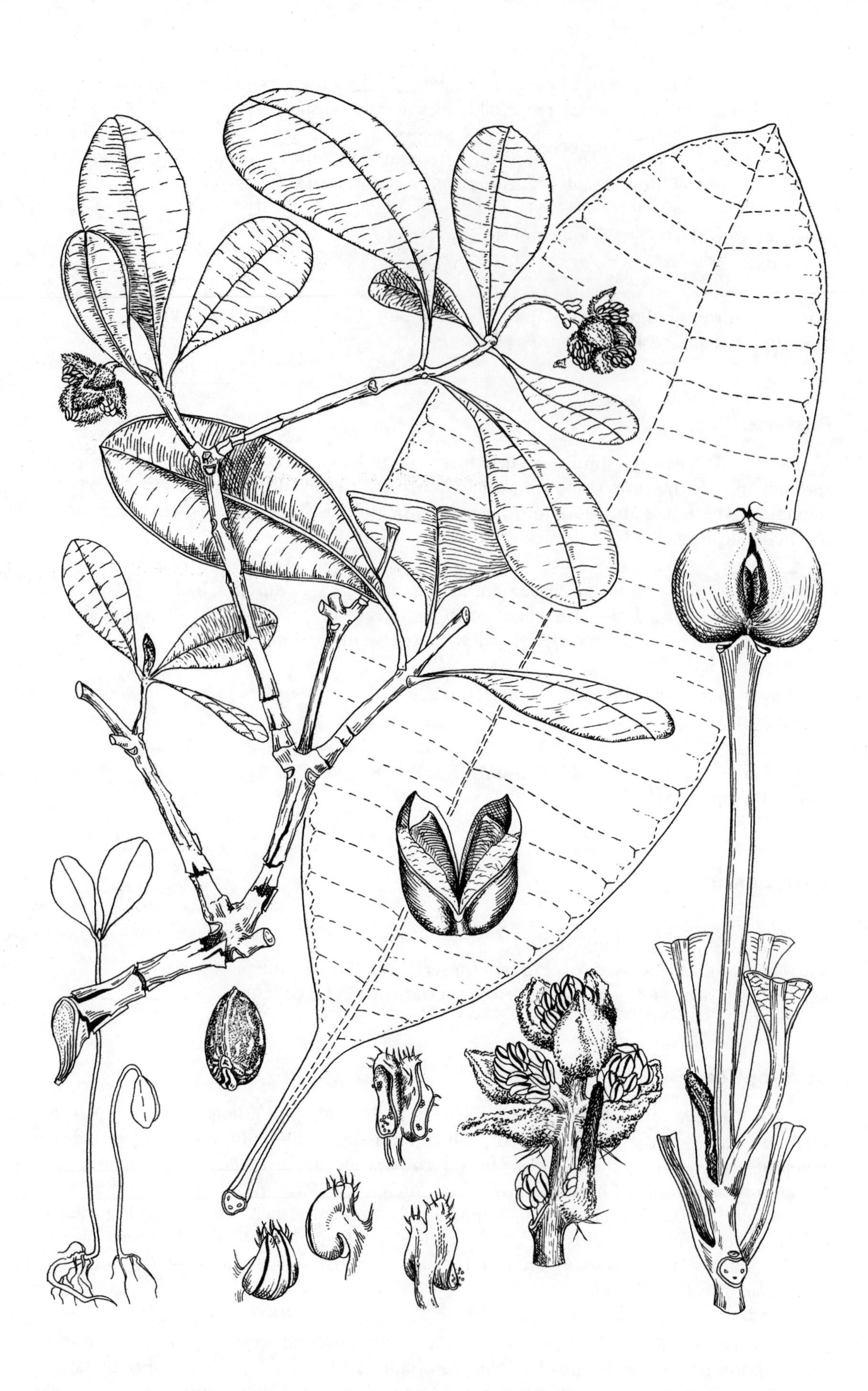

Borneodendron aenigmaticum Airy Shaw
Artist: Mary Grierson
Ic. Pl. 37(2): pl. 3633 (1967)
KEW ILLUSTRATIONS COLLECTION

combination *Botryophora geniculata* (Miq.) Beumée ex Airy Shaw along with a revised Latin description; range from Sumatra to Peninsular Malaysia and Borneo. An extensive discussion of affinities is presented, with the conclusion that the genus was most closely related to *Blumeodendron*.

Airy-Shaw, H. K. (1962). *Botryophora geniculata*. Ic. Pl. 36: pl. 3576. En. — Plant portrait with full description and documentation.

Botryophora Hook.f., Fl. Brit. India 5: 476 (1888).
S. Indo-China, W. Malesia. 41 42.

Botryophora geniculata (Miq.) Beumée ex Airy Shaw, Kew Bull. 3: 484 (1948 publ. 1949).
S. Burma, S. Thailand, Pen. Malaysia, Sumatera, Jawa, Borneo (Sabah, N. Kalimantan). 41 BMA THA 42 BOR JAW MLY SUM. Phan.
* *Sterculia geniculata* Miq., Fl. Ned. Ind., Eerste Bijv.: 400 (1861).
Botryophora kingii Hook.f., Fl. Brit. India 5: 476 (1888).

Synonyms:
Botryophora kingii Hook.f. === **Botryophora geniculata** (Miq.) Beumée ex Airy Shaw

Botryospora

An orthographic variant of *Botryophora*.

Botryphora

An orthographic variant of *Botryophora*.

Boutonia

Synonyms:
Boutonia Bojer ex Baill. === **Mallotus** Lour.
Boutonia Bojer === **Mallotus** Lour.
Boutonia acuminata Baill. === **Deuteromallotus acuminatus** (Baill.) Pax & K.Hoffm.
Boutonia mascareinensis Bojer === **Cordemoya integrifolia** (Willd.) Baill.

Brachystachys

Synonyms:
Brachystachys Klotzsch === **Croton** L.
Brachystachys hirta Klotzsch === **Croton glandulosus** L.

Bradleia

Synonyms:
Bradleia Banks ex Gaertn. === **Glochidion** J.R.Forst. & G.Forst.
Bradleia acuminata Wall. === **Glochidion triandrum** (Blanco) C.B.Rob.
Bradleia arborescens (Blume) Steud. === **Glochidion zeylanicum** var. **arborescens** (Blume) Chakrab. & M.G.Gangop.
Bradleia blumei Steud. === **Glochidion molle** Blume
Bradleia coccinea (Buch.-Ham.) Wall. === **Glochidion coccineum** (Buch.-Ham.) Müll.Arg.
Bradleia coriacea Wall. === **Glochidion coriaceum** Thwaites
Bradleia coronata Wall. === **Glochidion rubrum** Blume var. **rubrum**
Bradleia dioica (Schumach. & Thonn.) Gaertn. ex Vahl === **Flueggea virosa** (Roxb. ex Willd.) Voigt

Bradleia finlaysoniana Wall. === **Glochidion superbum** Baill. ex Müll.Arg.
Bradleia glauca Labill. === **Glochidion billardieri** Baill.
Bradleia glaucophylla Hassk. === **Glochidion obscurum** (Roxb. ex Willd.) Blume
Bradleia glochidion Gaertn. === **Glochidion ramiflorum** J.R.Forst. & G.Forst.
Bradleia hirsuta Roxb. === **Glochidion zeylanicum** var. **talbotii** (Hook.f.) Haines
Bradleia impuber Roxb. === **Glochidion impuber** (Roxb.) Govaerts
Bradleia kipareh Steud. === **Glochidion obscurum** (Roxb. ex Willd.) Blume
Bradleia laevigata Wall. === **Glochidion lutescens** Blume
Bradleia lanceolaria Roxb. === **Glochidion lanceolarium** (Roxb.) Voigt
Bradleia littorea (Blume) Steud. === **Glochidion littorale** Blume
Bradleia lucida (Blume) Steud. === **Glochidion lucidum** Blume
Bradleia lutescens (Blume) Steud. === **Glochidion lutescens** Blume
Bradleia macrocarpa (Blume) Steud. === **Glochidion macrocarpum** Blume
Bradleia macrophylla Labill. === **Glochidion macphersonii** Govaerts & Radcl.-Sm.
Bradleia mollis (Blume) Steud. === **Glochidion molle** Blume
Bradleia moluccana (Blume) Steud. === **Glochidion moluccanum** Blume
Bradleia multiloculare (Rottler ex Willd.) Spreng. === **Glochidion multiloculare** (Rottler ex Willd.) Voigt
Bradleia nephroia Steud. === **Cocculus trilobus** (Thunb.) DC. (Menispermaceae)
Bradleia nitida Roxb. === **Glochidion zeylanicum** (Gaertn.) A.Juss. var. **zeylanicum**
Bradleia obliqua (Willd.) Spreng. === **Glochidion zeylanicum** (Gaertn.) A.Juss. var. **zeylanicum**
Bradleia obtusa Wall. === **Glochidion littorale** Blume
Bradleia ovata Wall. === **Glochidion heyneanum** (Wight & Arn.) Wight
Bradleia philippensis Willd. === **Glochidion philippicum** (Cav.) C.B.Rob.
Bradleia philippica Cav. === **Glochidion philippicum** (Cav.) C.B.Rob.
Bradleia pinnata Roxb. === **Glochidion obscurum** (Roxb. ex Willd.) Blume
Bradleia pubera (L.) Roxb. === **Glochidion puberum** (L.) Hutch.
Bradleia rubra (Blume) Steud. === **Glochidion rubrum** Blume
Bradleia sinensis Siebold ex Miq. === **Glochidion obovatum** Siebold & Zucc.
Bradleia sinica Müll.Arg. === **Margaritaria nobilis** L.f.
Bradleia sinica Gaertn. === **Glochidion puberum** (L.) Hutch.
Bradleia timoriensis Steud. === **Glochidion zeylanicum** (Gaertn.) A.Juss. var. **zeylanicum**
Bradleia wightiana Wall. === **Glochidion ellipticum** Wight
Bradleia zeylanica Gaertn. === **Glochidion zeylanicum** (Gaertn.) A.Juss.
Bradleia zeylanica Labill. === **Glochidion caledonicum** Müll.Arg.

Bradleja

An orthographic variant of *Bradleia*.

Brexiopsis

Synonyms:
Brexiopsis H.Perrier === **Drypetes** Vahl
Brexiopsis aquifolia H.Perrier === **Drypetes bathiei** Capuron & Leandri

Breynia

35 species, Asia, Malesia, the Pacific and Australia; most strongly developed in W Malesia and SE Asia. Shrubs or small trees with 'plagiotropic' branching featuring spreading sprays of distichously arranged leaves. Some species may be from time to time deciduous (a series of observations was made on *B. cernua* in Java in 1916-21; see J.J. Smith in Ann. Jard. Bot. Buitenzorg 32: 97-102. 1923). Brunel (1987; see *Phyllanthus*) extends the genus to Africa, but

Lebrun and Stork in their *Énumération des plantes à fleurs d'Afrique tropicale* (vol. 1, 1991) do not list any native species. *B. disticha* 'Nivosa' (*B. nivosa* (W. Bull) Small), the 'Snow Bush', is widely cultivated. According to Brunel, however, *B. microphylla* var. *angustifolia* is possibly naturalised in Ivory Coast in West Africa. [The genus was never revised for *Pflanzenreich* and at present the species are not well understood. Malesian representatives have been under study by Wolfgang Stuppy as part of the Rijksherbarium/Hortus Botanicus programme for Malesian Euphorbiaceae.] (Phyllanthoideae)

Pax, F. & K. Hoffmann (1931). *Breynia*. In A. Engler (ed.), Die natürlichen Pflanzenfamilien, 2. Aufl., 19c: 59. Leipzig. Ge. — Description of genus; synoptic précis (20-30 species, most strongly developed in W Malesia and SE Asia but extending to the Pacific, Australia and S Asia).

Croizat, L. (1942b). Notes on the Euphorbiaceae, III. Bull. Bot. Gard. Buitenzorg, III, 17: 209-219. En. — Includes (among other topics) a proposal for conservation of *Breynia* over *Melanthesa*.

Airy-Shaw, H. K. (1971). Notes on Malesian and other Asiatic Euphorbiaceae, CXXII. A new *Breynia* from New Guinea. Kew Bull. 25: 488-489. En. — Description of *B. collaris* from the Highlands of Papua New Guinea.

Airy-Shaw, H. K. (1972). Notes on Malesian and other Asiatic Euphorbiaceae, CLIII: A new *Breynia* from New Guinea. Kew Bull. 27: 74. En. — Description of *B. platycalyx* from Irian Jaya (Kepala Burung).

Airy-Shaw, H. K. (1978). Notes on Malesian and other Asiatic Euphorbiaceae, CLXXXVIII. A new *Breynia* from New Guinea. Kew Bull. 32: 365-367. En. — Description of *B. podocarpa* from Long Island off the north coast of mainland Papua New Guinea. A similar plant was also recorded from the Northern Territory of Australia.

Radcliffe-Smith, A. (1981). A new status for the snow-bush. Kew Bull. 35: 498. En. — On *Breynia disticha* f. *nivosa*, based on *Phyllanthus nivosus*; a colour-form of *B. disticha*.

Brunel, J. F. (1987). *Breynia*. In *idem*, Sur le genre *Phyllanthus* L. et quelques genres voisins de la tribu des Phyllantheae Dumort. Strasbourg. [See also *Phyllanthus*.] Fr. — Description of the genus (pp. 417-418); *B. angustifolia* (=*B. microphylla* var. *angustifolia*) claimed as naturalised in Ivory Coast (West Africa), possibly through introduction by railway construction workers from Asia.

Breynia J.R.Forst. & G.Forst., Char. Gen. Pl.: 73 (1775), nom. cons.
Réunion, Trop. & Subtrop. Asia, Australia, Pacific. 29 36 38 40 41 42 50 60 (61).
Foersteria Scop., Intr. Hist. Nat.: 98 (1777).
Forsteria Steud., Nomencl. Bot. 1: 344 (1821).
Melanthesa Blume, Bijdr.: 590 (1826).
Melanthesopsis Müll.Arg., Linnaea 32: 74 (1863).

Breynia baudouinii Beille in H.Lecomte, Fl. Indo-Chine 5: 638 (1927).
Vietnam. 41 VIE. Nanophan.

Breynia cernua (Poir.) Müll.Arg. in A.P.de Candolle, Prodr. 15(2): 439 (1866).
Jawa, Borneo, Philippines, New Guinea, Northern Territory, Solomon Is. 42 BOR JAW NWG PHI 50 NTA 60 SOL. Nanophan. or phan.
Phyllanthus ruber Noronha, Verh. Batav. Genootsch. Kunsten 5(4): 22 (1790), nom. nud.
* *Phyllanthus cernuus* Poir. in J.B.A.M.de Lamarck, Encycl. 5: 298 (1804). *Melanthesa cernua* (Poir.) Decne., Nouv. Ann. Mus. Hist. Nat. 3: 483 (1834). *Breynia cernua* var. *genuina* Müll.Arg. in A.P.de Candolle, Prodr. 15(2): 439 (1866), nom. inval.
Melanthesa rubra Blume, Bijdr.: 591 (1826). *Phyllanthus blumei* Steud., Nomencl. Bot., ed. 2, 2: 326 (1841). *Breynia rubra* (Blume) Müll.Arg. in A.P.de Candolle, Prodr. 15(2): 438 (1866).
Melanthesa cernua var. *acutifolia* Müll.Arg., Linnaea 32: 74 (1863). *Breynia cernua* var. *acutifolia* (Müll.Arg.) Müll.Arg. in A.P.de Candolle, Prodr. 15(2): 439 (1866).
Breynia rumpens J.J.Sm., Nova Guinea 8: 227 (1910).

Breynia collaris Airy Shaw, Kew Bull. 25: 488 (1971).
Papua New Guinea. 42 NWG. Nanophan. or phan.

Breynia coriacea Beille in H.Lecomte, Fl. Indo-Chine 5: 638 (1927).
Vietnam. 41 VIE. Nanophan.

Breynia coronata Hook.f., Fl. Brit. India 5: 330 (1887).
Pen. Malaysia, S. Sumatera. 42 MLY SUM. Nanophan. or phan.
Phyllanthus agynus Hunter ex Ridl., J. Straits Branch Roy. Asiat. Soc. 53: 116 (1909).

Breynia discigera Müll.Arg. in A.P.de Candolle, Prodr. 15(2): 440 (1866).
S. Thailand, Pen. Malaysia, W. Sumatera. 41 THA 42 MLY SUM. Nanophan.
Melanthesa racemosa var. *pubescens* Müll.Arg., Linnaea 32: 73 (1863).
Melanthesa rhamnoides var. *pubescens* Müll.Arg., Linnaea 32: 74 (1863). *Breynia rhamnoides* var. *pubescens* (Müll.Arg.) Müll.Arg. in A.P.de Candolle, Prodr. 15(2): 441 (1866).

Breynia disticha J.R.Forst. & G.Forst., Char. Gen. Pl.: 73 (1775). *Breynia disticha* var. *genuina* Müll.Arg. in A.P.de Candolle, Prodr. 15(2): 439 (1866), nom. inval.
New Caledonia, Fiji, Vanuatu. 60 FIJ NWC VAN. Nanophan. or phan. – The variegated cultivar 'Nivosa' is widely grown.
Breynia axillaris Spreng., Pl. Min. Cogn. Pug. 2: 92 (1815).
Melanthesa neocaledonica Baill., Adansonia 2: 240 (1862). *Breynia disticha* var. *neocaledonica* (Baill.) Müll.Arg. in A.P.de Candolle, Prodr. 15(2): 439 (1866).
Melanthesa neocaledonica var. *forsteri* Müll.Arg., Linnaea 32: 74 (1863).
Phyllanthus nivosus W.Bull, Cat. New Pl.: 9 (1873). *Breynia nivosa* (W.Bull) Small, Bull. Torrey Bot. Club 37: 516 (1910). *Breynia disticha* f. *nivosa* (W.Bull) Croizat ex Radcl.-Sm., Kew Bull. 35: 498 (1980).
Phyllanthus sandwicensis var. *hypoglaucus* H.Lév., Repert. Spec. Nov. Regni Veg. 10: 124 (1911).

Breynia diversifolia Beille in H.Lecomte, Fl. Indo-Chine 5: 640 (1927).
S. Vietnam. 41 VIE. Nanophan.

Breynia fleuryi Beille in H.Lecomte, Fl. Indo-Chine 5: 634 (1927).
Vietnam. 41 VIE. Nanophan.

Breynia fruticosa (L.) Hook.f., Fl. Brit. India 5: 331 (1887).
S. China, Vietnam, Thailand. 36 CHC CHH CHS 41 THA VIE. Nanophan. to phan.
* *Andrachne fruticosa* L., Sp. Pl.: 1014 (1753).
Phyllanthus lucens Poir. in J.B.A.M.de Lamarck, Encycl. 5: 206 (1804).
Phyllanthus turbinatus Sims, Bot. Mag. 44: t. 1862 (1817).
Melanthesa chinensis Blume, Bijdr.: 592 (1826).
Phyllanthus introductus Steud., Nomencl. Bot., ed. 2, 2: 327 (1841).
Phyllanthus simsianus Wall., Numer. List: 7920 (1847), nom. inval.

Breynia glauca Craib, Bull. Misc. Inform. Kew 1911: 460 (1911).
Thailand, Burma. 41 BMA THA. Nanophan.
Breynia subterblanca (C.E.C.Fisch.) C.E.C.Fisch., Bull. Misc. Inform. Kew 1939: 98 (1938).

Breynia grandiflora Beille in H.Lecomte, Fl. Indo-Chine 5: 635 (1927).
Vietnam. 41 VIE. Nanophan.

Breynia heyneana J.J.Sm., Bull. Jard. Bot. Buitenzorg, III, 4: 236 (1922).
Lesser Sunda Is. 42 LSI.

Breynia hyposauropus Croizat, J. Arnold Arbor. 21: 493 (1940).
China (Yunnan, Guangxi). 36 CHC CHS. Nanophan.

Breynia indosinensis Beille in H.Lecomte, Fl. Indo-Chine 5: 634 (1927).
Vietnam, Laos. 41 LAO VIE. Phan.

Breynia massiei Beille in H.Lecomte, Fl. Indo-Chine 5: 640 (1927).
Laos. 41 LAO.

Breynia microphylla (Kurz ex Teijsm. & Binn.) Müll.Arg. in A.P.de Candolle, Prodr. 15(2): 442 (1866).
Burma, Thailand, W. & C. Malesia. 41 BMA THA 42 JAW MLY SUL SUM.
* *Melanthesa microphylla* Kurz ex Teijsm. & Binn., Tijdschr. Ned.-Indië 27: 49 (1864).

var. **angustifolia** (Hook.f.) Airy Shaw, Kew Bull. 36: 272 (1981).
Burma, Thailand, Pen. Malaysia, Sumatera. 41 BMA THA 42 MLY SUM.
* *Breynia angustifolia* Hook.f., Fl. Brit. India 5: 330 (1887).

var. **microphylla**
N. Sumatera, Jawa, S. Sulawesi. 42 JAW SUL SUM.

Breynia mollis J.J.Sm., Nova Guinea 8: 784 (1912).
New Guinea. 42 NWG. Nanophan.

Breynia oblongifolia (Müll.Arg.) Müll.Arg. in A.P.de Candolle, Prodr. 15(2): 440 (1866).
Papua New Guinea, N. & E. Australia. 42 NWG 50 NSW NTA QLD. (Cl.) nanophan. or phan.
* *Melanthesa rhamnoides* var. *oblongifolia* Müll.Arg., Linnaea 32: 73 (1863).

var. **oblongifolia**
Papua New Guinea, N. & E. Australia. 42 NWG 50 NSW NTA QLD. (Cl.) nanophan. or phan.
Melanthesa rhamnoides Decne., Nouv. Ann. Mus. Hist. Nat. 3: 483 (1834), nom. illeg.
Breynia cinerascens Baill., Adansonia 6: 344 (1866).

var. **suborbicularis** Airy Shaw, Kew Bull., Addit. Ser. 8: 40 (1980).
Papua New Guinea, Queensland. 42 NWG 50 QLD. Nanophan.

Breynia platycalyx Airy Shaw, Kew Bull. 27: 74 (1972).
Irian Jaya. 42 NWG. Nanophan.

Breynia podocarpa Airy Shaw, Kew Bull. 32: 365 (1978).
Papua New Guinea, Northern Territory. 42 NWG 50 NTA. Nanophan. or phan.

Breynia pubescens Merr., Philipp. J. Sci., C 11: 282 (1916).
Maluku. 42 MOL.

Breynia racemosa (Blume) Müll.Arg. in A.P.de Candolle, Prodr. 15(2): 441 (1866).
S. Thailand, Malesia. 41 THA 42 BIS JAW LSI MLY NWG PHI SUM. Nanophan. to phan.
* *Melanthesa racemosa* Blume, Bijdr.: 592 (1826). *Breynia racemosa* var. *genuina* Müll.Arg. in A.P.de Candolle, Prodr. 15(2): 441 (1866), nom. inval.

var. **aromatica** Airy Shaw, Kew Bull., Addit. Ser. 8: 41 (1980).
Papua New Guinea, Bismarck Archip. 42 BIS NWG. Nanophan. to phan.

var. **racemosa**
S. Thailand, Pen. Malaysia, Sumatera (incl. Bangka), Jawa, Lesser Sunda Is., Philippines. 41 THA 42 JAW LSI MLY PHI SUM. Nanophan. to phan.
Phyllanthus racemifer Steud., Nomencl. Bot., ed. 2, 2: 327 (1841).
Melanthesa acuminata Müll.Arg., Linnaea 32: 74 (1863). *Breynia acuminata* (Müll.Arg.) Müll.Arg. in A.P.de Candolle, Prodr. 15(2): 442 (1866).
Melanthesa rhamnoides var. *hypoglauca* Müll.Arg., Linnaea 32: 73 (1863). *Breynia rhamnoides* var. *hypoglauca* (Müll.Arg.) Müll.Arg. in A.P.de Candolle, Prodr. 15(2): 440 (1866).

Melanthesopsis lucens Müll.Arg., Linnaea 32: 75 (1863).
Breynia racemosa var. *concolor* Müll.Arg. in A.P.de Candolle, Prodr. 15(2): 441 (1866).
Melanthesopsis fruticosa Müll.Arg. in A.P.de Candolle, Prodr. 15(2): 437 (1866).
Phyllanthus rhamnoides Bojer ex Müll.Arg. in A.P.de Candolle, Prodr. 15(2): 435 (1866).

Breynia reclinata (Roxb.) Hook.f., Fl. Brit. India 5: 331 (1887).
S. Thailand, Pen. Malaysia (incl. Singapore), Sumatera, Jawa. 41 THA 42 JAW MLY SUM. Cl. nanophan.
* *Phyllanthus reclinatus* Roxb., Fl. Ind. ed. 1832, 3: 669 (1832). *Melanthesa reclinata* (Roxb.) Müll.Arg., Linnaea 32: 74 (1863).

Breynia retusa (Dennst.) Alston, Ann. Roy. Bot. Gard. (Peradeniya) 11: 204 (1929).
Mascarenes, India,to S. China. 29 REU 36 CHC CHS CHT 40 ASS IND NEP SRL 41 BMA THA VIE. Nanophan. or phan.
* *Phyllanthus retusus* Dennst., Schlüssel Hortus Malab.: 31 (1818). *Melanthesa retusa* (Dennst.) Kostel., Allg. Med.-Pharm. Fl. 5: 1771 (1836).
Phyllanthus pomaceus Moon, Cat. Pl. Ceylon: 65 (1824).
Phyllanthus patens Roxb., Fl. Ind. ed. 1832, 3: 667 (1832). *Melanthesopsis patens* (Roxb.) Müll.Arg. in A.P.de Candolle, Prodr. 15(2): 437 (1866). *Melanthesopsis patens* var. *vulgaris* Müll.Arg. in A.P.de Candolle, Prodr. 15(2): 437 (1866), nom. inval. *Breynia patens* (Roxb.) Rolfe, J. Bot. 20: 359 (1882).
Phyllanthus turbinatus K.D.Koenig ex Roxb., Fl. Ind. ed. 1832, 3: 666 (1832), nom. illeg. *Melanthesa turbinata* Oken, Allg. Naturgesch. 3(3): 1603 (1841). *Melanthesopsis patens* var. *turbinata* (Oken) Müll.Arg.in A.P.de Candolle, Prodr. 15(2): 437 (1866). *Breynia turbinata* (Oken) Cordem., Fl. Réunion: 348 (1895).
Phyllanthus suffultus Wall., Numer. List: 7939 (1847), nom. inval.
Melanthesa obliqua Wight, Icon. Pl. Ind. Orient. 5: t. 1898 (1852).
Melanthesopsis variabilis Müll.Arg., Linnaea 32: 75 (1863).
Phyllanthus naviluri Miq. ex Müll.Arg., Linnaea 32: 75 (1863), pro syn.
Melanthesopsis patens var. *oblongifolia* Müll.Arg. in A.P.de Candolle, Prodr. 15(2): 437 (1866).

Breynia rhynchocarpa Benth., Fl. Austral. 6: 114 (1873).
Northern Territory. 50 NTA. – Perhaps identical with B. cernua.

Breynia rostrata Merr., Philipp. J. Sci. 21: 346 (1922).
S. China, Hainan, Vietnam. 36 CHC CHH CHS 41 VIE. Nanophan. to phan.

Breynia septata Beille in H.Lecomte, Fl. Indo-Chine 5: 641 (1927).
Vietnam. 41 VIE. Nanophan. to phan.

Breynia stipitata Müll.Arg. in A.P.de Candolle, Prodr. 15(2): 442 (1866).
Northern Territory, Queensland. 50 NTA QLD. Nanophan. – Close to B. cernua.
Breynia muelleriana Baill., Adansonia 6: 344 (1866).

Breynia subangustifolia Thin, Euphorb. Vietnam: 50 (1995).
Vietnam. 41 VIE. Nanophan.

Breynia subindochinensis Thin, Euphorb. Vietnam: 50 (1995).
Vietnam. 41 VIE. Phan.

Breynia tonkinensis Beille in H.Lecomte, Fl. Indo-Chine 5: 636 (1927).
Vietnam. 41 VIE.

Breynia vestita Warb., Bot. Jahrb. Syst. 13: 354 (1891).
New Guinea, Maluku (Kai Is.). 42 MOL NWG. Nanophan. or phan.
Breynia ovalifolia J.J.Sm., Nova Guinea 8: 226 (1910).

Breynia virgata (Blume) Müll.Arg. in A.P.de Candolle, Prodr. 15(2): 441 (1866).
Jawa, N. Sumatera. 42 JAW SUM.
Phyllanthus turbinatus Noronha, Verh. Batav. Genootsch. Kunsten 5(4): 22 (1790), nom. nud.
* *Melanthesa virgata* Blume, Bijdr.: 592 (1826).
Phyllanthus sylvaticus Steud., Nomencl. Bot., ed. 2, 2: 327 (1841).

Breynia vitis-idaea (Burm.f.) C.E.C.Fisch., Bull. Misc. Inform. Kew 1932: 65 (1932).
Pakistan to Nansei-shoto and Sumatera. 36 CHC CHS 38 NNS TAI 40 IND PAK 41 BMA CBD THA VIE 42 MLY PHI SUM. Nanophan.
* *Rhamnus vitis-idaea* Burm.f., Fl. Indica: 61 (1768). *Phyllanthus vitis-idea* (Burm.f.) D.Koenig ex Roxb., Fl. Ind. ed. 1832, 3: 665 (1832).
Phyllanthus rhamnoides Retz., Observ. Bot. 5: 30 (1788). *Melanthesa rhamnoides* (Retz.) Blume, Bijdr.: 591 (1826). *Breynia rhamnoides* (Retz.) Müll.Arg. in A.P.de Candolle, Prodr. 15(2): 440 (1866). *Breynia rhamnoides* var. *genuina* Müll.Arg. in A.P.de Candolle, Prodr. 15(2): 440 (1886), nom. inval.
Phyllanthus tristis A.Juss., Euphorb. Gen.: 108 (1824).
Melanthesa ovalifolia Kostel., Allg. Med.-Pharm. Fl. 5: 1772 (1836).
Phyllanthus calycinus Wall., Numer. List: 7939B (1847), nom. inval.
Phyllanthus sepiarius Roxb. ex Wall., Numer. List: 7914 (1847), nom. inval.
Phyllanthus tinctorius Vahl ex Baill., Étude Euphorb.: 633 (1858).
Breynia officinalis Hemsl., J. Linn. Soc., Bot. 26: 427 (1894).
Breynia accrescens Hayata, J. Coll. Sci. Imp. Univ. Tokyo 20(3): 22 (1904). *Breynia officinalis* var. *accrescens* (Hayata) M.J.Deng & J.C.Wang, in Fl. Taiwan, ed. 2, 3: 430 (1993).
Breynia stipitata var. *formosana* Hayata, J. Coll. Sci. Imp. Univ. Tokyo 20(3): 22 (1904). *Breynia formosana* (Hayata) Hayata in ?, .
Breynia keithii Ridl., J. Straits Branch Roy. Asiat. Soc. 59: 174 (1911).
Breynia microcalyx Ridl., J. Fed. Malay States Mus. 10: 114 (1920).

Synonyms:
Breynia accrescens Hayata === **Breynia vitis-idaea** (Burm.f.) C.E.C.Fisch.
Breynia acuminata (Müll.Arg.) Müll.Arg. === **Breynia racemosa** (Blume) Müll.Arg. var. **racemosa**
Breynia angustifolia Hook.f. === **Breynia microphylla** var. **angustifolia** (Hook.f.) Airy Shaw
Breynia axillaris Spreng. === **Breynia disticha** J.R.Forst. & G.Forst.
Breynia cernua var. *acutifolia* (Müll.Arg.) Müll.Arg. === **Breynia cernua** (Poir.) Müll.Arg.
Breynia cernua var. *genuina* Müll.Arg. === **Breynia cernua** (Poir.) Müll.Arg.
Breynia cinerascens Baill. === **Breynia oblongifolia** Müll.Arg. var. **oblongifolia**
Breynia disticha var. *genuina* Müll.Arg. === **Breynia disticha** J.R.Forst. & G.Forst.
Breynia disticha var. *neocaledonica* (Baill.) Müll.Arg. === **Breynia disticha** J.R.Forst. & G.Forst.
Breynia disticha f. *nivosa* (W.Bull) Croizat ex Radcl.-Sm. === **Breynia disticha** J.R.Forst. & G.Forst.
Breynia formosana (Hayata) Hayata === **Breynia vitis-idaea** (Burm.f.) C.E.C.Fisch.
Breynia keithii Ridl. === **Breynia vitis-idaea** (Burm.f.) C.E.C.Fisch.
Breynia microcalyx Ridl. === **Breynia vitis-idaea** (Burm.f.) C.E.C.Fisch.
Breynia muelleriana Baill. === **Breynia stipitata** Müll.Arg.
Breynia nivosa (W.Bull) Small === **Breynia disticha** J.R.Forst. & G.Forst.
Breynia officinalis Hemsl. === **Breynia vitis-idaea** (Burm.f.) C.E.C.Fisch.
Breynia officinalis var. *accrescens* (Hayata) M.J.Deng & J.C.Wang === **Breynia vitis-idaea** (Burm.f.) C.E.C.Fisch.
Breynia ovalifolia J.J.Sm. === **Breynia vestita** Warb.
Breynia paniculata Spreng. === ?
Breynia patens (Roxb.) Rolfe === **Breynia retusa** (Dennst.) Alston
Breynia racemosa var. *concolor* Müll.Arg. === **Breynia racemosa** (Blume) Müll.Arg. var. **racemosa**
Breynia racemosa var. *genuina* Müll.Arg. === **Breynia racemosa** (Blume) Müll.Arg.
Breynia rhamnoides (Retz.) Müll.Arg. === **Breynia vitis-idaea** (Burm.f.) C.E.C.Fisch.

Breynia rhamnoides var. *genuina* Müll.Arg. === **Breynia vitis-idaea** (Burm.f.) C.E.C.Fisch.
Breynia rhamnoides var. *hypoglauca* (Müll.Arg.) Müll.Arg. === **Breynia racemosa** (Blume) Müll.Arg. var. **racemosa**
Breynia rhamnoides var. *pubescens* (Müll.Arg.) Müll.Arg. === **Breynia discigera** Müll.Arg.
Breynia rubra (Blume) Müll.Arg. === **Breynia cernua** (Poir.) Müll.Arg.
Breynia rumpens J.J.Sm. === **Breynia cernua** (Poir.) Müll.Arg.
Breynia stipitata var. *formosana* Hayata === **Breynia vitis-idaea** (Burm.f.) C.E.C.Fisch.
Breynia subterblanca (C.E.C.Fisch.) C.E.C.Fisch. === **Breynia glauca** Craib
Breynia turbinata (Oken) Cordem. === **Breynia retusa** (Dennst.) Alston

Breyniopsis

Synonyms:
Breyniopsis Beille === **Sauropus** Blume
Breyniopsis pierrei Beille === **Sauropus pierrei** (Beille) Croizat

Bridelia

49 species, Africa, Madagascar (2), and from the Arabian Peninsula across Asia and Malesia to Australia and the southwestern Pacific; erect or scandent shrubs or small to medium trees to 20 m (often 'plagiotropic' as in the African *B. micrantha*) or woody climbers, the twig-bases often persistent and becoming thorny. The older (and correct) spelling is *Briedelia* (Webster, Synopsis, 1994); a proposal for retention of the more familiar orthography was, however, made (Dressler 1996, in *Taxon*) and has since been accepted. Some species are common in secondary forest. The ovary is 2-carpelled and the fruits usually fleshy, in contrast to the closely related *Cleistanthus*. A revision for SE Asia and Malesia by Dressler (1996) accounts for 19 species; elsewhere, modern knowledge is less well consolidated. (Phyllanthoideae)

• Gehrmann, K. (1908). Vorarbeiten zu einer Monographie der Gattung *Bridelia* mit besonderer Berücksichtigung der afrikanischen Arten. Bot. Jahrb. Syst. 41, Beibl. 65: 1-42, illus., map. Ge. — Limits of genus and character survey; anatomical features; useful characters; classification and geographical distribution, with synopsis of sections and subsections (their ranges depicted on the accompanying map); general surveys of species in different regions; phylogeny (with phylograms); key to and synoptic treatment of genus with references, synonymy, descriptions of novelties, indication of distribution, and commentary; doubtful and excluded species. [43 species in 2 sections and 9 subsections accounted for. The greatest diversity was in Asia and Malesia, although subsect. *Micranthae* was wholly African.]

Jablonski, E. (as E. Jablonszky) (1915). *Bridelia*. In A. Engler (ed.), Das Pflanzenreich, IV 147 VIII (Euphorbiaceae-Phyllanthoideae-Bridelieae): 54-88, illus. Berlin. (Heft 65.) La/Ge. — 56 species, Africa, Madagascar, Asia, Malesia, and the Pacific; in 2 subgenera and 6 sections.

Léonard, J. (1955). Notulae systematicae XIX. Observations sur divers *Bridelia* africains (Euphorbiaceae). Bull. Jard. Bot. État 25: 359-374. Fr. — Novelties and notes with documentation and one partial key covering 6 species. [A precursor to the *Flore du Congo* treatment; see also Léonard (1959, 1961) under **Africa**.]

Leandri, J. (1958). *Bridelia*. Fl. Madag. Comores 111 (Euphorbiacées), I: 192-197. Paris. Fr. — Flora treatment (2 species).

Airy-Shaw, H. K. (1969). Notes on Malesian and other Asiatic Euphorbiaceae, CIV. New or noteworthy species of *Bridelia* Willd. Kew Bull. 23: 65-69. En. — 6 species, of which 2 new; some reductions.

Airy-Shaw, H. K. (1971). Notes on Malesian and other Asiatic Euphorbiaceae, CXXXIII. A new *Bridelia* from New Guinea. Kew Bull. 25: 512-514. En. — *B. macrocarpa* described.

Airy-Shaw, H. K. (1972). Notes on Malesian and other Asiatic Euphorbiaceae, CLVII: A new *Bridelia* from Malaya. Kew Bull. 27: 77-78. En. — *B. whitmorei* from Pahang, Peninsular Malaysia.

Airy-Shaw, H. K. (1978). Notes on Malesian and other Asiatic Euphorbiaceae, CXCVI. New species of *Bridelia* Willd. Kew Bull. 32: 383-387. En. — Descriptions of 3 species (two new); sections indicated.

Friis, I. & K. Vollesen (1980). The identity of the Ethiopian monotypic genus *Tzellemtinia* Chiov. Bot. Notis. 133: 347-349. En. — Reduction of genus to *Bridelia.*

Olowokudejo, J. D. & O. O. Bamgbowu (1993-94(1995?)). Leaf epidermal morphology in *Bridelia* (Euphorbiaceae) and its taxonomic significance. Bol. Soc. Brot. 66: 5-18, illus. En. — Leaf morphological study of 7 West African species, with particular reference to leaf surfaces and their identification; includes key (pp. 15-16). [Many species are of medicinal use in the region; in addition, the timber is of value for construction and fuel. *B. micrantha* and *B. ferruginea* also have edible sweet fruits.]

Dressler, S. (1996). *Bridelia* (Euphorbiaceae) in New Guinea with a description of a new species. Kew Bull. 51: 601-607, illus., maps. En. — Description of one new species (*B. erapensis* from northeastern mainland New Guinea); key to and distribution maps of 7 species (in 2 sections). [*B. erapensis* is related to *BB. macrocarpa* and *triplocarya*; all are allied to the west Malesian *B. stipularis*.]

Dressler, S. (1996). Proposal to conserve the name *Bridelia* (Euphorbiaceae) with a conserved spelling. Taxon 45: 337-338. En. — Nomenclatural; argument for retention of *Bridelia* against the etymologically more precise *Briedelia* originally used by Willdenow.

• Dressler, S. (1996). The genus *Bridelia* (Euphorbiaceae) in Malesia and Indochina: a regional revision. Blumea 41: 263-331, illus., maps. En. — Critical revision (19 species) with keys, descriptions, synonymy, references, citations, types, indication of distribution (with dot maps) and habitat, and commentary; literature, list of specimens seen and index at end. The general part includes a character analysis and biogeographical discussion.

Bridelia Willd., Sp. Pl. 4: 978 (1806), nom. & orth. cons.
Palaeotrop. 22 23 24 25 26 27 29 35 36 38 40 41 42 50 60 62 (63).
Candelabria Hochst., Flora 26: 79 (1843).
Pentameria Klotzsch ex Baill., Étude Euphorb.: 584 (1858).
Neogoetzea Pax, Bot. Jahrb. Syst. 28: 419 (1900).
Gentilia Beille, Compt. Rend. Hebd. Séances Acad. Sci. 145: 1294 (1907).
Tzellemtinia Chiov., Ann. Bot. (Rome) 9: 55 (1911).

Bridelia adusta Airy Shaw, Kew Bull., Addit. Ser. 4: 224 (1975).
Borneo (N. Sarawak, Sabah). 42 BOR. Phan.

Bridelia affinis Craib, Bull. Misc. Inform. Kew 1911: 456 (1911).
China (Yunnan), Hainan, Thailand. 36 CHC CHH 41 THA. Nanophan. or phan.
Bridelia henryana Jabl. in H.G.A.Engler, Pflanzenr., IV, 147, VIII: 62 (1915).
Bridelia colorata Airy Shaw, Kew Bull. 23: 66 (1969).

Bridelia alnifolia Griff., Not. Pl. Asiat. 4: 481 (1854).
Burma. 41 BMA. - Provisionally accepted.

Bridelia assamica Hook.f., Fl. Brit. India 5: 269 (1887).
Assam. 40 ASS. Nanophan. or phan.
Bridelia chartacea Kurz in ?, .

Bridelia atroviridis Müll.Arg., J. Bot. 2: 327 (1864).
Trop. Africa. 22 BEN GUI IVO NGA SIE TOG 23 CAF CMN EQG ZAI 24 ETH SUD 25 KEN TAN UGA 26 ANG ZAM ZIM. Phan.
Bridelia zenkeri Pax, Bot. Jahrb. Syst. 26: 327 (1899).

Bridelia brideliifolia (Pax) Fedde, Just's Bot. Jahresber. 36(11): 413 (1905).
Malawi, Tanzania, Zaire. 23 ZAI 25 TAN 26 MLW. Phan.
Bridelia brideliifolia subsp. *pubescentifolia* J.Léonard in ?, .
* *Neogoetzea brideliifolia* Pax, Bot. Jahrb. Syst. 28: 419 (1900). *Bridelia neogoetzea* Gehrm., Bot. Jahrb. Syst. 41(95): 40 (1908).
Bridelia ramiflora Gehrm. ex Jabl. in H.G.A.Engler, Pflanzenr., IV, 147, VIII: 83 (1915).

Bridelia cathartica Bertol., Mem. Reale Accad. Sci. Ist. Bologna 5: 476 (1854).
Ethiopia to S. Africa. 23 ZAI 24 ETH SOM SUD 25 KEN TAN 26 MLW MOZ ZAM ZIM 27 BOT CPV NAM NAT SWZ TVL. Nanophan. or phan.

subsp. **cathartica**
Mozambique, KwaZulu-Natal. 26 MOZ 27 NAT. Nanophan. or phan.
Bridelia schlechteri Hutch., Bull. Misc. Inform. Kew 1914: 249 (1914).

subsp. **melanthesoides** (Klotzsch ex Baill.) J.Léonard, Bull. Jard. Bot. État 25: 364 (1955).
Ethiopia to S. Africa. 23 ZAI 24 ETH SOM SUD 25 KEN TAN 26 MLW MOZ ZAM ZIM 27 BOT CPV NAM NAT SWZ TVL. Nanophan. or phan.
* *Pentameria melanthesoides* Klotzsch ex Baill., Étude Euphorb.: 584 (1858). *Bridelia melanthesoides* (Klotzsch ex Baill.) Klotzsch in W.C.H.Peters, Naturw. Reise Mossambique: 103 (1861). *Bridelia melanthesoides* var. *typica* Gehrm., Bot. Jahrb. Syst. 41(95): 34 (1908), nom. inval. *Bridelia cathartica* f. *melanthesoides* (Klotzsch ex Baill.) Radcl.-Sm., in Fl. Zambes. 9(4): 17 (1996). *Bridelia cathartica* var. *melanthesoides* (Klotzsch ex Baill.) Radcl.-Sm., in Fl. Zambes. 9(4): 16 (1996).
Bridelia fischeri Pax, Bot. Jahrb. Syst. 15: 531 (1893). *Bridelia cathartica* f. *fischeri* (Pax) Radcl.-Sm., Kew Bull. 51: 302 (1996).
Bridelia lingelsheimii Gehrm., Bot. Jahrb. Syst. 41(95): 36 (1908). *Bridelia fischeri* var. *lingelsheimii* (Gehrm.) Hutch. in D.Oliver, Fl. Trop. Afr. 6(1): 616 (1912). *Bridelia cathartica* var. *lingelsheimii* (Gehrm.) Radcl.-Sm., Kew Bull. 51: 302 (1996).
Bridelia melanthesoides var. *lanceolata* Gehrm., Bot. Jahrb. Syst. 41(95): 35 (1908).
Bridelia melanthesoides var. *ovata* Gehrm., Bot. Jahrb. Syst. 41(95): 35 (1908).
Bridelia niedenzui Gehrm., Bot. Jahrb. Syst. 41(95): 36 (1908). *Bridelia cathartica* f. *niedenzui* (Gehrm.) Radcl.-Sm., Kew Bull. 51: 302 (1996).
Bridelia niedenzui var. *njassae* Gehrm., Bot. Jahrb. Syst. 41(95): 37 (1908).
Bridelia niedenzui var. *pilosa* Gehrm., Bot. Jahrb. Syst. 41(95): 37 (1908).
Bridelia niedenzui var. *revoluta* Gehrm., Bot. Jahrb. Syst. 41(95): 37 (1908).
Bridelia cathartica f. *pubescens* Radcl.-Sm., Kew Bull. 51: 302 (1996).

Bridelia cinerascens Gehrm., Bot. Jahrb. Syst. 41(95): 30 (1908).
India. 40 IND.

Bridelia cinnamomea Hook.f., Fl. Brit. India 5: 273 (1887). *Bridelia griffithii* var. *cinnamomea* (Hook.f.) Gehrm., Bot. Jahrb. Syst. 41(95): 38 (1908).
Thailand, W. Malesia. 41 THA 42 BOR MLY SUM. (Cl.) nanophan. or phan.
Bridelia griffithii Hook.f., Fl. Brit. India 5: 272 (1887).
Bridelia gehrmannii Jabl. in H.G.A.Engler, Pflanzenr., IV, 147, VIII: 73 (1915).

Bridelia cuneata Gehrm., Bot. Jahrb. Syst. 41(95): 34 (1908).
Assam. 40 ASS. Nanophan. or phan.

Bridelia curtisii Hook.f., Fl. Brit. India 5: 273 (1887). *Bridelia ovata* var. *curtisii* (Hook.f.) Airy Shaw, Kew Bull. 26: 229 (1972).
Indo-China to N. Sumatera. 41 AND CBD NCB THA VIE 42 MLY SUM. (Cl.) nanophan. or phan.

Bridelia duvigneaudii J.Léonard, Bull. Jard. Bot. État 25: 365 (1955).
Zaire, Burundi, Tanzania, Zambia, Mozambique. 23 BUR ZAI 25 TAN 26 MOZ ZAM. Nanophan. or phan.

Bridelia eranalis J.Léonard, Bull. Jard. Bot. État 29: 385 (1959).
Zaire. 23 ZAI.

Bridelia erapensis S.Dressler, Kew Bull. 51: 601 (1996).
NE. New Guinea. 42 NWG. Phan.

Bridelia exaltata F.Muell., Fragm. 3: 32 (1862). *Bridelia ovata* var. *exaltata* (F.Muell.) Müll.Arg. in A.P.de Candolle, Prodr. 15(2): 495 (1866).
E. Australia. 50 NSW QLD. Phan.
Amanoa ovata Baill., Adansonia 6: 336 (1866).

Bridelia ferruginea Benth. in W.J.Hooker, Niger Fl.: 511 (1849). *Bridelia micrantha* var. *ferruginea* (Benth.) Müll.Arg. in A.P.de Candolle, Prodr. 15(2): 498 (1866).
Trop. & S. Africa. 22 BEN GUI MLI NGA SIE TOG 23 CAF CMN CON GAB ZAI 24 CHA SUD 25 UGA 26 ANG ZAM 27 NAM. Nanophan. or phan.
Bridelia speciosa var. *kourousensis* Beille, Bull. Soc. Bot. France 40(8): 68 (1908).

Bridelia fordii Hemsl., J. Linn. Soc., Bot. 26: 419 (1894).
S. China, Hainan. 36 CHC CHH CHS. Phan.

Bridelia glauca Blume, Bijdr.: 597 (1826).
S. China, Taiwan, Trop. Asia. 36 CHC CHS 38 TAI 40 ASS EHM NEP 41 BMA LAO THA 42 BIS BOR JAW MLY MOL NWG PHI SUL SUM. Phan.
Cleistanthus oblongifolius var. *scaber* Müll.Arg. in A.P.de Candolle, Prodr. 15(2): 506 (1866).
Bridelia multiflora Zipp. ex Scheff., Ann. Mus. Bot. Lugduno-Batavi 4: 119 (1869).
Bridelia pubescens Kurz, J. Asiat. Soc. Bengal, Pt. 2, Nat. Hist. 42(2): 241 (1874).
Cleistanthus myrianthoides C.B.Rob., Philipp. J. Sci., C 6: 325 (1911).
Bridelia acuminatissima Merr., Philipp. J. Sci., C 9: 473 (1914 publ. 1915). *Bridelia glauca* var. *acuminatissima* (Merr.) S.Dressler, Blumea 41: 314 (1996).
Bridelia laurifolia Elmer, Leafl. Philipp. Bot. 7: 2637 (1915). *Bridelia glauca* f. *laurifolia* (Elmer) Jabl. in H.G.A.Engler, Pflanzenr., IV, 147, VIII: 75 (1915).
Bridelia sosopodonica Airy Shaw, Kew Bull. 23: 67 (1969). *Bridelia glauca* var. *sosopodonica* (Airy Shaw) S.Dressler, Blumea 41: 315 (1996).
Bridelia nooteboomii Chakrab., J. Econ. Taxon. Bot. 5(4): 949 (1984).

Bridelia grandis Pierre ex Hutch. in D.Oliver, Fl. Trop. Afr. 6(1): 1042 (1913).
W. & WC. Trop. Africa. 22 BEN IVO LBR NIG SIE 23 CON GAB ZAI. Nanophan. or phan.

subsp. **grandis**
W. & WC. Trop. Africa. 22 BEN IVO LBR NIG SIE 23 GAB ZAI. Nanophan. or phan.
Bridelia aubrevillei Pellegr., Bull. Soc. Bot. France 78: 683 (1931 publ. 1932).

subsp. **puberula** J.Léonard, Bull. Jard. Bot. État 25: 396 (1955).
WC. Trop. Africa. 23 CON ZAI. Nanophan. or phan.

Bridelia hamiltoniana Wall. ex Müll.Arg., Linnaea 34: 77 (1865). *Bridelia hamiltoniana* var. *genuina* Müll.Arg. Linnaea 34: 77 (1865), nom. inval.
India. 40 IND. Nanophan.
Bridelia hamiltoniana var. *glabra* Müll.Arg., Linnaea 34: 77 (1865).

Bridelia harmandii Gagnep., Bull. Soc. Bot. France 70: 433 (1923).
Indo-China. 41 CBD LAO THA VIE. (Cl.) nanophan.

Bridelia insulana Hance, J. Bot. 15: 337 (1877).
Trop. & Subtrop. Asia to W. Pacific. 36 CHC CHH CHS 38 NNS TAI 41 BMA LAO NCB THA VIE 42 BIS BOR JAW LSI MLY MOL NWG PHI SUL SUM 50 QLD 60 SCZ SOL VAN 62 CRL. Phan.
Bridelia minutiflora Hook.f., Fl. Brit. India 5: 273 (1887).
Bridelia penangiana Hook.f., Fl. Brit. India 5: 272 (1887). *Bridelia griffithii* var. *penangiana* (Hook.f.) Gehrm., Bot. Jahrb. Syst. 41(95): 38 (1908).
Bridelia subnuda K.Schum. & Lauterb., Fl. Schutzgeb. Südsee: 393 (1900). *Bridelia penangiana* var. *subnuda* (K.Schum. & Lauterb.) Airy Shaw, Kew Bull., Addit. Ser. 8: 45 (1980). *Bridelia insulana* var. *subnuda* (K.Schum. & Lauterb.) S.Dressler, Blumea 41: 320 (1996).
Bridelia balansae Tutcher, J. Linn. Soc., Bot. 37: 66 (1905). – Perhaps a good species.
Bridelia kawakamii Hayata, J. Coll. Sci. Imp. Univ. Tokyo 22: 362 (1906), nom. nud.
Bridelia pachinensis Hayata, J. Coll. Sci. Imp. Univ. Tokyo 22: 362 (1906), nom. nud.
Bridelia griffithii var. *glabra* Gehrm., Bot. Jahrb. Syst. 41(95): 38 (1908).
Bridelia minutiflora var. *abbreviata* J.J.Sm. in S.H.Koorders & T.Valeton, Bijdr. Boomsoort. Java 12: 313 (1910).
Bridelia platyphylla Merr., Philipp. J. Sci., C 7: 384 (1912 publ. 1913).
Bridelia palauensis Kaneh. ex Kaneh. & Hatus., Bot. Mag. (Tokyo) 53: 152 (1939).
Bridelia morotaea Airy Shaw, Kew Bull. 37: 10 (1982).
Bridelia nicobarica Chakrab. & Vasudeva Rao, J. Econ. Taxon. Bot. 5(4): 945 (1984).

Bridelia leichhardtii Baill. ex Müll.Arg. in A.P.de Candolle, Prodr. 15(2): 499 (1866).
E. Australia. 50 NSW QLD.
Amanoa faginea Baill., Adansonia 6: 336 (1866). *Bridelia faginea* (Baill.) F.Muell. ex Benth., Fl. Austral. 6: 120 (1873).
Amanoa leichhardtii Baill., Adansonia 6: 336 (1866).
Bridelia melanthesoides var. *australis* Gehrm., Bot. Jahrb. Syst. 41(95): 35 (1908).

Bridelia macrocarpa Airy Shaw, Kew Bull. 25: 512 (1971).
Maluku, New Guinea. 42 MOL NWG. Phan.
Bridelia beguinii Airy Shaw, Kew Bull. 37: 10 (1982).

Bridelia micrantha (Hochst.) Baill., Adansonia 3: 164 (1863).
Trop. & S. Africa, Mascarenes. 22 BEN GAM GHA GUI IVO LBR MLI NGA SEN SIE TOG 23 BUR CAF CMN EQG GAB GGI RWA ZAI 24 ETH SUD 25 KEN TAN UGA 26 ANG MLW MOZ ZAM ZIM 27 CPP NAT SWZ TVL 29 REU. Phan.
* *Candelabria micrantha* Hochst., Flora 26: 79 (1843). *Bridelia micrantha* var. *genuina* Müll.Arg. in A.P.de Candolle, Prodr. 15(2): 498 (1866), nom. inval.

var. **gambicola** (Baill.) Müll.Arg. in A.P.de Candolle, Prodr. 15(2): 498 (1866).
W. Trop. Africa. 22 GAM SEN SIE. Phan.
* *Bridelia gambicola* Baill., Adansonia 1: 79 (1860).

var. **micrantha**
Trop. & S. Africa, Mascarenes. 22 BEN GAM GHA GUI IVO LBR MLI NGA SEN SIE TOG 23 BUR CAF CMN EQG GAB GGI RWA ZAI 24 ETH SUD 25 KEN TAN UGA 26 ANG MLW MOZ ZAM ZIM 27 CPP NAT SWZ TVL 29 REU. Phan.
Bridelia speciosa var. *trichoclada* Müll.Arg., J. Bot. 2: 327 (1864).
Bridelia stenocarpa Müll.Arg., Flora 47: 515 (1864).
Bridelia zanzibarensis Vatke & Pax, Bot. Jahrb. Syst. 15: 530 (1893).
Bridelia ferruginea var. *gambicola* Hiern, Cat. Afr. Pl. 1: 954 (1900).
Bridelia abyssinica Pax, Bot. Jahrb. Syst. 39: 630 (1907).
Bridelia abyssinica var. *densiflora* Gehrm., Bot. Jahrb. Syst. 41(95): 41 (1908).
Bridelia abyssinica var. *rosenii* Gehrm., Bot. Jahrb. Syst. 41(95): 41 (1908).
Bridelia mildbraedii Gehrm., Jahresber. Schles. Ges. Vaterl. Cult. 86: 2 (1908).
Bridelia zanzibarensis var. *sericea* Gehrm., Bot. Jahrb. Syst. 41(95): 40 (1908).

Bridelia microphylla Chiov., Result. Sci. Miss. Stefan.-Paoli Somal. Ital. 1: 160 (1916).
Somalia. 24 SOM.

Bridelia mollis Hutch., Bull. Misc. Inform. Kew 1912: 100 (1912).
SC. Trop. & S. Africa. 26 MLW MOZ ZAM ZIM 27 BOT NAM TVL. Phan.

Bridelia montana (Roxb.) Willd., Sp. Pl. 4: 978 (1806).
Nepal, Bhutan, Sikkim, Assam, China (SE. Yunnan). 36 CHC 40 ASS EHM NEP. Nanophan. or phan.
* *Clutia montana* Roxb., Pl. Coromandel 2: 38 (1802).
Andrachne elliptica Roth, Nov. Pl. Sp.: 364 (1821).

Bridelia moonii Thwaites, Enum. Pl. Zeyl.: 279 (1861).
Sri Lanka. 40 SRL. Phan.

Bridelia ndellensis Beille, Bull. Soc. Bot. France 55(8): 69 (1908).
Nigeria to Uganda. 22 NGA 23 CAF CMN ZAI UGA 24 CHA SUD 25 UGA. Nanophan. or phan.
Gentilia hygrophila Beille, Compt. Rend. Hebd. Séances Acad. Sci. 145: 70 (1908).
Bridelia ferruginea var. *orientalis* Hutch. in D.Oliver, Fl. Trop. Afr. 6(1): 620 (1912).

Bridelia oligantha Airy Shaw, Kew Bull. 32: 386 (1978).
Papua New Guinea (near Port Moresby). 42 NWG. Nanophan. or phan.

Bridelia ovata Decne., Nouv. Ann. Mus. Hist. Nat. 3: 484 (1834). *Bridelia ovata* var. *genuina* Müll.Arg. in A.P.de Candolle, Prodr. 15(2): 495 (1866), nom. inval.
Indo-China, W. Malesia. 41 AND BMA NCB THA 42 JAW LSI MLY. (Cl.) nanophan. or phan.
Bridelia lanceolata Kurz ex Teijsm. & Binn., Tijdschr. Ned.-Indië 27: 45 (1864).
Bridelia ovata var. *acutifolia* Müll.Arg. in A.P.de Candolle, Prodr. 15(2): 495 (1866).
Bridelia amoena Kurz, Forest Fl. Burma 2: 368 (1877), sensu auct.
Bridelia burmanica Hook.f., Fl. Brit. India 5: 269 (1887).
Bridelia kurzii Hook.f., Fl. Brit. India 5: 272 (1887).
Bridelia tomentosa var. *oblonga* Gehrm., Bot. Jahrb. Syst. 41(95): 32 (1908).
Bridelia pedicellata Ridl., J. Straits Branch Roy. Asiat. Soc. 59: 167 (1911).

Bridelia parvifolia Kuntze, Revis. Gen. Pl. 2: 594 (1891).
Hainan, Vietnam. 36 CHH 41 VIE. Nanophan.
Bridelia poilanei Gagnep., Bull. Soc. Bot. France 70: 434 (1923).

Bridelia pervilleana Baill., Adansonia 2: 38 (1861).
Madagascar. 29 MDG. Phan.
Bridelia berneriana Baill., Adansonia 2: 39 (1861).
Bridelia pervilleana var. *humbertii* Leandri, Notul. Syst. (Paris) 11: 154 (1944).

Bridelia pustulata Hook.f., Fl. Brit. India 5: 271 (1887).
Pen. Malaysia (incl. Singapore), N. Sumatera, Borneo, Philippines. 42 BOR MLY PHI SUM. Phan.

Bridelia retusa (L.) A.Juss., Euphorb. Gen.: 109 (1824). – FIGURE, p. 290.
S. China (WSW. Yunnan), India to W. Malesia. 36 CHC 40 ASS EHM IND SRL 41 BMA LAO THA VIE 42 MLY SUM. Nanophan. or phan.
* *Clutia retusa* L., Sp. Pl.: 1475 (1753). *Bridelia retusa* var. *genuina* Müll.Arg. in A.P.de Candolle, Prodr. 15(2): 493 (1866), nom. inval. *Bridelia airy-shawii* P.T.Li, Acta Phytotax. Sin. 20: 117 (1982), nom. illeg.

Bridelia retusa (L.) A. Juss. (as *Briedelia retusa*)
Artist: W.H. Fitch
Brandis, Forest Fl. N.W. India, Illustrations, pl. 55 (1874)
KEW ILLUSTRATIONS COLLECTION

Clutia squamosa Lam., Encycl. 2: 54 (1786). *Bridelia retusa* var. *squamosa* (Lam.) Müll.Arg. in A.P.de Candolle, Prodr. 15(2): 493 (1866). *Bridelia squamosa* var. *typica* Gehrm., Bot. Jahrb. Syst. 41(95): 30 (1908), nom. inval. *Bridelia squamosa* (Lam.) Gehrm., Bot. Jahrb. Syst. 41(95): 30 (1908).

Clutia spinosa Roxb., Pl. Coromandel 2: 38 (1802). *Bridelia spinosa* (Roxb.) Willd., Sp. Pl. 4: 979 (1806).

Bridelia fruticosa Pers., Syn. Pl. 2: 591 (1807).

Bridelia crenulata Roxb., Fl. Ind. ed. 1832, 3: 734 (1832). *Bridelia retusa* var. *roxburghiana* Müll.Arg. in A.P.de Candolle, Prodr. 15(2): 493 (1866). *Bridelia roxburghiana* (Müll.Arg.) Gehrm., Bot. Jahrb. Syst. 41(95): 30 (1908).
Andrachne doonkyboisca B.Heyne ex Wall., Numer. List: 7879 (1847), nom. inval.
Bridelia amoena Wall. ex Baill., Étude Euphorb.: 584 (1858).
Bridelia retusa var. *glauca* Hook.f., Fl. Brit. India 5: 268 (1887).
Bridelia retusa var. *glabra* Gehrm., Bot. Jahrb. Syst. 41(95): 30 (1908).
Bridelia retusa var. *pubescens* Gehrm., Bot. Jahrb. Syst. 41(95): 30 (1908).
Bridelia retusa var. *stipulata* Gehrm., Bot. Jahrb. Syst. 41(95): 30 (1908).
Bridelia squamosa var. *meeboldii* Gehrm., Bot. Jahrb. Syst. 41(95): 30 (1908).
Bridelia cambodiana Gagnep., Bull. Soc. Bot. France 70: 432 (1923).
Bridelia pierrei Gagnep., Bull. Soc. Bot. France 70: 434 (1923).
Bridelia chineensis Thin, J. Biol. (Vietnam) 9: 37 (1987).

Bridelia rhomboidalis Baill., Adansonia 2: 37 (1861).
Madagascar. 29 MDG. Nanophan. to phan. – Provisionally accepted.

Bridelia ripicola J.Léonard, Bull. Jard. Bot. État 25: 370 (1955).
Zaire. 23 ZAI. Nanophan. or phan.

Bridelia scleroneura Müll.Arg., Flora 47: 515 (1864).
Trop. Africa, Yemen. 22 BEN GHA NGA TOG 23 CAF ZAI 24 ETH 25 KEN TAN UGA 26 ANG ZAM ZIM 35 YEM. Nanophan. or phan.

subsp. **angolensis** (Welw. ex Müll.Arg.) Radcl.-Sm., Kew Bull. 51: 303 (1996).
Zaire, Angola. 23 ZAI 26 ANG.
* *Bridelia angolensis* Müll.Arg., J. Bot. 2: 327 (1864). *Bridelia angolensis* var. *typica* Gehrm., Bot. Jahrb. Syst. 41(95): 31 (1908), nom. inval.
Bridelia angolensis var. *welwitschii* Gehrm., Bot. Jahrb. Syst. 41(95): 31 (1908).

f. **habroneura** Radcl.-Sm., Kew Bull. 36: 222 (1981).
Yemen. 35 YEM. Nanophan. or phan.
* *Bridelia tomentosa* var. *glabrata* Schweinf., Bull. Herb. Boissier 7: 305 (1899).

subsp. **scleroneura**
Trop. Africa. 22 BEN GHA NGA TOG 23 CAF 24 ETH 25 KEN TAN UGA 26 ANG ZAM ZIM. Nanophan. or phan.
Bridelia scleroneuroides Pax, Bot. Jahrb. Syst. 15: 532 (1893). *Bridelia scleroneuroides* var. *typica* Gehrm., Bot. Jahrb. Syst. 41(95): 32 (1908), nom. inval.
Bridelia angolensis var. *nitida* Beille, Bull. Soc. Bot. France 40(8): 70 (1908).
Bridelia paxii Gehrm., Bot. Jahrb. Syst. 41(95): 31 (1908).
Bridelia scleroneura var. *barteri* Gehrm., Bot. Jahrb. Syst. 41(95): 31 (1908).
Bridelia scleroneura var. *togoensis* Gehrm., Bot. Jahrb. Syst. 41(95): 31 (1908).
Bridelia scleroneuroides var. *elliptica* Gehrm., Bot. Jahrb. Syst. 41(95): 32 (1908).
Tzellemtinia nervosa Chiov., Ann. Bot. (Rome) 9: 56 (1911).

Bridelia sikkimensis Gehrm., Bot. Jahrb. Syst. 41(95): 34 (1908).
Pakistan to Assam. 40 ASS BAN EHM NEP PAK. Nanophan. or phan.
Bridelia sikkimensis var. *macrophylla* Gehrm., Bot. Jahrb. Syst. 41(95): 34 (1908).
Bridelia sikkimensis var. *minuta* Gehrm., Bot. Jahrb. Syst. 41(95): 34 (1908).
Bridelia verrucosa Haines, J. Bot. 59: 189, 193 (1921).

Bridelia somalensis Hutch., Bull. Misc. Inform. Kew 1931: 413 (1931).
Somalia. 24 SOM.

Bridelia speciosa Müll.Arg., J. Bot. 2: 327 (1864).
Nigeria, Cameroon. 22 NGA 23 CMN. Nanophan. or phan.

Bridelia speciosa var. *psiloclada* Müll.Arg., J. Bot. 2: 327 (1864).
Bridelia perrotii Beille, Bull. Soc. Bot. France 55(8): 69 (1908).
Bridelia speciosa var. *medinanensis* Beille, Bull. Soc. Bot. France 55(8): 68 (1908).

Bridelia stipularis (L.) Blume, Bijdr.: 597 (1826).
Trop. & Subtrop. Asia. 36 CHC CHH CHS 38 TAI 40 ASS BAN EHM IND NEP WHM 41 BMA THA VIE 42 BOR JAW LSI MLY PHI SUM. (Cl.) nanophan. or phan.
* *Clutia stipularis* L., Mant. Pl.: 127 (1776). *Bridelia stipularis* var. *typica* Gehrm., Bot. Jahrb. Syst. 41(95): 29 (1908), nom. inval.
Clutia scandens Roxb., Pl. Coromandel 2: 39 (1802). *Bridelia scandens* (Roxb.) Willd., Sp. Pl. 4: 979 (1806).
Bridelia zollingeri Miq., Fl. Ned. Ind. 1(2): 364 (1859).
Bridelia dasycalyx Kurz, J. Asiat. Soc. Bengal, Pt. 2, Nat. Hist. 42(2): 241 (1873).
Bridelia dasycalyx var. *aridicola* Kurz, J. Asiat. Soc. Bengal, Pt. 2, Nat. Hist. 42(2): 241 (1873).
Bridelia stipularis var. *ciliata* Gehrm., Bot. Jahrb. Syst. 41(95): 29 (1908).
Bridelia montana Woodrow ex J.J.Sm. in S.H.Koorders & T.Valeton, Bijdr. Boomsoort. Java 12: 315 (1910), nom. nud.
Bridelia stipularis subsp. *philippinensis* Jabl. in H.G.A.Engler, Pflanzenr., IV, 147, VIII: 57 (1915).

Bridelia taitensis Vatke & Pax ex Pax, Bot. Jahrb. Syst. 15: 531 (1893).
Kenya. 25 KEN. Nanophan. or phan.
Bridelia nigricans Gehrm., Bot. Jahrb. Syst. 41(95): 35 (1908).

Bridelia tenuifolia Müll.Arg., J. Bot. 2: 328 (1864).
Angola, Namibia. 26 ANG 27 NAM. Nanophan. or phan.
Bridelia elegans Müll.Arg., J. Bot. 2: 327 (1864). *Bridelia tenuifolia* var. *elegans* (Müll.Arg.) Hutch. in D.Oliver, Fl. Trop. Afr. 6(1): 615 (1912).

Bridelia tomentosa Blume, Bijdr.: 597 (1826). *Bridelia tomentosa* var. *genuina* Müll.Arg. in A.P.de Candolle, Prodr. 15(2): 501 (1866), nom. inval.
Trop. & Subtrop. Asia, N. Australia. 36 CHC CHH CHS 38 TAI 40 ASS BAN EHM NEP WHM 41 AND BMA CBD NCB THA VIE 42 BOR JAW LSI MLY MOL NWG PHI SUL SUM 50 NTA QLD. (Cl.) nanophan. or phan.
Bridelia lanceifolia Roxb., Fl. Ind. ed. 1832, 3: 737 (1832). *Bridelia tomentosa* var. *lanceifolia* (Roxb.) Müll.Arg. in A.P.de Candolle, Prodr. 15(2): 502 (1866).
Bridelia loureirii Hook. & Arn., Bot. Beechey Voy.: 211 (1837).
Bridelia lancifolia Buch.-Ham. in N.Wallich, Numer. List: 7884 (1847), nom. inval.
Bridelia rhamnoides Griff., Not. Pl. Asiat. 4: 480 (1854). *Bridelia tomentosa* var. *rhamnoides* (Griff.) Müll.Arg. in A.P.de Candolle, Prodr. 15(2): 502 (1866).
Bridelia tomentosa var. *glabrescens* Benth., Hooker's J. Bot. Kew Gard. Misc. 6: 8 (1854).
Bridelia urticoides Griff., Not. Pl. Asiat. 4: 481 (1854).
Amanoa tomentosa Baill., Adansonia 6: 336 (1866).
Bridelia tomentosa var. *chinensis* Müll.Arg. in A.P.de Candolle, Prodr. 15(2): 501 (1866).
Bridelia tomentosa var. *trichadenia* Müll.Arg. in A.P.de Candolle, Prodr. 15(2): 501 (1866).
Cleistanthus lanceolatus Müll.Arg. in A.P.de Candolle, Prodr. 15(2): 507 (1866).
Phyllanthus loureirii Müll.Arg. in A.P.de Candolle, Prodr. 15(2): 435 (1866), sphalm.
Bridelia tomentosa var. *ovoidea* Benth., Fl. Austral. 6: 120 (1873).
Bridelia monoica Merr., Philipp. J. Sci., C 13: 142 (1918).
Bridelia phyllanthoides W.Fitzg., J. Roy. Soc. W. Australia 3: 163 (1918).
Bridelia glabrifolia Merr., Enum. Philipp. Fl. Pl. 2: 422 (1923). *Bridelia tomentosa* var. *glabrifolia* (Merr.) Airy Shaw, Kew Bull. 31: 383 (1976).
Bridelia tomentosa var. *eriantha* Airy Shaw, Kew Bull. 31: 384 (1976).
Bridelia nayarii P.Basu, J. Econ. Taxon. Bot. 7(3): 634 (1985 publ. 1986).

Bridelia triplocarya Airy Shaw, Kew Bull. 32: 385 (1978).
Papua New Guinea. 42 NWG. Phan.

Bridelia tulasneana Baill., Adansonia 2: 40 (1861).
E. & N. Madagascar. 29 MDG. Phan.
Bridelia coccolobifolia Baker, J. Linn. Soc., Bot. 21: 441 (1885).

Bridelia whitmorei Airy Shaw, Kew Bull. 27: 77 (1972).
Pen. Malaysia (Pahang). 42 MLY. Cl. phan.

Synonyms:

Bridelia abyssinica Pax === **Bridelia micrantha** (Hochst.) Baill. var. **micrantha**
Bridelia abyssinica var. *densiflora* Gehrm. === **Bridelia micrantha** (Hochst.) Baill. var. **micrantha**
Bridelia abyssinica var. *rosenii* Gehrm. === **Bridelia micrantha** (Hochst.) Baill. var. **micrantha**
Bridelia acuminata Wall. === **Glochidion triandrum** (Blanco) C.B.Rob.
Bridelia acuminatissima Merr. === **Bridelia glauca** Blume
Bridelia airy-shawii P.T.Li === **Bridelia retusa** (L.) A.Juss.
Bridelia amoena Wall. ex Baill. === **Bridelia retusa** (L.) A.Juss.
Bridelia amoena Kurz === **Bridelia ovata** Decne.
Bridelia angolensis Müll.Arg. === **Bridelia scleroneura** subsp. **angolensis** (Welw. ex Müll.Arg.) Radcl.-Sm.
Bridelia angolensis var. *nitida* Beille === **Bridelia scleroneura** Müll.Arg. subsp. **scleroneura**
Bridelia angolensis var. *typica* Gehrm. === **Bridelia scleroneura** subsp. **angolensis** (Welw. ex Müll.Arg.) Radcl.-Sm.
Bridelia angolensis var. *welwitschii* Gehrm. === **Bridelia scleroneura** subsp. **angolensis** (Welw. ex Müll.Arg.) Radcl.-Sm.
Bridelia attenuata Wall. ex Voigt === **Cleistanthus oblongifolius** (Roxb.) Müll.Arg.
Bridelia aubrevillei Pellegr. === **Bridelia grandis** Pierre ex Hutch. subsp. **grandis**
Bridelia balansae Tutcher === **Bridelia insulana** Hance
Bridelia beguinii Airy Shaw === **Bridelia macrocarpa** Airy Shaw
Bridelia berneriana Baill. === **Bridelia pervilleana** Baill.
Bridelia berryana Wall. === **Margaritaria indica** (Dalzell) Airy Shaw
Bridelia brideliifolia subsp. *pubescentifolia* J.Léonard === **Bridelia brideliifolia** (Pax) Fedde
Bridelia burmanica Hook.f. === **Bridelia ovata** Decne.
Bridelia buxifolia Baill. === **Cleistanthus stipitatus** (Baill.) Müll.Arg.
Bridelia cambodiana Gagnep. === **Bridelia retusa** (L.) A.Juss.
Bridelia cathartica f. *fischeri* (Pax) Radcl.-Sm. === **Bridelia cathartica** subsp. **melanthesoides** (Klotzsch ex Baill.) J.Léonard
Bridelia cathartica var. *lingelsheimii* (Gehrm.) Radcl.-Sm. === **Bridelia cathartica** subsp. **melanthesoides** (Klotzsch ex Baill.) J.Léonard
Bridelia cathartica var. *melanthesoides* (Klotzsch ex Baill.) Radcl.-Sm. === **Bridelia cathartica** subsp. **melanthesoides** (Klotzsch ex Baill.) J.Léonard
Bridelia cathartica f. *melanthesoides* (Klotzsch ex Baill.) Radcl.-Sm. === **Bridelia cathartica** subsp. **melanthesoides** (Klotzsch ex Baill.) J.Léonard
Bridelia cathartica f. *niedenzui* (Gehrm.) Radcl.-Sm. === **Bridelia cathartica** subsp. **melanthesoides** (Klotzsch ex Baill.) J.Léonard
Bridelia cathartica f. *pubescens* Radcl.-Sm. === **Bridelia cathartica** subsp. **melanthesoides** (Klotzsch ex Baill.) J.Léonard
Bridelia chartacea Wall. === **Cleistanthus oblongifolius** (Roxb.) Müll.Arg.
Bridelia chartacea Kurz === **Bridelia assamica** Hook.f.
Bridelia chineensis Thin === **Bridelia retusa** (L.) A.Juss.
Bridelia coccolobifolia Baker === **Bridelia tulasneana** Baill.
Bridelia collina (Roxb.) Hook. & Arn. === **Cleistanthus collinus** (Roxb.) Benth.
Bridelia colorata Airy Shaw === **Bridelia affinis** Craib
Bridelia crenulata Roxb. === **Bridelia retusa** (L.) A.Juss.

Bridelia dasycalyx Kurz === **Bridelia stipularis** (L.) Blume
Bridelia dasycalyx var. *aridicola* Kurz === **Bridelia stipularis** (L.) Blume
Bridelia diversifolia (Roxb.) Hook. & Arn. === **Cleistanthus diversifolius** (Roxb.) Müll.Arg.
Bridelia elegans Müll.Arg. === **Bridelia tenuifolia** Müll.Arg.
Bridelia faginea (Baill.) F.Muell. ex Benth. === **Bridelia leichardtii** Baill. ex Müll.Arg.
Bridelia ferruginea var. *gambicola* Hiern === **Bridelia micrantha** (Hochst.) Baill. var. **micrantha**
Bridelia ferruginea var. *orientalis* Hutch. === **Bridelia ndellensis** Beille
Bridelia fischeri Pax === **Bridelia cathartica** subsp. **melanthesoides** (Klotzsch ex Baill.) J.Léonard
Bridelia fischeri var. *lingelsheimii* (Gehrm.) Hutch. === **Bridelia cathartica** subsp. **melanthesoides** (Klotzsch ex Baill.) J.Léonard
Bridelia fruticosa Pers. === **Bridelia retusa** (L.) A.Juss.
Bridelia gambicola Baill. === **Bridelia micrantha** var. **gambicola** (Baill.) Müll.Arg.
Bridelia gehrmannii Jabl. === **Bridelia cinnamomea** Hook.f.
Bridelia glabrifolia Merr. === **Bridelia tomentosa** Blume
Bridelia glauca Wall. === **Glochidion rubrum** Blume var. **rubrum**
Bridelia glauca var. *acuminatissima* (Merr.) S.Dressler === **Bridelia glauca** Blume
Bridelia glauca f. *laurifolia* (Elmer) Jabl. === **Bridelia glauca** Blume
Bridelia glauca var. *sosopodonica* (Airy Shaw) S.Dressler === **Bridelia glauca** Blume
Bridelia griffithii Hook.f. === **Bridelia cinnamomea** Hook.f.
Bridelia griffithii var. *cinnamomea* (Hook.f.) Gehrm. === **Bridelia cinnamomea** Hook.f.
Bridelia griffithii var. *glabra* Gehrm. === **Bridelia insulana** Hance
Bridelia griffithii var. *penangiana* (Hook.f.) Gehrm. === **Bridelia insulana** Hance
Bridelia hamiltoniana var. *genuina* Müll.Arg. === **Bridelia hamiltoniana** Wall. ex Müll.Arg.
Bridelia hamiltoniana var. *glabra* Müll.Arg. === **Bridelia hamiltoniana** Wall. ex Müll.Arg.
Bridelia henryana Jabl. === **Bridelia affinis** Craib
Bridelia heterantha Wall. === **Glochidion glomerulatum** (Miq.) Boerl.
Bridelia horrida Dillwyn === **Scleropyrum pentandrum** (Dennst.) D.J.Mabb.
Bridelia insulana var. *subnuda* (K.Schum. & Lauterb.) S.Dressler === **Bridelia insulana** Hance
Bridelia kawakamii Hayata === **Bridelia insulana** Hance
Bridelia kurzii Hook.f. === **Bridelia ovata** Decne.
Bridelia lanceifolia Roxb. === **Bridelia tomentosa** Blume
Bridelia lanceolata Kurz ex Teijsm. & Binn. === **Bridelia ovata** Decne.
Bridelia lancifolia Buch.-Ham. === **Bridelia tomentosa** Blume
Bridelia laurifolia Elmer === **Bridelia glauca** Blume
Bridelia laurina Baill. === **Cleistanthus stipitatus** (Baill.) Müll.Arg.
Bridelia lingelsheimii Gehrm. === **Bridelia cathartica** subsp. **melanthesoides** (Klotzsch ex Baill.) J.Léonard
Bridelia loureirii Hook. & Arn. === **Bridelia tomentosa** Blume
Bridelia melanthesoides (Klotzsch ex Baill.) Klotzsch === **Bridelia cathartica** subsp. **melanthesoides** (Klotzsch ex Baill.) J.Léonard
Bridelia melanthesoides var. *australis* Gehrm. === **Bridelia leichardtii** Baill. ex Müll.Arg.
Bridelia melanthesoides var. *lanceolata* Gehrm. === **Bridelia cathartica** subsp. **melanthesoides** (Klotzsch ex Baill.) J.Léonard
Bridelia melanthesoides var. *ovata* Gehrm. === **Bridelia cathartica** subsp. **melanthesoides** (Klotzsch ex Baill.) J.Léonard
Bridelia melanthesoides var. *typica* Gehrm. === **Bridelia cathartica** subsp. **melanthesoides** (Klotzsch ex Baill.) J.Léonard
Bridelia micrantha var. *ferruginea* (Benth.) Müll.Arg. === **Bridelia ferruginea** Benth.
Bridelia micrantha var. *genuina* Müll.Arg. === **Bridelia micrantha** (Hochst.) Baill.
Bridelia mildbraedii Gehrm. === **Bridelia micrantha** (Hochst.) Baill. var. **micrantha**
Bridelia minutiflora Hook.f. === **Bridelia insulana** Hance
Bridelia minutiflora var. *abbreviata* J.J.Sm. === **Bridelia insulana** Hance
Bridelia monoica Merr. === **Bridelia tomentosa** Blume
Bridelia montana Woodrow ex J.J.Sm. === **Bridelia stipularis** (L.) Blume

Bridelia morotaea Airy Shaw === **Bridelia insulana** Hance
Bridelia multiflora Zipp. ex Scheff. === **Bridelia glauca** Blume
Bridelia nayarii P.Basu === **Bridelia tomentosa** Blume
Bridelia neogoetzea Gehrm. === **Bridelia brideliifolia** (Pax) Fedde
Bridelia nicobarica Chakrab. & Vasudeva Rao === **Bridelia insulana** Hance
Bridelia niedenzui Gehrm. === **Bridelia cathartica** subsp. **melanthesoides** (Klotzsch ex Baill.) J.Léonard
Bridelia niedenzui var. *njassae* Gehrm. === **Bridelia cathartica** subsp. **melanthesoides** (Klotzsch ex Baill.) J.Léonard
Bridelia niedenzui var. *pilosa* Gehrm. === **Bridelia cathartica** subsp. **melanthesoides** (Klotzsch ex Baill.) J.Léonard
Bridelia niedenzui var. *revoluta* Gehrm. === **Bridelia cathartica** subsp. **melanthesoides** (Klotzsch ex Baill.) J.Léonard
Bridelia nigricans Gehrm. === **Bridelia taitensis** Vatke & Pax ex Pax
Bridelia nooteboomii Chakrab. === **Bridelia glauca** Blume
Bridelia oblongifolius (Roxb.) Hook. & Arn. === **Cleistanthus oblongifolius** (Roxb.) Müll.Arg.
Bridelia ovata var. *acutifolia* Müll.Arg. === **Bridelia ovata** Decne.
Bridelia ovata var. *curtisii* (Hook.f.) Airy Shaw === **Bridelia curtisii** Hook.f.
Bridelia ovata var. *exaltata* (F.Muell.) Müll.Arg. === **Bridelia exaltata** F.Muell.
Bridelia ovata var. *genuina* Müll.Arg. === **Bridelia ovata** Decne.
Bridelia pachinensis Hayata === **Bridelia insulana** Hance
Bridelia palauensis Kaneh. ex Kaneh. & Hatus. === **Bridelia insulana** Hance
Bridelia patula (Roxb.) Hook. & Arn. === **Cleistanthus patulus** (Roxb.) Müll.Arg.
Bridelia paxii Gehrm. === **Bridelia scleroneura** Müll.Arg. subsp. **scleroneura**
Bridelia pedicellata Ridl. === **Bridelia ovata** Decne.
Bridelia penangiana Hook.f. === **Bridelia insulana** Hance
Bridelia penangiana var. *subnuda* (K.Schum. & Lauterb.) Airy Shaw === **Bridelia insulana** Hance
Bridelia perrotii Beille === **Bridelia speciosa** Müll.Arg.
Bridelia pervilleana var. *humbertii* Leandri === **Bridelia pervilleana** Baill.
Bridelia phyllanthoides W.Fitzg. === **Bridelia tomentosa** Blume
Bridelia pierrei Gagnep. === **Bridelia retusa** (L.) A.Juss.
Bridelia platyphylla Merr. === **Bridelia insulana** Hance
Bridelia poilanei Gagnep. === **Bridelia parvifolia** Kuntze
Bridelia polystachya (Hook.f. ex Planch.) Baill. === **Cleistanthus polystachyus** Hook.f. ex Planch.
Bridelia pubescens Kurz === **Bridelia glauca** Blume
Bridelia ramiflora Gehrm. ex Jabl. === **Bridelia brideliifolia** (Pax) Fedde
Bridelia retusa var. *genuina* Müll.Arg. === **Bridelia retusa** (L.) A.Juss.
Bridelia retusa var. *glabra* Gehrm. === **Bridelia retusa** (L.) A.Juss.
Bridelia retusa var. *glauca* Hook.f. === **Bridelia retusa** (L.) A.Juss.
Bridelia retusa var. *pubescens* Gehrm. === **Bridelia retusa** (L.) A.Juss.
Bridelia retusa var. *roxburghiana* Müll.Arg. === **Bridelia retusa** (L.) A.Juss.
Bridelia retusa var. *squamosa* (Lam.) Müll.Arg. === **Bridelia retusa** (L.) A.Juss.
Bridelia retusa var. *stipulata* Gehrm. === **Bridelia retusa** (L.) A.Juss.
Bridelia rhamnoides Griff. === **Bridelia tomentosa** Blume
Bridelia roxburghiana (Müll.Arg.) Gehrm. === **Bridelia retusa** (L.) A.Juss.
Bridelia rufa Hook.f. === **Cleistanthus rufus** (Hook.f.) Gehrm.
Bridelia rugosa Miq. === **Aporusa lunata** (Miq.) Kurz
Bridelia scandens (Roxb.) Willd. === **Bridelia stipularis** (L.) Blume
Bridelia schlechteri Hutch. === **Bridelia cathartica** Bertol. subsp. **cathartica**
Bridelia scleroneura var. *barteri* Gehrm. === **Bridelia scleroneura** Müll.Arg. subsp. **scleroneura**
Bridelia scleroneura var. *togoensis* Gehrm. === **Bridelia scleroneura** Müll.Arg. subsp. **scleroneura**
Bridelia scleroneuroides Pax === **Bridelia scleroneura** Müll.Arg. subsp. **scleroneura**

Bridelia scleroneuroides var. *elliptica* Gehrm. === **Bridelia scleroneura** Müll.Arg. subsp. **scleroneura**
Bridelia scleroneuroides var. *typica* Gehrm. === **Bridelia scleroneura** Müll.Arg. subsp. **scleroneura**
Bridelia sikkimensis var. *macrophylla* Gehrm. === **Bridelia sikkimensis** Gehrm.
Bridelia sikkimensis var. *minuta* Gehrm. === **Bridelia sikkimensis** Gehrm.
Bridelia sinica J.Graham === **Glochidion hohenackeri** (Müll.Arg.) Bedd. var. **hohenackeri**
Bridelia sosopodonica Airy Shaw === **Bridelia glauca** Blume
Bridelia speciosa var. *kourousensis* Beille === **Bridelia ferruginea** Benth.
Bridelia speciosa var. *medinanensis* Beille === **Bridelia speciosa** Müll.Arg.
Bridelia speciosa var. *psiloclada* Müll.Arg. === **Bridelia speciosa** Müll.Arg.
Bridelia speciosa var. *trichoclada* Müll.Arg. === **Bridelia micrantha** (Hochst.) Baill. var. **micrantha**
Bridelia spinosa (Roxb.) Willd. === **Bridelia retusa** (L.) A.Juss.
Bridelia spinosa DC. === **Damnacanthus** sp. (Rubiaceae)
Bridelia squamosa (Lam.) Gehrm. === **Bridelia retusa** (L.) A.Juss.
Bridelia squamosa var. *meeboldii* Gehrm. === **Bridelia retusa** (L.) A.Juss.
Bridelia squamosa var. *typica* Gehrm. === **Bridelia retusa** (L.) A.Juss.
Bridelia stenocarpa Müll.Arg. === **Bridelia micrantha** (Hochst.) Baill. var. **micrantha**
Bridelia stipitata Baill. === **Cleistanthus stipitatus** (Baill.) Müll.Arg.
Bridelia stipularis Hook. & Arn. === **Cleistanthus stipularis** (Kuntze) Müll.Arg.
Bridelia stipularis var. *ciliata* Gehrm. === **Bridelia stipularis** (L.) Blume
Bridelia stipularis subsp. *philippinensis* Jabl. === **Bridelia stipularis** (L.) Blume
Bridelia stipularis var. *typica* Gehrm. === **Bridelia stipularis** (L.) Blume
Bridelia subnuda K.Schum. & Lauterb. === **Bridelia insulana** Hance
Bridelia tenuifolia var. *elegans* (Müll.Arg.) Hutch. === **Bridelia tenuifolia** Müll.Arg.
Bridelia tomentosa var. *chinensis* Müll.Arg. === **Bridelia tomentosa** Blume
Bridelia tomentosa var. *eriantha* Airy Shaw === **Bridelia tomentosa** Blume
Bridelia tomentosa var. *genuina* Müll.Arg. === **Bridelia tomentosa** Blume
Bridelia tomentosa var. *glabrata* Schweinf. === **Bridelia scleroneura** f. **habroneura** Radcl.-Sm.
Bridelia tomentosa var. *glabrescens* Benth. === **Bridelia tomentosa** Blume
Bridelia tomentosa var. *glabrifolia* (Merr.) Airy Shaw === **Bridelia tomentosa** Blume
Bridelia tomentosa var. *lanceifolia* (Roxb.) Müll.Arg. === **Bridelia tomentosa** Blume
Bridelia tomentosa var. *oblonga* Gehrm. === **Bridelia ovata** Decne.
Bridelia tomentosa var. *ovoidea* Benth. === **Bridelia tomentosa** Blume
Bridelia tomentosa var. *rhamnoides* (Griff.) Müll.Arg. === **Bridelia tomentosa** Blume
Bridelia tomentosa var. *trichadenia* Müll.Arg. === **Bridelia tomentosa** Blume
Bridelia urticoides Griff. === **Bridelia tomentosa** Blume
Bridelia verrucosa Haines === **Bridelia sikkimensis** Gehrm.
Bridelia zanzibarensis Vatke & Pax === **Bridelia micrantha** (Hochst.) Baill. var. **micrantha**
Bridelia zanzibarensis var. *sericea* Gehrm. === **Bridelia micrantha** (Hochst.) Baill. var. **micrantha**
Bridelia zenkeri Pax === **Bridelia atroviridis** Müll.Arg.
Bridelia zollingeri Miq. === **Bridelia stipularis** (L.) Blume

Briedelia

The correct orthography for *Bridelia*; conservation of the commonly used (but incorrect) spelling has, however, been proposed (Dressler, 1996, in *Taxon* 45; see above) and subsequently accepted.

Bromfieldia

A Necker name, reduced (with doubt) to *Jatropha* in *Index Kewensis*.

Bruea

= *Laportea* (Urticaceae). Originally described by Gaudichaud (*Voyage ... sur les corvettes S.M. l'Uranie et La Physicienne par M. L. Freycinet, Botanique*: 511. 1826) in Urticaceae with a single species, *B. bengalensis,* but later referred to *Macaranga* (Baillon 1866). No mention of this genus is made by Webster (Synopsis, 1994).

Baillon, H. (1866). Sur le genre *Bruea*. Adansonia 7: 96-97. Fr. — Referred to *Macaranga* as *M. bengalensis*; range scattered from India to Australia (in the latter corresponding to *M. involucrata* var. *mallotoides*).

Brunsvia

Synonyms:
Brunsvia Neck. === **Croton** L.

Buraeavia

Conserved against *Bureava* Baill. (Combretaceae) but here merged with *Austrobuxus*.

Synonyms:
Buraeavia Baill. === **Austrobuxus** Miq.
Buraeavia carunculata (Baill.) Baill. === **Austrobuxus carunculatus** (Baill.) Airy Shaw
Buraeavia clusiaceus Baill. === **Austrobuxus clusiaceus** (Baill.) Airy Shaw
Buraeavia horneana A.C.Sm. === **Austrobuxus horneanus** (A.C.Sm.) Airy Shaw
Buraeavia rubiginosa Guillaumin === **Austrobuxus rubiginosus** (Guillaumin) Airy Shaw

Bureaua

An orthographic variant of *Buraeavia*, proposed by Kuntze.

Caelodepas

An orthographic variant of *Koilodepas*.

Synonyms:
Caelodepas Benth. & Hook.f. === **Koilodepas** Hassk.

Caletia

Synonyms:
Caletia Baill. === **Micrantheum** Desf.
Caletia divaricatissima Müll.Arg. === **Pseudanthus divaricatissimus** (Müll.Arg.) Benth.
Caletia divaricatissima var. *genuinus* Müll.Arg. === **Pseudanthus divaricatissimus** (Müll.Arg.) Benth.
Caletia divaricatissima var. *orbicularis* Müll.Arg. === **Pseudanthus divaricatissimus** (Müll.Arg.) Benth.
Caletia ericodes (Desf.) Kuntze === **Micrantheum ericoides** Desf.
Caletia hexandra (Hook.f.) Müll.Arg. === **Micrantheum hexandrum** Hook.f.
Caletia linearis Müll.Arg. === **Pseudanthus orientalis** F.Muell.
Caletia micrantheoides Baill. === **Micrantheum hexandrum** Hook.f.
Caletia orientalis (F.Muell.) Baill. === **Pseudanthus orientalis** F.Muell.
Caletia orientalis var. *orbicularis* Baill. === **Pseudanthus divaricatissimus** (Müll.Arg.) Benth.
Caletia ovalifolia (F.Muell.) Müll.Arg. === **Pseudanthus ovalifolius** F.Muell.
Caletia wilhelmii F.Muell. ex Müll.Arg. === **Pseudanthus ovalifolius** F.Muell.

Calococcus

Synonyms:
Calococcus Kurz ex Teijsm. & Binn. === **Margaritaria** L.f.
Calococcus sundaicus Kurz ex Teijsm. & Binn. === **Margaritaria indica** (Dalzell) Airy Shaw

Calpigyne

Synonyms:
Calpigyne Blume === **Koilodepas** Hassk.
Calpigyne frutescens Blume === **Koilodepas frutescens** (Blume) Airy Shaw
Calpigyne hainanensis Merr. === **Koilodepas hainanense** (Merr.) Croizat

Calycopeplus

5 species, Australia (W Australia, Northern Territory, NE Queensland). The genus comprises shrubs or small trees to 10 m in height with a seemingly leafless 'ephedroid' appearance, the twigs themselves being photosynthetic though not succulent and indeed somewhat woody. The largest species is the Cape York-distributed (and appropriately named!) *C. casuarinoides*. A recent critical revision is that of Forster (1995) who, however, draws no conclusions with regard to the position of the genus within the Euphorbieae — in contrast to Webster (1994; see **General**) who grouped it with *Neoguillauminia* into Croizat's subtribe Neoguillauminiinae. (Euphorbioideae (Euphorbieae))

Pax, F. & K. Hoffmann (1931). *Calycopeplus*. In A. Engler (ed.), Die natürlichen Pflanzenfamilien, 2. Aufl., 19c: 221. Leipzig. Ge. — Synopsis with description of genus; 3 species, W Australia. [Now out of date; see Forster 1995.]

• Forster, P. I. (1995). A taxonomic revision of *Calycopeplus* Planch. (Euphorbiaceae). Austrobaileya 4: 417-428, illus., map. En. — Descriptive treatment (5 species) with key, synonymy, references, types, localities with (sometimes selected) exsiccatae, indication of distribution and habitat (with references to 1-degree grid cells; see map, p. 423), commentary, and critical figures; references but no separate index.

Calycopeplus Planch., Bull. Soc. Bot. France 8: 30 (1861).
Australia. 50.

Calycopeplus casuarinoides L.S.Sm., Contr. Queensland Herb. 6: 4 (1969).
N. Queensland. 50 QLD. Nanophan. or phan.

Calycopeplus collinus P.I.Forst., Austrobaileya 4: 420 (1995).
N. Western Australia, Northern Territory. 50 NTA WAU. Nanophan.

Calycopeplus marginatus Benth., Fl. Austral. 6: 53 (1873).
Western Australia (Eyre). 50 WAU. Nanophan. or phan.

Calycopeplus oligandrus P.I.Forst., Austrobaileya 4: 426 (1995).
SW. Western Australia. 50 WAU. Nanophan.

Calycopeplus paucifolius (Klotzsch) Baill., Adansonia 6: 319 (1866).
S. Australia. 50 SOA WAU. Nanophan.
* *Euphorbia paucifolia* Klotzsch in J.G.C.Lehmann, Pl. Preiss. 1: 174 (1845).
Calycopeplus ephedroides Planch., Bull. Soc. Bot. France 8: 31 (1861).
Calycopeplus helmsii F.Muell. & Tate, Trans. Roy. Soc. South Australia 16: 341 (1896).

Synonyms:
Calycopeplus ephedroides Planch. === **Calycopeplus paucifolius** (Klotzsch) Baill.
Calycopeplus helmsii F.Muell. & Tate === **Calycopeplus paucifolius** (Klotzsch) Baill.

Calypteriopetalon

Synonyms:
Calypteriopetalon Hassk. === **Croton** L.

Calyptosepalum

Synonyms:
Calyptosepalum S.Moore === **Drypetes** Vahl
Calyptosepalum pacificum I.W.Bailey & A.C.Sm. === **Drypetes pacifica** (I.W.Bailey & A.C.Sm.) A.C.Sm.
Calyptosepalum sumatranum S.Moore === **Drypetes calyptosepala** Airy Shaw

Calyptriopetalum

Synonyms:
Calyptriopetalum Hassk. ex Müll.Arg. === **Croton** L.

Calyptroon

Synonyms:
Calyptroon Miq. === **Baccaurea** Lour.
Calyptroon sumatranum Miq. === **Baccaurea sumatrana** (Miq.) Müll.Arg.

Calyptrospatha

Synonyms:
Calyptrospatha Klotzsch ex Baill. === **Acalypha** L.
Calyptrospatha pubiflora Klotzsch === **Acalypha pubiflora** (Klotzsch) Baill.

Calyptrostigma

Synonyms:
Calyptrostigma Klotzsch === **Beyeria** Miq.
Calyptrostigma ledifolium Klotzsch === **Beyeria leschenaultii** (DC.) Baill.
Calyptrostigma oblongifolium Klotzsch === **Beyeria viscosa** (Labill.) Miq.

Camirium

Synonyms:
Camirium Gaertn. === **Aleurites** J.R.Forst. & G.Forst.
Camirium cordifolium Gaertn. === **Aleurites moluccana** (L.) Willd.

Canaca

This New Caledonia genus was accepted in *Families and genera of spermatophytes recognized by the Agricultural Research Service* (1992) but by McPherson & Tirel (1987) and Brummitt, *Vascular plant families and genera* (1992) reduced to *Austrobuxus* (g.v.). Some 7 species (nos. 9-15 of McPherson and Tirel) would seem to belong here if arguments for its segregation were accepted. (Oldfieldioideae)

Synonyms:
Canaca Guillaumin === **Austrobuxus** Miq.
Canaca vieillardii Guillaumin === **Austrobuxus vieillardii** (Guillaumin) Airy Shaw

Canariastrum

of Engler, invalidly published; = *Uapaca* Baill.

Candelabria

Synonyms:
Candelabria Hochst. === **Bridelia** Willd.
Candelabria micrantha Hochst. === **Bridelia micrantha** (Hochst.) Baill.
Candelabria polystachya (Hook.f. ex Planch.) Planch. === **Cleistanthus polystachyus** Hook.f. ex Planch.

Canschi

Synonyms:
Canschi Adans. === **Trewia** L.

Caoutchoua

Synonyms:
Caoutchoua J.F.Gmel. === **Hevea** Aubl.

Capellenia

Synonyms:
Capellenia Teijsm. & Binn. === **Endospermum** Benth.
Capellenia moluccana Teijsm. & Binn. === **Endospermum moluccanum** (Teijsm. & Binn.) Kurz

Caperonia

34 species, Madagascar, Africa and the Americas; sometimes prickly annual or perennial herbs or subshrubs with hollow stems usually found in wet places. *CC. stenophylla* and *angustissima* feature very narrow leaves, as illustrated by Pax. Of the 2 sections *Aculeolatae*, all perennial or suffrutescent, are entirely S American with the majority in Brazil. *C. palustris*, an herb with weedy tendencies widespread from the southern U.S.A. to Argentina, has become established in Guam. No full revision has appeared since 1912. (Acalyphoideae)

- Pax, F. (with K. Hoffmann) (1912). *Caperonia*. In A. Engler (ed.), Das Pflanzenreich, IV 147 VI (Euphorbiaceae-Acalypheae-Chrozophorinae): 27-49, illus. Berlin. (Heft 57.) La/Ge. — 33 species in 2 sections, Madagascar, Africa and the Americas; sect. *Aculeolatae* is entirely S American.

  Pax, F. & K. Hoffmann (1924). *Caperonia*. In A. Engler (ed.), Das Pflanzenreich, IV 147 XVII (Euphorbiaceae-Additamentum VII): 179. Berlin. (Heft 85.) La/Ge. — Additions.

  Léonard, J. (1956). Notulae systematicae XX. Contribution à l'étude des *Caperonia* africains (Euphorbiaceae). Bull. Jard. Bot. État 26: 313-320. Fr. — Treatment of 2 species with synonymy, selected exsiccatae, and notes; no key. [Features heavy reductions from the 12 names previously recorded.]

Caperonia A.St.-Hil., Hist. Pl. Remarq. Brésil: 244 (1825).
Mexico, S. America, Trop. & S. Africa, Madagascar. 22 23 24 25 26 27 29 (62) 79 80 81 82 83 84 85.
Cavanilla Vell., Fl. Flumin.: 226 (1829).
Meterana Raf., Sylva Tellur.: 65 (1838).

Lepidococca Turcz., Bull. Soc. Imp. Naturalistes Moscou 21(1): 588 (1848).
Androphoranthus H.Karst., Wochenschr. Gärtnerei Pflanzenk. 2: 5 (1859).
Acanthopyxis Miq. ex Lanj., Euphorb. Surinam: 128 (1931).

Caperonia aculeolata Müll.Arg., Linnaea 34: 152 (1865). *Argythamnia aculeolata* (Müll.Arg.) Kuntze, Revis. Gen. Pl. 2: 593 (1891).
SE. Brazil. 84 BZL. Cham.
Ditaxis polymorpha var. *brevifolia* Baill., Adansonia 4: 273 (1864). *Caperonia heteropetala* var. *brevifolia* (Baill.) Müll.Arg. in A.P.de Candolle, Prodr. 15(2): 752 (1866).
Caperonia heteropetala var. *elliptica* Müll.Arg., Linnaea 34: 152 (1865).
Caperonia spinosa Endl. ex Pax & K.Hoffm. in H.G.A.Engler, Pflanzenr., IV, 147, VI: 47 (1912), pro syn.

Caperonia angustissima Klotzsch, Hooker's J. Bot. Kew Gard. Misc. 2: 50 (1843). *Argythamnia angustissima* (Klotzsch) Kuntze, Revis. Gen. Pl. 2: 593 (1891).
Guyana. 82 GUY. Ther.

Caperonia bahiensis Müll.Arg. in C.F.P.von Martius, Fl. Bras. 11(2): 325 (1874). *Argythamnia bahiensis* (Müll.Arg.) Kuntze, Revis. Gen. Pl. 2: 593 (1891).
E. Brazil, Paraguay. 84 BZE 85 PAR. Ther.
Caperonia castaneifolia f. *succulenta* Wawra, Bot. Ergebn.: 32 (1866).
Caperonia bahiensis f. *angustior* Chodat & Hassl., Bull. Herb. Boissier, II, 5: 503 (1905).

Caperonia buettneriacea Müll.Arg. in C.F.P.von Martius, Fl. Bras. 11(2): 320 (1874). *Argythamnia buettneriacea* (Müll.Arg.) Kuntze, Revis. Gen. Pl. 2: 593 (1891).
SE. Brazil. 84 BZL. Cham.

Caperonia castaneifolia (L.) A.St.-Hil., Hist. Pl. Remarq. Brésil: 245 (1825).
Mexico, Trop. America. 79 MXG 80 COS NIC PAN 81 CUB DOM HAI JAM WIN 82 GUY 83 CLM 84 BZC BZN 85 PAR. Ther.
* *Croton castaneifolius* L., Sp. Pl.: 1005 (1753). *Ditaxis castaneifolia* (L.) Baill., Adansonia 4: 274 (1864). *Argythamnia castaneifolia* (L.) Kuntze, Revis. Gen. Pl. 2: 593 (1891).
Croton palustris Kunth in F.W.H.von Humboldt, A.J.A.Bonpland & C.S.Kunth, Nov. Gen. Sp. 2: 71 (1817), nom. illeg.
Caperonia paludosa Klotzsch, Hooker's J. Bot. Kew Gard. Misc. 2: 51 (1843). *Argythamnia paludosa* (Klotzsch) Kuntze, Revis. Gen. Pl. 2: 593 (1891).
Caperonia nervosa A.Rich. in R.de la Sagra, Hist. Fis. Cuba, Bot. 2: 213 (1850).
Croton nervosus Rich. ex A.Rich. in R.de la Sagra, Hist. Fis. Cuba, Bot. 2: 213 (1850).
Caperonia panamensis Klotzsch in B.Seemann, Bot. Voy. Herald: 103 (1853).
Caperonia cubensis M.R.Schomb. ex Pax & K.Hoffm. in H.G.A.Engler, Pflanzenr., IV, 147, VI: 31 (1912), pro syn.
Caperonia panamensis Pax & K.Hoffm. in H.G.A.Engler, Pflanzenr., IV, 147, VII: 424 (1914), nom. illeg. *Caperonia stenomeres* S.F.Blake, J. Wash. Acad. Sci. 14: 288 (1924).
Caperonia angusta S.F.Blake, J. Wash. Acad. Sci. 14: 188 (1924).

Caperonia castrobarrosiana Paula & Hamburgo, Rodriguésia 30(46): 164 (1978).
Brazil (Maranhão). 84 BZE.

Caperonia chiltepecensis Croizat, J. Wash. Acad. Sci. 33: 15 (1943).
Mexico (Oaxaca). 79 MXS.

Caperonia corchoroides Müll.Arg., Linnaea 34: 153 (1865). *Argythamnia corchorodes* (Müll.Arg.) Kuntze, Revis. Gen. Pl. 2: 594 (1891).
Guyana, Surinam. 82 GUY SUR. Cham.
Caperonia castaneifolia Miq., Linnaea 21: 478 (1848), nom. illeg.
Croton aculeatus Splitg. ex Lanj., Euphorb. Surinam: 127 (1931).

Caperonia cordata A.St.-Hil., Hist. Pl. Remarq. Brésil: 245 (1825). *Ditaxis cordata* (A.St.-Hil.) Baill., Adansonia 4: 272 (1864). *Argythamnia cordata* (A.St.-Hil.) Kuntze, Revis. Gen. Pl. 2: 593 (1891).
S. Brazil, Paraguay, Uruguay. 84 BZS 85 PAR URU. Hemicr.
Caperonia cordata var. *genuina* Pax & K.Hoffm. in H.G.A.Engler, Pflanzenr., IV, 147, VI: 43 (1911), nom. inval.
Caperonia cordata var. *mollis* Pax & K.Hoffm. in H.G.A.Engler, Pflanzenr., IV, 147, VI: 43 (1911).

Caperonia cubana Pax & K.Hoffm. in H.G.A.Engler, Pflanzenr., IV, 147, VI: 36 (1912).
W. & C. Cuba. 81 CUB. Ther.

Caperonia fistulosa Beille, Bull. Soc. Bot. France 55(8): 73 (1908).
Trop. & S. Africa. 22 MLI NGR 23 BUR ZAI 24 CHA ETH SOM SUD 25 KEN TAN 26 MLW ZAM 27 BOT NAM. Ther.
Caperonia hirtella Beille, Bull. Soc. Bot. France 55(8): 73 (1908).
Caperonia buchananii Baker, Bull. Misc. Inform. Kew 1912: 103 (1912).

Caperonia gardneri Müll.Arg. in C.F.P.von Martius, Fl. Bras. 11(2): 321 (1874). *Argythamnia muellerargoviana* Kuntze, Revis. Gen. Pl. 2: 594 (1891).
WC. Brazil. 84 BZC. Cham.

Caperonia glabrata Pax & K.Hoffm. in H.G.A.Engler, Pflanzenr., IV, 147, VI: 43 (1912).
Paraguay. 85 PAR. Hemicr.

Caperonia heteropetala Didr., Vidensk. Meddel. Dansk Naturhist. Foren. Kjøbenhavn 1857: 148 (1857). *Caperonia heteropetala* var. *genuina* Müll.Arg., Linnaea 34: 152 (1864), nom. inval. *Argythamnia heteropetala* (Didr.) Kuntze, Revis. Gen. Pl. 2: 593 (1891).
SE. Brazil. 84 BZL. Cham.
Croton lanceolatus Hornem. ex Didr., Vidensk. Meddel. Dansk Naturhist. Foren. Kjøbenhavn 1857: 148 (1857), nom. illeg.
Ditaxis polymorpha Baill., Adansonia 4: 273 (1864).
Caperonia heteropetala var. *lanceolata* Müll.Arg., Linnaea 34: 152 (1865).
Caperonia heteropetaloides Müll.Arg., Linnaea 34: 152 (1865). *Argythamnia heteropetalodes* (Müll.Arg.) Kuntze, Revis. Gen. Pl. 2: 594 (1891).
Caperonia heteropetala var. *major* Müll.Arg. in C.F.P.von Martius, Fl. Bras. 11(2): 320 (1874).
Caperonia heteropetala var. *oblongifolia* Müll.Arg. in C.F.P.von Martius, Fl. Bras. 11(2): 319 (1874).

Caperonia hystrix Pax & K.Hoffm. in H.G.A.Engler, Pflanzenr., IV, 147, VI: 41 (1912).
Brazil (Rio Grande do Sul). 84 BZS. Cham.

Caperonia langsdorffii Müll.Arg. in C.F.P.von Martius, Fl. Bras. 11(2): 319 (1874). *Argythamnia langsdorffii* (Müll.Arg.) Kuntze, Revis. Gen. Pl. 2: 594 (1891).
Brazil (São Paulo). 84 BZL. Cham.
Caperonia langsdorffii var. *lanceolata* Müll.Arg. in C.F.P.von Martius, Fl. Bras. 11(2): 319 (1874).
Caperonia langsdorffii var. *oblongifolia* Müll.Arg. in C.F.P.von Martius, Fl. Bras. 11(2): 319 (1874).

Caperonia latifolia Pax, Bol. Soc. Brot. 10: 159 (1892).
Nigeria to Tanzania. 22 NGA TOG 23 CMN 25 TAN. Ther.
Caperonia macrocarpa Pax & K.Hoffm. in H.G.A.Engler, Pflanzenr., IV, 147, VI: 39 (1912).

Caperonia latior (Chodat & Hassl.) Pax & K.Hoffm. in H.G.A.Engler, Pflanzenr., IV, 147, VI: 34 (1912).
Paraguay. 85 PAR. Ther.
* *Caperonia bahiensis* f. *latior* Chodat & Hassl., Bull. Herb. Boissier, II, 5: 503 (1905).

Caperonia linearifolia A.St.-Hil., Hist. Pl. Remarq. Brésil: 246 (1825). *Ditaxis linearifolia* (A.St.-Hil.) Baill., Adansonia 4: 273 (1864). *Argythamnia linearifolia* (A.St.-Hil.) Kuntze, Revis. Gen. Pl. 2: 593 (1891).
NE. Argentina, S. Brazil. 84 BZS 85 AGE. Cham.

Caperonia lutea Pax & K.Hoffm. in H.G.A.Engler, Pflanzenr., IV, 147, VI: 45 (1912).
Guyana. 82 GUY. Cham.

Caperonia multicostata Müll.Arg. in C.F.P.von Martius, Fl. Bras. 11(2): 23 (1873). *Argythamnia multicostata* (Müll.Arg.) Kuntze, Revis. Gen. Pl. 2: 593 (1891).
NE. Brazil. 84 BZE. Cham.

Caperonia neglecta G.L.Webster, Ann. Missouri Bot. Gard. 15: 192 (1967).
Panama to Venezuela. 80 PAN 82 VEN 83 CLM.

Caperonia palustris (L.) A.St.-Hil., Hist. Pl. Remarq. Brésil: 245 (1825).
SE. U.S.A., Mexico, Trop. America. (62) mrn 78 FLA 79 MXG MXS 80 COS 81 CUB DOM HAI LEE PUE TRT WIN 82 FRG GUY SUR VEN 83 CLM ECU PER 84 BZC BZL 85 AGE PAR URU. Ther.
* *Croton palustris* L., Sp. Pl.: 1004 (1753). *Argythamnia palustris* (L.) Kuntze, Revis. Gen. Pl. 2: 593 (1891).
Croton castaneifolius Kunth in F.W.H.von Humboldt, A.J.A.Bonpland & C.S.Kunth, Nov. Gen. Sp. 2: 70 (1817), nom. illeg.
Caperonia liebmanniana Didr. ex Pax & K.Hoffm. in H.G.A.Engler, Pflanzenr., IV, 147, VI: 33 (1912), pro syn.
Caperonia palustris var. *linearis* Standl. & L.O.Williams, Ceiba 1: 148 (1950).

Caperonia paraguayensis Pax & K.Hoffm. in H.G.A.Engler, Pflanzenr., IV, 147, VI: 35 (1912).
Paraguay. 85 PAR. Ther.

Caperonia pubescens S.F.Blake, Contr. U. S. Natl. Herb. 24: 12 (1922).
Guatemala. 80 GUA.

Caperonia regnellii Müll.Arg. in C.F.P.von Martius, Fl. Bras. 11(2): 321 (1874). *Argythamnia regnellii* (Müll.Arg.) Kuntze, Revis. Gen. Pl. 2: 593 (1891).
Brazil (Minas Gerais). 84 BZL. Cham.

Caperonia rutenbergii Müll.Arg., Bremen Abh. 7: 25 (1880). *Argythamnia rutenbergii* (Müll.Arg.) Kuntze, Revis. Gen. Pl. 2: 593 (1891).
W. Madagascar. 29 MDG. Ther.

Caperonia serrata (Turcz.) C.Presl, Epimel. Bot.: 213 (1851).
Trop. Africa. 22 BEN GAM GHA GUI IVO MLI NGA SEN SIE TOG 23 CAF CMN CON ZAI 24 ETH SUD 25 KEN UGA. Ther.
* *Lepidococca serrata* Turcz., Bull. Soc. Imp. Naturalistes Moscou 21(1): 589 (1848). *Croton serratus* (Turcz.) Hochst. ex Baill., Adansonia 1: 66 (1860).
Caperonia senegalensis Müll.Arg., Linnaea 34: 153 (1865). *Argythamnia senegalensis* (Müll.Arg.) Kuntze, Revis. Gen. Pl. 2: 593 (1891).
Argythamnia hochstetteri Kuntze, Revis. Gen. Pl. 2: 594 (1891).
Caperonia chevalieri Beille, Bull. Soc. Bot. France 55(8): 73 (1908).
Caperonia gallabatensis Pax & K.Hoffm. in H.G.A.Engler, Pflanzenr., IV, 147, VI: 39 (1912).

Caperonia similis Pax & K.Hoffm. in H.G.A.Engler, Pflanzenr., IV, 147, XIV: 6 (1919).
Brazil (Amazonas). 84 BZN.

Caperonia stenophylla Müll.Arg. in C.F.P.von Martius, Fl. Bras. 11(2): 326 (1874).
Argythamnia stenophylla (Müll.Arg.) Kuntze, Revis. Gen. Pl. 2: 594 (1891).
Guianas, Brazil (Minas Gerais). 82 FRG GUY SUR? 84 BZL. Ther.

Caperonia stuhlmannii Pax, Bot. Jahrb. Syst. 19: 81 (1894).
Tanzania (incl. Zanzibar), Mozambique, Malawi, Zambia, Zimbabwe, Natal. 25 TAN 26 MLW MOZ ZAM ZIM 27 NAT. Ther.

Caperonia subrotunda Chiov., Result. Sci. Miss. Stefan.-Paoli Somal. Ital. 1: 220 (1916).
Somalia. 24 SOM.

Caperonia velloziana Müll.Arg. in C.F.P.von Martius, Fl. Bras. 11(2): 323 (1874).
Brazil (Rio de Janeiro). 84 BZL. Cham.

Caperonia zaponzeta Mansf., Notizbl. Bot. Gart. Berlin-Dahlem 9: 265 (1925).
Peru. 83 PER.

Synonyms:
Caperonia acalyphifolia Griseb. === **Byttneria** sp. (Byttneriaceae)
Caperonia angusta S.F.Blake === **Caperonia castaneifolia** (L.) A.St.-Hil.
Caperonia bahiensis f. *angustior* Chodat & Hassl. === **Caperonia bahiensis** Müll.Arg.
Caperonia bahiensis f. *latior* Chodat & Hassl. === **Caperonia latior** (Chodat & Hassl.) Pax & K.Hoffm.
Caperonia buchananii Baker === **Caperonia fistulosa** Beille
Caperonia castaneifolia Miq. === **Caperonia corchoroides** Müll.Arg.
Caperonia castaneifolia f. *succulenta* Wawra === **Caperonia bahiensis** Müll.Arg.
Caperonia chevalieri Beille === **Caperonia serrata** (Turcz.) C.Presl
Caperonia cordata var. *genuina* Pax & K.Hoffm. === **Caperonia cordata** A.St.-Hil.
Caperonia cordata var. *mollis* Pax & K.Hoffm. === **Caperonia cordata** A.St.-Hil.
Caperonia cubensis M.R.Schomb. ex Pax & K.Hoffm. === **Caperonia castaneifolia** (L.) A.St.-Hil.
Caperonia gallabatensis Pax & K.Hoffm. === **Caperonia serrata** (Turcz.) C.Presl
Caperonia heteropetala var. *brevifolia* (Baill.) Müll.Arg. === **Caperonia aculeolata** Müll.Arg.
Caperonia heteropetala var. *elliptica* Müll.Arg. === **Caperonia aculeolata** Müll.Arg.
Caperonia heteropetala var. *genuina* Müll.Arg. === **Caperonia heteropetala** Didr.
Caperonia heteropetala var. *lanceolata* Müll.Arg. === **Caperonia heteropetala** Didr.
Caperonia heteropetala var. *major* Müll.Arg. === **Caperonia heteropetala** Didr.
Caperonia heteropetala var. *oblongifolia* Müll.Arg. === **Caperonia heteropetala** Didr.
Caperonia heteropetaloides Müll.Arg. === **Caperonia heteropetala** Didr.
Caperonia hirtella Beille === **Caperonia fistulosa** Beille
Caperonia langsdorffii var. *lanceolata* Müll.Arg. === **Caperonia langsdorffii** Müll.Arg.
Caperonia langsdorffii var. *oblongifolia* Müll.Arg. === **Caperonia langsdorffii** Müll.Arg.
Caperonia liebmanniana Didr. ex Pax & K.Hoffm. === **Caperonia palustris** (L.) A.St.-Hil.
Caperonia macrocarpa Pax & K.Hoffm. === **Caperonia latifolia** Pax
Caperonia nervosa A.Rich. === **Caperonia castaneifolia** (L.) A.St.-Hil.
Caperonia paludosa Klotzsch === **Caperonia castaneifolia** (L.) A.St.-Hil.
Caperonia palustris var. *linearis* Standl. & L.O.Williams === **Caperonia palustris** (L.) A.St.-Hil.
Caperonia panamensis Pax & K.Hoffm. === **Caperonia castaneifolia** (L.) A.St.-Hil.
Caperonia panamensis Klotzsch === **Caperonia castaneifolia** (L.) A.St.-Hil.
Caperonia senegalensis Müll.Arg. === **Caperonia serrata** (Turcz.) C.Presl
Caperonia spinosa Endl. ex Pax & K.Hoffm. === **Caperonia aculeolata** Müll.Arg.
Caperonia stenomeres S.F.Blake === **Caperonia castaneifolia** (L.) A.St.-Hil.

Carcia

Synonyms:
Carcia Raeusch. === **Garcia** Vahl ex Rohr

Carumbium

Synonyms:
Carumbium Reinw. === **Homalanthus** A.Juss.
Carumbium acuminatum Müll.Arg. === **Homalanthus acuminatus** (Müll.Arg.) Pax
Carumbium moerenhoutianum Müll.Arg. === **Homalanthus nutans** (G.Forst.) Guill.
Carumbium novoguineense Warb. === **Homalanthus novoguineensis** (Warb.) K.Schum.
Carumbium nutans (G.Forst.) Müll.Arg. === **Homalanthus nutans** (G.Forst.) Guill.
Carumbium polyandrum Hook.f. === **Homalanthus polyandrus** (Hook.f.) Cheeseman
Carumbium populneum (Geiseler) Müll.Arg. === **Homalanthus populneus** (Geiseler) Pax
Carumbium populneum var. *minus* Müll.Arg. === **Homalanthus populneus** (Geiseler) Pax

Carumbium

A later homonym of *Carumbium* Reinw.

Synonyms:
Carumbium Kurz === **Triadica** Lour.

Caryodendron

4 species, S. America (widely scattered from Panama to Colombia and Brazil); forest trees to 40 m with edible seeds. Pax (1914) suggested a habit similarity with *Sapium* but its relationships are with *Alchorneopsis* and *Discoglypremna*. (Acalyphoideae)

Pax, F. (with K. Hoffmann) (1914). *Caryodendron*. In A. Engler (ed.), Das Pflanzenreich, IV 147 VII (Euphorbiaceae-Acalypheae-Mercurialinae): 263-264. Berlin. (Heft 63.) La/Ge. — 2 species, S America.

Caryodendron H.Karst., Fl. Columb. 1: 91 (1860).
C. & S. Trop. America. 80 82 83 84.
Centrodiscus Müll.Arg. in C.F.P.von Martius, Fl. Bras. 11(2): 325 (1874).

Caryodendron amazonicum Ducke, Trop. Woods 76: 18 (1943).
Brazil (Amazonas). 84 BZN. Phan.

Caryodendron angustifolium Standl., Publ. Field Mus. Nat. Hist., Bot. Ser. 4: 217 (1929).
Costa Rica, W. Panama. 80 COS PAN. Phan.

Caryodendron janeirense Müll.Arg. in C.F.P.von Martius, Fl. Bras. 11(2): 707 (1874).
Brazil (Rio de Janeiro). 84 BZL. Phan.
Centrodiscus grandifolius Müll.Arg. in C.F.P.von Martius, Fl. Bras. 11(2): 325 (1874).
Caryodendron grandifolium (Müll.Arg.) Pax in H.G.A.Engler, Pflanzenr., IV, 147, V: 257 (1912).
Sapium macrophyllum Klotzsch ex Pax in H.G.A.Engler, Pflanzenr., IV, 147, V: 257 (1912), pro syn.

Caryodendron orinocense H.Karst., Fl. Columb. 1: 91 (1860).
Colombia, Venezuela. 82 VEN 83 CLM. Phan.

Synonyms:
Caryodendron grandifolium (Müll.Arg.) Pax **Caryodendron janeirense** Müll.Arg.

Casabitoa

= *Picramnia* (Simaroubaceae; Zanoni & Garcia, 1994).

Alain H. Liogier (1980). Novitates antillanae, VIII. Phytologia, 47: 167-198, illus. En. — Pp. 174-175 (with illustration) contain the protologue of *Casabitoa* and description of *C. perfae* from Hispaniola (using only pistillate material).
• Zanoni, T. A. & R. G. Garcia G. (1994). *Casabitoa perfae* (Euphorbiaceae): a new synonym of *Picramnia dictyoneura* (Simaroubaceae). Brittonia 46: 81-82. En. — Reduction and transfer.

Cascarilla

Synonyms:
Cascarilla Adans. === **Croton** L.

Castiglionia

Synonyms:
Castiglionia Ruiz & Pav. === **Jatropha** L.

Cataputia

Synonyms:
Cataputia Ludw. === **Ricinus** L.

Cathetus

Synonyms:
Cathetus Lour. === **Phyllanthus** L.
Cathetus fasciculata Lour. === **Phyllanthus cochinchinensis** Spreng.

Caturus

Synonyms:
Caturus L. === **Acalypha** L.
Caturus Lour. === **Alchornea** Sw.
Caturus scandens Lour. === **Alchornea scandens** (Lour.) Müll.Arg.
Caturus spiciflorus L. === **Acalypha caturus** Blume

Cavacoa

3 species, tropical and subtropical Africa south to KwaZulu-Natal; includes *Grossera* sect. *Racemiformes* of Pax & Hoffmann. The trees sometimes have fluted trunks (*C. quintasii*); the inflorescences are terminal and racemiform. The southern African *C. aurea* has been cultivated; it was long locally confused with *Heywoodia lucens*. (Crotonoideae)

Pax, F. (with K. Hoffmann) (1912). *Grossera*. In A. Engler (ed.), Das Pflanzenreich, IV 147 VI (Euphorbiaceae-Acalypheae-Chrozophorinae): 105-108. Berlin. (Heft 57.) La/Ge. — 3 species in 2 sections, Africa. [Sect. *Racemiformes* is now in *Cavacoa*.]

Léonard, J. (1955). À propos des genres africains *Grossera* Pax et *Cavacoa* J. Léonard. Bull. Jard. Bot. État 25: 315-324, 2 figl. (1 halftone). Fr. —*Grossera* of Pax divided into 2 genera, one of them *Cavacoa*, with 3 species; the species accounts include synonymy, selected localities with exsiccatae and commentary but no key. There are two illustrations, one of the fluted trunk, of *C. quintasii* which occurs within Congo.

Cavacoa J.Léonard, Bull. Jard. Bot. État 25: 320 (1955).
Trop. & S. Africa. 22 23 25 26 27.

Cavacoa aurea (Cavaco) J.Léonard, Bull. Jard. Bot. État 25: 323 (1955). – FIGURE, p. 308.
Kenya, Mozambique, Natal. 25 KEN 26 MOZ 27 NAT. Nanophan. or phan.
* *Grossera aurea* Cavaco, Bull. Mus. Natl. Hist. Nat., II, 21: 274 (1949).

Cavacoa baldwinii (Keay & Cavaco) J.Léonard, Bull. Jard. Bot. État 25: 324 (1955).
Sierra Leone, Liberia. 22 LBR SIE. Phan.
* *Grossera baldwinii* Keay & Cavaco, Portugaliae Acta Biol., Sér. B, Sist. 6: 1 (1955).

Cavacoa quintasii (Pax & K.Hoffm.) J.Léonard, Bull. Jard. Bot. État 25: 322 (1955).
Annobon, São Tomé, Zaire. 23 GGI ZAI. Phan.
* *Grossera quintasii* Pax & K.Hoffm. in H.G.A.Engler, Pflanzenr., IV, 147, VI: 108 (1912).

Cavanilla

Synonyms:
Cavanilla Vell. === **Caperonia** A.St.-Hil.

Cecchia

Synonyms:
Cecchia Chiov. === **Oldfieldia** Benth. & Hook.f.
Cecchia somalensis Chiov. === **Oldfieldia somalensis** (Chiov.) Milne-Redh.

Celaenodendron

Synonyms:
Celaenodendron Standl. === **Piranhea** Baill.
Celaenodendron mexicanum Standl. === **Piranhea mexicana** (Standl.) Radcl.-Sm.

Celianella

1 species, S America (Venezuelan Guayana); shrubs to 5 m with semisucculent leaves featuring a reddish margin. Allied to *Hieronyma* by Levin (1986; see General) on the basis of anatomical features. (Phyllanthoideae)

Jablonski, E. (1965). *Celianella*. Euphorbiaceae, Guayana Highland (Mem. New York Bot. Gard. 12(3)): 176-178. New York. En. — Monotypic, Venezuelan Guayana; very local (Cerro Yutajé).

Celianella Jabl., Mem. New York Bot. Gard. 12(3): 176 (1965).
Venezuela Highlands. 82. Nanophan.

Celianella montana Jabl., Mem. New York Bot. Gard. 12(3): 178 (1965).
S. Venezuela. 82 VEN. Nanophan.

Cavacoa aurea (Cavaco) J. Léonard
Artist: Stella Ross-Craig
Ic. Pl. 36(2): pl. 3561 (1956)
KEW ILLUSTRATIONS COLLECTION

Cenesmon

Synonyms:
Cenesmon Gagnep. === **Cnesmone** Blume
Cenesmon hainanense Merr. & Chun === **Cnesmone hainanensis** (Merr. & Chun) Croizat
Cenesmon laoticum Gagnep. === **Cnesmone laotica** (Gagnep.) Croizat
Cenesmon lineare Gagnep. === **Cnesmone linearis** (Gagnep.) Croizat
Cenesmon peltatum Gagnep. === **Cnesmone peltata** (Gagnep.) Croizat
Cenesmon poilanei Gagnep. === **Cnesmone poilanei** (Gagnep.) Croizat
Cenesmon tonkinense Gagnep. === **Cnesmone javanica** Blume var. **javanica**

Centrandra

Synonyms:
Centrandra H.Karst. === **Croton** L.
Centrandra hondensis H.Karst. === **Croton hondensis** (H.Karst.) G.L.Webster

Centrodiscus

Synonyms:
Centrodiscus Müll.Arg. === **Caryodendron** H.Karst.
Centrodiscus grandifolius Müll.Arg. === **Caryodendron janeirense** Müll.Arg.

Centrostylis

Synonyms:
Centrostylis Baill. === **Adenochlaena** Boivin ex Baill.
Centrostylis zeylanica Baill. === **Adenochlaena zeylanica** (Baill.) Thwaites

Cephalocroton

4 species, tropical to southern Africa; sometimes many-stemmed shrubs or subshrubs or woody perennials of open country. Related to *Adenochlaena* and *Cephalocrotonopsis*, now treated as separate genera in contrast to Radcliffe-Smith (1973). (Acalyphoideae)

Pax, F. (1910). *Cephalocroton*. In A. Engler (ed.), Das Pflanzenreich, IV 147 II (Euphorbiaceae-Adrianae): 7-12. Berlin. (Heft 44.) La/Ge. — 8 species, Africa. [Account superseded by Radcliffe-Smith 1973.]

Radcliffe-Smith, A. (1973). An account of the genus *Cephalocroton* Hochst. (Euphorbiaceae). Kew Bull. 28: 123-132. En. — Synoptic treatment with key, synonymy, references and citations, localities with exsiccatae, and commentary. [Genus here includes *Adenochlaena* and *Cephalocrotonopsis*, now separate.]

Cephalocroton Hochst., Flora 24: 370 (1841).
Trop. & S. Africa. 22 24 25 26 27.

Cephalocroton cordofanus Hochst., Flora 24: 370 (1841). *Cephalocroton cordofanus* var. *genuinus* Müll.Arg., Linnaea 34: 155 (1865), nom. inval.
Nigeria, E. Sudan, Ethiopia, Somalia, Kenya, NE. Tanzania. 22 NGA 24 ETH SOM SUD 25 KEN TAN. Cham. or nanophan.
Acalypha betulina Schweinf., Pl. Quaed. Nilot.: 13 (1862). *Cephalocroton cordofanus* var. *betulinus* (Schweinf.) Müll.Arg., Linnaea 34: 155 (1865).

Cephalocroton nudus Pax & K.Hoffm. in H.G.A.Engler, Pflanzenr., IV, 147, II: 10 (1910).
Cephalocroton scabridus Pax & K.Hoffm. in H.G.A.Engler, Pflanzenr., IV, 147, II: 9 (1910).
Cephalocroton velutinus Pax & K.Hoffm. in H.G.A.Engler, Pflanzenr., IV, 147, II: 10 (1910).

Cephalocroton incanus M.G.Gilbert, Kew Bull. 42: 365 (1987). – FIGURE, p. 311.
Nigeria, Ethiopia. 22 NGA 24 ETH.

Cephalocroton mollis Klotzsch in W.C.H.Peters, Naturw. Reise Mossambique: 99 (1861).
Tanzania, Zimbabwe, Malawi, Mozambique, Botswana, Namibia, Transvaal, Natal. 25 TAN 26 MLW MOZ ZIM 27 BOT NAM NAT TVL. Cham. or nanophan.
Cephalocroton pueschelii Pax, Bot. Jahrb. Syst. 43: 84 (1909).
Cephalocroton depauperatus Pax & K.Hoffm. in H.G.A.Engler, Pflanzenr., IV, 147, II: 12 (1910).

Cephalocroton polygynus Pax & K.Hoffm. in H.G.A.Engler, Pflanzenr., IV, 147, II: 10 (1910).
Somalia. 24 SOM. Nanophan. – Probably only a polygynous form of *C. cordofanus*.

Synonyms:
Cephalocroton albicans (Blume) Müll.Arg. === **Sumbaviopsis albicans** (Blume) J.J.Sm.
Cephalocroton albicans var. *virens* Müll.Arg. === **Cladogynos orientalis** Zipp. ex Span.
Cephalocroton cordifolius Baker === **Adenochlaena leucocephala** Baill.
Cephalocroton cordofanus var. *betulinus* (Schweinf.) Müll.Arg. === **Cephalocroton cordofanus** Hochst.
Cephalocroton cordofanus var. *genuinus* Müll.Arg. === **Cephalocroton cordofanus** Hochst.
Cephalocroton depauperatus Pax & K.Hoffm. === **Cephalocroton mollis** Klotzsch
Cephalocroton discolor Müll.Arg. === **Cladogynos orientalis** Zipp. ex Span.
Cephalocroton indicus Bedd. === **Epiprinus mallotiformis** (Müll.Arg.) Croizat
Cephalocroton leucocephalus (Baill.) Baill. === **Adenochlaena leucocephala** Baill.
Cephalocroton nudus Pax & K.Hoffm. === **Cephalocroton cordofanus** Hochst.
Cephalocroton orientalis (Zipp. ex Span) Miq. === **Cladogynos orientalis** Zipp. ex Span.
Cephalocroton pueschelii Pax === **Cephalocroton mollis** Klotzsch
Cephalocroton scabridus Pax & K.Hoffm. === **Cephalocroton cordofanus** Hochst.
Cephalocroton socotranus Balf.f. === **Cephalocrotonopsis socotranus** (Balf.f.) Pax
Cephalocroton velutinus Pax & K.Hoffm. === **Cephalocroton cordofanus** Hochst.
Cephalocroton zeylanicus (Baill.) Baill. === **Adenochlaena zeylanica** (Baill.) Thwaites

Cephalocrotonopsis

1 species, Africa (Socotra Is.); a sometimes dominant shrub or small tree with spiky interlacing branches and shoots. Sometimes included with *Cephalocroton* (Radcliffe-Smith, 1973). [*C. socotranus* illustrated in Balfour, *Botany of Socotra* (1887), pl. 94.] (Acalyphoideae)

Pax, F. (1910). *Cephalocrotonopsis*. In A. Engler (ed.), Das Pflanzenreich, IV 147 II (Euphorbiaceae-Adrianae): 15. Berlin. (Heft 44.) La/Ge. — 1 species, Yemen (Socotra I.). [Superseded by Radcliffe-Smith 1973.]

Radcliffe-Smith, A. (1973). An account of the genus *Cephalocroton* Hochst. (Euphorbiaceae). Kew Bull. 28: 123-132. En. — Synoptic treatment with key, synonymy, references and citations, localities with exsiccatae, and commentary. [*Cephalocrotonopsis*, pp. 131-132.]

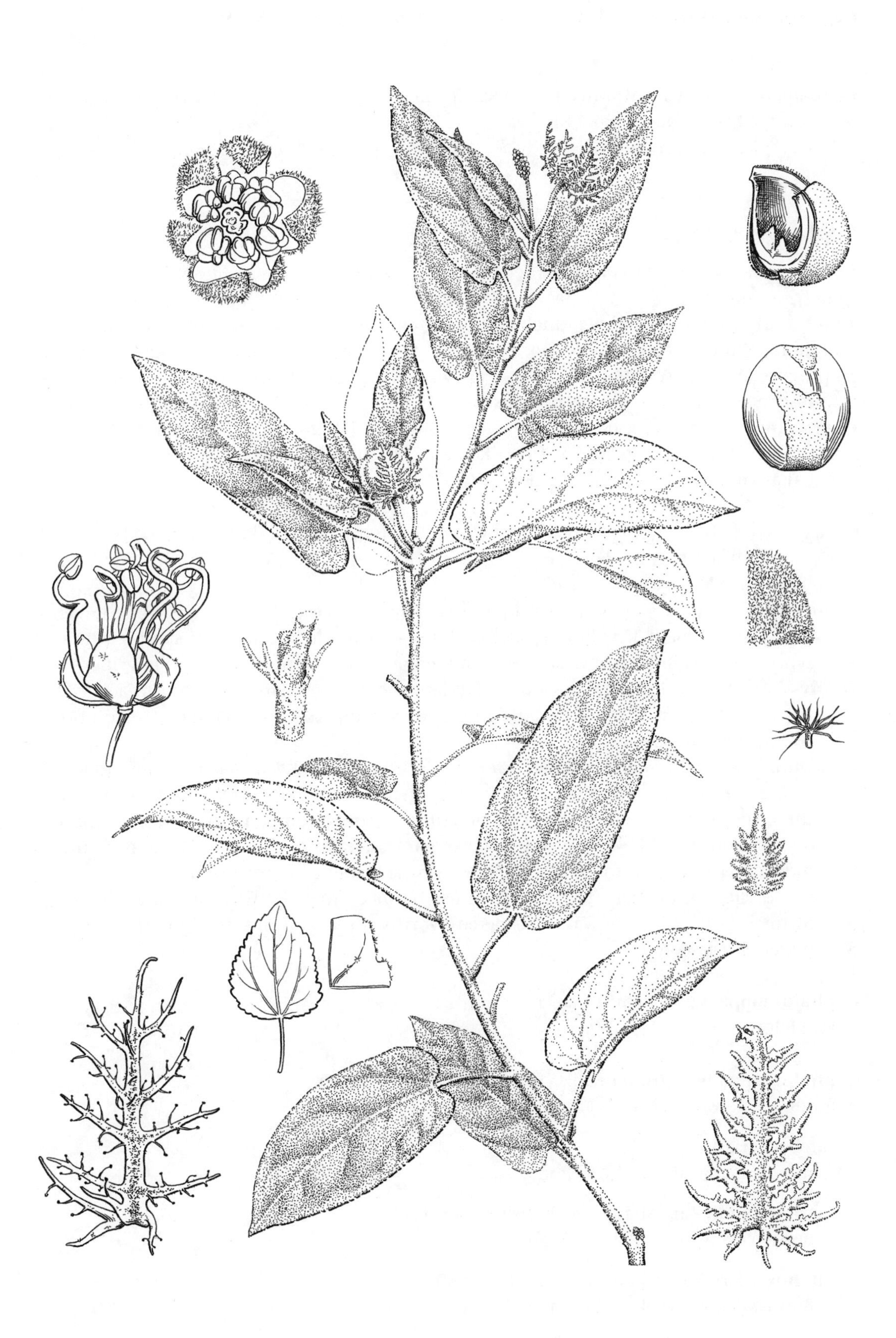

Cephalocroton incanus M.G. Gilbert
Artist: Eleanor Catherine
Kew Bull. 42(2): 366 (1987)
KEW ILLUSTRATIONS COLLECTION

Cephalocrotonopsis Pax in H.G.A.Engler, Pflanzenr., IV, 147, II: 15 (1910).
Socotra. 24.

Cephalocrotonopsis socotranus (Balf.f.) Pax in H.G.A.Engler, Pflanzenr., IV, 147, II: 15 (1910).
Socotra. 24 SOC. Nanophan.
* *Cephalocroton socotranus* Balf.f., Proc. Roy. Soc. Edinburgh 12: 95 (1884).

Cephalomappa

5 species, Malesia (Sumatera to Borneo); forest trees to 36 m (*C. penangensis*) with loosely spirally arranged to alternate leaves. The small flowers, in axillary inflorescences, are arranged in heads. *C. paludicola* inhabits freshwater peat-swamp forests; others are mainly in well-drained forest to 600 m. Webster (Synopsis, 1994) assigned it to its own subtribe within Epiprineae; Airy-Shaw (1960), however, previously suggested that in aspect the foliage, inflorescence and fruit were reminiscent of *Ptychopyxis* (in the neighboring tribe Pycnocomeae). A revision has recently been published (Widuri & van Welzen, 1998), with *Muricococcum* (*C. sinensis* (Chun & How) Kosterm.) excluded and *Koilodepas* (in Epiprininae) adopted as an outgroup. (Acalyphoideae)

Pax, F. (1910). *Cephalomappa*. In A. Engler (ed.), Das Pflanzenreich, IV 147 II (Euphorbiaceae-Adrianae): 16-17. Berlin. (Heft 44.) La/Ge. — 1 species, Borneo. [Superseded by Airy-Shaw 1960.]

Airy-Shaw, H. K. (1960). Notes on Malaysian Euphorbiaceae, X. A synopsis of the genus *Cephalomappa* Baill. Kew Bull. 14: 378-382. En. — Synoptic account of 5 species, with key, synonymy, localities with exsiccatae, commentary, and descriptions of two novelties.

Airy-Shaw, H. K. (1963). Notes on Malaysian and other Asiatic Euphorbiaceae, XXXV. Further notes on *Cephalomappa* Baillon. Kew Bull. 16: 353-354. En. — Additional records for four species.

• Widuri, R. & P. van Welzen (1998). A revision of the genus *Cephalomappa* (Euphorbiaceae) in Malesia. Reinwardtia 11: 153-184, illus., maps. En. — Revision (5 species, one with three additional varieties) with key, descriptions, brief commentary, and indication of distribution, habitat, ecology, vernacular names and uses, list of specimens seen and index at end. The introduction includes a phylogenetic analysis (rooted in *Koilodepas*) and discussion of characters. Particular use has been made of trichomes and epidermal features. *Muricococcum* was considered worthy of exclusion, though no authentic material was seen.

Cephalomappa Baill., Adansonia 11: 130 (1874).
W. Malesia. 42.

Cephalomappa beccariana Baill., Adansonia 11: 131 (1874).
Borneo (Sarawak). 42 BOR. Phan.

var. **beccariana**
Borneo (Sarawak). 42 BOR. Phan.

var. **havilandii** Airy Shaw, Kew Bull. 14: 380 (1960).
Borneo (Sarawak). 42 BOR. Phan.

var. **hosei** Airy Shaw, Kew Bull. 14: 380 (1960).
Borneo (Sarawak). 42 BOR. Phan.

var. **tenuifolia** Airy Shaw, Kew Bull., Addit. Ser. 4: 66 (1975).
Borneo (SW. Sarawak). 42 BOR. Phan.

Cephalomappa lepidotula Airy Shaw, Kew Bull. 14: 379 (1960).
S. Pen. Malaysia, Sumatera, N. & C. Borneo. 42 BOR MLY SUM. Phan.

Cephalomappa malloticarpa J.J.Sm., Bull. Jard. Bot. Buitenzorg, III, 6: 95 (1924).
Pen. Malaysia, Sumatera, Borneo (Sarawak, Sabah, N. & E. Kalimantan). 42 BOR MLY SUM. Phan.

Cephalomappa paludicola Airy Shaw, Kew Bull. 14: 380 (1960).
Borneo (Sarawak). 42 BOR. Phan.

Cephalomappa penangensis Ridl., Bull. Misc. Inform. Kew 1923: 368 (1923).
Pen. Malaysia. 42 MLY. Phan.

Synonyms:
Cephalomappa sinensis (Chun & How) Kosterm. === **Muricococcum sinense** Chun & F.C.How

Ceramanthus

Synonyms:
Ceramanthus Hassk. === **Phyllanthus** L.
Ceramanthus gracilis Hassk. === **Phyllanthus albidiscus** (Ridl.) Airy Shaw

Ceraselma

A Wittstein name, listed in *Vascular plant families and genera* (Brummitt 1992) = *Euphorbia.*

Ceratococcus

Synonyms:
Ceratococcus Meisn. === **Pterococcus** Hassk.

Ceratogynum

Synonyms:
Ceratogynum Wight === **Sauropus** Blume
Ceratogynum concolor Pritz. === **Ceratolobus concolor** Blume (Arecaceae)
Ceratogynum rhamnoides (Roxb.) Wight === **Sauropus quadrangularis** (Willd.) Müll.Arg.

Ceratophorus

Synonyms:
Ceratophorus Sond. === **Suregada** Roxb. ex Rottl.
Ceratophorus africanus Sond. === **Suregada africana** (Sond.) Müll.Arg.

Chaenotheca

Synonyms:
Chaenotheca Urb. === **Chascotheca** Urb.
Chaenotheca domingensis Urb. === **Chascotheca neopeltandra** (Griseb.) Urb.

Chaetocarpus

15 species, scattered in warmer regions (Sri Lanka and India to W. Malesia, Madagascar (1), WC. and S. tropical Africa (1), the Caribbean (5), and tropical S. America); includes *Mettenia*. Shrubs or small to large trees; *C. castanocarpus* reaches 45 m. The majority of the species (11) is American. The genus is related to *Trigonopleura* (van Welzen 1994). Its name is conserved against *Chaetocarpus* Schreb. (Sapotaceae). A revision of the entire genus has been undertaken by Marcus V. da Silva Alves (Museu Nacional, Rio de Janeiro, Brazil); a preliminary synopsis of American species appeared as Alves (1994). Capuron (1972, with added notes by Leandri) believed *C. rabaraba*, a hard-wooded tree of western Madagascar to 35 m, to add evidence for *Chaetocarpus* being representative 'de l'ancienne flore australe du globe'. (Acalyphoideae)

Pax, F. (with K. Hoffmann) (1912). *Chaetocarpus*. In A. Engler (ed.), Das Pflanzenreich, IV 147 IV (Euphorbiaceae-Gelonieae): 7-11. Berlin. (Heft 52.) La/Ge. — 7 species in 3 sections, tropics, scattered.

Pax, F. (with K. Hoffmann) (1912). *Mettenia*. In A. Engler (ed.), Das Pflanzenreich, IV 147 IV (Euphorbiaceae-Gelonieae): 11-12. Berlin. (Heft 52.) La/Ge. — Perhaps 2 species, West Indies. [Genus now merged with *Chaetocarpus*.]

Jablonski, E. (1967). *Chaetocarpus*. Euphorbiaceae, Guayana Highland (Mem. New York Bot. Gard. 17(1)): 161-162. New York. En. — 2 species, 1 poorly known.

Capuron, R. (1972). Contribution à l'étude de la flore forestière de Madagascar. Adansonia, II, 12: 205-211. Fr. — Comprises two parts, of which the second is a description of *Chaetocarpus rabaraba*, the first representative of the genus in Madagascar.

Alves, M. V. da Silva (1994). Novas informações sobre *Chaetocarpus* Thwaites (Euphorbiaceae). An. Jard. Bot. Madrid 51: 302-304. Pt. — Synopsis with descriptions of novelties, synonymy, references types, localities with exsiccatae, brief indication of distribution, and concise notes.

Welzen, P. C. van (1994). A taxonomic revision of S.E. Asian *Chaetocarpus* Thwaites (Euphorbiaceae). Rheedea 4: 93-101, illus., map. En. — Introduction, with background and evaluation of suspect records; descriptive treatment of the genus and the single SE Asian/Malesian species (*C. castanocarpus*) with synonymy, references and citations, typification, vernacular names, and extensive notes on distribution, habitat, ecology, uses, properties and variation; list of exsiccatae seen at end. [van Welzen takes here a broader view of species limits than does Philcox in *Revised Handbook to the Flora of Ceylon* 11 (1997); there, four species were accepted.]

Chaetocarpus Thwaites, Hooker's J. Bot. Kew Gard. Misc. 6: 300 (1854).
Trop. America, Trop. Africa, Madagascar, Trop. & Subtrop. Asia. 22 26 29 36 40 41 42 81 82 83 84.
Mettenia Griseb., Fl. Brit. W. I.: 43 (1859).
Regnaldia Baill., Adansonia 1: 187 (1861).
Gaedawakka L. ex Kuntze, Revis. Gen. Pl. 2: 606 (1891).
Neochevaliera A.Chev. & Beille, Compt. Rend. Hebd. Séances Acad. Sci. 145: 1294 (1907).

Chaetocarpus acutifolius (Britton & P.Wilson) Borhidi, Acta Bot. Acad. Sci. Hung. 25: 18 (1979).
Cuba (Sierra de Moa). 81 CUB. Nanophan. or phan.
* *Mettenia acutifolia* Britton & P.Wilson, Bull. Torrey Bot. Club 39: 9 (1912).

Chaetocarpus africanus Pax, Bot. Jahrb. Syst. 19: 113 (1894).
C. Trop. Africa. 23 CAB CON GAB ZAI 26 ANG ZAM. Nanophan.

Chaetocarpus castanocarpus (Roxb.) Thwaites, Enum. Pl. Zeyl.: 275 (1861).
China (S. Yunnan), Trop. Asia. 36 CHC 40 ASS SRL 41 AND BMA CBD LAO THA VIE 42 BOR MLY SUM. Phan.

* *Adelia castanocarpa* Roxb., Fl. Ind. ed. 1832, 3: 848 (1832). *Chaetocarpus castanocarpus* var. *genuinus* Müll.Arg. in A.P.de Candolle, Prodr. 15(2): 1122 (1866), nom. inval.
Chaetocarpus pungens Thwaites, Hooker's J. Bot. Kew Gard. Misc. 6: 301 (1854).

Chaetocarpus cordifolius (Urb.) Borhidi, Acta Bot. Acad. Sci. Hung. 25: 18 (1979).
E. Cuba, Jamaica, Hispaniola. 81 CUB DOM JAM. Nanophan. or phan.
* *Mettenia cordifolia* Urb., Symb. Antill. 9: 213 (1924).
Chaetocarpus domingensis Proctor, Moscosoa 2: 25 (1983).

Chaetocarpus coriaceus Thwaites, Enum. Pl. Zeyl.: 275 (1861).
Sri Lanka. 40 SRL. Phan.

Chaetocarpus cubensis Fawc. & Rendle, J. Bot. 57: 312 (1919).
W. Cuba. 81 CUB. Nanophan.
Mettenia humilis Ekman ex Urb., Repert. Spec. Nov. Regni Veg. 28: 229 (1930).
Chaetocarpus humilis (Ekman ex Urb.) Borhidi, Acta Bot. Acad. Sci. Hung. 25: 18 (1979).

Chaetocarpus echinocarpus (Baill.) Ducke, Arq. Serv. Florest. 1: 32 (1939).
Bolivia, Brazil. 83 BOL 84 BZC BZE BZN. Phan.
* *Pera echinocarpa* Baill., Adansonia 5: 223 (1865).
Chaetocarpus pohlii Müll.Arg. in C.F.P.von Martius, Fl. Bras. 11(2): 508 (1874).

Chaetocarpus ferrugineus Philcox, Kew Bull. 50: 121 (1995).
Sri Lanka. 40 SRL. Nanophan. or phan.

Chaetocarpus globosus (Sw.) Fawc. & Rendle, J. Bot. 57: 312 (1919).
Jamaica, Cuba. 81 CUB JAM. Phan.
* *Croton globosus* Sw., Prodr.: 100 (1788). *Ricinus globosus* (Sw.) Willd., Sp. Pl. 4: 567 (1805).

subsp. **globosus**
Jamaica, Cuba. 81 CUB JAM. Phan.
Mettenia lepidota Urb., Repert. Spec. Nov. Regni Veg. 28: 230 (1930). *Chaetocarpus globosus* var. *lepidotus* (Urb.) Borhidi, Acta Bot. Hung. 29: 184 (1983).
Chaetocarpus globosus f. *puberula* Borhidi, Acta Bot. Hung. 29: 184 (1983).

subsp. **oblongatus** (Alain) Borhidi, Acta Bot. Acad. Sci. Hung. 25: 17 (1979).
Cuba (Sierra de Moa). 81 CUB. Nanophan. or phan.
**Mettenia oblongata* Alain, Contr. Ocas. Mus. Hist. Nat. Colegio "De La Salle" 11: 10 (1952). *Chaetocarpus oblongatus* (Alain) Borhidi, Acta Bot. Acad. Sci. Hung. 29: 181 (1983).
Chaetocarpus oblongatus var. *monticola* Borhidi, Acta Bot. Hung. 29: 184 (1983).
Chaetocarpus oblongatus var. *subenervis* Borhidi, Acta Bot. Hung. 29: 184 (1983).

Chaetocarpus myrsinites Baill., Adansonia 11: 95 (1873).
N. South America, E. Brazil. 82 GUY VEN 84 BZE BZL. Phan.

var. **myrsinites**
E. Brazil. 84 BZE BZL. Phan.
Chaetocarpus blanchetii Müll.Arg. in C.F.P.von Martius, Fl. Bras. 11(2): 507 (1874).

var. **stipularis** (Gleason) M.V.da S. Alvez, Anal. Jard. Bot. Madrid 51: 303 (1994).
Venezuela (Bolívar), Guyana (Mt. Roraima). 82 GUY VEN. Phan.
* *Chaetocarpus stipularis* Gleason, Bull. Torrey Bot. Club 56: 397 (1929).

Chaetocarpus parvifolius Borhidi, Acta Bot. Hung. 29: 184 (1983).
E. Cuba. 81 CUB. Nanophan.

Chaetocarpus pearcei Rusby, Bull. New York Bot. Gard. 8: 102 (1912).
Bolivia. 83 BOL.
Chaetocarpus pearcei Rusby, Descr. S. Amer. Pl.: 49 (1920), nom. illeg.

Chaetocarpus pubescens (Thwaites) Hook.f., Fl. Brit. India 5: 461 (1887).
Sri Lanka. 40 SRL. Nanophan. or phan.
* *Chaetocarpus castanocarpus* var. *pubescens* Thwaites, Enum. Pl. Zeyl.: 275 (1861).

Chaetocarpus rabaraba Capuron, Adansonia, n.s. 12: 209 (1972).
Madagascar. 29 MDG. Nanophan. or phan.

Chaetocarpus schomburgkianus (Kuntze) Pax & K.Hoffm. in H.G.A.Engler, Pflanzenr., IV, 147, IV: 10 (1912).
Guianas, Venezuela, Brazil (Amazonas, Amapá). 82 FRG GUY SUR VEN 84 BZN. Phan.
* *Gaedawakka schomburgkiana* Kuntze, Revis. Gen. Pl. 2: 606 (1891).
Chaetocarpus williamsii Steyerm., Fieldiana, Bot. 28: 306 (1952).

Synonyms:
Chaetocarpus blanchetii Müll.Arg. === **Chaetocarpus myrsinites** Baill. var. **myrsinites**
Chaetocarpus castanocarpus var. *genuinus* Müll.Arg. === **Chaetocarpus castanocarpus** (Roxb.) Thwaites
Chaetocarpus castanocarpus var. *pubescens* Thwaites === **Chaetocarpus pubescens** (Thwaites) Hook.f.
Chaetocarpus domingensis Proctor === **Chaetocarpus cordifolius** (Urb.) Borhidi
Chaetocarpus globosus var. *lepidotus* (Urb.) Borhidi === **Chaetocarpus globosus** (Sw.) Fawc. & Rendle subsp. **globosus**
Chaetocarpus globosus f. *puberula* Borhidi === **Chaetocarpus globosus** (Sw.) Fawc. & Rendle subsp. **globosus**
Chaetocarpus humilis (Ekman ex Urb.) Borhidi === **Chaetocarpus cubensis** Fawc. & Rendle
Chaetocarpus oblongatus (Alain) Borhidi === **Chaetocarpus globosus** subsp. **oblongatus** (Alain) Borhidi
Chaetocarpus oblongatus var. *monticola* Borhidi === **Chaetocarpus globosus** subsp. **oblongatus** (Alain) Borhidi
Chaetocarpus oblongatus var. *subenervis* Borhidi === **Chaetocarpus globosus** subsp. **oblongatus** (Alain) Borhidi
Chaetocarpus pearcei Rusby === **Chaetocarpus pearcei** Rusby
Chaetocarpus pohlii Müll.Arg. === **Chaetocarpus echinocarpus** (Baill.) Ducke
Chaetocarpus pouteria J.F.Gmel. === **Pouteria guianensis** Aubl. (Sapotaceae)
Chaetocarpus pungens Thwaites === **Chaetocarpus castanocarpus** (Roxb.) Thwaites
Chaetocarpus stipularis Gleason === **Chaetocarpus myrsinites** var. **stipularis** (Gleason) M.V.da S. Alvez
Chaetocarpus williamsii Steyerm. === **Chaetocarpus schomburgkianus** (Kuntze) Pax & K.Hoffm.

Chamaesyce

For most Eastern Hemisphere workers outside Australia treated as part of *Euphorbia* but following work in the 1960s and 1970s by Webster and Burch (in North America) and Hassall (in Australia) becoming accepted in those continents as a genus in its own right. In *Euphorbia* it remains anagenetically the most distinctive subgenus but developmentally is highly specialised due to virtual suppression of the true vegetative part of its life-cycle; above-ground growths are morphologically part of the inflorescence. The largest species occur in Hawai'i; these present a reversion to woodiness largely lost elsewhere in the group.

Synonyms:

Chamaesyce Gray === **Euphorbia** L.
Chamaesyce abdita D.G.Burch === **Euphorbia abdita** (D.G.Burch) Radcl.-Sm.
Chamaesyce abramsiana (Wheeler) Koutnik === **Euphorbia abramsiana** Wheeler
Chamaesyce acuta Millsp. === **Euphorbia georgei** Oudejans
Chamaesyce adenoptera (Bertol.) Small === **Euphorbia adenoptera** Bertol.
Chamaesyce adenoptera subsp. *canescens* Proctor === **Euphorbia adenoptera** subsp. **canescens** (Proctor) Oudejans
Chamaesyce adhaerens Small === **Euphorbia deltoidea** var. **adhaerens** (Small) Oudejans
Chamaesyce adicioides Small === **Euphorbia garberi** Engelm. ex Chapm.
Chamaesyce aequata Lundell === **Euphorbia serpyllifolia** Pers. var. **serpyllifolia**
Chamaesyce albescens (Urb.) Millsp. === **Euphorbia turpinii** Boiss.
Chamaesyce albicaulis (Rydb.) Rydb. === **Euphorbia serpyllifolia** Pers. var. **serpyllifolia**
Chamaesyce albomarginata (Torr. & A.Gray) Small === **Euphorbia albomarginata** Torr. & A.Gray
Chamaesyce alsiniflora (Baill.) D.C.Hassall === **Euphorbia alsiniflora** Baill.
Chamaesyce ammannioides (Kunth) Small === **Euphorbia bombensis** Jacq.
Chamaesyce ammatotricha (Boiss.) Millsp. === **Euphorbia ammatotricha** Boiss.
Chamaesyce amoena (Klotzsch & Garcke) Millsp. === **Euphorbia adenoptera** Bertol. subsp. **adenoptera**
Chamaesyce amplexicaulis (Hook.f.) D.G.Burch === **Euphorbia amplexicaulis** Hook.f.
Chamaesyce andromedae (Millsp.) Millsp. === **Euphorbia torralbasii** Urb.
Chamaesyce anegadensis Millsp. === **Euphorbia turpinii** Boiss.
Chamaesyce angusta (Engelm.) Small === **Euphorbia angusta** Engelm.
Chamaesyce anisopetala Prokh. === **Euphorbia anisopetala** (Prokh.) Prokh.
Chamaesyce anthonyi (Brandegee) G.A.Levin === **Euphorbia anthonyi** Brandegee
Chamaesyce anychioides (Boiss.) Millsp. === **Euphorbia anychioides** Boiss.
Chamaesyce apatzingana (McVaugh) McVaugh === **Euphorbia apatzingana** McVaugh
Chamaesyce arabica (Hochst. & Steud. ex Anderson) Soják === **Euphorbia arabica** Hochst. & Steud. ex Anderson
Chamaesyce arenicola (Parish) Millsp. === **Euphorbia ocelleta** subsp. **arenicola** (Parish) Oudejans
Chamaesyce arequipensis Croizat === **Euphorbia peruviana** Wheeler
Chamaesyce arizonica (Engelm.) Arthur === **Euphorbia arizonica** Engelm.
Chamaesyce arnottiana (Endl.) O.Deg., I.Deg. & Croizat === **Euphorbia arnottiana** Endl.
Chamaesyce articulata (Burm.) Britton === **Euphorbia articulata** Burm.
Chamaesyce astyla (Engelm. ex Boiss.) Millsp. === **Euphorbia astyla** Engelm. ex Boiss.
Chamaesyce atoto (G.Forst.) Croizat === **Euphorbia atoto** G.Forst.
Chamaesyce atrococca (A.Heller) Croizat & O.Deg. === **Euphorbia atrococca** A.Heller
Chamaesyce aureola Millsp. === **Euphorbia melanadenia** Torr.
Chamaesyce auricularia (Boiss.) V.S.Raju & P.N.Rao === **Euphorbia glaucescens** Willd.
Chamaesyce australis (Boiss.) D.C.Hassall === **Euphorbia australis** Boiss.
Chamaesyce bahiensis (Klotzsch & Garcke) Dugand & Burch === **Euphorbia bahiensis** (Klotzsch & Garcke) Boiss.
Chamaesyce balbisii (Boiss.) Millsp. === **Euphorbia balbisii** Boiss.
Chamaesyce barberoana Croizat === **Euphorbia berteroana** Balb. ex Spreng.
Chamaesyce barbicarina Millsp. === **Euphorbia mendezii** Boiss.
Chamaesyce bartolomei (Greene) Millsp. === **Euphorbia bartolomei** Greene
Chamaesyce berteroana (Balb. ex Spreng.) Millsp. === **Euphorbia berteroana** Balb. ex Spreng.
Chamaesyce biconvexa (Domin) D.C.Hassall === **Euphorbia biconvexa** Domin
Chamaesyce bindloensis (Stewart) D.G.Burch === **Euphorbia punctulata** Andersson
Chamaesyce biramensis (Urb.) Alain === **Euphorbia serpens** Kunth
Chamaesyce blodgettii (Engelm. ex Hitchc.) Small === **Euphorbia blodgettii** Engelm. ex Hitchc.
Chamaesyce boliviana (Rusby) Croizat === **Euphorbia boliviana** Rusby
Chamaesyce bombensis (Jacq.) Dugand === **Euphorbia bombensis** Jacq.
Chamaesyce bracei (Millsp.) Millsp. === **Euphorbia cayensis** Millsp.

Chamaesyce brachypoda Small === **Euphorbia garberi** Engelm. ex Chapm.
Chamaesyce brandegeei (Millsp.) Millsp. === **Euphorbia brandegeei** Millsp.
Chamaesyce brasiliensis (Lam.) Small === **Euphorbia hyssopifolia** L.
Chamaesyce brittonii (Millsp.) Millsp. === **Euphorbia minutula** Boiss.
Chamaesyce bruntii Proctor === **Euphorbia bruntii** (Proctor) Oudejans
Chamaesyce bryophylla (Donn.Sm.) Millsp. === **Euphorbia dioeca** Kunth
Chamaesyce burmanica (Hook.f.) Soják === **Euphorbia burmanica** Hook.f.
Chamaesyce buxifolia (Lam.) Small === **Euphorbia mesembryanthemifolia** Jacq.
Chamaesyce caecorum (Mart. ex Boiss.) Croizat === **Euphorbia potentilloides** Boiss.
Chamaesyce camagueyensis Millsp. === **Euphorbia camagueyensis** (Millsp.) Urb.
Chamaesyce canescens (L.) Prokh. === **Euphorbia chamaesyce** L.
Chamaesyce canescens subsp. *glabra* (Roep.) Soják === **Euphorbia chamaesyce** L.
Chamaesyce canescens subsp. *massiliensis* (DC.) Soják === **Euphorbia chamaesyce** subsp. **massiliensis**
Chamaesyce capillaris (Gagnep.) Soják === **Euphorbia capillaris** Gagnep.
Chamaesyce capitellata (Engelm.) Millsp. === **Euphorbia capitellata** Engelm.
Chamaesyce carissoides (F.M.Bailey) D.C.Hassal ex P.I.Forst. & R.J.F.Hend. === **Euphorbia carissoides** F.M.Bailey
Chamaesyce carmenensis (N.E.Rose) Millsp. === **Euphorbia polycarpa** Benth. var. **polycarpa**
Chamaesyce carunculata (Waterf.) Shinners === **Euphorbia carunculata** Waterf.
Chamaesyce catamarcensis Croizat === **Euphorbia catamarcensis** (Croizat) Subils
Chamaesyce cayensis (Millsp.) Millsp. === **Euphorbia cayensis** Millsp.
Chamaesyce celastroides (Boiss.) Croizat & O.Deg. === **Euphorbia celastroides** Boiss.
Chamaesyce celastroides var. *laehiensis* (O.Deg., I.Deg. & Sherff) Koutnik === **Euphorbia celastroides** Boiss.
Chamaesyce celastroides var. *nelsonii* (H.St.John) V.S.Raju & P.N.Rao === **Euphorbia celastroides** Boiss.
Chamaesyce celastroides var. *tomentella* (Boiss.) Koutnik === **Euphorbia celastroides** Boiss.
Chamaesyce centralis (B.G.Thomson) P.I.Forst. & R.J.F.Hend. === **Euphorbia centralis** B.G.Thomson
Chamaesyce centunculoides (Kunth) Millsp. === **Euphorbia centunculoides** Kunth
Chamaesyce chaetocalyx (Boiss.) Wooton & Standl. === **Euphorbia chaetocalyx** (Boiss.) Tidestr.
Chamaesyce chaetocalyx var. *trigulata* (L.C.Wheeler) Mayfield === **Euphorbia chaetocalyx** (Boiss.) Tidestr.
Chamaesyce chalicophila (Weath.) Millsp. === **Euphorbia anychioides** Boiss.
Chamaesyce chamaecaula (Weath.) Millsp. === **Euphorbia chamaecaula** Weath.
Chamaesyce chamaerrhodos (Boiss.) Croizat === **Euphorbia chamaerrhodos** Boiss.
Chamaesyce chamaesyce (L.) Hurus. === **Euphorbia chamaesyce** L.
Chamaesyce chamaesycoides (B.Nord.) Koutnik === **Euphorbia chamaesycoides** B.Nord.
Chamaesyce chamissonis (Klotzsch & Garcke ex Klotzsch) F.C.Ho === **Euphorbia chamissonis** (Klotzsch & Garcke ex Klotzsch) Boiss.
Chamaesyce chiogenes Small === **Euphorbia chiogenes** (Small) Oudejans
Chamaesyce cinerascens (Engelm.) Small === **Euphorbia cinerascens** Engelm.
Chamaesyce clarkeana (Hook.f.) Soják === **Euphorbia clarkeana** Hook.f.
Chamaesyce clusiifolia (Hook. & Arn.) Arthur === **Euphorbia clusiifolia** Hook. & Arn.
Chamaesyce coccinea (B.Heyne ex Roth) Soják === **Euphorbia coccinea** B.Heyne ex Roth
Chamaesyce coghlanii (F.M.Bailey) D.C.Hassal ex P.I.Forst. & R.J.F.Hend. === **Euphorbia coghlanii** F.M.Bailey
Chamaesyce conferta Small === **Euphorbia conferta** (Small) B.E.Sm.
Chamaesyce conjuncta (Millsp.) Millsp. === **Euphorbia pediculifera** Engelm.
Chamaesyce consanguinea Millsp. === **Euphorbia serpyllifolia** Pers. var. **serpyllifolia**
Chamaesyce cordata (Klotzsch & Garcke) Arthur === **Euphorbia degeneri** Sherff
Chamaesyce cordifolia (Elliott) Small === **Euphorbia cordifolia** Elliott
Chamaesyce corrigioloides (Boiss.) Soják === **Euphorbia nodosa** Houtt.
Chamaesyce coudercii (Gagnep.) Soják === **Euphorbia coudercii** Gagnep.
Chamaesyce cowellii Millsp. ex Britton === **Euphorbia cowellii** (Millsp. ex Britton) Oudejans

Chamaesyce cozumelensis (Millsp.) Millsp. === **Euphorbia cozumelensis** Millsp.
Chamaesyce crassinodis (Urb.) Millsp. === **Euphorbia crassinodis** Urb.
Chamaesyce crepitata (L.C.Wheeler) Mayfield === **Euphorbia crepitata** L.C.Wheeler
Chamaesyce crepitata var. *longa* (M.C.Johnst.) Mayfield === **Euphorbia crepitata** L.C.Wheeler
Chamaesyce cristata (B.Heyne ex Roth) G.L.Webster === **Euphorbia cristata** B.Heyne ex Roth
Chamaesyce cumbrae (Boiss.) Millsp. === **Euphorbia cumbrae** Boiss.
Chamaesyce cumulicola Small === **Euphorbia cumulicola** (Small) Oudejans
Chamaesyce dallachyana (Baill.) D.C.Hassall === **Euphorbia dallachyana** Baill.
Chamaesyce degeneri (Sherff) Croizat & O.Deg. === **Euphorbia degeneri** Sherff
Chamaesyce deltoidea (Engelm. ex Chapm.) Small === **Euphorbia deltoidea** Engelm. ex Chapm.
Chamaesyce deltoidea subsp. *adhaerens* (Small) A.Herndon === **Euphorbia deltoidea** var. **adhaerens** (Small) Oudejans
Chamaesyce deltoidea subsp. *pinetorum* (Small) A.Herndon === **Euphorbia deltoidea** subsp. **pinetorum** (Small) Oudejans
Chamaesyce densiflora (Klotzsch) Millsp. === **Euphorbia densiflora** (Klotzsch) Klotzsch
Chamaesyce deppeana (Boiss.) Millsp. === **Euphorbia deppeana** Boiss.
Chamaesyce dioecia (Kunth) Millsp. === **Euphorbia dioeca** Kunth
Chamaesyce dorsiventralis (Urb.) Millsp. === **Euphorbia mendezii** Boiss.
Chamaesyce drummondii (Boiss.) Soják === **Euphorbia drummondii** Boiss.
Chamaesyce duckei Croizat === **Euphorbia duckei** (Croizat) Oudejans
Chamaesyce eichleri (Müll.Arg.) Croizat === **Euphorbia eichleri** Müll.Arg.
Chamaesyce eleanoriae M.E.Lawr. & W.L.Wagner === **Euphorbia eleanoriae** (M.E.Lawr. & W.L.Wagner) Govaerts
Chamaesyce elegans (Spreng.) Soják === **Euphorbia elegans** Spreng.
Chamaesyce emarginata (Klotzsch & Garcke) Croizat === **Euphorbia serpens** Kunth
Chamaesyce emodi (Hook.f.) Soják === **Euphorbia hispida** Boiss.
Chamaesyce engelmannii (Boiss.) Soják === **Euphorbia engelmannii** Boiss.
Chamaesyce erecta Lunell === **Euphorbia serpyllifolia** Pers. var. **serpyllifolia**
Chamaesyce erythroclada (Boiss.) Soják === **Euphorbia erythroclada** Boiss.
Chamaesyce exstipulata (Engelm.) Rydb. === **Euphorbia exstipulata** Engelm.
Chamaesyce exumensis Millsp. === **Euphorbia lecheoides** var. **exumensis** (Millsp.) Oudejans
Chamaesyce eylesii (Rendle) Koutnik === **Euphorbia eylesii** Rendle
Chamaesyce feddemae (McVaugh) McVaugh === **Euphorbia feddemae** McVaugh
Chamaesyce fendleri (Torr. & A.Gray) Small === **Euphorbia fendleri** Torr. & A.Gray
Chamaesyce festiva (Sherff) Croizat & O.Deg. === **Euphorbia deppeana** Boiss.
Chamaesyce filicaulis (Urb.) Alain === **Euphorbia filicaulis** Urb.
Chamaesyce filipes (Benth.) D.C.Hassall === **Euphorbia filipes** Benth.
Chamaesyce fimbriata (B.Heyne ex Roth) R.R.Rao & Razi === **Euphorbia laciniata** Panigrahi subsp. **laciniata**
Chamaesyce fimbriata subsp. *burmanica* (Panigrahi) K.S.Rao & M.N.V.Prasad === **Euphorbia laciniata** subsp. **burmanica** Panigrahi
Chamaesyce flagelliformis (Engelm.) Rydb. === **Euphorbia parryi** Engelm.
Chamaesyce floribunda (Engelm. ex Boiss.) Millsp. === **Euphorbia floribunda** Engelm. ex Boiss.
Chamaesyce florida (Engelm.) Millsp. === **Euphorbia florida** Engelm.
Chamaesyce forbesii (Sherff) Croizat & O.Deg. === **Euphorbia clusiifolia** Hook. & Arn.
Chamaesyce fosbergii Florence === **Euphorbia fosbergii** (Florence) Govaerts
Chamaesyce fruticulosa (Engelm. ex Boiss.) Millsp. === **Euphorbia fruticulosa** Engelm. ex Boiss.
Chamaesyce galapageia (B.L.Rob. & Greenm.) D.G.Burch === **Euphorbia galapageia** Rob. & Greenm.
Chamaesyce garanbiensis (Hayata) Hara === **Euphorbia garanbiensis** Hayata
Chamaesyce garberi (Engelm. ex Chapm.) Small === **Euphorbia garberi** Engelm. ex Chapm.
Chamaesyce garkeana (Boiss.) Millsp. === **Euphorbia garkeana** Boiss.
Chamaesyce gemella (Lag.) Small === **Euphorbia hirta** L.
Chamaesyce geyeri (Engelm. & A.Gray) Small === **Euphorbia geyeri** Engelm. & A.Gray

Chamaesyce geyeri var. *wheeleriana* (Warnock & M.C.Johnst.) Mayfield === **Euphorbia geyeri** Engelm. & A.Gray
Chamaesyce glanduligera (Pax) Koutnik === **Euphorbia glanduligera** Pax
Chamaesyce glaucophylla (Poir.) Croizat === **Euphorbia trinervia** Schumach. & Thonn.
Chamaesyce glomerifera Millsp. === **Euphorbia glomerifera** (Millsp.) Wheeler
Chamaesyce glyptosperma (Engelm.) Small === **Euphorbia glyptosperma** Engelm.
Chamaesyce golondrina (Wheeler) Shinners === **Euphorbia golondrina** Wheeler
Chamaesyce gooddingii Millsp. === **Euphorbia fendleri** Torr. & A.Gray
Chamaesyce gracillima (S.Watson) Millsp. === **Euphorbia gracillima** S.Watson
Chamaesyce grammata McVaugh === **Euphorbia grammata** (McVaugh) Oudejans
Chamaesyce granulata (Forssk.) Soják === **Euphorbia granulata** Forssk.
Chamaesyce granulata var. *dentata* (N.E.Br.) V.S.Raju & P.N.Rao === **Euphorbia inaegulatera** var. **dentata** (N.E.Br.) M.G.Gilbert
Chamaesyce granulata var. *glabrata* (Boiss.) V.S.Raju & P.N.Rao === **Euphorbia granulata** var. **glabrata** (Gay) Boiss.
Chamaesyce granulata var. *turcomanica* (Boiss.) V.S.Raju & P.N.Rao === **Euphorbia granulata** var. **turcomanica** (Boiss.) Hadidi
Chamaesyce greenei (Millsp.) Rydb. === **Euphorbia fendleri** Torr. & A.Gray
Chamaesyce grisea (Engelm. ex Boiss.) Millsp. === **Euphorbia grisea** Engelm. ex Boiss.
Chamaesyce gundlachii (Urb.) Alain === **Euphorbia adenoptera** subsp. **gundlachii** (Urb.) Oudejans
Chamaesyce gymnadenia (Urb.) Millsp. === **Euphorbia adenoptera** subsp. **pergamena** (Small) Oudejans
Chamaesyce halemanui (Sherff) Croizat & O.Deg. === **Euphorbia halemanui** Sherff
Chamaesyce halophila (Miq.) Croizat === **Euphorbia atoto** G.Forst.
Chamaesyce harmandii (Gagnep.) Soják === **Euphorbia harmandii** Gagnep.
Chamaesyce hartwegiana (Boiss.) Small === **Euphorbia albomarginata** Torr. & A.Gray
Chamaesyce hassleriana (Chodat) Soják === **Euphorbia selloi** (Klotzsch & Garcke) Boiss.
Chamaesyce helwigii (Urb. & Ekman) D.G.Burch === **Euphorbia helwigii** Urb. & Ekman
Chamaesyce hepatica (Urb. & Ekman) D.G.Burch === **Euphorbia hepatica** Urb. & Ekman
Chamaesyce heraldiana Millsp. === **Euphorbia heraldiana** (Millsp.) Oudejans
Chamaesyce herbstii W.L.Wagner === **Euphorbia herbstii** (W.L.Wagner) Oudejans
Chamaesyce heyneana (Spreng.) Soják === **Euphorbia heyneana** Spreng.
Chamaesyce heyneana subsp. *galioides* (Boiss.) V.S.Raju & P.N.Rao === **Euphorbia heyneana** subsp. **galioides** (Boiss.) Panigrahi
Chamaesyce heyneana subsp. *nilagirica* (Miq.) V.S.Raju & P.N.Rao === **Euphorbia heyneana** subsp. **nilagirica** (Miq.) Panigrahi
Chamaesyce hillebrandii (H.Lév.) Croizat & O.Deg. === **Euphorbia hillebrandii** H.Lév.
Chamaesyce hirsuta (Torr.) Arthur === **Euphorbia vermiculata** Raf.
Chamaesyce hirta (L.) Millsp. === **Euphorbia hirta** L.
Chamaesyce hirtella (Boiss.) Croizat === **Euphorbia hirtella** Boiss.
Chamaesyce hirtula (Engelm. ex S.Watson) Millsp. === **Euphorbia serpyllifolia** var. **hirtula** (Engelm. ex S.Watson) L.C.Wheeler
Chamaesyce hispida (Boiss.) V.S.Raju & P.N.Rao === **Euphorbia hispida** Boiss.
Chamaesyce hookeri (Steud.) Arthur === **Euphorbia arnottiana** Endl.
Chamaesyce hooveri (Wheeler) Koutnik === **Euphorbia hooveri** Wheeler
Chamaesyce hsinchuensis S.C.Lin & S.M.Chaw === **Euphorbia hsinchuensis** (S.C.Lin & Chaw) C.Y.Wu & J.S.Ma
Chamaesyce humifusa (Willd.) Prokh. === **Euphorbia humifusa** Willd.
Chamaesyce humistrata (Engelm. ex A.Gray) Small === **Euphorbia humistrata** Engelm. ex A.Gray
Chamaesyce hunzikeri (Subils) Holub === **Euphorbia hunzikeri** Subils
Chamaesyce hypericifolia (L.) Millsp. === **Euphorbia hypericifolia** L.
Chamaesyce hyssopifolia (L.) Small === **Euphorbia hyssopifolia** L.
Chamaesyce inaequalis (Klotzsch & Garcke) Millsp. === **Euphorbia adenoptera** Bertol. subsp. **adenoptera**

Chamaesyce inaequilatera (Sond.) Soják === **Euphorbia inaequilatera** Sond.
Chamaesyce inappendiculata (Domin) D.C.Hassall === **Euphorbia inappendiculata** Domin
Chamaesyce incerta (Brandegee) Millsp. === **Euphorbia incerta** Brandegee
Chamaesyce indica (Lam.) Croizat === **Euphorbia indica** Lam.
Chamaesyce indivisa (Engelm.) Millsp. === **Euphorbia indivisa** (Engelm.) Tidestr.
Chamaesyce ingallsii Small === **Euphorbia bombensis** Jacq.
Chamaesyce insulisalis Millsp. === **Euphorbia centunculoides** Kunth
Chamaesyce interaxillaris (Fernald) Millsp. === **Euphorbia stictospora** Engelm.
Chamaesyce intermixta (S.Watson) Millsp. === **Euphorbia polycarpa** Benth. var. **polycarpa**
Chamaesyce involuta (Millsp.) Millsp. === **Euphorbia pediculifera** Engelm.
Chamaesyce jejuna (M.C.Johnst. & Warnock) Shinners === **Euphorbia jejuna** M.C.Johnst. & Warnock
Chamaesyce jenningsii Millsp. ex Britton === **Euphorbia hyssopifolia** L.
Chamaesyce jodhpurensis (Blatt. & Hallb.) V.S.Raju & P.N.Rao === **Euphorbia clarkeana** Hook.f.
Chamaesyce johnstonii (Mayfield) Mayfield === **Euphorbia johnstonii** Mayfield
Chamaesyce jonesii (Millsp.) Millsp. === **Euphorbia hyssopifolia** L.
Chamaesyce jovetii (Huguet) Holub === **Euphorbia maculata** L.
Chamaesyce karwinskyi (Boiss.) Millsp. === **Euphorbia hirta** L.
Chamaesyce katrajensis (Gage) Soják === **Euphorbia katrajensis** Gage
Chamaesyce keyensis Small === **Euphorbia porteriana** (Small) Oudejans
Chamaesyce koerneriana (Allem & Irgang) V.S.Raju & P.N.Rao === **Euphorbia rochaensis** (Croizat) Alonso Paz & Marchesi
Chamaesyce kuriensis (Vierh.) Soják === **Euphorbia kuriensis** Vierh.
Chamaesyce kuwaleana (O.Deg. & Sherff) O.Deg. & I.Deg. === **Euphorbia kuwaleana** O.Deg. & Sherff
Chamaesyce laciniata (Panigrahi) V.S.Raju & P.N.Rao === **Euphorbia laciniata** Panigrahi
Chamaesyce laciniata subsp. *burmanica* (Panigrahi) V.S.Raju & P.N.Rao === **Euphorbia laciniata** subsp. **burmanica** Panigrahi
Chamaesyce lansingii Millsp. === **Euphorbia nutans** Lag.
Chamaesyce laredana (Millsp.) Small === **Euphorbia laredana** Millsp.
Chamaesyce lasiocarpa (Klotzsch) Arthur === **Euphorbia lasiocarpa** Klotzsch
Chamaesyce lata (Engelm.) Small === **Euphorbia lata** Engelm.
Chamaesyce lecheoides (Millsp.) Millsp. === **Euphorbia lecheoides** Millsp.
Chamaesyce leonardii D.G.Burch === **Euphorbia leonardii** (D.G.Burch) Radcl.-Sm.
Chamaesyce leucantha (Klotzsch & Garcke) Millsp. === **Euphorbia mendezii** Boiss.
Chamaesyce leucophylla (Benth.) Millsp. === **Euphorbia leucophylla** Benth.
Chamaesyce levis (Poir.) Croizat === **Euphorbia atoto** G.Forst.
Chamaesyce liliputiana (C.Wright ex Urb.) Millsp. === **Euphorbia minutula** Boiss.
Chamaesyce linearifolia Soják === **Euphorbia deccanensis** V.S.Raju
Chamaesyce linearifolia var. *nallamalayana* (J.L.Ellis) V.S.Raju & P.N.Rao === **Euphorbia deccanensis** V.S.Raju
Chamaesyce linearis (Retz.) Millsp. === **Euphorbia articulata** Burm.
Chamaesyce lineata (S.Watson) Millsp. === **Euphorbia lineata** S.Watson
Chamaesyce linguiformis (McVaugh) McVaugh === **Euphorbia linguiformis** McVaugh
Chamaesyce linguiformis var. *acutinadenia* McVaugh === **Euphorbia linguiformis** McVaugh
Chamaesyce lissosperma (S.Carter) Soják === **Euphorbia lissosperma** S.Carter
Chamaesyce liukiuensis (Hayata) Hara === **Euphorbia liukiuensis** Hayata
Chamaesyce livida (C.A.Mey. ex Boiss.) Koutnik === **Euphorbia livida** C.A.Mey. ex Boiss.
Chamaesyce longinsulicola (S.R.Hill) V.S.Raju & P.N.Rao === **Euphorbia longinsulicola** S.R.Hill
Chamaesyce longiramosa (S.Watson) Millsp. === **Euphorbia parryi** Engelm.
Chamaesyce longistyla (Boiss.) Croizat === **Euphorbia longistyla** Boiss.
Chamaesyce lorentzii (Müll.Arg.) Croizat === **Euphorbia hirtella** Boiss.
Chamaesyce lorifolia (A.Gray ex H.Mann) Croizat & O.Deg. === **Euphorbia celastroides** Boiss.
Chamaesyce luisensis Millsp. === **Euphorbia tomentella** Engelm. ex Boiss.
Chamaesyce lutulenta Croizat === **Euphorbia lutulenta** (Croizat) Oudejans

Chamaesyce luzoniensis (Merr.) Soják === **Euphorbia luzoniensis** Merr.
Chamaesyce macgillivrayi (Boiss.) D.C.Hassall === **Euphorbia macgillivrayi** Boiss.
Chamaesyce maculata (L.) Small === **Euphorbia maculata** L.
Chamaesyce magdalenae (Benth.) Millsp. === **Euphorbia magdalenae** Benth.
Chamaesyce makinoi (Hayata) Hara === **Euphorbia makinoi** Hayata
Chamaesyce malaca Small === **Euphorbia prostrata** Aiton
Chamaesyce mangletii (Urb.) Alain === **Euphorbia serpens** Kunth
Chamaesyce marayensis (Subils) Holub === **Euphorbia marayensis** Subils
Chamaesyce maritima Gray === **Euphorbia peplis** L.
Chamaesyce massiliensis (DC.) Galushko === **Euphorbia chamaesyce** L.
Chamaesyce mathewsii Small === **Euphorbia chamaesula** Boiss.
Chamaesyce mauritiana Comm. ex Denis === **Euphorbia thymifolia** L.
Chamaesyce melanadenia (Torr.) Millsp. === **Euphorbia melanadenia** Torr.
Chamaesyce mendezii (Boiss.) Millsp. === **Euphorbia mendezii** Boiss.
Chamaesyce mertonii (Fosberg) Soják === **Euphorbia mertonii** Fosberg
Chamaesyce mesembryanthemifolia (Jacq.) Dugand === **Euphorbia mesembryanthemifolia** Jacq.
Chamaesyce meyeniana (Klotzsch) Croizat === **Euphorbia meyeniana** Klotzsch
Chamaesyce micradenia (Boiss.) D.C.Hassall === **Euphorbia micradenia** Boiss.
Chamaesyce microcephala (Boiss.) Croizat === **Euphorbia microcephala** Boiss.
Chamaesyce microclada (Urb.) Alain === **Euphorbia serpens** Kunth
Chamaesyce micromera (Boiss. ex Engelm.) Wooton & Standl. === **Euphorbia micromera** Boiss. ex Engelm.
Chamaesyce microphylla (Lam.) Soják === **Euphorbia thymifolia** L.
Chamaesyce minbuensis (Gage) Soják === **Euphorbia minbuensis** Gage
Chamaesyce minutula (Boiss.) D.G.Burch === **Euphorbia minutula** Boiss.
Chamaesyce missurica (Raf.) Shinners === **Euphorbia missurica** Raf.
Chamaesyce missurica var. *calcicola* Shinners === **Euphorbia missurica** Raf.
Chamaesyce mitchelliana (Boiss.) D.C.Hassall === **Euphorbia mitchelliana** Boiss.
Chamaesyce monensis Millsp. === **Euphorbia adenoptera** subsp. **pergamena** (Small) Oudejans
Chamaesyce montana Alain === **Euphorbia alainii** Oudejans
Chamaesyce mosieri Small === **Euphorbia garberi** Engelm. ex Chapm.
Chamaesyce mossambicensis (Klotzsch & Garcke) Koutnik === **Euphorbia mossambicensis** (Klotzsch & Garcke) Boiss.
Chamaesyce multiformis (Gaudich. ex Hook. & Arn.) Croizat & O.Deg. === **Euphorbia multiformis** Gaudich. ex Hook. & Arn.
Chamaesyce multinodis (Urb.) Millsp. === **Euphorbia multinodis** Urb.
Chamaesyce myrtillifolia (L.) Millsp. === **Euphorbia myrtillifolia** L.
Chamaesyce myrtoides (Boiss.) D.C.Hassall === **Euphorbia myrtoides** Boiss.
Chamaesyce nana (Klotzsch & Garke) M.P.Simmons & W.J.Hayden === **Euphorbia chamaerrhodos** Boiss.
Chamaesyce nashii Small === **Euphorbia blodgettii** Engelm. ex Hitchc.
Chamaesyce neomexicana (Greene) Standl. === **Euphorbia serpyllifolia** Pers. var. **serpyllifolia**
Chamaesyce neopolycnemoides (Pax & K.Hoffm.) Koutnik === **Euphorbia neopolycnemoides** Pax & K.Hoffm.
Chamaesyce niqueroana (Urb.) Alain === **Euphorbia minutula** Boiss.
Chamaesyce nirurioides Millsp. === **Euphorbia hyssopifolia** L.
Chamaesyce notoptera (Boiss.) Soják === **Euphorbia notoptera** Boiss.
Chamaesyce nummularia (Hook.f.) D.G.Burch === **Euphorbia nummularia** Hook.f.
Chamaesyce nutans (Lag.) Small === **Euphorbia nutans** Lag.
Chamaesyce nuttallii (Engelm.) Small === **Euphorbia missurica** Raf.
Chamaesyce obliqua (F.A.Bauer ex Endl.) Florence === **Euphorbia obliqua** F.A.Bauer ex Endl.
Chamaesyce occidentalis (Drew) Millsp. === **Euphorbia serpyllifolia** Pers. var. **serpyllifolia**
Chamaesyce ocelleta (Durand & Hilg.) Millsp. === **Euphorbia ocelleta** Durand & Hilg.

Chamaesyce ocelleta subsp. *arenicola* (Parish) Thorne === **Euphorbia ocelleta** subsp. **arenicola** (Parish) Oudejans
Chamaesyce ocelleta subsp. *rattanii* (S.Watson) Koutnik === **Euphorbia ocelleta** var. **rattanii** (S.Watson) L.C.Wheeler
Chamaesyce olowaluana (Sherff) Croizat & O.Deg. === **Euphorbia olowaluana** Sherff
Chamaesyce ophthalmica (Pers.) D.G.Burch === **Euphorbia ophthalmica** Pers.
Chamaesyce oranensis Croizat === **Euphorbia oranensis** (Croizat) Subils
Chamaesyce orbiculata (Kunth) Soják === **Euphorbia orbiculata** Kunth
Chamaesyce orbiculata subsp. *galioides* (Boiss.) Soják === **Euphorbia heyneana** subsp. **galioides** (Boiss.) Panigrahi
Chamaesyce orbiculata subsp. *nilagirica* (Miq.) Soják === **Euphorbia heyneana** subsp. **nilagirica** (Miq.) Panigrahi
Chamaesyce orbifolia Alain === **Euphorbia orbifolia** (Alain) Oudejans
Chamaesyce ovalifolia (Engelm. ex Klotzsch) Croizat === **Euphorbia klotzschii** Oudejans
Chamaesyce pachypoda (Urb.) Alain === **Euphorbia centunculoides** Kunth
Chamaesyce parannaquensis (Blanco) Hara === **Euphorbia vachellii** Hook. & Arn.
Chamaesyce parciflora (Urb.) D.G.Burch === **Euphorbia parciflora** Urb.
Chamaesyce paredonensis Millsp. === **Euphorbia paredonensis** (Millsp.) Oudejans
Chamaesyce parryi (Engelm.) Rydb. === **Euphorbia parryi** Engelm.
Chamaesyce parviflora (L.) Soják === **Euphorbia parviflora** L.
Chamaesyce paucipila (Urb.) Millsp. === **Euphorbia minutula** Boiss.
Chamaesyce pediculifera (Engelm.) Rose & Standl. === **Euphorbia pediculifera** Engelm.
Chamaesyce peplis (L.) Prokh. === **Euphorbia peplis** L.
Chamaesyce perennans Shinners === **Euphorbia perennans** (Shinners) Warnock & M.C.Johnst.
Chamaesyce pergemena (Small) Small === **Euphorbia adenoptera** subsp. **pergamena** (Small) Oudejans
Chamaesyce pergracilis (P.G.Mey.) Koutnik === **Euphorbia pergracilis** P.G.Mey.
Chamaesyce perlignea (McVaugh) G.L.Webster === **Euphorbia perlignea** McVaugh
Chamaesyce petala (Ewart & L.R.Kerr) P.I.Forst. & R.J.F.Hend. === **Euphorbia petala** Ewart & L.R.Kerr
Chamaesyce petaloidea (Engelm.) Small === **Euphorbia missurica** Raf.
Chamaesyce petrina (S.Watson) Millsp. === **Euphorbia petrina** S.Watson
Chamaesyce picachensis (Brandegee) Millsp. === **Euphorbia picachensis** Brandegee
Chamaesyce pileoides (Millsp.) Millsp. === **Euphorbia serpens** Kunth
Chamaesyce pilosula (Engelm. ex Boiss.) Arthur === **Euphorbia anychioides** Boiss.
Chamaesyce pilulifera (L.) Small === **Euphorbia hirta** L.
Chamaesyce pinariona (Urb.) Alain === **Euphorbia camagueyensis** (Millsp.) Urb.
Chamaesyce pinetorum Small === **Euphorbia deltoidea** subsp. **pinetorum** (Small) Oudejans
Chamaesyce platysperma (Engelm. ex S.Watson) Shinners === **Euphorbia platysperma** Engelm. ex S.Watson
Chamaesyce podadenia (Boiss.) Millsp. === **Euphorbia podadenia** Boiss.
Chamaesyce polycarpa (Benth.) Millsp. === **Euphorbia polycarpa** Benth.
Chamaesyce polyclada (Boiss.) Small === **Euphorbia geyeri** Engelm. & A.Gray
Chamaesyce polycnemoides (Hochst. ex Boiss.) Soják === **Euphorbia polycnemoides** Hochst. ex Boiss.
Chamaesyce polygonifolia (L.) Small === **Euphorbia polygonifolia** L.
Chamaesyce pondii (Millsp.) Millsp. === **Euphorbia pondii** Millsp.
Chamaesyce porteriana Small === **Euphorbia porteriana** (Small) Oudejans
Chamaesyce portoricensis (Urb.) Millsp. === **Euphorbia turpinii** Boiss.
Chamaesyce portucasadiana Croizat === **Euphorbia portucasadiana** (Croizat) Subils
Chamaesyce portulana (S.Watson) Millsp. === **Euphorbia arizonica** Engelm.
Chamaesyce potentilloides (Boiss.) Croizat === **Euphorbia potentilloides** Boiss.
Chamaesyce potosina (Fernald) Arthur === **Euphorbia potosina** Fernald
Chamaesyce preslii (Guss.) Arthur === **Euphorbia nutans** Lag.
Chamaesyce prieuriana (Baill.) Soják === **Euphorbia convolvuloides** Hochst. ex Benth.

Chamaesyce proctorii D.G.Burch === **Euphorbia proctorii** (D.G.Burch) Correll
Chamaesyce prostrata (Aiton) Small === **Euphorbia prostrata** Aiton
Chamaesyce psammogeton (P.S.Green) P.I.Forst. & R.J.F.Hend. === **Euphorbia psammogeton** P.S.Green
Chamaesyce pseudonutans Thell. === **Euphorbia maculata** L.
Chamaesyce pseudoserpyllifolia (Millsp.) Millsp. === **Euphorbia micromera** Boiss. ex Engelm.
Chamaesyce puberula (Fernald) Millsp. === **Euphorbia bertereana** Balb. ex Spreng.
Chamaesyce punctulata (Andersson) D.G.Burch === **Euphorbia punctulata** Andersson
Chamaesyce purisimana (Millsp.) Millsp. === **Euphorbia arizonica** Engelm.
Chamaesyce pycnanthema (Engelm.) Millsp. === **Euphorbia capitellata** Engelm.
Chamaesyce pycnostegia (Boiss.) Soják === **Euphorbia pycnostegia** Boiss.
Chamaesyce radicans Millsp. === **Euphorbia serpens** Kunth
Chamaesyce radioloides (Boiss.) Millsp. === **Euphorbia radioloides** Boiss.
Chamaesyce rafinesquei (Greene) Small === **Euphorbia vermiculata** Raf.
Chamaesyce ramosa (Seaton) Millsp. === **Euphorbia ramosa** Seaton
Chamaesyce rattanii (S.Watson) Millsp. === **Euphorbia ocelleta** var. **rattanii** (S.Watson) L.C.Wheeler
Chamaesyce reconciliationis (Radcl.-Sm.) Soják === **Euphorbia reconciliationis** Radcl.-Sm.
Chamaesyce recurva (Hook.f.) D.G.Burch === **Euphorbia recurva** Hook.f.
Chamaesyce remyi (A.Gray ex Boiss.) Croizat & O.Deg. === **Euphorbia remyi** A.Gray ex Boiss.
Chamaesyce reniformis (Blume) Soják === **Euphorbia reniformis** Blume
Chamaesyce revoluta (Engelm.) Small === **Euphorbia revoluta** Engelm.
Chamaesyce rhytisperma (Klotzsch & Garcke) Soják === **Euphorbia rhytisperma** (Klotzsch & Garcke) Boiss.
Chamaesyce riebeckii (Pax) Soják === **Euphorbia riebeckii** Pax
Chamaesyce rochaensis Croizat === **Euphorbia rochaensis** (Croizat) Alonso Paz & Marchesi
Chamaesyce rockii (C.N.Forbes) Croizat & O.Deg. === **Euphorbia rockii** C.N.Forbes
Chamaesyce rockii var. *grandifolia* (Hillebr.) Koutnik === **Euphorbia herbstii** (W.L.Wagner) Oudejans
Chamaesyce rosea (Retz.) G.L.Webster === **Euphorbia rosea** Retz.
Chamaesyce rosei Millsp. === **Euphorbia hirta** L.
Chamaesyce rothrockii Millsp. === **Euphorbia rothrockii** (Millsp.) Oudejans
Chamaesyce rubida (Greenm.) Millsp. === **Euphorbia anychioides** Boiss.
Chamaesyce rubrosperma (Lotsy) Millsp. === **Euphorbia thymifolia** L.
Chamaesyce rugulosa (Engelm. ex Millsp.) Rydb. === **Euphorbia serpyllifolia** Pers. var. **serpyllifolia**
Chamaesyce ruizlealii (Subils) Holub === **Euphorbia ruizlealii** Subils
Chamaesyce rusbyi (Greene) Millsp. === **Euphorbia capitellata** Engelm.
Chamaesyce rutilis Millsp. === **Euphorbia rutilis** (Millsp.) Standl. & Steyerm.
Chamaesyce sachetiana Florence === **Euphorbia sachetiana** (Florence) Govaerts
Chamaesyce salsuginosa McVaugh === **Euphorbia salsuginosa** (McVaugh) Radcl.-Sm. & Govaerts
Chamaesyce sanmartensis (Rusby) Dugand === **Euphorbia sanmartensis** Rusby
Chamaesyce schlechteri (Pax) Koutnik === **Euphorbia schlechteri** Pax
Chamaesyce schultzii (Benth.) D.C.Hassall === **Euphorbia schultzii** Benth.
Chamaesyce scoparia Small === **Euphorbia porteriana** (Small) Oudejans
Chamaesyce scopulorum (Brandegee) Millsp. === **Euphorbia scopulorum** Brandegee
Chamaesyce scopulorum var. *inornata* (M.C.Johnst.) V.S.Raju & P.N.Rao === **Euphorbia scopulorum** Brandegee
Chamaesyce scopulorum var. *nuda* (M.C.Johnst.) V.S.Raju & P.N.Rao === **Euphorbia scopulorum** Brandegee
Chamaesyce scordiifolia (Jacq.) Croizat === **Euphorbia scordiifolia** Jacq.
Chamaesyce seleri (Donn.Sm.) Millsp. === **Euphorbia seleri** Donn.Sm.
Chamaesyce selloi (Klotzsch & Garcke) Croizat === **Euphorbia selloi** (Klotzsch & Garcke) Boiss.

Chamaesyce senguptae (N.P.Balakr. & Subr.) V.S.Raju & P.N.Rao === **Euphorbia senguptae** N.P.Balakr. & Subr.
Chamaesyce serpens (Kunth) Small === **Euphorbia serpens** Kunth
Chamaesyce serpyllifolia (Pers.) Small === **Euphorbia serpyllifolia** Pers.
Chamaesyce serpyllifolia subsp. *hirtula* (Engelm. ex S.Watson) Koutnik === **Euphorbia serpyllifolia** var. **hirtula** (Engelm. ex S.Watson) L.C.Wheeler
Chamaesyce serpyllum Small === **Euphorbia deltoidea** Engelm. ex Chapm. subsp. **deltoidea**
Chamaesyce serratifolia (S.Carter) Soják === **Euphorbia serratifolia** S.Carter
Chamaesyce serrula (Engelm.) Wooton & Standl. === **Euphorbia serrula** Engelm.
Chamaesyce setiloba (Engelm. ex Torr.) Norton === **Euphorbia setiloba** Engelm. ex Torr.
Chamaesyce setosa (Boiss.) M.P.Simmons & W.J.Hayden === **Euphorbia setosa** (Boiss.) Müll.Arg.
Chamaesyce simulans (L.C.Wheeler) Mayfield === **Euphorbia simulans** (L.C.Wheeler) Warnock & M.C.Johnst.
Chamaesyce sistanica (Dinelli & De Marco) Soják === **Euphorbia granulata** Forssk. var. **granulata**
Chamaesyce skottsbergii (Sherff) Croizat & O.Deg. === **Euphorbia skottsbergii** Sherff
Chamaesyce skottsbergii var. *vaccinioides* (Sherff) Koutnik === **Euphorbia skottsbergii** var. **vaccinioides** Sherff
Chamaesyce sparrmanii (Boiss.) Hurus. === **Euphorbia sparrmanii** Boiss.
Chamaesyce sparsiflora (A.Heller) Koutnik === **Euphorbia sparsiflora** A.Heller
Chamaesyce standleyi Millsp. === **Euphorbia standleyi** (Millsp.) Oudejans
Chamaesyce stanfieldii Small === **Euphorbia villifera** Scheele
Chamaesyce stictospora (Engelm.) Small === **Euphorbia stictospora** Engelm.
Chamaesyce stictospora var. *sublaevis* (M.C.Johnst.) V.S.Raju & P.N.Rao === **Euphorbia stictospora** Engelm.
Chamaesyce stoddartii (Fosberg) Soják === **Euphorbia stoddartii** Fosberg
Chamaesyce sulfurea Millsp. === **Euphorbia ocelleta** Durand & Hilg.
Chamaesyce supina (Raf.) Hara === **Euphorbia maculata** L.
Chamaesyce taihsiensis Chaw & Koutnik === **Euphorbia taihsiensis** (Chaw & Koutnik) Oudejans
Chamaesyce tamanduana (Boiss.) M.P.Simmons & W.J.Hayden === **Euphorbia tamanduana** Boiss.
Chamaesyce tamaulipasana Millsp. === **Euphorbia tamaulipasana** (Millsp.) Oudejans
Chamaesyce tashiroi (Hayata) Hara === **Euphorbia humifusa** Willd.
Chamaesyce tettensis (Klotzsch) Koutnik === **Euphorbia tettensis** Klotzsch
Chamaesyce theriaca (Wheeler) Shinners === **Euphorbia theriaca** Wheeler
Chamaesyce theriaca var. *spurca* (M.C.Johnst.) Mayfield === **Euphorbia theriaca** Wheeler
Chamaesyce thymifolia (L.) Millsp. === **Euphorbia thymifolia** L.
Chamaesyce tomentulosa (S.Watson) Millsp. === **Euphorbia tomentulosa** S.Watson
Chamaesyce tonsita Millsp. === **Euphorbia polycarpa** Benth. var. **polycarpa**
Chamaesyce torralbasii (Urb.) Millsp. === **Euphorbia torralbasii** Urb.
Chamaesyce trachysperma (Engelm.) Millsp. === **Euphorbia trachysperma** Engelm.
Chamaesyce tracyi Small === **Euphorbia maculata** L.
Chamaesyce trancapatae Croizat === **Euphorbia trancapatae** (Croizat) J.F.Macbr.
Chamaesyce tumistyla D.G.Burch === **Euphorbia tumistyla** (D.G.Burch) Radcl.-Sm.
Chamaesyce turcomanica (Boiss.) Prokh. === **Euphorbia granulata** var. **turcomanica** (Boiss.) Hadidi
Chamaesyce turpinii (Boiss.) Millsp. === **Euphorbia turpinii** Boiss.
Chamaesyce umbellulata (Engelm. ex Boiss.) Millsp. === **Euphorbia umbellulata** Engelm. ex Boiss.
Chamaesyce urbanii Millsp. === **Euphorbia dioeca** Kunth
Chamaesyce vachellii (Hook. & Arn.) Hara === **Euphorbia vachellii** Hook. & Arn.
Chamaesyce vaginulata (Griseb.) Millsp. === **Euphorbia vaginulata** Griseb.
Chamaesyce vahlii (Willd. ex Klotzsch & Garcke) P.Wilson === **Euphorbia articulata** Burm.

Chamaesyce vallismortuae Millsp. === **Euphorbia vallismortuae** (Millsp.) J.T.Howell
Chamaesyce velleriflora (Klotzsch & Garcke) Millsp. === **Euphorbia velleriflora** (Klotzsch & Garcke) Boiss.
Chamaesyce velligera (Schauer) Millsp. === **Euphorbia velligera** Schauer
Chamaesyce vermiculata (Raf.) House === **Euphorbia vermiculata** Raf.
Chamaesyce versicolor (Greene) Norton === **Euphorbia arizonica** Engelm.
Chamaesyce vestita (Boiss.) Millsp. === **Euphorbia vestita** Boiss.
Chamaesyce villifera (Scheele) Small === **Euphorbia villifera** Scheele
Chamaesyce villosior (Greenm.) Millsp. === **Euphorbia prostrata** Aiton
Chamaesyce viminea (Hook.f.) D.G.Burch === **Euphorbia viminea** Hook.f.
Chamaesyce viridula (Cordem. ex Radcl.-Sm.) Soják === **Euphorbia viridula** Cordem. ex Radcl.-Sm.
Chamaesyce viscoides (Boiss.) M.P.Simmons & W.J.Hayden === **Euphorbia viscoides** Boiss.
Chamaesyce vulgaris Prokh. === **Euphorbia chamaesyce** L.
Chamaesyce vulgaris subsp. *massiliensis* (DC.) Benedí & Orell === **Euphorbia chamaesyce** L.
Chamaesyce watsonii (Millsp.) Millsp. === **Euphorbia magdalenae** Benth.
Chamaesyce wheeleri (Baill.) D.C.Hassall === **Euphorbia wheeleri** Baill.
Chamaesyce wightiana V.S.Raju & P.N.Rao === **Euphorbia agowensis** Hochst. ex Boiss. var. **agowensis**
Chamaesyce wilsonii Millsp. === **Euphorbia lecheoides** var. **wilsonii** (Millsp.) Oudejans
Chamaesyce yayalesia (Urb.) Alain === **Euphorbia mesembryanthemifolia** Jacq.
Chamaesyce yucatanensis Millsp. === **Euphorbia yucatanensis** (Millsp.) Standl.
Chamaesyce zambesiana (Benth.) Koutnik === **Euphorbia zambesiana** Benth.
Chamaesyce zornioides (Boiss.) Soják === **Euphorbia pycnostegia** Boiss.
Chamaesyce zygophylloides (Boiss.) Small === **Euphorbia missurica** Raf.

Characias

Synonyms:
Characias Gray === **Euphorbia** L.

Charidia

Synonyms:
Charidia Baill. === **Blotia** Leandri

Chascotheca

2 species, West Indies (Cuba, Hispaniola), shrubs with spreading branches in the manner of *Breynia*. *C. neopeltandra* grows on limestone, at least in Hispaniola. The genus was never revised for *Pflanzenreich* (apart from *C. triplinervia*, included therein under *Drypetes* as no. 57) and in their *Pflanzenfamilien* account Pax & Hoffmann (1931) assigned the two then-known species (now united) to their *Securinega* sect. *Colmeiroa*, based on *S. buxifolia* from SW. Europe. Webster (1984; see *Flueggea*) removed this latter species (and sect. *Colmeiroa*) to *Flueggea*, limiting *Chascotheca* to the Antilles. (Phyllanthoideae)

Pax, F. & K. Hoffmann (1931). *Securinega*. In A. Engler (ed.), Die natürlichen Pflanzenfamilien, 2. Aufl., 19c: 60. Leipzig. Ge. — Synopsis with description of genus. [*Chascotheca* here treated as sect. *Colmeiroa*, with 1 species in SW Europe and 2 species in Cuba and Hispaniola (the latter mutually closely related and perhaps not distinct).]

Chascotheca Urb., Symb. Antill. 5: 14 (1904).
Cuba, Hispaniola. 81.
Chaenotheca Urb., Symb. Antill. 3: 284 (1902).

Chascotheca neopeltandra (Griseb.) Urb., Symb. Antill. 5: 14 (1904).
Cuba, Hispaniola. 81 CUB DOM HAI. Nanophan. or phan.
* *Phyllanthus neopeltandrus* Griseb., Nachr. Königl. Ges. Wiss. Georg-Augusts-Univ. 1: 167 (1865). *Securinega neopeltandra* (Griseb.) Urb. ex Pax & K.Hoffm. in H.G.A.Engler, Nat. Pflanzenfam. ed. 2, 19c: 60 (1931).
Chaenotheca domingensis Urb., Symb. Antill. 3: 285 (1902). *Chascotheca domingensis* (Urb.) Urb., Symb. Antill. 5: 14 (1904).

Chascotheca triplinervia (Müll.Arg.) G.L.Webster, J. Arnold Arbor. 48: 330 (1967).
Cuba. 81 CUB. Nanophan.
* *Drypetes triplinervia* Müll.Arg. in A.P.de Candolle, Prodr. 15(2): 456 (1866).

Synonyms:
Chascotheca domingensis (Urb.) Urb. === **Chascotheca neopeltandra** (Griseb.) Urb.

Cheilosa

1 species, W. Malesia and Philippines; small to medium trees of forest to 28 m with coarse foliage, often in steep places or along rivers where there is more light. *C. malayana* was recently reduced to *C. montana* (van Welzen et al., 1993). The genus is close to *Neoscortechinia* (Airy-Shaw 1972; see Asia); together they form tribe Cheiloseae in the Webster system. (Acalyphoideae)

Pax, F. (with K. Hoffmann) (1912). *Cheilosa*. In A. Engler (ed.), Das Pflanzenreich, IV 147 IV (Euphorbiaceae-Gelonieae): 12-14. Berlin. (Heft 52.) La/Ge. — 1 species, Java (*C. montana*).
Airy-Shaw, H. K. (1963). Notes on Malaysian and other Asiatic Euphorbiaceae, XLII. *Cheilosa* Blume in Borneo. Kew Bull. 16: 364-365. En. — Range extension and new name, *C. malayana*, with full synonymy, references and citations, and localities with exsiccatae outside Peninsular Malaysia and Philippines. [The name is now reduced to *C. montana*.]
Airy-Shaw, H. K. (1966). Notes on Malaysian and other Asiatic Euphorbiaceae, LXVI. *Cheilosa* Bl. in Sumatra and East Borneo. Kew Bull. 20: 49. En. — Range extensions.
Welzen, P.C. van, R. A. Banka & C. D. Leoncito (1993). A revision of the Malesian monotypic genus *Cheilosa* Blume (Euphorbiaceae). Blumea 38: 161-166, illus., map. En. — Revision (1 species) with generic and species descriptions, synonymy, types, field notes, indication of distribution, habitat and ecology, vernacular names, and critical remarks; excluded species, references, and identification list. *Cheilosa whiteana* is confirmed as excluded (to *Trigonostemon*).

Cheilosa Blume, Bijdr.: 613 (1826).
Malesia. 42.

Cheilosa montana Blume, Bijdr.: 614 (1826).
W. & NC. Malesia. 42 BOR JAW MLY PHI SUM. Phan.
Baliospermum malayanum Hook.f., Fl. Brit. India 5: 463 (1888). *Cheilosa malayana* (Hook.f.) Corner ex Airy Shaw, Kew Bull. 16: 364 (1963).
Baliospermum analayanum Hook.f. ex B.D.Jacks., Index Kew., Suppl. 1: (1896), nom. illeg.
Cheilosa homaliifolia Merr., Philipp. J. Sci., C 8: 379 (1913).
Cheilosa homaliifolia var. *grandifolia* Merr., Enum. Philipp. Fl. Pl. 2: 457 (1923).
Cheilosa montana var. *longifolia* S.Moore, J. Bot. 64(Suppl.): 104 (1925).

Synonyms:
Cheilosa homaliifolia Merr. === **Cheilosa montana** Blume
Cheilosa homaliifolia var. *grandifolia* Merr. === **Cheilosa montana** Blume

Cheilosa malayana (Hook.f.) Corner ex Airy Shaw === **Cheilosa montana** Blume
Cheilosa montana var. *longifolia* S.Moore === **Cheilosa montana** Blume
Cheilosa whiteana Croizat === **Trigonostemon whiteanus** (Croizat) Airy Shaw

Chiropetalum

22 species, Mexico (2) and S America (20, from Peru and Brazil to Argentina and Chile); herbs or subshrubs. Some authorities, among them Ingram (1980), have referred it to *Argythamnia*. *Aonikena* is included here following O'Donell & Lourteig and Webster (Synopsis, 1994); Ingram (1980) allocated it to neither of the larger genera. The genus is notable within the Americas for a disjunct bihemispheric distribution (for discussion, see Ingram 1980); several species are found in desert areas. (Acalyphoideae)

Pax, F. (with K. Hoffmann) (1912). *Aonikena*. In A. Engler (ed.), Das Pflanzenreich, IV 147 VI (Euphorbiaceae-Acalypheae-Chrozophorinae): 95-96. Berlin. (Heft 57.) La/Ge. — 1 species, southern S America (Patagonia). [Now included with *Chiropetalum*.]

Pax, F. (with K. Hoffmann) (1912). *Chiropetalum*. In A. Engler (ed.), Das Pflanzenreich, IV 147 VI (Euphorbiaceae-Acalypheae-Chrozophorinae): 86-95. Berlin. (Heft 57.) La/Ge. — 18 species, mainly in southern S America. [Genus reduced by Ingram to *Argythamnia*, a step first proposed by Croizat in 1945 but not upheld in recent reference works.]

• Ingram, J. (1980). A revision of *Argythamnia*, subgenus *Chiropetalum* (Euphorbiaceae). Gentes Herb. 11(7): 437-468, illus., maps. En. — Descriptive revision (20 species, some new) with key, synonymy, references, types, indication of distribution and localities with exsiccatae; list of references but no separate index. The general part includes sections on biogeography and characters but no general suggestions regarding evolutionary trends. [2 species are in Mexico and 18 in South America; many are in desert areas. Subgen. *Chiropetalum* was accorded generic rank by Pax and Hoffmann (1912) and in recent reference works, but the author has here reverted to Müller's circumscription.]

Ingram, J. (1980). The generic limits of *Argythamnia* (Euphorbiaceae) defined. Gentes Herb. 11(7): 426-436, illus., maps. En. — Biogeography (with postulated centre of origin and dispersal tracks); morphology and taxonomic relationships; revised generic description along with a key to and distribution maps of subgenera (*Argythamnia*, *Ditaxis* and *Chiropetalum*). [The reduction of *Ditaxis* by Croizat (1945) is accepted, along with the reduction (here, for the first time) of *Chiropetalum* to *Argythamnia*.]

Chiropetalum A.Juss., Ann. Sc. Nat. (Paris) 25: 21 (1832).
Mexico, Dry Trop. S. America. 79 83 84 85.
Desfontaena Vell., Fl. Flumin.: 95 (1829).
Chlorocaulon Klotzsch in S.L.Endlicher, Gen. Pl., Suppl. 4(3): 89 (1850).
Aonikena Speg., Anales Mus. Nac. Buenos Aires 7: 162 (1902).

Chiropetalum anisotrichum (Müll.Arg.) Pax & K.Hoffm. in H.G.A.Engler, Pflanzenr., IV, 147, VI: 93 (1912).
Brazil (Rio Grande do Sul). 84 BZS. Hemicr.
* *Argythamnia anisotricha* Müll.Arg. in C.F.P.von Martius, Fl. Bras. 11(2): 314 (1874).

Chiropetalum argentinense Skottsb., Lilloa 17: 304 (1949). *Argythamnia argentinense* (Skottsb.) Allem & Irgang, Revista Brasil. Biol. 36(2): 286 (1976).
N. Argentina. 85 AGE AGW. Hemicr.

Chiropetalum astroplethos (J.W.Ingram) Radcl.-Sm. & Govaerts, Kew Bull. 52: 478 (1997).
Mexico (Coahuila, Veracruz, Tamaulipas, San Luis Potosí, Nuevo León). 79 MXE MXG. Hemicr.

Argythamnia schiedeana var. *minor* Müll.Arg., Linnaea 34: 150 (1865). *Chiropetalum schiedeanum* var. *minus* (Müll.Arg.) Pax & K.Hoffm. in H.G.A.Engler, Pflanzenr., IV, 147, VI: 90 (1912).
* *Argythamnia astroplethos* J.W.Ingram, Gentes Herb. 11: 463 (1980).

Chiropetalum berterianum Schltdl., Linnaea 26: 637 (1853). *Argythamnia berteriana* (Schltdl.) Müll.Arg., Linnaea 34: 151 (1865).
N. & C. Chile. 85 CLN CLC. Hemicr.
* *Ditaxis chiropetala* Bertero, Bull. Ferussac 20: 110 (1830).

var. **berterianum**
N. & C. Chile. 85 CLN CLC. Hemicr.
Chiropetalum berterianum f. *macrantha* Skottsb., Acta Horti Gothob. 18: 78 (1949).

var. **psiladenium** Skottsb., Acta Horti Gothob. 18: 63 (1949). *Argythamnia berteriana* var. *psiladenia* (Skottsb.) J.W.Ingram, Gentes Herb. 11: 454 (1980).
Chile. 85 CLC. Hemicr.

Chiropetalum boliviense (Müll.Arg.) Pax & K.Hoffm. in H.G.A.Engler, Pflanzenr., IV, 147, VI: 94 (1912).
Bolivia, NW. Argentina. 83 BOL 85 AGW. Cham.
* *Argythamnia boliviensis* Müll.Arg., Linnaea 34: 149 (1865).
Chiropetalum triandrum Griseb., Abh. Königl. Ges. Wiss. Göttingen 24: 56 (1879).
Argythamnia triandra (Griseb.) Allem & Irgang, Revista Brasil. Biol. 36: 286 (1976).

Chiropetalum canescens Phil., Fl. Atacam.: 49 (1860). *Argythamnia canescens* (Phil.) F.Phil., Cat. Pl. Vasc. Chil.: 262 (1881).
N. Chile. 85 CLN. Hemicr.
Argythamnia sponiella Müll.Arg., Linnaea 34: 148 (1865). *Chiropetalum sponiella* (Müll.Arg.) Pax in H.G.A.Engler & K.A.E.Prantl, Nat. Pflanzenfam. 3(5): 45 (1890).
Chiropetalum gigouxii Espinosa, Revista Chilena Hist. Nat. 40: 190 (1936 publ. 1937).

Chiropetalum cremnophilum I.M.Johnst., Contr. Gray Herb. 85: 64 (1929). *Argythamnia cremnophila* (I.M.Johnst.) J.W.Ingram, Gentes Herb. 11: 449 (1980).
N. Chile. 85 CLN. Hemicr.

Chiropetalum foliosum (Müll.Arg.) Pax & K.Hoffm. in H.G.A.Engler, Pflanzenr., IV, 147, VI: 91 (1912).
S. Brazil (?). 84 BZS. Hemicr.
* *Argythamnia foliosa* Müll.Arg., Linnaea 34: 150 (1865).

Chiropetalum griseum Griseb., Abh. Königl. Ges. Wiss. Göttingen 24: 57 (1879). *Argythamnia grisea* (Griseb.) Allem & Irgang, Revista Brasil. Biol. 36: 286 (1976).
Paraguay, NE. Argentina. 85 AGE AGW PAR. Cham.
Chiropetalum cupreum Pax & K.Hoffm. in H.G.A.Engler, Pflanzenr., IV, 147, VI: 89 (1912).

Chiropetalum gymnadenium (Müll.Arg.) Pax & K.Hoffm. in H.G.A.Engler, Pflanzenr., IV, 147, VI: 91 (1912).
Brazil (Minas Gerais, São Paulo). 84 BZL. Cham. or nanophan.
* *Argythamnia gymnadenia* Müll.Arg. in C.F.P.von Martius, Fl. Bras. 11(2): 316 (1874).

Chiropetalum intermedium Pax & K.Hoffm. in H.G.A.Engler, Pflanzenr., IV, 147, VI: 91 (1912). *Argythamnia intermedia* (Pax & K.Hoffm.) Allem & Irgang, Revista Brasil. Biol. 36: 286 (1976).
Uruguay, Argentina (Buenos Aires). 85 AGE URU. Cham.

Chiropetalum molle (Baill.) Pax & K.Hoffm. in H.G.A.Engler, Pflanzenr., IV, 147, VI: 87 (1912).
Brazil (Rio Grande do Sul). 84 BZS. Cham. or nanophan.
* *Argythamnia mollis* Baill., Adansonia 4: 289 (1864).

Chiropetalum patagonicum (Speg.) O'Donell & Lourteig, Lilloa 8: 41 (1942).
Argentina (Córdoba, San Luís, La Pampa). 85 AGE AGW. Ther.
* *Aonikena patagonica* Speg., Anales Mus. Nac. Buenos Aires 7: 162 (1902).

Chiropetalum pavonianum (Müll.Arg.) Pax in H.G.A.Engler & K.A.E.Prantl, Nat. Pflanzenfam. 3(5): 45 (1890).
Peru. 83 PER. Cham.
* *Argythamnia pavoniana* Müll.Arg., Linnaea 34: 149 (1865).

Chiropetalum phalacradenium (J.W.Ingram) L.B.Sm. & Downs, Fl. Ilustr. Catar. 1(Euforbiac.): 155 (1988).
Brazil (Santa Catarina). 84 BZS. Hemicr.
* *Argythamnia phalacradenia* J.W.Ingram, Gentes Herb. 11: 458 (1980).

Chiropetalum pilosistylum (Allem & Irgang) Radcl.-Sm. & Govaerts, Kew Bull. 52: 478 (1997).
Brazil (Rio Grande do Sul). 84 BZS. Nanophan.
* *Argythamnia pilosistyla* Allem & Irgang, Revista Brasil. Biol. 36: 285 (1976).

Chiropetalum quinquecuspidatum (A.Juss.) Pax & K.Hoffm. in H.G.A.Engler, Pflanzenr., IV, 147, VI: 92 (1912).
Peru. 83 PER. Cham.
* *Croton quinquecuspidatus* A.Juss., Euphorb. Tent.: 110 (1828). *Argythamnia quinquecuspidata* (A.Juss.) Müll.Arg., Linnaea 34: 150 (1865).
Chiropetalum peruvianum A.Juss., Ann. Sc. Nat. (Paris) 25: 22 (1832), nom. illeg.

Chiropetalum ramboi (Allem & Irgang) Radcl.-Sm. & Govaerts, Kew Bull. 52: 478 (1997).
Brazil (Rio Grande do Sul). 84 BZS. Hemicr.
* *Argythamnia ramboi* Allem & Irgang, Revista Brasil. Biol. 36: 283 (1976).

Chiropetalum ruizianum (Müll.Arg.) Pax & K.Hoffm. in H.G.A.Engler, Pflanzenr., IV, 147, VI: 94 (1912).
Peru. 83 PER. Cham.
* *Argythamnia ruiziana* Müll.Arg., Linnaea 34: 151 (1865).
Croton striatus Ruiz ex Pax in H.G.A.Engler, Pflanzenr., IV, 147, VI: 94 (1912).

Chiropetalum schiedeanum (Müll.Arg.) Pax in H.G.A.Engler & K.A.E.Prantl, Nat. Pflanzenfam. 3(5): 45 (1890).
Mexico (México State, Hidalgo, Veracruz). 79 MXC MXE MXG. Hemicr.
* *Argythamnia schiedeana* Müll.Arg., Linnaea 34: 150 (1865). *Argythamnia schiedeana* var. *major* Müll.Arg., Linnaea 34: 150 (1865), nom. illeg.
Chiropetalum schiedeanum var. *major* Müll.Arg., Linnaea 34: 150 (1865).

Chiropetalum tricoccum (Vell.) Chodat & Hassl., Bull. Herb. Boissier, II, 5: 502 (1905).
S. Brazil, Uruguay, NE. Argentina. 84 BZS 85 AGE URU. Hemicr.
* *Desfontaena tricocca* Vell., Fl. Flumin.: 95 (1829). *Argythamnia tricocca* (Vell.) Müll.Arg., Linnaea 34: 150 (1865).
Chiropetalum lineatum Klotzsch, Arch. Naturgesch. 7: 199 (1841), nom. nud.
Argythamnia lineata (Klotzsch) Baill., Adansonia 4: 288 (1864), nom. illeg.
Chiropetalum tricoccum f. *latifolium* Chodat & Hassl., Bull. Herb. Boissier, II, 5: 502 (1905).

Chiropetalum tricuspidatum (Lam.) A.Juss., Ann. Sc. Nat. (Paris) 25: 22 (1832).
C. Chile. 85 CLC. Hemicr.

* *Croton tricuspidatus* Lam., Encycl. 2: 212 (1786). *Argythamnia tricuspidata* var. *genuina* Müll.Arg., Linnaea 34: 150 (1865), nom. inval. *Chiropetalum tricuspidatum* var. *genuina* (Müll.Arg.) Pax & K.Hoffm.in H.G.A.Engler, Pflanzenr., IV, 147, VI: 93 (1912), nom. inval. *Argythamnia tricuspidata* (Lam.) Müll.Arg., Linnaea 34: 150 (1865).
Croton lanceolatus Cav., Icon. 6: 38 (1800). *Chiropetalum lanceolatum* (Cav.) A.Juss., Ann. Sc. Nat. (Paris) 25: 21 (1832). *Argythamnia tricuspidata* var. *lanceolata* (Cav.) Müll.Arg., Linnaea 34: 150 (1865). *Chiropetalum tricuspidatum* var. *lanceolatum* (Cav.) Pax & K.Hoffm. in H.G.A.Engler, Pflanzenr., IV, 147, VI: 92 (1912).
Chiropetalum ovatum Phil., Linnaea 29: 42 (1858).

Synonyms:
Chiropetalum berterianum f. *macrantha* Skottsb. === **Chiropetalum berterianum** Schltdl. var. **berterianum**
Chiropetalum cupreum Pax & K.Hoffm. === **Chiropetalum griseum** Griseb.
Chiropetalum gigouxii Espinosa === **Chiropetalum canescens** Phil.
Chiropetalum lanceolatum (Cav.) A.Juss. === **Chiropetalum tricuspidatum** (Lam.) A.Juss.
Chiropetalum lineatum Klotzsch === **Chiropetalum tricoccum** (Vell.) Chodat & Hassl.
Chiropetalum ovatum Phil. === **Chiropetalum tricuspidatum** (Lam.) A.Juss.
Chiropetalum peruvianum A.Juss. === **Chiropetalum quinquecuspidatum** (A.Juss.) Pax & K.Hoffm.
Chiropetalum schiedeanum var. *major* Müll.Arg. === **Chiropetalum schiedeanum** (Müll.Arg.) Pax
Chiropetalum schiedeanum var. *minus* (Müll.Arg.) Pax & K.Hoffm. === **Chiropetalum astroplethos** (J.W.Ingram) Radcl.-Sm. & Govaerts
Chiropetalum sponiella (Müll.Arg.) Pax === **Chiropetalum canescens** Phil.
Chiropetalum triandrum Griseb. === **Chiropetalum boliviense** (Müll.Arg.) Pax & K.Hoffm.
Chiropetalum tricoccum f. *latifolium* Chodat & Hassl. === **Chiropetalum tricoccum** (Vell.) Chodat & Hassl.
Chiropetalum tricuspidatum var. *genuina* (Müll.Arg.) Pax & K.Hoffm. === **Chiropetalum tricuspidatum** (Lam.) A.Juss.
Chiropetalum tricuspidatum var. *lanceolatum* (Cav.) Pax & K.Hoffm. === **Chiropetalum tricuspidatum** (Lam.) A.Juss.

Chlamydojatropha

1 species, Africa (Cameroon); shrub of flood plains. Remains taxonomically unplaced as only the female was known and no further material appears to have been collected. Pax suggested an affinity with *Jatropha*. (Unplaced)

Pax, F. (with K. Hoffmann) (1912). *Chlamydojatropha*. In A. Engler (ed.), Das Pflanzenreich, IV 147 VI [Euphorbiaceae-Additamentum IV]: 125-126. Berlin. (Heft 57.) La/Ge. — 1 sp., *C. kamerunica*, Cameroun; only the female known. Based on *Ledermann 0884*.

Chlamydojatropha Pax & K.Hoffm. in H.G.A.Engler, Pflanzenr., IV, 147, VI: 125 (1912).
Cameroon. 23.

Chlamydojatropha kamerunica Pax & K.Hoffm. in H.G.A.Engler, Pflanzenr., IV, 147, VI: 125 (1912).
Cameroon. 23 CMN. Nanophan.

Chloradenia

Synonyms:
Chloradenia Baill. === **Cladogynos** Zipp. ex Span.

Chlorocaulon

Synonyms:
Chlorocaulon Klotzsch === **Chiropetalum** A.Juss.

Chloropatane

Synonyms:
Chloropatane Engl. === **Erythrococca** Benth.

Chondrostylis

2 species, SE Asia, W Malesia (southern Thailand to Sumatera and Borneo); trees to 15 m. Related to *Agrostistachys* but distinguished by branching inflorescences and apetalous flowers. Skeletally revised by Airy-Shaw (1960); more recently under study by Sofia Sevilla within the framework of the *Flora Malesiana* programme on Euphorbiaceae. [Illustration of *C. bancana* in Ic. Bog. 1: pl. 23. 1897.] (Acalyphoideae)

Pax, F. (with K. Hoffmann) (1914). *Chondrostylis*. In A. Engler (ed.), Das Pflanzenreich, IV 147 VII (Euphorbiaceae-Acalypheae-Mercurialinae): 15-16. Berlin. (Heft 63.) La/Ge. — 1 species, W Malesia (Banka). [Species now known from a greater range.]

Airy-Shaw, H. K. (1960). Notes on Malaysian Euphorbiaceae, VI. The genera *Chondrostylis* Boerl. and *Kunstlerodendron* Ridl. Kew Bull. 14: 358-362. En. —*Kunstlerodendron sublanceolata* reduced to *Chondrostylis* as *C. kunstleri*; key to it and *C. bancana* (p. 359) along with a synoptic treatment including synonymy and exsiccatae. [A second species of *Kunstlerodendron* is referred to *Mallotus*. *C. kunstleri* was previously treated as a *Mallotus* (no. 98 of Pax and Hoffmann's treatment).]

Airy-Shaw, H. K. (1963). Notes on Malaysian and other Asiatic Euphorbiaceae, XXVI. Further Siamese collections of *Chondrostylis kunstleri*. Kew Bull. 16: 345. En. — New records.

Airy-Shaw, H. K. (1966). Notes on Malaysian and other Asiatic Euphorbiaceae, LIX. *Chondrostylis kunstleri* in Sumatra. Kew Bull. 20: 27. En. — Range extension.

Airy-Shaw, H. K. (1966). Notes on Malaysian and other Asiatic Euphorbiaceae, LXXIV. *Chondrostylis* Boerl. in Sumatra. Kew Bull. 20: 398. En. — An inexplicable duplication of note LIX (1966).

Airy-Shaw, H. K. (1981). Notes on Asiatic, Malesian and Melanesian Euphorbiaceae, CCXLV. *Chondrostylis* Boerl. Kew Bull. 36: 605-606. En. — New material, of wild origin in south-central Borneo, of *C. bancana*; originally described from plants grown in Bogor but attributed to Banka.

Chondrostylis Boerl., Icon. Bogor.: t. 23 (1897).
Indo-China, Malesia. 41 42.
Kunstlerodendron Ridl., Fl. Malay Penins. 3: 283 (1924).

Chondrostylis bancana Boerl., Icon. Bogor.: t. 23 (1897).
Sumatera (Bangka), Borneo (S. Kalimantan). 42 BOR SUM. Phan.

Chondrostylis kunstleri (King ex Hook.f.) Airy Shaw, Kew Bull. 14: 359 (1960).
S. Thailand, Pen. Malaysia, Sumatera, Borneo (Sarawak). 41 THA 42 BOR MLY SUM. Phan.
* *Mallotus kunstleri* King ex Hook.f., Fl. Brit. India 5: 443 (1887).

Chonocentrum

1 species, S America (Brazil, upper Amazon Basin); a large shrub or forest tree uncertainly related to *Discocarpus*. Remains imperfectly known, however, Hayden & Hayden (1996; see *Discocarpus*) suggested an affinity with Antidesmeae. [An illustration was associated with the type collection, according to Pax & Hoffmann (1922).] (Phyllanthoideae)

Pax, F. & K. Hoffmann (1922). *Chonocentrum*. In A. Engler (ed.), Das Pflanzenreich, IV 147 XV (Euphorbiaceae-Phyllanthoideae-Phyllantheae): 205. Berlin. (Heft 81.) La/Ge. — 1 species, Brazil (upper Amazon basin).

Jablonski, E. (1967). *Chonocentrum*. Euphorbiaceae, Guayana Highland (Mem. New York Bot. Gard. 17(1)): 121. New York. En. — 1 species, northern S America; related to *Drypetes*. Only the type collection (male) known.

Chonocentrum Pierre ex Pax & K.Hoffm. in H.G.A.Engler, Pflanzenr., IV, 147, XV: 205 (1922).
NW. Brazil. 84.

Chonocentrum cyathophorum (Müll.Arg.) Pierre ex Pax & K.Hoffm. in H.G.A.Engler, Pflanzenr., IV, 147, XV: 205 (1922).
Brazil (Amazonas). 84 BZN. Phan.
* *Drypetes cyathophora* Müll.Arg. in A.P.de Candolle, Prodr. 15(2): 454 (1866).

Choriceras

2 species, N Australia (one also in New Guinea); shrubs or small slender trees of varied habitats with *C. majus* preferring vine-forest. Combined by Pax and Hoffmann (1922) with *Dissiliaria* but separated by Airy-Shaw and since upheld. The genus has been under study by Paul Forster (Queensland) along with other Australian Oldfieldioideae. (Oldfieldioideae)

Pax, F. & K. Hoffmann (1922). *Dissiliaria*. In A. Engler (ed.), Das Pflanzenreich, IV 147 XV (Euphorbiaceae-Phyllanthoideae-Phyllantheae): 291-292. Berlin. (Heft 81.) La/Ge. — 3 species. [*D. tricornis* now in *Choriceras*.]

Airy-Shaw, H. K. (1960). Notes on Malaysian Euphorbiaceae, IV. The genus *Choriceras* Baill. in New Guinea. Kew Bull. 14: 356-357. En. —*Dissiliaria tricornis* transferred to *Choriceras*; as *C. tricorne* distributed in Australia and the southern 'belly' of New Guinea.

Airy-Shaw, H. K. (1963). Notes on Malaysian and other Asiatic Euphorbiaceae, XXIV. Further collection of *Choriceras tricorne*. Kew Bull. 16: 344. En. — New record.

Airy-Shaw, H. K. (1980). *Choriceras*. Kew Bull. 35: 604-605. (Euphorbiaceae-Platylobeae of Australia.) En. — Treatment of 2 species.

Forster, P. I. & P. C. van Welzen (1999). The Malesian species of *Choriceras*, *Fontainea* and *Petalostigma* (Euphorbiaceae). Blumea 44: 99-107, illus. En. — *Choriceras*, pp. 100-101; treatment of 1 species (in Australia and southern New Guinea) with genus and species descriptions, synonymy, references, types, indication of distribution and ecology, and general notes; general references and list of specimens seen at end of paper.

Choriceras Baill., Adansonia 11: 119 (1874).
New Guinea, N. & NE. Australia. 42 50.

Choriceras majus Airy Shaw, Muelleria 4: 220 (1980).
Queensland (Cook). 50 QLD. Phan.

Choriceras tricorne (Benth.) Airy Shaw, Kew Bull. 14: 356 (1960).
Irian Jaya, Papua New Guinea, NE. Queensland, N. Northern Territory. 42 NWG 50 NTA QLD. Nanophan. or phan.
* *Dissiliaria tricornis* Benth., Fl. Austral. 6: 91 (1873).
Choriceras australiana Baill., Adansonia 11: 120 (1874).

Synonyms:
Choriceras australiana Baill. === **Choriceras tricorne** (Benth.) Airy Shaw

Choriophyllum

Synonyms:
Choriophyllum Benth. === **Austrobuxus** Miq.
Choriophyllum malayanum Benth. === **Austrobuxus nitidus** Miq. var. **nitidus**
Choriophyllum montanum Ridl. === **Austrobuxus nitidus** var. **montanus** (Ridl.) Whitmore

Chorisandra

A later homonym of *Chorisandra* R. Br. (1810) in Cyperaceae.

Synonyms:
Chorisandra Wight === **Phyllanthus** L.
Chorisandra orientalis Craib === **Phyllanthus orientalis** (Craib) Airy Shaw
Chorisandra pinnata Wight === **Phyllanthus pinnatus** (Wight) G.L.Webster

Chorisandrachne

1 species, SE. Asia (Thailand); shrub or small tree of light 'evergreen' forest. By Webster (Synopsis, 1994) reduced to *Leptopus*, an opinion also supported in USDA-GRIN (1992) but not by Brummitt (1992). Its separate recognition is supported by Stuppy (1995; see **Phyllanthoideae**). The plant has features of both *Leptopus* and *Phyllanthus*. (Phyllanthoideae)

Airy-Shaw, H. K. (1969). Notes on Malesian and other Asiatic Euphorbiaceae, C. A new phyllanthoid genus from Siam. Kew Bull. 23: 40-42. En. — Protologue of *Chorisandrachne* and description of *C. diplosperma*; considered to be of restricted distribution.

Chorisandrachne Airy Shaw, Kew Bull. 23: 40 (1969).
Indo-China. 41.

Chorisandrachne diplosperma Airy Shaw, Kew Bull. 23: 40 (1969). *Leptopus diplospermus* (Airy Shaw) G.L.Webster, Ann. Missouri Bot. Gard. 81: 40 (1994). – FIGURE, p. 335.
Thailand. 41 THA. Nanophan.

Chorizonema

Africa, Madagascar; a segregate of *Phyllanthus*. The name was intended by Brunel (1987) to replace *Chorisandra* Wight, a later homonym, but formally remains unpublished; it is not in *Index Kewensis*. [*Chorisandra* was included by Pax & Hoffmann (1931) as a section of *Phyllanthus* but Brunel recommended its re-instatement.] (Phyllanthoideae)

Brunel, J. F. (1987). *Chorizonema*. In *idem*, Sur le genre *Phyllanthus* L. et quelques genres voisins de la tribu des Phyllantheae Dumort: 254-257. Strasbourg. [See also *Phyllanthus*.] Fr. — Description of genus (the name replaces *Chorisandra* Wight, a later homonym). The 2 African species, here combined as *C. pinnata*, were previously included in *Phyllanthus*. [This name has not been validly published. *Chorisandra* has generally been included in *Phyllanthus*, following in the first instance Pax and Hoffmann wherein it appears as sect. *Chorisandra* (1931, p. 61).]

Chorizotheca

Synonyms:
Chorizotheca Müll.Arg. === **Pseudanthus** Sieber ex Spreng.
Chorizotheca macrophylla Heckel === **Uapaca bojeri** Baill.
Chorizotheca micrantheoides Müll.Arg. === **Pseudanthus virgatus** (Klotzsch) Müll.Arg.

Chorisandrachne diplosperma Airy Shaw
Artist: Mary Grierson
Ic. Pl. 38(1): pl. 3707 (1974)
KEW ILLUSTRATIONS COLLECTION

Chrozophora

8 species, Africa, S. Europe and across W. and S. Asia with *C. plicata* reaching Thailand and Java; annuals (sometimes prostrate) or subshrubs or shrubs in dry regions, some at least ruderal. The name is conserved against *Tournesol* Adans. Three sections were recognised by Pax (1912) while Prain (1918) accepted two. *C. tinctoria* (including *C. verbascifolia*) is a characteristic Mediterranean plant known since antiquity, particularly for its red dyestuff. It is likely that *C. gangetica*, published after Pax's and Prain's treatments, is a synonym of *C. plicata* as it falls well within the latter's range in India. The genus was thought not to have immediate close relatives but was seen in turn to be connected to the genera of the Asian and Malesian subtribe Doryxylinae, particularly *Thyrsanthera* of SE. Asia (Webster, Synopsis, 1994). Recent research by P. van Welzen, however, suggests that not only that genus but also others in Doryxylinae (e.g. *Doryxylon, Melanolepis*, and *Sumbaviopsis*) could well be combined with *Chrozophora*; the distinctions among them appear to be relatively weak. For practical reasons, though, it was expected that traditional circumscriptions would be maintained with respect to *Flora Malesiana*. (Acalyphoideae)

Pax, F. (with K. Hoffmann) (1912). *Chrozophora*. In A. Engler (ed.), Das Pflanzenreich, IV 147 VI (Euphorbiaceae-Acalypheae-Chrozophorinae): 17-27. Berlin. (Heft 57.) La/Ge. — 9 species in 3 sections.

• Prain, D. (1918). The genus *Chrozophora*. Bull. Misc. Inform. Kew 1918: 49-120. En/La. — Detailed treatment (11 species in 4 series, 2 in each of 2 sections) with keys (only within series), diagnostic descriptions (in Latin), indication of distribution and uses, localities with exsiccatae, and sometimes extensive commentary. The lengthy general part (pp. 49-89) includes a historical chapter with remarks on the approaches to the genus taken by numerous authors from the 16th century onwards, a statement of the characters of the genus and its subdivisions, and the author's own ideas along with a key (pp. 60-61); history of the species under *Ricinoides* and as part of *Croton* (representing opinions of some early authors); history of *Chrozophora* after 1824 (cf. de Jussieu, 1824, under General) through the work of Mueller in the *Prodromus*; history of the African, Indian, and SW Asian species after 1866; review of Pax's *Pflanzenreich* treatment in comparison with that of Müller.

Chrozophora Neck. ex A.Juss., Euphorb. Gen.: 27 (1824), nom. & typ. cons.
S. & E. Europe, Medit to Himalaya and SE. Asia, Africa. 12 13 14 20 21 22 23 24 25 26 27 29? 32 33 34 35 36 40 41 42.
Tournesol Adans., Fam. Pl. 2: 356 (1763), nom. rejic.
Tournesolia Nissole ex Scop., Intr. Hist. Nat.: 243 (1777), orth. var.
Ricinoides Tourn. ex Moench, Methodus: 286 (1794).
Crozophora A.Juss., Euphorb. Gen.: 27 (1824), orth. var.
Lepidocroton C.Presl, Epimel. Bot.: 213 (1851).

Chrozophora brocchiana (Vis.) Schweinf., Pl. Quaed. Nilot.: 9 (1862).
Egypt, Algeria, Cape Verde Is.,Senegal, Mauritania, Niger, Mali, Chad, Sudan. 20 ALG EGY 21 CVI 22 MLI MTN NGR SEN 24 CHA SUD. Cham. or nanophan.
* *Croton brocchianus* Vis., Pl. Aegypti: 39 (1836).
Chrozophora brocchiana var. *hartmannii* Schweinf., Pl. Quaed. Nilot.: 9 (1862).
Croton macrocalyx Ehrenb. ex Schweinf., Pl. Quaed. Nilot.: 9 (1862). *Chrozophora brocchiana* var. *hartmannii* Schweinf., Pl. Quaed. Nilot.: 9 (1862).
Chrozophora senegalensis var. *lanigera* Prain in D.Oliver, Fl. Trop. Afr. 6(1): 837 (1912). *Croton lanigerus* (Prain) Perr. ex Prain in D.Oliver, Fl. Trop. Afr. 6(1): 839 (1912).

Chrozophora gangetica Gand., Bull. Soc. Bot. France 66: 286 (1919 publ. 1920).
India. 40 IND. – Provisionally accepted.

Chrozophora mujunkumi Nasimova, Bot. Mater. Gerb. Inst. Bot. Akad. Nauk Uzbeksk. S.S.R. 20: 34 (1982).
C. Asia. 32 UZB. Ther.

Chrozophora oblongifolia (Delile) A.Juss. ex Spreng., Syst. Veg. 3: 850 (1826).
Sinai, Saudi Arabia, Yemen, Oman, Pakistan, Socotra, N. Somalia, Djibouti, Eritrea, E. Sudan. 24 DJI ETH SOC SOM SUD 29 COM? 35 OMA SAU SIN YEM. Cham. or nanophan.
* *Croton oblongifolius* Delile, Descr. Egypte, Hist. Nat.: 283 (1812).

Chrozophora plicata (Vahl) A.Juss. ex Spreng., Syst. Veg. 3: 850 (1826). *Chrozophora plicata* var. *genuina* Müll.Arg. in A.P.de Candolle, Prodr. 15(2): 747 (1866), nom. inval.
Trop. & Subtrop. Africa, Subtrop. Asia. 20 EGY 22 MLI MTN NGA NGR SEN 23 CAF CMN 24 CHA ETH SOM SUD 25 KEN TAN 26 MOZ ZAM ZIM 27 TVL 34 PAL SIN 35 SAU 40 ASS IND PAK 41 BMA THA 42 JAW. Ther.
Croton hastatus Burm.f., Fl. Indica: 305 (1768), nom. illeg.
Croton tinctorius Burm.f., Fl. Indica: 304 (1768), nom. illeg.
* *Croton plicatus* Vahl, Symb. Bot. 1: 78 (1790).
Croton moluccanus Willd., Sp. Pl. 4: 551 (1805).
Croton rottleri Geiseler, Croton. Monogr.: 54 (1807). *Chrozophora rottleri* (Geiseler) A.Juss. ex Spreng., Syst. Veg. 3: 850 (1826). *Chrozophora plicata* var. *rottleri* (Geiseler) Müll.Arg. in A.P.de Candolle, Prodr. 15(2): 747 (1866).
Chrozophora burmannii Spreng., Syst. Veg. 3: 851 (1826).
Croton asper K.D.Koenig ex Roxb., Fl. Ind. ed. 1832, 3: 681 (1832).
Croton obliquifolius Vis., Pl. Aegypti: 39 (1836). *Chrozophora obliquifolia* (Vis.) Baill., Étude Euphorb.: 322 (1858). *Chrozophora plicata* var. *obliquifolia* (Vis.) Prain in D.Oliver, Fl. Trop. Afr. 6(1): 835 (1912).
Chrozophora prostrata Dalzell & Gibson, Bombay Fl.: 233 (1861).
Chrozophora obliqua Schweinf., Pl. Quaed. Nilot.: 13 (1862), nom. illeg.
Chrozophora parvifolia Klotzsch ex Schweinf., Pl. Quaed. Nilot.: 11 (1862). *Chrozophora prostrata* var. *parvifolia* (Klotzsch ex Schweinf.) N.P.Balakr., Bull. Bot. Surv. India 15: 7 (1973 publ. 1976).
Croton lanuginosus K.Schum. ex Schweinf., Pl. Quaed. Nilot.: 10 (1862).

Chrozophora sabulosa Kar. & Kir., Bull. Soc. Imp. Naturalistes Moscou 15: 446 (1842).
Oman, Iran, Afghanistan, Pakistan, C. Asia, W. China. 32 KAZ KGZ TKM UZB 34 AFG IRN 35 OMA 36 CHX 40 PAK. Ther.
Chrozophora gracilis Fisch. & C.A.Mey. ex Ledeb., Fl. Ross. 3(2): 581 (1850).
Chrozophora pannosa Pazij, Bot. Mater. Gerb. Inst. Bot. Zool. Akad. Nauk Uzbeksk. S.S.R. 11: 25 (1948).

Chrozophora senegalensis (Lam.) A.Juss. ex Spreng., Syst. Veg. 3: 850 (1826).
Mauritania, Senegal, Gambia, Mali, Ghana, Benin, Niger. 22 BEN GAM GHA MLI MTN NGR SEN. Cham.
* *Croton senegalensis* Lam., Encycl. 2: 212 (1786).

Chrozophora tinctoria (L.) Raf., Chlor. Aetn.: 4 (1813).
Medit. to NW. India. 12 BAL FRA POR SAR COR SPA 13 ALB BUL GRC ITA KRI ROM YUG 14 KRY 20 ALG EGY LBY MOR TUN 24 SOC 32 KAZ TKM 34 AFG CYP IRN LBS PAL SIN TUR 35 SAU YEM 40 IND PAK. Ther.
* *Croton tinctorius* L., Sp. Pl.: 1004 (1753). *Chrozophora tinctoria* var. *genuina* Müll.Arg. in A.P.de Candolle, Prodr. 15(2): 749 (1866), nom. inval.
Croton argenteus Forssk., Fl. Aegypt.-Arab.: lxxv (1775), nom. illeg.
Croton obliquus Vahl, Symb. Bot. 1: 78 (1790). *Chrozophora obliqua* (Vahl) A.Juss. ex Spreng., Syst. Veg. 3: 850 (1826). *Chrozophora tinctoria* subsp. *obliqua* (Vahl) O.Bolòs & Vigo, Fl. Paisos Catalans 2: 547 (1990).

Croton verbascifolius Willd., Sp. Pl. 4: 539 (1805). *Chrozophora verbascifolia* (Willd.) A.Juss. ex Spreng., Syst. Veg. 3: 850 (1826). *Chrozophora tinctoria* var. *verbascifolia* (Willd.) Müll.Arg. in A.P.de Candolle, Prodr. 15(2): 748 (1866).
Croton patulus Lag., Gen. Sp. Pl.: 21 (1816).
Chrozophora hierosolymitana Spreng., Syst. Veg. 3: 850 (1826). *Chrozophora tinctoria* var. *hierosolymitana* (Spreng.) Müll.Arg. in A.P.de Candolle, Prodr. 15(2): 850 (1826).
Croton oblongifolius Sieber ex Spreng., Syst. Veg. 3: 850 (1826).
Chrozophora sieberi C.Presl, Abh. Königl. Böhm. Ges. Wiss., V, 3: 539 (1845).
Chrozophora integrifolia Bunge, Mém. Sav. Étr. Acad. Petersbourgh 7: 490 (1851).
Chrozophora tinctoria f. *brachypetala* Müll.Arg. in A.P.de Candolle, Prodr. 15(2): 749 (1866).
Chrozophora tinctoria var. *subplicata* Müll.Arg. in A.P.de Candolle, Prodr. 15(2): 749 (1866). *Chrozophora subplicata* (Müll.Arg.) Pax & K.Hoffm. in H.G.A.Engler, Pflanzenr., IV, 147, VI: 21 (1912).
Chrozophora warionii Coss. ex Batt. & Trab. in J.A.Battandier & al., Fl. Algérie, Dicot.: 804 (1888).
Chrozophora tinctoria var. *glabrata* Heldr., Parnassus Boicus: 277 (1899). *Chrozophora glabrata* (Heldr.) Pax & K.Hoffm. in H.G.A.Engler, Pflanzenr., IV, 147, VI: 24 (1912).
Chrozophora cordifolia Pazij, Bot. Mater. Gerb. Inst. Bot. Zool. Akad. Nauk Uzbeksk. S.S.R. 11: 23 (1948).
Chrozophora lepidocarpa Pazij, Bot. Mater. Gerb. Inst. Bot. Zool. Akad. Nauk Uzbeksk. S.S.R. 14: 23 (1954).

Synonyms:
Chrozophora brocchiana var. *hartmannii* Müll.Arg. === **Chrozophora brocchiana** (Vis.) Schweinf.
Chrozophora brocchiana var. *hartmannii* Schweinf. === **Chrozophora brocchiana** (Vis.) Schweinf.
Chrozophora burmannii Spreng. === **Chrozophora plicata** (Vahl) A.Juss. ex Spreng.
Chrozophora cordifolia Pazij === **Chrozophora tinctoria** (L.) Raf.
Chrozophora glabrata (Heldr.) === Pax & K.Hoffm. === **Chrozophora tinctoria** (L.) Raf.
Chrozophora gracilis Fisch. & C.A.Mey. ex Ledeb. === **Chrozophora sabulosa** Kar. & Kir.
Chrozophora hierosolymitana Spreng. === **Chrozophora tinctoria** (L.) Raf.
Chrozophora integrifolia Bunge === **Chrozophora tinctoria** (L.) Raf.
Chrozophora lepidocarpa Pazij === **Chrozophora tinctoria** (L.) Raf.
Chrozophora mollissima (Geiseler) A.Juss. === **Mallotus mollissimus** (Geiseler) Airy Shaw
Chrozophora obliqua (Vahl) A.Juss. ex Spreng. === **Chrozophora tinctoria** (L.) Raf.
Chrozophora obliqua Schweinf. === **Chrozophora plicata** (Vahl) A.Juss. ex Spreng.
Chrozophora obliquifolia Kotschy ex Pax & K.Hoffm. === **Chrozophora plicata** (Vahl) A.Juss. ex Spreng.
Chrozophora obliquifolia (Vis.) Baill. === **Chrozophora plicata** (Vahl) A.Juss. ex Spreng.
Chrozophora pannosa Pazij === **Chrozophora sabulosa** Kar. & Kir.
Chrozophora parvifolia Klotzsch ex Schweinf. === **Chrozophora plicata** (Vahl) A.Juss. ex Spreng.
Chrozophora peltata Labill. === **Codiaeum peltatum** (Labill.) P.S.Green
Chrozophora plicata var. *genuina* Müll.Arg. === **Chrozophora plicata** (Vahl) A.Juss. ex Spreng.
Chrozophora plicata var. *obliquifolia* (Vis.) Prain === **Chrozophora plicata** (Vahl) A.Juss. ex Spreng.
Chrozophora plicata var. *rottleri* (Geiseler) Müll.Arg. === **Chrozophora plicata** (Vahl) A.Juss. ex Spreng.
Chrozophora prostrata Dalzell & Gibson === **Chrozophora plicata** (Vahl) A.Juss. ex Spreng.
Chrozophora prostrata var. *parvifolia* (Klotzsch ex Schweinf.) N.P.Balakr. === **Chrozophora plicata** (Vahl) A.Juss. ex Spreng.
Chrozophora rottleri (Geiseler) A.Juss. ex Spreng. === **Chrozophora plicata** (Vahl) A.Juss. ex Spreng.
Chrozophora senegalensis var. *lanigera* Prain === **Chrozophora brocchiana** (Vis.) Schweinf.
Chrozophora sieberi C.Presl === **Chrozophora tinctoria** (L.) Raf.

Chrozophora subplicata (Müll.Arg.) Pax & K.Hoffm. === **Chrozophora tinctoria** (L.) Raf.
Chrozophora tinctoria f. *brachypetala* Müll.Arg. === **Chrozophora tinctoria** (L.) Raf.
Chrozophora tinctoria var. *genuina* Müll.Arg. === **Chrozophora tinctoria** (L.) Raf.
Chrozophora tinctoria var. *glabrata* Heldr. === **Chrozophora tinctoria** (L.) Raf.
Chrozophora tinctoria var. *hierosolymitana* (Spreng.) === Müll.Arg. **Chrozophora tinctoria** (L.) Raf.
Chrozophora tinctoria subsp. *obliqua* (Vahl) O.Bolòs & Vigo === **Chrozophora tinctoria** (L.) Raf.
Chrozophora tinctoria var. *subplicata* Müll.Arg. === **Chrozophora tinctoria** (L.) Raf.
Chrozophora tinctoria var. *verbascifolia* (Willd.) Müll.Arg. === **Chrozophora tinctoria** (L.) Raf.
Chrozophora verbascifolia (Willd.) A.Juss. ex Spreng. === **Chrozophora tinctoria** (L.) Raf.
Chrozophora warionii Coss. ex Batt. & Trab. === **Chrozophora tinctoria** (L.) Raf.

Chrysostemon

Synonyms:
Chrysostemon Klotzsch === **Pseudanthus** Sieber ex Spreng.
Chrysostemon virgatus Klotzsch === **Pseudanthus virgatus** (Klotzsch) Müll.Arg.

Chylogala

Synonyms:
Chylogala Fourr. === **Euphorbia** L.

Cicca

A segregate of *Phyllanthus*.

Synonyms:
Cicca L. === **Phyllanthus** L.
Cicca acida (L.) Merr. === **Phyllanthus acidus** (L.) Skeels
Cicca acidissima Blanco === **Phyllanthus acidus** (L.) Skeels
Cicca albizzioides Kurz === **Phyllanthus albizzioides** (Kurz) Hook.f.
Cicca anomala Baill. === **Margaritaria anomala** (Baill.) Fosberg
Cicca antillana A.Juss. === **Margaritaria nobilis** L.f.
Cicca antillana var. *glaucescens* Griseb. === **Margaritaria scandens** (Wright ex Griseb.) G.L.Webster
Cicca antillana var. *pedicellaris* Griseb. === **Margaritaria nobilis** L.f.
Cicca antillana var. *virens* Griseb. === **Margaritaria tetracocca** (Baill.) G.L.Webster
Cicca arborea C.Wright ex Griseb. === **Margaritaria indica** (Dalzell) Airy Shaw
Cicca brasiliensis (Aubl.) Baill. === **Phyllanthus brasiliensis** (Aubl.) Poir.
Cicca decandra Blanco === **Phyllanthus reticulatus** Poir. var. **reticulatus**
Cicca discoidea Baill. === **Margaritaria discoidea** (Baill.) G.L.Webster
Cicca disticha L. === **Phyllanthus acidus** (L.) Skeels
Cicca emblica (L.) Kurz === **Phyllanthus emblica** L.
Cicca flexuosa Siebold & Zucc. === **Phyllanthus flexuosus** (Siebold & Zucc.) Müll.Arg.
Cicca gaertneriana Baill. === **Margaritaria cyanosperma** (Gaertn.) Airy Shaw
Cicca guianensis (Klotzsch) Splitg. ex Lanj. === **Phyllanthus carolinensis subsp. guianensis**
Cicca leucopyrus (Willd.) Kurz === **Flueggea leucopyrus** Willd.
Cicca macrocarpa (Labill.) Kurz === **Glochidion macphersonii** Govaerts & Radcl.-Sm.
Cicca macrostachya (Müll.Arg.) Benth. === **Picramnia antidesma** Sw. (Picramniaceae; Simaroubaceae s.l.)
Cicca microcarpa Benth. === **Phyllanthus reticulatus** Poir. var. **reticulatus**
Cicca nodiflora Lam. === **Phyllanthus acidus** (L.) Skeels
Cicca obovata (Willd.) Kurz === **Flueggea virosa** (Roxb. ex Willd.) Voigt
Cicca pavoniana Baill. === **Margaritaria nobilis** L.f.
Cicca pentandra Blanco === **Flueggea virosa** (Roxb. ex Willd.) Voigt

Cicca racemosa Lour. === **Phyllanthus acidus** (L.) Skeels
Cicca reticulata (Poir.) Kurz === **Phyllanthus reticulatus** Poir.
Cicca rhomboidalis Baill. === **Margaritaria rhomboidalis** (Baill.) G.L.Webster
Cicca scandens C.Wright ex Griseb. === **Margaritaria scandens** (Wright ex Griseb.) G.L.Webster
Cicca sinica Baill. === **Margaritaria nobilis** L.f.
Cicca surinamensis Miq. === **Margaritaria nobilis** L.f.
Cicca virens (Griseb.) C.Wright ex Griseb. === **Margaritaria tetracocca** (Baill.) G.L.Webster

Cieca

Rejected against *Julocroton* Mart. (= *Croton* L.)

Synonyms:
Cieca Adans. === **Croton** L.
Cieca ackermanniana (Müll.Arg.) Kuntze === **Croton ackermannianus** (Müll.Arg.) G.L.Webster
Cieca argentea (L.) Kuntze === **Croton argenteus** L.
Cieca conspurcata (Schltdl.) Kuntze === **Croton conspurcatus** Schltdl.
Cieca decaloba (Müll.Arg.) Kuntze === **Croton decalobus** Müll.Arg.
Cieca doratophylla (Baill.) Kuntze === **Croton doratophyllus** Baill.
Cieca fuscescens (Spreng.) Kuntze === **Croton fuscescens** Spreng.
Cieca gardneri (Müll.Arg.) Kuntze === **Croton argentealbidus** Radcl.-Sm. & Govaerts
Cieca geraesensis (Baill.) Kuntze === **Croton geraesensis** (Baill.) G.L.Webster
Cieca hondensis (H.Karst.) Kuntze === **Croton hondensis** (H.Karst.) G.L.Webster
Cieca humilis (Didr.) Kuntze === **Croton didrichsenii** G.L.Webster
Cieca lanceolata (Klotzsch ex Müll.Arg.) Kuntze === **Croton lanceolaris** G.L.Webster
Cieca microcalyx (Müll.Arg.) Kuntze === **Croton microcalyx** (Müll.Arg.) G.L.Webster
Cieca montevidensis (Klotzsch ex Baill.) Kuntze === **Croton argenteus** L.
Cieca nervosa (Baill.) Kuntze === **Croton calonervosus** G.L.Webster
Cieca peruviana (Müll.Arg.) Kuntze === **Croton flavispicatus** Rusby
Cieca pycnophyllus (Schltdl. ex Müll.Arg.) Kuntze === **Croton salzmannii** (Baill.) G.L.Webster
Cieca riedeliana (Müll.Arg.) Kuntze === **Croton cordeiroae** G.L.Webster
Cieca solanacea (Müll.Arg.) Kuntze === **Croton solanaceus** (Müll.Arg.) G.L.Webster
Cieca triquetra (Lam.) Kuntze === **Croton triqueter** Lam.
Cieca verbascifolia (Klotzsch ex Baill.) Kuntze === **Croton verbascoides** G.L.Webster

Cinogasum

Synonyms:
Cinogasum Neck. === **Croton** L.

Cladodes

Synonyms:
Cladodes Lour. === **Alchornea** Sw.
Cladodes rugosa Lour. === **Alchornea rugosa** (Lour.) Müll.Arg.
Cladodes thozetiana Baill. === **Alchornea aquifolia** (Js.Sm.) Domin

Cladogelonium

1 species, Madagascar; related to *Suregada* (Leandri 1939). Shrubs to 4-5 m with flattened green branches, found in western parts. (**Euphorbioideae** (**except Euphorbieae**))

Leandri, J. (1939). Euphorbiacées malgaches nouvelles recoltées par M. H. Perrier de la Bâthie. Bull. Soc. Bot. France 85: 523-533, illus. Fr. — Includes description of *Cladogelonium*, related to *Suregada* and considered a 'neoendemic'.

Cladogelonium Leandri, Bull. Soc. Bot. France 85: 530 (1938 publ. 1939).
Madagascar. 29.

Cladogelonium madagascariense Leandri, Bull. Soc. Bot. France 85: 530 (1938 publ. 1939).
Madagascar. 29 MDG.

Cladogynos

1 species, SE Asia and Malesia, extending to Philippines, Timor and Maluku; shrubs to 2 m with a white indumentum on leaf undersurfaces. A preference is shown for more seasonal climates; in Java the plant has been recorded in forest and scrub over limestone. In Webster's system (Synopsis, 1994) the genus is most nearly related to *Cephalocrotonopsis* (Socotra) and *Cephalocroton* (Africa) with which are shared staminate flowers in capitula and non-glandular stipules. (Acalyphoideae)

Pax, F. (with K. Hoffmann) (1914). *Cladogynos*. In A. Engler (ed.), Das Pflanzenreich, IV 147 VII (Euphorbiaceae-Acalypheae-Mercurialinae): 264-266, illus. Berlin. (Heft 63.) La/Ge. — Monotypic, SE Asia and Malesia, extending to Timor and possibly Philippines.
Pax, F. (with K. Hoffmann) (1914). *Adenochlaena*. In A. Engler (ed.), Das Pflanzenreich, IV 147 VII [Euphorbiaceae-Additamentum V]: 401. Berlin. (Heft 63.) La/Ge. — 1 species, *A. siamensis*, Thailand and Malaysia; included with some doubt. [Now referable to *Cladogynos orientalis*.]

Cladogynos Zipp. ex Span., Linnaea 15: 349 (1841).
Malesia. 42.
Adenogynum Rchb.f. & Zoll., Acta Soc. Regiae Sci. Indo-Neerl. 1: 23 (1856).
Chloradenia Baill., Étude Euphorb.: 471 (1858).
Baprea Pierre ex Pax & K.Hoffm. in H.G.A.Engler, Pflanzenr., IV, 147, VII: 264 (1914), pro syn.

Cladogynos orientalis Zipp. ex Span., Linnaea 15: 349 (1841). *Cephalocroton orientalis* (Zipp. ex Span.) Miq., Ann. Mus. Bot. Lugduno-Batavi 4: 120 (1869).
N. Pen. Malaysia, Thailand, Vietnam, S. China, Philippines, Sulawesi, Maluku, Lesser Sunda Is. (Timor), Jawa. 36 CHS 41 MLY THA VIE 42 JAW LSI MOL PHI SUL. Nanophan.
Conceveiba tomentosa Span., Linnaea 15: 349 (1841).
Cephalocroton albicans var. *virens* Müll.Arg., Linnaea: 155 (1865). *Cladogynos orientalis* var. *virens* (Müll.Arg.) Pax & K.Hoffm. in H.G.A.Engler, Pflanzenr., IV, 147 VII: 266 (1914).
Cephalocroton discolor Müll.Arg. in A.P.de Candolle, Prodr. 15(2): 761 (1866).
Adenochlaena siamensis Ridl., J. Straits Branch Roy. Asiat. Soc. 59: 180 (1911).
Baprea bicolor Pierre ex Pax & K.Hoffm. in H.G.A.Engler, Pflanzenr., IV, 147, VII: 265 (1914), pro syn.
Cladogynos orientalis var. *grossedentata* Pax & K.Hoffm. in H.G.A.Engler, Pflanzenr., IV, 147 VII: 266 (1914).

Synonyms:
Cladogynos orientalis var. *grossedentata* Pax & K.Hoffm. === **Cladogynos orientalis** Zipp. ex Span.
Cladogynos orientalis var. *virens* (Müll.Arg.) Pax & K.Hoffm. === **Cladogynos orientalis** Zipp. ex Span.

Clambus

Synonyms:
Clambus Miers === **Phyllanthus** L.

Claoxylon

113 species, Madagascar (10), Mascarenes, Asia, Malesia, Australia and Pacific Is. (to the Society Islands and Hawaii); somewhat nondescript shrubs or trees of primary or secondary forest. In addition to the standard keys, the revision of Pax (1914), the most recent available but now quite out of date, features a geographically arranged key to species. Apart from *Claoxylopsis*, the most closely related genus is the African *Mareya*. (Acalyphoideae)

- Pax, F. (with K. Hoffmann) (1914). *Claoxylon*. In A. Engler (ed.), Das Pflanzenreich, IV 147 VII (Euphorbiaceae-Acalypheae-Mercurialinae): 100-131, illus. Berlin. (Heft 63.) La/Ge. — 57 species in 12 sections, Madagascar, Asia, Malesia, Australia and Pacific Is. Additions in ibid., XIV (Additamentum VI): 15-16 (1919).

Airy-Shaw, H. K. (1966). Notes on Malaysian and other Asiatic Euphorbiaceae, LXII. New or noteworthy species of *Claoxylon* from New Guinea. Kew Bull. 20: 31-38. En. — Novelties (6 new species) and notes. New Guinea revealed as a secondary centre for speciation in *Claoxylon* and, in addition, several species were shown to have more than 100 stamens.

Airy-Shaw, H. K. (1966). Notes on Malaysian and other Asiatic Euphorbiaceae, LXXV. New species of *Claoxylon* Juss. Kew Bull. 20: 398-405. En. — Descriptions of novelties; key to *CC. tumidum* and *ledermanii* and varieties of the latter (p. 405).

Airy-Shaw, H. K. (1968). Notes on Malesian and other Asiatic Euphorbiaceae, LXXXIX. New species of *Claoxylon* Juss. Kew Bull. 21: 375-379. En. — 3 species in 3 different sections, Borneo and New Guinea.

Airy-Shaw, H. K. (1969). Notes on Malesian and other Asiatic Euphorbiaceae, CVI. New species of *Claoxylon* Juss. Kew Bull. 23: 77-79. En. — Descriptions of 2 new species, both in sect. *Indica*.

Airy-Shaw, H. K. (1971). Notes on Malesian and other Asiatic Euphorbiaceae, CXXXVI. New or noteworthy species of *Claoxylon* Juss. Kew Bull. 25: 519-524. En. — 5 novelties and a range extension for *C. carolinianum*.

Airy-Shaw, H. K. (1974). Notes on Malesian and other Asiatic Euphorbiaceae, CLXXX. New Bornean species of *Claoxylon* Juss. Kew Bull. 29: 313-315. En. — Two novelties described.

Airy-Shaw, H. K. (1978). Notes on Malesian and other Asiatic Euphorbiaceae, CCXV. *Claoxylon* Juss. Kew Bull. 33: 61-62. En. — Description of *C. carrii* from southeastern New Guinea.

Airy-Shaw, H. K. (1978). Notes on Malesian and other Asiatic Euphorbiaceae, CXCVIII. New species of *Claoxylon* Juss. Kew Bull. 32: 389-400. En. — 9 novelties (mostly from New Guinea and the SW. Pacific); sections not indicated.

Airy-Shaw, H. K. (1979). Notes on Malesian and other Asiatic Euphorbiaceae, CCXXV. *Claoxylon* Juss. Kew Bull. 33: 532-534. En. — Descriptions of 2 species, Solomon Is. and Bougainville.

Airy-Shaw, H. K. (1981). Notes on Asiatic, Malesian and Melanesian Euphorbiaceae, CCXLVI. *Claoxylon* Juss. Kew Bull. 36: 606. En. — Description of *C. lambiricum* from Sarawak.

Claoxylon A.Juss., Euphorb. Gen.: 43 (1824).
Madagascar, Trop. & Subtrop. Asia to W. & S. Pacific. 29 36 38 40 41 42 50 51 60 62 63.
Erythrochilus Reinw., Flora 8(1): 103 (1825).
Quadrasia Elmer, Leafl. Philipp. Bot. 7: 2656 (1915).

Claoxylon abbreviatum J.J.Sm. ex Koord. & Valeton, Meded. Dept. Landb. Ned.-Indië 10: 376 (1910).
Jawa, Lesser Sunda Is. (Flores), Sulawesi. 42 JAW LSI SUL.

Claoxylon affine Zoll. & Moritzi, Natuur.-Geneesk. Arch. Ned.-Indië 2: 582 (1845).
Jawa, Lesser Sunda Is. 42 JAW LSI.
Claoxylon graciliflorum Miq., Fl. Ned. Ind. 1(2): 386 (1859).

Claoxylon albicans (Blanco) Merr., Sp. Blancoan.: 220 (1918).
Philippines. 42 PHI.
* *Prockia albicans* Blanco, Fl. Filip.: 430 (1837).
Claoxylon rubescens var. *cumingianum* Müll.Arg. in A.P.de Candolle, Prodr. 15(2): 788 (1866).
Claoxylon elongatum Merr., Philipp. J. Sci. 1(Suppl.): 204 (1906).

Claoxylon albiflorum Pax & K.Hoffm. in H.G.A.Engler, Pflanzenr., IV, 147, VII: 121 (1914).
Bismarck Archip. 42 BIS. Nanophan.

Claoxylon angustifolium Müll.Arg., Linnaea 34: 165 (1865). *Mercurialis angustifolia* (Müll.Arg.) Baill., Adansonia 6: 322 (1866).
Queensland. 50 QLD. Nanophan.

Claoxylon anomalum Hook.f., Fl. Brit. India 5: 412 (1887).
SW. India (Idukki, Courtallam). 40 IND. Nanophan.

Claoxylon arboreum Elmer, Leafl. Philipp. Bot. 2: 486 (1908).
Philippines. 42 PHI.
Claoxylon pedicellare Pax & K.Hoffm. in H.G.A.Engler, Pflanzenr., IV, 147, VII: 114 (1914), nom. illeg.
Claoxylon grandifolium Elmer, Leafl. Philipp. Bot. 7: 2638 (1915), nom. illeg.

Claoxylon attenuatum Airy Shaw, Kew Bull. 29: 314 (1974).
Borneo (Sabah). 42 BOR. Nanophan.

Claoxylon australe Baill. ex Müll.Arg. in A.P.de Candolle, Prodr. 15(2): 788 (1866).
Queensland, New South Wales. 50 NSW QLD. Phan.
Mercurialis australis Baill., Adansonia 6: 322 (1866).
Claoxylon australe var. *dentata* Benth., Fl. Austral. 6: 131 (1873).
Claoxylon australe var. *laxiflora* Benth., Fl. Austral. 6: 131 (1873).

Claoxylon bicarpellatum K.Schum. & Lauterb., Fl. Schutzgeb. Südsee: 393 (1900).
NE. Papua New Guinea. 42 NWG. Nanophan.

Claoxylon biciliatum Guillaumin, J. Linn. Soc., Bot. 51: 560 (1938).
Vanuatu. 60 VAN.

Claoxylon brachyandrum Pax & K.Hoffm. in H.G.A.Engler, Pflanzenr., IV, 147, VII: 115 (1914).
Borneo (Sabah), Philippines. 42 BOR PHI. Phan.

Claoxylon capillipes Airy Shaw, Kew Bull. 20: 34 (1966). – FIGURE, p. 344.
Lesser Sunda Is. (Flores) ?, NE. Papua New Guinea. 42 LSI? NWG. Nanophan. or phan.

Claoxylon carinatum Airy Shaw, Kew Bull. 20: 401 (1966).
Borneo (E. Kalimantan). 42 BOR. Nanophan. or phan.

Claoxylon carolinianum Pax & K.Hoffm. in H.G.A.Engler, Pflanzenr., IV, 147, XIV: 16 (1919).
Caroline Is. 62 CRL.

Claoxylon carrii Airy Shaw, Kew Bull. 33: 61 (1978).
Papua New Guinea. 42 NWG. Nanophan. or phan.

Claoxylon centenarium Koidz., Bot. Mag. (Tokyo) 33: 119 (1919).
Ogasawara-shoto. 62 OGA.

Claoxylon capillipes Airy Shaw
Artist: Ann Davies
Airy-Shaw, Euphorbiaceae of New Guinea, pl. 5, no. 4 (1980)
KEW ILLUSTRATIONS COLLECTION

Claoxylon colfsii Airy Shaw, Kew Bull. 25: 520 (1971).
Lesser Sunda Is. (Sumbawa). 42 LSI.

Claoxylon collenettei Riley, Bull. Misc. Inform. Kew 1926: 55 (1926).
Tubuai Is. (Rapa I.). 61 TUB. Phan.

Claoxylon coriaceolanatum Airy Shaw, Kew Bull. 20: 37 (1966).
NE. Papua New Guinea. 42 NWG. Phan.

Claoxylon crassipes Pax & K.Hoffm. in H.G.A.Engler, Pflanzenr., IV, 147, VII: 121 (1914).
Philippines. 42 PHI.

Claoxylon crassivenium Pax & K.Hoffm. in H.G.A.Engler, Pflanzenr., IV, 147, VII: 120 (1914).
Philippines. 42 PHI.

Claoxylon cuneatum J.J.Sm., Nova Guinea 8: 232 (1910).
Irian Jaya. 42 NWG. Phan.
Claoxylon tumidum J.J.Sm., Nova Guinea 8: 233 (1910).

Claoxylon decaryanum Leandri, Notul. Syst. (Paris) 9: 174 (1941).
Madagascar. 29 MDG.

Claoxylon dolichostachyum Cordem., Fl. Réunion: 341 (1895).
Réunion. 29 REU. Nanophan.

Claoxylon echinospermum Müll.Arg. in A.P.de Candolle, Prodr. 15(2): 787 (1866).
Fiji. 60 FIJ. Nanophan. or phan.
Claoxylon archboldianum Croizat, Sargentia 1: 50 (1942).
Claoxylon sitibundum Croizat, Sargentia 1: 51 (1942).

Claoxylon ellipticum Merr., Philipp. J. Sci. 16: 553 (1920).
Philippines. 42 PHI.

Claoxylon erythrophyllum Miq., Fl. Ned. Ind. 1(2): 387 (1859).
Sulawesi, Maluku (Bacan), Lesser Sunda Is., Irian Jaya. 42 LSI MOL NWG SUL. Nanophan.

var. **brassii** Airy Shaw, Kew Bull., Addit. Ser. 8: 51 (1980).
Irian Jaya. 42 NWG. Nanophan.

var. **erythrophyllum**
Sulawesi, Maluku (Bacan), Lesser Sunda Is. 42 LSI MOL SUL. Nanophan.
Claoxylon coriaceum Baill., Étude Euphorb.: 493 (1858), nom. nud.

Claoxylon euphorbioides (Elmer) Merr., Enum. Philipp. Fl. Pl. 2: 430 (1923).
Philippines. 42 PHI.
* *Quadrasia euphorbioides* Elmer, Leafl. Philipp. Bot. 7: 2656 (1915).

Claoxylon extenuatum Airy Shaw, Kew Bull. 25: 523 (1971).
Solomon Is. (Bougainville). 60 SOL.

Claoxylon fallax Müll.Arg. in A.P.de Candolle, Prodr. 15(2): 780 (1866).
Fiji, Tonga. 60 FIJ TON. Nanophan. or phan.
Claoxylon parviflorum Seem., Bonplandia 9: 258 (1861), nom. illeg.
Claoxylon parvicoccum Croizat, Sargentia 1: 49 (1942).

Claoxylon flavum Scott-Elliot, J. Linn. Soc., Bot. 29: 49 (1891).
Madagascar. 29 MDG. Nanophan.
Claoxylon scottianum Baill., Bull. Mens. Soc. Linn. Paris 2: 996 (1892).

Claoxylon fulvescens Airy Shaw, Kew Bull. 32: 390 (1978).
Sulawesi ?, Lesser Sunda Is. (Flores) ?, NE. Papua New Guinea. 42 LSI? NWG SUL?

Claoxylon gillisonii Airy Shaw, Kew Bull. 32: 399 (1978).
Vanuatu. 60 VAN.

Claoxylon glabrifolium Miq., Fl. Ned. Ind. 1(2): 387 (1859).
Jawa, Lesser Sunda Is. 42 JAW LSI. Nanophan.

var. **glabrifolium**
Jawa. 42 JAW. Nanophan.
Claoxylon indicum f. *angustifolium* J.J.Sm., Meded. Dept. Landb. Ned.-Indië 10: 371, 374 (1912).

var. **integrifolium** Pax & K.Hoffm. in H.G.A.Engler, Pflanzenr., IV, 147, XIV: 16 (1919).
Lesser Sunda Is. (Sumbawa). 42 LSI.

Claoxylon glandulosum Boivin ex Baill., Adansonia 1: 281 (1861).
Mauritius. 29 MAU. Nanophan. or phan.

Claoxylon goodenoviense Airy Shaw, Kew Bull. 32: 391 (1978).
New Guinea (D'Entrecasteaux Is.). 42 NWG. Phan.

Claoxylon grandifolium (Poir.) Müll.Arg., Linnaea 34: 163 (1865).
Réunion. 29 REU.
* *Acalypha grandifolia* Poir. in J.B.A.M.de Lamarck, Encycl. 6: 204 (1804). *Claoxylon grandifolium* var. *genuinum* Müll.Arg., Linnaea 34: 163 (1865), nom. inval.
Claoxylon metzieri Bouton ex Bojer, Hortus Maurit.: 285 (1837), nom. nud.
Claoxylon crassifolium Baill., Adansonia 1: 279 (1861).
Claoxylon grandifolium var. *submembranaceum* Müll.Arg., Linnaea 34: 163 (1865). *Claoxylon submembranaceum* (Müll.Arg.) Pax & K.Hoffm. in H.G.A.Engler, Pflanzenr., IV, 147, VII: 122 (1914).

Claoxylon gymnadenum Airy Shaw, Kew Bull. 25: 522 (1971).
Solomon Is. (Bougainville). 60 SOL.

Claoxylon hainanense Pax & K.Hoffm. in H.G.A.Engler, Pflanzenr., IV, 147, VII: 128 (1914).
Hainan. 36 CHH.
Mercurialis indica Lour., Fl. Cochinch.: 628 (1790).

Claoxylon hirsutellum Airy Shaw, Kew Bull. 20: 402 (1966).
Borneo (E. Kalimantan). 42 BOR. Nanophan.

Claoxylon hosei (Merr.) Airy Shaw, Kew Bull. 14: 391 (1960).
Borneo (NE. Sarawak: Baram Reg.). 42 BOR. Nanophan.
* *Koilodepas hosei* Merr., Philipp. J. Sci., C 11: 66 (1916).

Claoxylon humbertii Leandri, Notul. Syst. (Paris) 9: 175 (1941).
Madagascar. 29 MDG.

Claoxylon indicum (Reinw. ex Blume) Hassk., Cat. Hort. Bot. Bogor.: 235 (1844).
S. China, Indo-China to New Guinea. 36 CHS 41 AND BMA NCB THA VIE 42 BOR JAW NWG PHI SUL SUM XMS. Nanophan. or phan.

Croton pigmentarius Noronha, Verh. Batav. Genootsch. Kunsten 5(4): 13 (1790), nom. nud.
Croton tabacifolius Geiseler, Croton. Monogr.: 26 (1807). Provisional synonym.
* *Erythrochilus indicus* Reinw. ex Blume, Bijdr.: 615 (1826). *Claoxylon indicum* var. *genuinum* Müll.Arg. in A.P.de Candolle, Prodr. 15(2): 782 (1866), nom. inval.
Erythrochilus minor Blume, Bijdr.: 616 (1826). *Claoxylon minus* (Blume) Hassk., Pl. Jav. Rar.: 251 (1848).
Erythrochilus mollis Blume, Bijdr.: 615 (1826). *Claoxylon molle* (Blume) Miq., Fl. Ned. Ind. 1(2): 386 (1859).
Croton halecum Roxb., Fl. Ind. ed. 1832, 3: 683 (1832).
Claoxylon macrophyllum Bojer, Hortus Maurit.: 284 (1837), nom. nud.
Claoxylon parviflorum Hook. & Arn., Bot. Beechey Voy.: 212 (1837), nom. illeg.
Claoxylon macrophyllum Hassk., Pl. Jav. Rar.: 251 (1848). *Claoxylon indicum* var. *macrophyllum* (Hassk.) Müll.Arg. in A.P.de Candolle, Prodr. 15(2): 782 (1866).
Claoxylon indicum var. *spathulatum* Müll.Arg. in A.P.de Candolle, Prodr. 15(2): 782 (1866).
Claoxylon caerulescens Ridl., J. Straits Branch Roy. Asiat. Soc. 45: 223 (1906).
Claoxylon indicum f. *gracilius* J.J.Sm., Meded. Dept. Landb. Ned.-Indië 10: 369 (1910).
Claoxylon polot Merr., Interpr. Herb. Amboin.: 200 (1917).

Claoxylon insigne Airy Shaw, Kew Bull. 29: 313 (1974).
Borneo (Sarawak). 41 BOR. Nanophan.

Claoxylon insulanum Müll.Arg., Linnaea 34: 164 (1865).
New Caledonia (incl. Loyalty Is.). 60 NWC. Nanophan. or phan.
Claoxylon affine Baill., Adansonia 2: 227 (1862), nom. illeg.
Claoxylon brachybotryon Müll.Arg. ex Sebert & Pancher in H.Sebert & J.A.I.Pancher, Not. Bois Nouv. Caléd. 41: 212 (1874).
Claoxylon indicum var. *neocaledonicum* Schltr., Bot. Jahrb. Syst. 39: 149 (1906).

Claoxylon kaievskii Airy Shaw, Kew Bull. 33: 532 (1979).
Solomon Is., Santa Cruz Is. 60 SCZ SOL.

Claoxylon khasianum Hook.f., Fl. Brit. India 5: 411 (1887).
Assam, Bangladesh. 40 ASS BAN. Nanophan. or phan.
Claoxylon khasianum var. *serratum* Hook.f., Fl. Brit. India 5: 411 (1887).

Claoxylon kinabaluense Airy Shaw, Kew Bull. 20: 401 (1966).
Borneo (Sabah). 42 BOR. Nanophan. or phan.

Claoxylon kingii Hook.f. ex Ridl., Fl. Malay Penins. 3: 272 (1924).
S. Thailand, Pen. Malaysia (Perak). 41 THA 42 MLY. Nanophan.

Claoxylon kotoense Hayata, Icon. Pl. Formosan. 9: 101 (1920).
Taiwan. 38 TAI.

Claoxylon lambiricum Airy Shaw, Kew Bull. 36: 606 (1981).
Borneo (Sarawak). 42 BOR.

Claoxylon ledermannii Airy Shaw, Kew Bull. 20: 404 (1966).
Papua New Guinea. 42 NWG. Nanophan. or phan.

var. **intermedium** Airy Shaw, Kew Bull. 20: 404 (1966).
Papua New Guinea. 42 NWG. Nanophan. or phan.

var. **ledermannii**
Papua New Guinea. 42 NWG. Nanophan. or phan.

Claoxylon linostachys Baill., Adansonia 1: 281 (1861).
Mauritius. 29 MAU. Nanophan.

subsp. **brachyphyllum** (Croizat) Coode, Kew Bull. 34: 46 (1979).
Mauritius (Pétrin). 29 MAU.
* *Claoxylon brachyphyllum* Croizat, Trop. Woods 77: 15 (1944).

subsp. **linostachys**
Mauritius. 29 MAU. Nanophan.
Claoxylon scabrum Bojer, Hortus Maurit.: 285 (1837), nom. nud.
Acalypha spiciflora Thouars ex Baill., Adansonia 1: 281 (1861).

subsp. **pedicellare** Coode, Kew Bull. 34: 47 (1979).
Mauritius. 29 MAU. Nanophan.

Claoxylon longifolium (Blume) Endl. ex Hassk., Cat. Hort. Bot. Bogor.: 235 (1844).
Trop. Asia, Caroline Is. 40 ASS IND 41 THA VIE 42 BOR JAW MLY MOL NWG PHI SUL SUM 62 CRL. Nanophan. or phan.
* *Erythrochilus longifolius* Blume, Bijdr.: 616 (1826).

var. **longifolium**
Indo-China, Malesia, Caroline Is. 41 AND BMA NCB THA VIE 42 BOR JAW MLY MOL NWG PHI SUL SUM 62 CRL. Nanophan. or phan.
Claoxylon longifolium subsp. *glabrum* Müll.Arg. in A.P.de Candolle, Prodr. 15(2): 781 (1866).
Claoxylon elegans Airy Shaw, Kew Bull. 20: 398 (1966).
Claoxylon papyraceum Airy Shaw, Kew Bull. 23: 77 (1969).

var. **rugifrux** Airy Shaw, Kew Bull., Addit. Ser. 4: 72 (1975).
Borneo (Sarawak, E. Kalimantan). 42 BOR. Phan.

Claoxylon longipetiolatum Kurz, J. Asiat. Soc. Bengal, Pt. 2, Nat. Hist. 42(2): 244 (1873).
Sikkim to Burma. 40 ASS EHM 41 AND? BMA. Nanophan. or phan.

Claoxylon longiracemosum Hosok., Trans. Nat. Hist. Soc. Taiwan 25: 25 (1935).
Caroline Is. (Palau). 62 CRL.

Claoxylon lutescens Pax & K.Hoffm. in H.G.A.Engler, Pflanzenr., IV, 147, XIV: 15 (1919).
New Guinea. 42 NWG. Phan.

Claoxylon macranthum Müll.Arg., Linnaea 34: 164 (1865).
Madagascar, Mauritius. 29 MAU MDG.

Claoxylon mananarense Leandri, Notul. Syst. (Paris) 9: 176 (1941).
Madagascar. 29 MDG.

Claoxylon marianum Müll.Arg. in A.P.de Candolle, Prodr. 15(2): 783 (1866).
Marianas. 62 MRN. Nanophan. or phan.

Claoxylon medullosum Baill., Adansonia 1: 283 (1861).
Madagascar. 29 MDG.

Claoxylon microcarpum Airy Shaw, Kew Bull. 20: 403 (1966).
Papua New Guinea. 42 NWG. Phan.

Claoxylon monoicum Baill., Adansonia 1: 283 (1861).
Madagascar. 29 MDG.

Claoxylon muscisilvae Airy Shaw, Kew Bull. 21: 377 (1968).
NE. Papua New Guinea. 42 NWG. Phan.

Claoxylon neoebudicum (Guillaumin) Airy Shaw, Kew Bull. 32: 398 (1978).
Vanuatu, Santa Cruz Is., Fiji ? 60 FIJ? SCZ VAN. Phan.
* *Claoxylon taitense* var. *neoebudicum* Guillaumin, J. Arnold Arbor. 13: 92 (1932).

Claoxylon nervosum Pax & K.Hoffm. in H.G.A.Engler, Pflanzenr., IV, 147, VII: 106 (1914).
NE. Papua New Guinea. 42 NWG. Phan.

Claoxylon nigtanig Airy Shaw, Kew Bull. 33: 533 (1979).
Solomon Is. (Bougainville). 60 SOL.

Claoxylon nubicola Airy Shaw, Kew Bull. 20: 36 (1966).
Papua New Guinea. 42 NWG. Nanophan. or phan.

Claoxylon oblanceolatum (Merr.) Pax & K.Hoffm. in H.G.A.Engler, Pflanzenr., IV, 147, VII: 115 (1914).
Philippines. 42 PHI.
* *Claoxylon rubescens* var. *oblanceolatum* Merr., Philipp. J. Sci. 1(Suppl.): 79 (1906).

Claoxylon oliganthum Airy Shaw, Kew Bull. 32: 71 (1977).
S. Thailand. 41 THA. Nanophan.

Claoxylon ooumuense Fosberg & Sachet, Smithsonian Contr. Bot. 47: 6 (1981).
Marquesas (Nuku Hiva). 61 MRQ. Nanophan. or phan.

Claoxylon papuae Airy Shaw, Kew Bull. 32: 392 (1978).
Papua New Guinea. 42 NWG. Phan.

Claoxylon parviflorum A.Juss., Euphorb. Gen.: 114 (1824).
Réunion, Rodrigues. 29 REU ROD. Nanophan.
Acalypha spiciflora Poir. in J.B.A.M.de Lamarck, Encycl. 6: 206 (1804), nom. illeg.
Acalypha scabra Vahl ex Baill., Adansonia 1: 280 (1861).

Claoxylon paucinerve Airy Shaw, Kew Bull. 20: 33 (1966).
Papua New Guinea. 42 NWG. Phan.

Claoxylon perrieri Leandri, Notul. Syst. (Paris) 9: 176 (1941).
Madagascar. 29 MDG.

Claoxylon physocarpum Airy Shaw, Kew Bull. 20: 399 (1966).
Sumatera. 42 SUM. Phan.

Claoxylon platyphyllum Airy Shaw, Kew Bull. 32: 390 (1978).
Papua New Guinea. 42 NWG. Nanophan. or phan.

Claoxylon porphyrostemon Airy Shaw, Kew Bull. 20: 32 (1966).
NE. Papua New Guinea. 42 NWG. Nanophan. or phan.

Claoxylon praetermissum Airy Shaw, Kew Bull. 23: 78 (1969).
Borneo (Sabah). 42 BOR. Phan. – Perhaps identical with *C. pseudoinsulanum*.

Claoxylon pseudoinsulanum Pax & K.Hoffm. in H.G.A.Engler, Pflanzenr., IV, 147, VII: 113 (1914).
Borneo (E. Kalimantan). 42 BOR. Phan.

Claoxylon psilogyne Airy Shaw, Kew Bull. 32: 398 (1978).
Vanuatu. 60 VAN.

Claoxylon pubescens Quisumb., Philipp. J. Sci. 76(3): 42 (1944).
Philippines. 42 PHI.

Claoxylon purpureum Merr., Philipp. J. Sci. 1(Suppl.): 204 (1906).
Philippines. 42 PHI.

Claoxylon putii Airy Shaw, Kew Bull. 25: 519 (1971).
NW. Thailand. 41 THA.

Claoxylon racemiflorum Baill., Étude Euphorb.: 493 (1858).
Réunion (Basse Vallée). 29 REU.
Claoxylon grandidentatum Boivin ex Baill., Adansonia 1: 283 (1861).

Claoxylon raymondianum Leandri, Notul. Syst. (Paris) 9: 177 (1941).
Madagascar. 29 MDG.

Claoxylon rostratum Airy Shaw, Kew Bull. 32: 389 (1978).
Andaman Is., Nicobar Is., Burma. 41 AND BMA NCB. Nanophan. or phan.

Claoxylon rubescens Miq., Fl. Ned. Ind. 1(2): 387 (1859).
Lesser Sunda Is. (Lombok), Philippines. 42 LSI PHI.
Claoxylon rubrinerve Baill., Étude Euphorb.: 498 (1858), nom. nud.

Claoxylon rubrivenium Pax & K.Hoffm. in H.G.A.Engler, Pflanzenr., IV, 147, XVII: 182 (1924).
Philippines (Luzon). 42 PHI.

Claoxylon salicinum Airy Shaw, Kew Bull. 21: 376 (1968).
Borneo (Sabah). 42 BOR. Nanophan. or phan.

Claoxylon salomonense Airy Shaw, Kew Bull. 32: 393 (1978).
Papua New Guinea (incl. D'Entrecasteaux Is., Louisiade Archip.), Solomon Is. 42 NWG 60 SOL. Phan.

Claoxylon samoense Pax & K.Hoffm. in H.G.A.Engler, Pflanzenr., IV, 147, VII: 111 (1914).
Samoa. 60 SAM. Phan.

Claoxylon sanctae-crucis Airy Shaw, Kew Bull. 32: 396 (1978).
Santa Cruz Is. 60 SCZ.

Claoxylon sandwicense Müll.Arg., Linnaea 34: 165 (1865).
Hawaiian Is. 63 HAW. Nanophan. or phan.
Claoxylon sandwicense f. *glabrescens* Sherff in ?, *Claoxylon sandwicense* var. *glabrescens* (Sherff) Sherff in ?, .
Claoxylon sandwicense var. *degeneri* Sherff in ?
Claoxylon sandwicense var. *hillebrandtii* Sherff in ?
Claoxylon sandwicense var. *magnifolium* Sherff in ?
Claoxylon sandwicense var. *tomentosum* Hillebr., Fl. Hawaiian Isl.: 399 (1888). *Claoxylon sandwicense* f. *tomentosum* (Hillebr.) Sherff in ?, *Claoxylon tomentosum* (Hillebr.) A.Heller, Minnesota Bot. Stud. 1: 843 (1897).
Claoxylon insigne Baill. ex Drake, Ill. Fl. Ins. Pacif.: 291 (1892), nom. nud.
Claoxylon helleri Sherff, Amer. J. Bot. 24: 88 (1937).

Claoxylon sarasinorum Pax & K.Hoffm. in H.G.A.Engler, Pflanzenr., IV, 147, XVII: 182 (1924).
Sulawesi. 42 SUL.

Claoxylon scabratum (Pax & K.Hoffm.) Airy Shaw, Kew Bull. 20: 31 (1966).
Irian Jaya, NE. Papua New Guinea. 42 NWG. Phan.
* *Claoxylon indicum* var. *scabratum* Pax & K.Hoffm. in H.G.A.Engler, Pflanzenr., IV, 147, VII: 110 (1914).

Claoxylon setosum Coode, Kew Bull. 34: 45 (1979).
Réunion. 29 REU. Nanophan. or phan.

Claoxylon spathulatum Pax & K.Hoffm. in H.G.A.Engler, Pflanzenr., IV, 147, VII: 120 (1914).
Philippines. 42 PHI.

Claoxylon stapfianum Airy Shaw, Kew Bull. 20: 400 (1966).
Borneo (N. Sarawak, Sabah). 42 BOR. Phan.
* *Claoxylon pauciflorum* Stapf, Trans. Linn. Soc. London, Bot. 4: 225 (1894), nom. illeg.

Claoxylon subbullatum Airy Shaw, Kew Bull. 21: 375 (1968).
Borneo (Sabah). 42 BOR. Nanophan.

Claoxylon subsessiliflorum Croizat, J. Arnold Arbor. 23: 506 (1942).
Vietnam, China (Yunnan). 36 CHC 41 VIE.

Claoxylon subviride Elmer, Leafl. Philipp. Bot. 3: 907 (1910).
Philippines. 42 PHI.

Claoxylon taitense Müll.Arg., Linnaea 34: 165 (1865).
Society Is. 61 SCI. Phan.

Claoxylon tenerifolium (Baill.) F.Muell., Fragm. 6: 183 (1868).
Lesser Sunda Is., New Guinea (incl. Aru Is.), Northern Territory, Queensland. 42 LSI NWG 50 NTA QLD. Nanophan. or phan.
* *Mercurialis tenerifolia* Baill., Adansonia 6: 323 (1866).
Claoxylon australe var. *latifolia* Benth., Fl. Austral. 6: 131 (1873).
Claoxylon hillii Benth., Fl. Austral. 6: 131 (1873).
Claoxylon indicum var. *novoguineense* J.J.Sm. ex Valeton, Bull. Dép. Agric. Indes Néerl. 10: 26 (1907).
Claoxylon delicatum Airy Shaw, Kew Bull. 20: 32 (1966).

Claoxylon tenuiflorum Airy Shaw, Kew Bull. 36: 278 (1981).
Sumatera (Atjeh). 42 SUM. Nanophan. or phan.

Claoxylon tetracoccum Airy Shaw, Kew Bull. 25: 521 (1971).
Irian Jaya. 42 NWG. Phan.

Claoxylon tsaratananae Leandri, Notul. Syst. (Paris) 9: 173 (1941).
Madagascar. 29 MDG.

Claoxylon velutinum J.J.Sm., Bull. Jard. Bot. Buitenzorg, III, 1: 395 (1920).
Borneo (E. Kalimantan). 42 BOR.

Claoxylon vitiense Gillespie, Bernice P. Bishop Mus. Bull. 91: 13 (1932).
Fiji. 60 FIJ. Nanophan. or phan.

Claoxylon wallichianum Müll.Arg. in A.P.de Candolle, Prodr. 15(2): 781 (1866).
Pen. Malaysia. 42 MLY. Phan.

Claoxylon warburgianum Pax & K.Hoffm. in H.G.A.Engler, Pflanzenr., IV, 147, XIV: 16 (1919). Sulawesi. 42 SUL.

Claoxylon winkleri Pax & K.Hoffm. in H.G.A.Engler, Pflanzenr., IV, 147, VII: 106 (1914). Borneo (SE. Kalimantan). 42 BOR.

Synonyms:

Claoxylon affine Baill. === **Claoxylon insulanum** Müll.Arg.
Claoxylon africanum (Baill.) Müll.Arg. === **Erythrococca africana** (Baill.) Prain
Claoxylon angolense Müll.Arg. === **Erythrococca angolensis** (Müll.Arg.) Prain
Claoxylon archboldianum Croizat === **Claoxylon echinospermum** Müll.Arg.
Claoxylon atrovirens Pax === **Erythrococca atrovirens** (Pax) Prain
Claoxylon australe var. *dentata* Benth. === **Claoxylon australe** Baill. ex Müll.Arg.
Claoxylon australe var. *latifolia* Benth. === **Claoxylon tenerifolium** (Baill.) F.Muell.
Claoxylon australe var. *laxiflora* Benth. === **Claoxylon australe** Baill. ex Müll.Arg.
Claoxylon bakerianum Baill. === **Lobanilia bakeriana** (Baill.) Radcl.-Sm.
Claoxylon barteri Hook.f. === **Erythrococca africana** (Baill.) Prain
Claoxylon beddomei Hook.f. === **Micrococca beddomei** (Hook.f.) Prain
Claoxylon brachybotryon Müll.Arg. ex Sebert & Pancher === **Claoxylon insulanum** Müll.Arg.
Claoxylon brachyphyllum Croizat === **Claoxylon linostachys** subsp. **brachyphyllum** (Croizat) Coode
Claoxylon caerulescens Ridl. === **Claoxylon indicum** (Reinw. ex Blume) Hassk.
Claoxylon capense Baill. === **Micrococca capensis** (Baill.) Prain
Claoxylon chevalieri Beille === **Erythrococca chevalieri** (Beille) Prain
Claoxylon columnare Müll.Arg. === **Erythrococca columnaris** (Müll.Arg.) Prain
Claoxylon cordifolium Benth. === **Mallotus oppositifolius** (Geiseler) Müll.Arg. var. **oppositifolius**
Claoxylon coriaceum Baill. === **Claoxylon erythrophyllum** Miq. var. **erythrophyllum**
Claoxylon crassifolium Baill. === **Claoxylon grandifolium** (Poir.) Müll.Arg.
Claoxylon deflersii Schweinf. ex Pax & K.Hoffm. === **Erythrococca abyssinica** Pax
Claoxylon delicatum Airy Shaw === **Claoxylon tenerifolium** (Baill.) F.Muell.
Claoxylon dewevrei Pax=== **Erythrococca dewevrei** (Pax) Prain
Claoxylon digynum Wight === **Macaranga digyna** (Wight) Müll.Arg.
Claoxylon elegans Airy Shaw === **Claoxylon longifolium** (Blume) Endl. ex Hassk. var. **longifolium**
Claoxylon elongatum Merr. === **Claoxylon albicans** (Blanco) Merr.
Claoxylon flaccidum Pax === **Erythrococca atrovirens** var. **flaccida** (Pax) Radcl.-Sm.
Claoxylon graciliflorum Miq. === **Claoxylon affine** Zoll. & Moritzi
Claoxylon grandidentatum Boivin ex Baill. === **Claoxylon racemiflorum** Baill.
Claoxylon grandifolium Elmer === **Claoxylon arboreum** Elmer
Claoxylon grandifolium var. *genuinum* Müll.Arg. === **Claoxylon grandifolium** (Poir.) Müll.Arg.
Claoxylon grandifolium var. *submembranaceum* Müll.Arg. === **Claoxylon grandifolium** (Poir.) Müll.Arg.
Claoxylon helleri Sherff === **Claoxylon sandwicense** Müll.Arg.
Claoxylon hexandrum Müll.Arg. === **Discoclaoxylon hexandrum** (Müll.Arg.) Pax & K.Hoffm.
Claoxylon hillii Benth. === **Claoxylon tenerifolium** (Baill.) F.Muell.
Claoxylon hirsutum Hook.f. === **Micrococca wightii** var. **hirsuta** (Hook.f.) Prain
Claoxylon hirtellum Baill. === **Lobanilia hirtella** (Baill.) Radcl.-Sm.
Claoxylon hispidum Pax === **Erythrococca hispida** (Pax) Prain
Claoxylon holstii Pax === **Micrococca holstii** (Pax) Prain
Claoxylon humblotianum Baill. === **Micrococca humblotiana** (Baill.) Prain
Claoxylon inaequilaterum Pax === **Erythrococca atrovirens** (Pax) Prain var. **atrovirens**
Claoxylon indicum f. *angustifolium* J.J.Sm. === **Claoxylon glabrifolium** Miq. var. **glabrifolium**
Claoxylon indicum var. *genuinum* Müll.Arg. === **Claoxylon indicum** (Reinw. ex Blume) Hassk.
Claoxylon indicum f. *gracilius* J.J.Sm. === **Claoxylon indicum** (Reinw. ex Blume) Hassk.

Claoxylon indicum var. *macrophyllum* (Hassk.) Müll.Arg. === **Claoxylon indicum** (Reinw. ex Blume) Hassk.
Claoxylon indicum var. *neocaledonicum* Schltr. === **Claoxylon insulanum** Müll.Arg.
Claoxylon indicum var. *novoguineense* J.J.Sm. ex Valeton === **Claoxylon tenerifolium** (Baill.) F.Muell.
Claoxylon indicum var. *scabratum* Pax & K.Hoffm. === **Claoxylon scabratum** (Pax & K.Hoffm.) Airy Shaw
Claoxylon indicum var. *spathulatum* Müll.Arg. === **Claoxylon indicum** (Reinw. ex Blume) Hassk.
Claoxylon insigne Baill. ex Drake === **Claoxylon sandwicense** Müll.Arg.
Claoxylon khasianum var. *serratum* Hook.f. === **Claoxylon khasianum** Hook.f.
Claoxylon kirkii Müll.Arg. === **Erythrococca kirkii** (Müll.Arg.) Prain
Claoxylon lancifolium (Prain) Leandri === **Micrococca lancifolia** Prain
Claoxylon lasiococcum Pax === **Erythrococca trichogyne** (Müll.Arg.) Prain var. **trichogyne**
Claoxylon leucocarpum Kurz === **Mallotus leucocarpus** (Kurz) Airy Shaw
Claoxylon longifolium Baill. === **Micrococca oligandra** (Müll.Arg.) Prain
Claoxylon longifolium subsp. *glabrum* Müll.Arg. === **Claoxylon longifolium** (Blume) Endl. ex Hassk. var. **longifolium**
Claoxylon luteobrunneum (Baker) Baill. === **Lobanilia luteobrunnea** (Baker) Radcl.-Sm.
Claoxylon macrophyllum Bojer === **Claoxylon indicum** (Reinw. ex Blume) Hassk.
Claoxylon macrophyllum Prain === **Erythrococca macrophylla** Prain
Claoxylon macrophyllum Hassk. === **Claoxylon indicum** (Reinw. ex Blume) Hassk.
Claoxylon mannii Hook.f. === **Erythrococca mannii** (Hook.f.) Prain
Claoxylon membranaceum Müll.Arg. === **Erythrococca membranacea** (Müll.Arg.) Prain
Claoxylon menyharthii Pax === **Erythrococca menyharthii** (Pax) Prain
Claoxylon mercurialis (L.) Thwaites === **Micrococca mercurialis** (L.) Benth.
Claoxylon metzieri Bouton ex Bojer === **Claoxylon grandifolium** (Poir.) Müll.Arg.
Claoxylon mildbraedii Pax === **Erythrococca trichogyne** (Müll.Arg.) Prain var. **trichogyne**
Claoxylon minus (Blume) Hassk. === **Claoxylon indicum** (Reinw. ex Blume) Hassk.
Claoxylon molle (Blume) Miq. === **Claoxylon indicum** (Reinw. ex Blume) Hassk.
Claoxylon molleri Pax === **Erythrococca molleri** (Pax) Prain
Claoxylon muricatum Wight === **Mallotus resinosus** (Blanco) Merr.
Claoxylon neraudianum Baill. === **Orfilea neraudiana** (Baill.) G.L.Webster
Claoxylon occidentale Müll.Arg. === **Discoclaoxylon occidentale** (Müll.Arg.) Pax & K.Hoffm.
Claoxylon oleraceum Prain === **Erythrococca atrovirens** var. **flaccida** (Pax) Radcl.-Sm.
Claoxylon oligandrum Müll.Arg. === **Micrococca oligandra** (Müll.Arg.) Prain
Claoxylon ovale Baill. === **Lobanilia ovalis** (Baill.) Radcl.-Sm.
Claoxylon papyraceum Airy Shaw === **Claoxylon longifolium** (Blume) Endl. ex Hassk. var. **longifolium**
Claoxylon parvicoccum Croizat === **Claoxylon fallax** Müll.Arg.
Claoxylon parviflorum Hook. & Arn. === **Claoxylon indicum** (Reinw. ex Blume) Hassk.
Claoxylon parviflorum Seem. === **Claoxylon fallax** Müll.Arg.
Claoxylon patulum Prain === **Erythrococca patula** (Prain) Prain
Claoxylon pauciflorum Müll.Arg. === **Erythrococca pauciflora** (Müll.Arg.) Prain
Claoxylon pauciflorum Stapf === **Claoxylon stapfianum** Airy Shaw
Claoxylon pedicellare Pax & K.Hoffm. === **Claoxylon arboreum** Elmer
Claoxylon pedicellare Müll.Arg. === **Discoclaoxylon pedicellare** (Müll.Arg.) Pax & K.Hoffm.
Claoxylon poggei Prain === **Erythrococca poggei** (Prain) Prain
Claoxylon polot Merr. === **Claoxylon indicum** (Reinw. ex Blume) Hassk.
Claoxylon polyandrum Pax & K.Hoffm. === **Erythrococca polyandra** (Pax & K.Hoffm.) Prain
Claoxylon preussii Pax === **Discoclaoxylon hexandrum** (Müll.Arg.) Pax & K.Hoffm.
Claoxylon purpurascens Beille === **Erythrococca molleri** (Pax) Prain
Claoxylon remyi Sherff === **Platydesma remyi** (Sherff) O.Deg. et al. (Rutaceae)
Claoxylon reticulatum (Poir.) Bojer === **Acalypha filiformis** Poir. subsp. **filiformis**
Claoxylon rivulare Müll.Arg. === **Erythrococca rivularis** (Müll.Arg.) Prain
Claoxylon rubescens var. *cumingianum* Müll.Arg. === **Claoxylon albicans** (Blanco) Merr.

Claoxylon rubescens var. *oblanceolatum* Merr. === **Claoxylon oblanceolatum** (Merr.) Pax & K.Hoffm.
Claoxylon rubrinerve Baill. === **Claoxylon rubescens** Miq.
Claoxylon sandwicense var. *degeneri* Sherff === **Claoxylon sandwicense** Müll.Arg.
Claoxylon sandwicense var. *glabrescens* (Sherff) Sherff === **Claoxylon sandwicense** Müll.Arg.
Claoxylon sandwicense f. *glabrescens* Sherff === **Claoxylon sandwicense** Müll.Arg.
Claoxylon sandwicense var. *hillebrandtii* Sherff === **Claoxylon sandwicense** Müll.Arg.
Claoxylon sandwicense var. *magnifolium* Sherff === **Claoxylon sandwicense** Müll.Arg.
Claoxylon sandwicense f. *tomentosum* (Hillebr.) Sherff === **Claoxylon sandwicense** Müll.Arg.
Claoxylon sandwicense var. *tomentosum* Hillebr. === **Claoxylon sandwicense** Müll.Arg.
Claoxylon scabrum Bojer === **Claoxylon linostachys** Baill. subsp. **linostachys**
Claoxylon schweinfurthii Pax === **Erythrococca atrovirens** (Pax) Prain var. **atrovirens**
Claoxylon scottianum Baill. === **Claoxylon flavum** Scott-Elliot
Claoxylon sitibundum Croizat === **Claoxylon echinospermum** Müll.Arg.
Claoxylon sphaerocarpum Kuntze === **Croton sylvaticus** Hochst.
Claoxylon spiciflorum (Burm.f.) Baill. === **Cleidion spiciflorum** (Burm.f.) Merr.
Claoxylon spiciflorum (Burm.f.) A.Juss. === **Cleidion spiciflorum** (Burm.f.) Merr.
Claoxylon stipulosum Rchb.f. ex Zoll. === **Mallotus dispar** (Blume) Müll.Arg.
Claoxylon submembranaceum (Müll.Arg.) Pax & K.Hoffm. === **Claoxylon grandifolium** (Poir.) Müll.Arg.
Claoxylon taitense var. *neoebudicum* Guillaumin === **Claoxylon neoebudicum** (Guillaumin) Airy Shaw
Claoxylon tomentosum (Hillebr.) A.Heller === **Claoxylon sandwicense** Müll.Arg.
Claoxylon trichogyne Müll.Arg. === **Erythrococca trichogyne** (Müll.Arg.) Prain
Claoxylon triste Müll.Arg. === **Erythrococca tristis** (Müll.Arg.) Prain
Claoxylon tumidum J.J.Sm. === **Claoxylon cuneatum** J.J.Sm.
Claoxylon ubanghense A.Chev.?
Claoxylon virens N.E.Br. === **Erythrococca menyharthii** (Pax) Prain
Claoxylon volkensii Pax === **Micrococca volkensii** (Pax) Prain
Claoxylon welwitschianum Müll.Arg. === **Erythrococca welwitschiana** (Müll.Arg.) Prain
Claoxylon wightii Hook.f. === **Micrococca wightii** (Hook.f.) Prain
Claoxylon wightii var. *angustatum* S.R.M.Susila Rani & N.P.Balakr. === **Micrococca wightii** var. **angustata** (S.R.M.Susila Rani & N.P.Balakr.) Radcl.-Sm. & Govaerts
Claoxylon wightii var. *glabratum* S.R.M.Susila Rani & N.P.Balakr. === **Micrococca wightii** var. **glabrata** (S.R.M.Susila Rani & N.P.Balakr.) Radcl.-Sm. & Govaerts
Claoxylon wightii var. *hirsutum* (Hook.f.) S.R.M.Susila Rani & Balakr. === **Micrococca wightii** var. **hirsuta** (Hook.f.) Prain

Claoxylopsis

3 species, Madagascar; shrubs or small trees of wet submontane forest (500-2000 m) with *C. andapensis* scandent. Closely related to *Claoxylon* from which it differs mainly in having 15 stamens/flower rather than 20 or more and by its plumose styles. (Acalyphoideae)

Leandri, J. (1939). Euphorbiacées malgaches nouvelles récoltées par M. H. Perrier de la Bâthie. Bull. Soc. Bot. France 85: 523-533. Fr. — Includes protologue of *Claoxylopsis* (related to *Claoxylon*) and description of *C. perrieri*.

Radcliffe-Smith, A. (1988). Notes on Madagascan Euphorbiaceae, I. On the identity of *Paragelonium* and on the affinities of *Benoistia* and *Claoxylopsis*. Kew Bull. 43: 625-647, illus., maps. En. — Includes (pp. 642-646) descriptions of 2 new species of *Claoxylopsis* (both illustrated) for a total of 3; additional records for *C. perrieri* (one collected by its original author!) on p. 646; no key.

Claoxylopsis Leandri, Bull. Soc. Bot. France 85: 526 (1938 publ. 1939).
Madagascar. 29.

Claoxylopsis andapensis Radcl.-Sm., Kew Bull. 43: 642 (1988).
N. Madagascar. 29 MDG. Nanophan.

Claoxylopsis perrieri Leandri, Bull. Soc. Bot. France 85: 526 (1938 publ. 1939).
E. Madagascar. 29 MDG. Nanophan.

Claoxylopsis purpurascens Radcl.-Sm., Kew Bull. 43: 645 (1988).
N. & NC. Madagascar. 29 MDG. (Cl.) nanophan. or phan.

Clarorivinia

Synonyms:
Clarorivinia Pax & K.Hoffm. === **Ptychopyxis** Miq.
Clarorivinia chrysantha (K.Schum.) Pax & K.Hoffm. === **Ptychopyxis chrysantha** (K.Schum.) Airy Shaw
Clarorivinia grandifolia Pax & K.Hoffm. === **Ptychopyxis chrysantha** (K.Schum.) Airy Shaw

Clavipodium

Synonyms:
Clavipodium Desv. ex Grüning === **Beyeria** Miq.

Clavistylus

Synonyms:
Clavistylus J.J.Sm. ex Koord. & Valeton === **Megistostigma** Hook.f.
Clavistylus peltatus J.J.Sm. === **Megistostigma peltatum** (J.J.Sm.) Croizat

Cleidiocarpon

2 species, Asia (China and SE. Asia); small or medium to large trees to 30 m in more or less seasonal evergreen forest. The wide scattering of localities suggested to Airy-Shaw (1981) that the genus was an old one in process of decline. It was not at the time, however, possible to be confident about the number of species; again, one was faced with a puzzle without yet all the pieces to hand. The genus is related to *Epiprinus*. (Acalyphoideae)

Airy-Shaw, H. K. (1965). Notes on Malaysian and other Asiatic Euphorbiaceae, LIV. A new genus from Burma and western China. Kew Bull. 19: 313-314. En. — Protologue of genus, description of *C. laurinum*, and new combination for *Baccaurea cavaleriei*.

Tsiang, Y. (Tsiang Ying) (1973). (*Sinopimeleodendron*, a new genus of Euphorbiaceae from Kwangsi.) Acta Bot. Sin. 15: 131-135, illus. Ch. — Protologue of genus and description of *S. kwangsiense* with specimens seen; illustration. [Genus now reduced to *Cleidiocarpon*.]

Airy-Shaw, H. K. (1978). Notes on Malesian and other Asiatic Euphorbiaceae, CCII. On the affinities and identity of the genus *Sinopimelodendron* Tsiang. Kew Bull. 32: 408-410. En. — Discussion of *Sinopimelodendron kwangsiense*; reduction to *Cleidiocarpon cavaleriei* following a suggestion of C.Y. Wu.

Airy-Shaw, H. K. (1981). Notes on Asiatic, Malesian and Melanesian Euphorbiaceaee, CCXLVIII. *Cleidiocarpon* Airy Shaw. Kew Bull. 36: 608-609. En. — Summary of accumulated knowledge; the wide scattering of localities suggested to the author that the genus was an old one in process of decline. No satisfactory answer was possible respecting the number of taxa to be formally recognised.

Cleidiocarpon Airy Shaw, Kew Bull. 19: 313 (1965).
Burma, S. China. 36 41.
Sinopimelodendron Tsiang, Acta Bot. Sin. 15: 132 (1973).

Cleidiocarpon cavaleriei (H.Lév.) Airy Shaw, Kew Bull. 19: 314 (1965).
S. China (Guizhou, Guangxi). 36 CHC CHS. Phan.
* *Baccaurea cavaleriei* H.Lév., Fl. Kouy-Tchéou: 159 (1914).

Cleidiocarpon laurinum Airy Shaw, Kew Bull. 19: 313 (1965).
Burma. 41 BMA. Phan.

Cleidion

35 species, tropics (S. China, India to N. Australia and SW. Pacific (most); West and West-Central tropical Africa (1); Madagascar (1), Mesoamerica and tropical South America (5)). Shrubs or small to medium trees. In the SW. Pacific 12 species are represented in New Caledonia. The genus is well-marked with only *Wetria* closely related. [Lolita Bulalacao (Manila) has taken an interest in this genus within the framework of the Malesian Euphorbiaceae programme of the Rijksherbarium/Hortus Botanicus (Leiden).] (Acalyphoideae)

- Pax, F. (with K. Hoffmann) (1914). *Cleidion*. In A. Engler (ed.), Das Pflanzenreich, IV 147 VII (Euphorbiaceae-Acalypheae-Mercurialinae): 288-298. Berlin. (Heft 63.) La/Ge. — 17 species, tropics; the greater number in the Asia-Pacific region, but only 1 in Africa. Additions: ibid., XIV (Additamentum VI): 23-24.

Airy-Shaw, H. K. (1969). Notes on Malesian and other Asiatic Euphorbiaceae, CX. Notes on *Cleidion* Bl. in the Solomon Is. and New Hebrides. Kew Bull. 23: 85-88. En. — Treatment of 3 species (2 new). *C. papuanum* not recollected in New Guinea, but evidently frequent in the Solomons.

Airy-Shaw, H. K. (1971). Notes on Malesian and other Asiatic Euphorbiaceae, CXL. A misplaced Philippine *Cleidion*. Kew Bull. 25: 528-529. En. — Formerly known as *Actephila megistophylla*; only fruiting material available.

Leandri, J. (1972). Le genre *Cleidion* (Euphorbiacées) à Madagascar. Adansonia, II, 12: 193-196. Fr. — Genus recorded for the first time from Madagascar with description of *C. capuronii*; extensive discussion. [The species was, however, in the 19th century apparently introduced into botanical gardens outside Madagascar.]

McPherson, G. & C. Tirel (1987). *Cleidion*. Fl. Nouvelle-Calédonie, 14 (Euphorbiacées, I): 143-169. Paris. Fr. — Flora treatment (12 species) with key.

Cleidion Blume, Bijdr.: 612 (1826).
Trop. America, W. & WC. Trop. Africa, Madagascar, S. China to SW. Pacific. 22 23 36 40 41 42 50 60 80 83 84.
Psilostachys Turcz., Bull. Soc. Imp. Naturalistes Moscou 16: 58 (1843).
Redia Casar., Nov. Stirp. Brasil.: 51 (1843).
Lasiostyles C.Presl, Abh. Königl. Böhm. Ges. Wiss., V, 3: 579 (1845).
Tetraglossa Bedd., Madras J. Lit. Sci., II, 22: 70 (1861).

Cleidion alongense Bennet & Subh.Chandra, Indian Forester 111: 846 (1985).
India. 40 IND.

Cleidion amazonicum Ule, Verh. Bot. Vereins Prov. Brandenburg 50: 76 (1908 publ. 1909).
Brazil (Amazonas), Bolivia. 83 BOL 84 BZN. Phan.
Cleidion tricoccum Rusby ex Pax & K.Hoffm. in H.G.A.Engler, Pflanzenr., IV, 147, VII: 294 (1914), pro syn.

Cleidion bishnui Chakrab. & M.G.Gangop., J. Econ. Taxon. Bot. 12: 473 (1988 publ. 1989).
Burma. 41 BMA.

Cleidion bracteosum Gagnep., Bull. Soc. Bot. France 71: 569 (1924).
Vietnam. 41 VIE.

Cleidion brevipetiolatum Pax & K.Hoffm. in H.G.A.Engler, Pflanzenr., IV, 147, VII: 292 (1914).
Vietnam. 41 VIE. Nanophan.

Cleidion capuronii Leandri, Adansonia, n.s., 12: 196 (1972).
N. Madagascar. 29 MDG. Phan.

Cleidion castaneifolium Müll.Arg., Linnaea 34: 184 (1865).
Mexico to Peru. 79 MXT 80 COS HON 83 ECU PER. Phan.
Alchornea oblongifolia Standl., Publ. Carnegie Inst. Wash. 461: 66 (1935). *Cleidion oblongifolium* (Standl.) Croizat, J. Arnold Arbor. 24: 166 (1943).

Cleidion claoxyloides Müll.Arg., Linnaea 34: 184 (1865).
New Caledonia. 60 NWC. Nanophan. or phan.
Cleidion obovatum S.Moore, J. Linn. Soc., Bot. 45: 406 (1921).

Cleidion gabonicum Baill., Adansonia 11: 129 (1874).
Ghana, Cameroon, Gabon. 22 GHA 23 CMN GAB. Phan.

Cleidion lanceolatum Merr., Philipp. J. Sci., C 9: 474 (1914 publ. 1915).
Philippines. 42 PHI.

Cleidion lasiophyllum Pax & K.Hoffm. in H.G.A.Engler, Pflanzenr., IV, 147, VII: 296 (1914).
SE. New Caledonia. 60 NWC. Nanophan. or phan.
Cleidion panduratum S.Moore, J. Linn. Soc., Bot. 45: 405 (1921).

Cleidion lemurum McPherson, in Fl. N. Caled. & Depend. 14: 146 (1987).
New Caledonia (Grottes de Hienghène). 60 NWC. Nanophan.

Cleidion leptostachyum (Müll.Arg.) Pax & K.Hoffm. in H.G.A.Engler, Pflanzenr., IV, 147, VII: 293 (1914).
Fiji. 60 FIJ. Nanophan. or phan.
* *Mappa leptostachya* Müll.Arg., Linnaea 34: 198 (1865). *Macaranga leptostachya* (Müll.Arg.) Müll.Arg. in A.P.de Candolle, Prodr. 15(2): 1007 (1866).
Cleidion vieillardii var. *vitiensis* Müll.Arg. in A.P.de Candolle, Prodr. 15(2): 1007 (1866).
Cleidion degeneri Croizat, Sargentia 1: 51 (1942).

Cleidion lochmion McPherson, in Fl. N. Caled. & Depend. 14: 144 (1987).
WC. New Caledonia. 60 NWC. Nanophan.

Cleidion macarangoides Guillaumin, Mém. Mus. Natl. Hist. Nat., B, Bot. 8: 257 (1962).
C. & NW. New Caledonia. 60 NWC. Phan.

Cleidion macrophyllum Baill., Adansonia 2: 219 (1862). *Cleidion vieillardii* var. *macrophyllum* (Baill.) Müll.Arg. in A.P.de Candolle, Prodr. 15(2): 985 (1866).
New Caledonia. 60 NWC. Nanophan. or phan.
Cleidion viridiflorum S.Moore, J. Linn. Soc., Bot. 45: 403 (1921).

Cleidion marginatum McPherson, in Fl. N. Caled. & Depend. 14: 149 (1987).
EC. New Caledonia. 60 NWC. Nanophan.

Cleidion megistophyllum (Quisumb. & Merr.) Airy Shaw, Kew Bull. 25: 528 (1971).
Philippines (Luzon). 42 PHI.
* *Actephila megistophylla* Quisumb. & Merr., Philipp. J. Sci. 37: 158 (1928).

Cleidion membranaceum Pax & K.Hoffm. in H.G.A.Engler, Pflanzenr., IV, 147, XIV: 23 (1919).
Panama, Venezuela. 80 PAN 82 VEN. Phan.
Cleidion woodsonianum Croizat, J. Arnold Arbor. 24: 167 (1943). *Adenophaedra woodsoniana* (Croizat) Croizat, Trop. Woods 88: 31 (1946).

Cleidion microcarpum Merr., Philipp. J. Sci. 30: 404 (1926).
Philippines (Tawitawi). 42 PHI.

Cleidion minahassae Pax & K.Hoffm. in H.G.A.Engler, Pflanzenr., IV, 147, XIV: 23 (1919).
Sulawesi. 42 SUL.

Cleidion neoebudicum Airy Shaw, Kew Bull. 23: 87 (1969).
Vanuatu. 60 VAN.

Cleidion nitidum (Müll.Arg.) Thwaites ex Kurz, J. Asiat. Soc. Bengal, Pt. 2, Nat. Hist. 42(2): 245 (1873).
Andaman Is., Sri Lanka. 40 SRL 41 AND. Phan.
* *Mallotus nitidus* Müll.Arg. in A.P.de Candolle, Prodr. 15(2): 979 (1866).

Cleidion papuanum Lauterb. in K.M.Schumann & C.A.G.Lauterbach, Fl. Schutzgeb. Südsee, Nachtr.: 296 (1905).
NE. Papua New Guinea, Bismarck Archip., Solomon Is. 42 BIS NWG 60 SOL. Nanophan. or phan.
Trigonostemon oliganthus K.Schum. in K.M.Schumann & C.A.G.Lauterbach, Fl. Schutzgeb. Südsee, Nachtr.: 298 (1905).

Cleidion ramosii (Merr.) Merr., Philipp. J. Sci. 20: 400 (1922).
Philippines. 42 PHI.
* *Mallotus ramosii* Merr., Philipp. J. Sci., C 8: 401 (1913).
Mallotus samarensis Merr., Philipp. J. Sci., C 9: 488 (1914 publ. 1915).

Cleidion salomonis Airy Shaw, Kew Bull. 23: 87 (1969).
Solomon Is. 60 SOL.

Cleidion sessile Kaneh. & Hatus., Bot. Mag. (Tokyo) 53: 151 (1939).
Caroline Is. (Palau). 62 CRL.

Cleidion spathulatum Baill., Adansonia 2: 221 (1862).
New Caledonia. 60 NWC. Nanophan. or phan.
Cleidion paucidentatum S.Moore, J. Linn. Soc., Bot. 45: 406 (1921).

Cleidion spiciflorum (Burm.f.) Merr., Interpr. Herb. Amboin.: 322 (1917).
SE. China, Trop. Asia, Queensland (Cape York Pen.), Solomon Is., Vanuatu. 36 CHS 40 BHU IND 41 THA 42 BIS BOR JAW LSI MLY NWG PHI SUL SUM 50 QLD 60 SOL VAN. Phan.
* *Acalypha spiciflora* Burm.f., Fl. Indica: 203 (1768). *Claoxylon spiciflorum* (Burm.f.) A.Juss., Euphorb. Gen.: 43 (1824). *Claoxylon spiciflorum* (Burm.f.) Baill., Étude Euphorb., Atlas: 37 (1858), nom. illeg.

var. **moniliflorum** (Airy Shaw) Radcl.-Sm. & Govaerts, Kew Bull. 52: 478 (1997).
Bismarck Archip. (New Britain). 42 BIS. Phan.
* *Cleidion moniliflorum* Airy Shaw, Kew Bull. 34: 591 (1980). *Cleidion javanicum* var. *moniliflorum* (Airy Shaw) Chakrab. & M.G.Gangop., J. Econ. Taxon. Bot. 12: 493 (1988 publ. 1989).

var. **spiciflorum**
SE. China, Trop. Asia, Queensland (Cape York Pen.), Solomon Is., Vanuatu. 36 CHS 40 BHU IND 41 THA 42 BIS BOR JAW LSI MLY NWG PHI SUL SUM 50 QLD 60 SOL VAN. Phan.
Tragia filiformis Poir. in J.B.A.M.de Lamarck, Encycl. 7: 727 (1806).
Cleidion javanicum Blume, Bijdr.: 613 (1826).
Acalypha spicigera Klotzsch in B.Seemann, Bot. Voy. Herald: 101 (1853).
Acalypha acuminata Vahl ex Baill., Adansonia 1: 267 (1861), nom. illeg.
Acalypha acuminata Baill. in A.Grandidier, Hist. Phys. Madagascar, Atlas: 188 (1891), nom. illeg.
Macaranga tamiana K.Schum., Notizbl. Bot. Gart. Berlin-Dahlem 1: 52 (1895).
Mallotus geloniifolius Müll.Arg. ex Pax & K.Hoffm. in H.G.A.Engler, Pflanzenr., IV, 147, VII: 290 (1914), pro syn.
Cleidion javanicum var. *longipedicellatum* Chakrab. & M.G.Gangop., J. Econ. Taxon. Bot. 12: 491 (1988 publ. 1989).

Cleidion tavnguavenense Thin, Euphorb. Vietnam: 47 (1995).
Vietnam. 41 VIE. Phan.

Cleidion tricoccum (Casar.) Baill., Adansonia 4: 370 (1864).
E. Brazil. 84 BZE BZL. Nanophan. or phan.
* *Reidia tricocca* Casar., Nov. Stirp. Brasil. 6: 52 (1843).

Cleidion veillonii McPherson, in Fl. N. Caled. & Depend. 14: 154 (1987).
NW. New Caledonia. 60 NWC. Nanophan. or phan.

Cleidion velutinum McPherson, in Fl. N. Caled. & Depend. 14: 164 (1987).
NW. New Caledonia. 60 NWC. Nanophan.

Cleidion verticillatum Baill., Adansonia 2: 221 (1862).
NW. & SE. New Caledonia (incl. Ile des Pins, Loyalty Is.). 60 NWC. Nanophan.

Cleidion vieillardii Baill., Adansonia 2: 220 (1862). *Cleidion vieillardii* var. *genuinum* Müll.Arg. in A.P.de Candolle, Prodr. 15(2): 986 (1866), nom. inval.
New Caledonia. 60 NWC. Nanophan. or phan.

var. **mareense** Guillaumin in F.Sarasin & J.Roux, Nova Caledonia, Bot. 1(2): 166 (1920).
New Caledonia (I. des Pins, Loyalty Is.). 60 NWC. Nanophan. or phan.

var. **vieillardii**
New Caledonia (incl. I. des Pins). 60 NWC. Nanophan. or phan.
Cleidion vieillardii var. *acutifolium* Müll.Arg. in A.P.de Candolle, Prodr. 15(2): 986 (1866).
Cleidion baillonii Vieill. ex Guillaumin, Ann. Inst. Bot.-Géol. Colon. Marseille, II, 9: 228 (1911), nom. illeg.
Cleidion angustifolium Pax & K.Hoffm. in H.G.A.Engler, Pflanzenr., IV, 147, VII: 293 (1914).
Cleidion comptonii S.Moore, J. Linn. Soc., Bot. 45: 405 (1921).
Cleidion sylvestre S.Moore, J. Linn. Soc., Bot. 45: 404 (1921).

Synonyms:
Cleidion angustifolium Pax & K.Hoffm. === **Cleidion vieillardii** Baill. var. **vieillardii**
Cleidion baillonii Vieill. ex Guillaumin === **Cleidion vieillardii** Baill. var. **vieillardii**
Cleidion cafcaf Croizat === **Orfilea neraudiana** (Baill.) G.L.Webster
Cleidion comptonii S.Moore === **Cleidion vieillardii** Baill. var. **vieillardii**
Cleidion coriaceum Baill. === **Macaranga coriacea** (Baill.) Müll.Arg.
Cleidion degeneri Croizat === **Cleidion leptostachyum** (Müll.Arg.) Pax & K.Hoffm.
Cleidion denticulatum Standl. === **Adenophaedra grandifolia** (Klotzsch) Müll.Arg.
Cleidion javanicum Blume === **Cleidion spiciflorum** (Burm.f.) Merr. var. **spiciflorum**

Cleidion javanicum var. *longipedicellatum* Chakrab. & M.G.Gangop. === **Cleidion spiciflorum** (Burm.f.) Merr. var. **spiciflorum**
Cleidion javanicum var. *moniliflorum* (Airy Shaw) Chakrab. & M.G.Gangop. === **Cleidion spiciflorum** var. **moniliflorum** (Airy Shaw) Radcl.-Sm. & Govaerts
Cleidion lutescens Pax & Lingelsh. === **Macaranga lutescens** (Pax & Lingelsh.) Pax
Cleidion mannii Baker === **Tetracarpidium conophorum** (Müll.Arg.) Hutch. & Dalziel
Cleidion moniliflorum Airy Shaw === **Cleidion spiciflorum** var. **moniliflorum** (Airy Shaw) Radcl.-Sm. & Govaerts
Cleidion nicaraguense Hemsl. === **Acidoton nicaraguensis** (Hemsl.) G.L.Webster
Cleidion oblongifolium (Standl.) Croizat === **Cleidion castaneifolium** Müll.Arg.
Cleidion obovatum S.Moore === **Cleidion claoxyloides** Müll.Arg.
Cleidion panduratum S.Moore === **Cleidion lasiophyllum** Pax & K.Hoffm.
Cleidion paucidentatum S.Moore === **Cleidion spathulatum** Baill.
Cleidion platystigma Schltr. === **Bocquillonia codonostylis** (Baill.) Airy Shaw
Cleidion populifolium Zipp. ex Span.?
Cleidion prealtum Croizat === **Adenophaedra prealta** (Croizat) Croizat
Cleidion preussii (Pax) Baker === **Tetracarpidium conophorum** (Müll.Arg.) Hutch. & Dalziel
Cleidion sylvestre S.Moore === **Cleidion vieillardii** Baill. var. **vieillardii**
Cleidion tenuispica Schltr. === **Macaranga vieillardii** (Müll.Arg.) Müll.Arg.
Cleidion tricoccum Rusby ex Pax & K.Hoffm. === **Cleidion amazonicum** Ule
Cleidion ulmifolium Müll.Arg. === **Alchornea ulmifolia** (Müll.Arg.) Hurus.
Cleidion vieillardii var. *acutifolium* Müll.Arg. === **Cleidion vieillardii** Baill. var. **vieillardii**
Cleidion vieillardii var. *genuinum* Müll.Arg. === **Cleidion vieillardii** Baill.
Cleidion vieillardii var. *macrophyllum* (Baill.) Müll.Arg. === **Cleidion macrophyllum** Baill.
Cleidion vieillardii var. *vitiensis* Müll.Arg. === **Cleidion leptostachyum** (Müll.Arg.) Pax & K.Hoffm.
Cleidion viridiflorum S.Moore === **Cleidion macrophyllum** Baill.
Cleidion woodsonianum Croizat === **Cleidion membranaceum** Pax & K.Hoffm.
Cleidion xyphophylloides Croizat === **Trigonostemon xyphophylloides** (Croizat) L.K.Dai & T.L.Wu

Cleistanthus

148 species, Africa, Madagascar (c. 6), Asia, Malesia, Australia and western Pacific Is.; includes *Schistostigma*. Shrubs or small to large trees to 20 m with more or less distichously arranged, reticulately veined leaves, often in forest understorey; a very characteristic element in Malesia. Jablonski (1915) recognised 10 sections partly on single characters. Most species are evidently local; however, *CC. myrianthus* and *sumatranus* are widely distributed in Asia and Malesia and *C. polystachyus* similarly in Africa. Full, though not recent, revisions have been published for Africa and Madagascar. Its rather larger representation in Asia and Malesia has, however, not been examined in full since 1915 although many additions were published by Airy-Shaw from 1967 through the early 1980s. A third more species are now recognised when compared with Jablonski's treatment, with, for example, 29 recorded in Peninsular Malaysia and considered adequately known (*Tree Flora of Malaya*, 2: 80. 1972). Léonard (1960) believed the genus to be 'en pleine évolution et loin d'être fixée'. It is related to *Bridelia* but the ovary is usually tricarpellate and the fruits generally capsular; however, further study of its delimitation against that genus along with the relationship of both to *Amanoa* is needed (Webster, Synopsis, 1994). Suggestions have also been made that *Cleistanthus* is heterogeneous. (Phyllanthoideae)

Pax, F. (with K. Hoffmann) (1911). *Schistostigma*. In A. Engler (ed.), Das Pflanzenreich, IV 147 III (Euphorbiaceae-Cluytieae): 84-85. Berlin. (Heft 47.) La/Ge. — 1 species, New Guinea (Papua New Guinea). [Now part of *Cleistanthus*.]

- Jablonski, E. (as E. Jablonszky) (1915). *Cleistanthus*. In A. Engler (ed.), Das Pflanzenreich, IV 147 VIII (Euphorbiaceae-Phyllanthoideae-Bridelieae): 8-54. Berlin. (Heft 65.) La/Ge. — 106 species in 10 sections, Africa, Madagascar, Asia, Malesia, Australia and the Pacific.

Leandri, J. (1957). Les espèces malgaches du genre *Cleistanthus* Hook. f. (Euphorbiacées). Naturaliste Malgache 9: 41-47, illus. Fr. — Introductory note with background; illustrated, briefly descriptive treatment (any novelties being more fully described, in Latin) with synonymy, references, phenolgy, distribution and habitat, localities with exsiccatae, and notes (covering 6 species); key (p. 47).

Leandri, J. (1958). *Cleistanthus*. Fl. Madag. Comores 111 (Euphorbiacées), I: 181-192. Paris. Fr. — Flora treatment (6 species); key.

• Léonard, J. (1960). Notulae systematicae XXIX. Révision des *Cleistanthus* d'Afrique continentale (Euphorbiacées). Bull. Jard. Bot. État 30: 421-461, 3 maps. Fr. — Synoptic treatment, with descriptions of novelties but no keys, of 23 species; geographical review, pp. 452-456, followed by conclusions (among them that the genus was, along with others, 'en pleine évolution et loin d'être fixée'.). [The main division (p. 422) was into species with glabrous or externally somewhat pubescent sepals vs. those with sepals externally densely hairy.]

Airy-Shaw, H. K. (1966). Notes on Malaysian and other Asiatic Euphorbiaceae, LXXI. New species of *Cleistanthus* Hook. f. Kew Bull. 20: 389-393. En. — 3 novelties, Malesia.

Airy-Shaw, H. K. (1968). Notes on Malesian and other Asiatic Euphorbiaceae, LXXXVII. New or noteworthy species of *Cleistanthus* Hook. f. ex Planch. Kew Bull. 21: 362-374. En. — 10 species covered of which 8 new; in Borneo, Peninsular Malaysia, Singapore and SE Asia.

Airy-Shaw, H. K. (1969). Notes on Malesian and other Asiatic Euphorbiaceae, CIII. New or noteworthy species of *Cleistanthus* Hook. f. ex Planch. Kew Bull. 23: 62-65. En. — 3 species, of which 1 new; some reductions. Sections are indicated.

Airy-Shaw, H. K. (1971). Notes on Malesian and other Asiatic Euphorbiaceae, CXXXII. New species of *Cleistanthus* Hook. f. ex Planch. Kew Bull. 25: 510-512. En. — 2 species from W Malesia; in different sections.

Airy-Shaw, H. K. (1972). Notes on Malesian and other Asiatic Euphorbiaceae, CLVI: Notes on *Cleistanthus* Hook. f. Kew Bull. 27: 76-77. En. — 2 species, one of them with a new variety.

Airy-Shaw, H. K. (1974). Notes on Malesian and other Asiatic Euphorbiaceae, CLXXVI. A new *Cleistanthus* from Malaya. Kew Bull. 29: 309-310. En. — Description of *C. longinervis*.

Airy-Shaw, H. K. (1978). Notes on Malesian and other Asiatic Euphorbiaceae, CXCV. New or noteworthy species of *Cleistanthus* Hook. f. Kew Bull. 32: 382-383. En. — Description of *C. jacobsianus* from Borneo (Sarawak, on limestone); range extension for *C. erycibifolius*. Sections are indicated.

Airy-Shaw, H. K. (1978). Notes on Malesian and other Asiatic Euphorbiaceae, CCXIII. *Cleistanthus* Hook. f. ex Planch. Kew Bull. 33: 43-54. En. — 13 spp., various sections; includes partial key to *C. rufescens* and relatives (sect. *Ferruginosi*, nos. 47-71 in Jablonski 1915).

• Dressler, S. (1999). Revision of the genus *Cleistanthus* in the Philippines. Blumea 44: 109-148, illus., maps. En. — Revision of 16 species with key, descriptions, synonymy, references, types, indication of distribution and ecology, illustrations, maps and commentary; general references, list of specimens seen and index to names at end. One additional species from (Palawan) is imperfectly known. [The author notes that no overall classification of the genus has succeeded that of Jablonski in 1915. The choice of the Philippines was designed to complement the studies of Airy-Shaw for *Cleistanthus* in other parts of Malesia. Reductions are substantial, but could not be incorporated in time for the present book.]

Cleistanthus Hook.f. ex Planch., Hooker's Icon. Pl. 8: t. 779 (1848).

Trop. & S. Africa, Madagascar, S. China, Trop. Asia, W. Pacific. 22 23 25 26 27 29 36 40 41 42 50 60 62. Nanophan. or phan.

Nanopetalum Hassk., Verslagen Meded. Afd. Natuurk. Kon. Akad. Wetensch. 4: 140 (1855).
Lebidiera Baill., Étude Euphorb., Atlas: 50 (1858).
Stenonia Baill., Étude Euphorb.: 578 (1858).
Leiopyxis Miq., Fl. Ned. Ind., Eerste Bijv.: 445 (1861).
Lebidieropsis Müll.Arg., Linnaea 32: 79 (1863).
Kaluhaburunghos L. ex Kuntze, Revis. Gen. Pl. 2: 607 (1891).

Stenoniella Kuntze in T.E.von Post & C.E.O.Kuntze, Lex. Gen. Phan.: 535 (1903).
Schistostigma Lauterb. in K.M.Schumann & C.A.G.Lauterbach, Fl. Schutzgeb. Südsee, Nachtr.: 299 (1905).
Godefroya Gagnep., Bull. Soc. Bot. France 70: 435 (1923).
Paracleisthus Gagnep., Bull. Soc. Bot. France 70: 499 (1923).

Cleistanthus acuminatissimus Merr., Univ. Calif. Publ. Bot. 15: 155 (1929).
Borneo (Sabah). 42 BOR. Nanophan. or phan. – Close to *C. beccarianus*.

Cleistanthus acuminatus (Thwaites) Müll.Arg. in A.P.de Candolle, Prodr. 15(2): 508 (1866).
Sri Lanka. 40 SRL.
**Amanoa acuminata* Thwaites, Enum. Pl. Zeyl.: 428 (1864).

Cleistanthus angularis Kaneh., Bot. Mag. (Tokyo) 45: 289 (1931).
Caroline Is. (Palau). 62 CRL.

Cleistanthus angustifolius Merr., Philipp. J. Sci., C 7: 386 (1912 publ. 1913).
Philippines (Luzon). 42 PHI. Nanophan. or phan.

Cleistanthus annamensis Gagnep., Bull. Soc. Bot. France 72: 461 (1925).
Vietnam. 41 VIE.

Cleistanthus apiculatus C.B.Rob., Philipp. J. Sci., C 3: 189 (1908).
Philippines (Mindanao). 42 PHI. Phan.

Cleistanthus apodus Benth., Fl. Austral. 6: 122 (1873).
NE. Queensland, Papua New Guinea. 42 NWG 50 QLD. Phan.

Cleistanthus bakonensis Airy Shaw, Kew Bull. 21: 369 (1968).
Borneo (Sarawak, Sabah). 42 BOR. Phan.

Cleistanthus balakrishnanii Chakrab., J. Econ. Taxon. Bot. 5(4): 951 (1984).
Nicobar Is. 41 NCB.

Cleistanthus baramicus Jabl. in H.G.A.Engler, Pflanzenr., IV, 147, VIII: 24 (1915).
Borneo. 42 BOR. Phan.
Cleistanthus glaucus Jabl. in H.G.A.Engler, Pflanzenr., IV, 147, VIII: 25 (1915), nom. illeg.

Cleistanthus beccarianus Jabl. in H.G.A.Engler, Pflanzenr., IV, 147, VIII: 19 (1915).
Borneo (Sarawak). 42 BOR. Phan.

Cleistanthus bipindensis Pax, Bot. Jahrb. Syst. 33: 282 (1903).
Cameroon, Central African Rep., Zaïre. 23 CAF CMN ZAI. Phan.

Cleistanthus boivinianus (Baill.) Müll.Arg. in A.P.de Candolle, Prodr. 15(2): 505 (1866).
Madagascar (incl. Nosy Bé I.). 29 MDG. Phan.
**Amanoa boiviniana* Baill., Étude Euphorb.: 582 (1858).
Cleistanthus boivinianus f. *humbertii* Leandri, Naturaliste Malgache 9: 42 (1957).
Cleistanthus boivinianus f. *occidentalis* Leandri, Naturaliste Malgache 9: 42 (1957).
Cleistanthus boivinianus f. *orientalis* Leandri, Naturaliste Malgache 9: 42 (1957).

Cleistanthus bracteosus Jabl. in H.G.A.Engler, Pflanzenr., IV, 147, VIII: 41 (1915).
Pen. Malaysia. 42 MLY. Phan.

Cleistanthus brideliifolius C.B.Rob., Philipp. J. Sci., C 3: 191 (1908).
Philippines, Borneo, Sumatera, Jawa, Pen. Malaysia ? 42 BOR JAW MLY? PHI SUM. Phan.

var. **brideliifolius**
Philippines, Borneo, Sumatera, Jawa, Pen. Malaysia ? 42 BOR JAW MLY? PHI SUM. Phan.
Cleistanthus pallidus var. *subcordatus* J.J.Sm. in S.H.Koorders & T.Valeton, Bijdr. Boomsoort. Java 12: 304 (1910). *Cleistanthus subcordatus* (J.J.Sm.) Jabl. in H.G.A.Engler, Pflanzenr., IV, 147, VIII: 22 (1915).

var. **calcicola** Airy Shaw, Kew Bull. 33: 51 (1978).
Borneo. 42 BOR.

Cleistanthus camerunensis J.Léonard, Bull. Jard. Bot. État 30: 430 (1960).
Cameroon. 23 CMN. Phan.

Cleistanthus capuronii Leandri, Naturaliste Malgache 9: 45 (1957).
E. Madagascar. 29 MDG. Phan.

Cleistanthus carolinianus Jabl. in H.G.A.Engler, Pflanzenr., IV, 147, VIII: 40 (1915).
Caroline Is. (Palau). 62 CRL. – Close to *C. myrianthus*.

Cleistanthus caudatus Pax ex De Wild. & T.Durand, Ann. Mus. Congo Belge, Bot., II, 1: 49 (1899).
Zaire. 23 ZAI. Nanophan.

Cleistanthus celebicus Jabl. in H.G.A.Engler, Pflanzenr., IV, 147, VIII: 22 (1915).
Borneo (Sabah), Sulawesi. 42 BOR SUL. Phan.

Cleistanthus chlorocarpus Airy Shaw, Kew Bull. 33: 45 (1978).
Borneo (Sarawak). 42 BOR. Nanophan. or phan.

Cleistanthus collinus (Roxb.) Benth. in G.Bentham & J.D.Hooker, Gen. Pl. 3: 268 (1880).
Pakistan, India, Bangladesh, Sri Lanka. 40 BAN IND PAK SRL. Nanophan. or phan.
* *Clutia collina* Roxb., Pl. Coromandel 2: 37 (1802). *Bridelia collina* (Roxb.) Hook. & Arn., Bot. Beechey Voy.: 211 (1837). *Amanoa collina* (Roxb.) Baill., Étude Euphorb.: 582 (1858). *Lebidieropsis collina* (Roxb.) Müll.Arg., Linnaea 32: 80 (1863). *Lebidieropsis orbiculata* var. *collina* (Roxb.) Müll.Arg. in A.P.de Candolle, Prodr. 15(2): 509 (1866), nom. illeg.
Andrachne orbiculata Roth, Nov. Pl. Sp.: 364 (1821). *Lebidieropsis orbiculata* (Roth) Müll.Arg. in A.P.de Candolle, Prodr. 15(2): 509 (1866).
Emblica palasis Buch.-Ham., Trans. Linn. Soc. London 13: 507 (1822).
Andrachne cadishaco Roxb. ex Wall., Numer. List: 7877 (1847), nom. inval.
Lebidieropsis orbiculata var. *lambertii* Müll.Arg. in A.P.de Candolle, Prodr. 15(2): 510 (1866).

Cleistanthus concinnus Croizat, J. Arnold Arbor. 23: 41 (1942).
Hainan, Vietnam. 36 CHH 41 VIE. Nanophan.
Phyllanthus dongfangensis P.T.Li, Acta Phytotax. Sin. 25: 382 (1987). *Cleistanthus dongfangensis* (P.T.Li) H.S.Kiu, Acta Phytotax. Sin. 27: 454 (1989).

Cleistanthus contractus Airy Shaw, Kew Bull. 25: 511 (1971).
Pen. Malaysia, Borneo (Sabah). 42 BOR MLY. Phan.

Cleistanthus coriaceus Airy Shaw, Kew Bull. 21: 362 (1968).
Borneo (Sarawak). 42 BOR. Phan.

Cleistanthus cunninghamii (Müll.Arg.) Müll.Arg. in A.P.de Candolle, Prodr. 15(2): 506 (1866).
Queensland, New South Wales. 50 NSW QLD. Nanophan. or phan.
* *Lebidiera cunninghamii* Müll.Arg., Linnaea 32: 80 (1863). *Amanoa cunninghamii* (Müll.Arg.) Baill., Adansonia 6: 335 (1866).

Cleistanthus curtisii Jabl. in H.G.A.Engler, Pflanzenr., IV, 147, VIII: 22 (1915).
Pen. Malaysia. 42 MLY.

Cleistanthus dallachyanus (Baill.) Benth., Fl. Austral. 6: 122 (1873).
Northern Territory, Queensland. 50 NTA QLD. Phan.
* *Amanoa dallachyana* Baill., Adansonia 6: 335 (1866).

Cleistanthus decipiens C.B.Rob., Philipp. J. Sci., C 3: 195 (1908).
Philippines. 42 PHI. Nanophan. or phan.

Cleistanthus decurrens Hook.f., Fl. Brit. India 5: 278 (1887).
Pen. Malaysia, Borneo (Sarawak), Philippines (Palawan). 42 BOR MLY PHI. Phan.
Cleistanthus mattangensis Jabl. in H.G.A.Engler, Pflanzenr., IV, 147, VIII: 33 (1915).

Cleistanthus denudatus Airy Shaw, Kew Bull. 21: 367 (1968).
Thailand. 41 THA. Nanophan. or phan.

Cleistanthus discolor Summerh., Bull. Misc. Inform. Kew 1928: 144 (1928).
Queensland. 50 QLD.

Cleistanthus diversifolius (Roxb.) Müll.Arg. in A.P.de Candolle, Prodr. 15(2): 509 (1866).
? 42 +. – Provisionally accepted.
* *Clutia diversifolia* Roxb., Fl. Ind. ed. 1832, 3: 731 (1832). *Bridelia diversifolia* (Roxb.) Hook. & Arn., Bot. Beechey Voy.: 211 (1837).

Cleistanthus dolichophyllus Airy Shaw, Kew Bull. 33: 47 (1978).
Pen. Malaysia. 42 MLY. Nanophan.

Cleistanthus duvipermaniorum J.Léonard, Bull. Jard. Bot. État 30: 447 (1960).
Zaire, Angola. 23 ZAI 26 ANG. Nanophan. or phan.

Cleistanthus eberhardtii (Gagnep.) Croizat, J. Arnold Arbor. 23: 39 (1942).
Vietnam. 41 VIE.
* *Paracleisthus eberhardtii* Gagnep., Bull. Soc. Bot. France 70: 499 (1923).

Cleistanthus ellipticus Hook.f., Fl. Brit. India 5: 281 (1887).
Pen. Malaysia. 42 MLY. Phan.

Cleistanthus elongatus Jabl. in H.G.A.Engler, Pflanzenr., IV, 147, VIII: 23 (1915).
Borneo (SW. Sarawak). 42 BOR. Phan.

Cleistanthus erycibifolius Airy Shaw, Kew Bull. 20: 389 (1966 publ. 1967).
Pen. Malaysia, Borneo (Kalimantan). 42 BOR MLY. Phan.

Cleistanthus everettii C.B.Rob., Philipp. J. Sci., C 3: 194 (1908).
Philippines (Negros). 42 PHI. Nanophan. or phan.

Cleistanthus evrardii J.Léonard, Bull. Jard. Bot. État 30: 446 (1960).
Zaire. 23 ZAI. Phan.

Cleistanthus ferrugineus (Thwaites) Müll.Arg. in A.P.de Candolle, Prodr. 15(2): 507 (1866).
Sri Lanka. 40 SRL. Phan.
* *Amanoa ferruginea* Thwaites, Enum. Pl. Zeyl.: 280 (1861).

Cleistanthus flavescens Jabl. in H.G.A.Engler, Pflanzenr., IV, 147, VIII: 19 (1915).
Pen. Malaysia. 42 MLY. Phan.

Cleistanthus floricola Airy Shaw, Kew Bull. 33: 52 (1978).
Lesser Sunda Is. (Flores). 42 LSI. Phan.

Cleistanthus gabonensis Hutch., Bull. Misc. Inform. Kew 1912: 332 (1912).
Congo, Gabon, Zaire ? 23 CON GAB ZAI? Phan.

Cleistanthus glaber Airy Shaw, Kew Bull. 21: 366 (1968).
Pen. Malaysia, Borneo (Sarawak, Sabah), Philippines. 42 BOR MLY PHI. Phan.

Cleistanthus glabratus Jabl. in H.G.A.Engler, Pflanzenr., IV, 147, VIII: 40 (1915).
Borneo (SW. Sarawak). 42 BOR. Phan.

Cleistanthus glandulosus Jabl. in H.G.A.Engler, Pflanzenr., IV, 147, VIII: 18 (1915).
Pen. Malaysia. 42 MLY.

Cleistanthus gracilis Hook.f., Fl. Brit. India 5: 277 (1887).
S. Thailand, Pen. Malaysia, Borneo, Philippines? 41 THA 42 BOR MLY PHI? Phan.
Cleistanthus dasyphyllus F.N.Williams, Bull. Herb. Boissier, II, 5: 31 (1904).
Cleistanthus blancoi f. *dubius* Jabl. in H.G.A.Engler, Pflanzenr., IV, 147, VIII: 13 (1915).
Cleistanthus gracilis var. *parvifolia* Ridl., Fl. Malay Penins. 3: 190 (1924).

Cleistanthus helferi Hook.f., Fl. Brit. India 5: 280 (1887).
Burma, Thailand. 41 BMA THA. Phan.
Cleistanthus meeboldii Jabl. in H.G.A.Engler, Pflanzenr., IV, 147, VIII: 20 (1915).

Cleistanthus hirsutipetalus Gage, Bull. Misc. Inform. Kew 1914: 239 (1914).
Pen. Malaysia (incl. Singapore). 42 MLY.

Cleistanthus hirsutulus Hook.f., Fl. Brit. India 5: 278 (1887).
Vietnam, Pen. Malaysia, S. Thailand, Borneo (Sabah, NE. Sarawak), Sumatera. 41 THA VIE 42 BOR MLY SUM. Phan.
Cleistanthus siamensis Craib, Bull. Misc. Inform. Kew 1913: 71 (1913). *Paracleisthus siamensis* (Craib) Gagnep., Bull. Soc. Bot. France 70: 497 (1923).
Cleistanthus cochinchinae Jabl. in H.G.A.Engler, Pflanzenr., IV, 147, VIII: 21 (1915).
Cleistanthus penangensis Jabl. in H.G.A.Engler, Pflanzenr., IV, 147, VIII: 21 (1915).

Cleistanthus hylandii Airy Shaw, Kew Bull. 31: 379 (1976).
Queensland. 50 QLD.

Cleistanthus indochinensis Merr. ex Croizat, J. Arnold Arbor. 23: 40 (1942).
Vietnam. 41 VIE.

Cleistanthus inglorius Airy Shaw, Kew Bull. 33: 52 (1978).
Papua New Guinea. 42 NWG. Phan.

Cleistanthus insignis Airy Shaw, Kew Bull. 20: 391 (1966 publ. 1967).
Papua New Guinea (Morobe). 42 NWG. Phan.

Cleistanthus insularis Kaneh., Bot. Mag. (Tokyo) 47: 673 (1933).
Caroline Is. (Palau). 62 CRL.

Cleistanthus inundatus J.Léonard, Bull. Jard. Bot. État 30: 442 (1960).
Zaire, Angola. 23 ZAI 26 ANG. Phan.
Cleistanthus inundatus var. *velutinus* J.Léonard, Bull. Jard. Bot. État 30: 445 (1960).

Cleistanthus isabellinus Elmer, Leafl. Philipp. Bot. 3: 911 (1910).
Philippines (Sibuyan). 42 PHI. Phan.

Cleistanthus itsoghensis Pellegr., Bull. Mus. Natl. Hist. Nat. 34: 467 (1928).
Equatorial Guinea, Gabon, Zaire. 23 EQG GAB ZAI. Phan.

Cleistanthus jacobsianus Airy Shaw, Kew Bull. 32: 382 (1978).
Borneo (Sarawak). 42 BOR. Phan.

Cleistanthus kasaiensis J.Léonard, Bull. Jard. Bot. État 30: 425 (1960).
Cameroon, Gabon, Zaire, Kenya ? 23 CMN GAB ZAI 25 KEN? Phan.
Cleistanthus kasaiensis var. *paulopubescens* J.Léonard, Bull. Jard. Bot. État 30: 427 (1960).

Cleistanthus kingii Jabl. in H.G.A.Engler, Pflanzenr., IV, 147, VIII: 40 (1915).
Pen. Malaysia. 42 MLY. Nanophan. or phan. – Close to *C. myrianthus*.

Cleistanthus kwangensis J.Léonard, Bull. Jard. Bot. État 30: 450 (1960).
Zaire, Angola. 23 ZAI 26 ANG. Phan.

Cleistanthus langkawiensis Airy Shaw, Kew Bull. 33: 47 (1978).
Pen. Malaysia (Langkawi I.). 42 MLY. Phan.

Cleistanthus lanuginosus Jabl. in H.G.A.Engler, Pflanzenr., IV, 147, VIII: 26 (1915).
Pen. Malaysia (Mt. Ophir). 42 MLY. Phan.

Cleistanthus letouzeyi J.Léonard, Adansonia, n.s., 3: 65 (1963).
Cameroon. 23 CMN.

Cleistanthus libericus N.E.Br., J. Linn. Soc., Bot. 37: 113 (1905).
W. & WC. Trop. Africa. 22 GHA IVO LBR NGA SIE 23 GAB GGI ZAI.

Cleistanthus longinervis Airy Shaw, Kew Bull. 29: 309 (1974).
Pen. Malaysia. 42 MLY. Nanophan. or phan.

Cleistanthus macrophyllus Hook.f., Fl. Brit. India 5: 278 (1887).
S. China (Yunnan), W. Malesia. 36 CHC 42 BOR MLY. Phan.

Cleistanthus maingayi Hook.f., Fl. Brit. India 5: 280 (1887).
Pen. Malaysia, Borneo (SW. Sarawak). 42 BOR MLY. Phan.
Cleistanthus borneensis Jabl. in H.G.A.Engler, Pflanzenr., IV, 147, VIII: 25 (1915).

Cleistanthus major Airy Shaw, Kew Bull. 25: 510 (1971).
Pen. Malaysia. 42 MLY. Phan.

Cleistanthus malabaricus (Müll.Arg.) Müll.Arg. in A.P.de Candolle, Prodr. 15(2): 508 (1866).
W. India. 40 IND. Phan.
* *Lebidiera malabarica* Müll.Arg., Linnaea 32: 81 (1863).

Cleistanthus malaccensis Hook.f., Fl. Brit. India 5: 277 (1887).
Pen. Malaysia. 42 MLY. Phan.

Cleistanthus megacarpus C.B.Rob., Philipp. J. Sci., C 6: 323 (1911).
Borneo, Philippines. 42 BOR PHI. Phan.

Cleistanthus membranaceus Hook.f., Fl. Brit. India 5: 278 (1887).
Pen. Malaysia. 42 MLY. Phan.
Cleistanthus stipulatus Hook.f., Fl. Brit. India 5: 281 (1887).

Cleistanthus michelsonii J.Léonard, Bull. Jard. Bot. État 25: 282 (1955).
Zaire. 23 ZAI. Phan.

Cleistanthus micranthus Croizat, J. Arnold Arbor. 26: 98 (1945).
Fiji. 60 FIJ.

Cleistanthus mildbraedii Jabl. in H.G.A.Engler, Pflanzenr., IV, 147, VIII: 51 (1915).
Zaire. 23 ZAI. Phan.

Cleistanthus minahassae Koord., Meded. Lands Plantentuin 19: 625 (1898).
Sulawesi. 42 SUL. Phan.

Cleistanthus mindanaensis C.B.Rob., Philipp. J. Sci., C 6: 324 (1911).
Philippines (Mindanao). 42 PHI. Phan.

Cleistanthus misamisensis C.B.Rob., Philipp. J. Sci., C 6: 325 (1911).
Philippines (Mindanao). 42 PHI. Phan.

Cleistanthus monocarpus R.I.Milne, Kew Bull. 49: 450 (1994).
Borneo (Sabah). 42 BOR.

Cleistanthus monoicus (Lour.) Müll.Arg. in A.P.de Candolle, Prodr. 15(2): 508 (1866).
China (Guangdong). 36 CHS. Nanophan. or phan. – Provisionally accepted.
* *Clutia monoica* Lour., Fl. Cochinch.: 638 (1790).

Cleistanthus morii Kaneh., Bot. Mag. (Tokyo) 46: 456 (1932).
Caroline Is. (Truk). 62 CRL.

Cleistanthus myrianthus (Hassk.) Kurz, Prelim. Rep., App. A: cx (1875).
Trop. Asia to Solomon Is. 40 IND 41 AND 42 BOR JAW MLY NWG PHI SUM 50 QLD 60 SOL. Nanophan. or phan.
* *Nanopetalum myrianthum* Hassk., Verslagen Meded. Afd. Natuurk. Kon. Akad. Wetensch. 4: 141 (1855).
Cleistanthus cupreus Vidal, Revis. Pl. Vasc. Filip.: 235 (1886). *Cleistanthus myrianthus* subsp. *cupreus* (Vidal) Jabl. in H.G.A.Engler, Pflanzenr., IV, 147, VIII: 39 (1915).
Cleistanthus pseudocanescens Elmer, Leafl. Philipp. Bot. 3: 910 (1910).
Cleistanthus pseudocinereus Elmer, Leafl. Philipp. Bot. 3: 910 (1910).
Cleistanthus myrianthus f. *scortechinii* Jabl. in H.G.A.Engler, Pflanzenr., IV, 147, VIII: 39 (1915).
Cleistanthus pseudomyrianthus Jabl. in H.G.A.Engler, Pflanzenr., IV, 147, VIII: 41 (1915).
Cleistanthus castus S.Moore, J. Bot. 63(Suppl.): 93 (1925).
Cleistanthus myrianthus var. *concinnus* Airy Shaw, Kew Bull. 33: 44 (1978).
Cleistanthus myrianthus var. *spicatus* Airy Shaw, Kew Bull. 33: 44 (1978).

Cleistanthus namatanaiensis Jabl. in H.G.A.Engler, Pflanzenr., IV, 147, VIII: 40 (1915).
Bismarck Archip. 42 BIS. Phan.

Cleistanthus ngounyensis Pellegr., Bull. Mus. Natl. Hist. Nat. 34: 467 (1928).
Gabon. 23 GAB.

Cleistanthus nitidus Hook.f., Fl. Brit. India 5: 280 (1887).
Pen. Malaysia (incl. Singapore). 42 MLY. Phan.

Cleistanthus normanbyanus Airy Shaw, Kew Bull. 33: 54 (1978).
New Guinea (D'Entrecasteaux Is.). 42 NWG. Phan.

Cleistanthus oblongatus Airy Shaw, Kew Bull. 33: 50 (1978).
Borneo (Sarawak). 42 BOR. Phan.

Cleistanthus oblongifolius (Roxb.) Müll.Arg. in A.P.de Candolle, Prodr. 15(2): 506 (1866).
Bangladesh. 40 BAN. Nanophan.
* *Clutia oblongifolia* Roxb., Fl. Ind. ed. 1832, 3: 730 (1832). *Bridelia oblongifolius* (Roxb.) Hook. & Arn., Bot. Beechey Voy.: 212 (1837).
Bridelia attenuata Wall. ex Voigt, Hort. Suburb. Calcutt.: 156 (1845), nom. nud.
Bridelia chartacea Wall., Numer. List: 7881 (1847), nom. inval.
Amanoa chartacea Baill. ex Müll.Arg. in A.P.de Candolle, Prodr. 15(2): 507 (1866), pro syn.
Cleistanthus chartaceus Müll.Arg. in A.P.de Candolle, Prodr. 15(2): 507 (1866).
Cleistanthus myrianthus subsp. *attenuatus* Jabl. in H.G.A.Engler, Pflanzenr., IV, 147, VIII: 39 (1915).

Cleistanthus occidentalis (Leandri) Leandri, Naturaliste Malgache 9: 45 (1957).
W. Madagascar. 29 MDG. Phan.
* *Cleistanthus stenonia* var. *occidentalis* Leandri, Notul. Syst. (Paris) 11: 153 (1944).

Cleistanthus orgyialis (Blanco) Merr., Publ. Bur. Sci. Gov. Lab. 27: 75 (1905).
Philippines (Luzon). 42 PHI. Nanophan.
* *Gluta orgyialis* Blanco, Fl. Filip. ed. 2: 451 (1845).

Cleistanthus ovatus C.B.Rob., Philipp. J. Sci., C 3: 194 (1908).
Philippines (Camiguin). 42 PHI. Nanophan. or phan.

Cleistanthus pallidus (Thwaites) Müll.Arg. in A.P.de Candolle, Prodr. 15(2): 508 (1866).
Sri Lanka. 40 SRL. Phan.
* *Amanoa pallida* Thwaites, Enum. Pl. Zeyl.: 280 (1861).
Cleistanthus pallidus var. *subglaucus* Trimen, Syst. Cat. Fl. Pl. Ceylon: 78 (1885).
Cleistanthus subglaucus Thwaites ex Trimen, Syst. Cat. Fl. Pl. Ceylon: 78 (1885).
Cleistanthus subpallidus Thwaites ex Hook.f., Fl. Brit. India 5: 279 (1887).

Cleistanthus papuanus (Lauterb.) Jabl. in H.G.A.Engler, Pflanzenr., IV, 147, VIII: 42 (1915).
Papua New Guinea. 42 NWG. Nanophan. or phan.
* *Schistostigma papuanum* Lauterb. in K.M.Schumann & C.A.G.Lauterbach, Fl. Schutzgeb. Südsee, Nachtr.: 299 (1905).

Cleistanthus papyraceus Airy Shaw, Kew Bull. 21: 368 (1968).
Burma, Thailand. 41 BMA THA. Nanophan. or phan.

Cleistanthus parvifolius Hook.f., Fl. Brit. India 5: 281 (1887).
Pen. Malaysia. 42 MLY. Phan.

Cleistanthus patulus (Roxb.) Müll.Arg. in A.P.de Candolle, Prodr. 15(2): 505 (1866).
E. India, Sri Lanka. 40 IND SRL. Phan.
* *Clutia patula* Roxb., Pl. Coromandel 2: 37 (1802). *Bridelia patula* (Roxb.) Hook. & Arn., Bot. Beechey Voy.: 212 (1837). *Amanoa patula* (Roxb.) Thwaites, Enum. Pl. Zeyl.: 280 (1861).
Amanoa indica Wight, Icon. Pl. Ind. Orient. 5: t. 1911 (1852).
Amanoa indica f. *minor* Thwaites, Enum. Pl. Zeyl.: 428 (1864).

Cleistanthus paxii Jabl. in H.G.A.Engler, Pflanzenr., IV, 147, VIII: 27 (1915).
Borneo (SW. Sarawak). 42 BOR. Nanophan. or phan.

Cleistanthus pedicellatus Hook.f., Fl. Brit. India 5: 281 (1887).
S. China (Guangxi), Pen. Malaysia, Borneo (Sarawak, Sabah), Philippines, New Guinea. 36 CHS 42 BOR MLY NWG PHI. Phan.

Cleistanthus integer C.B.Rob., Philipp. J. Sci., C 3: 196 (1908).
Cleistanthus quadrifidus C.B.Rob., Philipp. J. Sci., C 3: 197 (1908).
Cleistanthus dichotomus J.J.Sm., Nova Guinea 8: 786 (1912).

Cleistanthus peninsularis Airy Shaw & B.Hyland, Muelleria 4: 222 (1980).
Papua New Guinea, Queensland. 42 NWG 50 QLD.

Cleistanthus perrieri Leandri, Naturaliste Malgache 9: 44 (1957).
E. Madagascar. 29 MDG. Nanophan. or phan.
Cleistanthus perrieri var. *minor* Leandri, Naturaliste Malgache 9: 44 (1957).

Cleistanthus petelotii Merr. ex Croizat, J. Arnold Arbor. 23: 40 (1942).
N. Vietnam, S. China (W. Guangxi). 36 CHS 41 VIE. Phan.

Cleistanthus pierlotii J.Léonard, Bull. Jard. Bot. État 30: 427 (1960).
Zaire. 23 ZAI. Phan.

Cleistanthus pierrei (Gagnep.) Croizat, J. Arnold Arbor. 23: 39 (1942).
Vietnam. 41 VIE.
* *Paracleisthus pierrei* Gagnep., Bull. Soc. Bot. France 70: 500 (1923).

Cleistanthus pilosus C.B.Rob., Philipp. J. Sci., C 6: 326 (1911).
Philippines (Basilan). 42 PHI. Phan.

Cleistanthus podocarpus Hook.f., Fl. Brit. India 5: 281 (1887).
Pen. Malaysia. 42 MLY. Phan.

Cleistanthus podopyxis Airy Shaw, Kew Bull. 21: 363 (1968).
Borneo (Sabah, E. Kalimantan). 42 BOR. Phan.

Cleistanthus polyneurus Airy Shaw, Kew Bull. 33: 48 (1978).
Pen. Malaysia. 42 MLY. Phan.

Cleistanthus polyphyllus F.N.Williams, Bull. Herb. Boissier, II, 5: 31 (1904).
Pen. Malaysia. 42 MLY. Phan.
Cleistanthus trichocarpa Ridl., J. Straits Branch Roy. Asiat. Soc. 59: 167 (1911).

Cleistanthus polystachyus Hook.f. ex Planch., Hooker's Icon. Pl. 8: t. 779 (1848).
Candelabria polystachya (Hook.f. ex Planch.) Planch., Ann. Sci. Nat., Bot., IV, 2: 264 (1854). *Bridelia polystachya* (Hook.f. ex Planch.) Baill., Étude Euphorb.: 584 (1858).
Trop. Africa. 22 IVO LBR SIE 23 CMN ZAI 25 KEN TAN UGA 26 ANG MLW MOZ ZAM ZIM. Nanophan. or phan.

subsp. **milleri** (Dunkley) Radcl.-Sm., Kew Bull. 51: 303 (1996).
SC. Trop. Africa to Zaire. 23 ZAI 26 MLW MOZ ZAM ZIM. Nanophan. or phan.
Cleistanthus apetalus S.Moore, J. Linn. Soc., Bot. 40: 191 (1911).
* *Cleistanthus milleri* Dunkley, Bull. Misc. Inform. Kew 1937: 468 (1937).

subsp. **polystachyus**
Trop. Africa. 22 IVO LBR SIE 23 CMN ZAI 25 KEN TAN UGA 26 ANG. Nanophan. or phan.
Cleistanthus angolensis Welw. ex Müll.Arg., J. Bot. 2: 339 (1864).
Cleistanthus amaniensis Jabl. in H.G.A.Engler, Pflanzenr., IV, 147, VIII: 46 (1915).

Cleistanthus praetermissus Gage, Bull. Misc. Inform. Kew 1914: 240 (1914).
S. Thailand, Pen. Malaysia, Borneo (Sabah). 41 THA 42 BOR MLY. Phan.

Cleistanthus pseudopallidus Jabl. in H.G.A.Engler, Pflanzenr., IV, 147, VIII: 23 (1915).
Sumatera. 42 SUM.

Cleistanthus pseudopodocarpus Jabl. in H.G.A.Engler, Pflanzenr., IV, 147, VIII: 29 (1915).
Borneo (Sarawak, Sabah), Sumatera, Pen. Malaysia. 42 BOR MLY SUM. Phan.

var. **leptopus** Airy Shaw, Kew Bull. 27: 76 (1972).
Borneo (Sarawak). 42 BOR. Nanophan. or phan.

var. **pseudopodocarpus**
Borneo (Sarawak, Sabah), Sumatera, Pen. Malaysia. 42 BOR MLY SUM. Phan.
Cleistanthus rufescens Jabl. in H.G.A.Engler, Pflanzenr., IV, 147, VIII: 30 (1915).

Cleistanthus pubens Airy Shaw, Kew Bull. 21: 365 (1968).
Borneo. 42 BOR. Nanophan. or phan.

Cleistanthus pyrrhocarpus Airy Shaw, Kew Bull. 21: 370 (1968).
Borneo (Sarawak, Brunei). 42 BOR. Phan.

Cleistanthus racemosus Pierre ex Hutch. in D.Oliver, Fl. Trop. Afr. 6(1): 1048 (1913).
Congo, Gabon. 23 CON GAB. Phan. – Close to *C. polystachyus*.

Cleistanthus ripicola J.Léonard, Bull. Jard. Bot. État 30: 438 (1960).
W. & WC. Trop. Africa. 22 GHA IVO NGA 23 CAF ZAI. Phan.

Cleistanthus robinsonii Elmer, Leafl. Philipp. Bot. 3: 909 (1910).
Philippines (Sibuyan). 42 PHI. Phan.

Cleistanthus robustus (Thwaites) Müll.Arg. in A.P.de Candolle, Prodr. 15(2): 504 (1866).
Sri Lanka. 40 SRL.
Amanoa indica Thwaites, Enum. Pl. Zeyl.: 428 (1864), nom. illeg.
* *Amanoa robusta* Thwaites, Enum. Pl. Zeyl.: 428 (1864).

Cleistanthus rotundatus Jabl. in H.G.A.Engler, Pflanzenr., IV, 147, VIII: 21 (1915).
Godefroya rotundata (Jabl.) Gagnep., Bull. Soc. Bot. France 70: 435 (1923).
Cambodia. 41 CBD. Phan.

Cleistanthus rufus (Hook.f.) Gehrm., Bot. Jahrb. Syst. 41(95): 2 (1908).
Pen. Malaysia. 42 MLY. Nanophan. or phan.
* *Bridelia rufa* Hook.f., Fl. Brit. India 5: 273 (1887).

Cleistanthus salicifolius Airy Shaw, Kew Bull. 33: 46 (1978).
Borneo (Sarawak). 42 BOR. Phan.

Cleistanthus samarensis Merr., Philipp. J. Sci., C 9: 475 (1914 publ. 1915).
Philippines (Samar). 42 PHI.

Cleistanthus sankunnianus Sivar. & Balach., Kew Bull. 40: 121 (1985).
India. 40 IND.

Cleistanthus sarawakensis Jabl. in H.G.A.Engler, Pflanzenr., IV, 147, VIII: 34 (1915).
Borneo (Sarawak). 42 BOR.

Cleistanthus schlechteri (Pax) Hutch. in W.H.Harvey, Fl. Cap. 5(2): 382 (1915).
Kenya to KwaZulu-Natal. 25 KEN TAN 26 MLW MOZ ZIM 27 NAT TVL.
* *Securinega schlechteri* Pax, Bot. Jahrb. Syst. 28: 18 (1899).

var. **pubescens** (Hutch.) J.Léonard, Bull. Jard. Bot. État 30: 423 (1960).
Kenya to Mozambique. 25 KEN TAN 26 MOZ ZIM.
* *Cleistanthus johnsonii* var. *pubescens* Hutch., Bull. Misc. Inform. Kew 1909: 380 (1909).
Cleistanthus holtzii var. *pubescens* (Hutch.) Hutch. in D.Oliver, Fl. Trop. Afr. 6(1): 623 (1912).

var. **schlechteri**
Kenya to KwaZulu-Natal. 25 KEN TAN 26 MLW MOZ ZIM 27 NAT TVL.
Cleistanthus holtzii Pax, Bot. Jahrb. Syst. 43: 77 (1909).
Cleistanthus johnsonii Hutch., Bull. Misc. Inform. Kew 1909: 380 (1909).

Cleistanthus semiopacus F.Muell. ex Benth., Fl. Austral. 6: 123 (1873).
Queensland. 50 QLD. Phan.
Cleistanthus semiopacus var. *curvaminis* Airy Shaw, Kew Bull. 31: 380 (1976).

Cleistanthus stenonia (Baill.) Jabl. in H.G.A.Engler, Pflanzenr., IV, 147, VIII: 50 (1915).
Madagascar, Comoros (Mayotte). 29 COM MDG. Phan.
* *Amanoa stenonia* Baill. in A.Grandidier, Hist. Phys. Madagascar, Atlas: 211 (1892).

Cleistanthus stenophyllus Kurz, J. Asiat. Soc. Bengal, Pt. 2, Nat. Hist. 42(2): 242 (1874).
Burma. 41 BMA.
Cleistanthus lancifolius Hook.f., Fl. Brit. India 5: 277 (1887).

Cleistanthus stipitatus (Baill.) Müll.Arg. in A.P.de Candolle, Prodr. 15(2): 506 (1866).
New Caledonia. 60 NWC. Phan.
Bridelia buxifolia Baill., Adansonia 2: 230 (1862). *Cleistanthus stipitatus* f. *buxifolia* (Baill.) Guillaumin, Arch. Bot. Mém. 2(3): 28 (1929).
Bridelia laurina Baill., Adansonia 2: 229 (1862). *Cleistanthus stipitatus* var. *laurinus* (Baill.) Müll.Arg. in A.P.de Candolle, Prodr. 15(2): 507 (1866).
* *Bridelia stipitata* Baill., Adansonia 2: 229 (1862).
Cleistanthus stipitatus var. *hypoleucus* Müll.Arg. in A.P.de Candolle, Prodr. 15(2): 507 (1866).
Cleistanthus stipitatus f. *subcanescens* Jabl. in H.G.A.Engler, Pflanzenr., IV, 147, VIII: 36 (1915).

Cleistanthus stipularis (Kuntze) Müll.Arg. in A.P.de Candolle, Prodr. 15(2): 509 (1866).
W. India. 40 IND. – Provisionally accepted.
* *Bridelia stipularis* Hook. & Arn., Bot. Beechey Voy.: 211 (1837), nom. illeg.
Kaluhaburunghos stipularis Kuntze, Revis. Gen. Pl. 2: 607 (1891).

Cleistanthus striatus Airy Shaw, Kew Bull. 23: 63 (1969).
Borneo (Sarawak, Sabah). 42 BOR. Phan.

Cleistanthus suarezensis Leandri, Naturaliste Malgache 9: 46 (1957).
NW. Madagascar. 29 MDG. Phan.

Cleistanthus sumatranus (Miq.) Müll.Arg. in A.P.de Candolle, Prodr. 15(2): 504 (1866).
China, Trop. Asia. 36 CHC CHH CHS 41 CBD THA VIE 42 BOR JAW LSI MLY MOL NWG PHI SUM. Nanophan. or phan.
* *Leiopyxis sumatrana* Miq., Fl. Ned. Ind., Eerste Bijv.: 446 (1861).
Cleistanthus ferrugineus Fern.-Vill. in F.M.Blanco, Fl. Filip., ed. 3, 4(21A): 187 (1880), nom. illeg.
Cleistanthus blancoi Rolfe, J. Linn. Soc., Bot. 21: 315 (1884).
Cleistanthus heterophyllus Hook.f., Fl. Brit. India 5: 276 (1887).
Cleistanthus laevis Hook.f., Fl. Brit. India 5: 277 (1887).
Cleistanthus vidalii C.B.Rob., Philipp. J. Sci., C 3: 193 (1908).

Cleistanthus laevigatus Jabl. in H.G.A.Engler, Pflanzenr., IV, 147, VIII: 12 (1915).
Cleistanthus oligophlebius Merr., Philipp. J. Sci., C 13: 80 (1918).
Cleistanthus saichikii Merr., Philipp. J. Sci. 23: 248 (1923).
Paracleisthus subgracilis Gagnep., Bull. Soc. Bot. France 70: 500 (1923).
Cleistanthus euphlebius Merr., Pap. Michigan Acad. Sci. 24: 78 (1938 publ. 1939).

Cleistanthus tenerifolius Airy Shaw, Kew Bull. 37: 380 (1982).
Sumatera, Borneo (Kalimantan). 42 BOR SUM.

Cleistanthus tomentosus Hance, J. Bot. 15: 337 (1877).
Thailand, Vietnam, China (Guangdong, Hainan). 36 CHH CHS 41 THA VIE. Phan.
Cleistanthus eburneus Gagnep., Bull. Soc. Bot. France 70: 501 (1923).
Cleistanthus eburneus var. *sordidus* Gagnep., Bull. Soc. Bot. France 70: 502 (1923).

Cleistanthus tonkinensis Jabl. in H.G.A.Engler, Pflanzenr., IV, 147, VIII: 16 (1915).
Paracleisthus tonkinensis (Jabl.) Gagnep., Bull. Soc. Bot. France 70: 499 (1923).
Vietnam, S. China. 36 CHC CHS 41 VIE. Phan.

Cleistanthus travancorensis Jabl. in H.G.A.Engler, Pflanzenr., IV, 147, VIII: 21 (1915).
India. 40 IND. Phan.

Cleistanthus venosus C.B.Rob., Philipp. J. Sci., C 3: 192 (1908).
Pen. Malaysia, Borneo (Sabah), Sumatera, Philippines (Mindanao). 42 BOR MLY PHI SUM. Phan.
Cleistanthus elmeri Merr., Univ. Calif. Publ. Bot. 15: 154 (1929).

Cleistanthus vestitus Jabl. in H.G.A.Engler, Pflanzenr., IV, 147, VIII: 32 (1915).
Pen. Malaysia, Borneo, Sumatera, Philippines. 42 BOR MLY PHI SUM. Nanophan. or phan.
Cleistanthus vestitus f. *perakensis* Jabl. in H.G.A.Engler, Pflanzenr., IV, 147, VIII: 32 (1915). *Cleistanthus perakensis* (Jabl.) Gage ex Ridl., Fl. Malay Penins. 3: 195 (1924).
Cleistanthus barrosii Merr., Philipp. J. Sci. 20: 400 (1922).

Cleistanthus willmannianus J.Léonard, Bull. Jard. Bot. État 30: 430 (1960).
Gabon. 23 GAB.

Cleistanthus winkleri Jabl. in H.G.A.Engler, Pflanzenr., IV, 147, VIII: 12 (1915).
Borneo (Sarawak, Brunei). 42 BOR. Phan.

Cleistanthus xanthopus Airy Shaw, Kew Bull. 20: 392 (1966 publ. 1967).
E. Papua New Guinea. 42 NWG. Phan.

Cleistanthus xerophilus Domin, Biblioth. Bot. 89: 325 (1927).
Queensland. 50 QLD.
Cleistanthus densiflorus C.T.White, Proc. Roy. Soc. Queensland 55: 82 (1944).

Cleistanthus zenkeri Jabl. in H.G.A.Engler, Pflanzenr., IV, 147, VIII: 49 (1915).
S. Cameroon to Cabinda. 23 CAB CMN GAB ZAI. Phan.

Synonyms:
Cleistanthus albidiscus Ridl. === **Phyllanthus albidiscus** (Ridl.) Airy Shaw
Cleistanthus amaniensis Jabl. === **Cleistanthus polystachyus** Hook.f. ex Planch. subsp. **polystachyus**
Cleistanthus angolensis Welw. ex Müll.Arg. === **Cleistanthus polystachyus** Hook.f. ex Planch. subsp. **polystachyus**
Cleistanthus apetalus S.Moore === **Cleistanthus polystachyus** subsp. **milleri** (Dunkley) Radcl.-Sm.
Cleistanthus barrosii Merr. === **Cleistanthus vestitus** Jabl.

Cleistanthus blancoi Rolfe === **Cleistanthus sumatranus** (Miq.) Müll.Arg.
Cleistanthus blancoi f. *dubius* Jabl. === **Cleistanthus gracilis** Hook.f.
Cleistanthus boivinianus f. *humbertii* Leandri === **Cleistanthus boivinianus** (Baill.) Müll.Arg.
Cleistanthus boivinianus f. *occidentalis* Leandri === **Cleistanthus boivinianus** (Baill.) Müll.Arg.
Cleistanthus boivinianus f. *orientalis* Leandri === **Cleistanthus boivinianus** (Baill.) Müll.Arg.
Cleistanthus borneensis Jabl. === **Cleistanthus maingayi** Hook.f.
Cleistanthus castus S.Moore === **Cleistanthus myrianthus** (Hassk.) Kurz
Cleistanthus chartaceus Müll.Arg. === **Cleistanthus oblongifolius** (Roxb.) Müll.Arg.
Cleistanthus cochinchinae Jabl. === **Cleistanthus hirsutulus** Hook.f.
Cleistanthus cupreus Vidal === **Cleistanthus myrianthus** (Hassk.) Kurz
Cleistanthus dasyphyllus F.N.Williams === **Cleistanthus gracilis** Hook.f.
Cleistanthus densiflorus C.T.White === **Cleistanthus xerophilus** Domin
Cleistanthus dichotomus J.J.Sm. === **Cleistanthus pedicellatus** Hook.f.
Cleistanthus dongfangensis (P.T.Li) H.S.Kiu === **Cleistanthus concinnus** Croizat
Cleistanthus dubius Ridl. === **Phyllanthus roseus** (Craib & Hutch.) Beille
Cleistanthus eburneus Gagnep. === **Cleistanthus tomentosus** Hance
Cleistanthus eburneus var. *sordidus* Gagnep. === **Cleistanthus tomentosus** Hance
Cleistanthus elmeri Merr. === **Cleistanthus venosus** C.B.Rob.
Cleistanthus euphlebius Merr. === **Cleistanthus sumatranus** (Miq.) Müll.Arg.
Cleistanthus ferrugineus Fern.-Vill. === **Cleistanthus sumatranus** (Miq.) Müll.Arg.
Cleistanthus glaucus Hiern === **Pseudolachnostylis maprouneifolia** var. **glabra** (Pax) Brenan
Cleistanthus glaucus Jabl. === **Cleistanthus baramicus** Jabl.
Cleistanthus gracilis var. *parvifolia* Ridl. === **Cleistanthus gracilis** Hook.f.
Cleistanthus heterophyllus Hook.f. === **Cleistanthus sumatranus** (Miq.) Müll.Arg.
Cleistanthus holtzii Pax === **Cleistanthus schlechteri** (Pax) Hutch. var. **schlechteri**
Cleistanthus holtzii var. *pubescens* (Hutch.) Hutch. === **Cleistanthus schlechteri** var. **pubescens** (Hutch.) J.Léonard
Cleistanthus integer C.B.Rob. === **Cleistanthus pedicellatus** Hook.f.
Cleistanthus inundatus var. *velutinus* J.Léonard === **Cleistanthus inundatus** J.Léonard
Cleistanthus johnsonii Hutch. === **Cleistanthus schlechteri** (Pax) Hutch. var. **schlechteri**
Cleistanthus johnsonii var. *pubescens* Hutch. === **Cleistanthus schlechteri** var. **pubescens** (Hutch.) J.Léonard
Cleistanthus kasaiensis var. *paulopubescens* J.Léonard === **Cleistanthus kasaiensis** J.Léonard
Cleistanthus laevigatus Jabl. === **Cleistanthus sumatranus** (Miq.) Müll.Arg.
Cleistanthus laevis Hook.f. === **Cleistanthus sumatranus** (Miq.) Müll.Arg.
Cleistanthus lanceolatus Müll.Arg. === **Bridelia tomentosa** Blume
Cleistanthus lancifolius Hook.f. === **Cleistanthus stenophyllus** Kurz
Cleistanthus longipedicellatus Merr. === **Actephila longipedicellata** (Merr.) Croizat
Cleistanthus mattangensis Jabl. === **Cleistanthus decurrens** Hook.f.
Cleistanthus meeboldii Jabl. === **Cleistanthus helferi** Hook.f.
Cleistanthus milleri Dunkley === **Cleistanthus polystachyus** subsp. **milleri** (Dunkley) Radcl.-Sm.
Cleistanthus minutiflorus Ridl. === **Phyllanthus ridleyanus** Airy Shaw
Cleistanthus myrianthoides C.B.Rob. === **Bridelia glauca** Blume
Cleistanthus myrianthus subsp. *attenuatus* Jabl. === **Cleistanthus oblongifolius** (Roxb.) Müll.Arg.
Cleistanthus myrianthus var. *concinnus* Airy Shaw === **Cleistanthus myrianthus** (Hassk.) Kurz
Cleistanthus myrianthus subsp. *cupreus* (Vidal) Jabl. === **Cleistanthus myrianthus** (Hassk.) Kurz
Cleistanthus myrianthus f. *scortechinii* Jabl. === **Cleistanthus myrianthus** (Hassk.) Kurz
Cleistanthus myrianthus var. *spicatus* Airy Shaw === **Cleistanthus myrianthus** (Hassk.) Kurz
Cleistanthus oblongifolius var. *scaber* Müll.Arg. === **Bridelia glauca** Blume
Cleistanthus oligophlebius Merr. === **Cleistanthus sumatranus** (Miq.) Müll.Arg.
Cleistanthus pallidus var. *subcordatus* J.J.Sm. === **Cleistanthus brideliifolius** C.B.Rob. var. **brideliifolius**
Cleistanthus pallidus var. *subglaucus* Trimen === **Cleistanthus pallidus** (Thwaites) Müll.Arg.

Cleistanthus penangensis Jabl. === **Cleistanthus hirsutulus** Hook.f.
Cleistanthus perakensis (Jabl.) Gage ex Ridl. === **Cleistanthus vestitus** Jabl.
Cleistanthus perrieri var. *minor* Leandri === **Cleistanthus perrieri** Leandri
Cleistanthus pseudocanescens Elmer === **Cleistanthus myrianthus** (Hassk.) Kurz
Cleistanthus pseudocinereus Elmer === **Cleistanthus myrianthus** (Hassk.) Kurz
Cleistanthus pseudomyrianthus Jabl. === **Cleistanthus myrianthus** (Hassk.) Kurz
Cleistanthus quadrifidus C.B.Rob. === **Cleistanthus pedicellatus** Hook.f.
Cleistanthus rufescens Jabl. === **Cleistanthus pseudopodocarpus** Jabl. var. **pseudopodocarpus**
Cleistanthus saichikii Merr. === **Cleistanthus sumatranus** (Miq.) Müll.Arg.
Cleistanthus semiopacus var. *curvaminis* Airy Shaw === **Cleistanthus semiopacus** F.Muell. ex Benth.
Cleistanthus siamensis Craib === **Cleistanthus hirsutulus** Hook.f.
Cleistanthus stenonia var. *occidentalis* Leandri === **Cleistanthus occidentalis** (Leandri) Leandri
Cleistanthus stipitatus f. *buxifolia* (Baill.) Guillaumin === **Cleistanthus stipitatus** (Baill.) Müll.Arg.
Cleistanthus stipitatus var. *hypoleucus* Müll.Arg. === **Cleistanthus stipitatus** (Baill.) Müll.Arg.
Cleistanthus stipitatus var. *laurinus* (Baill.) Müll.Arg. === **Cleistanthus stipitatus** (Baill.) Müll.Arg.
Cleistanthus stipitatus f. *subcanescens* Jabl. === **Cleistanthus stipitatus** (Baill.) Müll.Arg.
Cleistanthus stipulatus Hook.f. === **Cleistanthus membranaceus** Hook.f.
Cleistanthus subcordatus (J.J.Sm.) Jabl. === **Cleistanthus brideliifolius** C.B.Rob. var. **brideliifolius**
Cleistanthus subglaucus Thwaites ex Trimen === **Cleistanthus pallidus** (Thwaites) Müll.Arg.
Cleistanthus subpallidus Thwaites ex Hook.f. === **Cleistanthus pallidus** (Thwaites) Müll.Arg.
Cleistanthus trichocarpa Ridl. === **Cleistanthus polyphyllus** F.N.Williams
Cleistanthus vestitus f. *perakensis* Jabl. === **Cleistanthus vestitus** Jabl.
Cleistanthus vidalii C.B.Rob. === **Cleistanthus sumatranus** (Miq.) Müll.Arg.

Cleodora

Synonyms:
Cleodora Klotzsch === **Croton** L.

Cleopatra

Synonyms:
Cleopatra Pancher ex Croizat === **Neoguillauminia** Croizat

Clistanthus

An orthographic variant of *Cleistanthus*.

Clistranthus

Synonyms:
Clistranthus Poit. ex Baill. === **Pera** Mutis

Clonostachys

Synonyms:
Clonostachys Klotzsch === **Sebastiania** Spreng.

Clonostylis

1 species, Malesia (Sumatera). Reduced to *Spathiostemon* by Airy-Shaw (1978) where long confused with *S. moniliformis* but restored by van Welzen (1998). (Acalyphoideae)

Airy-Shaw, H. K. (1978). Notes on Malesian and other Asiatic Euphorbiaceae, CC. *Clonostylis* S. Moore reduced to *Spathiostemon* Bl. Kew Bull. 32: 407-408. En. — Reduction of genus and transfer of *C. forbesii*; key to 3 species.

- Welzen, P. C. van (1998). Revisions and phylogenies of Malesian Euphorbiaceae subtribe Lasiococcinae (*Homonoia, Lasiococca, Spathiostemon*) and *Clonostylis, Ricinus* and *Wetria*. Blumea 43: 131-164. (*Lasiococca* with Nguyen Nghia Thin & Vu Hoai Duc.) En. — General introduction, with history of studies; a note of caution on the use as a character of monoecy vs. dioecy given the imperfect state of much material; phylogeny with character table and cladogram; revision of *Clonostylis* (pp. 150-151; 1 species) with description, synonymy, type, literature citations, indication of distribution, ecology and habitat, and other notes; identification list at end. [The single species is still known only from the type, *Forbes 3027*, from the upper Palembang basin in Sumatera Selatan. Its reduction to *Spathiostemon moniliformis* by Airy-Shaw was unfortuntely wrong.]

Clonostylis S.Moore, J. Bot. 63 (Suppl.): 101 (1925).
W. Malesia. 42.

Clonostylis forbesii S.Moore, J. Bot. 63 (Suppl.): 101 (1925). *Spathiostemon forbesii* (S.Moore) Airy Shaw, Kew Bull. 32: 407 (1978). – FIGURE, p. 376.
Sumatera. 42 SUM. Phan. – Known only from the type.

Clusiophyllum

Synonyms:
Clusiophyllum Müll.Arg. === **Micrandra** Benth.
Clusiophyllum sprucei Müll.Arg. === **Micrandra sprucei** (Müll.Arg.) R.E.Schult.

Clutia

55 species, Africa (2 also in Arabian Peninsula) but in addition; well developed in S Africa, 24 species occur in tropical Africa. Together southern and eastern Africa have all eight sections. Shrubs or small trees of various sizes with some in habit resembling *Bridelia* and *Phyllanthus* as well as *Chrozophora*. Some, such as the variable *C. alaternoides*, are 'Cape' heaths. Most treatments since 1911-19 have been floristic; however, there has been a fair amount of reduction of names in more recent decades. No study of Southern African species has, however, appeared since that of N.E. Brown in *Flora capensis* (1920)The genus has since the time of Mueller (1866; see **General**) been recognised as somewhat distinctive and by Webster (Synopsis, 1994) assigned uniquely to Clutieae. To quote Radcliffe-Smith (1992) in his review of the genus, 'its proper placement within the Euphorbiaceae is somewhat problematical'. (Acalyphoideae)

- Pax, F. (with K. Hoffmann) (1911). *Cluytia*. In A. Engler (ed.), Das Pflanzenreich, IV 147 III (Euphorbiaceae-Cluytieae): 50-83, illus. Berlin. (Heft 47.) La/Ge. — 48 species in 8 sections, 'fere omnes africanae'. The distribution of the sections is mapped (p. 52).

Pax, F. (with K. Hoffmann) (1914). *Clutia*. In A. Engler (ed.), Das Pflanzenreich, IV 147 VII [Euphorbiaceae-Additamentum V]: 404-406. Berlin. (Heft 63.) La/Ge. — 10 species. Further additions in ibid., XIV (Additamentum VI): 39-40 (1919).

Léonard, J. (1961). Notulae systematicae XXXII. Observations sur des espèces africaines de *Clutia, Ricinodendron* et *Sapium* (Euphorbiacées). Bull. Jard. Bot. État 31: 391-406. Fr. —*Clutia*, pp. 391-396, with one new species; no keys.

Clonostylis forbesii S. Moore (as *Spathiostemon forbesii* (S. Moore) Airy Shaw)
Artist: Mary Grierson
Ic. Pl. 38(1): pl. 3720 (1974)
KEW ILLUSTRATIONS COLLECTION

Mtotomwena, K. & R. L. A. Mahunnah (1985). Pollen grain studies on the genus *Clutia* L. (Euphorbiaceae) in Tanzania with reference to infrageneric taxonomy. J. Econ. Taxon. Bot. 7: 153-166, illus. En. — Introduction and prior studies; list of species studied; results. A proposal is made that *C. abyssinica* s.l. should be split into three species; a new combination is made for one of them. No formal taxonomic account is presented.

Radcliffe-Smith, A. (1992). Notes on African Euphorbiaceae, XXVII. *Clutia*. Kew Bull. 47: 111-119. En. — Novelties and notes precursory to *Flora Zambesiaca*, with one new species; includes a brief history of the genus.

Clutia Boerh. ex L., Sp. Pl.: 1042 (1753).
Trop. & S. Africa, Arabian Pen. 22 23 24 25 27 28 35.
Altora Adans., Fam. Pl. 2: 356 (1763).
Cratochwilia Neck., Elem. Bot. 2: 339 (1790).
Middelbergia Schinz ex Pax in H.G.A.Engler, Pflanzenr., IV, 147, III: 66 (1911).

Clutia abyssinica Jaub. & Spach, Ill. Pl. Orient. 5: 57 (1855).
Ethiopia to S. Africa. 23 BUR RWA ZAI 24 ETH SOM SUD 25 KEN TAN UGA 26 ANG MLW MOZ ZAM ZIM 27 CPP NAT. Cham. or phan.

var. **abyssinica** – FIGURE, p. 378.
Sudan to S. Africa. 23 ZAI 24 ETH SOM SUD 25 KEN TAN UGA 26 ANG MLW MOZ ZAM ZIM 27 CPP NAT. Cham. or phan.
Clutia lanceolata var. *glabra* A.Rich., Tent. Fl. Abyss. 2: 253 (1850).
Clutia abyssinica var. *deserticola* Volkens, Bot. Jahrb. Syst. 23: 531 (1897).
Clutia glabrescens Knauf, Bot. Jahrb. Syst. 30: 340 (1901).
Clutia abyssinica var. *calvescens* Pax & K.Hoffm. in H.G.A.Engler, Pflanzenr., IV, 147, III: 57 (1911).
Clutia abyssinica var. *firma* Pax & K.Hoffm. in H.G.A.Engler, Pflanzenr., IV, 147, III: 57 (1911).
Clutia abyssinica var. *glabra* Pax in H.G.A.Engler, Pflanzenr., IV, 147, III: 57 (1911).
Clutia anomala Pax & K.Hoffm. in H.G.A.Engler, Pflanzenr., IV, 147, VII: 405 (1914).

var. **pedicellaris** (Pax) Pax in H.G.A.Engler, Pflanzenr., IV, 147, III: 57 (1911).
Ethiopia to Malawi. 23 BUR RWA ZAI 24 ETH 25 KEN TAN UGA 26 MLW. Cham. or phan.
* *Clutia richardiana* var. *pedicellaris* Pax, Bot. Jahrb. Syst. 23: 531 (1897). *Clutia pedicellaris* (Pax) Hutch. in D.Oliver, Fl. Trop. Afr. 6(1): 806 (1912).

var. **usambarica** Pax & K.Hoffm. in H.G.A.Engler, Pflanzenr., IV, 147, III: 57 (1911).
E. Trop. Africa. 25 KEN TAN. Cham. or phan.
Clutia leuconeura Pax, Bot. Jahrb. Syst. 19: 112 (1894).
Clutia mollis Pax, Bot. Jahrb. Syst. 19: 112 (1894).
Clutia rotundifolia Pax, Bot. Jahrb. Syst. 43: 85 (1909).
Clutia abyssinica var. *ovalifolia* Pax & K.Hoffm. in H.G.A.Engler, Pflanzenr., IV, 147, III: 57 (1911).

Clutia affinis Sond., Linnaea 23: 126 (1850). *Clutia affinis* var. *genuina* Müll.Arg. in A.P.de Candolle, Prodr. 15(2): 1050 (1866), nom. inval.
S. Africa. 27 CPP NAT SWZ TVL. Nanophan.
Clutia hirsuta Eckl. & Zeyh. ex Sond., Linnaea 23: 126 (1850), pro syn.
Clutia pubescens Eckl. & Zeyh. ex Sond., Linnaea 23: 126 (1850), nom. illeg.
Clutia phyllanthifolia Baill., Adansonia 3: 153 (1863). *Clutia affinis* var. *phyllanthifolia* (Baill.) Müll.Arg. in A.P.de Candolle, Prodr. 15(2): 1051 (1866).

Clutia africana Poir. in J.B.A.M.de Lamarck, Encycl., Suppl. 2: 302 (1811).
Cape Prov. 27 CPP.
Clutia alaternoides var. *major* Krauss, Flora 28: 82 (1845).
Clutia alaternoides var. *latifolia* Sond., Linnaea 23: 127 (1850).

Clutia abyssinica Jaub. & Spach var. *abyssinica*
Artist: Pat Halliday
Fl. Trop. East Africa, Euphorbiaceae, 1: 335, fig. 63 (1987)
KEW ILLUSTRATIONS COLLECTION

Clutia alaternoides L., Sp. Pl.: 1042 (1753). *Clutia alaternoides* var. *genuina* Müll.Arg. in A.P.de Candolle, Prodr. 15(2): 1048 (1866), nom. inval.
Cape Prov. 27 CPP. Cham.
Clutia polygalifolia Salisb., Prodr. Stirp. Chap. Allerton: 390 (1796).
Clutia alaternoides f. *lanceolata* Sond., Linnaea 23: 128 (1850). *Clutia alaternoides* var. *lanceolata* (Sond.) Müll.Arg. in A.P.de Candolle, Prodr. 15(2): 1048 (1866).
Clutia floribunda Baill., Étude Euphorb., Atlas: 30 (1858).
Clutia alaternoides f. *brevifolia* Müll.Arg. in A.P.de Candolle, Prodr. 15(2): 1048 (1866).
Clutia alaternoides f. *imbricata* Müll.Arg. in A.P.de Candolle, Prodr. 15(2): 1048 (1866).
Clutia alaternoides f. *leptophylla* Müll.Arg. in A.P.de Candolle, Prodr. 15(2): 1048 (1866).
Clutia alaternoides f. *longifolia* Müll.Arg. in A.P.de Candolle, Prodr. 15(2): 1048 (1866).
Clutia alaternoides f. *obovata* Müll.Arg. in A.P.de Candolle, Prodr. 15(2): 1048 (1866).
Clutia alaternoides lusus *acutangula* Müll.Arg. in A.P.de Candolle, Prodr. 15(2): 1048 (1866).
Clutia alaternoides lusus *floribunda* Müll.Arg. in A.P.de Candolle, Prodr. 15(2): 1047 (1866).
Clutia alaternoides lusus *oxygona* Müll.Arg. in A.P.de Candolle, Prodr. 15(2): 1048 (1866).
Clutia alaternoides var. *angustifolia* Müll.Arg. in A.P.de Candolle, Prodr. 15(2): 1048 (1866).
Clutia glauca Pax ex Zahlbr., Ann. K. K. Naturhist. Hofmus. 15: 50 (1900 publ. 1901).
Clutia alaternoides f. *glauca* (Pax ex Zahlbr.) Pax in H.G.A.Engler, Pflanzenr., IV, 147, III: 70 (1911).
Clutia angulata Burch. ex Prain, Bull. Misc. Inform. Kew 1913: 396 (1913).
Clutia myrtifolia Burch. ex Prain, Bull. Misc. Inform. Kew 1913: 396 (1913), nom. illeg.

Clutia alpina Prain, Bull. Misc. Inform. Kew 1913: 403 (1913).
Cape Prov. 27 CPP.

Clutia angustifolia Knauf, Bot. Jahrb. Syst. 30: 340 (1901).
Burundi, Zaire (Shaba), Tanzania, N. Zambia, Malawi, Mozambique. 23 BUR ZAI 25 TAN 26 MLW MOZ ZAM. Nanophan.
Clutia lasiococca Pax & K.Hoffm. in H.G.A.Engler, Pflanzenr., IV, 147, VII: 405 (1914).

Clutia benguelensis Müll.Arg., J. Bot. 2: 337 (1864).
Angola. 26 ANG. Hemicr.

Clutia brassii Brenan, Mem. New York Bot. Gard. 9: 71 (1954).
Malawi. 26 MLW.

Clutia brevifolia Sond., Linnaea 23: 125 (1850). *Clutia polifolia* var. *brevifolia* (Sond.) Müll.Arg. in A.P.de Candolle, Prodr. 15(2): 1049 (1866).
Cape Prov. 27 CPP.

Clutia conferta Hutch. in D.Oliver, Fl. Trop. Afr. 6(1): 805 (1912).
Malawi. 26 MLW. Cham.

Clutia cordata Bernh. ex Krauss, Flora 28: 81 (1845).
S. Africa. 27 CPP NAT. Cham.
Clutia heterophylla Sond., Linnaea 23: 128 (1850), nom. illeg.

Clutia daphnoides Lam., Encycl. 2: 54 (1786). *Clutia daphnoides* var. *genuina* Müll.Arg. in A.P.de Candolle, Prodr. 15(2): 1050 (1866), nom. inval.
Cape Prov. 27 CPP.
Clutia polygonoides Thunb., Prodr. Pl. Cap.: 53 (1794), nom. illeg.
Clutia tomentosa Thunb., Fl. Cap., ed. 2: 271 (1823), nom. illeg.
Clutia daphnoides var. *incana* Sond., Linnaea 23: 126 (1850).
Clutia daphnoides var. *glabrata* Müll.Arg. in A.P.de Candolle, Prodr. 15(2): 1050 (1866).
Clutia cinerea Burm. ex Prain, Bull. Misc. Inform. Kew 1913: 412 (1913), nom. illeg.

Clutia disceptata Prain, Bull. Misc. Inform. Kew 1913: 410 (1913).
Cape Prov., KwaZulu-Natal. 27 CPP NAT.

Clutia dregeana Scheele, Linnaea 25: 583 (1852).
Cape Prov. 27 CPP.
Clutia sonderiana Müll.Arg. in A.P.de Candolle, Prodr. 15(2): 1051 (1866).
Clutia sonderiana var. *glabra* Müll.Arg. in A.P.de Candolle, Prodr. 15(2): 1051 (1866).
Clutia sonderiana var. *pubescens* Müll.Arg. in A.P.de Candolle, Prodr. 15(2): 1051 (1866).
Clutia sonderiana var. *ovalifolia* Pax in H.G.A.Engler, Pflanzenr., IV, 147, III: 73 (1911).

Clutia eckloniana Müll.Arg. in A.P.de Candolle, Prodr. 15(2): 1054 (1866).
Cape Prov. 27 CPP. Cham.

Clutia ericoides Thunb., Prodr. Pl. Cap.: 53 (1794).
Cape Prov. 27 CPP.
Clutia tenuifolia Willd., Sp. Pl. 4: 880 (1806).
Clutia ericoides var. *minor* Krauss, Flora 28: 82 (1845).
Clutia ericoides var. *tenuis* Sond., Linnaea 23: 122 (1850).
Clutia gracilis Baill., Adansonia 3: 151 (1863).
Clutia ambigua Pax & K.Hoffm. in H.G.A.Engler, Pflanzenr., IV, 147, III: 82 (1911).
Clutia pachyphylla Spreng. ex Prain, Bull. Misc. Inform. Kew 1913: 385 (1913), nom. illeg.

Clutia galpinii Pax, Bull. Herb. Boissier 6: 736 (1898).
S. Africa. 27 BOT TVL. Cham.
Clutia pulchella var. *ovalis* Müll.Arg. in A.P.de Candolle, Prodr. 15(2): 1046 (1866). *Clutia pulchella* f. *ovalis* (Müll.Arg.) Pax in H.G.A.Engler, Pflanzenr., IV, 147, III: 55 (1911).

Clutia govaertsii Radcl.-Sm., Kew Bull. 52: 478 (1997).
SW. Cape Prov. 27 CPP. Cham.
* *Clutia thunbergii* var. *vaccinioides* Pax & K.Hoffm. in H.G.A.Engler, Pflanzenr., IV, 147, III: 76 (1911). *Clutia vaccinioides* (Pax & K.Hoffm.) Prain, Bull. Misc. Inform. Kew 1913: 413 (1913), nom. illeg.

Clutia heterophylla Thunb., Prodr. Pl. Cap.: 53 (1794).
Cape Prov. 27 CPP.
Phyllanthus vaccinioides Scheele, Linnaea 25: 585 (1852).
Clutia similis Müll.Arg. in A.P.de Candolle, Prodr. 15(2): 1046 (1866).
Clutia vaccinioides Müll.Arg. in A.P.de Candolle, Prodr. 15(2): 436 (1866).
Clutia dumosa Cooper ex Pax in H.G.A.Engler, Pflanzenr., IV, 147, III: 66 (1911), pro syn.
Clutia dumosa Harv. ex Prain, Bull. Misc. Inform. Kew 1913: 411 (1913), nom. inval.

Clutia hirsuta (Sond.) Müll.Arg. in A.P.de Candolle, Prodr. 15(2): 1046 (1866).
Zimbabwe, Mozambique, S. Africa. 26 MOZ ZIM 27 CPP LES NAT OFS TVL.
* *Clutia heterophylla* var. *hirsuta* Sond., Linnaea 23: 129 (1850).

var. **hirsuta**
Zimbabwe, Mozambique, S. Africa. 26 MOZ ZIM 27 CPP LES NAT OFS TVL.
Clutia krookii Pax, Ann. K. K. Naturhist. Hofmus. 15: 49 (1900 publ. 1901).
Clutia schlechteri Pax, Bot. Jahrb. Syst. 34: 373 (1904).
Clutia inyangensis Hutch. in D.Oliver, Fl. Trop. Afr. 6(1): 804 (1912).
Clutia volubilis Hutch. in D.Oliver, Fl. Trop. Afr. 6(1): 809 (1912).

var. **robusta** Prain, Bull. Misc. Inform. Kew 1913: 409 (1913).
S. Africa. 27 NAT OFS TVL.

Clutia × **hybrida** Pax & K.Hoffm. in H.G.A.Engler, Pflanzenr., IV, 147, III: 60 (1911). C. hirsuta var. hirsuta × C. pulchella var. pulchella.
KwaZulu-Natal. 27 NAT. Nanophan.

Clutia imbricata E.Mey. ex Prain, Bull. Misc. Inform. Kew 1913: 398 (1913).
Cape Prov. 27 CPP.

Clutia impedita Prain, Bull. Misc. Inform. Kew 1913: 402 (1913).
Cape Prov. 27 CPP.

Clutia jaubertiana Müll.Arg. in A.P.de Candolle, Prodr. 15(2): 1044 (1866).
Yemen. 35 YEM. Cham.
* *Clutia lanceolata* Jaub. & Spach, Ill. Pl. Orient. 5: 467 (1855), nom. illeg.

Clutia kamerunica Pax, Bot. Jahrb. Syst. 45: 238 (1910).
Cameroon. 23 CMN. Nanophan. or phan.

Clutia katharinae Pax in H.G.A.Engler, Pflanzenr., IV, 147, III: 58 (1911).
Cape Prov., KwaZulu-Natal. 27 CPP NAT.

Clutia kilimandscharica Engl. in L.R.von Höhnel, Rudolph-Stephanie See, App.: 5 (1892). *Clutia robusta* var. *kilimandscharica* (Engl.) Pax in H.G.A.Engler, Pflanzenr., IV, 147, III: 61 (1911).
Ethiopia to Zimbabwe. 24 ETH SOM 25 KEN TAN UGA 26 ZIM. Nanophan. or phan.
Clutia richardiana var. *trichophora* Müll.Arg. in A.P.de Candolle, Prodr. 15(2): 1044 (1866).
Clutia robusta Pax in H.G.A.Engler, Pflanzenw. Ost-Afrikas C: 241 (1895). *Clutia robusta* var. *genuina* Pax in H.G.A.Engler, Pflanzenr., IV, 147, III: 61 (1911), nom. inval. *Clutia lanceolata* subsp. *robusta* (Pax) M.G.Gilbert, Nordic J. Bot. 12: 401 (1992).
Clutia robusta var. *rhododendroides* Pax, Bot. Jahrb. Syst. 43: 85 (1909).
Clutia brachyadenia Volkens ex Pax in H.G.A.Engler, Pflanzenr., IV, 147, III: 61 (1911).
Clutia robusta var. *acutifolia* Pax in H.G.A.Engler, Pflanzenr., IV, 147, III: 61 (1911).
Clutia robusta var. *polyphylla* Pax in H.G.A.Engler, Pflanzenr., IV, 147, III: 61 (1911).
Clutia robusta var. *salicifolia* Pax in H.G.A.Engler, Pflanzenr., IV, 147, III: 61 (1911).
Clutia stenophylla Pax & K.Hoffm. in H.G.A.Engler, Pflanzenr., IV, 147, III: 63 (1911).

Clutia lanceolata Forssk., Fl. Aegypt.-Arab.: 170 (1775).
Yemen. 35 YEM.

Clutia laxa Eckl. ex Sond., Linnaea 23: 128 (1850).
S. Africa. 27 CPP NAT SWZ TVL.

Clutia marginata E.Mey. ex Sond., Linnaea 23: 130 (1850). *Clutia tomentosa* var. *marginata* (E.Mey. ex Sond.) Müll.Arg. in A.P.de Candolle, Prodr. 15(2): 1053 (1866).
Cape Prov. 27 CPP.
Clutia incanescens Prain, Bull. Misc. Inform. Kew 1913: 388 (1913), nom. illeg.

Clutia monticola S.Moore, J. Linn. Soc., Bot. 40: 197 (1911).
Zimbabwe, Mozambique, S. Africa. 26 MOZ ZIM 27 CPP NAT OFS SWZ TVL. Hemicr.

var. **monticola**
Zimbabwe, Mozambique, S. Africa. 26 MOZ ZIM 27 CPP NAT OFS SWZ TVL. Hemicr.

var. **stelleroides** (S.Moore) Radcl.-Sm., Kew Bull. 47: 114 (1992).
Zimbabwe. 26 ZIM. Hemicr.
* *Clutia stelleroides* S.Moore, J. Linn. Soc., Bot. 40: 198 (1911).

Clutia myricoides Jaub. & Spach, Ill. Pl. Orient. 5: 73 (1855).
Ethiopia, Yemen. 24 ETH 35 YEM. Cham.
Clutia lanceolata var. *angustifolia* A.Rich., Tent. Fl. Abyss. 2: 253 (1850).

Clutia nana Prain, Bull. Misc. Inform. Kew 1913: 386 (1913).
S. Africa. 27 CPP LES NAT.

Clutia natalensis Bernh., Flora 28: 81 (1845). *Clutia natalensis* var. *genuina* Müll.Arg. in A.P.de Candolle, Prodr. 15(2): 1052 (1866), nom. inval.
S. Africa. 27 CPP LES NAT OFS TVL.
Clutia natalensis var. *glabrata* Sond., Linnaea 23: 127 (1850).

Clutia ovalis Sond., Linnaea 23: 129 (1850).
Cape Prov. 27 CPP.

Clutia paxii Knauf, Bot. Jahrb. Syst. 30: 341 (1901).
Rwanda to Mozambique. 23 BUR RWA ZAI 25 TAN 26 MLW MOZ ZAM ZIM. Nanophan.
Clutia phyllanthoides S.Moore, J. Linn. Soc., Bot. 40: 198 (1911).
Clutia gracilis Hutch. in D.Oliver, Fl. Trop. Afr. 6(1): 809 (1912), nom. illeg.

Clutia pentheriana Gand., Bull. Soc. Bot. France 66: 286 (1919 publ. 1920).
Cape Prov. 27 CPP.

Clutia platyphylla Pax & K.Hoffm. in H.G.A.Engler, Pflanzenr., IV, 147, III: 74 (1911).
Cape Prov. 27 CPP.

Clutia polifolia Jacq., Pl. Hort. Schoenbr. 2: 67 (1797). *Clutia polifolia* var. *genuina* Müll.Arg. in A.P.de Candolle, Prodr. 15(2): 1049 (1866), nom. inval.
Cape Prov. 27 CPP.
Clutia acuminata E.Mey. ex Sond., Linnaea 23: 125 (1850), nom. illeg.
Clutia teretifolia Sond., Linnaea 23: 124 (1850). *Clutia polifolia* var. *teretifolia* (Sond.) Müll.Arg. in A.P.de Candolle, Prodr. 15(2): 1049 (1866).
Clutia meyeriana Müll.Arg. in A.P.de Candolle, Prodr. 15(2): 1055 (1866).
Clutia polifolia var. *cinerascens* Müll.Arg. in A.P.de Candolle, Prodr. 15(2): 1049 (1866).

Clutia polyadenia Pax, Bot. Jahrb. Syst. 43: 84 (1909).
Tanzania. 25 TAN.

Clutia polygonoides L., Sp. Pl. ed. 2: 1475 (1763). *Clutia polygonoides* var. *genuina* Müll.Arg. in A.P.de Candolle, Prodr. 15(2): 1054 (1866), nom. inval.
Cape Prov. 27 CPP.
Clutia ericoides Krebs, Flora 28: 82 (1845), nom. illeg.
Clutia diosmoides Sond., Linnaea 23: 122 (1850).
Clutia tabularis Eckl. & Zeyh. ex Sond., Linnaea 23: 122 (1850).
Clutia daphnoides Eckl. & Zeyh. ex Müll.Arg. in A.P.de Candolle, Prodr. 15(2): 1054 (1866), pro syn.

Clutia pterogona Müll.Arg. in A.P.de Candolle, Prodr. 15(2): 1048 (1866).
Cape Prov. 27 CPP.
Clutia polygonoides var. *heterophylla* Krauss, Flora 28: 82 (1845). *Clutia pterogona* var. *heterophylla* (Krauss) Müll.Arg. in A.P.de Candolle, Prodr. 15(2): 1049 (1866).
Clutia alaternoides f. *revoluta* Sond., Linnaea 23: 128 (1850). *Clutia pterogona* var. *revoluta* (Sond.) Müll.Arg. in A.P.de Candolle, Prodr. 15(2): 1049 (1866).
Clutia polifolia Sond., Linnaea 23: 124 (1850), nom. illeg.
Clutia lavandulifolia Rchb. ex Pax in H.G.A.Engler, Pflanzenr., IV, 147, III: 78 (1911), nom. illeg.

Clutia pterogona var. *angustifolia* (Krauss) Pax in H.G.A.Engler, Pflanzenr., IV, 147, III: 78 (1911).

Clutia pubescens Thunb., Prodr. Pl. Cap.: 53 (1794).
Cape Prov. 27 CPP.
Clutia humilis Bernh., Flora 28: 81 (1845).
Clutia pubescens var. *glabrata* Sond., Linnaea 23: 124 (1850). *Clutia glabrata* (Sond.) Pax in H.G.A.Engler, Pflanzenr., IV, 147, III: 80 (1911).
Clutia rustii Knauf, Geogr. Verbr. Cluytia: 49, 54 (1903).
Clutia fallacina Pax & K.Hoffm. in H.G.A.Engler, Pflanzenr., IV, 147, III: 80 (1911).
Clutia intertexta Pax & K.Hoffm. in H.G.A.Engler, Pflanzenr., IV, 147, III: 80 (1911).

Clutia pulchella L., Sp. Pl.: 1042 (1753). *Clutia pulchella* var. *genuina* Müll.Arg. in A.P.de Candolle, Prodr. 15(2): 1045 (1866), nom. inval. *Clutia pulchella* f. *genuina* (Müll.Arg.) Paxin H.G.A.Engler, Pflanzenr., IV, 147, III: 54 (1911), nom. inval.
Zimbabwe to S. Africa. 26 MOZ ZIM 27 BOT CPP LES NAT OFS TVL.

var. **franksiae** Prain, Bull. Misc. Inform. Kew 1913: 405 (1913).
KwaZulu-Natal. 27 NAT.

var. **obtusata** (Sond.) Müll.Arg. in A.P.de Candolle, Prodr. 15(2): 1046 (1866).
Zimbabwe to S. Africa. 26 MOZ ZIM 27 BOT CPP LES NAT OFS TVL.
* *Clutia pulchella* f. *obtusata* Sond., Linnaea 23: 129 (1850).
Clutia pulchella f. *macrophylla* Müll.Arg. in A.P.de Candolle, Prodr. 15(2): 1045 (1866).
Clutia pulchella f. *microphylla* Pax, Ann. K. K. Naturhist. Hofmus. 15: 49 (1900).

var. **pulchella**
Mozambique, Zimbabwe, S. Africa. 26 MOZ ZIM 27 BOT LES NAT TVL.
Clutia cotinifolia Salisb., Prodr. Stirp. Chap. Allerton: 390 (1796).

Clutia punctata Wild, Kirkia 4: 142 (1964).
Zimbabwe. 26 ZIM.

Clutia richardiana Müll.Arg. in A.P.de Candolle, Prodr. 15(2): 1044 (1866).
Eritrea, Ethiopia, Yemen. 24 ERI ETH 35 YEM. Nanophan.

var. **pubescens** (Rich.) Müll.Arg. in A.P.de Candolle, Prodr. 15(2): 1044 (1866).
Ethiopia. 24 ETH.
* *Clutia lanceolata* var. *pubescens* A.Rich., Tent. Fl. Abyss. 2: 253 (1850).

var. **richardiana**
Eritrea, Ethiopia, Yemen. 24 ERI ETH 35 YEM. Nanophan.

Clutia rubricaulis Eckl. ex Sond., Linnaea 23: 128 (1850).
Cape Prov. 27 CPP.

var. **grandifolia** (Krauss) Prain, Bull. Misc. Inform. Kew 1913: 400 (1913).
Cape Prov. 27 CPP.
* *Clutia polygonoides* var. *grandifolia* Krauss, Flora 28: 82 (1845). *Clutia alaternoides* f. *grandifolia* (Krauss) Pax in H.G.A.Engler, Pflanzenr., IV, 147, III: 68 (1911).
Clutia alaternoides f. *elliptica* Müll.Arg. in A.P.de Candolle, Prodr. 15(2): 1048 (1866).
Clutia alaternoides f. *oblongata* Müll.Arg. in A.P.de Candolle, Prodr. 15(2): 1048 (1866).

var. **microphylla** (Müll.Arg.) Prain, Bull. Misc. Inform. Kew 1913: 400 (1913).
Cape Prov. 27 CPP.
Clutia alaternoides var. *brevifolia* E.Mey. ex Sond., Linnaea 23: 128 (1850), pro syn.
* *Clutia alaternoides* var. *microphylla* Müll.Arg. in A.P.de Candolle, Prodr. 15(2): 1048 (1866). *Clutia alaternoides* f. *typica* Pax & K.Hoffm. in H.G.A.Engler, Pflanzenr., IV, 147, III: 70 (1911), nom. inval.

Clutia gnidioides Willd. ex Prain, Bull. Misc. Inform. Kew 1913: 400 (1913), pro syn.
Clutia microphylla Burch. ex Prain, Bull. Misc. Inform. Kew 1913: 400 (1913), pro syn.

var. **rubricaulis**
Cape Prov. 27 CPP.
Clutia polygonoides Willd., Hort. Berol.: 51 (1816), nom. illeg.
Clutia curvata E.Mey. ex Sond., Linnaea 23: 121 (1850). *Clutia polygonoides* var. *curvata* (E.Mey. ex Sond.) Sond., Linnaea 23: 123 (1850).
Clutia alaternoides f. *typica* Pax & K.Hoffm. in H.G.A.Engler, Pflanzenr., IV, 147, III: 70 (1911), nom. inval.

var. **tenuifolia** Prain, Bull. Misc. Inform. Kew 1913: 400 (1913).
Cape Prov. 27 CPP.
Clutia alaternoides f. *brachyphylla* Müll.Arg. in A.P.de Candolle, Prodr. 15(2): 1048 (1866).
Clutia thymifolia Willd. ex Prain, Bull. Misc. Inform. Kew 1913: 400 (1913), pro syn.

Clutia sericea Müll.Arg. in A.P.de Candolle, Prodr. 15(2): 1053 (1866).
Cape Prov. 27 CPP.

Clutia sessilifolia Radcl.-Sm., Kew Bull. 47: 115 (1992).
E. Zimbabwe. 26 ZIM. Nanophan.

Clutia stuhlmannii Pax, Bot. Jahrb. Syst. 19: 112 (1894).
Tanzania, Burundi. 23 BUR 25 TAN. Hemicr. or cham.

Clutia swynnertonii S.Moore, J. Linn. Soc., Bot. 40: 197 (1911).
Mozambique, Zimbabwe. 26 MOZ ZIM. Nanophan.

Clutia thunbergii Sond., Linnaea 23: 130 (1850). *Clutia daphnoides* var. *thunbergii* (Sond.) Müll.Arg. in A.P.de Candolle, Prodr. 15(2): 1050 (1866).
Cape Prov. 27 CPP.
Clutia pubescens Willd., Sp. Pl. 4: 881 (1806), nom. illeg.
Clutia crassifolia Pax, Bull. Herb. Boissier 6: 736 (1898).
Clutia karreensis Schltr. ex Pax in H.G.A.Engler, Pflanzenr., IV, 147, III: 76 (1911), pro syn.
Clutia thunbergii var. *canescens* Pax & K.Hoffm. in H.G.A.Engler, Pflanzenr., IV, 147, III: 76 (1911).

Clutia timpermaniana J.Léonard, Bull. Jard. Bot. État 31: 391 (1961).
Congo. 23 CON.

Clutia tomentosa L., Mant. Pl.: 299 (1767). *Clutia tomentosa* var. *genuina* Müll.Arg. in A.P.de Candolle, Prodr. 15(2): 1053 (1866), nom. inval.
Cape Prov. 27 CPP.
Clutia tomentosa var. *elliptica* Müll.Arg. in A.P.de Candolle, Prodr. 15(2): 1053 (1866).

Clutia virgata Pax & K.Hoffm. in H.G.A.Engler, Pflanzenr., IV, 147, III: 71 (1911).
S. Africa. 27 CPP NAT SWZ TVL.

Clutia whytei Hutch. in D.Oliver, Fl. Trop. Afr. 6(1): 806 (1912).
Tanzania, Malawi, Zambia. 25 TAN 26 MLW ZAM. Hemicr. or cham.

var. **monticoloides** Radcl.-Sm., Kew Bull. 47: 117 (1992).
N. Zambia. 26 ZAM. Hemicr. or cham.

var. **whytei**
Tanzania, Malawi, Zambia. 25 TAN 26 MLW ZAM. Hemicr. or cham.
Clutia densifolia Gilli, Ann. Naturhist. Mus. Wien 74: 437 (1970 publ. 1971).

Synonyms:

Clutia abyssinica var. *calvescens* Pax & K.Hoffm. === **Clutia abyssinica** Jaub. & Spach var. **abyssinica**

Clutia abyssinica var. *deserticola* Volkens === **Clutia abyssinica** Jaub. & Spach var. **abyssinica**

Clutia abyssinica var. *firma* Pax & K.Hoffm. === **Clutia abyssinica** Jaub. & Spach var. **abyssinica**

Clutia abyssinica var. *glabra* Pax === **Clutia abyssinica** Jaub. & Spach var. **abyssinica**

Clutia abyssinica var. *ovalifolia* Pax & K.Hoffm. === **Clutia abyssinica** var. **usambarica** Pax & K.Hoffm.

Clutia acuminata Thunb. === **Lachnostylis hirta** (L.f.) Müll.Arg.

Clutia acuminata E.Mey. ex Sond. === **Clutia polifolia** Jacq.

Clutia acuminata L.f. === **Lachnostylis hirta** (L.f.) Müll.Arg.

Clutia affinis var. *genuina* Müll.Arg. === **Clutia affinis** Sond.

Clutia affinis var. *phyllanthifolia* (Baill.) Müll.Arg. === **Clutia affinis** Sond.

Clutia alaternoides lusus *acutangula* Müll.Arg. === **Clutia alaternoides** L.

Clutia alaternoides var. *angustifolia* Müll.Arg. === **Clutia alaternoides** L.

Clutia alaternoides f. *brachyphylla* Müll.Arg. === **Clutia rubricaulis** var. **tenuifolia** Prain

Clutia alaternoides var. *brevifolia* E.Mey. ex Sond. === **Clutia rubricaulis** var. **microphylla** (Müll.Arg.) Prain

Clutia alaternoides f. *brevifolia* Müll.Arg. === **Clutia alaternoides** L.

Clutia alaternoides f. *elliptica* Müll.Arg. === **Clutia rubricaulis** var. **grandifolia** (Krauss) Prain

Clutia alaternoides lusus *floribunda* Müll.Arg. === **Clutia alaternoides** L.

Clutia alaternoides var. *genuina* Müll.Arg. === **Clutia alaternoides** L.

Clutia alaternoides f. *glauca* (Pax ex Zahlbr.) Pax === **Clutia alaternoides** L.

Clutia alaternoides f. *grandifolia* (Krauss) Pax === **Clutia rubricaulis** var. **grandifolia** (Krauss) Prain

Clutia alaternoides f. *imbricata* Müll.Arg. === **Clutia alaternoides** L.

Clutia alaternoides f. *lanceolata* Sond. === **Clutia alaternoides** L.

Clutia alaternoides var. *lanceolata* (Sond.) Müll.Arg. === **Clutia alaternoides** L.

Clutia alaternoides var. *latifolia* Sond. === **Clutia africana** Poir.

Clutia alaternoides f. *leptophylla* Müll.Arg. === **Clutia alaternoides** L.

Clutia alaternoides f. *longifolia* Müll.Arg. === **Clutia alaternoides** L.

Clutia alaternoides var. *major* Krauss === **Clutia africana** Poir.

Clutia alaternoides var. *microphylla* Müll.Arg. === **Clutia rubricaulis** var. **microphylla** (Müll.Arg.) Prain

Clutia alaternoides f. *oblongata* Müll.Arg. === **Clutia rubricaulis** var. **grandifolia** (Krauss) Prain

Clutia alaternoides f. *obovata* Müll.Arg. === **Clutia alaternoides** L.

Clutia alaternoides lusus *oxygona* Müll.Arg. === **Clutia alaternoides** L.

Clutia alaternoides f. *revoluta* Sond. === **Clutia pterogona** Müll.Arg.

Clutia alaternoides f. *typica* Pax & K.Hoffm. === **Clutia rubricaulis** Eckl. ex Sond. var. **rubricaulis**

Clutia ambigua Pax & K.Hoffm. === **Clutia ericoides** Thunb.

Clutia androgyna L. === **Sauropus androgynus** (L.) Merr.

Clutia angulata Burch. ex Prain === **Clutia alaternoides** L.

Clutia anomala Pax & K.Hoffm. === **Clutia abyssinica** Jaub. & Spach var. **abyssinica**

Clutia berberifolia Pax === **Phyllanthus calycinus** Labill.

Clutia berteriana Sieber ex Pax === **Ditaxis polygama** (Jacq.) Wheeler

Clutia brachyadenia Volkens ex Pax === **Clutia kilimandscharica** Engl.

Clutia cascarilla L. === **Croton cascarilla** (L.) L.

Clutia cinerea Burm. ex Prain === **Clutia daphnoides** Lam.

Clutia collina Roxb. === **Cleistanthus collinus** (Roxb.) Benth.

Clutia cotinifolia Salisb. === **Clutia pulchella** L. var. **pulchella**

Clutia crassifolia Pax === **Clutia thunbergii** Sond.

Clutia curvata E.Mey. ex Sond. === **Clutia rubricaulis** Eckl. ex Sond. var. **rubricaulis**

Clutia daphnoides Eckl. & Zeyh. ex Müll.Arg. === **Clutia polygonoides** L.

Clutia daphnoides var. *genuina* Müll.Arg. === **Clutia daphnoides** Lam.

Clutia daphnoides var. *glabrata* Müll.Arg. === **Clutia daphnoides** Lam.
Clutia daphnoides var. *incana* Sond. === **Clutia daphnoides** Lam.
Clutia daphnoides var. *thunbergii* (Sond.) Müll.Arg. === **Clutia thunbergii** Sond.
Clutia decandra Crantz === **Croton eluteria** (L.) W.Wright
Clutia densifolia Gilli === **Clutia whytei** Hutch. var. **whytei**
Clutia diosmoides Sond. === **Clutia polygonoides** L.
Clutia diversifolia Roxb. === **Cleistanthus diversifolius** (Roxb.) Müll.Arg.
Clutia dumosa Harv. ex Prain === **Clutia heterophylla** Thunb.
Clutia dumosa Cooper ex Pax === **Clutia heterophylla** Thunb.
Clutia eluteria L. === **Croton eluteria** (L.) W.Wright
Clutia ericoides Krebs === **Clutia polygonoides** L.
Clutia ericoides var. *minor* Krauss === **Clutia ericoides** Thunb.
Clutia ericoides var. *tenuis* Sond. === **Clutia ericoides** Thunb.
Clutia fallacina Pax & K.Hoffm. === **Clutia pubescens** Thunb.
Clutia floribunda Baill. === **Clutia alaternoides** L.
Clutia glabrata (Sond.) Pax === **Clutia pubescens** Thunb.
Clutia glabrescens Knauf === **Clutia abyssinica** Jaub. & Spach var. **abyssinica**
Clutia glauca Pax ex Zahlbr. === **Clutia alaternoides** L.
Clutia gnidioides Willd. ex Prain === **Clutia rubricaulis** var. **microphylla** (Müll.Arg.) Prain
Clutia gracilis Baill. === **Clutia ericoides** Thunb.
Clutia gracilis Hutch. === **Clutia paxii** Knauf
Clutia heterophylla Sond. === **Clutia cordata** Bernh. ex Krauss
Clutia heterophylla var. *hirsuta* Sond. === **Clutia hirsuta** (Sond.) Müll.Arg.
Clutia hirsuta Eckl. & Zeyh. ex Sond. === **Clutia affinis** Sond.
Clutia hirta Vahl === **Lachnostylis hirta** (L.f.) Müll.Arg.
Clutia hirta L.f. === **Lachnostylis hirta** (L.f.) Müll.Arg.
Clutia humilis Bernh. === **Clutia pubescens** Thunb.
Clutia incanescens Prain === **Clutia marginata** E.Mey. ex Sond.
Clutia intertexta Pax & K.Hoffm. === **Clutia pubescens** Thunb.
Clutia inyangensis Hutch. === **Clutia hirsuta** (Sond.) Müll.Arg. var. **hirsuta**
Clutia karreensis Schltr. ex Pax === **Clutia thunbergii** Sond.
Clutia krookii Pax === **Clutia hirsuta** (Sond.) Müll.Arg. var. **hirsuta**
Clutia lanceolata Jaub. & Spach === **Clutia jaubertiana** Müll.Arg.
Clutia lanceolata var. *angustifolia* A.Rich. === **Clutia myricoides** Jaub. & Spach
Clutia lanceolata var. *glabra* A.Rich. === **Clutia abyssinica** Jaub. & Spach var. **abyssinica**
Clutia lanceolata var. *pubescens* A.Rich. === **Clutia richardiana** var. **pubescens** (Rich.) Müll.Arg.
Clutia lanceolata subsp. *robusta* (Pax) M.G.Gilbert === **Clutia kilimandscharica** Engl.
Clutia lasiococca Pax & K.Hoffm. === **Clutia angustifolia** Knauf
Clutia lavandulifolia Rchb. ex Pax === **Clutia pterogona** Müll.Arg.
Clutia leuconeura Pax === **Clutia abyssinica** var. **usambarica** Pax & K.Hoffm.
Clutia meyeriana Müll.Arg. === **Clutia polifolia** Jacq.
Clutia microphylla Burch. ex Prain === **Clutia rubricaulis** var. **microphylla** (Müll.Arg.) Prain
Clutia mollis Pax === **Clutia abyssinica** var. **usambarica** Pax & K.Hoffm.
Clutia monoica Lour. === **Cleistanthus monoicus** (Lour.) Müll.Arg.
Clutia montana Roxb. === **Bridelia montana** (Roxb.) Willd.
Clutia myrtifolia Burch. ex Prain === **Clutia alaternoides** L.
Clutia natalensis var. *genuina* Müll.Arg. === **Clutia natalensis** Bernh.
Clutia natalensis var. *glabrata* Sond. === **Clutia natalensis** Bernh.
Clutia oblongifolia Roxb. === **Cleistanthus oblongifolius** (Roxb.) Müll.Arg.
Clutia ovalis (E.Mey. ex Sond.) Scheele === **Andrachne ovalis** (E.Mey. ex Sond.) Müll.Arg.
Clutia pachyphylla Spreng. ex Prain === **Clutia ericoides** Thunb.
Clutia patula Roxb. === **Cleistanthus patulus** (Roxb.) Müll.Arg.
Clutia pedicellaris (Pax) Hutch. === **Clutia abyssinica** var. **pedicellaris** (Pax) Pax
Clutia phyllanthifolia Baill. === **Clutia affinis** Sond.
Clutia phyllanthoides S.Moore === **Clutia paxii** Knauf

Clutia polifolia Sond. === **Clutia pterogona** Müll.Arg.
Clutia polifolia var. *brevifolia* (Sond.) Müll.Arg. === **Clutia brevifolia** Sond.
Clutia polifolia var. *cinerascens* Müll.Arg. === **Clutia polifolia** Jacq.
Clutia polifolia var. *genuina* Müll.Arg. === **Clutia polifolia** Jacq.
Clutia polifolia var. *teretifolia* (Sond.) Müll.Arg. === **Clutia polifolia** Jacq.
Clutia polygalifolia Salisb. === **Clutia alaternoides** L.
Clutia polygonoides Thunb. === **Clutia daphnoides** Lam.
Clutia polygonoides Willd. === **Clutia rubricaulis** Eckl. ex Sond. var. **rubricaulis**
Clutia polygonoides var. *curvata* (E.Mey. ex Sond.) Sond. === **Clutia rubricaulis** Eckl. ex Sond. var. **rubricaulis**
Clutia polygonoides var. *genuina* Müll.Arg. === **Clutia polygonoides** L.
Clutia polygonoides var. *grandifolia* Krauss === **Clutia rubricaulis** var. **grandifolia** (Krauss) Prain
Clutia polygonoides var. *heterophylla* Krauss === **Clutia pterogona** Müll.Arg.
Clutia pterogona var. *angustifolia* (Krauss) Pax === **Clutia pterogona** Müll.Arg.
Clutia pterogona var. *heterophylla* (Krauss) Müll.Arg. === **Clutia pterogona** Müll.Arg.
Clutia pterogona var. *revoluta* (Sond.) Müll.Arg. === **Clutia pterogona** Müll.Arg.
Clutia pubescens Willd. === **Clutia thunbergii** Sond.
Clutia pubescens Eckl. & Zeyh. ex Sond. === **Clutia affinis** Sond.
Clutia pubescens var. *glabrata* Sond. === **Clutia pubescens** Thunb.
Clutia pulchella f. *genuina* (Müll.Arg.) Pax === **Clutia pulchella** L.
Clutia pulchella var. *genuina* Müll.Arg. === **Clutia pulchella** L.
Clutia pulchella f. *macrophylla* Müll.Arg. === **Clutia pulchella** var. **obtusata** (Sond.) Müll.Arg.
Clutia pulchella f. *microphylla* Pax === **Clutia pulchella** var. **obtusata** (Sond.) Müll.Arg.
Clutia pulchella f. *obtusata* Sond. === **Clutia pulchella** var. **obtusata** (Sond.) Müll.Arg.
Clutia pulchella f. *ovalis* (Müll.Arg.) Pax === **Clutia galpinii** Pax
Clutia pulchella var. *ovalis* Müll.Arg. === **Clutia galpinii** Pax
Clutia retusa L. === **Bridelia retusa** (L.) A.Juss.
Clutia richardiana var. *pedicellaris* Pax === **Clutia abyssinica** var. **pedicellaris** (Pax) Pax
Clutia richardiana var. *trichophora* Müll.Arg. === **Clutia kilimandscharica** Engl.
Clutia robusta Pax === **Clutia kilimandscharica** Engl.
Clutia robusta var. *acutifolia* Pax === **Clutia kilimandscharica** Engl.
Clutia robusta var. *genuina* Pax === **Clutia kilimandscharica** Engl.
Clutia robusta var. *kilimandscharica* (Engl.) Pax === **Clutia kilimandscharica** Engl.
Clutia robusta var. *polyphylla* Pax === **Clutia kilimandscharica** Engl.
Clutia robusta var. *rhododendroides* Pax === **Clutia kilimandscharica** Engl.
Clutia robusta var. *salicifolia* Pax === **Clutia kilimandscharica** Engl.
Clutia rotundifolia Pax === **Clutia abyssinica** var. **usambarica** Pax & K.Hoffm.
Clutia rustii Knauf === **Clutia pubescens** Thunb.
Clutia scandens Roxb. === **Bridelia stipularis** (L.) Blume
Clutia schlechteri Pax === **Clutia hirsuta** (Sond.) Müll.Arg. var. **hirsuta**
Clutia semperflorens Roxb. === **Trigonostemon semperflorens** (Roxb.) Müll.Arg.
Clutia sempervirens Müll.Arg. === **Trigonostemon semperflorens** (Roxb.) Müll.Arg.
Clutia similis Müll.Arg. === **Clutia heterophylla** Thunb.
Clutia sonderiana Müll.Arg. === **Clutia dregeana** Scheele
Clutia sonderiana var. *glabra* Müll.Arg. === **Clutia dregeana** Scheele
Clutia sonderiana var. *ovalifolia* Pax === **Clutia dregeana** Scheele
Clutia sonderiana var. *pubescens* Müll.Arg. === **Clutia dregeana** Scheele
Clutia spinosa Roxb. === **Bridelia retusa** (L.) A.Juss.
Clutia squamosa Lam. === **Bridelia retusa** (L.) A.Juss.
Clutia stelleroides S.Moore === **Clutia monticola** var. **stelleroides** (S.Moore) Radcl.-Sm.
Clutia stenophylla Pax & K.Hoffm. === **Clutia kilimandscharica** Engl.
Clutia stipularis L. === **Bridelia stipularis** (L.) Blume
Clutia tabularis Eckl. & Zeyh. ex Sond. === **Clutia polygonoides** L.
Clutia tenuifolia Willd. === **Clutia ericoides** Thunb.
Clutia teretifolia Sond. === **Clutia polifolia** Jacq.
Clutia thunbergii var. *canescens* Pax & K.Hoffm. === **Clutia thunbergii** Sond.

Clutia thunbergii var. *vaccinioides* Pax & K.Hoffm. === **Clutia govaertsii** Radcl.-Sm.
Clutia thymifolia Willd. ex Prain === **Clutia rubricaulis** var. **tenuifolia** Prain
Clutia tomentosa Thunb. === **Clutia daphnoides** Lam.
Clutia tomentosa var. *elliptica* Müll.Arg. === **Clutia tomentosa** L.
Clutia tomentosa var. *genuina* Müll.Arg. === **Clutia tomentosa** L.
Clutia tomentosa var. *marginata* (E.Mey. ex Sond.) Müll.Arg. === **Clutia marginata** E.Mey. ex Sond.
Clutia vaccinioides Müll.Arg. === **Clutia heterophylla** Thunb.
Clutia vaccinioides (Pax & K.Hoffm.) Prain === **Clutia govaertsii** Radcl.-Sm.
Clutia volubilis Hutch. === **Clutia hirsuta** (Sond.) Müll.Arg. var. **hirsuta**

Cluytia

An orthographic variant of *Clutia*.

Cluytiandra

Synonyms:
Cluytiandra Müll.Arg. === **Meineckia** Baill.
Cluytiandra baronii Hutch. === **Meineckia baronii** (Hutch.) G.L.Webster
Cluytiandra capillariformis (Vatke & Pax ex Pax) Pax & K.Hoffm. === **Meineckia phyllanthoides** subsp. **capillariformis** (Vatke & Pax ex Pax) G.L.Webster
Cluytiandra decaryi Leandri === **Meineckia decaryi** (Leandri) Brunel ex Radcl.-Sm.
Cluytiandra decaryi var. *occidentalis* Leandri === **Meineckia decaryi** var. **occidentalis** (Leandri) Radcl.-Sm.
Cluytiandra engleri Pax === **Meineckia fruticans** var. **engleri** (Pax) G.L.Webster
Cluytiandra fruticans Pax === **Meineckia fruticans** (Pax) G.L.Webster
Cluytiandra madagascariensis Leandri ex Humbert === **Meineckia madagascariensis** (Leandri) G.L.Webster
Cluytiandra orientalis Leandri === **Meineckia orientalis** (Leandri) G.L.Webster
Cluytiandra peltata Hutch. === **Meineckia peltata** (Hutch.) G.L.Webster
Cluytiandra perrieri Leandri ex Humbert === ?
Cluytiandra schinzii Pax === **Phyllanthus pinnatus** (Wight) G.L.Webster
Cluytiandra somalensis Pax === **Meineckia phyllanthoides** subsp. **somalensis** (Pax) G.L.Webster
Cluytiandra trichopoda Müll.Arg. === **Meineckia phyllanthoides** subsp. **trichopoda** (Müll.Arg.) G.L.Webster

Clytia

An orthographic variant of *Clutia*.

Cnemidostachys

Synonyms:
Cnemidostachys Mart. === **Microstachys** A.Juss.
Cnemidostachys acalyphoides Mart. === **Microstachys corniculata** (Vahl) Griseb.
Cnemidostachys bidentata Klotzsch ex Baill. === **Microstachys bidentata** (Mart. & Zucc.) Esser
Cnemidostachys bidentata Mart. === **Microstachys bidentata** (Mart. & Zucc.) Esser
Cnemidostachys coriacea Mart. === **Microstachys marginata** (Mart.) Klotzsch ex Müll.Arg.
Cnemidostachys crotonoides Mart. === **Microstachys** sp.
Cnemidostachys daphnoides Mart. === **Microstachys daphnoides** (Mart.) Müll.Arg.
Cnemidostachys ditassoides Didr. === **Microstachys ditassoides** (Didr.) Esser
Cnemidostachys dubia Wawra === **Microstachys corniculata** (Vahl) Griseb.
Cnemidostachys glabrata Mart. === **Gymnanthes multiramea** (Klotzsch) Müll.Arg.
Cnemidostachys glandulosa Mart. === **Microstachys corniculata** (Vahl) Griseb.

Cnemidostachys hispida Mart. === **Microstachys corniculata** (Vahl) Griseb.
Cnemidostachys longifolia Mart. === **Microstachys corniculata** (Vahl) Griseb.
Cnemidostachys marginata Mart. === **Microstachys marginata** (Mart.) Klotzsch ex Müll.Arg.
Cnemidostachys myrtilloides Mart. === **Microstachys daphnoides** (Mart.) Müll.Arg.
Cnemidostachys oleoides Mart. === **Microstachys daphnoides** (Mart.) Müll.Arg.
Cnemidostachys patula Mart. === **Microstachys corniculata** (Vahl) Griseb.
Cnemidostachys patula f. *dubia* (Wawra) Wawra === **Microstachys corniculata** (Vahl) Griseb.
Cnemidostachys prostrata Mart. === **Microstachys corniculata** (Vahl) Griseb.
Cnemidostachys salicifolia Mart. === **Microstachys corniculata** (Vahl) Griseb.
Cnemidostachys scoparia Mart. === **Microstachys bidentata** (Mart. & Zucc.) Esser
Cnemidostachys serrulata Mart. === **Microstachys serrulata** (Mart.) Müll.Arg.
Cnemidostachys tragioides Mart. === **Microstachys corniculata** (Vahl) Griseb.

Cnesmone

11 species, SE. Asia and W. Malesia to Bali and the Philippines; small shrubs or slender, slightly woody twiners with stinging hairs. *C. javanica* is easily the most widely distributed. No good modern revision is available, that by Pax & Hoffmann (1919) being obsolete given the many later additions in E. and SE. Asia. Most species remain imperfectly known. Present generic limits follow Croizat (1941; see Asia); among related genera are the Asian/Malesian *Megistostigma* and the African *Sphaerostylis* and *Tragiella*. (Acalyphoideae)

Pax, F. & K. Hoffmann (1919). *Cnesmone*. In A. Engler (ed.), Das Pflanzenreich, IV 147 IX (Euphorbiaceae-Acalypheae-Plukenetiinae): 102-103. Berlin. (Heft 68.) La/Ge. — 1 species, SE. Asia, W. Malesia. [*C. javanica* here too widely defined.]

Gagnepain, F. (1925). Quelques genres nouveaux d'Euphorbiacées. Bul. Soc. Bot. France 71: 864-879. Fr. — Includes (pp. 864-870) protologue of *Cenesmon* and accounts of 6 new species from Yunnan (China), Vietnam, and Laos. The genus was considered related to *Cnesmone* (with which it is now merged).

Airy-Shaw, H. K. (1969). Notes on Malesian and other Asiatic Euphorbiaceae, CXII. Notes on the subtribe Plukenetiinae Pax. Kew Bull. 23: 114-121. En. —*Cnesmone*, pp. 117-119; notes on 4 species.

Cnesmone Blume, Bijdr.: 630 (1826).
S. China, Trop. Asia. 36 40 41 42.
Cenesmon Gagnep., Bull. Soc. Bot. France 71: 864 (1924 publ. 1925).

Cnesmone anisopetala (Merr. & Chun) Croizat, J. Arnold Arbor. 22: 429 (1941).
Hainan. 36 CHH.
* *Tragia anisopetala* Merr. & Chun, Sunyatsenia 2: 261 (1935).

Cnesmone hainanensis (Merr. & Chun) Croizat, J. Arnold Arbor. 22: 430 (1941).
Hainan. 36 CHH.
* *Cenesmon hainanense* Merr. & Chun, Sunyatsenia 5: 94 (1940).

Cnesmone javanica Blume, Bijdr.: 630 (1826).
Assam, Indo-China, Sumatera, Jawa, Lesser Sunda Is. (Bali), Borneo (Sabah, SE. Kalimantan). 40 ASS 41 AMD BMA MLY THA VIE 42 BOR JAW LSI SUM. Cl. cham. or cl. nanophan.

var. **glabriuscula** N.P.Balakr. & N.G.Nair, Gard. Bull. Singapore 31: 49 (1978).
Andaman Is. 41 AND.

var. **javanica**
Assam, Indo-China, Sumatera, Jawa, Lesser Sunda Is. (Bali), Borneo (Sabah, SE. Kalimantan). 40 ASS 41 BMA MLY THA VIE 42 BOR JAW LSI SUM. Cl. cham. or cl. nanophan.

Tragia macrophylla Wall., Numer. List: 7796 (1847), nom. inval.
Tragia rugosa Wall., Numer. List: 7794 (1847), nom. inval.
Tragia hastata Reinw. ex Hassk., Pl. Jav. Rar.: 245 (1868).
Cenesmon tonkinense Gagnep., Bull. Soc. Bot. France 71: 869 (1924 publ. 1925).
Cnesmone tonkinensis (Gagnep.) Croizat, J. Arnold Arbor. 22: 429 (1941).

Cnesmone laevis (Ridl.) Airy Shaw, Kew Bull. 23: 118 (1969).
S. Thailand, Pen. Malaysia (Lankawi I.). 41 THA 42 MLY. Cl. nanophan.
* *Tragia laevis* Ridl., Bull. Misc. Inform. Kew 1923: 368 (1923).

Cnesmone laotica (Gagnep.) Croizat, J. Arnold Arbor. 22: 428 (1941).
N. Thailand, Laos, Cambodia. 41 CBD LAO THA. Cham. or nanophan.
* *Cenesmon laoticum* Gagnep., Bull. Soc. Bot. France 71: 867 (1924 publ. 1925).

Cnesmone linearis (Gagnep.) Croizat, J. Arnold Arbor. 22: 428 (1941).
Vietnam. 41 VIE.
* *Cenesmon lineare* Gagnep., Bull. Soc. Bot. France 71: 867 (1924 publ. 1925).

Cnesmone mairei (H.Lév.) Croizat, J. Arnold Arbor. 22: 429 (1941).
SC. China (Yunnan). 36 CHC.
* *Alchornea mairei* H.Lév., Cat. Pl. Yun-Nan: 94 (1916). *Tragia mairei* (H.Lév.) Rehder, J. Arnold Arbor. 18: 214 (1937).

Cnesmone peltata (Gagnep.) Croizat, J. Arnold Arbor. 22: 428 (1941).
Vietnam. 41 VIE.
* *Cenesmon peltatum* Gagnep., Bull. Soc. Bot. France 71: 868 (1924 publ. 1925).

Cnesmone philippinensis (Merr.) Airy Shaw, Kew Bull. 23: 119 (1969).
Philippines. 42 PHI.
* *Tragia philippinensis* Merr., Philipp. J. Sci. 16: 565 (1920).

Cnesmone poilanei (Gagnep.) Croizat, J. Arnold Arbor. 22: 428 (1941).
Vietnam. 41 VIE.
* *Cenesmon poilanei* Gagnep., Bull. Soc. Bot. France 71: 869 (1924 publ. 1925).

Cnesmone subpeltata Ridl., Bull. Misc. Inform. Kew 1923: 368 (1923).
Pen. Malaysia. 42 MLY.

Synonyms:
Cnesmone glabrata Kurz === **Megistostigma malaccense** Hook.f.
Cnesmone tonkinensis (Gagnep.) Croizat === **Cnesmone javanica** Blume var. **javanica**

Cnesmosa

An orthographic variant of *Cnesmone*.

Cnidoscolus

67 species, Americas; perennial herbs with a woody base or shrubs or small trees with characteristic palmate foliage. The plants are covered with hypodermic needle-hairs which produce painful stings; this is reflected in the generic name which translates as 'nettle-spine'. The genus is closely related to *Jatropha* and in the past was combined with it, e.g. by Pax (1910). It was definitively separated by Pax and Hoffmann (1931) and Miller & Webster (1962); opinion now places the genus nearer *Manihot* (Webster, Synopsis, 1994). Infrageneric groups were revised by McVaugh (1944) who recognised 5 sections. More recent systematic work in the genus has been largely floristic or at infrageneric level. Some species are cultivated (cf. Ingram 1957), and *C. elasticus* (Mexico) yields a rubber-forming latex. (Crotonoideae)

Pax, F. (1910). *Jatropha*. In A. Engler (ed.), Das Pflanzenreich, IV 147 [I] (Euphorbiaceae-Jatropheae): 21-113. Berlin. (Heft 42.) La/Ge. — 156 species, in 3 subgenera and 14 sections (the last subgenus, with 6 sections, comprising *Cnidoscolus* (pp. 86-110)). See also Addenda (pp. 133-134) with 2 additional species and, in IV 147 II, Additamentum I (1910) for one further species. [The authors separated the two genera in their 1931 survey.]

Pax, F. & K. Hoffmann (1931). *Cnidoscolus*. In A. Engler (ed.), Die natürlichen Pflanzenfamilien, 2. Aufl., 19c: 164-167. Leipzig. Ge. — Synopsis with description of genus; c. 50 species in 7 sections with representative species listed. [The genus has here been separated from *Jatropha* in contrast to the authors' *Pflanzenreich* treatment.]

Croizat, L. (1943). New and critical Euphorbiaceae of Brazil. Tropical Woods 76: 11-14. En. — Includes discussion of mutual relationships among *Jatropha*, *Cnidoscolus* and *Manihot* (with a belief that they are interrelated and that *Cnidoscolus* merits only subgeneric rank within *Jatropha*).

McVaugh, R. (1944). The genus *Cnidoscolus*: generic limits and intergeneric groups. Bull. Torrey Bot. Club 71: 457-474. En. — General review of genus and key to 5 sections; supraspecific treatment with synonymy, type species, and commentary along with descriptions of new species; nomenclator of accepted species (pp. 472-473); list of references at end. [One less section recognised than in Pax and Hoffmann.]

Lundell, C. L. (1945). The genus *Cnidoscolus* in Mexico: new species and critical notes. Bull. Torrey Bot. Club 72: 319-334. En. — Additions to Pax and Hoffmann's treatments; 12 new species described with notes on others but no key.

Ingram, J. (1957). Notes on the cultivated Euphorbiaceae, 2. *Cnidoscolus* and *Jatropha*. Baileya 5: 110-117. En. — Includes keys, descriptions and illustrations; covers 4 species of *Cnidoscolus* and 6 of *Jatropha*. General features of the genera are covered in an introduction on p. 110.

Miller, K. I. & G. L. Webster (1962). Systematic position of *Cnidoscolus* and *Jatropha*. Brittonia 14: 174-180. En. — Study of additional evidence (anatomy, karology, palynology); support for separation of genera (which furthermore appear not to be that closely related). No key is furnished. [A range of species was chosen from each genus.]

Johnston, M. C. & B. H. Warnock (1963). The species of *Cnidoscolus* and *Jatropha* (Euphorbiaceae) in far western Texas. Southwestern Nat. 8: 121-126. En. — Treatment of *C. texanus* with distribution map and key contrasting it with the two reported species of *Jatropha*; no exsiccatae.

Jablonski, E. (1967). *Cnidoscolus*. Euphorbiaceae, Guayana Highland (Mem. New York Bot. Gard. 17(1)): 156. New York. En. — 1 species, *C. urens*.

Breckon, G. J. (1975). *Cnidoscolus*, sect. *Calyptrosolen* (Euphorbiaceae) in Mexico and Central America. 463 pp. Davis, Calif. (Unpubl. Ph.D. dissertation, Univ. of California, Davis.) En. — An unpublished revision; corresponds to section IV of Pax and Hoffmann (1931).

Breckon, G. J. (1979). Studies in *Cnidoscolus* (Euphorbiaceae), 1. *Jatropha tubulosa*, *J. liebmannii* and allied taxa from Central Mexico. Brittonia 31: 125-148, illus., graph, map. En. — Transfer and redefinition of *J. liebmannii* and *J. tubulosa* and revision of these and related Mexican species of sect. *Calyptrosolen* (sect. III.4 in the revision of *Jatropha* by Pax (1910). The formal treatment covers 4 species with keys, descriptions, synonymy, references, types, localities with exsiccatae, indication of distribution and habitat, and commentary and citations. 1 new species and 2 new subspecies are described.]

Levin, G. A. (1995). Euphorbiaceae, spurge family: 1. *Acalypha* and *Cnidoscolus*. J. Ariz.-Nev. Acad. Sci. 29: 18-24, illus., maps. En. — Revision of Arizonan species of *Cnidoscolus* precursory to treatment of the genus in a new state flora.

Cnidoscolus Pohl, Pl. Bras. Icon. Descr. 1: 56 (1827).

Trop. & Subtrop. America. 74 76 77 78 79 80 81 82 83 84 85.

Jussieuia Houst., Reliq. Houstoun.: 6 (1781).

Bivonea Raf., Fl. Ludov.: 138 (1817).

Victorinia Léon, Mem. Soc. Cub. Hist. Nat. "Felipe Poey" 15: 242 (1941).

Cnidoscolus aconitifolius (Mill.) I.M.Johnst., Contr. Gray Herb. 68: 86 (1923).
Mexico, C. America. 79 MXG MXS MXT 80 COS HON NIC (83) per. Nanophan. or phan.
* *Jatropha aconitifolia* Mill., Gard. Dict. ed. 8: 6 (1768). *Jatropha aconitifolia* var. *genuina* Müll.Arg. in A.P.de Candolle, Prodr. 15(2): 1100 (1866), nom. inval.

subsp. **aconitifolius**
Mexico (Veracruz, Yucatán), C. America. 79 MXG MXT 80 COS HON NIC (83) per. Nanophan. or phan.
Jatropha papaya Medik., Bot. Beob. 1782: 194 (1782). *Jatropha aconitifolia* var. *papaya* (Medik.) Pax in H.G.A.Engler, Pflanzenr., IV, 147, I: 101 (1910).
Jatropha napeifolia Desr. in J.B.A.M.de Lamarck, Encycl. 4: 15 (1797). *Cnidoscolus napeifolius* (Desr.) Pohl, Pl. Bras. Icon. Descr. 1: 63 (1827).
Jatropha palmata Willd., Sp. Pl. 4: 562 (1805). *Cnidoscolus palmatus* (Willd.) Pohl, Pl. Bras. Icon. Descr. 1: 63 (1827). *Jatropha aconitifolia* var. *palmata* (Willd.) Müll.Arg. in A.P.de Candolle, Prodr. 15(2): 1100 (1866).
Jatropha aconitifolia var. *multipartita* Müll.Arg. in A.P.de Candolle, Prodr. 15(2): 1100 (1866).

subsp. **polyanthus** (Pax & K.Hoffm.) Breckon, Syst. Bot. 9: 22 (1984).
Mexico (Michoacán, Guerrero). 79 MXS. Nanophan.
* *Jatropha polyantha* Pax & K.Hoffm. in H.G.A.Engler, Pflanzenr., IV, 147, I: 105 (1910). *Cnidoscolus polyanthus* (Pax & K.Hoffm.) I.M.Johnst., Contr. Gray Herb. 68: 86 (1923).

Cnidoscolus acrandrus (Urb.) Pax & K.Hoffm. in H.G.A.Engler, Nat. Pflanzenfam. ed. 2, 19c: 167 (1931).
Dominican Rep. 81 DOM.
* *Jatropha acrandra* Urb., Symb. Antill. 7: 515 (1913).

Cnidoscolus albidus Lundell, Bull. Torrey Bot. Club 72: 320 (1945).
Mexico (Hidalgo). 79 MXE. Nanophan.

Cnidoscolus albomaculatus (Pax) I.M.Johnst., Contr. Gray Herb. 68: 86 (1923).
Paraguay. 85 PAR. Cham.
Jatropha vitifolia f. *nana* Chodat & Hassl., Bull. Herb. Boissier, II, 5: 613 (1905). *Jatropha albomaculata* var. *nana* (Chodat & Hassl.) Pax in H.G.A.Engler, Pflanzenr., IV, 147, I: 91 (1910).
Jatropha vitifolia f. *stimulosissima* Chodat & Hassl., Bull. Herb. Boissier, II, 5: 613 (1905). *Jatropha albomaculata* var. *stimulosissima* (Chodat & Hassl.) Pax in H.G.A.Engler, Pflanzenr., IV, 147, I: 91 (1910).
* *Jatropha albomaculata* Pax in H.G.A.Engler, Pflanzenr., IV, 147, I: 90 (1910).
Jatropha albomaculata var. *subcuneata* Pax in H.G.A.Engler, Pflanzenr., IV, 147, I: 91 (1910).

Cnidoscolus angustidens Torr. in W.H.Emory, Rep. U.S. Mex. Bound. 2(1): 198 (1858).
Jatropha angustidens (Torr.) Müll.Arg. in A.P.de Candolle, Prodr. 15(2): 1102 (1866).
Arizona, W. Mexico. 76 ARI 79 MXN MXS. Hemicr. or tuber geophyte.
Cnidoscolus pringlei I.M.Johnst., Contr. Gray Herb. 68: 85 (1923). *Jatropha pringlei* (I.M.Johnst.) Standl., Contr. U. S. Natl. Herb. 23: 1670 (1926).

Cnidoscolus appendiculatus (Pax & K.Hoffm.) Pax & K.Hoffm. in H.G.A.Engler, Nat. Pflanzenfam. ed. 2, 19c: 164 (1931).
Paraguay. 85 PAR. Cham.
* *Jatropha appendiculata* Pax & K.Hoffm. in H.G.A.Engler, Pflanzenr., IV, 147, I: 92 (1910).

Cnidoscolus autlanensis Breckon, Contr. Univ. Michigan Herb. 20: 201 (1995).
Mexico (Jalisco, Colima). 79 MXS.

Cnidoscolus bahianus (Ule) Pax & K.Hoffm. in H.G.A.Engler, Nat. Pflanzenfam. ed. 2, 19c: 164 (1931).
Brazil (Bahia, Pernambuco). 84 BZE. Nanophan. or phan.
**Jatropha bahiana* Ule, Bot. Jahrb. Syst. 42: 220 (1908). *Jatropha bahiana* var. *genuina* Pax in H.G.A.Engler, Pflanzenr., IV, 147, I: 89 (1910), nom. inval.
Jatropha bahiana var. *rupestris* Ule, Bot. Jahrb. Syst. 42: 220 (1908).

Cnidoscolus basiacanthus (Pax & K.Hoffm.) J.F.Macbr., Publ. Field Mus. Nat. Hist., Bot. Ser. 13(3A1): 164 (1951).
Peru. 83 PER. Nanophan. or phan.
**Jatropha basiacantha* Pax & K.Hoffm. in H.G.A.Engler, Pflanzenr., IV, 147, I: 90 (1910).

Cnidoscolus bellator (Ekman ex Urb.) Léon, Mem. Soc. Cub. Hist. Nat. "Felipe Poey" 15: 237 (1941).
W. Cuba. 81 CUB.
**Jatropha bellatrix* Ekman ex Urb., Repert. Spec. Nov. Regni Veg. 28: 228 (1930).

Cnidoscolus calyculatus (Pax & K.Hoffm.) I.M.Johnst., Contr. Gray Herb. 68: 86 (1923).
Mexico (Michoacán). 79 MXS. Hemicr. or cham.
**Jatropha calyculata* Pax & K.Hoffm. in H.G.A.Engler, Pflanzenr., IV, 147, I: 97 (1910).

Cnidoscolus campanulatus (Pax) Pax in H.G.A.Engler, Nat. Pflanzenfam. ed. 2, 19c: 164 (1931).
S. Paraguay, Argentina (Tucumán). 85 AGW PAR. Nanophan.
**Jatropha campanulata* Pax in H.G.A.Engler, Pflanzenr., IV, 147, I: 91 (1910).

Cnidoscolus chaya Lundell, Bull. Torrey Bot. Club 72: 321 (1945).
Mexico (Yucatán). 79 MXT. Nanophan.

Cnidoscolus chayamansa McVaugh, Bull. Torrey Bot. Club 72: 466 (1944).
Mexico (Yucatán), Belize. 79 MXT 80 BLZ. Succ. nanophan. – Young leaves and shoots edible.
Jatropha urens var. *inermis* Calvino, Revista Agric. Comércio Trab. 2: 364 (1919).

Cnidoscolus cnicodendron Griseb., Abh. Königl. Ges. Wiss. Göttingen 24: 53 (1879).
Jatropha vitifolia var. *cnicodendron* (Griseb.) Pax in H.G.A.Engler, Pflanzenr., IV, 147, I: 88 (1910).
Brazil (Goiás, Bahia), Argentina (Salta). 84 BZC BZE 85 AGW. Nanophan. or phan.
Jatropha vitifolia var. *maritima* Müll.Arg. in C.F.P.von Martius, Fl. Bras. 11(2): 498 (1874).
Cnidoscolus vitifolius var. *repandus* Griseb., Abh. Königl. Ges. Wiss. Göttingen 24: 53 (1879). *Jatropha vitifolia* var. *repanda* (Griseb.) Pax in H.G.A.Engler, Pflanzenr., IV, 147, I: 88 (1910).
Jatropha vitifolia var. *grisebachii* Pax in H.G.A.Engler, Pflanzenr., IV, 147, I: 88 (1910).

Cnidoscolus diacanthus (Pax & K.Hoffm.) J.F.Macbr., Publ. Field Mus. Nat. Hist., Bot. Ser. 13(3A1): 164 (1951).
Peru. 83 PER.
**Jatropha diacantha* Pax & K.Hoffm. in H.G.A.Engler, Pflanzenr., IV, 147, VII: 399 (1914).

Cnidoscolus elasticus Lundell, Field & Lab. 12: 33 (1944).
Mexico (Sinaloa, Durango). 79 MXE MXN. Phan.

Cnidoscolus fragrans (Kunth) Pohl, Pl. Bras. Icon. Descr. 1: 63 (1827).
Cuba (La Habana). 81 CUB. Phan.
**Jatropha fragrans* Kunth in F.W.H.von Humboldt, A.J.A.Bonpland & C.S.Kunth, Nov. Gen. Sp. 2: 105 (1817).

Cnidoscolus hamosus Pohl, Pl. Bras. Icon. Descr. 1: 57 (1827). *Jatropha hamosa* (Pohl) Müll.Arg. in A.P.de Candolle, Prodr. 15(2): 1097 (1866).
Brazil (Minas Gerais). 84 BZL. Nanophan.

Cnidoscolus hasslerianus (Pax) Pax in H.G.A.Engler, Nat. Pflanzenfam. ed. 2, 19c: 164 (1931).
Paraguay. 85 PAR. Cham. or nanophan.
* *Jatropha hassleriana* Pax in H.G.A.Engler, Pflanzenr., IV, 147, I: 91 (1910).

Cnidoscolus horridus (Müll.Arg.) Pax & K.Hoffm. in H.G.A.Engler, Nat. Pflanzenfam. ed. 2, 19c: 164 (1931).
SE. Brazil. 84 BZL. Hemicr.
* *Jatropha horrida* Müll.Arg., Linnaea 34: 210 (1865).
Jatropha ferox Müll.Arg. in C.F.P.von Martius, Fl. Bras. 11(2): 497 (1874).

Cnidoscolus hypoleucus (Pax) Pax in H.G.A.Engler, Nat. Pflanzenfam. ed. 2, 19c: 164 (1931).
Peru. 83 PER. Nanophan.
* *Jatropha hypoleuca* Pax in H.G.A.Engler, Pflanzenr., IV, 147, I: 96 (1910).

Cnidoscolus infestus Pax & K.Hoffm. in H.G.A.Engler, Pflanzenr., IV, 147, XVI: 193 (1924).
Brazil (Paraíba). 84 BZE.

Cnidoscolus jaenensis (Pax & K.Hoffm.) J.F.Macbr., Publ. Field Mus. Nat. Hist., Bot. Ser. 13(3A1): 165 (1951).
Peru. 83 PER.
* *Jatropha jaenensis* Pax & K.Hoffm. in H.G.A.Engler, Pflanzenr., IV, 147, VII: 400 (1914).

Cnidoscolus kunthianus (Müll.Arg.) Pax & K.Hoffm. in H.G.A.Engler, Nat. Pflanzenfam. ed. 2, 19c: 165 (1931).
Colombia, Guyana, Venezuela. 82 GUY VEN 83 CLM. Nanophan.
Cnidoscolus quinquelobus Pohl, Pl. Bras. Icon. Descr. 1: 63 (1827), nom. illeg.
* *Jatropha kunthiana* Müll.Arg., Linnaea 34: 211 (1865).

Cnidoscolus leuconeurus (Pax & K.Hoffm.) Pax & K.Hoffm. in H.G.A.Engler, Nat. Pflanzenfam. ed. 2, 19c: 164 (1931).
Paraguay. 85 PAR. Cham.
* *Jatropha leuconeura* Pax & K.Hoffm. in H.G.A.Engler, Pflanzenr., IV, 147, I: 94 (1910).

Cnidoscolus liebmannii (Müll.Arg.) Lundell, Bull. Torrey Bot. Club 72: 324 (1945).
Mexico (Puebla). 79 MXC. Nanophan.
* *Jatropha liebmannii* Müll.Arg., Linnaea 34: 212 (1865).
Cnidoscolus armatus Lundell, Bull. Torrey Bot. Club 72: 320 (1945).

Cnidoscolus loasoides (Pax) I.M.Johnst., Contr. Gray Herb. 68: 86 (1923).
Argentina (Corrientes). 85 AGE. Cham.
* *Jatropha loasoides* Pax in H.G.A.Engler, Pflanzenr., IV, 147, I: 92 (1910).

Cnidoscolus loefgrenii (Pax & K.Hoffm.) Pax & K.Hoffm. in H.G.A.Engler, Nat. Pflanzenfam. ed. 2, 19c: 166 (1931).
Brazil (São Paulo). 84 BZL. Nanophan.
* *Jatropha loefgrenii* Pax & K.Hoffm. in H.G.A.Engler, Pflanzenr., IV, 147, I: 107 (1910).

Cnidoscolus longipedunculatus (Brandegee) Pax & K.Hoffm. in H.G.A.Engler, Pflanzenr., IV, 147, XVI: 193 (1924).
Mexico (Veracruz). 79 MXG.
Jatropha longipedunculata Brandegee, Univ. Calif. Publ. Bot. 7: 328 (1920).
* *Jatropha urens* var. *longipedunculata* Brandegee, Univ. Calif. Publ. Bot. 7: 368 (1920).

Cnidoscolus longipes (Pax) I.M.Johnst., Contr. Gray Herb. 68: 86 (1923).
Colombia. 83 CLM. Nanophan.
* *Jatropha longipes* Pax in H.G.A.Engler, Pflanzenr., IV, 147, I: 106 (1910).

Cnidoscolus macrandrus Lundell, Bull. Torrey Bot. Club 72: 324 (1945).
Mexico (Veracruz). 79 MXG. Nanophan.

Cnidoscolus maculatus (Brandegee) Pax & K.Hoffm. in H.G.A.Engler, Pflanzenr., IV, 147, XVI: 193 (1924).
Mexico (Baja California Sur). 79 MXN. Hemicr. or tuber geophyte.
* *Jatropha maculata* Brandegee, Univ. Calif. Publ. Bot. 6: 359 (1916).

Cnidoscolus maracayensis (Chodat & Hassl.) Pax & K.Hoffm. in H.G.A.Engler, Nat. Pflanzenfam. ed. 2, 19c: 164 (1931).
Paraguay. 85 PAR. Cham.
* *Jatropha maracayensis* Chodat & Hassl., Bull. Herb. Boissier, II, 5: 613 (1905).

Cnidoscolus matosii Léon, Mem. Soc. Cub. Hist. Nat. "Felipe Poey" 15: 238 (1941).
E. Cuba. 81 CUB. Phan.

Cnidoscolus multilobus (Pax) I.M.Johnst., Contr. Gray Herb. 68: 86 (1923).
Mexico (Veracruz). 79 MXG. Nanophan.
* *Jatropha multiloba* Pax in H.G.A.Engler, Pflanzenr., IV, 147, I: 107 (1910).

Cnidoscolus oligandrus (Müll.Arg.) Pax in H.G.A.Engler, Nat. Pflanzenfam. ed. 2, 19c: 166 (1931).
Brazil (Rio de Janeiro). 84 BZL. Nanophan. or phan.
* *Jatropha oligandra* Müll.Arg. in C.F.P.von Martius, Fl. Bras. 11(2): 502 (1874).
Cnidoscolus oligandrus var. *xerophilus* A.Mattos & al. in ?, (1992).

Cnidoscolus orbiculatus Lundell, Bull. Torrey Bot. Club 72: 325 (1945).
Mexico (Morelos). 79 MXC. Cham.

Cnidoscolus palmeri (S.Watson) Rose, Contr. U. S. Natl. Herb. 12: 282 (1909).
Mexico (Baja California Sur, Sonora). 79 MXN. Nanophan.
* *Jatropha palmeri* S.Watson, Proc. Amer. Acad. Arts 24: 76 (1889).

Cnidoscolus parviflorus Lundell, Bull. Torrey Bot. Club 72: 327 (1945).
Mexico (San Luis Potosí). 79 MXE. Nanophan.

Cnidoscolus paucistamineus (Pax) Pax in H.G.A.Engler, Nat. Pflanzenfam. ed. 2, 19c: 166 (1931).
Brazil (Mato Grosso). 84 BZC. Phan.
* *Jatropha paucistaminea* Pax in H.G.A.Engler, Pflanzenr., IV, 147, I: 110 (1910).

Cnidoscolus peruvianus (Müll.Arg.) Pax & K.Hoffm. in H.G.A.Engler, Nat. Pflanzenfam. ed. 2, 19c: 164 (1931).
Peru. 83 PER.
* *Jatropha peruviana* Müll.Arg., Linnaea 34: 210 (1865).

Cnidoscolus pubescens Pohl, Pl. Bras. Icon. Descr. 1: 62 (1827). *Jatropha obtusifolia* var. *pubescens* (Pohl) Müll.Arg. in A.P.de Candolle, Prodr. 15(2): 1097 (1866).
Brazil (Pernambuco, Bahia). 84 BZE. Nanophan. or phan.
Cnidoscolus obtusifolius Pohl ex Baill., Adansonia 4: 275 (1864). *Jatropha obtusifolia* var. *genuina* Müll.Arg. in A.P.de Candolle, Prodr. 15(2): 1097 (1866), nom. inval. *Jatropha obtusifolia* (Pohl ex Baill.) Müll.Arg. in A.P.de Candolle, Prodr. 15(2): 1097 (1866).

Cnidoscolus pyrophorus (Pax) J.F.Macbr., Publ. Field Mus. Nat. Hist., Bot. Ser. 13(3A1): 166 (1951).
Peru. 83 PER. Nanophan.
* *Jatropha pyrophora* Pax in H.G.A.Engler, Pflanzenr., IV, 147, I: 101 (1910).

Cnidoscolus quercifolius Pohl, Pl. Bras. Icon. Descr. 1: 62 (1827). *Jatropha phyllacantha* var. *quercifolia* (Pohl) Müll.Arg. in A.P.de Candolle, Prodr. 15(2): 1098 (1866).
Brazil (Piauí, Pernambuco, Bahia, São Paulo). 84 BZE BZL. Phan.
Cnidoscolus lobatus Pohl, Pl. Bras. Icon. Descr. 1: 62 (1827). *Jatropha phyllacantha* var. *lobata* (Pohl) Müll.Arg. in A.P.de Candolle, Prodr. 15(2): 1098 (1866).
Cnidoscolus repandus Pohl, Pl. Bras. Icon. Descr. 1: 62 (1827). *Jatropha phyllacantha* var. *repanda* (Pohl) Müll.Arg. in A.P.de Candolle, Prodr. 15(2): 1098 (1866).
Jatropha phyllacantha Müll.Arg. in A.P.de Candolle, Prodr. 15(2): 1098 (1866). *Cnidoscolus phyllacanthus* (Müll.Arg.) Pax & K.Hoffm. in H.G.A.Engler, Nat. Pflanzenfam. ed. 2, 19c: 165 (1931).

Cnidoscolus quinquelobatus (Mill.) Leon, Mem. Soc. Cub. Hist. Nat. "Felipe Poey" 15: 236 (1941).
Cuba (La Habana). 81 CUB. Nanophan.
* *Jatropha quinquelobata* Mill., Gard. Dict. ed. 8: 2 (1768).
Jatropha quinqueloba Cerv., Supl. Gaz. Lit. Mexico: 4 (2 July 1794).

Cnidoscolus rangel (M.Gómez) McVaugh, Bull. Torrey Bot. Club 71: 464 (1944).
Cuba (Pinar del Río). 81 CUB. Nanophan. or phan.
* *Jatropha rangel* M.Gómez, Mem. Real Soc. Esp. Hist. Nat. 23: 51 (1894).
Jatropha platyandra Pax in H.G.A.Engler, Pflanzenr., IV, 147, I: 110 (1910). *Cnidoscolus platyandrus* (Pax) I.M.Johnst., Contr. Gray Herb. 68: 86 (1923).

Cnidoscolus regina (Léon) Radcl.-Sm. & Govaerts, Kew Bull. 52: 183 (1997).
Cuba (Sierra Sagua Baracoa). 81 CUB. Nanophan. or phan.
* *Jatropha regina* Léon, Mem. Soc. Cub. Hist. Nat. "Felipe Poey" 12: 352 (1938).

Cnidoscolus rostratus Lundell, Bull. Torrey Bot. Club 72: 328 (1945).
Mexico. 79 MXC MXS. Nanophan. or phan.

subsp. **glabratus** Breckon, Brittonia 31: 139 (1979).
Mexico (Guerrero, México State, Michoacán, Morelos). 79 MXC MXS. Nanophan.

subsp. **hintonii** Breckon, Brittonia 31: 139 (1979).
Mexico (Guerrero, Michoacán). 79 MXS. Nanophan.

subsp. **rostratus**
Mexico (Oaxaca, Puebla). 79 MXC MXS. Nanophan. or phan.

Cnidoscolus rotundifolius (Müll.Arg.) McVaugh, Bull. Torrey Bot. Club 71: 466 (1944).
Mexico (Tamaulipas). 79 MXE. Nanophan.
* *Jatropha rotundifolia* Müll.Arg., Linnaea 34: 211 (1865).
Cnidoscolus inermiflorus I.M.Johnst., Contr. Gray Herb. 68: 85 (1923).
Jatropha inermiflora Standl., Contr. U. S. Natl. Herb. 23: 1670 (1926).

Cnidoscolus sellowianus Klotzsch ex Pax in H.G.A.Engler, Pflanzenr., IV, 147: 90 (1910). *Jatropha sellowiana* (Klotzsch ex Pax) Pax & K.Hoffm. in H.G.A.Engler, Pflanzenr., IV, 147, I: 90 (1910).
Brazil (Minas Gerais). 84 BZL. Nanophan.

Cnidoscolus serrulatus (Pax & K.Hoffm.) Pax & K.Hoffm. in H.G.A.Engler, Nat. Pflanzenfam. ed. 2, 19c: 164 (1931).

Paraguay. 85 PAR. Cham.
Jatropha serrulata Pax & K.Hoffm. in H.G.A.Engler, Pflanzenr., IV, 147, I: 94 (1910).

Cnidoscolus shrevei I.M.Johnst., J. Arnold Arbor. 21: 260 (1940).
Mexico (Durango). 79 MXE. Nanophan.

Cnidoscolus souzae McVaugh, Bull. Torrey Bot. Club 71: 468 (1944).
Mexico (Yucatán, Campeche), Belize. 79 MXT 80 BLZ. Cham. or nanophan.

Cnidoscolus spinosus Lundell, Bull. Torrey Bot. Club 72: 329 (1945).
Mexico (Jalisco). 79 MXS. Nanophan. or phan.

Cnidoscolus subinteger (Pax & K.Hoffm.) Pax & K.Hoffm. in H.G.A.Engler, Nat. Pflanzenfam. ed. 2, 19c: 164 (1931).
Paraguay. 85 PAR. Cham.
Jatropha vitifolia f. *subintegra* Chodat & Hassl., Bull. Herb. Boissier, II, 5: 613 (1905).
Jatropha subintegra (Pax & K.Hoffm.) Pax & K.Hoffm. in H.G.A.Engler, Pflanzenr., IV, 147, I: 92 (1910).

Cnidoscolus tehuacanensis Breckon, Brittonia 31: 140 (1979).
Mexico (Puebla, Oaxaca). 79 MXC MXS. Nanophan.

Cnidoscolus tenuifolius (Pax & K.Hoffm.) I.M.Johnst., Contr. Gray Herb. 68: 86 (1923).
Paraguay. 85 PAR. Nanophan.
Jatropha tenuifolia Pax & K.Hoffm. in H.G.A.Engler, Pflanzenr., IV, 147, I: 107 (1910), nom. illeg.

Cnidoscolus tenuilobus Lundell, Bull. Torrey Bot. Club 72: 330 (1945).
Mexico (Guerrero). 79 MXS. Nanophan.

Cnidoscolus tetracyclus Pax & K.Hoffm. in H.G.A.Engler, Pflanzenr., IV, 147, XVI: 193 (1924).
Paraguay. 85 PAR. Cham.

Cnidoscolus texanus (Müll.Arg.) Small, Fl. S.E. U.S.: 706 (1903).
Texas, Oklahoma. 74 OKL 77 TEX. Hemicr.
Jatropha texana Müll.Arg., Linnaea 34: 211 (1865). *Bivonea texana* (Müll.Arg.) House, Bull. New York State Mus. Nat. Hist. 233-234: 61 (1920 publ. 1921).

Cnidoscolus tubulosus (Müll.Arg.) I.M.Johnst., Contr. Gray Herb. 68: 86 (1923).
Mexico to Peru. 79 MXS MXT 80 COS ELS NIC GUA PAN 83 PER. Nanophan. or phan.
Jatropha tubulosa Müll.Arg., Linnaea 34: 212 (1865), nom. illeg.
Jatropha tubulosa var. *quinqueloba* Müll.Arg., Linnaea 34: 212 (1865).
Jatropha tubulosa var. *septemloba* Müll.Arg., Linnaea 34: 212 (1865).
Jatropha tubulosa var. *triloba* Müll.Arg., Linnaea 34: 212 (1865).
Jatropha jurgensenii Briq., Annuaire Conserv. Jard. Bot. Genève 4: 229 (1900).
Cnidoscolus jurgensenii (Briq.) Lundell, Bull. Torrey Bot. Club 72: 324 (1945).
Jatropha tepiquensis Costantin & Gallaud, Rev. Gén. Bot. 18: 391 (1906). *Cnidoscolus tepiquensis* (Costantin & Gallaud) Lundell, Field & Lab. 12: 36 (1944).
Jatropha cordifolia Pax in H.G.A.Engler, Pflanzenr., IV, 147, I: 107 (1910). *Cnidoscolus cordifolius* (Pax) I.M.Johnst., Contr. Gray Herb. 68: 86 (1923).
Cnidoscolus hernandezii Lundell, Bull. Torrey Bot. Club 72: 323 (1945).
Cnidoscolus tomentosus Lundell, Bull. Torrey Bot. Club 72: 332 (1945).

Cnidoscolus ulei (Pax) Pax in H.G.A.Engler, Nat. Pflanzenfam. ed. 2, 19c: 164 (1931).
Brazil (Bahia). 84 BZE. Nanophan. or phan.
Jatropha ulei Pax in H.G.A.Engler, Pflanzenr., IV, 147, I: 88 (1910).

Cnidoscolus urens (L.) Arthur, Torreya 21: 11 (1921).
SE. U.S.A., Mexico, Trop. America. 78 ALA FLA NCA SCA VRG 79 MXG MXT 80 COS GUA PAN 81 DOM LEE TRT VNA WIN 82 FRG GUY SUR VEN 83 CLM 84 BZC BZE BZL 85 AGE PAR. Nanophan.
* *Jatropha urens* L., Sp. Pl.: 1007 (1753). *Manihot urens* (L.) Crantz, Inst. Rei Herb. 1: 167 (1766). *Jatropha urens* var. *genuina* Müll.Arg. in A.P.de Candolle, Prodr. 15(2): 1100 (1866), nom. inval. *Bivonea urens* (L.) Arthur, Torreya 22: 30 (1922).

var. **stimulosus** (Michx.) Govaerts in R.Govaerts, D.G.Frodin & A.Radcliffe-Smith, World Checklist Bibliogr. Euphorbiaceae: 398 (2000).
SE. U.S.A. 78 ALA FLA NCA SCA VRG. Cham. or nanophan.
* *Jatropha stimulosa* Michx., Fl. Bor.-Amer. 2: 216 (1803). *Bivonea stimulosa* (Michx.) Raf., Specchio Sci. 1: 156 (1814). *Cnidoscolus michauxii* Pohl, Pl. Bras. Icon. Descr. 1: 58 (1827), nom. illeg. *Cnidoscolus stimulosus* (Michx.) Engelm. & A.Gray, Boston J. Nat. Hist. 5: 234 (1845). *Jatropha urens* var. *stimulosa* (Michx.) Müll.Arg. in A.P.de Candolle, Prodr. 15(2): 1101 (1866).
Jatropha stipulosa Steud., Nomencl. Bot., ed. 2, 1: 800 (1840).

var. **urens**
Mexico, Trop. America. 79 MXG MXT 80 COS GUA PAN 81 DOM LEE TRT VNA WIN 82 FRG GUY SUR VEN 83 CLM 84 BZC BZE BZL 85 AGE PAR. Nanophan.
Jatropha herbacea L., Sp. Pl.: 1007 (1753). *Manihot herbacea* (L.) Crantz, Inst. Rei Herb. 1: 167 (1766). *Jatropha urens* var. *herbacea* (L.) Müll.Arg. in A.P.de Candolle, Prodr. 15(2): 1101 (1866). *Cnidoscolus herbaceus* (L.) I.M.Johnst., Contr. Gray Herb. 68: 86 (1923).
Cnidoscolus marcgravii Pohl, Pl. Bras. Icon. Descr. 1: 58 (1827). *Jatropha urens* var. *marcgravii* (Pohl) Müll.Arg. in A.P.de Candolle, Prodr. 15(2): 1101 (1866).
Cnidoscolus neglectus Pohl, Pl. Bras. Icon. Descr. 1: 60 (1827). *Jatropha neglecta* (Pohl) Houst. ex Baill., Étude Euphorb.: 304 (1858).
Cnidoscolus osteocarpus Pohl, Pl. Bras. Icon. Descr. 1: 63 (1827). *Jatropha urens* f. *osteocarpa* (Pohl) Müll.Arg. in A.P.de Candolle, Prodr. 15(2): 1101 (1866). *Jatropha urens* var. *osteocarpa* (Pohl) Müll.Arg. in C.F.P.von Martius, Fl. Bras. 11(2): 500 (1874).
Jatropha urens var. *brachyloba* Müll.Arg. in C.F.P.von Martius, Fl. Bras. 11(2): 500 (1874).
Cnidoscolus mexicanus Klotzsch ex Pax in H.G.A.Engler, Pflanzenr., IV, 147: 98 (1910).
Jatropha osteocarpa Schott ex Pax in H.G.A.Engler, Pflanzenr., IV, 147, I: 98 (1910), pro syn.
Jatropha adenophila Pax & K.Hoffm. in H.G.A.Engler, Pflanzenr., IV, 147, VII: 400 (1914). *Cnidoscolus adenophilus* (Pax & K.Hoffm.) Pax & K.Hoffm. in H.G.A.Engler, Nat. Pflanzenfam. ed. 2, 19c: 166 (1931). *Cnidoscolus urens* subsp. *adenophilus* (Pax & K.Hoffm.) Breckon, Ann. Missouri Bot. Gard. 75: 1114 (1988).

Cnidoscolus urnigerus (Pax) Pax in H.G.A.Engler, Nat. Pflanzenfam. ed. 2, 19c: 166 (1931).
Brazil (Bahia). 84 BZE. Nanophan.
* *Jatropha urnigera* Pax in H.G.A.Engler, Pflanzenr., IV, 147, I: 104 (1910).

Cnidoscolus velutinus Lundell, Bull. Torrey Bot. Club 72: 333 (1945).
Mexico (Guerrero). 79 MXS. Nanophan. or phan.

Cnidoscolus vitifolius (Mill.) Pohl, Pl. Bras. Icon. Descr. 1: 61 (1827).
Mexico (near Cartagena). 79 + 84 BZE? Nanophan.
* *Jatropha vitifolia* Mill., Gard. Dict. ed. 8: 5 (1768). *Jatropha vitifolia* var. *genuina* Müll.Arg. in A.P.de Candolle, Prodr. 15(2): 1097 (1866), nom. inval.
Jatropha vitifolia var. *obtusifolia* Müll.Arg. in A.P.de Candolle, Prodr. 15(2): 1097 (1866).

Synonyms:

Cnidoscolus adenophilus (Pax & K.Hoffm.) Pax & K.Hoffm. === **Cnidoscolus urens** (L.) Arthur var. **urens**
Cnidoscolus armatus Lundell === **Cnidoscolus liebmannii** (Müll.Arg.) Lundell
Cnidoscolus cordifolius (Pax) I.M.Johnst. === **Cnidoscolus tubulosus** (Müll.Arg.) I.M.Johnst.
Cnidoscolus herbaceus (L.) I.M.Johnst. === **Cnidoscolus urens** (L.) Arthur var. **urens**
Cnidoscolus hernandezii Lundell === **Cnidoscolus tubulosus** (Müll.Arg.) I.M.Johnst.
Cnidoscolus inermiflorus I.M.Johnst. === **Cnidoscolus rotundifolius** (Müll.Arg.) McVaugh
Cnidoscolus jurgensenii (Briq.) Lundell === **Cnidoscolus tubulosus** (Müll.Arg.) I.M.Johnst.
Cnidoscolus lobatus Pohl === **Cnidoscolus quercifolius** Pohl
Cnidoscolus marcgravii Pohl === **Cnidoscolus urens** (L.) Arthur var. **urens**
Cnidoscolus mexicanus Klotzsch ex Pax === **Cnidoscolus urens** (L.) Arthur var. **urens**
Cnidoscolus michauxii Pohl === **Cnidoscolus urens** var. **stimulosus** (Michx.) Govaerts
Cnidoscolus napeifolius (Desr.) Pohl === **Cnidoscolus aconitifolius** (Mill.) I.M.Johnst. subsp. **aconitifolius**
Cnidoscolus neglectus Pohl === **Cnidoscolus urens** (L.) Arthur var. **urens**
Cnidoscolus obtusifolius Pohl ex Baill. === **Cnidoscolus pubescens** Pohl
Cnidoscolus oligandrus var. *xerophilus* A.Mattos & al. === **Cnidoscolus oligandrus** (Müll.Arg.) Pax
Cnidoscolus osteocarpus Pohl === **Cnidoscolus urens** (L.) Arthur var. **urens**
Cnidoscolus palmatus (Willd.) Pohl === **Cnidoscolus aconitifolius** (Mill.) I.M.Johnst. subsp. **aconitifolius**
Cnidoscolus phyllacanthus (Müll.Arg.) Pax & K.Hoffm. === **Cnidoscolus quercifolius** Pohl
Cnidoscolus platyandrus (Pax) I.M.Johnst. === **Cnidoscolus rangel** (M.Gómez) McVaugh
Cnidoscolus polyanthus (Pax & K.Hoffm.) I.M.Johnst. === **Cnidoscolus aconitifolius** subsp. **polyanthus** (Pax & K.Hoffm.) Breckon
Cnidoscolus pringlei I.M.Johnst. === **Cnidoscolus angustidens** Torr.
Cnidoscolus pubescens (Pax) Pax === **Jatropha froesii** Croizat
Cnidoscolus quinquelobus Pohl === **Cnidoscolus kunthianus** (Müll.Arg.) Pax & K.Hoffm.
Cnidoscolus repandus Pohl === **Cnidoscolus quercifolius** Pohl
Cnidoscolus stimulosus (Michx.) Engelm. & A.Gray === **Cnidoscolus urens** var. **stimulosus** (Michx.) Govaerts
Cnidoscolus surinamensis Miq. === **Croton lobatus** L.
Cnidoscolus tepiquensis (Costantin & Gallaud) Lundell === **Cnidoscolus tubulosus** (Müll.Arg.) I.M.Johnst.
Cnidoscolus tomentosus Lundell === **Cnidoscolus tubulosus** (Müll.Arg.) I.M.Johnst.
Cnidoscolus urens subsp. *adenophilus* (Pax & K.Hoffm.) Breckon === **Cnidoscolus urens** (L.) Arthur var. **urens**
Cnidoscolus vitifolius var. *repandus* Griseb. === **Cnidoscolus cnicodendron** Griseb.

Coccoceras

Synonyms:

Coccoceras Miq. === **Mallotus** Lour.
Coccoceras anisopodum Gagnep. === **Mallotus anisopodus** (Gagnep.) Airy Shaw
Coccoceras borneense (Müll.Arg.) J.J.Sm. === **Mallotus muticus** (Müll.Arg.) Airy Shaw
Coccoceras muticum Müll.Arg. === **Mallotus muticus** (Müll.Arg.) Airy Shaw
Coccoceras muticum var. *pedicellatum* Hook.f. === **Mallotus leucodermis** Hook.f.
Coccoceras plicatum Müll.Arg. === **Mallotus plicatus** (Müll.Arg.) Airy Shaw
Coccoceras sumatranum Miq. === **Mallotus sumatranus** (Miq.) Airy Shaw

Coccoglochidion

A segregate of *Glochidion*.

Synonyms:
Coccoglochidion K.Schum. === **Glochidion** J.R.Forst. & G.Forst.
Coccoglochidion erythrococcus K.Schum. === **Glochidion philippicum** (Cav.) C.B.Rob.

Coccomelia

Synonyms:
Coccomelia Reinw. ex Blume === **Baccaurea** Lour.
Coccomelia racemosa Reinw. ex Blume === **Baccaurea racemosa** (Reinw. ex Blume) Müll.Arg.

Cocconerion

2 species, New Caledonia (including Loyalty Is.); shrubs or trees from 2-25 m with a red latex in maquis or forests at lower elevations. The leaves are finely laterally veined. *C. minus* is the more widespread species. The most closely related genus is *Myricanthe*, also of New Caledonia; more distant is *Bertya* (Webster, Synopsis, 1994). An alliance with *Borneodendron* has also been suggested. The genus was never revised in *Pflanzenreich*; the modern treatment of McPherson & Tirel (1987) must be considered definitive. (Crotonoideae)

Airy-Shaw, H. K. (1971). Notes on Malesian and other Asiatic Euphorbiaceae, CXXX. The male flower of *Cocconerion* Bl. Kew Bull. 25: 503-506. En. — Amplification for *C. minus* from New Caledonia.

Airy-Shaw, H. K. (1978). Notes on Malesian and other Asiatic Euphorbiaceae, CXCIV. The male flower of *Cocconerion balansae*. Kew Bull. 32: 382. En. — Addition to existing knowledge.

McPherson, G. & C. Tirel (1987). *Cocconerion*. Fl. Nouvelle-Calédonie, 14 (Euphorbiacées, I): 38-43. Paris. Fr. — Flora treatment (2 species; genus endemic); key.

Cocconerion Baill., Adansonia 11: 87 (1873).
New Caledonia. 60.

Cocconerion balansae Baill., Adansonia 11: 88 (1873).
SE. New Caledonia. 60 NWC. Phan.

Cocconerion minus Baill., Adansonia 11: 88 (1873).
New Caledonia. 60 NWC. Nanophan. or phan.

Codiaeum

17 species, Malesia (Jawa, Lesser Sunda Is., Borneo, and Philippines to New Guinea), Australia (Queensland) and Pacific Islands; shrubs or small understorey trees. In Borneo and Banggi the genus is limited to limestone. Revisionary studies have been undertaken by Esti Munawaroh (Bogor). The genus is notable for the many 'fancy' forms of *C. variegatum*, collectively termed 'croton'. These arose in Melanesia and are widely cultivated there and elsewhere in warm regions (cf. Brown 1995). They were also popular in Europe in the second half of the nineteenth century (particularly from the 1860s through the 1880s), and many cultivars were named and offered as species by commercial nurserymen. From a somewhat later date they were brought into North America. Their listing by Nicholson and other contemporary horticultural writers, as given here, may not reflect actual dates of publication. (Crotonoideae)

Pax, F. (with K. Hoffmann) (1911). *Codiaeum*. In A. Engler (ed.), Das Pflanzenreich, IV 147 III (Euphorbiaceae-Cluytieae): 23-30. Berlin. (Heft 47.) La/Ge. — 6 species, C and E Malesia and Pacific Is.; *C. variegatum* with numerous named forms, of which 7 recognized here.

Airy-Shaw, H. K. (1978). Notes on Malesian and other Asiatic Euphorbiaceae, CCXX. *Codiaeum* Juss. Kew Bull. 33: 74-77. En. — Descriptions of 2 new species from New Guinea (one with 2 additional varieties).
Green, P. S. (1986). New combinations in *Baloghia* and *Codiaeum* (Euphorbiaceae). Kew Bull. 41: 1026. En. — Contributions to Australasian island floras.
McPherson, G. & C. Tirel (1987). *Codiaeum*. Fl. Nouvelle-Calédonie, 14 (Euphorbiacées, I): 95-101. Paris. Fr. — Flora treatment (2 native species); key.
Brown, B. F. (1995). A *Codiaeum* encyclopedia: crotons of the world. 136 pp., text-figs., numerous col. pls. Valkaria, Florida: Valkaria Tropical Garden. En. — History of introductions, especially to Europe and North America; modern developments; usage of the plants in SE. Asia; growth, propagation, and (pp. 101 and 104) classification; pictorial index. [Written, it would appear, without serious advice and badly organised. All the colour plates are interleaved with the text and arranged in seemingly random or whimsical fashion.]
Kiew, R. & P. van Welzen (1998). *Codiaeum variegatum* var. *cavernicola* var. nov. (Euphorbiaceae), the second *Codiaeum* from Borneo. *Gardens Bull. Singapore* 50: 31-34. — Description of novelty; note on the genus in Borneo.

Codiaeum Rumph. ex A.Juss., Euphorb. Gen.: 33 (1824), nom. & typ. cons.
Malesia, Queensland, SW. Pacific. 42 50 60.
Phyllaurea Lour., Fl. Cochinch.: 575 (1790), nom. rejic.
Crozophyla Raf., Sylva Tellur.: 64 (1838).
Synaspisma Endl., Gen. Pl.: 1110 (1840).
Junghuhnia Miq., Fl. Ned. Ind. 1(2): 412 (1859).

Codiaeum affine Merr., Philipp. J. Sci. 29: 385 (1926).
Borneo (Sabah: Banggi). 42 BOR. Nanophan.

Codiaeum bractiferum (Roxb.) Merr., Interpr. Herb. Amboin.: 47 326 (1917).
Lesser Sunda Is., Maluku. 42 LSI MOL. Nanophan.
* *Croton bractiferus* Roxb., Fl. Ind. ed. 1832, 3: 680 (1832).
Codiaeum brevistylum Pax & K.Hoffm. in H.G.A.Engler, Pflanzenr., IV, 147, III: 28 (1911).

Codiaeum ciliatum Merr., Philipp. J. Sci. 16: 571 (1920).
Philippines. 42 PHI.

Codiaeum finisterrae Pax & K.Hoffm. in H.G.A.Engler, Pflanzenr., IV, 147, V: 284 (1912).
NE. New Guinea. 42 NWG. Nanophan.

Codiaeum hirsutum Merr., Philipp. J. Sci., C 9: 476 (1914 publ. 1915).
Philippines. 42 PHI.

Codiaeum ludovicianum Airy Shaw, Kew Bull. 33: 75 (1978).
New Guinea (Louisiade Archip.). 42 NWG. Nanophan.
Codiaeum ludovicianum var. *praeruptorum* Airy Shaw, Kew Bull. 33: 76 (1978).
Codiaeum ludovicianum var. *rheophyticum* Airy Shaw, Kew Bull. 33: 76 (1978).

Codiaeum luzonicum Merr., Philipp. J. Sci. 1(Suppl.): 81 (1906).
Philippines. 42 PHI.
Codiaeum cuneifolium Pax & K.Hoffm. in H.G.A.Engler, Pflanzenr., IV, 147, III: 28 (1911), nom. illeg.
Codiaeum irosinense Elmer ex Merr., Enum. Philipp. Fl. Pl. 2: 453 (1923), pro syn.

Codiaeum macgregorii Merr., Philipp. J. Sci. 16: 573 (1920).
Philippines. 42 PHI.

Codiaeum megalanthum Merr., Philipp. J. Sci. 16: 570 (1920).
Philippines. 42 PHI.

Codiaeum membranaceum S.Moore, J. Linn. Soc., Bot. 45: 219 (1920).
Queensland (Cape York Pen.). 50 QLD. Phan.

Codiaeum oligogynum McPherson, in Fl. N. Caled. & Depend. 14: 100 (1987).
NW. New Caledonia. 60 NWC. Nanophan.

Codiaeum palawanense Elmer, Leafl. Philipp. Bot. 4: 1283 (1911).
Philippines (Palawan). 42 PHI.

Codiaeum peltatum (Labill.) P.S.Green, Kew Bull. 41: 1026 (1986). – FIGURE, p. 403.
New Caledonia (incl. Loyalty Is.). 60 NWC. Nanophan. or phan.
* *Chrozophora peltata* Labill., Sert. Austro-Caledon.: 74 (1825).

Codiaeum stellingianum Warb., Bot. Jahrb. Syst. 13: 353 (1891).
New Guinea, Maluku (Kai Is.). 42 MOL NWG. Nanophan.

Codiaeum tenerifolium Airy Shaw, Kew Bull. 33: 74 (1978).
New Guinea. 42 NWG. Phan.

Codiaeum trichocalyx Merr., Philipp. J. Sci. 16: 570 (1920).
Philippines (Luzon). 42 PHI.

Codiaeum variegatum (L.) Blume, Bijdr.: 606 (1826).
Jawa, Lesser Sunda Is., Sulawesi, Maluku, Philippines, New Guinea, N. Queensland, Vanuatu, Fiji. 42 BIS BOR JAW LSI MOL NWG PHI SUL 50 QLD 60 FIJ VAN. Nanophan. or phan. – Numerous cultivars are widely grown; leaves chewed with salt as a contraceptive.
* *Croton variegatus* L., Sp. Pl.: 1199 (1753). *Codiaeum chrysosticton* Rumph. ex Spreng., Syst. Veg. 3: 866 (1826), nom. illeg. *Codiaeum variegatum* var. *genuinum* Müll.Arg. in A.P.de Candolle, Prodr. 15(2): 1119 (1866), nom. inval.

var. **cavernicola** Kiew & Welzen, Gardens Bull Singapore 50: 32 (1998).
Borneo (Sabah). 42 BOR. Nanophan. – Limited to limestone cave areas.

var. **variegatum**
Jawa, Lesser Sunda Is., Sulawesi, Maluku, Philippines, New Guinea, N. Queensland, Vanuatu, Fiji. 42 BIS BOR JAW LSI MOL NWG PHI SUL 50 QLD 60 FIJ VAN. Nanophan. or phan. – Numerous cultivars are widely grown; leaves chewed with salt as a contraceptive.
Ricinus pictus Noronha, Verh. Batav. Genootsch. Kunsten 5(4): 2 (1790).
Croton incanus Blume, Catalogus: 104 (1823).
Codiaeum timorense A.Juss., Euphorb. Gen.: 34 (1824), nom. nud.
Croton pictus Lodd., Bot. Cab.: t. 870 (1824). *Codiaeum pictum* (Lodd.) Hook., Bot. Mag. 58: t. 3051 (1831). *Codiaeum variegatum* var. *pictum* (Lodd.) Müll.Arg. in A.P.de Candolle, Prodr. 15(2): 1119 (1866).
Codiaeum moluccanum Decne., Nouv. Ann. Mus. Hist. Nat. 3: 485 (1834). *Codiaeum variegatum* var. *moluccanum* (Decne.) Müll.Arg. in A.P.de Candolle, Prodr. 15(2): 1119 (1866).
Codiaeum cuneifolium Zipp. ex Span., Linnaea 15: 348 (1841).
Croton baliospermum Span., Linnaea 15: 348 (1841).
Codiaeum obovatum Zoll. & Moritzi, Natuur.-Geneesk. Arch. Ned.-Indië 2: 582 (1845).
Codiaeum medium Baill., Adansonia 1: 251 (1861).
Codiaeum chrysosticton var. *angustifolium* Rumph. ex Müll.Arg. in A.P.de Candolle, Prodr. 15(2): 1120 (1866), nom. illeg.
Codiaeum chrysosticton var. *latifolium* Rumph. ex Müll.Arg. in A.P.de Candolle, Prodr. 15(2): 1119 (1866), nom. illeg.

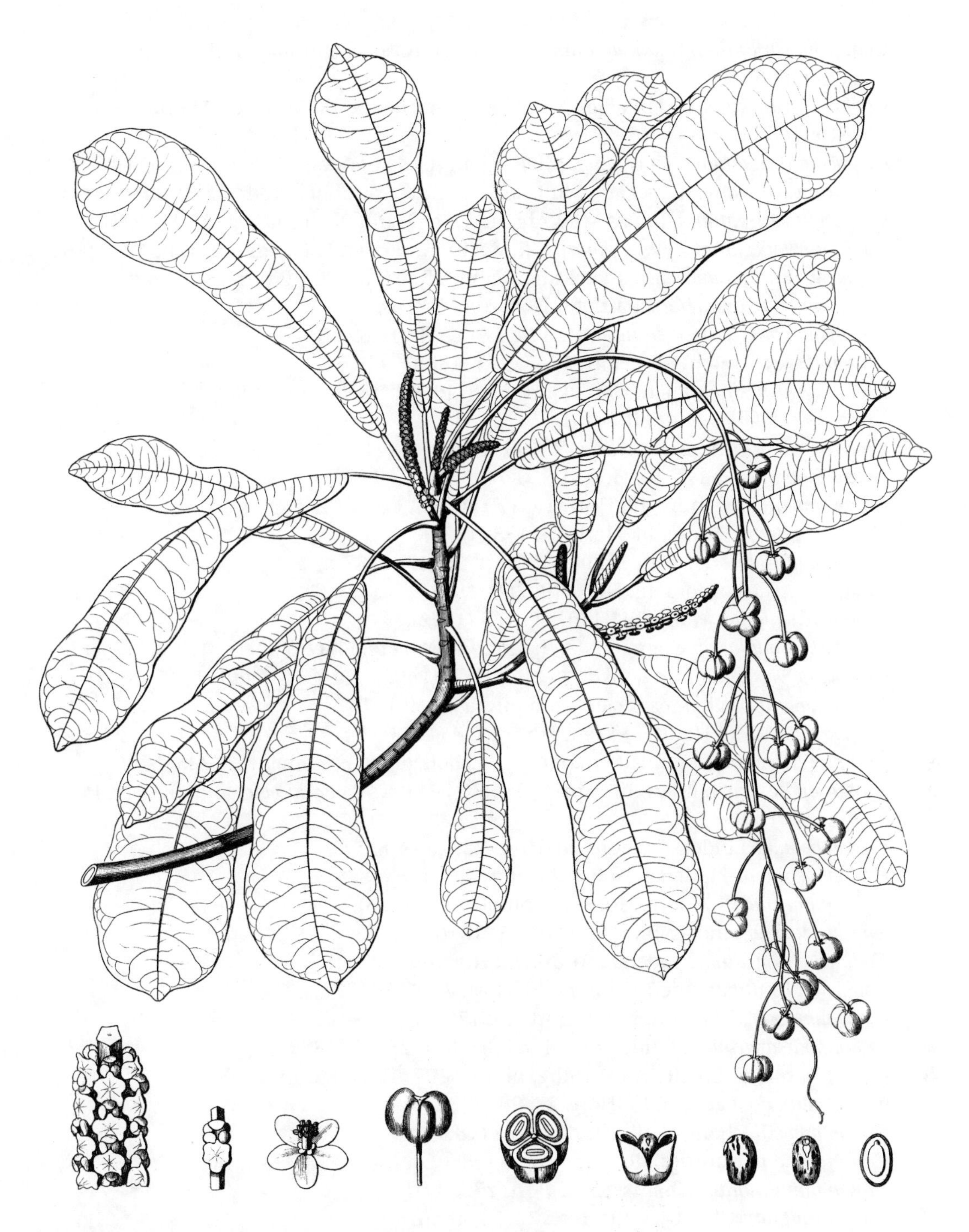

Codiaeum peltatum (Labill.) P.S. Green (as *Crozophora peltata* Labill.)
Artist: P.J.F. Turpin
Labilliardière, Sert. Austro-Caledon.: pl. 75 (1825)
KEW ILLUSTRATIONS COLLECTION

Codiaeum chrysosticton var. *medium* Rumph. ex Müll.Arg. in A.P.de Candolle, Prodr. 15(2): 1120 (1866), nom. illeg.
Codiaeum crispum Rumph. ex Müll.Arg. in A.P.de Candolle, Prodr. 15(2): 1120 (1866).
Codiaeum sylvestre Rumph. ex Müll.Arg. in A.P.de Candolle, Prodr. 15(2): 1120 (1866).
Codiaeum taeniosum Rumph. ex Müll.Arg. in A.P.de Candolle, Prodr. 15(2): 1120 (1866).
Codiaeum variegatum f. *angustifolium* Müll.Arg. in A.P.de Candolle, Prodr. 15(2): 1120 (1866).
Codiaeum variegatum f. *crispum* Müll.Arg. in A.P.de Candolle, Prodr. 15(2): 1120 (1866).

Codiaeum variegatum f. *lanceolatum* Müll.Arg. in A.P.de Candolle, Prodr. 15(2): 1119 (1866).
Codiaeum variegatum f. *latifolium* Müll.Arg. in A.P.de Candolle, Prodr. 15(2): 1119 (1866).
Codiaeum variegatum f. *longifolium* Müll.Arg. in A.P.de Candolle, Prodr. 15(2): 1119 (1866).
Codiaeum variegatum f. *medium* Müll.Arg. in A.P.de Candolle, Prodr. 15(2): 1120 (1866).
Codiaeum variegatum f. *minus* Müll.Arg. in A.P.de Candolle, Prodr. 15(2): 1119 (1866).
Codiaeum variegatum f. *parvifolium* Müll.Arg. in A.P.de Candolle, Prodr. 15(2): 1120 (1866).
Codiaeum variegatum f. *taeniosum* Müll.Arg. in A.P.de Candolle, Prodr. 15(2): 1120 (1866).
Codiaeum maximum Verschaff., Ill. Hort. 14: 534 (1867).
Codiaeum variegatum f. *cornutum* André, Nouv. Hort.: 69 (1867).
Croton hookeri Veitch, Rev. Hort. 1867: 295 (1867).
Croton angustissimus auct., Gard. Chron. 1871: 612 (1871), nom. illeg.
Croton aucubifolius Andr., Ill. Hort. 19: 327 (1872).
Croton cornutus André, Ill. Hort. 19: 188 (1872).
Croton hillianus Veitch, Ill. Hort. 19: 326 (1872).
Croton interruptus André, Ill. Hort. 19: 170 (1872).
Croton irregularis André, Ill. Hort. 19: 135 (1872).
Croton multicolor Veitch ex Andre, Ill. Hort. 19: 120 (1872).
Croton undulatus André, Ill. Hort. 19: 265 (1872).
Croton veitchianus André, Ill. Hort. 19: 135 (1872).
Codiaeum undulatum Cogn. & Marchand in A.Dallière, Pl. Ornament. 1: 15 (1873).
Codiaeum weismannii Cogn. & Marchand in A.Dallière, Pl. Ornament. 1: 22 (1873).
Croton johannis Veitch ex Regel, Gartenflora 22: 251 (1873).
Croton lacteus Van Houtte, J. Gén. Hort. 19: 8 (1873).
Croton weismannii Cogn. & Marchand in A.Dallière, Pl. Ornament. 1: 22 (1873).
Croton andreanus Linden, Ill. Hort. 22: 56 (1875). *Croton andreeus* Linden, Rev. Hort.: 87 (1877).
Croton bellulus Linden & André, Ill. Hort. 22: 104 (1875).
Croton hastiferus Linden & André, Ill. Hort. 22: 136 (1875).
Croton imperialis T.Moore, Florist & Pomol.: 209 (1876).
Croton vervaetii Linden, Ill. Hort. 23: 145 (1876).
Codiaeum elongatum Linden & André, Ill. Hort. 24: 187 (1877).
Codiaeum lyratum Linden & André, Ill. Hort. 24: 155 (1877).
Codiaeum roseopictum André, Ill. Hort. 26: 170 (1879).
Croton massangeanum Linden ex André, Ill. Hort. 26: 77 (1879).
Croton bergmanii Chantrier ex André, Ill. Hort. 27: 102 (1880).
Croton carrieri Chantrier, Ill. Hort. 27: 90 (1880).
Croton drouetii Chantrier, Ill. Hort. 27: 73 (1880).
Croton duvalii Chantrier, Ill. Hort. 27: 73 (1880).
Croton latimaculatus Chantrier, Ill. Hort. 27: 73 (1880).
Croton truffautii Chantrier, Ill. Hort. 27: 73 (1880).
Codiaeum elegantissimum W.Bull, Ill. Hort. 29: 175 (1882).
Codiaeum magnificum Linden, Ill. Hort. 24: 57 (1882).
Croton musaicus André, Rev. Hort. 1882: 240 (1882).
Codiaeum vanoosterzeei Rodigin, Ill. Hort. 30: 173 (1883).
Codiaeum variegatum f. *appendiculatum* Celak., Abh. Böhm. Ges. Wiss. 6: 21 (1884).
Codiaeum albicans G.Nicholson, Ill. Dict. Gard. 1: 350 (1885).
Codiaeum angustifolium G.Nicholson, Ill. Dict. Gard. 1: 350 (1885).
Codiaeum burtonii G.Nicholson, Ill. Dict. Gard. 1: 351 (1885).
Codiaeum chelsonii G.Nicholson, Ill. Dict. Gard. 1: 351 (1885).
Codiaeum chrysophyllum G.Nicholson, Ill. Dict. Gard. 1: 351 (1885).
Codiaeum cooperi G.Nicholson, Ill. Dict. Gard. 1: 351 (1885).
Codiaeum dodgonae G.Nicholson, Ill. Dict. Gard. 1: 352 (1885).
Codiaeum eburneum G.Nicholson, Ill. Dict. Gard. 1: 352 (1885).
Codiaeum elegans G.Nicholson, Ill. Dict. Gard. 1: 352 (1885).

Codiaeum evansianum G.Nicholson, Ill. Dict. Gard. 1: 352 (1885).
Codiaeum fucatum G.Nicholson, Ill. Dict. Gard. 1: 352 (1885).
Codiaeum goldiei G.Nicholson, Ill. Dict. Gard. 1: 353 (1885).
Codiaeum grande G.Nicholson, Ill. Dict. Gard. 1: 353 (1885).
Codiaeum hanburyanum G.Nicholson, Ill. Dict. Gard. 1: 353 (1885).
Codiaeum henryanum G.Nicholson, Ill. Dict. Gard. 1: 353 (1885).
Codiaeum illustre G.Nicholson, Ill. Dict. Gard. 1: 354 (1885).
Codiaeum imperiale G.Nicholson, Ill. Dict. Gard. 1: 355 (1885).
Codiaeum insigne G.Nicholson, Ill. Dict. Gard. 1: 355 (1885).
Codiaeum jamesii G.Nicholson, Ill. Dict. Gard. 1: 355 (1885).
Codiaeum lancifolium G.Nicholson, Ill. Dict. Gard. 1: 355 (1885).
Codiaeum macfarlanei G.Nicholson, Ill. Dict. Gard. 1: 355 (1885).
Codiaeum maculatum G.Nicholson, Ill. Dict. Gard. 1: 355 (1885).
Codiaeum majesticum W.Bull ex Nicholson, Ill. Dict. Gard. 1: 355 (1885).
Codiaeum multicolor G.Nicholson, Ill. Dict. Gard. 1: 355 (1885).
Codiaeum mutabile G.Nicholson, Ill. Dict. Gard. 1: 355 (1885).
Codiaeum nevilliae G.Nicholson, Ill. Dict. Gard. 1: 355 (1885).
Codiaeum pilgrimii G.Nicholson, Ill. Dict. Gard. 1: 355 (1885).
Codiaeum recurvifolium G.Nicholson, Ill. Dict. Gard. 1: 355 (1885).
Codiaeum spirale G.Nicholson, Ill. Dict. Gard. 1: 355 (1885).
Codiaeum stewartii G.Nicholson, Ill. Dict. Gard. 1: 355 (1885).
Codiaeum superbiens G.Nicholson, Ill. Dict. Gard. 1: 355 (1885).
Codiaeum trilobum G.Nicholson, Ill. Dict. Gard. 1: 355 (1885).
Codiaeum triumphans G.Nicholson, Ill. Dict. Gard. 1: 355 (1885).
Codiaeum triumphans var. *harwoodianum* W.Bull ex Nicholson, Ill. Dict. Gard. 1: 355 (1885).
Codiaeum volutum G.Nicholson, Ill. Dict. Gard. 1: 355 (1885).
Codiaeum warrenii G.Nicholson, Ill. Dict. Gard. 1: 355 (1885).
Codiaeum williamsii G.Nicholson, Ill. Dict. Gard. 1: 355 (1885).
Codiaeum wilsonii G.Nicholson, Ill. Dict. Gard. 1: 355 (1885).
Codiaeum youngii G.Nicholson, Ill. Dict. Gard. 1: 355 (1885).
Croton aigburtensis auct., Gard. Chron., II, 26: 749 (1886), nom. illeg.
Croton philippsii B.S.Williams, Cat.: 24 (1886).
Croton wigmannii B.S.Williams, Cat.: 24 (1886).
Croton newmannii auct., Bull Catal.: c, 10 (1887).
Croton picturatum André, Rev. Hort. 60: 423 (1888).
Codiaeum variegatum f. *perringii* K.Schum. in K.M.Schumann & U.M.Hollrung, Fl. Kais. Wilh. Land: 75 (1889).
Croton russelli auct., Gard. Chron., III, 1893: 629 (1893).
Croton thomsonii auct., Gard. Chron., III, 1893: 461 (1893).
Croton evansianum auct., Rev. Hort. 1894: 534 (1894).
Croton regelii auct., Rev. Hort. 1894: 534 (1894).
Codiaeum variegatum f. *ambiguum* Pax in H.G.A.Engler, Pflanzenr., IV, 147, III: 25 (1911).
Codiaeum variegatum f. *lobatum* Pax in H.G.A.Engler, Pflanzenr., IV, 147, III: 26 (1911).
Codiaeum variegatum f. *platyphyllum* Pax in H.G.A.Engler, Pflanzenr., IV, 147, III: 24 (1911).
Croton mirus Domin, Biblioth. Bot. 89: 328 (1927).
Codiaeum interruptum Baum, Oesterr. Bot. Z. 99: 422 (1952).

Synonyms:

Codiaeum albicans G.Nicholson === **Codiaeum variegatum** (L.) Blume
Codiaeum alternifolium (Baill.) Müll.Arg. === **Baloghia alternifolia** Baill.
Codiaeum andamanicum Kurz === **Blachia andamanica** (Kurz) Hook.f.
Codiaeum angustifolium G.Nicholson === **Codiaeum variegatum** (L.) Blume
Codiaeum aurantiacum (Kurz ex Teijsm. & Binn.) Müll.Arg. === **Trigonostemon aurantiacus** (Kurz ex Teijsm. & Binn.) Boerl.

Codiaeum balansae Baill. === **Baloghia balansae** (Baill.) Pax
Codiaeum brevistylum Pax & K.Hoffm. === **Codiaeum bractiferum** (Roxb.) Merr.
Codiaeum brongniartii Baill. === **Baloghia brongniartii** (Baill.) Pax
Codiaeum bureavii Baill. === **Baloghia bureavii** (Baill.) Schltr.
Codiaeum burtonii G.Nicholson === **Codiaeum variegatum** (L.) Blume
Codiaeum carunculatum (Baill.) Müll.Arg. === **Austrobuxus carunculatus** (Baill.) Airy Shaw
Codiaeum chelsonii G.Nicholson === **Codiaeum variegatum** (L.) Blume
Codiaeum chrysophyllum G.Nicholson === **Codiaeum variegatum** (L.) Blume
Codiaeum chrysosticton Rumph. ex Spreng. === **Codiaeum variegatum** (L.) Blume
Codiaeum chrysosticton var. *angustifolium* Rumph. ex Müll.Arg. === **Codiaeum variegatum** (L.) Blume
Codiaeum chrysosticton var. *latifolium* Rumph. ex Müll.Arg. === **Codiaeum variegatum** (L.) Blume
Codiaeum chrysosticton var. *medium* Rumph. ex Müll.Arg. === **Codiaeum variegatum** (L.) Blume
Codiaeum cooperi G.Nicholson === **Codiaeum variegatum** (L.) Blume
Codiaeum crispum Rumph. ex Müll.Arg. === **Codiaeum variegatum** (L.) Blume
Codiaeum cuneifolium Zipp. ex Span. === **Codiaeum variegatum** (L.) Blume
Codiaeum cuneifolium Pax & K.Hoffm. === **Codiaeum luzonicum** Merr.
Codiaeum deplanchei Baill. === **Baloghia deplanchei** (Baill.) Pax
Codiaeum dodgonae G.Nicholson === **Codiaeum variegatum** (L.) Blume
Codiaeum drimiflorum Baill. === **Baloghia drimiflora** (Baill.) Schltr.
Codiaeum eburneum G.Nicholson === **Codiaeum variegatum** (L.) Blume
Codiaeum elegans G.Nicholson === **Codiaeum variegatum** (L.) Blume
Codiaeum elegantissimum W.Bull === **Codiaeum variegatum** (L.) Blume
Codiaeum elongatum Linden & André === **Codiaeum variegatum** (L.) Blume
Codiaeum evansianum G.Nicholson === **Codiaeum variegatum** (L.) Blume
Codiaeum fucatum G.Nicholson === **Codiaeum variegatum** (L.) Blume
Codiaeum goldiei G.Nicholson === **Codiaeum variegatum** (L.) Blume
Codiaeum grande G.Nicholson === **Codiaeum variegatum** (L.) Blume
Codiaeum hanburyanum G.Nicholson === **Codiaeum variegatum** (L.) Blume
Codiaeum henryanum G.Nicholson === **Codiaeum variegatum** (L.) Blume
Codiaeum illustre G.Nicholson === **Codiaeum variegatum** (L.) Blume
Codiaeum imperiale G.Nicholson === **Codiaeum variegatum** (L.) Blume
Codiaeum inophyllum (G.Forst.) Müll.Arg. === **Baloghia inophylla** (G.Forst.) P.S.Green
Codiaeum insigne G.Nicholson === **Codiaeum variegatum** (L.) Blume
Codiaeum interruptum Baum === **Codiaeum variegatum** (L.) Blume
Codiaeum irosinense Elmer ex Merr. === **Codiaeum luzonicum** Merr.
Codiaeum jamesii G.Nicholson === **Codiaeum variegatum** (L.) Blume
Codiaeum lancifolium G.Nicholson === **Codiaeum variegatum** (L.) Blume
Codiaeum lucidum (Endl.) Müll.Arg. === **Baloghia inophylla** (G.Forst.) P.S.Green
Codiaeum ludovicianum var. *praeruptorum* Airy Shaw === **Codiaeum ludovicianum** Airy Shaw
Codiaeum ludovicianum var. *rheophyticum* Airy Shaw === **Codiaeum ludovicianum** Airy Shaw
Codiaeum lutescens Kurz === **Sphyranthera lutescens** (Kurz) Pax & K.Hoffm.
Codiaeum lyratum Linden & André === **Codiaeum variegatum** (L.) Blume
Codiaeum macfarlanei G.Nicholson === **Codiaeum variegatum** (L.) Blume
Codiaeum maculatum G.Nicholson === **Codiaeum variegatum** (L.) Blume
Codiaeum magnificum Linden === **Codiaeum variegatum** (L.) Blume
Codiaeum majesticum W.Bull ex Nicholson === **Codiaeum variegatum** (L.) Blume
Codiaeum maximum Verschaff. === **Codiaeum variegatum** (L.) Blume
Codiaeum medium Baill. === **Codiaeum variegatum** (L.) Blume
Codiaeum moluccanum Decne. === **Codiaeum variegatum** (L.) Blume
Codiaeum montanum (Müll.Arg.) Baill. === **Baloghia montana** (Müll.Arg.) Pax
Codiaeum multicolor G.Nicholson === **Codiaeum variegatum** (L.) Blume

Codiaeum mutabile G.Nicholson === **Codiaeum variegatum** (L.) Blume
Codiaeum nevilliae G.Nicholson === **Codiaeum variegatum** (L.) Blume
Codiaeum obovatum Zoll. & Moritzi === **Codiaeum variegatum** (L.) Blume
Codiaeum pancheri (Baill.) Müll.Arg. === **Fontainea pancheri** (Baill.) Heckel
Codiaeum pentzii Müll.Arg. === **Blachia pentzii** (Müll.Arg.) Benth.
Codiaeum pictum (Lodd.) Hook. === **Codiaeum variegatum** (L.) Blume
Codiaeum pilgrimii G.Nicholson === **Codiaeum variegatum** (L.) Blume
Codiaeum recurvifolium G.Nicholson === **Codiaeum variegatum** (L.) Blume
Codiaeum roseopictum André === **Codiaeum variegatum** (L.) Blume
Codiaeum spirale G.Nicholson === **Codiaeum variegatum** (L.) Blume
Codiaeum stewartii G.Nicholson === **Codiaeum variegatum** (L.) Blume
Codiaeum superbiens G.Nicholson === **Codiaeum variegatum** (L.) Blume
Codiaeum sylvestre Rumph. ex Müll.Arg. === **Codiaeum variegatum** (L.) Blume
Codiaeum taeniosum Rumph. ex Müll.Arg. === **Codiaeum variegatum** (L.) Blume
Codiaeum timorense A.Juss. === **Codiaeum variegatum** (L.) Blume
Codiaeum trilobum G.Nicholson === **Codiaeum variegatum** (L.) Blume
Codiaeum triumphans G.Nicholson === **Codiaeum variegatum** (L.) Blume
Codiaeum triumphans var. *harwoodianum* W.Bull ex Nicholson === **Codiaeum variegatum** (L.) Blume
Codiaeum umbellatum (Willd.) Müll.Arg. === **Blachia umbellata** (Willd.) Baill.
Codiaeum undulatum Cogn. & Marchand === **Codiaeum variegatum** (L.) Blume
Codiaeum vanoosterzeei Rodigin === **Codiaeum variegatum** (L.) Blume
Codiaeum variegatum f. *ambiguum* Pax === **Codiaeum variegatum** (L.) Blume
Codiaeum variegatum f. *angustifolium* Müll.Arg. === **Codiaeum variegatum** (L.) Blume
Codiaeum variegatum f. *appendiculatum* Celak. === **Codiaeum variegatum** (L.) Blume
Codiaeum variegatum f. *cornutum* André === **Codiaeum variegatum** (L.) Blume
Codiaeum variegatum f. *crispum* Müll.Arg. === **Codiaeum variegatum** (L.) Blume
Codiaeum variegatum var. *genuinum* Müll.Arg. === **Codiaeum variegatum** (L.) Blume
Codiaeum variegatum f. *lanceolatum* Müll.Arg. === **Codiaeum variegatum** (L.) Blume
Codiaeum variegatum f. *latifolium* Müll.Arg. === **Codiaeum variegatum** (L.) Blume
Codiaeum variegatum f. *lobatum* Pax === **Codiaeum variegatum** (L.) Blume
Codiaeum variegatum f. *longifolium* Müll.Arg. === **Codiaeum variegatum** (L.) Blume
Codiaeum variegatum f. *medium* Müll.Arg. === **Codiaeum variegatum** (L.) Blume
Codiaeum variegatum f. *minus* Müll.Arg. === **Codiaeum variegatum** (L.) Blume
Codiaeum variegatum var. *moluccanum* (Decne.) Müll.Arg. === **Codiaeum variegatum** (L.) Blume
Codiaeum variegatum f. *parvifolium* Müll.Arg. === **Codiaeum variegatum** (L.) Blume
Codiaeum variegatum f. *perringii* K.Schum. === **Codiaeum variegatum** (L.) Blume
Codiaeum variegatum var. *pictum* (Lodd.) Müll.Arg. === **Codiaeum variegatum** (L.) Blume
Codiaeum variegatum f. *platyphyllum* Pax === **Codiaeum variegatum** (L.) Blume
Codiaeum variegatum f. *taeniosum* Müll.Arg. === **Codiaeum variegatum** (L.) Blume
Codiaeum volutum G.Nicholson === **Codiaeum variegatum** (L.) Blume
Codiaeum warrenii G.Nicholson === **Codiaeum variegatum** (L.) Blume
Codiaeum weismannii Cogn. & Marchand === **Codiaeum variegatum** (L.) Blume
Codiaeum williamsii G.Nicholson === **Codiaeum variegatum** (L.) Blume
Codiaeum wilsonii G.Nicholson === **Codiaeum variegatum** (L.) Blume
Codiaeum youngii G.Nicholson === **Codiaeum variegatum** (L.) Blume

Codonocalyx

Synonyms:
Codonocalyx Klotzsch ex Baill. === **Croton** L.

Coelebogyne

Synonyms:
Coelebogyne Js.Sm. === **Alchornea** Sw.
Coelebogyne aquifolium (Js.Sm.) Domin === **Alchornea aquifolia** (Js.Sm.) Domin
Coelebogyne ilicifolia Js.Sm. === **Alchornea aquifolia** (Js.Sm.) Domin
Coelebogyne thozetiana (Baill.) Pax & K.Hoffm. === **Alchornea aquifolia** (Js.Sm.) Domin

Coelobogyne

An orthographic variant of *Coelebogyne*.

Coelodepas

An orthographic variant of *Koilodepas*.

Synonyms:
Coelodepas Hassk. === **Koilodepas** Hassk.

Coelodiscus

A segregate of *Mallotus*.

Synonyms:
Coelodiscus Baill. === **Mallotus** Lour.
Coelodiscus cambodianus Gagnep. === **Mallotus cambodianus** (Gagnep.) Airy Shaw
Coelodiscus coudercii Gagnep. === **Mallotus glabriusculus** (Kurz) Pax & K.Hoffm.
Coelodiscus glabriusculus Kurz === **Mallotus glabriusculus** (Kurz) Pax & K.Hoffm.
Coelodiscus hirsutulus Kurz === **Mallotus longipes** Müll.Arg.
Coelodiscus lanceolatus Gagnep. === **Mallotus lanceolatus** (Gagnep.) Airy Shaw
Coelodiscus lauterbachianus Pax & K.Hoffm. === **Mallotus lauterbachianus** (Pax & K.Hoffm.) Pax & K.Hoffm.
Coelodiscus longipes Kurz === **Mallotus myanmaricus** P.T.Li
Coelodiscus montanus Müll.Arg. === **Mallotus montanus** (Müll.Arg.) Airy Shaw
Coelodiscus pierrei Gagnep. === **Mallotus pierrei** (Gagnep.) Airy Shaw
Coelodiscus speciosus Müll.Arg. === **Sumbaviopsis albicans** (Blume) J.J.Sm.
Coelodiscus subcuneatus Gage === **Mallotus subcuneatus** (Gage) Airy Shaw
Coelodiscus thorelii Gagnep. === **Mallotus nanus** Airy Shaw
Coelodiscus thunbergianus Müll.Arg. === **Mallotus thunbergianus** (Müll.Arg.) Pax & K.Hoffm.
Coelodiscus ustulatus Gagnep. === **Mallotus ustulatus** (Gagnep.) Airy Shaw

Coilmeroa

Synonyms:
Coilmeroa Endl. === **Flueggea** Willd.

Collenucia

Synonyms:
Collenucia Chiov. === **Jatropha** L.
Collenucia paradoxa Chiov. === **Jatropha paradoxa** (Chiov.) Chiov.

Colliguaja

5 species, southern S. America (southern Brazil and Paraguay southwards) with a particular development in Chile; willow-like or leathery-leaved shrubs or small trees, the leaves opposite or alternate and inflorescences terminal. The genus is quite distinctive; Esser (1994; see **Euphorbioideae (except Euphorbieae)**) argues that it is possibly related only to *Gymnanthes* among other Hippomaneae. No overall revision has appeared since 1912, and a new account is needed. The leaves are, according to Esser, variable and fewer species should perhaps be accepted. Moreover, the position of *Colliguaja patagonica* Speg., a largely leafless stem-succulent, is problematic; in Esser's view, it requires either a new genus or placement within, and redefinition of, *Stillingia*. We have assigned it to **Unplaced taxa**. (Euphorbioideae (except Euphorbieae))

Pax, F. (with K. Hoffmann) (1912). *Colliguaya*. In A. Engler (ed.), Das Pflanzenreich, IV 147 V (Euphorbiaceae-Hippomaneae): 265-268. Berlin. (Heft 52.) La/Ge. — 5 species, S South America (well represented in Chile).

Colliguaja Molina, Sag. Stor. Nat. Chili: 158, 354 (1782).
Brazil, S. South America. 84 85.

Colliguaja brasillensis Klotzsch ex Baill., Étude Euphorb.: 535 (1858).
Paraguay, S. Brazil, Uruguay. 84 BZS 85 PAR URU. Nanophan. or phan.

Colliguaja dombeyana A.Juss., Ann. Sc. Nat. (Paris) 25: 23 (1832).
SC. Chile. 85 CLC. Nanophan.

Colliguaja integerrima Gillies & Hook., Bot. Misc. 1: 140 (1830).
SC. Chile, Argentina (Neuquén, Rio Negro, Chubut). 85 AGS CLN CLC. Nanophan.
Colliguaja bridgesii Müll.Arg., Linnaea 32: 125 (1863).

Colliguaja odorifera Molina, Sag. Stor. Nat. Chili: 158, 354 (1782). *Croton colliguaja* Spreng., Syst. Veg. 3: 875 (1826). – FIGURE, p. 410.
C. Chile. 85 CLC. Nanophan.
Colliguaja triquetra Gillies & Hook., Bot. Misc. 1: 142 (1830).
Colliguaja obtusa Regel, Index Seminum (LE) 1855: 18 (1855).

Colliguaja salicifolia Gillies & Hook., Bot. Misc. 1: 141 (1830).
C. Chile. 85 CLC. Nanophan.
Colliguaja dombeyana C.Gay, Fl. Chil. 5: 342 (1851), nom. illeg.

Synonyms:
Colliguaja bridgesii Müll.Arg. === **Colliguaja integerrima** Gillies & Hook.
Colliguaja dombeyana C.Gay === **Colliguaja salicifolia** Gillies & Hook.
Colliguaja obtusa Regel === **Colliguaja odorifera** Molina
Colliguaja patagonica Speg. === ?
Colliguaja triquetra Gillies & Hook. === **Colliguaja odorifera** Molina

Colmeiroa

Synonyms:
Colmeiroa Reut. === **Flueggea** Willd.
Colmeiroa buxifolia Reut. === **Flueggea tinctoria** (L.) G.L.Webster

Comatocroton

Synonyms:
Comatocroton H.Karst. === **Croton** L.

Colliguaja odorifera Molina
Artist: Eulalia Delile
Delessert, Icon. Sel. Pl. 3: pl. 88 (1838)
KEW ILLUSTRATIONS COLLECTION

Cometia

With respect to the *Pflanzenreich* treatment of *Cometia* by Pax & Hoffmann (1922: 78-79; see *Drypetes*) Leandri (1958; see **Malagassia**) transferred the type species, *C. thouarsii*, to *Drypetes* and treated *C. lucida* as *Thecacoris cometia*. (Phyllanthoideae)

Synonyms:
Cometia Thouars ex Baill. === **Drypetes** Vahl
Cometia lucida Baill. === **Thecacoris cometia** Leandri
Cometia thouarsii Baill. === **Drypetes thouarsii** (Baill.) Leandri

Commia

Excoecaria

Synonyms:
Commia Lour. === **Excoecaria** L.

Comopyena

Pycnocoma

Synonyms:
Comopyena Kuntze === **Pycnocoma** Benth.

Conami

Phyllanthus

Synonyms:
Conami Aubl. === **Phyllanthus** L.
Conami brasiliensis Aubl. === **Phyllanthus brasiliensis** (Aubl.) Poir.
Conami conami (Sw.) Britton === **Phyllanthus brasiliensis** (Aubl.) Poir.
Conami ovalifolia Britton === **Phyllanthus cinctus** Urb.
Conami portoricensis (Kuntze) Britton === **Flueggea virosa** (Roxb. ex Willd.) Voigt

Conceveiba

12 species, Middle and northern S. America (Costa Rica to the Guianas, Brazil and Peru); West Central Africa (Cameroon, Gabon). Medium-sized forest trees with conspicuous paniculate male inflorescences. to Includes *Conceveibastrum*, *Conceveibum* and *Veconcibea* following study by Jablonski (1967). Allies include *Gavarretia* and *Polyandra* (Webster, Synopsis, 1994). A thorough study of these as well as the former segregates of *Conceveiba* is needed for a better understanding of relationships. (Acalyphoideae)

Pax, F. (with K. Hoffmann) (1914). *Conceveiba*. In A. Engler (ed.), Das Pflanzenreich, IV 147 VII (Euphorbiaceae-Acalypheae-Mercurialinae): 214-217. Berlin. (Heft 63.) La/Ge. — 4 species, S America (Amazon Basin, Guayana).

Pax, F. (with K. Hoffmann) (1914). *Conceveibastrum*. In A. Engler (ed.), Das Pflanzenreich, IV 147 VII (Euphorbiaceae-Acalypheae-Mercurialinae): 217-218. Berlin. (Heft 63.) La/Ge. — Monotypic, S America (Amazon Basin). [More recently reduced to *Conceveiba*.]

Pax, F. (with K. Hoffmann) (1914). *Veconcibea*. In A. Engler (ed.), Das Pflanzenreich, IV 147 VII (Euphorbiaceae-Acalypheae-Mercurialinae): 218-219. Berlin. (Heft 63.) La/Ge. — 2 species, C and S America. [Now referred to *Conceveiba*.]

Jablonski, E. (1967). *Conceveiba*. Euphorbiaceae, Guayana Highland (Mem. New York Bot. Gard. 17(1)): 131-135. New York. En. — 6 species, including 1 new comb.

Thomas, D. W. (1990). *Conceveiba* Aublet (Euphorbiaceae) new to Africa. Ann. Missouri Bot. Gard. 77: 856-858, illus. En. — First report of genus for Africa and description of a new species, *C. africana* (=*C. macrostachys*), from Gabon. [A part of the rich Mayombe flora of West Central Africa, shown to have links with that of tropical South America.]

Breteler, F J. (1993). Novitates gabonenses, 11. The distribution of two noteworthy Gaboneae Euphorbiaceae: *Conceveiba macrostachys* and *Pogonophora letouzeyi*. Bull. Jard. Bot. Natl. Belg. 62: 191-195. En. — Includes required renaming of *C. africana* as *C. macrostachys* with discussion of distribution.

Breteler, F. J. (with A.M.W. Mennega) (1994). Novitates gabonenses, 17. *Conceveiba leptostachys*, a new Eup[horbiaceae from Gabon and Cameroon. Bull. Jard. Bot. Natl. Belg. 63: 209-217, illus., map. En. — Includes description of *C. leptostachys* (Gabon and Cameroon), with key to the 2 African species, and a study (by Mennega) of the wood structure.

Conceveiba Aubl., Hist. Pl. Guiane 2: 923 (1775).
S. America, WC. Trop. Africa. 23 81 82 83 84.
Conceveibastrum (Müll.Arg.) Pax & K.Hoffm. in H.G.A.Engler, Pflanzenr., IV, 147, VII: 217 (1914).
Veconcibea Pax & K.Hoffm. in H.G.A.Engler, Pflanzenr., IV, 147, VII: 218 (1914).

Conceveiba guianensis Aubl., Hist. Pl. Guiane 2: 924 (1775).
Venezuela, Guianas, N. & W. Brazil, Peru (Loreto). 82 FRG GUY SUR VEN 83 PER 84 BZC BZN. Phan.
Conceveiba ovata Rich. ex A.Juss., Euphorb. Gen.: 43 (1824).
Conceveiba krukoffii Steyerm., Publ. Field Mus. Nat. Hist., Bot. Ser. 17: 414 (1938).
Conceveiba simulata Steyerm., Publ. Field Mus. Nat. Hist., Bot. Ser. 17: 416 (1938).

Conceveiba hostmanii Benth., Hooker's J. Bot. Kew Gard. Misc. 6: 332 (1854).
Surinam, Brazil (Amazonas). 82 SUR 84 BZN. Phan.

Conceveiba latifolia Benth., Hooker's J. Bot. Kew Gard. Misc. 6: 332 (1854).
Brazil (Amazonas: Rio Uaupés). 84 BZN. Phan.

Conceveiba leptostachys Breteler, Bull. Jard. Bot. Belg. 63: 210 (1994).
Cameroon, Gabon. 23 CMN GAB.

Conceveiba macrostachys Breteler, Bull. Jard. Bot. Belg. 63: 192 (1993).
Gabon. 23 GAB.
* *Conceveiba africana* D.W.Thomas, Ann. Missouri Bot. Gard. 77: 856 (1990), nom. illeg.

Conceveiba magnifica Steyerm., Publ. Field Mus. Nat. Hist., Bot. Ser. 17: 414 (1938).
Brazil (Amazonas). 84 BZN. Phan.

Conceveiba martiana Baill., Adansonia 5: 221 (1865). *Alchornea martiana* (Baill.) Müll.Arg. in C.F.P.von Martius, Fl. Bras. 11(2): 375 (1874).
Guiana, Venezuela (Amazonas), Brazil (Amazonas), Peru (Loreto). 82 FRG VEN 83 PER 84 BZN. Phan.
Conceveiba megalophylla Müll.Arg., Linnaea 34: 167 (1865).

Conceveiba parvifolia McPherson, Novon 5: 287 (1995).
Panama, Colombia. 80 PAN 83 CLM. Phan.

Conceveiba pleiostemona Donn.Sm., Bot. Gaz. 54: 243 (1912).
Costa Rica. 80 COS. Phan.

Conceveiba ptariana (Steyerm.) Jabl., Mem. New York Bot. Gard. 17: 134 (1967).
Venezuela (Amazonas, Bolívar). 82 VEN.
* *Conceveibastrum ptarianum* Steyerm., Fieldiana, Bot. 28: 308 (1952).

Conceveiba rhytidocarpa Müll.Arg. in C.F.P.von Martius, Fl. Bras. 11(2): 372 (1874).
Peru. 83 PER. Phan.

Conceveiba trigonocarpa Müll.Arg. in C.F.P.von Martius, Fl. Bras. 11(2): 372 (1874).
Brazil (Amazonas: lake Tefé). 84 BZN. Phan.

Synonyms:
Conceveiba africana Müll.Arg. === **Neoboutonia mannii** Benth. & Hook.f.
Conceveiba africana D.W.Thomas === **Conceveiba macrostachys** Breteler
Conceveiba cordata A.Juss. === **Aparisthmium cordatum** (A.Juss.) Baill.
Conceveiba javanensis Blume === **Alchornea rugosa** (Lour.) Müll.Arg.
Conceveiba krukoffii Steyerm. === **Conceveiba guianensis** Aubl.
Conceveiba latifolia Zipp. ex Span. === **Alchornea rugosa** (Lour.) Müll.Arg.
Conceveiba macropylla (Mart.) Klotzsch ex Benth. === **Aparisthmium cordatum** (A.Juss.) Baill.
Conceveiba megalophylla Müll.Arg. === **Conceveiba martiana** Baill.
Conceveiba ovata Rich. ex A.Juss. === **Conceveiba guianensis** Aubl.
Conceveiba poeppigiana Klotzsch ex Pax & K.Hoffm. === **Aparisthmium cordatum** (A.Juss.) Baill.
Conceveiba pubescens Britton === **Alchornea iricurana** Casar.
Conceveiba simulata Steyerm. === **Conceveiba guianensis** Aubl.
Conceveiba terminalis (Baill.) Müll.Arg. === **Gavarretia terminalis** Baill.
Conceveiba tomentosa Span. === **Cladogynos orientalis** Zipp. ex Span.

Conceveibastrum

Synonyms:
Conceveibastrum (Müll.Arg.) Pax & K.Hoffm. === **Conceveiba** Aubl.
Conceveibastrum ptarianum Steyerm. === **Conceveiba ptariana** (Steyerm.) Jabl.

Conosapium

1 species, Madagascar. Included in *Sapium* by Pax (1912) but now restored to generic rank by Esser (1994) and Kruijt (1996; for both, see **Euphorbiaceae (except Euphorbieae)**). Tall glabrous shrubs with oblanceolate leaves to 20 cm long and inflorescences of up to similar length. The genus is allegedly related to *Taeniosapium* (Kruijt) but the overall picture appears more complicated. (Euphorbioideae (except Euphorbieae))

Pax, F. (with K. Hoffmann) (1912). *Sapium*. In A. Engler (ed.), Das Pflanzenreich, IV 147 V (Euphorbiaceae-Hippomaneae): 199-258. Berlin. (Heft 52.) La/Ge. — No. 92 (p. 255), *S. madagascariense*, is the type species of *Conosapium* (and subgen. *Conosapium*). [The other species here in subgen. *Conosapium* is now in *Taeniosapium*.]

Conosapium Müll.Arg., Linnaea 32: 87 (1863).
Madagascar. 29.

Conosapium madagascariensis Müll.Arg., Linnaea 32: 87 (1863). *Sapium madagascariense* (Müll.Arg.) Pax in H.G.A.Engler & K.A.E.Prantl, Nat. Pflanzenfam. 3(5): 98 (1891).
NW. Madagascar. 29 MDG.

Cordemoya

1 species, Mascarene Is. (Mauritius, Réunion). Tree to 15 m (sometimes only a shrub) of wet forest and scrub with acuminate, penninerved leaves featuring many cross-veins. For a modern treatment (by M.J.E. Coode) see *Flore des Mascareignes* 160: 50-53 (1982). The genus is close to *Malllotus*. (Acalyphoideae)

Pax, F. (with K. Hoffmann) (1914). *Cordemoya*. In A. Engler (ed.), Das Pflanzenreich, IV 147 VII (Euphorbiaceae-Acalypheae-Mercurialinae): 208-209. Berlin. (Heft 63.) La/Ge. — Monotypic, Mascarenes.

Cordemoya Baill., Adansonia 1: 255 (1861).
Mascarenes. 29.

Cordemoya integrifolia (Willd.) Baill., Adansonia 1: 255 (1861).
Réunion, Mauritius. 29 MAU REU. Phan.
* *Ricinus integrifolius* Willd., Sp. Pl. 4: 567 (1805). *Mallotus integrifolius* (Willd.) Müll.Arg., Linnaea 34: 186 (1865).
Ricinus dioicus Chev. ex Steud., Nomencl. Bot., ed. 2, 2: 459 (1841).
Boutonia mascareinensis Bojer, Trav. Soc. Hist. Nat. Ile Maurice 1842-1846: 151 (1846).
Ricinus lanceolatus Thouars ex Baill., Adansonia 1: 255 (1861).

Synonyms:
Cordemoya acuminata (Baill.) Baill. === **Deuteromallotus acuminatus** (Baill.) Pax & K.Hoffm.

Corythea

Synonyms:
Corythea S.Watson === **Acalypha** L.
Corythea filipes S.Watson === **Acalypha filipes** (S.Watson) McVaugh
Corythea multiflora Standl. === **Acalypha multiflora** (Standl.) Radcl.-Sm.

Coulejia

Synonyms:
Coulejia Dennst. === **Antidesma** L.

Cratochwilia

Synonyms:
Cratochwilia Neck. === **Clutia** Boerh. ex L.

Cremophyllum

Synonyms:
Cremophyllum Scheidw. === **Dalechampia** Plum. ex L.
Cremophyllum spathulatum Scheidw. === **Dalechampia spathulata** (Scheidw.) Baill.

Crepidaria

Synonyms:
Crepidaria Haw. === **Pedilanthus** Neck.
Crepidaria subcarinata Haw. === **Pedilanthus tithymaloides** (L.) Poit. subsp. **tithymaloides**

Croizatia

4 species, Americas (Panama to Peru and Venezuela); now includes *Pseudosagotia*, described by Secco (1985). Shrubs or small trees with fairly stout twigs, oblanceolate leaves and conspicuous and distinctive axillary inflorescences. A collection from Ecuador tentatively assigned to *C. panamensis* by Webster et al. (1987) may prove to be a distinct species; in addition, material possibly referable to *C. neotropica* has been reported from Peru.The presence of petals renders the genus unique in the subfamily. Webster et al. (1987) argued for a placement in the Phyllanthoideae but acknowledged that it was very close to Oldfieldioideae. In 1994 Webster did assign it to its own tribe at the beginning of Oldfieldioideae but allied it to tribe Amanoeae in Phyllanthoideae. A strong resemblance to *Actephila* has also been reported (Webster et al. 1987). Still imperfectly known, the genus appears to be a connecting link between the subfamilies but is somewhat isolated from other Oldfieldioideae. Within it, the species are mutually rather closely related with delimitation 'provisional' (Webster et al. 1987). (Oldfieldioideae)

Secco, R. de S. (1985). *Pseudosagotia*: um novo gênero de Euphorbiaceae para a Amazônia Venezuelana. Bol. Mus. Paraense 'Emilio Goeldi', Bot., 2(1): 23-27, illus. Pt. — 1 species, *P. brevipetiolata*, western Venezuela; based on *Steyermark et Rabe 97322*. [The genus is now united with *Croizatia*.]

• Webster, G., L. J. Gillespie & J. A. Steyermark (1987). Systematics of *Croizatia* (Euphorbiaceae). Syst. Bot. 12: 1-8. En. — Introduction, character review, and comparisons with related genera; revision of 3 species with key, descriptions, synonymy, references, types, new exsiccatae, and commentary; previous papers cited. A comparative table of characters of *Croizatia* with Phyllanthoideae and Oldfieldioideae is furnished.

Croizatia Steyerm., Fieldiana, Bot. 28: 308 (1952).
Panama to Ecuador. 80 (81) 82 83. Nanophan. or phan.
Pseudosagotia Secco, Bol. Mus. Paraense Emílio Goeldi, Bot 2(1): 23 (1985).

Croizatia brevipetiolata (Secco) Govaerts in R.Govaerts & D.G.Frodin, World Checklist Bibliogr. Euphorbiaceae: 415 (2000).
Venezuela (Amazonas). 82 VEN. Phan.
Pseudosagotia brevipetiolata Secco, Bol. Mus. Paraense Emílio Goeldi, Bot 2(1): 24 (1985).

Croizatia naiguatensis Steyerm., Brittonia 30: 40 (1978).
Venezuela (Vargas). 82 VEN. Phan.

Croizatia neotropica Steyerm., Fieldiana, Bot. 28: 309 (1952).
Venezuela (Anzoategui). 82 VEN. Phan.

Croizatia panamensis G.L.Webster, Syst. Bot. 12: 7 (1987).
Panama, Colombia (Chocó), Ecuador? 80 PAN 83 CLM ECU? Nanophan. or phan. – The collection from Ecuador may prove to be a different species.

Crossophora

An orthographic variant of *Chrozophora*.